Eukaryotic Microorganisms as Sources of Bioproducts

Related Titles:

Microbiology of Green Fuels
(ISBN: 978-1-00-317115-7)

Biofuels from Algae
(ISBN: 978-0-444-59558-4)

Advances in Yeast Biotechnology for Biofuels and Sustainability
(ISBN: 978-0-323-95449-5)

Microalgae-Based Biofuels and Bioproducts
(ISBN: 978-0-08-101023-5)

Woodhead Publishing Series in Energy

Eukaryotic Microorganisms as Sources of Bioproducts

From Microalgae and Yeast Cultivation to End-Products

Second Edition

Edited by

Cristina González-Fernández

Elia Tomás-Pejó

ELSEVIER

WP
WOODHEAD PUBLISHING
An imprint of Elsevier

Woodhead Publishing is an imprint of Elsevier
50 Hampshire Street, 5th Floor, Cambridge, MA 02139, United States
125 London Wall, London EC2Y 5AS, United Kingdom

ISBN: 978-0-443-30188-9 (print)
ISBN: 978-0-443-30189-6 (online)

For information on all Woodhead Publishing publications
visit our website at https://www.elsevier.com/books-and-journals

Publisher: Megan Ball
Acquisitions Editor: Peter Adamson
Editorial Project Manager: Teddy A. Lewis
Production Project Manager: Erragounta Saibabu Rao
Cover Designer: Mark Rogers

Typeset by MPS Limited, Chennai, India

Contents

Section 2 Some key microalgae and yeast species for bioproduction

Section 3 Up-scaling, industrial application and process assessments

List of contributors

Jorge Aburto Bioenergy Consultant, Mexico City, Mexico

Filiz Alemdar Pak Gıda Üretim ve Pazarlama A.Ş., Kocaeli, Turkey

Rodrigo Ledesma Amaro Department of Bioengineering, Imperial College London, London, United Kingdom

Roxana Ángeles Institute of Sustainable Processes, University of Valladolid, Valladolid, Spain; Department of Chemical Engineering and Environmental Technology, University of Valladolid, Valladolid, Spain; Associate Laboratory i4HB—Institute for Health and Bioeconomy, NOVA School of Science and Technology, NOVA University Lisbon, Caparica, Portugal; UCIBIO—Applied Molecular Biosciences Unit, Department of Chemistry, NOVA School of Science and Technology, NOVA University Lisbon, Caparica, Portugal

Bahar Aslanbay Guler Bioengineering Department, Ege University, Izmir, Turkey

Manuel Becerra CICA – Interdisciplinary Center for Chemistry and Biology, Departament of Biology, Faculty of Sciences, Biomedical Research Institute A Coruña (INIBIC), University of A Coruña, A Coruña, Spain

Paola Branduardi Department of Biotechnology and Biosciences, University of Milano-Bicocca, Milan, Italy

Anna Bzducha-Wróbel Department of Food Biotechnology and Microbiology, Institute of Food Sciences, Warsaw University of Life Sciences, Warsaw, Poland

Elisabete Coelho LAQV-REQUIMTE, "GlycoFoodChem" Group, Department of Chemistry, University of Aveiro, Aveiro, Portugal

Marta de Vicente Biotechnological Processes Unit, IMDEA Energy, Móstoles (Madrid), Spain

María-Eugenia DeCastro CICA – Interdisciplinary Center for Chemistry and Biology, Departament of Biology, Faculty of Sciences, Biomedical Research Institute A Coruña (INIBIC), University of A Coruña, A Coruña, Spain

Ümit Cem Derman Pak Gıda Üretim ve Pazarlama A.Ş., Kocaeli, Turkey

Joana C. Fradinho Associate Laboratory i4HB—Institute for Health and Bioeconomy, NOVA School of Science and Technology, NOVA University Lisbon, Caparica, Portugal; UCIBIO—Applied Molecular Biosciences Unit, Department of Chemistry, NOVA School of Science and Technology, NOVA University Lisbon, Caparica, Portugal

André Freches Associate Laboratory i4HB—Institute for Health and Bioeconomy, NOVA School of Science and Technology, NOVA University Lisbon, Caparica, Portugal; UCIBIO—Applied Molecular Biosciences Unit, Department of Chemistry, NOVA School of Science and Technology, NOVA University Lisbon, Caparica, Portugal

Octavio García-Depraect Institute of Sustainable Processes, University of Valladolid, Valladolid, Spain; Department of Chemical Engineering and Environmental Technology, University of Valladolid, Valladolid, Spain

Cristina González-Fernández Department of Chemical Engineering and Environmental Technology, School of Industrial Engineering, University of Valladolid, Valladolid, Spain; Institute of Sustainable Processes, Valladolid, Spain; Biotechnological Processes Unit, IMDEA Energy, Móstoles (Madrid), Spain

María-Isabel González-Siso CICA – Interdisciplinary Center for Chemistry and Biology, Departament of Biology, Faculty of Sciences, Biomedical Research Institute A Coruña (INIBIC), University of A Coruña, A Coruña, Spain

Silvia Greses Department of Chemical Engineering, University of Valencia, Valencia, Spain

Timea I. Hajnal-Jafari Department of Field and Vegetable Crops, Faculty of Agriculture, University of Novi Sad, Novi Sad, Serbia

Esra Imamoglu Bioengineering Department, Ege University, Izmir, Turkey

Christian Kennes Chemical Engineering Laboratory, Faculty of Sciences and Interdisciplinary Centre of Chemistry and Biology, Centro Interdisciplinar de Química y Biología (CICA), BIOENGIN Group, University of La Coruña, A Coruña, Spain

Raquel Lebrero Institute of Sustainable Processes, University of Valladolid, Valladolid, Spain; Department of Chemical Engineering and Environmental Technology, University of Valladolid, Valladolid, Spain

Marcela Levío-Raimán Department of Chemical Engineering and Environmental Technology, School of Industrial Engineering, University of Valladolid, Valladolid, Spain; Institute of Sustainable Processes, Valladolid, Spain

Fakhra Liaqat Center of Environmental and Biological Studies, Arabian Gulf University, College of Education, Administrative and Technical Sciences, Manama, Bahrain

Mihai Lucian Lomanar Department of Chemical Engineering and Environmental Technology, School of Industrial Engineering, University of Valladolid, Valladolid, Spain; Institute of Sustainable Processes, Valladolid, Spain

Carlos Lucena Department of Agronomy, ceiA3 Campus of International Excellence in Agri-Food of Rabanales, María de Maeztu Unit of Excellence 2021-2024, University of Córdoba, Córdoba, Spain

Valeria Mapelli Department of Biotechnology and Biosciences, University of Milano-Bicocca, Milan, Italy

Alejandra Martorell-Múgica Biotechnological Processes Unit, IMDEA Energia, Móstoles, Spain

Gavin Peter McStay School of Pharmacy and Biomolecular Sciences, Liverpool John Moores University, Liverpool, United Kingdom

Aleksandra Maria Mirończuk Laboratory for Biosustainability, Institute of Environmental Biology, Wrocław University of Environmental and Life Sciences, Wroclaw, Poland

Antonio David Moreno Advanced Biofuels and Bioproducts Unit, Department of Energy, Research Center for Energy, Environment and Technology (CIEMAT), Madrid, Spain

Bettina Müller Department of Molecular Sciences, Swedish University of Agricultural Sciences, Uppsala, Sweden

Raúl Muñoz Institute of Sustainable Processes, University of Valladolid, Valladolid, Spain; Department of Chemical Engineering and Environmental Technology, University of Valladolid, Valladolid, Spain

Samina Naseeh School of Biological Sciences, The University of Manchester, Manchester, United Kingdom

Cecilia Naveira-Pazos Chemical Engineering Laboratory, Faculty of Sciences and Interdisciplinary Centre of Chemistry and Biology, Centro Interdisciplinar de Química y Biología (CICA), BIOENGIN Group, University of La Coruña, A Coruña, Spain

María José Negro Advanced Biofuels and Bioproducts Unit, Department of Energy, Research Center for Energy, Environment and Technology (CIEMAT), Madrid, Spain

Volkmar Passoth Department of Molecular Sciences, Swedish University of Agricultural Sciences, Uppsala, Sweden

José Ramos Department of Agriculture Chemistry, Edaphology and Microbiology, ceiA3 Campus of International Excellence in Agri-Food of Rabanales, University of Córdoba, Córdoba, Spain

Maria A.M. Reis Associate Laboratory i4HB—Institute for Health and Bioeconomy, NOVA School of Science and Technology, NOVA University Lisbon, Caparica, Portugal; UCIBIO—Applied Molecular Biosciences Unit, Department of Chemistry, NOVA School of Science and Technology, NOVA University Lisbon, Caparica, Portugal

Yadira Rodríguez Institute of Sustainable Processes, University of Valladolid, Valladolid, Spain; Department of Chemical Engineering and Environmental Technology, University of Valladolid, Valladolid, Spain; Associate Laboratory i4HB—Institute for Health and Bioeconomy, NOVA School of Science and Technology, NOVA University Lisbon, Caparica, Portugal; UCIBIO—Applied Molecular Biosciences Unit, Department of Chemistry, NOVA School of Science and Technology, NOVA University Lisbon, Caparica, Portugal

Kyler Roy Department of Bioengineering, Imperial College London, London, United Kingdom

Dorota Alicja Rzechonek Department of Life Sciences (LIFE), University of Technology, Göteborg, Sweden

Almudena Saavedra-Bouza CICA – Interdisciplinary Center for Chemistry and Biology, Departament of Biology, Faculty of Sciences, Biomedical Research Institute A Coruña (INIBIC), University of A Coruña, A Coruña, Spain

Vittorio Senatore Department of Biotechnology and Biosciences, University of Milano-Bicocca, Milan, Italy

Rebecca Serna-García Department of Chemical Engineering, University of Valencia, Valencia, Spain

Immacolata Serra Department of Biotechnology and Biosciences, University of Milano-Bicocca, Milan, Italy

Dragana Stamenov Department of Field and Vegetable Crops, Faculty of Agriculture, University of Novi Sad, Novi Sad, Serbia

Mateusz Szczepańczyk Laboratory for Biosustainability, Institute of Environmental Biology, Wrocław University of Environmental and Life Sciences, Wroclaw, Poland

Rudolphus Antonius Timmers Area of Circular Economy, CARTIF Technology Centre, Boecillo, Valladolid, Spain; Circular Economy Area, CARTIF Technology Centre, Valladolid, Spain

Elia Tomás-Pejó Biotechnological Processes Unit, IMDEA Energy, Móstoles (Madrid), Spain

Mustafa Türker Pak Gıda Üretim ve Pazarlama A.Ş., Kocaeli, Turkey

María C. Veiga Chemical Engineering Laboratory, Faculty of Sciences and Interdisciplinary Centre of Chemistry and Biology, Centro Interdisciplinar de Química y Biología (CICA), BIOENGIN Group, University of La Coruña, A Coruña, Spain

Aysun Yurdunuseven Yildiz Department of Food Engineering, Pamukkale University, Denizli, Turkey

Arzu Yıldırım Bioengineering Department, Ege University, Izmir, Turkey

Acknowledgments

First and foremost, we would like to express our sincere gratitude to the European Commission through the COST (European Cooperation in Science and Technology) program for providing all the authors with the opportunity to collaborate and network during the preparation of this book. COST is a funding organization that facilitates the creation of research networks known as COST Actions. These networks offer an open platform for collaboration among scientists from across Europe (and beyond), driving research advancements and fostering innovation.

This publication is based on the work carried out under COST Action CA18220: Non-conventional Yeasts for the Production of Bioproducts (YEAST4BIO, (https://yeast4bio.eu/), and serves as a second edition of the previous book Microalgae-Based Biofuels and Bioproducts, which was the result of collaborative efforts within COST Action ES1408: European Network for Algal-Bioproducts (EUALGAE).

YEAST4BIO brings together an innovative group of researchers, combining expertise and experience to explore how nonconventional yeasts can be successfully integrated into the biotechnology industry. Moreover, YEAST4BIO unites leading scientists in the field, making it a key pillar globally. With the support of the Action, European scientists are shaping the future standards for research on nonconventional yeasts.

It was through the work of YEAST4BIO that the idea for this book emerged, and all the authors embraced the project with enthusiasm. Their dedication and hard work have culminated in this final product. We extend our warm thanks to all the contributing authors for their talent and effort in creating high-quality chapters and ensuring the reliability of the information provided. This publication also benefited from the significant contributions of renowned experts who were not directly involved in the project.

Finally, we would like to express our gratitude to the Editorial Team at Elsevier for their invaluable assistance and guidance during the preparation and editing of this book.

Cristina González-Fernández
Elia Tomás-Pejó

Section 1

Microalgae and yeast-based bioprocesses

Mitochondrial function in eukaryotic microbes for sustainable chemical and fuel production

1

Gavin Peter McStay[1], *and Samina Naseeb*[2]
[1]School of Pharmacy and Biomolecular Sciences, Liverpool John Moores University, Liverpool, United Kingdom, [2]School of Biological Sciences, The University of Manchester, Manchester, United Kingdom

Mitochondria are considered the powerhouse of the cell. These powerhouse functions relate to the ability of mitochondria to produce energy in the form of adenosine triphosphate (ATP) as well as their role in critical metabolic pathways. Mitochondria also function in the pathway of programed cell death and responses to stress. Therefore, it can be seen that mitochondria play a pivotal role in the fitness of a cell that could be used to generate useful bioproducts. Eukaryotes used in bioproduction all contain mitochondria which provide essential functions to these cells. However, the dependence on mitochondria varies in different species which is related to favored metabolic substrates and environmental conditions.

1.1 Fundamental biology of mitochondria

1.1.1 Mitochondrial structure

Mitochondria exist as a network within cells that are constantly undergoing fusion and fragmentation depending on cellular physiology. The mitochondrial ultrastructure is made up of two membranes—an outer membrane and an inner membrane. Within these membranes are two compartments known as the intermembrane space (IMS) and the matrix (Fig. 1.1).

These membrane barriers and submitochondrial compartments are crucial for proper mitochondrial function. The outer mitochondrial membrane (OMM) is in direct contact with the cellular cytoplasm and allows for the passage of small molecules from the cytoplasm to the IMS. Proteins residing in the OMM allow the passage of small molecules up to a molecular mass of approximately 8000 kDa (Zalman et al., 1980). The proteins responsible for this passage of molecules are generally known as Porins. The IMS is somewhat continuous with the cytoplasm in the sense that small molecules are able to pass freely through the OMM; however, larger molecules such as proteins and nucleic acids are unable to pass through the OMM and rely on the action of specific transport proteins. The specific proteins

Eukaryotic Microorganisms as Sources of Bioproducts. DOI: https://doi.org/10.1016/B978-0-443-30188-9.00001-9

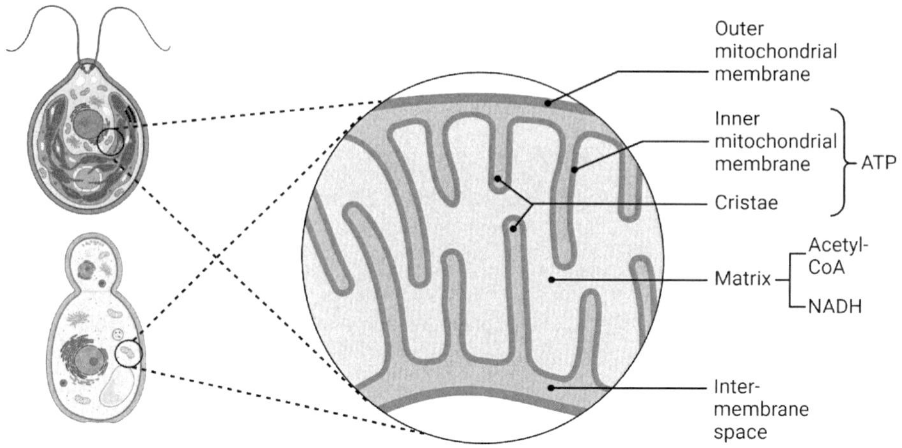

Figure 1.1 Ultrastructure of mitochondria in unicellular eukaryotes. Mitochondria are double-membrane organelles. The outer mitochondrial membrane surrounds the organelle and is in contact with the cytoplasm. The inner mitochondrial membrane is encased within the outer mitochondrial membrane. The inner mitochondrial membrane has a much larger surface area compared to the outer mitochondrial membrane and is highly involuted forming invaginations, known as cristae. The intermembrane space lies between the outer mitochondrial membrane and the inner mitochondrial membrane. The matrix lies within the inner mitochondrial membrane. The location of production sites of important substrates and cofactors in the mitochondria is indicated for Acetyl-CoA, ATP, and NADH.
Source: Created in BioRender.

localized to the IMS therefore provide specific functions to the IMS, which are also vital for mitochondrial function. The inner mitochondrial membrane (IMM) is a highly specific and generally impermeable membrane. The IMM has a large number of specific transport proteins that allow for the passage of specific molecules into and out of the mitochondrial matrix. This means that the IMM is the main barrier for the passage of important substrates into and out of mitochondria for engineered metabolic processes. The IMM is much larger in surface area compared to the OMM and to fit within these constraints the IMM is highly folded resulting in invaginations known as cristae. These invaginations are densely packed with oxidative phosphorylation (OXPHOS) complexes and are specialized sites for this purpose. The mitochondrial matrix is also a very distinct compartment which allows for specific metabolic pathways to occur in this space. The matrix is the location of almost all of the metabolic pathways of interest that have been engineered to increase production of important molecules due to the presence of useful substrates and cofactors, such as acetyl-CoA, ATP, and NADH. The matrix also houses mitochondrial DNA (mtDNA), this is most typically a circular DNA molecule that exists in multiple copies, but in fungi, mtDNA can be linear in organization and range in size from 10 to 350 kb (Tang et al., 2024). The fungal mtDNA encodes 14 core genes involved in oxidative phosphorylation (OXPHOS), as well as RNA genes required for translation (Tang et al., 2024). These genes are responsible for the expression of proteins required for ATP production in mitochondria. The presence of DNA within mitochondria indicates the endosymbiotic origin of mitochondria, but this also leads to difficulties with the manipulation and engineering of mtDNA for biotechnological purposes.

1.1.2 Mitochondrial function

The most well-known function of mitochondria is the production of ATP. This is vital for cellular activity, but crucial for cells with engineered metabolic pathways to maintain production and viability. ATP production is performed by the action of OXPHOS, which involves the electron transport chain (ETC) and ATP synthase. The ETC makes use of electrons released from metabolic pathways such as glycolysis, tricarboxylic acid cycle, fatty acid oxidation, and other pathways. The electrons are used in the ETC to promote proton pumping from the matrix into the IMS. This buildup of protons across the IMM results in an electrochemical gradient that can then be used to produce ATP, but also other important mitochondrial processes such as protein translocation, metabolite transport, and mitochondrial dynamics and turnover. The membrane potential results in a more alkaline pH within the matrix of 7.5 compared to pH 7.2 in the cytosol, which alters depending on nutrient availability and type (Orij et al., 2009). Within the IMM, ATP synthase will allow the passage of protons from the IMS to the matrix while harnessing the energy released from the electrochemical gradient to produce ATP. The correct function of mitochondria is required for many engineered pathways to function successfully to ensure suitable energy availability, localization of targeted mitochondrial proteins, and proper control of oxidation and reduction status within the matrix (Fig. 1.2).

To fuel the ETC, electron donors are generated from various metabolic pathways. The metabolic pathways within mitochondria are therefore important for the production of ATP. Within the mitochondrial matrix, metabolic pathways make use of electron carriers such as nicotinamide adenine dinucleotide (NAD) and flavin adenine dinucleotide (FAD) to capture electrons and convert them into energy in the form of ATP. In fungi, the main source of electron carriers is the tricarboxylic acid (TCA) cycle, while β-oxidation of fatty acids is a less commonly found metabolic pathway in the mitochondria of fungi. Fatty acid β-oxidation occurs mostly in peroxisomes in fungi. The other metabolic pathways include the synthesis and breakdown of amino acids, nucleotides, lipids, and other small molecules. It is here within the mitochondrial matrix where these metabolic pathways are located and the importance of mitochondria in the generation of bioproducts should be appreciated. Metabolic pathways within the mitochondria are interconnected with and supported by other cellular pathways, forming an extensive metabolic network that functions within and beyond the mitochondria.

An important integration point of these metabolic pathways is in aroma generation in alcoholic beverages. Mitochondria are pivotal in the metabolism of alcohols and amino acids, which are the precursors to aromatic molecules such as fatty acids, esters, acids, and alcohols. Enzymes within mitochondria, including

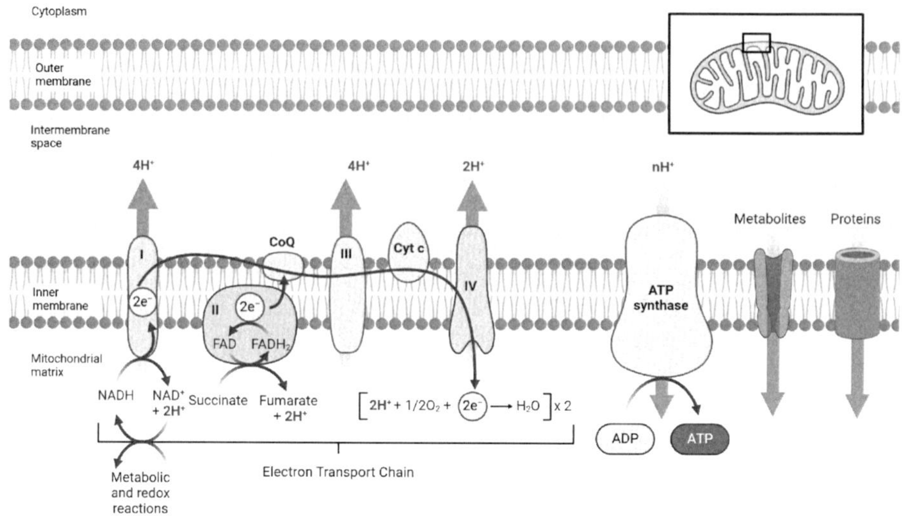

Figure 1.2 The process of oxidative phosphorylation and its role in membrane transport and redox regulation in mitochondria. Oxidative phosphorylation is made up of four complexes in the electron transport chain and ATP synthase. Electrons are passed from carriers to Complex I and Complex II then to Co-enzyme Q to Complex III. Electrons are passed to Cytochrome c and then to Complex IV before finally being accepted by molecular oxygen. Protons are pumped from the matrix to the intermembrane space through some of the electron transport complexes which is species dependent. The protons then pass through ATP synthase back to the matrix with the concommitant production of ATP. The membrane potential is required for the translocation of polypeptides and metabolites into mitochondria. NADH transfers electrons into the electron transport chain generating NAD^+ and maintaining redox regulation of the mitochondrial matrix.
Source: Created in BioRender.

alcohol dehydrogenases, branched-chain amino acid transferases, isoprene synthetic enzymes, acetyl/acyl-CoA transferases, synthases, decarboxylases, electron transport chain components, and NADPH dehydrogenases are involed in biosynthetic pathways resulting in aromatic products. Extensive deletion analysis in *S. cerevisiae* has identified genes responsible for generation of these aromatic molecules (Styger et al., 2013). These include mitochondrial enzymes alcohol dehydrogenase (ADH3), branched amino acid amino transferase (BAT1), ethanol acetly transferase (EAT1), α-ketoisovalerate synthesis pathway enzymes (ILV2, 3, 5), and α-isopropylmalate synthesis pathway enzymes (LEU4, 9) (Holt et al., 2019).

1.1.3 Mitochondrial life cycle

Mitochondria can only be generated from preexisting mitochondria—no cellular process for de novo generation has been identified. A mitochondrion will undergo growth through incorporation of lipids to increase membrane mass, increased protein import of nuclear-encoded genes, increased protein synthesis of mitochondrial encoded genes, and replication of mtDNA. Once a certain biomass is achieved, the mitochondrion will undergo fission to form two mitochondria. A reverse process exists where two separate mitochondria can fuse together to form a larger mitochondrion (Fig. 1.3).

These processes are dynamic and highly regulated depending on cellular states. This process of mitochondrial dynamics is crucial for both biogenesis and also degradation of mitochondria. Once mitochondria become dysfunctional, either functionally or through damage to mitochondrial macromolecules, a signal to specifically remove these mitochondria, such as mitophagy, or mitochondrial subcompartments, such as mitochondrial-derived vesicles is activated. This pathway includes several types of degradative pathways, such as mitophagy, a mitochondria specific form of autophagy, a cellular self-degradation system (Fig. 1.4).

This pathway has been extensively characterized in *S. cerevisiae* and is conserved in other fungal species (Schuster & Okamoto, 2022). Recently, more compartment specific degradative pathways have been identified that include mitochondria derived vesicles of both the OMM and IMM, which can be targeted to the autophagic pathway or targeted for extracellular release (Hughes et al., 2016). These processes ensure mitochondrial function is maintained suitably for cellular requirements by removing damaged components along with the recycling of salvageable building blocks of these mitochondrial components. Mitochondrial degradation can also be activated by nutrient limitation to provide a recycled

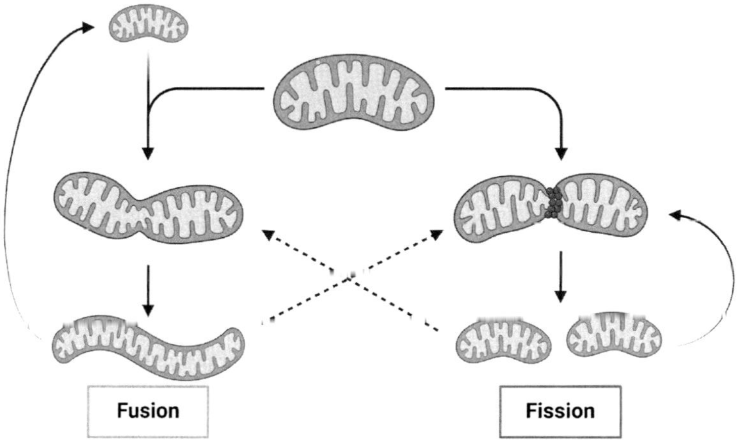

Figure 1.3 Mitochondrial fission and fusion. A mitochondrion (top center) can undergo either fission (right pathway) or fusion (left pathway). After fission, two mitochondria will be generated, after fusion, one larger mitochondrion will be generated. These mitochondria can then undergo further fission or fusion.
Source: Created using BioRender.

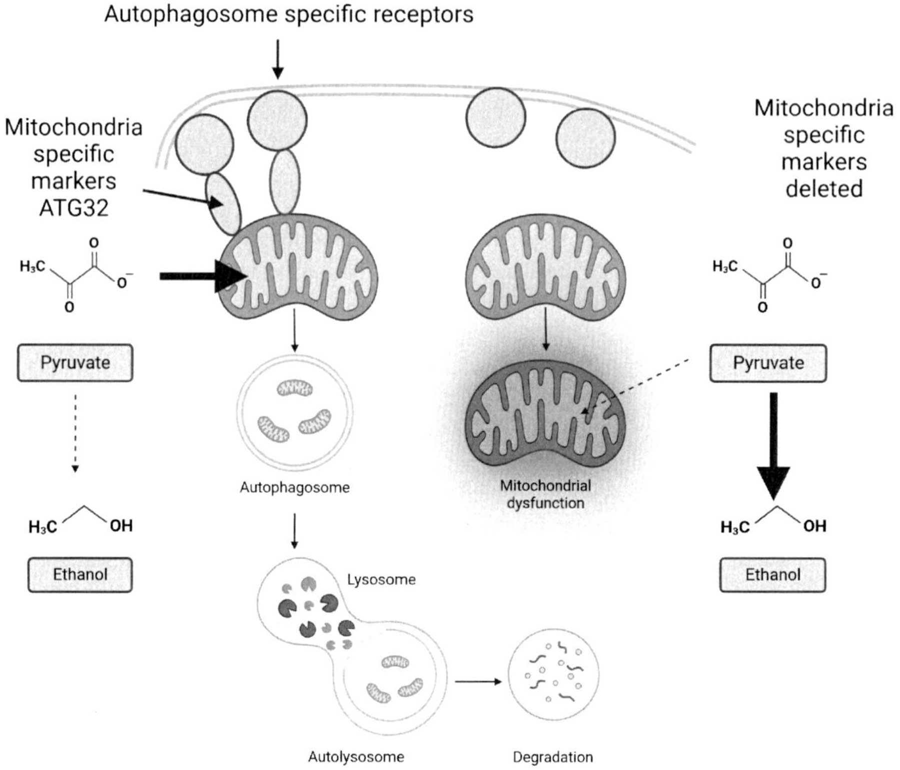

Figure 1.4 Role of mitophagy in bioproduction. Mitochondria interact with the developing autophagosome through specific molecular interactions, involving mitochondrial specific markers such as ATG32. The mitochondria become enveloped by the autophagosome before fusing with the lysosome. Degradative enzymes now breakdown macromolecules within the lysosome for reuse. In the case of ATG32 deficiency mitochondria are not turned over leading to increased mitochondrial dysfunction impacting on the metabolism of pyruvate in the mitochondria and instead is channeled into ethanol production and other cytosolic metabolic pathways.
Source: Created in BioRender.

source of building blocks that can be repurposed for cellular survival processes until favorable conditions return. If nutrient limiting conditions are prolonged, cells can no longer sustain vital functions and die. Therefore, it can be seen if limitations in growth conditions are experienced, then a decrease in bioproduction would result — therefore maintenance of mitochondrial function through sufficient nutrient supply is essential to achieve maximal yields. Several studies have investigated the role of mitophagy in bioproduction. In Sake strains of *S. cerevisiae,* loss of the mitophagy specific receptor ATG32 caused an increase in ethanol production. Mitophagy in many yeast species is activated under nutrient limiting conditions, such as those found in brewing, allowing for carbon captured in mitochondrial structures to be utilized for survival. By interrupting mitophagy under these conditions, carbon is not released from degraded mitochondria and the carbon is instead channeled for fermentation resulting in increased ethanol production (Shiroma et al., 2014). However, this increase in ethanol production was not seen in other situations, where sugarcane was used as a carbon source, in three different yeast strains (Eliodório et al., 2022). Deletion of ATG32 in engineered *S. cerevisiae* has also resulted in increased production of other useful molecules, such as 2, 3-butanediol, on the order of 20-fold compared to wild-type (Morita et al., 2019). These approaches disrupt pyruvate metabolism entry into mitochondria, through mitophagy inhibition, leading to increased dysfunction of mitochondria, and pyruvate can be channeled into cytosolic metabolic pathways in engineered strains.

1.1.4 Mitochondrial segregation during cell division

To distribute mitochondria in a dividing yeast cell, the cytoskeleton and membranes within the cell are used as tracks and anchors, respectively. The successful segregation of mitochondria is important as defects in this process results in the presence of dysfunctional mitochondria in the daughter cells and potential generation of cells lacking the ability to respire on respiratory substrates. Dysfunction can occur due to inappropriate segregation of mtDNA between dividing mitochondria. This results in a daughter cell with compromised function and risk to the culture. Inappropriate segregation results from genetic or physiological deficiencies of mitochondrial dynamics or segregation. The generation of respiration deficient daughter cells can result from missegregation of mitochondria during cell division in certain strains of yeast. Mitochondrial deficiency can result in a petite phenotype, as colonies are smaller than respiration competent clones. This has been extensively studied in *S. cerevisiae* (Leite et al., 2023). The smaller colony size represents a limitation in the ability to grow on glucose as a carbon source, as ethanol produced by the yeast cannot be further metabolized. In respiration competent cells, glucose is metabolized to ethanol through fermentation, and the ethanol can be metabolized further using oxidative phosphorylation in the mitochondria. More recent analysis has determined that amino acid metabolism is also impaired resulting in deficiencies in glutamate, glutamine, leucine, and arginine (Vowinckel et al., 2021) some of which are important for bioproduction and aroma generation in alcoholic beverages.

The yeast life cycle follows a general growth phase followed by a division stage resulting in a mother and daughter cell. This growth and division phase can happen rapidly making yeast an ideal cell-based system for bioproduction. Yeast growth follows typical eukaryotic cell cycle stages to grow, duplicate DNA and cell division. These cell cycle events can result in division processes resulting in cellular budding or binary fission, and different yeast species will use one or the other. Mitochondria will also grow, segregate, and be distributed between the mother and daughter cells. Evidence suggests that healthier mitochondria will be segregated into the daughter (thus younger) cell, while less healthy mitochondria will remain with the mother (thus older) cell (McFaline-Figueroa et al., 2011).

1.1.5 Yeast mating and mitochondrial DNA

Fungi can exist either as haploid or diploid states. Haploid cells harbor single copies of the nuclear genes, while diploid cells contain two alleles of each gene. In addition to the nuclear genome, mitochondria harbor their own DNA which exists most frequently in a circular form and in multiple copies. Mitochondria in yeast tend to exist as a homoplasmic population, meaning all copies of mtDNA are the same. This occurs through preferential propagation of one mitochondrial haplotype, known as vegetative segregation, through the DNA recombinase enzyme Mhr1p (Ling & Shibata, 2004). This happens most frequently when two haploid cells conjugate resulting in a cell where the two mitochondrial haplotypes now exist. Selection of the mitochondrial haplotype occurs based on the growth advantage provided by one haplotype over another. This selection can be directed to the alternative haplotype under different growth conditions. This selection indicates that mitochondrial haplotypes favor different growth conditions that could be used to provide improved growth properties for the yeast in question (Hewitt et al., 2020).

1.1.6 Interorganelle communication

As mitochondria are the main site of ATP production in a cell their function can be seen to have an indirect contribution to the function of other energy consuming compartments in a cell. In addition to this, recent work has identified direct and specific interactions between mitochondria and other cellular compartments such as the endoplasmic reticulum, lysosome/vacuole, nucleus, peroxisome, and plasma membrane (Kakimoto et al., 2018). These are mediated by large protein complexes that span between the two organelles, including Endoplasmic Reticulum-Mitochondria Encounter Structure (ERMES) and vaCuoLe And Mitochondria Patch (vCLAMP), Mitochondria-Endoplasmic reticulum Cortex Anchor (MECA) along with other less well characterized structures (Eisenberg-Bord et al., 2021; Elbaz-Alon et al., 2014; Lackner et al., 2013). These interactions are required for channeling of metabolites for efficient metabolic pathways, calcium signaling, mitochondrial dynamics, and mitochondrial segregation during cell division. These studies have revealed that mitochondria do not work in isolation within a cell and communicate with other cellular components that are required for the quality control, metabolism, and maintenance — all crucial aspects for mitochondrial processes involved in bioproduction.

1.2 Fungi and yeast used for bioproduction

The fungal kingdom has been used for centuries by humans for the generation of numerous molecules including ethanol and carbon dioxide that can be used to make food, drink, and other useful products. With this long-term use came a desire to understand fungi which led to the use of several species within this kingdom as laboratory models for eukaryotic biology. One of the first to be used and then fully characterized is the yeast species *S. cerevisiae*. The use of *Saccharomyces* species

Table 1.1 Characterization status of different types of yeast used for bioproduction.

Species	Genome	Transcriptome	Proteome	Metabolome	Database	Bioproducts
Saccharomyces cerevisiae	Goffeau et al. (1996)	David et al. (2006)	Ghaemmaghami et al. (2003)	Ramirez-Gaona et al. (2017)	Saccharomyces genome database yeastgenome.org	Ethanol Organic acids Enzymes Pharmaceuticals Carotenoids
Pichia pastoris	De Schutter et al. (2009); Sturmberger et al. (2016)	Love et al. (2016)	Hou et al. (2022)	Chung et al. (2010)		Protein
Kluyveromyces lactis	Chuzel et al. (2017)	Suleau et al. (2006)	Secretome (Swaim et al., 2008)	Dias et al. (2014)		Enzymes Bioethanol Organic acids
Kluyveromyces marxianus	Inokuma et al. (2016); Jeong et al. (2012); Suzuki et al. (2014)	Lertwattanasakul et al. (2015)	Li, Fu, et al. (2019)	Zhang et al. (2022)		Bioethanol
Aspergillus niger	Baker (2006)	Delmas et al. (2012)	Takagi et al. (2020)	Sun et al. (2007)		Citric acid Enzymes
Scheffersomyces stipitis (*Pichia* stipitis)	Jeffries et al. (2007)	Zhu et al. (2019)	Hao et al. (2013)	Caspeta et al. (2012)		Ethanol Recombinant proteins and enzymes Pharmaceuticals
Candida utilis	Tomita et al. (2012)	He et al. (2023)	Buerth et al. (2011)	Kieliszek et al. (2020)		Single-cell protein Yeast extract
Candida tropicalis	O'Brien et al. (2021)					Biodiesel
Penicillium chrysogenum	Van Den Berg et al. (2008)	Van Den Berg et al. (2008)	Douma et al. (2011)	Douma et al. (2011)		Penicillin
Yarrowia lipolytica	Pomraning & Baker (2015) Liu & Alper (2014)	Sabra et al. (2017)	da Silva et al. (2020)	Sabra et al. (2017)		Biofuesls Citric acid Lipids Proteins

dates back centuries to the production of leavened breads and fermented beverages. Within the last 100 years industrial use of *Saccharomyces* species and adoption of *S. cerevisiae* as a laboratory model have meant that these are very well characterized at the molecular level, revealing the genome, transcriptome, proteome, and metabolome. This broad understanding of *S. cerevisiae* biology has led to the ability to manipulate yeast function with the goal of generating a desired bioproduct. As the use of yeast for bioproduction increases, more species of yeast are being used for these efforts, and with this comes the desire to understand the molecular make-up of these particular species (Table 1.1). Mitochondrial metabolism is required for the generation of many types of molecules that can be grouped into alcohols, organic acids, fatty acids, flavorings, and isoprene derived products. Other types of molecules and products also require mitochondrial function, such as recombinant proteins, yeast extract, and single cell protein. Yeast provides an alternative to prokaryotic microbial cell factories due to compartmentalization of metabolism while minimizing threats to cultures from bacteriophages infection and reducing costs that are associated with gene induction systems that require complex media and carefully controlled growth conditions.

1.3 Mitochondrial functions involved in bioproduction

The abundance of fungal species existing on our planet is enormous and yeast are found naturally in almost every explored environmental niche. This means the diversity of yeast is incredible resulting in species specific for a wide variety of growth conditions—including carbon source, temperature, environment, and others. This means that there is an almost indefinite amount of fungal species that could be used and adapted for bioproduction. Fungi can naturally use simple carbohydrates, polysaccharides, and lipids as carbon sources. These fungi can be found in both cool and warm climates on earth, demonstrating that metabolism and cellular vitality can be optimally achieved in these environments. The natural variation in growth environments and fuel requirements are two characteristics that the bioproduction industry desires to optimize to produce useful molecules of increased purity, cheaper cultivation costs, and increased yield. These optimizations can be achieved through selection, by laboratory-based evolution approaches, such as evolutionary engineering/adaptive laboratory evolution, but also through directed genetic manipulation. Laboratory-based evolution approaches can be achieved in straightforward and cost-effective manner because of the fast and definable growth profiles of fungi using selection pressure. This results in optimized laboratory strains that can then be characterized by whole genome sequencing to identify genetic changes providing the optimized growth profile. The evolved strain can then be used for the bioproduction process or identified genetic changes can be introduced into a wild-type fungus of known growth characteristics to use the genetic advantage. Earlier attempts to evolve laboratory strains made use of random mutagenesis, using chemical or physical mutagens, that could be screened for the desired phenotype. Both of these approaches have advantages and disadvantages—laboratory-based evolution will generate strains that are specific to the desired phenotype but have a lower chance of success as genetic changes rely on the inherent error rates of DNA polymerases. Mutagenesis approaches nonspecifically generate genome-wide variation, that is likely to impact on a wide variety of cellular processes, that may include changes relevant to the desired phenotype. Strains derived using mutagens therefore need to be carefully characterized to ensure no detrimental phenotypes impacting on growth, culture condition tolerance, stability of phenotype or stress resistance are also selected. However, both of these approaches can be used to inform more directed genetic modification. Here, the identified mutants can be introduced in isolation into the natural host species or another useful strain. These directed genetic approaches can also be informed through computational analysis, such as in silico modeling of proteins to determine more productive or temperature tolerant enzymes (Mans et al., 2018).

1.3.1 Alternative mitochondrial functions

Due to the diversity of the fungal kingdom, essential mitochondrial functions are not all accomplished through orthologous genes. The overall reactions of the mitochondria can be achieved by alternative proteins, enzymes, and electron carriers. This again ensures a large molecular foundation that can be exploited for directed evolution or genetic manipulation for improved bioproduction. In the electron transport chain, Complex I is conventionally thought of as a large multisubunit complex comprised of subunits derived from both the nuclear and mitochondrial genome. This complex is able to accept electrons from NADH derived from the TCA cycle, lipid β-oxidation, and other metabolic pathways. The electron passage through Complex I results in energy transfer, causing conformational alterations in proton shuttling domains, allowing the movement of protons from the matrix to the IMS against the concentration gradient. In *S. cerevisiae,* the large multisubunit and dual origin Complex I does not exist and is instead replaced by two single subunit proteins, Ndi1p and Nde1/2p, which exist on the matrix face and IMS face of the IMM, respectively. These three proteins are functionally redundant and yeast lacking each gene can grow on nonfermentable carbon sources (Saladi et al., 2020). Ndi1p is a dimeric protein located on the matrix face of the IMM and accepts electrons from NADH derived from matrix and matrix facing reactions, such as the TCA cycle and fatty acid β oxidation. Ndi1p is associated with one FAD cofactor to pass electrons from NADH to ubiquinone (Iwata et al., 2012). Nde1/2p is tethered to the IMM and the NADH dehydrogenase domain exists in the IMS (Saladi et al., 2020). Nde1/2p accept electrons from NADH derived from cytosolic pathways, such as glycolysis and fermentation. These NADH dehydrogenase enzymes do not contain the full complement of cofactors, lacking iron-sulfur clusters, but contain flavin based co-factors. The extremely simplified system for NADH channeling into the ETC has been exploited in conditions where Complex I activity is compromised and Ndi1p has been proposed to be a genetic therapy for diseases where Complex I activity may be deficient, such as inherited genetic disorders and Parkinson's Disease. Introduction of Ndi1p restores mitochondrial function to cell and animal models of chemical-induced

Parkinson's Disease (Li et al., 2022) and in neuronal Complex I deficiencies (Granat et al., 2023). *Aspergillus niger* has an alternative NADH dehydrogenase and chemical inhibition results in increased growth rate and biomass (Voulgaris et al., 2012). *Yarrowia lipolytica* has both a multisubunit Complex I, with 42 subunits with flavin and iron-sulfur co-factors (Grba & Hirst, 2020), and an alternative external NADH Dehydrogenase (NDH2). Some fungal species have more than one complex for each step of the ETC. Alternative oxidases exist which provide an alternative terminal electron acceptor mechanism apart from molecular oxygen used by Cytochrome c oxidase (Complex IV). The alternative oxidases bypass the need for Complex III and Complex IV entirely. *Y. lipolytica* is one of the species with an alternative oxidase (AOX), which is not able to pump protons across the IMM, providing an alternative terminal electron acceptor without the need to contribute to mitochondrial membrane potential and can contribute to heat production, energy dissipation or reactive oxygen species generation prevention. Metabolic flux simulations predicted citrate production would be enhanced by AOX inhibition. Addition of a chemical inhibitor of AOX to *Y. lipolytica* cultures doubled the production of citrate, demonstrating that manipulation of mitochondrial function can enhance desired product generation (da Veiga Moreira et al., 2021). These alternatives to ETC and OXPHOS may benefit cells during the stationary stage of growth, when the reducing power in the cell is high (Guerrero-Castillo et al., 2009; Guerrero-Castillo et al., 2012). Interestingly, there is a correlation in yeast species that have Complex I also containing alternative oxidase, while those lacking Complex I lack alternative oxidase (Veiga et al., 2003).

1.3.2 Mitochondrial functions as protective responses

Mitochondrial function is primarily for the generation of ATP to support energy demanding processes. In addition to this, mitochondrial functions can be altered to enable a protective response against inhibitors or toxins, by increasing ATP levels to support increased energy demands, or to generate reductive power, in the form of NADH. Furfural and acetic acid are products of lignocellulosic material acid hydrolysis and are potent inhibitors of yeast growth. Attempts to generate yeast tolerant to furfural have been successful in *S. cerevisiae* and *P. stipitis,* and it has been shown through transcriptomics and proteomics that TCA cycle genes are upregulated, along with other pathways. This protective response caused by upregulated TCA cycle genes is proposed to maintain production of NADH to enable the reduction of furfural and render nontoxic (Hao et al., 2013; Lin et al., 2009). Further investigations into the mechanisms of tolerance using transcriptomics in *S. cerevisiae* demonstrated that overexpression of HAP4, a gene involved in activating the respiratory gene response, also provided tolerance to furfural, acetic acid, and a combination of both, presumably by increasing expression of TCA cycle genes (Li et al., 2021).

1.4 Retargeting functions and pathways to mitochondria

Approximately, 1000 proteins exist in mitochondria based on large-scale proteomic studies in *Saccharomyces cerevisiae* (Reinders et al., 2006). These proteins are encoded by both nuclear genes and mitochondrial genes. Nuclear genes bound for the mitochondria have amino acid signal sequences that target the protein to the organelle and then to the subcompartment, known as the mitochondrial localization sequence (MLS). These signal sequences can be generalized into those in proteins destined for membrane compartments of those destined for the matrix and IMS compartments. Soluble proteins in these compartments often have an amino terminal sequence that is removed by mitochondrial proteases to reveal the mature functional protein. Membrane bound proteins often have internal signal sequences that are inherent to the overall structure of the protein and are not removed. If a protein is to be engineered to relocate to the mitochondria, one of these signal sequences needs to be introduced into the coding region of the gene. These approaches work extremely well for the soluble proteins using all types of signal sequences. One of the most frequently used sequences is that of the Cytochrome c oxidase subunit, COX4 (Hurt et al., 1985). A broader analysis of MLS was recently performed to determine alternatives to the COX4 sequence. Many of the tested sequences were competent for exogenous protein import into mitochondria and were able to support relocalisation of the α-santalene biosynthetic pathway into mitochondria and increased production by 3.7-fold (Dong et al., 2021). These sequences could be used to expand the toolbox for mitochondrial retargeting, which could help with more efficient expression. The use of different MLSs will result in the use of different import pathways and processing enzymes in relocalised multigene pathways that will minimize competition and saturation. However, the retargeting of exogenously targetted membrane localized proteins to mitochondria has not been reported.

Once a protein has been relocated to the mitochondria, it is important to determine functionality of that protein within the new environment. Within the mitochondrial matrix, the pH is more alkaline than the cytosol and so this could impact protein structure and function. Phenotypic screening in deficient mutants can often determine whether the relocated protein is functioning like wild-type or better. However, a more systematic biochemical analysis is required to understand if function is retained or improved. The overall goal of relocation to specific compartments is to enhance bioproduction, so this measurement alone may be sufficient to determine if relocation has been successful. In addition to the new mitochondrial environment not being ideal for the protein, other factors could impact on whether relocation will be successful or not. Many relocated proteins are enzymes that are part of a metabolic pathway, so the substrates and products need to be accessible to the enzyme. Substrate availability is governed by membrane transporters allowing access of the substrate or precursor to the mitochondrial compartment. This can be alleviated by the introduction of a specific substrate or precursor transporter being introduced by relocation. The product of the enzyme may also need to leave the mitochondria, which depends on the type of molecule being produced, so specific efflux transporters may also need to be introduced. Substrates are also often part of multiple metabolic pathways that compete for the substrate based on cellular physiology. This can be addressed by inactivation of the competing pathway through genetic ablation or mutation of key enzymes in the metabolic pathway. However, this can result in growth deficiencies if the pathway is important. Examples exist where whole metabolic pathways have been relocated to the mitochondria to ensure the substrates and products are available for all of the enzymes of this pathway.

Relocation can result in improper expression levels of proteins that may introduce stress to the cell. Relocation to mitochondria could cause an overload of the mitochondrial import systems, which results in a feedback mechanism to reconcile overload and decrease mitochondrial import and protein synthesis (Maruszczak et al., 2023). Relocation of proteins within the mitochondria could also cause internal mitochondrial stress due to inherent properties of the protein such as aggregation, altered half-life or improper interactions resulting in activation of processes such as the mitochondrial unfolded protein response (Wrobel et al., 2015) or retrograde signaling pathway (da Cunha et al., 2015) or even cell death (Wang & Chen, 2015).

The origin of the retargeted sequence also needs to be considered. It is possible for any sequence to be used for retargeting—this could mean the endogenous gene is retargeted by introduction of a mitochondrial targeting sequence or an exogenous gene from another species.

1.4.1 Examples of mitochondrial retargeting for enhanced bioproduction

Retargeting of enzymes to mitochondria has been used to make a variety of products, including fusel alcohols, isoprene, geraniol, and amorpha-4,11-diene. Fusel alcohol production was increased by targeting both α-ketoacid decarboxylase (α-KDC) and alcohol dehydrogenase (ADH) to mitochondria in *S. cerevisiae.* α-KDC was from *Lactobacillus lactis* or *S. cerevisiae* with gene expression driven by a TDH3 promoter. ADH was from *S. cerevisiae, Escherichia coli, or Lactobacillus lactis* and driven by the TEF1 promoter. Both of these proteins were targeted to the mitochondria using the COX4 mitochondrial targeting sequence. Highest isobutanol yields were seen with *Lactobacillus lactis* α-KDC and *S. cerevisiae* ARO10 with all three ADH enzymes acting equally (Avalos et al., 2013). More recent works resulted in the successful bioconversion of xylose to isobutanol in *S. cerevisiae* through introduction of the same isobutanol metabolic pathway in mitochondria. Xylose is a carbon source that does not repress mitochondrial biogenesis and therefore improved isobutanol yield. However, maximal production of isobutanol was delayed as the engineered yeast grew slower on xylose compared to glucose (Fig. 1.5) (Lane et al., 2020).

Organic acids are generated through mitochondrial metabolism and several valuable types of organic acids have had yield increased through manipulation of biosynthetic pathways. 3-hydroxypropionate (3HP) is a three carbon molecule that can be used as starting material for acrylic acid, which is used in plastic, pharmaceutical, and agricultural industries. 3HP production can be enhanced by heterologous expression of Malonoyl-CoA reductase (MCR) from *Chloroflexus aurantiacus* in mitochondria compared to overexpression in the cytosol. The enhanced production of 3HP was predicted using metabolic modeling. MCR requires NADPH to reduce Malonoyl-CoA to 3HP; however, Malonoyl-CoA and NADPH are limiting for this reaction to occur. Acetyl-CoA conversion to Malonoyl-CoA

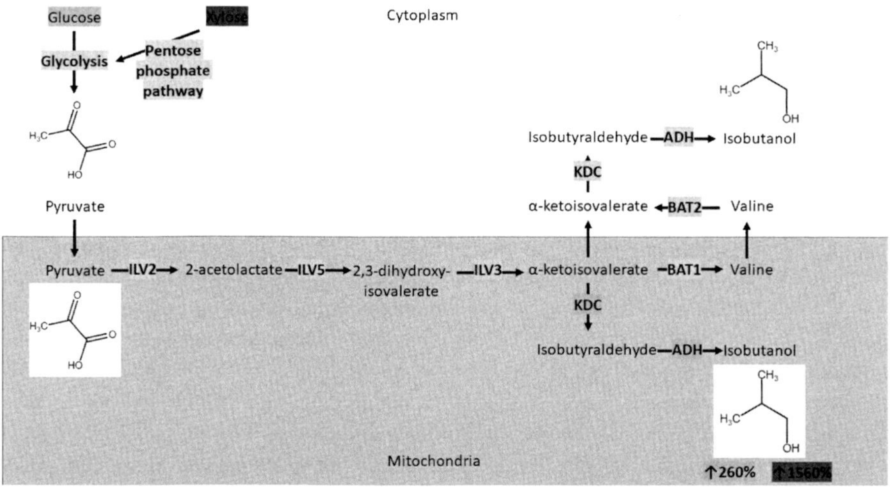

Figure 1.5 Engineering of mitochondrial metabolic pathways to increase isobutanol production. Isobutanol is endogenously produced through a metabolic pathway requiring steps in the cytoplasm and mitochondria (blue highlighted enzymes). By retargetting of ADH1 to the mitochondria conversion of isobutyraldehyde to isobutanol enhances production by 260% using glucose as a carbon source. If a nonrepressive sugar, such as xylose, is used, the production of isobutanol increases by 1560%.
Source: Created in BioRender.

conversion was directed by expression of modified Acetyl-CoA carboxylase to enhance Hfa1p activity, an endogenous mitochondrial-localized Acetyl-CoA carboxylase. Mitochondrial NADPH concentrations were elevated by overexpression of Pos5p, a NAD kinase, and Idp1p, which catalyzes the $NADP^+$-dependent oxidation of isocitrate to α-ketoglutarate, producing NADPH as a by-product (Zhang et al., 2023). Itaconic acid is a renewable nonpetroleum-based monomer that can be used in the production of many commercially valuable polymers, resins, coatings, adhesives, and plasticisers. Itaconic acid is generated through decarboxylation of *cis*-aconitate catalyzed by the cytosolic enzyme *cis*-aconitate decarboxylase. An engineered cytosolic pathway in *Neurospora crassa* converting malic acid to cis-aconitate enhanced itaconic acid production, and this was further enhanced by the over-expression of the transporter MTTA, a mitochondrial tricarboxylate transporter, from *Aspergillus terreus* (Zhao et al., 2024).

Isoprene is a product of the mevalonate pathway and is used for a range of industrial applications. The mevalonate pathway is a cytosolic process that channels products to the endoplasmic reticulum. By relocating the entire mevalonate pathway to mitochondria of *S. cerevisiae* isoprene production was increased to greater than 2 g per litre of culture. The seven enzymes of the mevalonate pathway were relocalised to the mitochondria using the mitochondria localization sequence of COX4. Optimal isoprene production was acheived using combinations of both cytosolic and mitochondria engineered steps along with optimized culture conditions (Lv et al., 2016). Cytoplasmic engineering combined with mitochondrial engineering enabled the pool of acetyl-CoA localized in both compartments to be exploited while minimizing the generation of side-products, such as squalene. This production of isoprene was further enhanced by optimization of enzymes in the pathway alongside engineering of the pathway in the mitochondria (Fig. 1.6) (Yao et al., 2018).

The mevalonate pathway has also been engineered for the production of plant terpenoids, which lead to the generation of useful products, such as the antimalarial—arteminisin. The precursor molecule farnesyl diphosphate is channeled into many pathways for the production of important cellular biomolecules. Through metabolic pathway engineering, including overexpression, deletion, heterologous expression, and mitochondrial retargeting, farnesyl disphosphate was directed to form terpenoids, such as Valencene—an aroma constituent, and Amorphadiene—an Arteminisin precursor, from competing pathways (Farhi et al., 2011). Modification of the mevalonate pathway has also been used to increase bioproduction of geraniol, a plant-based molecule, that is a precursor to strictosidine, the universal precursor to monoterpene indole alkaloids, which have many medicinal properties, such as the antimalarial quinine and anticancer agent vinblastine (Yee et al., 2019). However, not all retargeting approaches result in success. Carotenoids are used as preservatives and as supplements due to their color and antioxidant activity. These are endogenously found in plants but cannot be easily scaled for mass production. Other species can produce carotenoids, such as carotenogenic fungi, but the use of these for human consumption is questioned. *S. cerevisiae* is suitable

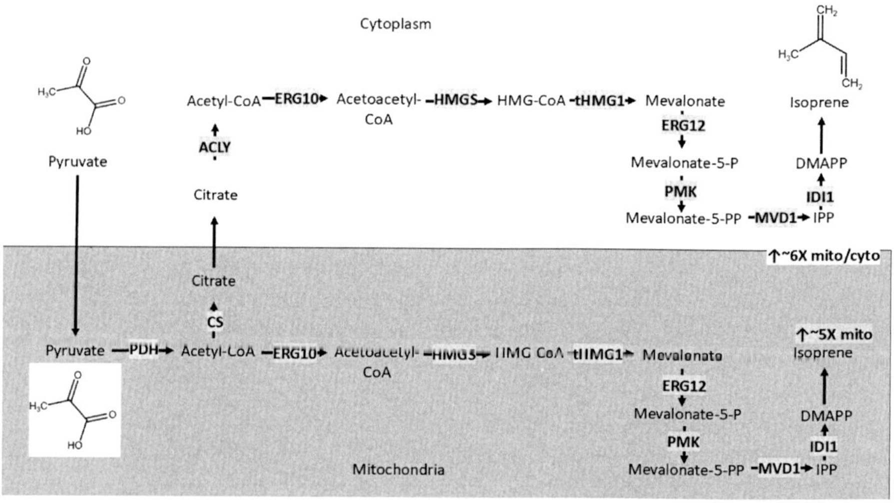

Figure 1.6 Modified pathways for enhanced isoprene production. The isoprene biosynthetic pathway is outlined with endogenous localization of relevant enzymes (blue highlight). Modified pathways after relocalisation in mitochondria (green highlight) improve isoprene yield by fivefold compared to cytoplasmic localization. A combined strategy of both cytoplasmic and mitochondrial localization increases yield approximately sixfold.
Source: Created in BioRender.

and safe for human consumption and the pathway to generate β-carotene from the precursor phytoene has been introduced. The enzymes naturally exist in the endoplasmic reticulum and relocalisation of the enzyme responsible for phytoene production, CrtYB, into the mitochondria increased phytoene and β-carotene production. However, the relocalisation of the next enzyme in phytoene metabolism to β-carotene, CrtI, caused a decrease in β-carotene production. This was due to inhibition of CrtI in the mitochondrial environment (Arhar et al., 2024).

1.5 Mitochondrial responses in bioproducing yeast strains

For cells to produce optimal amounts of the bioproduct, different media or different adapted strains are most commonly used. This enables favored production to ensure cost-effective purification and downstream use. All species of yeast have undergone adaptation to generate an optimal strain for bioproduction. This goes hand in hand with optimization of the media used for fermentation. *Y. lipolytica* has been identified as a promising microbial cell factory for citric acid production. An adapted strain was used to determine the response to different media used for biomass increase compared to citric acid production. Nitrogen limitation is required for citric acid production in *Y. lipolytica*; however, this results in the unwanted by-product isocitric acid. This is generated through the action of the enzyme aconitase, which may be a further enzyme of interest to favor citric acid production (da Silva et al., 2020).

Introduction or retargeting of proteins and metabolic pathways can lead to overload and depletion of metabolites that may cause cellular responses, through changes in gene expression or regulation of critical processes resulting in altered phenotypes. This has been seen in the case of *S. cerevisiae* engineered to metabolize xylose—a major component of hemicellulose. Xylose fermentation occurs in certain species of yeast, such as *Scheffersomyces* stipitis (*P.* stipitis), but requires the presence of oxygen for growth. The same is true for the xylose-fermenting engineered strains of *S. cerevisiae*. Metabolism of xylose is a respiratory process as mitochondrial gene expression is upregulated, this results in low rates of ethanol production. However, ethanol production can be increased by making *S. cerevisiae* respiration incompetent (Jin et al., 2004). This approach provides further options for the transformation of biomass to biofuels.

Accumulation of metabolites greater than in native conditions can result in toxicity. This has been seen when the mevalonate pathway was introduced into mitochondria to produce terpenes. Phosphorylated mevalonate intermediates are toxic when they accumulate in mitochondria by inhibiting oxidative phosphorylation (Zhu et al., 2021). This toxicity can be relieved by retargeting some of the metabolic steps outside of the mitochondria. Improved production of squalene was measured as success of the retargeting of the mevalonate pathway to the cytosol.

1.6 Yeast hybridization in evolution and biotechnology

Hybridization, a process recognized for its evolutionary significance since the early 20th century, has become increasingly appreciated in recent decades due to advances in sequencing technologies. These technological advancements have revealed that hybrids are prevalent across many species, contributing significantly to genetic diversity and adaptive evolution (Abbott et al., 2013). In both natural and domesticated settings, hybrids are common, with an estimated 25% of plant species and 10% of animal species engaging in natural hybridization (Mallet, 2007). In the microbial world, particularly among fungi, hybridization has played a crucial role in generating the diversity observed today (Dujon & Louis, 2017; Taylor & Larson, 2019).

1.6.1 *The* Saccharomyces *genus: a model for hybridization studies*

The yeast genus *Saccharomyces* is of particular interest due to its ability for hybridization, both in the laboratory and in natural environments. Despite significant genetic divergence (10%–20% nucleotide differences, comparable to that between humans and mice), species within this genus can produce viable hybrid offspring (Shen et al., 2018). This ability makes *Saccharomyces* an ideal model for studying hybridization's role in adaptive evolution (Botstein & Fink, 2011; Hittinger, 2013).

Recent discoveries have expanded our understanding of the evolutionary outcomes of hybridization within the *Saccharomyces* genus, with five new species identified in the last two decades. These findings underscore the dynamic nature of this field, driven by rapid and cost-effective genome sequencing.

1.6.2 *Mitochondrial inheritance in yeast hybrids*

In yeast, mtDNA is inherited biparentally but typically becomes homoplasmic, where only one parental mtDNA type is retained. This phenomenon is particularly evident in industrially relevant yeast hybrids, such as *Saccharomyces pastorianus*, a hybrid used in lager brewing, which retains mtDNA exclusively from the cryotolerant parent *S. eubayanus* (Dunn & Sherlock, 2008; Peris et al., 2014; Rainieri et al., 2008). The process of mtDNA retention in yeast hybrids is influenced by various genetic and environmental factors, rather than being a random occurrence.

During hybridization, environmental conditions, including temperature and the type of carbon source available, play a significant role in determining which parent's mtDNA is retained (Hsu & Chou, 2017). This retention is further shaped by cellular mechanisms governing mtDNA inheritance. Yeast cells initially inherit 30–100 copies of mtDNA from both parents, but the heteroplasmic state in progeny is transient due to mitotic segregation. New buds receive relatively few mtDNA copies, leading to rapid homoplasmy (Thrailkill & Birky, 1980). This transition may result from the random selection of an mtDNA template for replication and segregation, which becomes fixed through successive replication cycles. Furthermore, Albertin et al. (2013) and Solieri et al. (2008) demonstrated that in hybrids between *S. cerevisiae* and *S. uvarum*, mtDNA showed uniparental inheritance in all hybrids after a given number of generations, further highlighting the consistent shift toward homoplasmy in yeast hybrids.

Hybrids between *S. cerevisiae* and *S. paradoxus* demonstrated that mtDNA inheritance is uniparental and strongly influenced by environmental factors and the specific strains involved in the cross. Fitness differences between hybrids carrying different mtDNA genotypes were observed, and fitness competition assays revealed that the selective retention of mtDNA is often driven by the genotype's fitness advantage under specific stress conditions (Hsu & Chou, 2017). The inherited mtDNA confers adaptive advantages depending on the parental lineage and the environment, which is particularly crucial in determining hybrid vigor (Hewitt et al., 2020).

Selective pressures, such as the number of origin of replication (ori) sequences, can also influence mtDNA competition (Macalpine et al., 2001). For example, *S. cerevisiae* mtDNA, with eight ori sequences, outcompetes *S. bayanus* mtDNA, which has only four ori sequences (Marinoni et al., 1999; Piškur et al., 1998). However, in hybrids of *S. cerevisiae* and *S. paradoxus*, which have similar ori numbers (seven in *S. paradoxus*), biased mtDNA inheritance cannot be explained by ori number alone.

Interestingly, while all hybrids maintain respiratory competency, mitochondrial activity alone does not fully explain the fitness advantages observed under stress, indicating a complex interplay between genetic and environmental factors in shaping mtDNA inheritance and its impact on hybrid vigor.

1.6.3 Mitochondrial influence on yeast hybrid phenotypes

The genus *Saccharomyces* encompasses eight species of yeasts, known for their significance in scientific research. Hybrids of these *Saccharomyces* species serve as models for studying polyploidy, which aids in understanding how organisms adapt to new environments within the Eukaryotic domain. The retention of mtDNA following yeast hybridization is crucial for determining population fitness, hence the study of mitochondrial inheritance and its influence on hybrid phenotypes is very important.

Mitochondria play a critical role in various cellular processes, extending beyond their traditional role in OXPHOS. They are involved in nucleotide synthesis, calcium-ion homeostasis, amino acid metabolism, and apoptosis, among other functions (Spinelli & Haigis, 2018; Zardoya, 2020). *Saccharomyces* yeasts possess the ability to either respire, which depends on functional mitochondria, or to ferment, a process that does not require them. They can still survive in the absence of functional mitochondria, although they develop into smaller "petite" colonies. The mitochondrial genome plays a significant role on various phenotypic characteristics among *Saccharomyces* species.

1.6.4 Yeast thermotolerance and bioproduction

In yeast hybrids, mitochondrial function has been linked to thermotolerance, a trait of significant industrial importance. For instance, the mtDNA inherited from the parental strain was identified as a key factor influencing significant variations in the phenotypes of *S. cerevisiae* x *S. eubayanus* synthetic and industrial hybrids, particularly across a wide range of temperatures and in nonfermentable media (Baker et al., 2019). *S. cerevisiae* x *S. uvarum* hybrids have also been shown to exhibit different levels of thermotolerance depending on the origin of their mtDNA, with notable effects on their transcriptome and fitness under various environmental conditions (Baker et al., 2019; Hewitt et al., 2020; Li, Peris, et al., 2019). A genome-wide screen comparing *S. cerevisiae* and *Saccharomyces uvarum* hybrids revealed that while nuclear genes contribute moderately to thermal tolerance, the mitotype exerts a substantial influence on both heat and cold tolerance. These studies have shown that specific mtDNA inherited in yeast hybrids significantly influences the expression of nuclear genes, particularly those involved in mitochondrial function and energy metabolism. Genes encoding proteins required for oxidative phosphorylation, mitochondrial biogenesis, glycolysis, gluconeogenesis, and the TCA cycle were shown to be differentially expressed, reflecting the direct impact of mtDNA on energy production and mitochondrial processes. Moreover, stress response genes, including those for heat shock proteins and antioxidant systems, also showed altered gene expression, suggesting that the cells adapted to the functional differences in mitochondria caused by the specific mtDNA inherited (Hewitt et al., 2020).

A more recent study, Naseeb et al. (2021) demonstrated that the type of mitochondria inherited by yeast hybrids significantly alters their Quantitative Trait Loci (QTL) landscapes, resulting in distinct variations in traits such as temperature tolerance and stress resistance. This study showed that while nuclear genes contribute to phenotypic traits, the mitochondrial genome can exert an even more substantial influence, especially in hybrid backgrounds where new QTLs, not present in the parent species, emerge (Solieri et al., 2008). Indeed, interactions between mitochondrial and nuclear genomes have been shown to play a significant role in phenotypic variation in both intra-species and interspecies yeast hybrids. These interactions influence various biological processes, including nuclear transcription (Hewitt et al., 2020), respiration (Piatkowska et al., 2013), fermentation (Hewitt et al., 2020; Solieri et al., 2008), phenotypic fitness (Zeyl, 2006), and reproductive isolation (Lee et al., 2008). Moreover, these mitochondrial–nuclear interactions have a genome-wide impact and are crucial in the context of evolution, as demonstrated by multiple studies in yeast (Lee et al., 2008; Wolters et al., 2018) and other organisms (Ellison et al., 2008; Mossman et al., 2016; Wernick et al., 2019).

The parental origin of mtDNA in yeast hybrids can also impact resistance to environmental stresses and the production of specific aroma profiles in fermented products. For example, in wine fermentation, *Saccharomyces* hybrids tend to inherit aroma profiles more closely aligned with the parent from which they inherited their mtDNA. This bias in mtDNA inheritance has been observed under various environmental conditions and is influenced by factors such as sugar availability and osmotic stress (Manousaki et al., 2022; Picazo et al., 2015). Moreover, it has been shown that mtDNA also plays a significant role in determining the phenotypic traits of lager brewing hybrids between *S. cerevisiae* and *S. bayanus*. These hybrids consistently inherit mtDNA from the *S. bayanus* parent. This specific mitochondrial inheritance is likely a result of human-driven selection, favoring strains with optimal fermentation characteristics, particularly at the low temperatures typical for lager brewing (Rainieri et al., 2008).

The study of yeast hybrids and their mitochondrial function is a rapidly evolving field with significant implications for bioproduction. Understanding the complex interactions between nuclear and mitochondrial genomes in hybrids can unlock new possibilities for strain development, particularly in improving traits such as thermotolerance, fermentation efficiency, and stress resistance. As research continues to expand our knowledge of yeast mitogenomics and mitonuclear interactions, these insights will be invaluable in harnessing the full potential of yeast hybrids for industrial applications.

1.7 Conclusion

Mitochondria play a very important role in the function of unicellular eukaryotes, from the generation of energy providing ATP, metabolism of different molecules and tolerance to stress. The basic understanding of mitochondria has grown through the study of model unicellular eukaryote cells, which has largely focused on *S. cerevisiae*. However, harnessing this knowledge so it can be applied to the improved production of useful molecules is still underway. Further large-scale characterization of industrially important unicellular eukaryotes, followed by molecular and phenotypic characterization is required to identify both conserved and novel functions that may provide major boosts to cellular production processes. The immense variety of unicellular species, growth conditions, metabolic pathways, and genetic manipulation that can be modified for each bioproduct will require a vast amount of computational power. Predictive metabolic modeling, experimental design and analysis using the design of experiments approach, and scale-up optimization will be needed. These approaches will enable efficient engineering of unicellular eukaryotes with the end goal of generating useful bioproducts.

Acknowledgments

Figures were created in BioRender.com.

References

Abbott, R., Albach, D., Ansell, S., Arntzen, J. W., Baird, S. J. E., Bierne, N., Boughman, J., Brelsford, A., Buerkle, C. A., Buggs, R., Butlin, R. K., Dieckmann, U., Eroukhmanoff, F., Grill, A., Cahan, S. H., Hermansen, J. S., Hewitt, G., Hudson, A. G., Jiggins, C., ... Zinner, D. (2013). Hybridization and speciation. *Journal of Evolutionary Biology, 26*(2), 229–246. https://doi.org/10.1111/j.1420-9101.2012.02599.x.

Albertin, W., da Silva, T., Rigoulet, M., Salin, B., Masneuf-Pomarede, I., de Vienne, D., Sicard, D., Bely, M., & Marullo, P. (2013). The Mitochondrial Genome Impacts Respiration but Not Fermentation in Interspecific Saccharomyces Hybrids. *PLoS ONE, 8*(9). https://doi.org/10.1371/journal.pone.0075121.

Arhar, S., Pfaller, R., Athenstaedt, K., Lins, T., Gogg-Fassolter, G., Züllig, T., & Natter, K. (2024). Retargeting of heterologous enzymes results in improved β-carotene synthesis in Saccharomyces cerevisiae. *Journal of Applied Microbiology, 135*(9). https://doi.org/10.1093/jambio/lxae224.

Avalos, J. L., Fink, G. R., & Stephanopoulos, G. (2013). Compartmentalization of metabolic pathways in yeast mitochondria improves the production of branched-chain alcohols. *Nature Biotechnology, 31*(4), 335–341. https://doi.org/10.1038/nbt.2509.

Baker, E. C. P., Peris, D., Moriarty, R. V., Li, X. C., Fay, J. C., & Hittinger, C. T. (2019). Mitochondrial DNA and temperature tolerance in lager yeasts. *Science Advances, 5*(1). https://doi.org/10.1126/sciadv.aav1869.

Baker, S. E. (2006). Aspergillus niger genomics: Past, present and into the future. *Medical Mycology, 44*(SUPPL. 1), 17–21. https://doi.org/10.1080/13693780600921037.

Botstein, D., & Fink, G. R. (2011). Yeast: an experimental organism for 21st Century biology. *Genetics, 189*(3), 695–704. https://doi.org/10.1534/genetics.111.130765.

Buerth, C., Heilmann, C. J., Klis, F. M., de Koster, C. G., Ernst, J. F., & Tielker, D. (2011). Growth-dependent secretome of Candida utilis. *Microbiology, 157*(9), 2493–2503. https://doi.org/10.1099/mic.0.049320-0.

Caspeta, L., Shoaie, S., Agren, R., Nookaew, I., & Nielsen, J. (2012). Genome-scale metabolic reconstructions of Pichia stipitis and Pichia pastoris and in silico evaluation of their potentials. *BMC Systems Biology, 6*. https://doi.org/10.1186/1752-0509-6-24.

Chung, B. K. S., Selvarasu, S., Andrea, C., Ryu, J., Lee, H., Ahn, J., Lee, H., & Lee, D.-Y. (2010). Genome-scale metabolic reconstruction and in silico analysis of methylotrophic yeast Pichia pastoris for strain improvement. *Microbial Cell Factories, 9*. https://doi.org/10.1186/1475-2859-9-50.

Chuzel, L., Ganatra, M. B., Schermerhorn, K. M., Gardner, A. F., Anton, B. P., & Taron, C. H. (2017). Complete genome sequence of Kluyveromyces lactis strain GG799, a common yeast host for heterologous protein expression. *Genome Announcements, 5*(30). https://doi.org/10.1128/genomeA.00623-17.

da Cunha, F. M., Torelli, N. Q., & Kowaltowski, A. J. (2015). Mitochondrial Retrograde Signaling: Triggers, Pathways, and Outcomes. *Oxidative Medicine and Cellular Longevity, 2015*. https://doi.org/10.1155/2015/482582.

David, L., Huber, W., Granovskaia, M., Toedling, J., Palm, C. J., Bofkin, L., Jones, T., Davis, R. W., & Steinmetz, L. M. (2006). A high-resolution map of transcription in the yeast genome. *Proceedings of the National Academy of Sciences of the United States of America, 103*(14), 5320–5325.

De Schutter, K., Lin, Y.-C., Tiels, P., Van Hecke, A., Glinka, S., Weber-Lehmann, J., Rouzé, P., Van De Peer, Y., & Callewaert, N. (2009). Genome sequence of the recombinant protein production host Pichia pastoris. *Nature Biotechnology, 27*(6), 561–566. https://doi.org/10.1038/nbt.1544.

Delmas, S., Pullan, S. T., Gaddipati, S., Kokolski, M., Malla, S., Blythe, M. J., Ibbett, R., Campbell, M., Liddell, S., Aboobaker, A., Tucker, G. A., & Archer, D. B. (2012). Uncovering the Genome-Wide Transcriptional Responses of the Filamentous Fungus Aspergillus niger to Lignocellulose Using RNA Sequencing. *PLoS Genetics, 8*(8). https://doi.org/10.1371/journal.pgen.1002875.

Dias, O., Pereira, R., Gombert, A. K., Ferreira, E. C., & Rocha, I. (2014). iOD907, the first genome-scale metabolic model for the milk yeast Kluyveromyces lactis. *Biotechnology Journal, 9*(6), 776–790. https://doi.org/10.1002/biot.201300242.

Dong, C., Shi, Z., Huang, L., Zhao, H., Xu, Z., & Lian, J. (2021). Cloning and characterization of a panel of mitochondrial targeting sequences for compartmentalization engineering in Saccharomyces cerevisiae. *Biotechnology and Bioengineering, 118*(11), 4269–4277. https://doi.org/10.1002/bit.27896.

Douma, R. D., Batista, J. M., Touw, K. M., Kiel, J. A. K. W., Krikken, A. M., Zhao, Z., Veiga, T., Klaassen, P., Bovenberg, R. A. L., Daran, J.-M., Heijnen, J. J., & van Gulik, W. M. (2011). Degeneration of penicillin production in ethanol-limited chemostat cultivations of Penicillium chrysogenum: A systems biology approach. *BMC Systems Biology, 5*. https://doi.org/10.1186/1752-0509-5-132.

Dujon, B. A., & Louis, E. J. (2017). Genome Diversity and Evolution in the Budding Yeasts (Saccharomycotina). *Genetics, 206*(2), 717–750. https://doi.org/10.1534/genetics.116.199216.

Dunn, B., & Sherlock, G. (2008). Reconstruction of the genome origins and evolution of the hybrid lager yeast Saccharomyces pastorianus. *Genome Research, 18*(10), 1610–1623. https://doi.org/10.1101/gr.076075.108.

Eisenberg-Bord, M., Zung, N., Collado, J., Drwesh, L., Fenech, E. J., Fadel, A., Dezorella, N., Bykov, Y. S., Rapaport, D., Fernandez-Busnadiego, R., & Schuldiner, M. (2021). CNM1 mediates nucleus–mitochondria contact site formation in response to phospholipid levels. *Journal of Cell Biology, 220*(11). https://doi.org/10.1083/jcb.202104100.

Elbaz-Alon, Y., Rosenfeld-Gur, E., Shinder, V., Futerman, A. H., Geiger, T., & Schuldiner, M. (2014). A dynamic interface between vacuoles and mitochondria in yeast. *Developmental Cell, 30*(1), 95–102.

Eliodório, K. P., de Gois e Cunha, G. C., White, B. A., Patel, D. H. M., Zhang, F., Hettema, E. H., Basso, T. O., Gombert, A. K., & Raghavendran, V. (2022). Blocking Mitophagy Does Not Significantly Improve Fuel Ethanol Production in Bioethanol Yeast Saccharomyces cerevisiae. *Applied and Environmental Microbiology, 88*(5). https://doi.org/10.1128/aem.02068-21.

Ellison, Niehuis, & Gadau (2008). Hybrid breakdown and mitochondrial dysfunction in hybrids of Nasonia parasitoid wasps. *Journal of Evolutionary Biology, 21*(6), 1844–1851. https://doi.org/10.1111/j.1420-9101.2008.01608.x.

Farhi, M., Marhevka, E., Masci, T., Marcos, E., Eyal, Y., Ovadis, M., Abeliovich, H., & Vainstein, A. (2011). Harnessing yeast subcellular compartments for the production of plant terpenoids. *Metabolic Engineering, 13*(5), 474–481. https://doi.org/10.1016/j.ymben.2011.05.001.

Ghaemmaghami, S., Huh, W.-K., Bower, K., Howson, R. W., Belle, A., Dephoure, N., O'Shea, E. K., & Weissman, J. S. (2003). Global analysis of protein expression in yeast. *Nature, 425*(6959), 737–741.

Goffeau, A., Barrell, B. G., Bussey, H., Davis, R. W., Dujon, B., Feldmann, H., Galibert, F., Hoheisel, J. D., Jacq, C., Johnston, M., Louis, E. J., Mewes, H. W., Murakami, Y., Philippsen, P., Tettelin, H., & Oliver, S. G. (1996). Life with 6000 genes. *Science (New York, N.Y.). 274*(5287).

Granat, L., Knorr, D. Y., Ranson, D. C., Hamer, E. L., Chakrabarty, R. P., Mattedi, F., Fort-Aznar, L., Hirth, F., Sweeney, S. T., Vagnoni, A., Chandel, N. S., & Bateman, J. M. (2023). Yeast NDI1 reconfigures neuronal metabolism and prevents the unfolded protein response in mitochondrial complex I deficiency. *PLoS Genetics, 19*(7). https://doi.org/10.1371/journal.pgen.1010793.

Grba, D. N., & Hirst, J. (2020). Mitochondrial complex I structure reveals ordered water molecules for catalysis and proton translocation. *Nature Structural and Molecular Biology, 27*(10), 892–900. https://doi.org/10.1038/s41594-020-0473-x.

Guerrero-Castillo, S., Vázquez-Acevedo, M., González-Halphen, D., & Uribe-Carvajal, S. (2009). In Yarrowia lipolytica mitochondria, the alternative NADH dehydrogenase interacts specifically with the cytochrome complexes of the classic respiratory pathway. *Biochimica et Biophysica Acta, 1787*(2), 75–85. https://doi.org/10.1016/j.bbabio.2008.10.008.

Guerrero-Castillo, S., Cabrera-Orefice, A., Vázquez-Acevedo, M., González-Halphen, D., & Uribe-Carvajal, S. (2012). During the stationary growth phase, Yarrowia lipolytica prevents the overproduction of reactive oxygen species by activating an uncoupled mitochondrial respiratory pathway. *Biochimica et Biophysica Acta, 1817*(2), 353–362. https://doi.org/10.1016/j.bbabio.2011.11.007.

Hao, X.-C., Yang, X.-S., Wan, P., & Tian, S. (2013). Comparative proteomic analysis of a new adaptive Pichia Stipitis strain to furfural, a lignocellulosic inhibitory compound. *Biotechnology for Biofuels, 6*(1). https://doi.org/10.1186/1754-6834-6-34.

He, Q., Gong, G., Wan, T., Hu, H., & Yu, P. (2023). An integrated transcriptomic and metabolic phenotype analysis to uncover the metabolic characteristics of a genetically engineered Candida utilis strain expressing δ-zein gene. *Frontiers in Microbiology, 14*. https://doi.org/10.3389/fmicb.2023.1241462.

Hewitt, S. K., Duangrattanalert, K., Burgis, T., Zeef, L. A. H., Naseeb, S., & Delneri, D. (2020). Plasticity of mitochondrial DNA inheritance and its impact on nuclear gene transcription in yeast hybrids. *Microorganisms, 8*(4). https://doi.org/10.3390/microorganisms8040494.

Hittinger, C. T. (2013). Saccharomyces diversity and evolution: a budding model genus. *Trends in genetics: TIG, 29*(5), 309–317.

Holt, S., Miks, M. H., De Carvalho, B. T., Foulquié-Moreno, M. R., & Thevelein, J. M. (2019). The molecular biology of fruity and floral aromas in beer and other alcoholic beverages. *FEMS Microbiology Reviews, 43*(3), 193–222. https://doi.org/10.1093/femsre/fuy041.

Hou, R., Gao, L., Liu, J., Liang, Z., Zhou, Y. J., Zhang, L., & Zhang, Y. (2022). Comparative proteomics analysis of Pichia pastoris cultivating in glucose and methanol. *Synthetic and Systems Biotechnology, 7*(3), 862–868. https://doi.org/10.1016/j.synbio.2022.04.005.

Hsu, Y.-Y., & Chou, J.-Y. (2017). Environmental factors can influence mitochondrial inheritance in the saccharomyces yeast hybrids. *PLoS ONE, 12*(1). https://doi.org/10.1371/journal.pone.0169953.

Hughes, A. L., Hughes, C. E., Henderson, K. A., Yazvenko, N., & Gottschling, D. E. (2016). Selective sorting and destruction of mitochondrial membrane proteins in aged yeast. *eLife, 5*(APRIL 2016). https://doi.org/10.7554/eLife.13943.

Hurt, Pesold-Hurt, Suda, Oppliger, & Schatz (1985). The first twelve amino acids (less than half of the pre-sequence) of an imported mitochondrial protein can direct mouse cytosolic dihydrofolate reductase into the yeast mitochondrial matrix. *The EMBO Journal, 4*(8), 2061–2068. https://doi.org/10.1002/j.1460-2075.1985.tb03892.x.

Inokuma, K., Ishii, J., Hara, K. Y., Mochizuki, M., Hasunuma, T., & Kondo, A. (2016). Complete genome sequence of Kluyveromyces marxianus NBRC1777, a nonconventional thermotolerant yeast. *Genome Announcements, 3*(2). https://doi.org/10.1128/genomeA.00389-15.

Iwata, M., Lee, Y., Yamashita, T., Yagi, T., Iwata, S., Cameron, A. D., & Maher, M. J. (2012). The structure of the yeast NADH dehydrogenase (Ndi1) reveals overlapping binding sites for water- and lipid-soluble substrates. *Proceedings of the National Academy of Sciences of the United States of America, 109*(38), 15247–15252. https://doi.org/10.1073/pnas.1210059109.

Jeffries, T. W., Grigoriev, I. V., Grimwood, J., Laplaza, J. M., Aerts, A., Salamov, A., Schmutz, J., Lindquist, E., Dehal, P., Shapiro, H., Jin, Y.-S., Passoth, V., & Richardson, P. M. (2007). Genome sequence of the lignocellulose-bioconverting and xylose-fermenting yeast Pichia stipitis. *Nature Biotechnology, 25*(3), 319–326.

Jeong, H., Lee, D.-H., Kim, S. H., Kim, H.-J., Lee, K., Song, J. Y., Kim, B. K., Sung, B. H., Park, J. C., Sohn, J. H., Koo, H. M., & Kim, J. F. (2012). Genome sequence of the thermotolerant yeast Kluyveromyces marxianus var. marxianus KCTC 17555. *Eukaryotic Cell, 11*(12), 1584–1585. https://doi.org/10.1128/EC.00260-12.

Jin, Y.-S., Laplaza, J. M., & Jeffries, T. W. (2004). Saccharomyces cerevisiae engineered for xylose metabolism exhibits a respiratory response. *Applied and Environmental Microbiology, 70*(11), 6816–6825.

Kakimoto, Y., Tashiro, S., Kojima, R., Morozumi, Y., Endo, T., & Tamura, Y. (2018). Visualizing multiple inter-organelle contact sites using the organelle-targeted split-GFP system. *Scientific Reports, 8*(1). https://doi.org/10.1038/s41598-018-24466-0.

Kieliszek, M., Bierla, K., Jiménez-Lamana, J., Kot, A. M., Alcántara-Durán, J., Piwowarek, K., Błażejak, S., & Szpunar, J. (2020). Metabolic response of the yeast candida utilis during enrichment in Selenium. *International Journal of Molecular Sciences, 21*(15), 1–18. https://doi.org/10.3390/ijms21155287.

Lackner, L. L., Ping, H., Graef, M., Murley, A., & Nunnari, J. (2013). Endoplasmic reticulum-associated mitochondria-cortex tether functions in the distribution and inheritance of mitochondria. *Proceedings of the National Academy of Sciences of the United States of America, 110*(6). https://doi.org/10.1073/pnas.1215232110.

Lane, S., Zhang, Y., Yun, E. J., Ziolkowski, L., Zhang, G., Jin, Y.-S., & Avalos, J. L. (2020). Xylose assimilation enhances the production of isobutanol in engineered Saccharomyces cerevisiae. *Biotechnology and Bioengineering, 117*(2), 372–381. https://doi.org/10.1002/bit.27202.

Lee, H.-Y., Chou, J.-Y., Cheong, L., Chang, N.-H., Yang, S.-Y., & Leu, J.-Y. (2008). Incompatibility of Nuclear and Mitochondrial Genomes Causes Hybrid Sterility between Two Yeast Species. *Cell, 135*(6), 1065–1073. https://doi.org/10.1016/j.cell.2008.10.047.

Leite, A. C., Costa, V., & Pereira, C. (2023). Mitochondria and the cell cycle in budding yeast. *International Journal of Biochemistry and Cell Biology, 161*. https://doi.org/10.1016/j.biocel.2023.106444.

Lertwattanasakul, N., Kosaka, T., Hosoyama, A., Suzuki, Y., Rodrussamee, N., Matsutani, M., Murata, M., Fujimoto, N., Suprayogi, Tsuchikane, K., Limtong, S., Fujita, N., & Yamada, M. (2015). Genetic basis of the highly efficient yeast Kluyveromyces marxianus: Complete genome sequence and transcriptome analyses. *Biotechnology for biofuels, 8*(1). https://doi.org/10.1186/s13068-015-0227-x.

Li, B., Wang, L., Wu, Y.-J., Xia, Z.-Y., Yang, B.-X., & Tang, Y.-Q. (2021). Improving Acetic Acid and Furfural Resistance of Xylose-Fermenting Saccharomyces cerevisiae Strains by Regulating Novel Transcription Factors Revealed via Comparative Transcriptomic Analysis. *Applied and Environmental Microbiology, 87*(10), 1–15. https://doi.org/10.1128/AEM.00158-21.

Li, H., Sun, B., Huang, Y., Zhang, J., Xu, X., Shen, Y., Chen, Z., Yang, J., Chen, L., Hu, Y., & Gu, H. (2022). Gene therapy of yeast NDI1 on mitochondrial complex I dysfunction in rotenone-induced Parkinson's disease models in vitro and vivo. *Molecular medicine (Cambridge, Mass.). 28*(1). https://doi.org/10.1186/s10020-022-00456-x.

Li, P., Fu, X., Chen, M., Zhang, L., & Li, S. (2019). Proteomic profiling and integrated analysis with transcriptomic data bring new insights in the stress responses of Kluyveromyces marxianus after an arrest during high-temperature ethanol fermentation. *Biotechnology for biofuels, 12*(1). https://doi.org/10.1186/s13068-019-1390-2.

Li, X. C., Peris, D., Hittinger, C. T., Sia, E. A., & Fay, J. C. (2019). Mitochondria-encoded genes contribute to evolution of heat and cold tolerance in yeast. *Science Advances, 5*(1). https://doi.org/10.1126/sciadv.aav1848.

Lin, F.-M., Qiao, B., & Yuan, Y.-J. (2009). Comparative proteomic analysis of tolerance and adaptation of ethanologenic Saccharomyces cerevisiae to furfural, a lignocellulosic inhibitory compound. *Applied and Environmental Microbiology, 75*(11), 3765–3776. https://doi.org/10.1128/AEM.02594-08.

Ling, F., & Shibata, T. (2004). Mhr1p-dependent concatemeric mitochondrial DNA formation for generating yeast mitochondrial homoplasmic cells. *Molecular Biology of the Cell, 15*(1), 310–322.

Liu, L., & Alper, H. S. (2014). Draft genome sequence of the oleaginous yeast Yarrowia lipolytica PO1f, a commonly used metabolic engineering host. *Genome Announcements, 2*(4). https://doi.org/10.1128/genomea.00652-14.

Love, K. R., Shah, K. A., Whittaker, C. A., Wu, J., Bartlett, M. C., Ma, D., Leeson, R. L., Priest, M., Borowsky, J., Young, S. K., & Love, J. C. (2016). Comparative genomics and transcriptomics of Pichia pastoris. *BMC Genomics, 17*(1). https://doi.org/10.1186/s12864-016-2876-y.

Lv, X., Wang, F., Zhou, P., Ye, L., Xie, W., Xu, H., & Yu, H. (2016). Dual regulation of cytoplasmic and mitochondrial acetyl-CoA utilization for improved isoprene production in Saccharomyces cerevisiae. *Nature Communications, 7*. https://doi.org/10.1038/ncomms12851.

Macalpine, D. M., Kolesar, J., Okamoto, K., Butow, R. A., & Perlman, P. S. (2001). Replication and preferential inheritance of hypersuppressive petite mitochondrial DNA. *EMBO Journal, 20*(7), 1807–1817. https://doi.org/10.1093/emboj/20.7.1807.

Mallet, J. (2007). Hybrid speciation. *Nature, 446*(7133), 279–283.

Manousaki, A., Bagnall, J., Spiller, D., Balarezo-Cisneros, L. N., White, M., & Delneri, D. (2022). Quantitative Characterisation of Low Abundant Yeast Mitochondrial Proteins Reveals Compensation for Haplo-Insufficiency in Different Environments. *International Journal of Molecular Sciences, 23*(15). https://doi.org/10.3390/ijms23158532.

Mans, R., Daran, J.-M. G., & Pronk, J. T. (2018). Under pressure: evolutionary engineering of yeast strains for improved performance in fuels and chemicals production. *Current Opinion in Biotechnology, 50*, 47–56.

Marinoni, G., Manuel, M., Petersen, R. F., Hvidtfeldt, J., & Sulo, P. (1999). Jure Piškur, Horizontal transfer of genetic material among Saccharomyces yeasts. *Journal of Bacteriology, 181*(20), 6488–6496. https://doi.org/10.1128/JB.181.20.6488-6496.1999.

Maruszczak, K. K., Ayyamperumal, S., & Chacinska, A. (2023). Defective mitochondrial import as a challenge for cellular protein homeostasis. *FEBS Letters, 597*(12), 1555–1568. https://doi.org/10.1002/1873-3468.14677.

McFaline-Figueroa, J. R., Vevea, J., Swayne, T. C., Zhou, C., Liu, C., Leung, G., Boldogh, I. R., & Pon, L. A. (2011). Mitochondrial quality control during inheritance is associated with lifespan and mother-daughter age asymmetry in budding yeast. *Aging Cell, 10*(5), 885–895. https://doi.org/10.1111/j.1474-9726.2011.00731.x.

Morita, K., Matsuda, F., Okamoto, K., Ishii, J., Kondo, A., & Shimizu, H. (2019). Repression of mitochondrial metabolism for cytosolic pyruvate-derived chemical production in Saccharomyces cerevisiae. *Microbial Cell Factories, 18*(1). https://doi.org/10.1186/s12934-019-1226-6.

Mossman, J. A., Tross, J. G., Li, N., Wu, Z., & Rand, D. M. (2016). Mitochondrial-nuclear interactions mediate sex-specific transcriptional profiles in Drosophila. *Genetics, 204*(2), 613–630. https://doi.org/10.1534/genetics.116.192328.

Naseeb, S., Visinoni, F., Hu, Y., Hinks Roberts, A. J., Maslowska, A., Walsh, T., Smart, K. A., Louis, E. J., & Delneri, D. (2021). Restoring fertility in yeast hybrids: Breeding and quantitative genetics of beneficial traits. *Proceedings of the National Academy of Sciences of the United States of America, 118*(38), e2101242118. https://doi.org/10.1073/pnas.2101242118.

O'Brien, C. E., Oliveira-Pacheco, J., Cinnéide, E.Ó., Haase, M. A. B., Hittinger, C. T., Rogers, T. R., Zaragoza, O., Bond, U., & Butler, G. (2021). Population genomics of the pathogenic yeast Candida tropicalis identifies hybrid isolates in environmental samples. *PLoS pathogens, 17*(3). https://doi.org/10.1371/journal.ppat.1009138.

Orij, R., Postmus, J., Ter Beek, A., Brul, S., & Smits, G. J. (2009). In vivo measurement of cytosolic and mitochondrial pH using a pH-sensitive GFP derivative in Saccharomyces cerevisiae reveals a relation between intracellular pH and growth. *Microbiology (Reading, England), 155*(Pt 1), 268–278. https://doi.org/10.1099/mic.0.022038-0.

Peris, D., Sylvester, K., Libkind, D., Gonçalves, P., Sampaio, J. P., Alexander, W. G., & Hittinger, C. T. (2014). Population structure and reticulate evolution of Saccharomyces eubayanus and its lager-brewing hybrids. *Molecular Ecology, 23*(8), 2031–2045. https://doi.org/10.1111/mec.12702.

Piatkowska, E. M., Naseeb, S., Knight, D., & Delneri, D. (2013). Chimeric Protein Complexes in Hybrid Species Generate Novel Phenotypes. *PLoS genetics, 9*(10). https://doi.org/10.1371/journal.pgen.1003836.

Picazo, Gamero-Sandemetrio, Orozco, Albertin, Marullo, Matallana, & Aranda (2015). Mitochondria inheritance is a key factor for tolerance to dehydration in wine yeast production. *Letters in Applied Microbiology, 60*(3), 217–222. https://doi.org/10.1111/lam.12369.

Piškur, J., Smole, S., Groth, C., Petersen, R. F., & Pedersen, M. B. (1998). Structure and genetic stability of mitochondrial genomes vary among yeasts of the genus Saccharomyces. *International Journal of Systematic Bacteriology, 48*(3), 1015–1024. https://doi.org/10.1099/00207713-48-3-1015.

Pomraning, K. R., & Baker, S. E. (2015). Draft genome sequence of the dimorphic yeast Yarrowia lipolytica strain W29. *Genome Announcements, 3*(6). https://doi.org/10.1128/genomea.01211-15.

Rainieri, S., Kodama, Y., Nakao, Y., Pulvirenti, A., & Giudici, P. (2008). The inheritance of mtDNA in lager brewing strains. *FEMS Yeast Research, 8*(4), 586–596. https://doi.org/10.1111/j.1567-1364.2008.00363.x.

Ramirez-Gaona, M., Marcu, A., Pon, A., Guo, A. C., Sajed, T., Wishart, N. A., Karu, N., Djoumbou Feunang, Y., Arndt, D., & Wishart, D. S. (2017). YMDB 2.0: A significantly expanded version of the yeast metabolome database. *Nucleic Acids Research, 45*(D1). https://doi.org/10.1093/nar/gkw1058.

Reinders, J., Zahedi, R. P., Pfanner, N., Meisinger, C., & Sickmann, A. (2006). Toward the complete yeast mitochondrial proteome: multidimensional separation techniques for mitochondrial proteomics. *Journal of Proteome Research, 5*(7), 1543–1554.

Sabra, W., Bommareddy, R. R., Maheshwari, G., Papanikolaou, S., & Zeng, A.-P. (2017). Substrates and oxygen dependent citric acid production by Yarrowia lipolytica: Insights through transcriptome and fluxome analyses. *Microbial Cell Factories, 16*(1). https://doi.org/10.1186/s12934-017-0690-0.

Saladi, S. D., Boos, F., Poglitsch, M., Meyer, H., Sommer, F., Mühlhaus, T., Schroda, M., Schuldiner, M., Madeo, F., & Herrmann, J. M. (2020). The NADH Dehydrogenase Nde1 Executes Cell Death after Integrating Signals from Metabolism and Proteostasis on the Mitochondrial Surface. *Molecular Cell, 77*(1). https://doi.org/10.1016/j.molcel.2019.09.027.

Schuster, R., & Okamoto, K. (2022). An overview of the molecular mechanisms of mitophagy in yeast. *Biochimica et biophysica acta. General subjects, 1866*(11).

Shen, X.-X., Opulente, D. A., Kominek, J., Zhou, X., Steenwyk, J. L., Buh, K. V., Haase, M. A. B., Wisecaver, J. H., Wang, M., Doering, D. T., Boudouris, J. T., Schneider, R. M., Langdon, Q. K., Ohkuma, M., Endoh, R., Takashima, M., Manabe, R.-I., Čadež, N., Libkind, D., ... Rokas, A. (2018). Tempo and Mode of Genome Evolution in the Budding Yeast Subphylum. *Cell, 175*(6).

Shiroma, S., Jayakody, L. N., Horie, K., Okamoto, K., & Kitagaki, H. (2014). Enhancement of ethanol fermentation in Saccharomyces cerevisiae sake yeast by disrupting mitophagy function. *Applied and Environmental Microbiology, 80*(3), 1002–1012. https://doi.org/10.1128/AEM.03130-13.

da Silva, L. V., Coelho, M. A. Z., da Silva, M. R. S., & Amaral, P. F. F. (2020). Investigation of mitochondrial protein expression profiles of Yarrowia lipolytica in response to citric acid production. *Bioprocess and Biosystems Engineering, 43*(9), 1703–1715. https://doi.org/10.1007/s00449-020-02363-z.

Solieri, L., Antúnez, O., Pérez-Ortín, J. E., Barrio, E., & Giudici, P. (2008). Mitochondrial inheritance and fermentative: oxidative balance in hybrids between Saccharomyces cerevisiae and Saccharomyces uvarum. *Yeast (Chichester, England), 25*(7), 485–500. https://doi.org/10.1002/yea.1600.

Spinelli, J. B., & Haigis, M. C. (2018). The multifaceted contributions of mitochondria to cellular metabolism. *Nature Cell Biology, 20*(7), 745–754. https://doi.org/10.1038/s41556-018-0124-1.

Sturmberger, L., Chappell, T., Geier, M., Krainer, F., Day, K. J., Vide, U., Trstenjak, S., Schiefer, A., Richardson, T., Soriaga, L., Darnhofer, B., Birner-Gruenberger, R., Glick, B. S., Tolstorukov, I., Cregg, J., Madden, K., & Glieder, A. (2016). Refined Pichia pastoris reference genome sequence. *Journal of Biotechnology, 235*, 121–131. https://doi.org/10.1016/j.jbiotec.2016.04.023.

Styger, G., Jacobson, D., Prior, B. A., & Bauer, F. F. (2013). Genetic analysis of the metabolic pathways responsible for aroma metabolite production by Saccharomyces cerevisiae. *Applied Microbiology and Biotechnology, 97*(10), 4429–4442. https://doi.org/10.1007/s00253-012-4522-1.

Suleau, A., Gourdon, P., Reitz-Ausseur, J., & Casaregola, S. (2006). Transcriptomic analysis of extensive changes in metabolic regulation in Kluyveromyces lactis strains. *Eukaryotic Cell, 5*(8), 1360–1370. https://doi.org/10.1128/EC.00087-06.

Sun, J., Lu, X., Rinas, U., & Zeng, A. P. (2007). Metabolic peculiarities of Aspergillus niger disclosed by comparative metabolic genomics. *Genome Biology, 8*(9). https://doi.org/10.1186/gb-2007-8-9-r182.

Suzuki, T., Hoshino, T., & Matsushika, A. (2014). Draft genome sequence of Kluyveromyces marxianus strain DMB1, isolated from sugarcane bagasse hydrolysate. *Genome Announcements, 2*(4). https://doi.org/10.1128/genomeA.00733-14.

Swaim, C. L., Anton, B. P., Sharma, S. S., Taron, C. H., & Benner, J. S. (2008). Physical and computational analysis of the yeast Kluyveromyces lactis secreted proteome. *Proteomics, 8*(13), 2714–2723. https://doi.org/10.1002/pmic.200700764.

Takagi, S., Kojima, K., & Ohashi, S. (2020). Proteomic analysis on Aspergillus strains that are useful for industrial enzyme production. *Bioscience, Biotechnology, and Biochemistry,* 2241–2252. https://doi.org/10.1080/09168451.2020.1794784.

Tang, J., Zhang, L., Su, J., Ye, Q., Li, Y., Liu, D., Cui, H., Zhang, Y., & Ye, Z. (2024). Insights into Fungal Mitochondrial Genomes and Inheritance Based on Current Findings from Yeast-like Fungi. *Journal of Fungi, 10*(7). https://doi.org/10.3390/jof10070441.

Taylor, S. A., & Larson, E. L. (2019). Insights from genomes into the evolutionary importance and prevalence of hybridization in nature. *Nature Ecology & Evolution, 3*(2), 170–177. https://doi.org/10.1038/s41559-018-0777-y.

Thrailkill, K. M., & Birky, C. W. (1980). Intracellular population genetics: evidence for random drift of mitochondrial allele frequencies in Saccharomyces cerevisiae and Schizosaccharomyces pombe. *Genetics, 96*(1), 237–262.

Tomita, Y., Ikeo, K., Tamakawa, H., Gojobori, T., & Ikushima, S. (2012). Genome and transcriptome analysis of the food-yeast candida utilis. *PLoS ONE, 7*(5). https://doi.org/10.1371/journal.pone.0037226.

Van Den Berg, M. A., Albang, R., Albermann, K., Badger, J. H., Daran, J.-M., M Driessen, A. J., Garcia-Estrada, C., Fedorova, N. D., Harris, D. M., Heijne, W. H. M., Joardar, V., W Kiel, J. A. K., Kovalchuk, A., Martín, J. F., Nierman, W. C., Nijland, J. G., Pronk, J. T., Roubos, J. A., Van Der Klei, I. J., ... Bovenberg, R. A. L. (2008). Genome sequencing and analysis of the filamentous fungus Penicillium chrysogenum. *Nature Biotechnology, 26*(10), 1161–1168. https://doi.org/10.1038/nbt.1498.

Veiga, A., João, Arrabaça, D., Sansonetty, F., & Ludovico, P. (2003). Manuela Côrte-Real, Maria C Loureiro-Dias, Energy conversion coupled to cyanide-resistant respiration in the yeasts Pichia membranifaciens and Debaryomyces hansenii. *FEMS Yeast Research, 3*(2), 141–148.

da Veiga Moreira, J., Jolicoeur, M., Schwartz, L., & Peres, S. (2021). Fine-tuning mitochondrial activity in Yarrowia lipolytica for citrate overproduction. *Scientific Reports, 11*(1). https://doi.org/10.1038/s41598-020-79577-4.

Voulgaris, I., O'Donnell, A., Harvey, L. M., & McNeil, B. (2012). Inactivating alternative NADH dehydrogenases: enhancing fungal bioprocesses by improving growth and biomass yield? *Scientific Reports, 2*. https://doi.org/10.1038/srep00322.

Vowinckel, J., Hartl, J., Marx, H., Kerick, M., Runggatscher, K., Keller, M. A., Mülleder, M., Day, J., Weber, M., Rinnerthaler, M., Yu, J. S. L., Aulakh, S. K., Lehmann, A., Mattanovich, D., Timmermann, B., Zhang, N., Dunn, C. D., MacRae, J. I., Breitenbach, M., & Ralser, M. (2021). The metabolic growth limitations of petite cells lacking the mitochondrial genome. *Nature Metabolism, 3*(11), 1521–1535. https://doi.org/10.1038/s42255-021-00477-6.

Wang, X., & Chen, X. J. (2015). A cytosolic network suppressing mitochondria-mediated proteostatic stress and cell death. *Nature, 524*(7566), 481–484. https://doi.org/10.1038/nature14859.

Wernick, R. I., Christy, S. F., Howe, D. K., Sullins, J. A., Ramirez, J. F., Sare, M., Penley, M. K. J., Morran, L. T., Denver, D. R., & Estes, S. (2019). Sex and mitonuclear adaptation in experimental caenorhabditis elegans populations. *Genetics, 211*(3), 1045–1058. https://doi.org/10.1534/genetics.119.301935

Wolters, J. F., Charron, G., Gaspary, A., Landry, C. R., Fiumera, A. C., & Fiumera, H. L. (2018). Mitochondrial recombination reveals mito–mito epistasis in yeast. *Genetics, 209*(1), 307–319. https://doi.org/10.1534/genetics.117.300660.

Wrobel, L., Topf, U., Bragoszewski, P., Wiese, S., Sztolsztener, M. E., Oeljeklaus, S., Varabyova, A., Lirski, M., Chroscicki, P., Mroczek, S., Januszewicz, E., Dziembowski, A., Koblowska, M., Warscheid, B., & Chacinska, A. (2015). Mistargeted mitochondrial proteins activate a proteostatic response in the cytosol. *Nature, 524*(7566), 485–488. https://doi.org/10.1038/nature14951.

Yao, Z., Zhou, P., Su, B., Su, S., Ye, L., & Yu, H. (2018). Enhanced Isoprene Production by Reconstruction of Metabolic Balance between Strengthened Precursor Supply and

Improved Isoprene Synthase in Saccharomyces cerevisiae. *ACS Synthetic Biology, 7*(9), 2308–2316. https://doi.org/10.1021/acssynbio.8b00289.

Yee, D. A., DeNicola, A. B., Billingsley, J. M., Creso, J. G., Subrahmanyam, V., & Tang, Y. (2019). Engineered mitochondrial production of monoterpenes in Saccharomyces cerevisiae. *Metabolic Engineering, 55*, 76–84.

Zalman, L. S., Nikaido, H., & Kagawa, Y. (1980). Mitochondrial outer membrane contains a protein producing nonspecific diffusion channels. *The Journal of Biological Chemistry, 255*(5), 1771–1774.

Zardoya, R. (2020). Recent advances in understanding mitochondrial genome diversity. *F1000Research, 9*. https://doi.org/10.12688/f1000research.21490.1.

Zeyl, C. (2006). Experimental evolution with yeast. *FEMS yeast research, 6*(5), 685–691. https://doi.org/10.1111/j.1567-1364.2006.00061.x.

Zhang, Y., Su, M., Chen, Y., Wang, Z., Nielsen, J., & Liu, Z. (2023). Engineering yeast mitochondrial metabolism for 3-hydroxypropionate production. *Biotechnology for Biofuels and Bioproducts, 16*(1). https://doi.org/10.1186/s13068-023-02309-z.

Zhang, Z., Lan, Q., Yu, Y., Zhou, J., & Lu, H. (2022). Comparative metabolome and transcriptome analyses of the properties of Kluyveromyces marxianus and Saccharomyces yeasts in apple cider fermentation. *Food Chemistry: Molecular Sciences, 4*. https://doi.org/10.1016/j.fochms.2022.100095.

Zhao, C., Zhao, J., Han, J., Mei, Y., & Fang, H. (2024). Improved consolidated bioprocessing for itaconic acid production by simultaneous optimization of cellulase and metabolic pathway of Neurospora crass. *Biotechnology for Biofuels and Bioproducts, 17*(1). https://doi.org/10.1186/s13068-024-02505-5.

Zhu, Y., Wu, L., Zhu, J., Xu, Y., & Yu, S. (2019). Transcriptome and metabolome analysis of Pichia stipitis to three representative lignocellulosic inhibitors. *Archives of Microbiology, 201*(5), 581–589. https://doi.org/10.1007/s00203-018-1600-5.

Zhu, Z.-T., Du, M.-M., Gao, B., Tao, X.-Y., Zhao, M., Ren, Y.-H., Wang, F.-Q., & Wei, D.-Z. (2021). Metabolic compartmentalization in yeast mitochondria: Burden and solution for squalene overproduction. *Metabolic Engineering, 68*, 232–245.

Genome editing techniques for microalgae and yeast

2

Arzu Yıldırım[1], *Kyler Roy*[2], *Bahar Aslanbay Guler*[1], *Esra Imamoglu*[1], *and Rodrigo Ledesma Amaro*[2]
[1]Bioengineering Department, Ege University, Izmir, Turkey, [2]Department of Bioengineering, Imperial College London, London, United Kingdom

2.1 Introduction

Through several years of genetic research, understanding and modifying the complex mechanisms of organisms has been extensively studied to develop or enhance their specific characteristics. In general, a series of conventional gene modification techniques, including homologous recombination (HR), natural mutagenesis, mutagenesis induced by a mutagen (i.e., chemicals, X-rays, ion beam, and many more), and insertional mutagenesis by T-DNA insertions, have been used to alter genome sequences. Despite their considerable efficiency, these methods are usually limited by several drawbacks, including the random nature of the induced mutations, high cost, and their intensive labor and time requirements (Nishida & Kondo, 2021; Zhang, Raboanatahiry et al., 2017; Zhang, Zhang et al., 2017). Currently, advancements in genetic engineering have opened the door to sophisticated genome editing methods, which enable researchers to precisely modify an organism's DNA. These approaches cover novel technologies that enable targeted modifications of the genome, including mutagenesis or site-specific insertion, deletion, or substitution, at precise locations through the activity of nucleases (Duensing et al., 2018; Wani et al., 2023). Some of the most common nuclease families performed are zinc finger nucleases (ZFNs), transcription activator-like effector nucleases (TALENs), and clustered regularly interspaced short palindromic repeats-associated proteins (CRISPR-Cas) (Castro et al., 2021; Venezia & Creasey Krainer, 2021). As a precise and programmable tool for gene modification, nuclease-mediated genome editing has been effectively employed to modify genes in various organisms, including humans, animals, plants, bacteria, microalgae, and yeast.

Microalgae and yeast are used as cell factories to produce numerous valuable bioproducts across a wide range of sectors, including pharmaceuticals, cosmetics, food, feed, energy, and agriculture. Both organisms offer unique advantages as model systems and production platforms owing to their relatively simple genetics, rapid growth rates, and ease of cultivation. Maximizing the production of their metabolites and enhancing their quality pose a significant challenge for industrial bioprocesses. Therefore, genome editing in microalgae and yeast represents a cutting-edge approach for improving industrially important traits by facilitating the optimization

Eukaryotic Microorganisms as Sources of Bioproducts. DOI: https://doi.org/10.1016/B978-0-443-30188-9.00018-4

of metabolic pathways, introducing novel functionalities, and studying gene functions with unprecedented accuracy (Fajardo et al., 2020; Ng et al., 2020; Yang & Blenner, 2020). To date, many microalgae species, including *Chlamydomonas reinhardtii, Phaeodactylum tricornutum, Nannochloropsis sp.*, and yeasts such as *Saccharomyces cerevisiae, Yarrowia lipolitica, Pichia pastoris, Schizosaccharomyces pombe, Starmerella bombicola,* and *Candida tropicalis*, have been commonly subjected to genome editing due to their well-known genome sequences. While these studies show immense potential of the genome editing tools for the targeted microorganisms, several challenges persist, such as the development of efficient delivery techniques, the minimization of off-target effects, and navigating regulatory frameworks (Aljabali et al., 2024; Jeong et al., 2023; Sproles et al., 2021; Yang & Blenner, 2020). Overcoming these obstacles is essential for fully unlocking the capabilities of genome editing in harnessing these microorganisms for sustainable biotechnological processes.

This chapter provides a concise overview of novel genome editing techniques, including TALENs, ZFNs, and CRISPR-Cas, applied to microalgae and yeast (Fig. 2.1). This review discusses the main principles of these techniques and their applications across a wide variety of microalgae and yeast strains as well as recent case studies. Additionally, this chapter addresses the main challenges that hinder the widespread use of editing tools across diverse species.

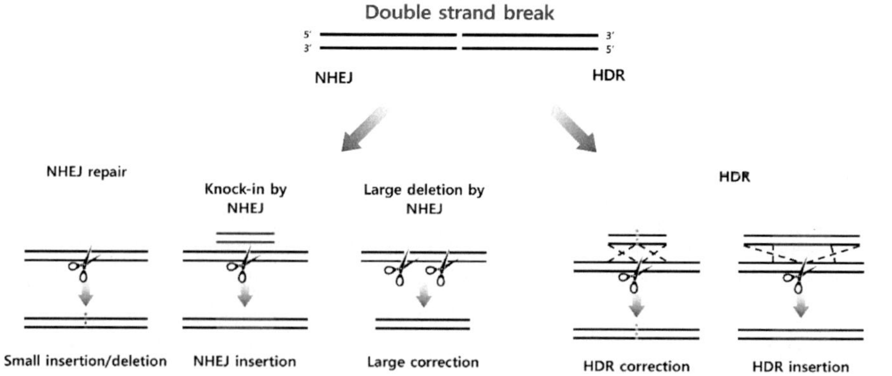

Figure 2.1 Novel genome editing techniques for microalgae and yeast.
Source: Adapted from Jeong, B., Jang, J. & Jin, E. (2023). Genome engineering via gene editing technologies in microalgae. *Bioresource Technology*, *373*, 128701. https://doi.org/10.1016/j.biortech.2023.128701.

2.2 Genome editing techniques

2.2.1 ZFNs gene editing technique

ZFNs are initial genome editing tools that have played a critical role in both the advancement and implementation of genome engineering technologies. They are one of the gene editing methods consisting of a chimeric nuclease that contains a DNA-binding zinc finger domain and a nonspecific FOKl endonuclease domain (Ahmar et al., 2020). The FokI endonuclease domain, derived from the *Flavobacterium okeanokoites*, is responsible for DNA cleavage and is capable of forming a dimer to precisely target sequences within the genome for efficient gene editing. On the other hand, zinc finger modules consist of up to six eukaryotic proteins that comprise approximately 30 amino acids arranged in a ββα configuration, forming a finger-shaped structure due to the bonding between a Zn^{2+} atom and Cys_2–His_2 residue. This Cys_2-His_2 residue, together with the zinc finger protein, is the most prevalent DNA-binding motif among eukaryotic transcription factors (TFs). Each protein in the zinc finger domain identify three base pairs along a single DNA strand, providing precise and specific binding properties. These characteristics of the engineered zinc-finger domain guide the ZFN to target an exact location within the genome (Bao et al., 2020; Li et al., 2020; Mushtaq et al., 2019). Since the FokI nuclease requires dimerization for DNA cleavage, two ZFN molecules are typically required to bind the target site with the correct orientation. These two zinc-finger proteins bind to complementary strands of DNA, facilitating the assembly of the fused FokI endonuclease into an active dimeric form, which then cleaves the DNA at specific target sites (Li et al., 2020; Zhang et al., 2019).

The ZFNs gene editing technique has widespread applications due to its specificity and simplicity. Over the past decade, it has been employed to target genes in numerous plants and animals, including *Arabidopsis thaliana*, maize, soybean, tobacco, catfish, rainbow trout, rats, frogs, silk worms, pigs, *Drosophila melanogaster*, and some mammalian cells (Bao et al., 2020; Singh, 2021; Yang et al., 2022). However, it has some disadvantages and technical complexities, including low efficiency, the ability to target only a single site at a time, and being time-consuming. In addition, they are expensive because of the high cost of the module assembly process (Bao et al., 2020; Mushtaq et al., 2018).

2.2.1.1 Application of ZFNs in microalgae

Despite the widespread application of ZFNs genome editing techniques across various organisms, their utilization in microalgae species remains limited. Very few studies focus on the utilization of ZFNs in microalgae species, and all of them have been conducted on *C. reinhardtii* (Table 2.1). In the first report, a mutant *C. reinhardtii* strain was designed by inserting a nonfunctional aminoglycoside 3'-phosphotransferase VIII (aphVIII) marker interspaced with a short COP3 target sequence into the nuclear genome of the microalgae. Engineered ZFNs, targeting short COP3 gene, were cotransformed into the nonfunctional aphVIII strain using a partial aphVIII DNA template. When ZFNs were active, they triggered a double-

Table 2.1 Genome editing via TALENs, ZFNs, and CRISPR in microalgae.

Editing method	Microorganism	Target gene(s)	Result	References
ZFN	*Chlamydomonas reinhardtii*	COP3 gene that encodes light-activated ion channel channelrhodopsin-1	Repair of a defective aphVIII gene by the activity of ZFN	Sizova et al. (2013)
ZFN	*Chlamydomonas reinhardtii*	Photoreceptor genes including, COP1/2, COP3, COP4, COP5, PHOT, UVR8, VGCC, MAT3, and aCRY	Successfully editing of target photoreceptor genes, which led to a better understanding of phototropism in cells	Greiner et al. (2017)
TALEN	*Pseudochoricystis ellipsoidea*	Uridine monophosphate synthase gene (*PeUMPS*)	Development of a self-cloning system	Kasai et al. (2015)
Platinum TALE-N	*Nannochloropsis oceanica*	Nitrate reductase and acyltransferase genes	Highly efficient genome-editing with no off-target mutations	Kurita et al. (2020)
Platinum TALE-N	*Nannochloropsis oceanica*	Nitrate reductase gene (*NoNR*)	Development of removable gene editing tool	Kurita et al. (2022)
TALEN	*Phaeodactylum tricornutum*	Acyl-CoA thioesterase gene	Modified lipid metabolism	Hao et al. (2018)
TALEN	*Microchloropsis gaditana*	Acyl-CoA synthetases (*ACS*)	Characterization of ACS on lipid production	Billey et al. (2021)
CRISPR-Cas9	*Chlamydomonas reinhardtii*	Lycopene ε-cyclase (*LCY-E*)	The carotenoid pathway shifted towards the β-carotene branch, which was then directed towards astaxanthin production	Kneip et al. (2024)
CRISPR-Cas9	*Chlamydomonas reinhardtii*	Lonely Guy (*LOG*) and isopentenyl transferase (*IPT*) genes related to the stimulation of cytokinin	The role of *LOG* and *IPT* genes of microalga on the cytokinin-related plant defense system was revealed	Sandor et al. (2024)

CRISPR-Cas9	*Phaeodactylum tricornutum*	Zeaxanthin epoxidase genes, *ZEP2* and *ZEP3*	A mutant strain capable of producing diatoxanthin	Græsholt et al. (2024)
CRISPR-Cas9	*Nannochloropsis oceanica*	*Nolsmt* gene, a homolog of Rubisco large-subunit methyltransferase (*LSMT*)	Identification of the function of LSMT on fructose-1,6-bisphosphate aldolase (FBA) trimethilation	Liang et al. (2024)
CRISPR-Cas9	*Phaeodactylum tricornutum*	Cobalamin acquisition protein 1 (*CBA1*)	The relation between vitamin B12 uptake and CBA1 was identified.	Sayer et al. (2024)
CRISPR-Cas9	*Phaeodactylum tricornutum*	*Phaeodactylum tricornutum* acyl-CoA binding protein (*PtACBP*)	Pt ACBP was found to be active in the intracellular acyl transport in the lipid metabolism	Leyland et al. (2024)
CRISPR acti-vation	*Chlorella sorokiniana*	Glutamate synthase (*GS*)	Stimulation of the endogenous GS expression to enhance the growth and high-value molecule accumulation.	Teng and Ng (2023)
CRISPR-Cas9	*Phaeodactylum tricornutum*	Stramenopile-type lipid droplet protein (*StLDP*) gene	Decrease in the number of lipid droplets per cell	Yoneda et al. (2023)
CRISPR-Cas9	*Phaeodactylum tricornutum*	The gene of chloroplast signal recognition particle receptor (*CpFTSY*)	CpFTSY does not show activity in targeting the thylakoid membrane in diatoms.	Nymark et al. (2023)
CRISPR-Cas9	*Phaeodactylum tricornutum*	PhoD-type alkaline phosphatase (*AP*)	Higher fatty acid accumulation with an increased content of PUFAs	Zhang et al. (2023)
CRISPR-Cas9	*Phaeodactylum tricornutum*	*P. tricornutum* fucosyltransferase 1 (*PtFucT1*)	PtFucT1 functions in the noncore α-1,4-fucose modification of N-glycans	Xie et al. (2023)
CRISPR-Cas9	*Chlamydomonas reinhardtii*	Bacterial phytase gene	Successful transformation of the nuclear genome by homolog recombination	Zadabbas Shahabadi et al. (2023)
CRISPR-Cas9	*Chlamydomonas reinhardtii*	Ornithine decarboxylase 1 (*ODC1*) and amine oxidase 2 (*AMX2*)	Increase in intracellular putrescine accumulation	Freudenberg et al. (2022)

stranded break in the COP3 sequence located within the inactive aphVIII marker. This break allows for the repair of aphVIII through HR using the partial-aphVIII DNA template (Sizova et al., 2013).

In another study, the ZFNs editing strategy was optimized for *C. reinhardtii* by targeting numerous photoreceptor genes including COP1/2, COP3, COP4, COP5, PHOT, UVR8, VGCC, MAT3, and aCRY. ZFNs technology was found to be highly successful in the disruption of target photoreceptor genes and related proteins. It was reported that ZFNs were utilized to edit genes in a dependable and predictable manner through HR. However, it is challenging to create new ZFNs pairs because of the restricted number of effective target sites within a gene (Greiner et al., 2017), resulting in an increasing trend in the development and application of novel editing technologies.

2.2.1.2 Applications of ZFNs in yeast

Although ZFNs have not been widely used for bioproduction in yeast, yeasts remain crucial for the screening and validation of these nucleases. Using a *S. cerevisiae* mating assay, researchers can evaluate the cleavage efficiency and dimerization of ZFN pairs in vitro before their introduction into more complex biological systems. This assay has enabled scientists to not only confirm the functionality of ZFNs in organisms such as zebrafish, plants, and mammals but also to develop novel ZFNs that exhibit reduced off-target cleavage by suppressing homodimerization (Doyon et al., 2011; Townsend et al., 2009; Wang et al., 2013).

2.2.2 TALEN-based genome editing technique

TALENs are one of the gene editing tools used to modify a target DNA region. It is mainly composed of two components, one of which is the DNA binding domain, which recognizes the specific region of DNA, and the other is the functional domain (in this case an endonuclease), which acts as a cleavage unit to create a double-strand break (DSB) at the target site. The origin of the DNA binding domain comes from a plant pathogenic bacterium, namely *Xanthomonas sp*. Bacteria secrete TAL effector proteins to invade the host DNA and act as a TF to manipulate the genes related to the disease susceptibility in the plant cell (Ji et al., 2022; Scholze & Boch, 2010). This component consists of approximately 17 repetitive modules of 34 amino acids carrying repeat variable di-amino acid residues (RVDs) at positions 12 and 13, which can only bind to specific nucleotides (Teper et al., 2023). The specificity of these RVDs in binding to one of each nucleotide is the key to the TALEN technology to design elements for specific gene editing. The nuclease activity of this system is created by fusing the TALEs domain to FokI endonuclease (Hiroyuki & Susumu, 1981), which does not require a specific restriction site, and can cleave virtually any combination of bases creating a sticky end. FokI is a monomeric enzyme, and its catalytic part needs to be dimerized for the proper cutting of both DNA strands (Halford et al., 2011). Therefore, to utilize a TALEN system for gene editing, one must design each of the corresponding arms of the DNA on opposite sides within the range of a spacer. The spacer length between the two TALEN units may vary from 12 to 23 bp according to the length of the TALE component

(Becker & Boch, 2021). The DSB can be repaired by either nonhomologous end-joining (NHEJ) or HR, with the former being the most likely. Accordingly, the type of repair can result in mutations that result in gene knockouts, leading to gene silencing. In addition to TALENs being particularly effective in understanding gene functions, TALE proteins can also be employed as regulators by fusion to an activator or repressor molecule (Bhardwaj & Nain, 2021).

2.2.2.1 Application of TALENs in microalgae

Although their number is limited compared to that of the CRISPR system, TALENs have also been utilized to edit the genome of microalgae, generally with a focus on lipid metabolism (Table 2.1). Kasai et al. (2015) used TALEN technology to create uracil auxotrophic mutants in the unicellular green alga *Pseudochoricystis ellipsoidea*, a suitable strain for biodiesel production (Kasai et al., 2015). In this way, they introduced a system for self-cloning and positive selection of this strain to prevent large-scale production of genetically modified organisms. Similarly, another lipid accumulating strain, *Nannochloropsis oceanica*, was used to develop a highly efficient genome-editing tool. Kurita et al. (2020) engineered *N. oceanica* genome using Platinum TALENs, where the 4th and 32nd amino acid residues of the TALE repeats are exchanged to enhance the activity (Kurita et al., 2020). The researchers were able to express the nitrate reductase gene at the *NoNR* locus and acyltransferase gene at the *LPAT1* locus simultaneously without the loss of function. The same research group made additional improvements to this system by designing a removable vector for Platinum TALEN, which comprised a yeast centromere and autonomous replication sequence (CEN/ARS). This vector operated similarly to episomes in *Nannochloropsis species* (Kurita et al., 2022). This highly efficient removable gene-editing tool has also been suggested for the large-scale outdoor cultivation of genetically modified microalgae.

Hao et al. (2018) used TALEN system to modify the lipid metabolism in the diatom *P. tricornutum*. The researchers mutated acyl-CoA thioesterase gene at a region located in the 4HBT_2 domain of ptTES1 using TALEN as a tool and obtained a significant increase in triacylglycerols without changing the level of other phosphoglycerolipids and galactoglycerolipids, demonstrating the effectiveness of the TALEN technique for designing diatoms to manufacture biofuels and other high-value fatty acid-derived compounds (Hao et al., 2018). Billey et al. (2021) applied the TALEN genome engineering technique to *Microchloropsis gaditana* (formerly known as *Nannochloropsis gaditana*) Bubblegum subfamily to modify fatty acid metabolism by creating mutations on acyl-CoA synthetases (ACS) (Billey et al., 2021). Instead of gene knockout, the researchers observed point mutations affecting the enzyme activities and supported the information and characterization of ACSs on lipid production and profiling in this strain.

2.2.2.2 Applications of TALENs in yeast

Due to their high specificity, TALENs have been utilized for genome editing in various yeast species, such as *S. cerevisiae* and *Y. lipolytica*, for bioproduction purposes (Table 2.2) (Gan et al., 2018; Rigouin et al., 2017; Zhang et al., 2015). Using a TALENs-

Table 2.2 Applications of genome editing in yeast for bioproduction.

Editing method	Microorganism	Target Gene(s)	Product	References
TALEN	*Saccharyomyces cerevisiae*	Generated a mutant library for enhanced stress tolerance	Heat and ethanol tolerance	Gan et al. (2018)
	Yarrowia lipolitica	Ketoacyl synthase (KS)	Myristic acid	Rigouin et al. (2017)
CRIPSR-Cas	*Yarrowia lipolitica*	Overexpression of ARO4, ARO7, and HPD1 Deletions to ARO8, ARO9, AAT1, AAT2, and HMG2	Homogentisic acid	Yuzbashev et al. (2023)
		β-glucosidase	β-glucosidase	Swietalski et al. (2020)
	Pichia pastoris	Pyruvate decarboxylase (PDC1)	Isopentanol	Siripong et al. (2020)
	Candida tropicalis	Carotenoid producing genes: carB, carRP	β-carotene	Zhang et al. (2020)
	Schizosaccharomyces pombe	Disrupted two pyruvate decarboxylases (PDCs), an L-lactic acid dehydrogenes (L-LDH) and an alcohol dehydrogenase (SPBC337.11) Introduced heterologous acetaldehyde dehydrogenases (MhpF and EutE)	Lactic acid	Ozaki et al. (2017)
	Starmerella bombicola	Overexpression of NADPH cytochrome P450 reductase (CPR) and cytochrome P450 oxidoreductase (CYP52M1), Disruption of a peroxisome membrane transporter (PXA1)	Acid-type sophoro-lipids	Shi et al. (2022)
	Saccharomyces cerevisiae	Naringenin producing genes: PAL2, C4H/ATR2, 4CL2, CHS, CHI, TSC13	Naringenin	Vanegas et al. (2017)
		38 heterologous enzymes	QS-21	Martin et al. (2024)
		11 genes in central metabolism	Succinic acid	Shaw et al. (2022)

assisted multiplexed editing (TAME) toolkit—a recently developed platform that enables the rapid and precise genomic editing of *S. cerevisiae*—researchers were able to rapidly engineer *S. cerevisiae* strains with enhanced industrial phenotypes, including increased ethanol tolerance and heat-shock resistance (Gan et al., 2018; Zhang et al., 2015). Furthermore, through TALEN-mediated directed mutagenesis of *Y. lipolytica* ketoacyl-synthase (YALI0B19382p), researchers enhanced the bioproduction of the medium-chain fatty acid myristic acid (Rigouin et al., 2017).

2.2.3 CRISPR gene editing technique

Initially discovered as a mechanism for adaptive immune responses in archaea and bacteria, clustered regularly interspaced short palindromic repeats (CRISPR) have since been repurposed to provide a simple, yet robust tool for genome engineering across a wide range of organisms (Barrangou et al., 2007; Bolotin et al., 2005; Garneau et al., 2010; Ishino et al., 1987; Jinek et al., 2012). In its native context, CRISPR acts as a defense mechanism against invading DNA by storing snippets of DNA from previous infections within genomic repeat-spacer arrays. Upon reinfection, these snippets are transcribed into CRISPR RNA (crRNA), which, together with a cascade of CRISPR-associated (Cas) proteins, recognizes and degrades the foreign DNA (Barrangou et al., 2007; Bolotin et al., 2005; Garneau et al., 2010). Although these repeat-spacer clusters were initially discovered in the 1980s, it was not until 2012 that scientists fully realized the potential of this system to target virtually any DNA sequence (Gasiunas et al., 2012; Ishino et al., 1987; Jinek et al., 2012). By simply modifying the spacer sequence within the crRNA, CRISPR-Cas enables the precise targeting of virtually any genetic sequence without the significant protein modifications required by earlier genome editing tools such as ZFNs and TALENs (Ishino et al., 1987; Jinek et al., 2012). Furthermore, the CRISPR-Cas system can be multiplexed, allowing the concurrent expression of several guide RNAs (gRNAs) to edit multiple genomic loci simultaneously (Cong et al., 2013; McCarty et al., 2020; Shaw et al., 2022; Zetsche et al., 2017).

A remarkably simple system, genome editing with CRISPR requires only the coexpression of two components: a guide RNA (gRNA) to target a specific site of the genome to edit and a Cas endonuclease to initiate a DSB (Gasiunas et al., 2012; Jinek et al., 2012). The gRNA consists of approximately 20 nucleotides, known as the protospacer, which targets a specific genomic sequence, and a "repeat" scaffold that facilitates Cas protein recognition. This configuration enables the recruitment of the ribonucleoprotein complex to a specific genomic locus, which, if proximal to a protospacer-adjacent motif (PAM), facilitates the formation of a DSB. Although various Cas orthologs differ in gRNA size, structure, and PAM requirements, their primary function is to guide Cas endonucleases to precise genomic locations (Jinek et al., 2012; Zetsche et al., 2015, 2017).

The Cas family is a diverse group of proteins, with each member playing a specialized role in the adaptive immune response, from the acquisition and processing of CRISPR arrays to the detection and cleavage of invading DNA. To date, two major classes and six types of Cas proteins have been identified, with the class II endonucleases Cas9 and Cas12 (also known as Cpf1) being the most characterized for genome editing (Koonin et al., 2017; Makarova et al., 2020). Cas9, a dual RNA-guided endonuclease, works in tandem with a crRNA and a trans-activating RNA (tracrRNA) to form a functional ribonucleoprotein complex. Together, this complex guides Cas9 to a targeted genomic site, where upon localization, introduces a DSB using its HNH and RuvC-like nuclease domains (Nishimasu et al., 2014). For genome engineering, this process has been simplified by fusing the crRNA and tracrRNA to form a chimeric single-guide RNA (sgRNA) (Jinek et al., 2012). More recently, Cas12a has been adapted for genome editing. A compact type-V CRISPR system, Cas12a is a single-RNA guided endonuclease that mediates a DSB at the genomic site using a single RuvC domain (Zetsche et al., 2015, 2017). Regardless of the Cas ortholog used, its expression—whether through genomic integration, plasmid transformation, electroporation, or as part of a ribonucleoprotein complex—results in the formation of a targeted DSB. These breaks are then repaired by endogenous mechanisms such as NHEJ and homology-directed repair (HDR), leading to deletions or insertions at the site of the DSB.

As the field of genetic engineering has evolved, so too has genome editing with CRISPR-Cas, leading to the development of Cas variants with enhanced functionalities. By mutating one or both nuclease domains of the Cas enzyme, researchers can create either Cas nickases or nuclease-null (dCas) enzymes. When fused with effector domains, such as those from base editors, prime editors, and TFs, these modified Cas proteins enable precise genetic modifications without generating a DSB (Gaudelli et al., 2017; Komor et al., 2016; Shaw et al., 2022; Anzalone 2019). Base editors, such as adenine or cytosine deaminases, facilitate the deamination of DNA bases, allowing for the precise editing of individual nucleotides to create point mutations (Gaudelli et al., 2017). Prime editing extends these capabilities further by enabling precise deletions and insertions through the fusion of a reverse transcription domain, such as that from the Moloney murine leukemia virus, to a Cas9 nickase. When coexpressed with a prime editing gRNA (pegRNA)—a gRNA that also serves as a template for the reverse transcription—this system can synthesize the desired insertion or deletion sequence directly at the target site to reduce off-target effects, enhance specificity, and eliminate the requirement for donor DNA (Anzalone et al., 2019). Finally, using CRISPR-Cas, genomic landscapes can be manipulated without introducing any genetic changes by employing CRISPR activation (CRISPRa) and CRISPR inhibition (CRISPRi) for transcriptional regulation. By fusing dCas9 to a TF domain, such as a VP-activation or Mxi1-repression domain, this approach allows the recruitment of the ribonucleoprotein-TF fusion to a promoter of interest to up- or down-regulate gene expression as desired (Shaw et al., 2022).

Because of its simplicity and adaptability to reprograming across a wide range of hosts, genome engineering using CRISPR-Cas has become a powerful tool for enhancing bioproduction in diverse organisms, including yeast, bacteria, microalgae, plants, and mammalian cells.

2.2.3.1 Application of CRISPR-Cas in microalgae

The applications of the CRISPR-Cas system in microalgae up to 2022 have recently been reviewed by several research groups (Dhokane et al., 2023; Hassanien et al.,

2023; Jeong et al., 2023; Lee et al., 2023; Zhang et al., 2019). The use of CRISPR system has accelerated over the years, compared to ZNF or TALEN tools due to ease and effectiveness of the technique. CRISPR has been used in many studies ranging from the identification of the activity of genes and metabolic pathways to the production of valuable metabolites (Table 2.1). Kneip et al. (2024); Lee et al. (2023) manipulated the carotenoid pathway by knocking out the gene for lycopene ε-cyclase, one of the key enzymes responsible for the conversion of lycopene into α-carotene, using CRISPR-Cas 9 tool in *C. reinhardtii*. The mutation redirected the carotenoid pathway to the β-carotene branch, and subsequent metabolic engineering resulted in the accumulation of astaxanthin in cells.

Sandor et al. (2024) used the CRISPR-Cas9 system to knock out Lonely Guy (LOG) and isopentenyl transferase (IPT) genes responsible for the biosynthesis of cytokinin, a hormone that stimulates the plant defense against pathogens, in *C. reinhardtii* CC125 strain. The results of this study confirmed that cytokinins produced by *C. reinhardti* provide defense against plant pathogens in tobacco.

Græsholt et al. (2024) evaluated the CRISPR-Cas9 method for converting carotenoid metabolism in the diatom *P. tricornutum*. The researchers identified that the ZEP3 gene is effective in the conversion of diatoxhantin to diatinoxhanthin in the light-to-dark cycle shift during the growth of *P. tricornutum*. They silenced ZEP3 using the CRISPR-Cas9 tool and obtained a mutant strain that can further be used for commercial applications of diatoxanthin production.

Liang et al. (2024) investigated the function of Rubisco large-subunit methyltransferase (LSMT) on the regulation of carbon fixation by knocking out the homolog version of this gene (*nolsmt*) exists in the model algae *Nannochloropsis oceanica*. Disruption of this gene prevents the trimethylation of fructose-1,6-bisphosphate aldolase (FBA) rather than Rubisco, which in turn increases biomass accumulation by improving carbon fixation under high-light culture conditions (Liang et al., 2024).

Sayer et al. (2024) identified that the vitamin B12 uptake in algae is controlled by the cobalamin acquisition protein 1 (CBA1) gene by showing the loss of function when knocking out the gene by CRISPR-Cas9 in *P. tricornutum*.

Leyland et al. (2024) investigated the function of a novel protein from *P. tricornutum* on lipid metabolism. The researchers observed that the lipid droplet and triacylglycerol degradations were disturbed during the shift from stress to nonstress growth conditions if the gene for *P. tricornutum* acyl-CoA binding protein (PtACBP) was knocked out.

Teng and Ng (2023) used the CRISPR activation system (CRISPR-a) to stimulate the expression of the endogenous glutamate synthase in *Chlorella sorokiniana* and developed a cost-effective system, where microalgae can be produced with high efficiency in the salt water with the additional advantage of low-carbon emissions and high-value metabolite accumulation.

Yoneda et al. (2023) identified the function of stramenopile-type lipid droplet protein (StLDP) by knocking out the gene responsible for CRISPR-Cas9 and by reintroducing a recognition site-modified version of the *stldp* in *P. tricornutum*. They revealed that StLDP functions as a scaffold protein to shape lipid droplets in this microalga (Yoneda et al., 2023).

Nymark et al. (2023) investigated the role of the chloroplast signal recognition particle receptor (CpFTSY), which is generally responsible for targeting light-harvesting complex proteins (LHCPs) to the thylakoid membranes of chloroplasts in the plants and algae. The researchers created mutants lacking this gene using CRISPR-Cas9 in *P. tricornutum*. They discovered that this protein is not involved in the insertion of LHCPs into thylakoid membranes, suggesting that the posttranslational phase of the CpSRP pathway is not active in diatoms.

Zhang et al. (2023) created a mutant strain with the CRISPR-Cas9 tool to observe the effect of PhoD-type alkaline phosphatase (AP) on lipid metabolism and composition in *P. tricornutum*. The mutation resulted in higher fatty acid accumulation, with varying fatty acid contents suitable for biomedical and biotechnological applications.

Fucosyltransferase (FucT) is one of the glycotranferase enzymes responsible for N-glycan formation in the Golgi apparatus. Xie et al. (2023) knocked out the PtFucT1 enzyme gene, one of the three glycosyltransferases acting in the Golgi of *P. tricornutum*, using CRISPR-Cas9. They shed light on the mechanism of N-glycosylation of proteins in microalgae, showing the potential of the microalgal platform to produce biopharmaceuticals.

Zadabbas Shahabadi et al. (2023) used the CRISPR-Cas9 system to knock in a gene into the nuclear genome of *C. reinhardtii* by HR to avoid positional effects on the transgene, which could lead to low expression or even gene silencing. They designed a ribonucleoprotein (RNP) complex that includes the Cas9 enzyme and a sgRNA carrying the bacterial phytase gene, which was targeted for the nitrate reductase gene exon 2. They reached about 15% editing efficiency, showing the potential of the CRISPR tool for effective nuclear transformation of the microalgae platform.

Freudenberg et al. (2022) applied a metabolic engineering approach for enhanced production of putrescine in *C. reinhardtii*. CRISPR-Cas9 tool contributed to the understanding of the function of key enzymes of the pathway, such as ornithine decarboxylase 1 (ODC1) and amine oxidase 2 (AMX2), which are responsible for putrescine degradation (Freudenberg et al., 2022). Knocking out this gene combined with the overexpression of arginases resulted in a 10-fold increase in putrescine accumulation in *C. reinhardtii*.

2.2.3.2 Applications of CRISPR in yeast

Owing to its simplicity in genetic engineering, CRISPR has been utilized extensively to engineer yeasts for the sustainable bioproduction of diverse products such as pigments, fragrances, bioplastics, pharmaceuticals, and biofuels (Table 2.2). For example, researchers have recently used CRISPR to integrate 38 pathway enzymes from six species into *S. cerevisiae*, enabling the production of the vaccine adjuvant QS-21 (Martin et al., 2024). Additionally, genome editing with CRISPR-Cas has proven effective in both introducing novel functionalities to chassis hosts and enhancing bioproduction by disrupting genes from competing pathways. Using multiplexed CRISPR, researchers have simultaneously disrupted two genes, PXA1 and SBLE, in *Starmerella bombicola* to increase sophorolipid production, which is a biosurfactant used in many cleaning products and cosmetics (Shi et al., 2022). Furthermore, the development of novel CRISPR toolkits has enabled the use of nonconventional hosts, such as the oleaginous yeast *Y. lipolytica*. Utilizing a novel *Yarrowia* CRISPR toolkit, *Yalicraft*, researchers have engineered a strain to enhance the production of homogentisic acid by knocking out four genes and overexpressing three (Yuzbashev et al., 2023). Finally, using an inducible multiplexed CRISPRai system, researchers were able to simultaneously manipulate the expression of 11 genes within the central metabolism of *S. cerevisiae* to improve the bioproduction of succinic acid—a valuable precursor to bioplastics—by 45-fold (Shaw et al., 2022). Additional examples of CRISPR-Cas applications in yeast bioproduction are summarized in Table 2.2 as well as in several review articles (Jiao et al., 2022; Mitsui et al., 2019; Patra et al., 2021).

2.3 Challenges of genome editing techniques

Although ZFNs, TALENs, and CRISPR have provided excellent tools for genome editing in microalgae and yeasts, many challenges remain, particularly in reducing off-target effects, minimizing cytotoxicity, delivery of the ribonucleoprotein complex, and enhancing genome editing capacity.

Despite their high specificity, ZFNs, TALENs, and CRISPR, all have the potential to bind and cleave unintended sites within the genome (Kim et al., 2012; Ran et al., 2013). To mitigate this, several algorithms have been designed to help create genome editing tools that reduce undesired off-target effects, including ZiFiT for ZFNs, E-TALEN for TALENs, and CHOP-CHOP for CRISPR (Heigwer et al., 2013; Montague et al., 2014; Sander et al., 2010). Furthermore, by utilizing dual Cas-nickases, one targeting each strand, editing can become more specific, as genome editing at a particular locus requires the targeting of two Cas complexes to initiate a DSB and reduce off-target cleavage by up to 1000-fold (Ran et al., 2013). More recently, researchers have trained a large language model (LLM) on over 1 million CRISPR operons to generate OPENCRISPR-1, the first AI-generated gene editor. This new tool exhibits a 95% reduction in off-target editing compared to wild-type Cas9 from *Streptococcus pyogenes* (Ruffolo et al., 2024).

In addition to off targets, genome editing tools have been found to exhibit various cytotoxic effects within hosts. For example, the overexpression of Cas endonucleases has been found to be toxic to many cell types, including microalgae and yeast, and its prolonged expression can lead to increased off-target editing (Jiang et al., 2014). To reduce these effects, many strategies have been developed to deliver the Cas9 protein into cells via episomal plasmids or ribonucleoprotein (RNP) complexes without genome integration (Kelterborn et al., 2022; Poliner et al., 2018; Zhang et al., 2019). Additionally, controlling the expression of Cas endonucleases via an inducible promoter, regulated by factors such as heat, chemical, or light, allows for the spatial expression of the editing system, thereby mitigating potential burden (Patel et al., 2023). In addition, the generation of DSBs has also been found to be toxic in many cell types. By introducing ZFN, TALEN, and Cas-nickases, researchers have been able to reduce genome editing-associated cytotoxicity by generating only single-stranded DNA breaks and circumventing the DSB immune response (Kim et al., 2012; Ran et al., 2013; Wu et al., 2014).

Due to the large size of TALEN and CRISPR proteins, an additional challenge in genome editing is the delivery of the ribonucleoprotein complex. To circumvent this, researchers have been searching for miniature Cas orthologs as well as engineering novel Cas variants through rational mutagenesis (Wu et al., 2021; Xu et al., 2021). Additionally, TALENs often stand more expensive than other gene-editing technologies, in terms of designing, constructing, and validating since both DNA strands should be considered.

Finally, one of the biggest challenges in genome editing is increasing the editing efficiency at a desired locus. In nonconventional hosts, a significant issue is nuclear transgene expression when a gene cassette is inserted into random positions in the genome, resulting in variable expression levels and transgene instability (Cerutti et al., 1997; Kim et al., 2015; Perozeni & Baier, 2023; Zhang et al., 2014). Due to this, many strategies have been employed to increase the gene editing efficiencies and stability of transgene expression by increasing HDR proficiency. By fusing Cas9 to Brex27, a protein involved in human HDR, researchers were able to increase the quadruple integration of heterologous genes by up to 78% in *S. cerevisiae* (Meng et al., 2023). Additionally, using the lowered indel nuclease system enabling accurate repair (LINEAR) CRISPR system, scientists were able to increase precise genome editing efficiencies by 67–100% in the NHEJ-proficient yeast *S. stipites* (Ploessl et al., 2022). Likewise, by grafting Rad52 from *S. cerevisiae* into *Y. lipolytica*, researchers achieved a 6.5-fold increase in gene editing efficiency over the wild-type strain (Ji et al., 2020). Similar efforts to enhance HDR in microalgae have also been explored. Using an RNP complex, researchers successfully integrated a bacterial phytase into the genome of *C. reinhardtii*, mitigating gene silencing and nuclear positional effects typically seen with random (Zadabbas Shahabadi et al., 2023). Finally, an additional factor limiting CRISPR-Cas on-target editing is the stringent requirement for a protospacer adjacent motif (PAM) sequence proximal to the intended editing site. To circumvent this, researchers have engineered novel Cas orthologs with relaxed PAM requirements, allowing for a broader range of genome editing capabilities without PAM limitations (Leenay & Beisel, 2017).

2.4 Conclusion

In conclusion, genome editing techniques have profoundly transformed genetic engineering capabilities, particularly in microalgae and yeast. Methods like ZFNs, TALENs, and CRISPR-Cas have revolutionized the precision with which DNA modifications can be made, offering unparalleled advantages over traditional methods. These technologies enable targeted modifications to optimize metabolic pathways, introduce new functionalities, and investigate gene functions crucial for enhancing bioproduct synthesis in these organisms. TALENs, specifically, have been instrumental in generating auxotrophic mutants and implementing innovative selection systems in microalgae, showcasing their versatility and potential for biotechnological applications. However, the field faces challenges such as delivery efficiency, minimizing unintended genetic alterations, and navigating regulatory landscapes, which are pivotal to fully realizing the benefits of genome editing in industrial settings. Addressing these hurdles is essential for advancing sustainable biotechnological processes utilizing microalgae and yeast, paving the way for future innovations and applications in diverse sectors ranging from biofuels to pharmaceuticals. As research continues to refine these technologies, their integration promises to further accelerate progress towards efficient and environmentally friendly bioproduction systems.

References

Ahmar, S., Saeed, S., Khan, M. H. U., Khan, S. U., Mora-Poblete, F., Kamran, M., Faheem, A., Maqsood, A., Rauf, M., Saleem, S., Hong, W. J., & Jung, K. H. (2020). A revolution toward gene-editing technology and its application to crop improvement. *International Journal of Molecular Sciences, 21*(16), 1–28. https://doi.org/10.3390/ijms21165665, https://www.mdpi.com/1422-0067/21/16/5665/pdf.

Aljabali, A. A. A., El-Tanani, M., & Tambuwala, M. M. (2024). Principles of CRISPR-Cas9 technology: Advancements in genome editing and emerging trends in drug delivery. *Journal of Drug Delivery Science and Technology, 92*. https://doi.org/10.1016/j.jddst.2024.105338, https://www.sciencedirect.com/science/journal/17732247.

Anzalone, A.V., Randolph, P.B., Davis, J.R., Sousa, A.A., Koblan, L.W., Levy, J.M., Chen, P.J., Wilson, C., Newby, G.A., Raguram, A., & Liu, D.R. (2019). Search-and-replace genome editing without double-strand breaks or donor DNA. Nature, 576(7785), 149–157. https://doi.org/10.1038/s41586-019-1711-4.

Bao, A., Zhang, C., Huang, Y., Chen, H., Zhou, X., & Cao, D. (2020). Genome editing technology and application in soybean improvement. *Oil Crop Science, 5*(1), 31–40. https://doi.org/10.1016/j.ocsci.2020.03.001, https://www.sciencedirect.com/journal/oil-crop-science.

Barrangou, R., Fremaux, C., Deveau, H., Richards, M., Boyaval, P., Moineau, S., Romero, D. A., & Horvath, P. (2007). CRISPR provides acquired resistance against viruses in prokaryotes. *Science (New York, N.Y.), 315*(5819), 1709–1712. https://doi.org/10.1126/science.1138140.

Becker, S., & Boch, J. (2021). TALE and TALEN genome editing technologies. *Gene and Genome Editing, 2*, 100007. https://doi.org/10.1016/j.ggedit.2021.100007.

Bhardwaj, A., & Nain, V. (2021). TALENs—an indispensable tool in the era of CRISPR: A mini review. *Journal of Genetic Engineering and Biotechnology, 19*.

Billey, E., Magneschi, L., Leterme, S., Bedhomme, M., Andres-Robin, A., Poulet, L., Michaud, M., Finazzi, G., Dumas, R., Crouzy, S., Laueffer, F., Fourage, L., Rébeillé, F., Amato, A., Collin, S., Jouhet, J., & Maréchal, E. (2021). Characterization of the Bubblegum acyl-CoA synthetase of Microchloropsis gaditana. *Plant Physiology, 185*(3), 815–835. https://doi.org/10.1093/plphys/kiaa110.

Bolotin, A., Quinquis, B., Sorokin, A., & Ehrlich, S. D. (2005). Clustered regularly interspaced short palindrome repeats (CRISPRs) have spacers of extrachromosomal origin. *Microbiology (Reading, England), 151*(8), 2551–2561. https://doi.org/10.1099/mic.0.28048-0.

Castro, N. G., Bjelic, J., Malhotra, G., Huang, C., & Alsaffar, S. H. (2021). Comparison of the feasibility, efficiency, and safety of genome editing technologies. *International Journal of Molecular Sciences, 22*(19). https://doi.org/10.3390/ijms221910355, https://www.mdpi.com/1422-0067/22/19/10355/pdf.

Cerutti, H., Johnson, A. M., Gillham, N. W., & Boynton, J. E. (1997). Epigenetic silencing of a foreign gene in nuclear transformants of Chlamydomonas. *The Plant Cell, 9*(6), 925–945. https://doi.org/10.1105/tpc.9.6.925, http://www.plantcell.org/.

Cong, L., Ran, F. A., Cox, D., Lin, S., Barretto, R., Habib, N., Hsu, P. D., Wu, X., Jiang, W., Marraffini, L. A., & Zhang, F. (2013). Multiplex genome engineering using CRISPR/Cas systems. *Science (New York, N.Y.), 339*(6121), 819–823. https://doi.org/10.1126/science.1231143, http://www.sciencemag.org/content/339/6121/819.full.pdf.

Dhokane, D., Shaikh, A., Yadav, A., Giri, N., Bandyopadhyay, A., Dasgupta, S., & Bhadra, B. (2023). CRISPR-based bioengineering in microalgae for production of industrially important biomolecules. *Frontiers in Bioengineering and Biotechnology, 11*. https://doi.org/10.3389/fbioe.2023.1267826, http://journal.frontiersin.org/journal/bioengineering-and-biotechnology#archive.

Doyon, Y., Vo, T. D., Mendel, M. C., Greenberg, S. G., Wang, J., Xia, D. F., Miller, J. C., Urnov, F. D., Gregory, P. D., & Holmes, M. C. (2011). Enhancing zinc-finger-nuclease activity with improved obligate heterodimeric architectures. *Nature Methods, 8*(1), 74–79. https://doi.org/10.1038/nmeth.1539.

Duensing, N., Sprink, T., Parrott, W. A., Fedorova, M., Lema, M. A., Wolt, J. D., & Bartsch, D. (2018). Novel features and considerations for ERA and regulation of crops produced by genome editing. *Frontiers in Bioengineering and Biotechnology, 9*. https://doi.org/10.3389/fbioe.2018.00079, https://www.frontiersin.org/articles/10.3389/fbioe.2018.00079/full.

Fajardo, C., De Donato, M., Carrasco, R., Martínez-Rodríguez, G., Mancera, J. M., & Fernández-Acero, F. J. (2020). Advances and challenges in genetic engineering of microalgae. *Reviews in Aquaculture, 12*(1), 365–381. https://doi.org/10.1111/raq.12322, http://onlinelibrary.wiley.com/journal/10.1111/(ISSN)1753-5131.

Freudenberg, R. A., Wittemeier, L., Einhaus, A., Baier, T., & Kruse, O. (2022). Advanced pathway engineering for phototrophic putrescine production. *Plant Biotechnology Journal, 20*(10), 1968–1982. https://doi.org/10.1111/pbi.13879, http://onlinelibrary.wiley.com/journal/10.1111/(ISSN)1467-7652.

Gan, Y., Lin, Y., Guo, Y., Qi, X., & Wang, Q. (2018). Metabolic and genomic characterisation of stress-tolerant industrial Saccharomyces cerevisiae strains from TALENs-assisted multiplex editing. *FEMS Yeast Research, 18*(5). https://doi.org/10.1093/femsyr/foy045, http://femsyr.oxfordjournals.org/.

Garneau, J. E., Dupuis, M. E., Villion, M., Romero, D. A., Barrangou, R., Boyaval, P., Fremaux, C., Horvath, P., Magadán, A. H., & Moineau, S. (2010). The CRISPR/cas bacterial immune system cleaves bacteriophage and plasmid DNA. *Nature, 468*(7320), 67–71. https://doi.org/10.1038/nature09523.

Gasiunas, G., Barrangou, R., Horvath, P., & Siksnys, V. (2012). Cas9-crRNA ribonucleoprotein complex mediates specific DNA cleavage for adaptive immunity in bacteria. *Proceedings of the National Academy of Sciences of the United States of America, 109*(39), E2579. https://doi.org/10.1073/pnas.1208507109, http://www.pnas.org/content/109/39/E2579.full.pdf+html. Lithuania.

Gaudelli, N. M., Komor, A. C., Rees, H. A., Packer, M. S., Badran, A. H., Bryson, D. I., & Liu, D. R. (2017). Programmable base editing of T to G C in genomic DNA without DNA cleavage. *Nature, 551*(7681), 464–471. https://doi.org/10.1038/nature24644, http://www.nature.com/nature/index.html.

Græsholt, C., Brembu, T., Volpe, C., Bartosova, Z., Serif, M., Winge, P., & Nymark, M. (2024). Zeaxanthin epoxidase 3 Knockout Mutants of the Model Diatom Phaeodactylum tricornutum Enable Commercial Production of the Bioactive Carotenoid Diatoxanthin. *Marine Drugs, 22*(4), 185. https://doi.org/10.3390/md22040185.

Greiner, A., Kelterborn, S., Evers, H., Kreimer, G., Sizova, I., & Hegemann, P. (2017). Targeting of photoreceptor genes in Chlamydomonas reinhardtii via zinc-finger nucleases and CRISPR/Cas9. *The Plant Cell, 29*(10), 2498–2518. https://doi.org/10.1105/tpc.17.00659, http://www.plantcell.org/content/29/10/2498.full.pdf.

Halford, S. E., Catto, L. E., Pernstich, C., Rusling, D. A., & Sanders, K. L. (2011). The reaction mechanism of FokI excludes the possibility of targeting zinc finger nucleases to unique DNA sites. *Biochemical Society Transactions, 39*(2), 584–588. https://doi.org/10.1042/BST0390584, http://www.biochemsoctrans.org/bst/039/0584/0390584.pdf. United Kingdom.

Hao, X., Luo, L., Jouhet, J., Rébeillé, F., Maréchal, E., Hu, H., Pan, Y., Tan, X., Chen, Z., You, L., Chen, H., Wei, F., & Gong, Y. (2018). Enhanced triacylglycerol production in the diatom Phaeodactylum tricornutum by inactivation of a Hotdog-fold thioesterase gene using TALEN-based targeted mutagenesis. *Biotechnology for Biofuels, 11*(1). https://doi.org/10.1186/s13068-018-1309-3.

Hassanien, A., Saadaoui, I., Schipper, K., Al-Marri, S., Dalgamouni, T., Aouida, M., Saeed, S., & Al-Jabri, H. M. (2023). Genetic engineering to enhance microalgal-based produced water treatment with emphasis on CRISPR/Cas9: A review. *Frontiers in Bioengineering and Biotechnology, 10*. https://doi.org/10.3389/fbioe.2022.1104914, http://journal.frontiersin.org/journal/bioengineering-and-biotechnology#archive.

Heigwer, F., Kerr, G., Walther, N., Glaeser, K., Pelz, O., Breinig, M., & Boutros, M. (2013). E-TALEN: A web tool to design TALENs for genome engineering. *Nucleic Acids Research, 41*(20), e190. https://doi.org/10.1093/nar/gkt789.

Hiroyuki, S., & Susumu, K. (1981). New restriction endonucleases from Flavobacterium okeanokoites (FokI) and Micrococcus luteus (MluI). *Gene, 16*(1-3), 73–78. https://doi.org/10.1016/0378-1119(81)90062-7.

Ishino, Y., Shinagawa, H., Makino, K., Amemura, M., & Nakata, A. (1987). Nucleotide sequence of the iap gene, responsible for alkaline phosphatase isozyme conversion in Escherichia coli, and identification of the gene product. *Journal of Bacteriology, 169*(12), 5429–5433. https://doi.org/10.1128/jb.169.12.5429-5433.1987.

Jeong, B.-ryool, Jang, J., & Jin, E. S. (2023). Genome engineering via gene editing technologies in microalgae. *Bioresource Technology, 373*, 128701. https://doi.org/10.1016/j.biortech.2023.128701.

Ji, Q., Mai, J., Ding, Y., Wei, Y., Ledesma-Amaro, R., & Ji, X.-J. (2020). Improving the homologous recombination efficiency of Yarrowia lipolytica by grafting heterologous component from Saccharomyces cerevisiae. *Metabolic Engineering Communications, 11*, e00152. https://doi.org/10.1016/j.mec.2020.e00152.

Ji, Z., Guo, W., Chen, X., Wang, C., & Zhao, K. (2022). Plant executor genes. *International Journal of Molecular Sciences, 23*(3), 1524. https://doi.org/10.3390/ijms23031524.

Jiang, W., Brueggeman, A. J., Horken, K. M., Plucinak, T. M., & Weeks, D. P. (2014). Successful transient expression of Cas9 and single guide RNA genes in Chlamydomonas reinhardtii. *Eukaryotic Cell, 13*(11), 1465–1469. https://doi.org/10.1128/EC.00213-14, http://ec.asm.org/content/13/11/1465.full.pdf.

Jiao, X., Gu, Y., Zhou, P., Yu, H., & Ye, L. (2022). Recent advances in construction and regulation of yeast cell factories. *World Journal of Microbiology and Biotechnology, 38*(4). https://doi.org/10.1007/s11274-022-03241-4, http://www.wkap.nl/journalhome.htm/0959-3993.

Jinek, M., Chylinski, K., Fonfara, I., Hauer, M., Doudna, J. A., & Charpentier, E. (2012). A programmable dual-RNA-guided DNA endonuclease in adaptive bacterial immunity. *Science (New York, N.Y.), 337*(6096), 816–821. https://doi.org/10.1126/science.1225829, http://www.sciencemag.org/content/337/6096/816.full.pdf.

Kasai, Y., Oshima, K., Ikeda, F., Abe, J., Yoshimitsu, Y., & Harayama, S. (2015). Construction of a self-cloning system in the unicellular green alga Pseudochoricystis ellipsoidea. *Biotechnology for Biofuels, 8*(1). https://doi.org/10.1186/s13068-015-0277-0, http://www.biotechnologyforbiofuels.com/.

Kelterborn, S., Boehning, F., Sizova, I., Baidukova, O., Evers, H., & Hegemann, P. (2022). Gene Editing in Green Alga Chlamydomonas reinhardtii via CRISPR-Cas9 Ribonucleoproteins. *Methods in Molecular Biology, 2379*, 45–65. https://doi.org/10.1007/978-1-0716-1791-5_3, http://www.springer.com/series/7651.

Kim, E., Kim, S., Kim, D. H., Choi, B. S., Choi, I. Y., & Kim, J. S. (2012). Precision genome engineering with programmable DNA-nicking enzymes. *Genome Research, 22*(7), 1327–1333. https://doi.org/10.1101/gr.138792.112, http://genome.cshlp.org/content/22/7/1327.full.pdf+html. South Korea.

Kim, E. J., Ma, X., & Cerutti, H. (2015). Gene silencing in microalgae: Mechanisms and biological roles. *Bioresource Technology, 184*, 23–32. https://doi.org/10.1016/j.biortech.2014.10.119, http://www.elsevier.com/locate/biortech.

Kneip, J. S., Kniepkamp, N., Jang, J., Mortaro, M. G., Jin, E. S., Kruse, O., & Baier, T. (2024). CRISPR/Cas9-mediated knockout of the lycopene ε-cyclase for efficient astaxanthin production in the green microalga chlamydomonas reinhardtii. *Plants, 13*(10). https://doi.org/10.3390/plants13101393, http://www.mdpi.com/journal/plants.

Komor, A. C., Kim, Y. B., Packer, M. S., Zuris, J. A., & Liu, D. R. (2016). Programmable editing of a target base in genomic DNA without double-stranded DNA cleavage. *Nature, 533*, 420–424. https://doi.org/10.1038/nature17946, http://www.nature.com/nature/index.html.

Koonin, E. V., Makarova, K. S., & Zhang, F. (2017). Diversity, classification and evolution of CRISPR-Cas systems. *Current Opinion in Microbiology, 37*, 67–78. https://doi.org/10.1016/j.mib.2017.05.008, http://www.elsevier.com/locate/mib.

Kurita, T., Iwai, M., Moroi, K., Okazaki, K., Nomura, S., Saito, F., Maeda, S., Takami, A., Sakamoto, A., Ohta, H., Sakuma, T., & Yamamoto, T. (2022). Genome editing with removable TALEN vectors harboring a yeast centromere and autonomous replication sequence in oleaginous microalga. *Scientific Reports, 12*(1). https://doi.org/10.1038/s41598-022-06495-y, http://www.nature.com/srep/index.html.

Kurita, T., Moroi, K., Iwai, M., Okazaki, K., Shimizu, S., Nomura, S., Saito, F., Maeda, S., Takami, A., Sakamoto, A., Ohta, H., Sakuma, T., & Yamamoto, T. (2020). Efficient and multiplexable genome editing using Platinum TALENs in oleaginous microalga, Nannochloropsis oceanica NIES-2145. *Genes to Cells, 25*(10), 695–702. https://doi.org/10.1111/gtc.12805, http://onlinelibrary.wiley.com/journal/10.1111/(ISSN)1365-2443.

Lee, T. M., Lin, J. Y., Tsai, T. H., Yang, R. Y., & Ng, I. S. (2023). Clustered regularly interspaced short palindromic repeats (CRISPR) technology and genetic engineering strategies for microalgae towards carbon neutrality: A critical review. *Bioresource Technology, 368*. https://doi.org/10.1016/j.biortech.2022.128350, http://www.elsevier.com/locate/biortech.

Leenay, R. T., & Beisel, C. L. (2017). Deciphering, Communicating, and Engineering the CRISPR PAM. *Journal of Molecular Biology, 429*(2), 177–191. https://doi.org/10.1016/j.jmb.2016.11.024, https://www.journals.elsevier.com/journal-of-molecular-biology.

Leyland, B., Novichkova, E., Dolui, A. K., Jallet, D., Daboussi, F., Legeret, B., Li, Z., Li-Beisson, Y., Boussiba, S., & Khozin-Goldberg, I. (2024). Acyl-CoA binding protein is required for lipid droplet degradation in the diatom Phaeodactylum tricornutum. *Plant Physiology, 194*(2), 958–981. https://doi.org/10.1093/plphys/kiad525, https://academic.oup.com/plphys/article/194/2/958/7293255.

Li, H., Yang, Y., Hong, W., Huang, M., Wu, M., & Zhao, X. (2020). Applications of genome editing technology in the targeted therapy of human diseases: mechanisms, advances and prospects. *Signal Transduction and Targeted Therapy, 5*(1). https://doi.org/10.1038/s41392-019-0089-y, http://www.nature.com/sigtrans.

Liang, W., Wei, L., Wang, Q., You, W., Poetsch, A., Du, X., Lv, N., & Xu, J. (2024). Knocking out chloroplastic aldolases/rubisco lysine methyltransferase enhances biomass accumulation in nannochloropsis oceanica under high-light stress. *International Journal of Molecular Sciences, 25*(7), 3756. https://doi.org/10.3390/ijms25073756.

Makarova, K. S., Wolf, Y. I., Iranzo, J., Shmakov, S. A., Alkhnbashi, O. S., Brouns, S. J. J., Charpentier, E., Cheng, D., Haft, D. H., Horvath, P., Moineau, S., Mojica, F. J. M., Scott, D., Shah, S. A., Siksnys, V., Terns, M. P., Venclovas, M. F., White, A. F., Yakunin, W., ... Koonin (2020). Evolutionary classification of CRISPR–Cas systems: a burst of class 2 and derived variants. *Nature Reviews. Microbiology, 18*(2), 67–83. https://doi.org/10.1038/s41579-019-0299-x, http://www.nature.com/nrmicro/index.html.

Martin, L. B. B., Kikuchi, S., Rejzek, M., Owen, C., Reed, J., Orme, A., Misra, R. C., El-Demerdash, A., Hill, L., Hodgson, H., Liu, Y., Keasling, J. D., Field, R. A., Truman, A. W., & Osbourn, A. (2024). Complete biosynthesis of the potent vaccine adjuvant QS-21. *Nature Chemical Biology, 20*(4), 493–502. https://doi.org/10.1038/s41589-023-01538-5, https://www.nature.com/nchembio.

McCarty, N. S., Graham, A. E., Studená, L., & Ledesma-Amaro, R. (2020). Multiplexed CRISPR technologies for gene editing and transcriptional regulation. *Nature Communications, 11*(1). https://doi.org/10.1038/s41467-020-15053-x, http://www.nature.com/ncomms/index.html.

Meng, J., Qiu, Y., Zhang, Y., Zhao, H., & Shi, S. (2023). CMI: CRISPR/Cas9 Based Efficient Multiplexed Integration in Saccharomyces cerevisiae. *ACS Synthetic Biology, 12*(5), 1408–1414. https://doi.org/10.1021/acssynbio.2c00591.

Mitsui, R., Yamada, R., & Ogino, H. (2019). CRISPR system in the yeast Saccharomyces cerevisiae and its application in the bioproduction of useful chemicals. *World Journal of Microbiology and Biotechnology, 35*(7). https://doi.org/10.1007/s11274-019-2688-8, http://www.wkap.nl/journalhome.htm/0959-3993.

Montague, T. G., Cruz, J. M., Gagnon, J. A., Church, G. M., & Valen, E. (2014). CHOPCHOP: A CRISPR/Cas9 and TALEN web tool for genome editing. *Nucleic Acids Research, 42*(1), W401. https://doi.org/10.1093/nar/gku410, http://nar.oxfordjournals.org/.

Mushtaq, M., Bhat, J. A., Mir, Z. A., Sakina, A., Ali, S., Singh, A. K., Tyagi, A., Salgotra, R. K., Dar, A. A., & Bhat, R. (2018). CRISPR/Cas approach: A new way of looking at plant-abiotic interactions. *Journal of Plant Physiology, 224-225*, 156–162. https://doi.org/10.1016/j.jplph.2018.04.001, http://www.urbanfischer.de/journals/jpp/p_physio.htm.

Mushtaq, M., Sakina, A., Wani, S. H., Shikari, A. B., Tripathi, P., Zaid, A., Galla, A., Abdelrahman, M., Sharma, M., Singh, A. K., & Salgotra, R. K. (2019). Harnessing genome editing techniques to engineer disease resistance in plants. *Frontiers in Plant Science, 10*. https://doi.org/10.3389/fpls.2019.00550, https://www.frontiersin.org/articles/10.3389/fpls.2019.00550/pdf.

Ng, I. S., Keskin, B. B., & Tan, S. I. (2020). A critical review of genome editing and synthetic biology applications in metabolic engineering of microalgae and cyanobacteria. *Biotechnology Journal, 15*(8). https://doi.org/10.1002/biot.201900228, http://onlinelibrary.wiley.com/journal/10.1002/(ISSN)1860-7314.

Nishida, K., & Kondo, A. (2021). CRISPR-derived genome editing technologies for metabolic engineering. *Metabolic Engineering, 63*, 141–147. https://doi.org/10.1016/j.ymben.2020.12.002, http://www.elsevier.com/inca/publications/store/6/2/2/9/1/3/index.htt.

Nishimasu, H., Ran, F. A., Hsu, P. D., Konermann, S., Shehata, S. I., Dohmae, N., Ishitani, R., Zhang, F., & Nureki, O. (2014). Crystal structure of Cas9 in complex with guide RNA and target DNA. *Cell, 156*(5), 935–949. https://doi.org/10.1016/j.cell.2014.02.001, https://www.sciencedirect.com/journal/cell.

Nymark, M., Finazzi, G., Volpe, C., Serif, M., de Miranda Fonseca, D., Sharma, A., Sanchez, N., Sharma, A. K., Ashcroft, F., Kissen, R., Winge, P., & Bones, A. M. (2023). Loss of CpFTSY reduces photosynthetic performance and affects insertion of PsaC of PSI in diatoms. *Plant and Cell Physiology, 64*(6), 583–603. https://doi.org/10.1093/pcp/pcad014, http://pcp.oxfordjournals.org/.

Ozaki, A., Konishi, R., Otomo, C., Kishida, M., Takayama, S., Matsumoto, T., Tanaka, T., & Kondo, A. (2017). Metabolic engineering of Schizosaccharomyces pombe via CRISPR-Cas9 genome editing for lactic acid production from glucose and cellobiose. *Metabolic Engineering Communications, 5*, 60–67. https://doi.org/10.1016/j.meteno.2017.08.002, http://www.journals.elsevier.com/metabolic-engineering-communications/.

Patel, V. K., Das, A., Kumari, R., & Kajla, S. (2023). Recent progress and challenges in CRISPR-Cas9 engineered algae and cyanobacteria. *Algal Research, 71*, 103068. https://doi.org/10.1016/j.algal.2023.103068.

Patra, P., Das, M., Kundu, P., & Ghosh, A. (2021). Recent advances in systems and synthetic biology approaches for developing novel cell-factories in non-conventional yeasts. *Biotechnology Advances, 47*, 107695. https://doi.org/10.1016/j.biotechadv.2021.107695.

Perozeni, F., & Baier, T. (2023). Current nuclear engineering strategies in the green microalga chlamydomonas reinhardtii. *Life (Chicago, Ill.: 1978), 13*(7), 1566. https://doi.org/10.3390/life13071566.

Ploessl, D., Zhao, Y., Cao, M., Ghosh, S., Lopez, C., Sayadi, M., Chudalayandi, S., Severin, A., Huang, L., Gustafson, M., & Shao, Z. (2022). A repackaged CRISPR platform increases homology-directed repair for yeast engineering. *Nature Chemical Biology, 18*(1), 38–46. https://doi.org/10.1038/s41589-021-00893-5, http://www.nature.com/nchembio.

Poliner, E., Takeuchi, T., Du, Z. Y., Benning, C., & Farré, E. M. (2018). Nontransgenic Marker-Free Gene Disruption by an Episomal CRISPR System in the Oleaginous Microalga, Nannochloropsis oceanica CCMP1779. *ACS Synthetic Biology, 7*(4), 962–968. https://doi.org/10.1021/acssynbio.7b00362, http://pubs.acs.org/journal/asbcd6.

Ran, F. A., Hsu, P. D., Lin, C. Y., Gootenberg, J. S., Konermann, S., Trevino, A. E., Scott, D. A., Inoue, A., Matoba, S., Zhang, Y., & Zhang, F. (2013). XDouble nicking by RNA-guided CRISPR cas9 for enhanced genome editing specificity. *Cell, 154*(6). https://doi.org/10.1016/j.cell.2013.08.021, https://www.sciencedirect.com/journal/cell.

Rigouin, C., Gueroult, M., Croux, C., Dubois, G., Borsenberger, V., Barbe, S., Marty, A., Daboussi, F., André, I., & Bordes, F. (2017). Production of medium chain fatty acids by yarrowia lipolytica: Combining molecular design and TALEN to engineer the fatty acid synthase. *ACS Synthetic Biology, 6*(10), 1870–1879. https://doi.org/10.1021/acssynbio.7b00034.

Ruffolo, J. A., Nayfach, S., Gallagher, J., Bhatnagar, A., Beazer, J., Hussain, R., Russ, J., Yip, J., Hill, E., Pacesa, M., Meeske, A. J., Cameron, P., & Madani, A. (2024). Design of highly functional genome editors by modeling the universe of CRISPR-Cas sequences. *bioRxiv*. https://doi.org/10.1101/2024.04.22.590591, https://www.biorxiv.org.

Sander, J. D., Maeder, M. L., Reyon, D., Voytas, D. F., Joung, J. K., & Dobbs, D. (2010). ZiFiT (Zinc Finger Targeter): An updated zinc finger engineering tool. *Nucleic Acids Research, 38*(2), W462. https://doi.org/10.1093/nar/gkq319.

Sandor, R., Wagh, S. G., Kelterborn, S., Großkinsky, D. K., Novak, O., Olsen, N., Paul, B., Petřík, I., Wu, S., Hegemann, P., Strnad, M., Červený, J., & Roitsch, T. (2024). Cytokinin-deficient C hlamydomonas reinhardtii CRISPR-Cas9 mutants show reduced ability to prime resistance of tobacco against bacterial infection. *Physiologia Plantarum, 176*.

Sayer, A. P., Llavero-Pasquina, M., Geisler, K., Holzer, A., Bunbury, F., Mendoza-Ochoa, G. I., Lawrence, A. D., Warren, M. J., Mehrshahi, P., & Smith, A. G. (2024). Conserved cobalamin acquisition protein 1 is essential for vitamin B12 uptake in both Chlamydomonas and Phaeodactylum. *Plant Physiology, 194*(2), 698–714. https://doi.org/10.1093/plphys/kiad564, https://academic.oup.com/plphys/article/194/2/698/7326690.

Scholze, H., & Boch, J. (2010). TAL effector-DNA specificity. *Virulence, 1*(5), 428–432. https://doi.org/10.4161/viru.1.5.12863, http://www.landesbioscience.com/journals/virulence/ScholzeVIRU1-5.pdf.

Shaw, W. M., Studená, L., Roy, K., Hapeta, P., McCarty, N. S., Graham, A. E., Ellis, T., & Ledesma-Amaro, R. (2022). Inducible expression of large gRNA arrays for multiplexed CRISPRai applications. *Nature Communications, 13*(1). https://doi.org/10.1038/s41467-022-32603-7, https://www.nature.com/ncomms/.

Shi, Y., Zhang, L., Zhang, M., Chu, J., Xia, Y., Yang, H., Liu, L., & Chen, X. (2022). A CRISPR–Cas9 system-mediated genetic disruption and multi-fragment assembly in starmerella bombicola. *ACS Synthetic Biology, 11*(4), 1497–1509. https://doi.org/10.1021/acssynbio.1c00582.

Singh, V. (2021). An introduction to CRISPR-Cas systems for reprogramming the genome of mammalian cells. *Progress in Molecular Biology and Translational Science, 181*, 1–13. https://doi.org/10.1016/bs.pmbts.2021.01.010, http://www.elsevier.com/books/book-series/progress-in-molecular-biology-and-translational-science#.

Siripong, W., Angela, C., Tanapongpipat, S., & Runguphan, W. (2020). Metabolic engineering of Pichia pastoris for production of isopentanol (3-Methyl-1-butanol). *Enzyme and Microbial Technology, 138*, 109557. https://doi.org/10.1016/j.enzmictec.2020.109557.

Sizova, I., Greiner, A., Awasthi, M., Kateriya, S., & Hegemann, P. (2013). Nuclear gene targeting in C hlamydomonas using engineered zinc-finger nucleases. *The Plant Journal, 73*(5), 873–882. https://doi.org/10.1111/tpj.12066.

Sproles, A. E., Fields, F. J., Smalley, T. N., Le, C. H., Badary, A., & Mayfield, S. P. (2021). Recent advancements in the genetic engineering of microalgae. *Algal Research, 53*. https://doi.org/10.1016/j.algal.2020.102158, http://www.sciencedirect.com/science/journal/aip/22119264

Swietalski, P., Hetzel, F., Seitl, I., & Fischer, L. (2020). Secretion of a low and high molecular weight β-glycosidase by Yarrowia lipolytica. *Microbial Cell Factories, 19*(1). https://doi.org/10.1186/s12934-020-01358-5, http://www.microbialcellfactories.com/home/.

Teng, C. S., & Ng, I. S. (2023). Optimization of 4-aminobutyric acid feeding strategy and clustered regularly interspaced short palindromic repeats activation for enhanced value-added chemicals in halophilic Chlorella sorokiniana. *Bioresource Technology, 387*. https://doi.org/10.1016/j.biortech.2023.129599, http://www.elsevier.com/locate/biortech.

Teper, D., White, F. F., & Wang, N. (2023). The dynamic transcription activator-like effector family of xanthomonas. *Phytopathology, 113*(4), 651–666. https://doi.org/10.1094/PHYTO-10-22-0365-KD, https://apsjournals.apsnet.org/doi/epdf/10.1094/PHYTO-10-22-0365-KD.

Townsend, J. A., Wright, D. A., Winfrey, R. J., Fu, F., Maeder, M. L., Joung, J. K., & Voytas, D. F. (2009). High-frequency modification of plant genes using engineered zinc-finger nucleases. *Nature, 459*(7245), 442–445. https://doi.org/10.1038/nature07845.

Vanegas, K. G., Lehka, B. J., & Mortensen, U. H. (2017). Switch: A dynamic CRISPR tool for genome engineering and metabolic pathway control for cell factory construction in Saccharomyces cerevisiae. *Microbial Cell Factories, 16*(1). https://doi.org/10.1186/s12934-017-0632-x, http://www.microbialcellfactories.com/home/.

Venezia, M., & Creasey Krainer, K. M. (2021). Current advancements and limitations of gene editing in orphan crops. *Frontiers in Plant Science, 12*. https://doi.org/10.3389/fpls.2021.742932, https://www.frontiersin.org/journals/plant-science.

Wang, L., Lin, J., Zhang, T., Xu, K., Ren, C., Zhang, Z., & Isalan, M. (2013). Simultaneous screening and validation of effective zinc finger nucleases in yeast. *PLoS One, 8*(5), e64687. https://doi.org/10.1371/journal.pone.0064687.

Wani, A. K., Akhtar, N., Singh, R., Prakash, A., Raza, S. H. A., Cavalu, S., Chopra, C., Madkour, M., Elolimy, A., & Hashem, N. M. (2023). Genome centric engineering using ZFNs, TALENs and CRISPR-Cas9 systems for trait improvement and disease control in animals. *Veterinary Research Communications, 47*(1), 1–16. https://doi.org/10.1007/s11259-022-09967-8, https://www.springer.com/journal/11259.

Wu, Y., Gao, T., Wang, X., Hu, Y., Hu, X., Hu, Z., Pang, J., Li, Z., Xue, J., Feng, M., Wu, L., & Liang, D. (2014). TALE nickase mediates high efficient targeted transgene integration at the human multi-copy ribosomal DNA locus. *Biochemical and Biophysical Research Communications, 446*(1), 261–266. https://doi.org/10.1016/j.bbrc.2014.02.099, http://www.elsevier.com/inca/publications/store/6/2/2/7/9/0/index.htt.

Wu, Z., Zhang, Y., Yu, H., Pan, D., Wang, Y., Wang, Y., Li, F., Liu, C., Nan, H., Chen, W., & Ji, Q. (2021). Programmed genome editing by a miniature CRISPR-Cas12f nuclease. *Nature Chemical Biology, 17*(11), 1132–1138. https://doi.org/10.1038/s41589-021-00868-6.

Xie, X., Yang, J., Du, H., Chen, J., Sanganyado, E., Gong, Y., Du, H., Chen, W., Liu, Z., & Liu, X. (2023). Golgi fucosyltransferase 1 reveals its important role in α-1,4-fucose modification of N-glycan in CRISPR/Cas9 diatom Phaeodactylum tricornutum. *Microbial Cell Factories, 22*(1). https://doi.org/10.1186/s12934-022-02000-2.

Xu, X., Chemparathy, A., Zeng, L., Kempton, H. R., Shang, S., Nakamura, M., & Qi, L. S. (2021). Engineered miniature CRISPR-Cas system for mammalian genome regulation and editing. *Molecular Cell, 81*(20), 4333. https://doi.org/10.1016/j.molcel.2021.08.008, http://www.molecule.org.

Yang, Z., & Blenner, M. (2020). Genome editing systems across yeast species. *Current Opinion in Biotechnology, 66*, 255–266. https://doi.org/10.1016/j.copbio.2020.08.011, http://www.elsevier.com/locate/copbio.

Yang, Z., Yu, Y., Tay, Y. X., & Yue, G. H. (2022). Genome editing and its applications in genetic improvement in aquaculture. *Reviews in Aquaculture, 14*(1), 178–191. https://doi.org/10.1111/raq.12591, http://onlinelibrary.wiley.com/journal/10.1111/(ISSN)1753-5131.

Yoneda, K., Oishi, R., Yoshida, M., Matsuda, Y., & Suzuki, I. (2023). Stramenopile-type lipid droplet protein functions as a lipid droplet scaffold protein in the marine diatom phaeodactylum tricornutum. *Plant and Cell Physiology, 64*(7), 803–813. https://doi.org/10.1093/pcp/pcad040, http://pcp.oxfordjournals.org/.

Yuzbashev, T., Yuzbasheva, E., Melkina, O., Patel, D., Bubnov, D., Dietz, H., & Ledesma-Amaro, R. (2023). A DNA assembly toolkit to unlock the CRISPR/Cas9 potential for metabolic engineering. *Research Square*. https://doi.org/10.21203/rs.3.rs-2738543/v1, https://www.researchsquare.com/browse.

Zadabbas Shahabadi, H., Akbarzadeh, A., Ofoghi, H., & Kadkhodaei, S. (2023). Site-specific gene knock-in and bacterial phytase gene expression in Chlamydomonas reinhardtii via Cas9 RNP-mediated HDR. *Frontiers in Plant Science, 14*. https://doi.org/10.3389/fpls.2023.1150436, https://www.frontiersin.org/journals/plant-science.

Zetsche, B., Gootenberg, J. S., Abudayyeh, O. O., Slaymaker, I. M., Makarova, K. S., Essletzbichler, P., Volz, S. E., Joung, J., Van Der Oost, J., Regev, A., Koonin, E. V., & Zhang, F. (2015). Cpf1 Is a single RNA-guided endonuclease of a class 2 CRISPR-cas system. *Cell, 163*(3), 759–771. https://doi.org/10.1016/j.cell.2015.09.038, https://www.sciencedirect.com/journal/cell.

Zetsche, B., Heidenreich, M., Mohanraju, P., Fedorova, I., Kneppers, J., Degennaro, E. M., Winblad, N., Choudhury, S. R., Abudayyeh, O. O., Gootenberg, J. S., Wu, W. Y., Scott, D. A., Severinov, K., Van Der Oost, J., & Zhang, F. (2017). Multiplex gene editing by CRISPR-Cpf1 using a single crRNA array. *Nature Biotechnology, 35*(1), 31–34. https://doi.org/10.1038/nbt.3737, http://www.nature.com/nbt/index.html.

Zhang, G., Lin, Y., Qi, X., Li, L., Wang, Q., & Ma, Y. (2015). TALENs-assisted multiplex editing for accelerated genome evolution to improve yeast phenotypes. *ACS Synthetic Biology, 4*(10), 1101–1111. https://doi.org/10.1021/acssynbio.5b00074, http://pubs.acs.org/journal/asbcd6.

Zhang, H., Zhang, J., Lang, Z., Botella, J. R., & Zhu, J.-K. (2017). Genome editing—principles and applications for functional genomics research and crop improvement. *Critical Reviews in Plant Sciences, 36*(4), 291–309. https://doi.org/10.1080/07352689.2017.1402989.

Zhang, H. X., Zhang, Y., & Yin, H. (2019). Genome editing with mRNA encoding ZFN, TALEN, and Cas9. *Molecular Therapy, 27*(4), 735–746. https://doi.org/10.1016/j.ymthe.2019.01.014, https://www.journals.elsevier.com/molecular-therapy.

Zhang, K., Raboanatahiry, N., Zhu, B., & Li, M. (2017). Progress in genome editing technology and its application in plants. *Frontiers in Plant Science, 8*. https://doi.org/10.3389/fpls.2017.00177, http://journal.frontiersin.org/article/10.3389/fpls.2017.00177/full.

Zhang, K., Li, J., Cheng, J., & Lin, S. (2023). Alkaline phosphatase PhoD mutation induces fatty acid and long-chain polyunsaturated fatty acid (LC-PUFA)-bound phospholipid production in the model diatom phaeodactylum tricornutum. *Marine Drugs, 21*(11), 560. https://doi.org/10.3390/md21110560.

Zhang, L., Zhang, H., Liu, Y., Zhou, J., Shen, W., Liu, L., Li, Q., & Chen, X. (2020). A CRISPR–Cas9 system for multiple genome editing and pathway assembly in Candida tropicalis. *Biotechnology and Bioengineering, 117*(2), 531–542. https://doi.org/10.1002/bit.27207.

Zhang, R., Patena, W., Armbruster, U., Gang, S. S., Blum, S. R., & Jonikas, M. C. (2014). High-throughput genotyping of green algal mutants reveals random distribution of mutagenic insertion sites and endonucleolytic cleavage of transforming DNA. *The Plant Cell, 26*(4), 1398–1409. https://doi.org/10.1105/tpc.114.124099, http://www.plantcell.org/content/26/4/1398.full.pdf.

Zhang, Y.T., Jiang, J.Y., Shi, T.Q., Sun, X.M., Zhao, Q.Y., Huang, H., & Ren, L.J. (2019). Application of the CRISPR/Cas system for genome editing in microalgae. *Applied Microbiology and Biotechnology, 103*(8), 3239–3248. https://doi.org/10.1007/s00253-019-09726-x, https://link.springer.com/content/pdf/10.1007/s00253-019-09726-x.pdf.

Yeasts as cell factories for the heterologous production of thermophilic enzymes

3

Almudena Saavedra-Bouza, María-Eugenia DeCastro, Manuel Becerra, and María-Isabel González-Siso
CICA – Interdisciplinary Center for Chemistry and Biology, Departament of Biology, Faculty of Sciences, Biomedical Research Institute A Coruña (INIBIC), University of A Coruña, A Coruña, Spain

3.1 Introduction

The global industrial enzymes market is growing sharply. Thermozymes, those enzymes that perform optimally and are stable at high temperatures, are specially demanded by the industry because they are robust, resistant to other denaturing agents besides heat, and allow advantages derived of catalysis at high temperature such as increased reaction rate and decreased microbial contamination risk, lower viscosity and higher solubility of substrates and products, and many more. Thermophilic microorganisms, living above 55°C, are the natural source of thermozymes. Therefore, during the last years, abundant bioprospecting has been made in thermophilic environments, both by culture-based and culture-independent (metagenomics) approaches, looking for novel biocatalysts. Culturable thermophilic microorganisms show very low growth rates and enzyme productivities in laboratory cultures, while metagenomics allows access to genes from unknown, nonculturable microorganisms. In both approaches, the production of thermozymes is achieved, through expression in heterologous mesophilic hosts with well-established recombinant DNA tools and culture conditions. Although *Escherichia coli* is a common host, several yeast strains have been taking this role too. In Table 3.1 there is a recompilation of cases of thermozymes expressed from yeasts during the last 10 years, including both those of thermophilic origin and those whose thermostability was improved by protein engineering techniques. Next here, the main expression systems that allow to use yeasts as hosts for thermozymes production are explained, inferred mostly from the articles cited in Table 3.1. The history since the first reported case of a recombinant yeast producing a thermophilic enzyme is also briefly described. Up to our knowledge, this is the first study that reviews the performance of yeasts as hosts for specifically thermozymes expression.

3.2 The beginnings

The first publications about the suitability of the yeast *Saccharomyces cerevisiae* as host for the heterologous expression of thermophilic enzymes are dated in 1988

Eukaryotic Microorganisms as Sources of Bioproducts. DOI: https://doi.org/10.1016/B978-0-443-30188-9.00002-0

Table 3.1 Thermozymes heterologously expressed from yeasts in the last 10 years. The bibliographic search was made during May 2024 using "thermophilic or thermostable and yeast expression" as keywords.

	Year	References	Yeast host	Thermozyme	Expression system	Aim—Application
1	2024	Serra et al. (2024)	*Komagataella phaffii* (*Pichia pastoris*) X33	Polygalacturonase (TL-PG1) from the thermophilic fungus *Thermomyces lanuginosus*	pPICZαA vector AOX1 promoter His tag	Citrus pulp hydrolysis Textile bioscouring
2	2024	Han et al. (2024)	*Pichia pastoris* GS115	Acidic exochitinase from the thermophilic fungus *Paecilomyces thermophila* J18	pPIC9K vector α-factor secretion signal AOX1 promoter	Degrading the cell walls of *Ganoderma lucidum* spores for the extraction of bioactive components
3	2024	Han et al. (2024)	*Saccharomyces cerevisiae* BJ3505	Metagenomic thermophilic esterase	YEpFLAG-1 episomal vector α-factor secretion signal *ADH2* promoter	Characterization novel enzyme from nonculturable microorganism
4	2024	Ahmed et al. (2024)	*Pichia pastoris* BG10	Peptidyl-lys metalloendopeptidase (Tc-LysN) from *Tramates coccinea* BRFM310	pBSY2S1Z vector α-factor secretion signal *AOX1* promoter	Proteomics workflows: protein digestion and mass spectrometry fingerprinting
5	2024	Song et al. (2024)	*Saccharomyces cerevisiae* 1447 ura3Δ	Thermostable laccase Lcc1 from *Trametes trogii*	Original signal peptide *PGK1* promoter Integration using CRISPR-Cas9 method	Decolorization and detoxication of malachite green pollutants
6	2023	Dai et al. (2023)	*Pichia pastoris* C1	Thermophilic endoglucanase from *Thermothielavioides terrestris* (TtCel45A)	CRISPR-Cas9-meditated chromosomal integration in three copies *AOX1* promoter His tag	Cost effective enzyme and SCP production

7	2023	Javanmard et al. (2023)	*Pichia pastoris* GS115	Thermophilic cellulases: cgxA gene (from *Ampullaria crossean*, with endo- and exoglucanase activities*), cglT gene (from *Thermoanaerobacter brockii*, with β-glucosidase activity) *These activities were not detected	pPICZαA vector α-factor secretion signal MF4I secretion signal *AOX1* promoter	Cost-effective cellulase enzymes production
8	2023	Zhou et al. (2023)	*Pichia pastoris GS115*	Thermostable endoglucanase (MtEG5-1) from *Myceliophthora thermophila*	pPIC9K vector α-factor secretion signal *AOX1* promoter His tag	Degradation of corn straw
9	2023	Elgharbi et al. (2023)	*Pichia pastoris* SMD1168H	Xylanase from *Aspergillus niger* US368	pGAPZαB expression vector α-factor secretion signal *GAP* promoter His tag	To investigate the difference observed toward copper between two recombinant forms of the xylanase
10	2022	Li et al. (2022)	*Pichia pastoris* GS115	β-mannanase (McMan5B) from the thermophilic fungus *Malbranchea cinnamomea* S168	pPIC9K vector α-factor secretion signal *AOX1* promoter	To produce partially hydrolyzed konjac powder as a a fiber supplement for steamed bread (Mantou) preparation
11	2022	Beklemishev et al. (2022)	*Pichia pastoris* X33	1,3-specific thermostable lipase from *Thermomyces lanuginosus*	pPICZα-A vector α-factor secretion signal *AOX1* promoter His tag	Immobilization for esterification of fatty acids with aliphatic alcohols in organic solvents
12	2022	Simatupang et al. (2022)	*Pichia pastoris* GS115	Thermostable lipases Itb1.1 and Lk3	pPICZα-A vector α-factor secretion signal *AOX1* promoter His tag	Comparison of the recombinant proteins from *P. pastoris* and *E.coli*

(*Continued*)

Table 3.1 (Continued)

	Year	References	Yeast host	Thermozyme	Expression system	Aim—Application
13	2022	Karim and Tasnim et al. (2022)	*Pichia pastoris* GS115	Thermostable glucoamylase from *Aspergillus flavus* NSH9	pPICZαC, pPICZB vectors α-factor secretion signal Native secretion signal *AOX1* promoter FLAG tag His tag	To measure the secretion levels of recombinant glucoamylase from *Pichia pastoris*, by using the α-mating factor secretion signal peptide and the native secretion signal
14	2022	Escuder-Rodríguez et al. (2022)	*Saccharomyces cerevisiae* BJ3505	Metagenomic thermophilic endoglucanase	YEpFLAG-1 vector α-factor secretion signal *ADH2* promoter His tag	Characterization of a novel enzyme from nonculturable microorganism and comparison of the recombinant proteins from *S.cerevisiae* and *E.coli*
15	2021	Bunterngsook et al. (2021)	*Pichia pastoris* KM71	Metagenomic lytic polysaccharide monooxygenase (BgAA9)	pPICZαA vector α-factor secretion signal *AOX1* promoter	Enhancement of commercial cellulase for biomass conversion industry
16	2021	Navone et al. (2021)	*Pichia pastoris* BG11	Thermostable variant of *E. coli* AppA phytase	pD912 vector α-factor secretion signal Hybrid Suc2, Pho1, and Ost1 secretion signals (with the α-factor pro-region at the 3′ end) *AOX1* promoter His tag	Protein engineering for increasing thermostability to withstand the high temperatures during industrial steam pelleting

17	2021	Kashyap & Gupta (2021)	*Pichia pastoris* X33	Lipase Lip11 from oleaginous yeast *Yarrowia lipolytica* MSR80	pPICZA, pPICZαA α-factor secretion signal Native secretion signal *AOX1* promoter His tag	Protein engineering to increase thermostability and catalytic efficiency
18	2021	Dwyer et al. (2021)	*Pichia pastoris* X33	Two GH18 chitinases from the thermophilic fungus *Rasamsonia emersonii* (previously *Talaromyces emersonii*)	pPICZαA vector α-factor secretion signal *AOX1* promoter His tag	Potential use of chitinases in crustacean chitin hydrolysis
19	2021	Hou et al. (2021)	*Pichia pastoris* GS115	Thermostable and alkaline feruloyl esterase from thermophilic *Chaetomium thermophilum*	pPIC9K vector α-factor secretion signal *AOX1* promoter His tag	Biobleaching (delignification) of wheat straw pulp
20	2021	Liu et al. (2021)	*Pichia pastoris* X33	Thermophilic and acidophilic β-mannanase (ManAK) from *Aspergillus kawachii* IFO 4308	pPICZαA vector α-factor secretion signal *AOX1* promoter His tag	Animal feed additive Development of a directed evolution (to enhance specific activity) and metabolic engineering (to increase enzyme productivity) tool
21	2021	Abedi et al. (2021)	*Pichia pastoris* pPink strain 4	Xylanase 10 from fungus *Bispora* sp. MEY-1 (X10), xylanase 11 from bacterium *Thermobacillus xylanilyticus* (X11), and their chimera.	pPinkα-HC vector α-factor secretion signal *AOX1* promoter	Creating a chimeric enzyme through protein engineering that can substitute two separate enzymes and therefore minimize the production costs
22	2021	Gao et al. (2021)	*Pichia pastoris* KM71	Thermostable endo-β-1,3-glucanase from *Trichoderma harzianum*	pPIC9K vector α-factor secretion signal AOX1 promoter	Production of oligoglucosides, through curdlan hydrolysis, used as a functional food additive

(*Continued*)

Table 3.1 (Continued)

	Year	References	Yeast host	Thermozyme	Expression system	Aim—Application
23	2021	Agrawal et al. (2021)	*Pichia pastoris* X33	Feruloyl esterase D (CE1) and β-xylosidase (GH43) from thermophilic fungus *Scytalidium thermophilum*	pPICZαA vector α-factor secretion signal *AOX1* promoter His tag	To improve commercial lignocellulolytic enzyme cocktails
24	2021	Li et al. (2021)	*Pichia pastoris* GS115	Thermophilic chitinase from the marine bacterium *Microbulbifer* sp. BN3	pPIC9K vector α-factor secretion signal *AOX1* promoter	Potential application in producing $(GlcNAc)_2$ and GlcNAc
25	2020	Agrawal et al. (2020)	*Pichia pastoris* X33	Two thermostable lytic polysaccharide monooxygenases from thermophilic fungi *Scytalidium thermophilum* and *Malbranchea cinnamomea*	pPICZαA vector *AOX1* promoter	Saccharification of acid/alkali pretreated sugarcane bagasse and rice straw in enzyme cocktails
26	2020	Zhang et al. (2020)	*Pichia pastoris* X33	Thermostable endo-1,5-α-l-arabinanase from *Bacillus thermodenitrificans* TS-3	pPICZαB vector α-factor secretion signal *AOX1* promoter His tag	To extract feruloylated arabino-oligosaccharides from sugar beet pulp
27	2020	Qin et al. (2020)	*S. cerevisiae* (Mav203)	Tyrosyl-tRNA synthetase from thermophilic bacterium *Geobacillus stearothermophilus*	pESC vector, pCMV vector Flag tag, His tag	Genetic code expansion in eukaryotic cells
28	2020	Liu et al. (2020)	*Pichia pastoris* X33	Engineered thermophilic and acidophilic β-mannanase from *Aspergillus kawachii* IFO 4308	pPICZαA vector α-factor secretion signal *AOX1* promoter His tag	Degradation of locust bean gum, konjac gum, and guar gum into small molecular mannooligosaccharides (<2000 Da)
29	2020	Xu et al. (2020)	*Pichia pastoris* X33	*Candida rugosa* lipase 1	pPICZαA vector α-factor secretion signal AOX1 promoter	Docosahexaenoic acid enrichment in algae oil

30	2020	Nimpiboon et al. (2020)	*Saccharomyces cerevisiae* FGY217	Amylomaltase from *Thermus filiformis*	pDDGFP-2 vector *GAL1* promoter His tag	To make starches with new functionality
31	2020	Yang et al. (2020)	*Pichia pastoris* X33	*Endo-β-1,4-xylanase from Thermotoga maritima, TmxB mutants*	pPICZαA vector α-factor secretion signal *AOX1* promoter	Rational design to enhance specific activity and catalytic efficiency
32	2020	Ronghua et al. (2020)	*Pichia pastoris* GS115	Chitosanase from *Aspergillus oryzae* NKY2017	*AOX1* promoter	To convert chitosan into oligosaccharides
33	2020	Chen et al. (2020)	*Pichia pastoris* GS115	The GH5 cellulase from *Aspergillus glaucus* CCHA	pPIC9K vector α-factor secretion signal *AOX1* promoter	Corn and rice straw conversion into reducing sugars
34	2020	Tan et al. (2020)	*Pichia pastoris* GS115	Endo-polygalacturonase II B from *Aspergillus luchuensis*	pPICZαA vector α-factor secretion signal *AOX1* promoter	Juice clarification
35	2019	Duan et al. (2019)	*Pichia pastoris* GS115	Lipase B from the thermophilic fungus *Malbranchea cinnamomea*	pPIC9K vector Native secretion sequence *AOX1* promoter	Potential for production of lipolyzed milkfat flavor and biodegradation of phthalate esters
36	2019	Schmidt et al. (2019)	*Pichia pastoris* GS115	Directed evolution library of *Aspergillus awamori* glucoamylase	pPEHα episomal vector α-factor secretion signal *AOX1* promoter	To increase the thermostability of the glucoamylase
37	2019	Brito e Cunha et al. (2019)	*Pichia pastoris X33*	Lipase B from *Candida antarctica*	pPGKΔ3_PRO vector α-factor secretion signal *PGK* constitutive promoter	To study the influence of structure on thermostability

(*Continued*)

Table 3.1 (Continued)

	Year	References	Yeast host	Thermozyme	Expression system	Aim—Application
38	2019	Hu et al. (2019)	*Pichia pastoris*	*Bacillus licheniformis* α-amylase	pPIC9K vector α-factor secretion signal *AOX1* promoter	To investigate the effects of glycosylation on α-amylase
39	2019	Menéndez et al. (2019)	*Pichia pastoris* X33	Thermostable β–fructosidase from *Thermotoga maritima*	pGAPZαC vector α-factor secretion signal Constitutive GAP promoter His tag	Rational design of the β-fructosidase for the continuous production of FOS-containing invert syrup at pasteurization temperatures
40	2018	Yang et al. (2018)	*Pichia pastoris* GS115	Mutants of the cellulase (TaCel45) from the thermophilic fungus *Thielavia arenaria* XZ7	pPIC9 vector	To investigate the effects of each disulfide bond on the structure, activity, and stability of TaCel45
41	2018	Sun et al. (2018)	*Pichia pastoris* GS115	Aspartic protease from the thermophilic fungus *Rhizomucor miehei* CAU432	pPIC9K vector α-factor secretion signal *AOX1* promoter	Potential application in food industries
42	2018	Jain et al. (2018)	*Pichia pastoris* X33	Thermostable endoglucanase from *Thermoascus aurantiacus* RCKK	pPICZαC vector α-factor secretion signal *AOX1* promoter His tag	Potential applicability in pretreatment of lignocellulosic biomass
43	2018	Han et al. (2018)	*Pichia pastoris* GS115	Mutants of thermostable cellobiohydrolase CtCel6 from *Chaetomium thermophilum*	pPIC9K vector α-factor secretion signal *AOX1* promoter His tag	Enzyme redesign to increase catalytic activity and thermostability

44	2018	Cha et al. (2018)	*Pichia pastoris* GS115	Endoglucanase from the brown rot fungus *Fomitopsis palustris*	pPICZαC vector α-factor secretion signal *AOX1* promoter His tag	To expand knowledge and understanding of the catalytic mechanism of GH45 subfamily C enzymes
45	2018	Ionata et al. (2018)	*Saccharomyces cerevisiae* SoKY117	Thermostable beta-glycosidase from *Sulfolobus solfataricus*	pYMM2 vector Intracellular expression	Immobilization to create functional foods for low lactose diets
46	2018	Wang et al. (2018)	*Pichia pastoris* GS115	Bovine enterokinase light chain and deglycosylated mutants	pPIC9K vector α-factor secretion signal *AOX1* promoter	To study the effects of N-glycosylation on the biochemical properties of the enzyme
47	2018	Niderhaus et al. (2018)	*Pichia pastoris* GS115	Endo-β-1,4 xylanase from the white-rot basidiomycete *Pycnoporus sanguineus* BAFC 2126	pPIC9 backbone vector α-factor secretion signal *AOX1* promoter His tag	To complement commercial cellulases in lignocellulose bioconversion to fermentable sugars
48	2017	Burén et al. (2017)	*Saccharomyces cerevisiae* W303-1a	O_2-sensitive S-adenosyl methionine –radical enzyme from *Azotobacter vinelandii* and *Methanocaldococcus infernus*	pESC vectors Intracellular expression GAL promoters SU9 leader sequences	To engineer nitrogenase in eukaryotic organisms (e.g., plants)
49	2017	Luo et al. (2017)	*Pichia pastoris* X33	Endo β-mannanase from thermophilic *Bacillus subtilis* (TBS2)	pPICZαA vector α-factor secretion signal *AOX1* promoter	The production of manno-oligosaccharides and animal feed
50	2017	Berezina et al. (2017)	*Pichia pastoris* GS115	Xyloglucanase from the thermophilic fungus *Myceliophthora thermophila* VKPM F-244	pPH93 vector α-factor pro-MEKR secretory leader sequence *AOX1* promoter	Saccharification of lignocellulosic biomass

(*Continued*)

Table 3.1 (Continued)

	Year	References	Yeast host	Thermozyme	Expression system	Aim—Application
51	2017	Karnaouri et al. (2017)	*Pichia pastoris* X33	Endoglucanase from the thermophilic fungus *Myceliophthora thermophila* ATCC 42464	pPICZαC vector α-factor secretion signal *AOX1* promoter His tag	To study the synergy between endoglucanase and lytic polysaccharide monooxygenase for the hydrolysis of lignocellulosic substrates
52	2017	Gao et al. (2017)	*Pichia pastoris*	Endoglucanase from the thermophilic fungus *Theilavia terrestris*	pPICZαA vector α-factor secretion signal *AOX1* promoter	To solve high-resolution crystal structure of the enzyme and to understand the catalytic mechanism
53	2017	Nisha & Satyanarayana (2017)	*Pichia pastoris* X33	3′-deleted amylopullulanase from the extreme thermophile *Geobacillus thermoleovorans* NP33	pPICZαA vector α-factor secretion signal *AOX1* and GAP promoters His tag	To compare methanol inducible and constitutive expression, and expression from *E. coli* and *P. pastoris*, for using the enzyme in starch saccharification
54	2017	Ranjan et al. (2017)	*Pichia pastoris* GS115	Cyanate hydratase from the thermophilic fungus *Thermomyces lanuginosus* SSBP	pPICZαA vector α-factor secretion signal *AOX1* and GAP promoters His tag	Detoxification of cyanate
55	2017	Dua et al. (2017)	*Pichia pastoris*	Chitinase from the thermophilic mold *Myceliopthora thermophila* BJA	pGAPZαA vector α-factor secretion signal GAP constitutive promoter Tween-20 in the medium to favor secretion	To inhibit the growth of phytopathogenic fungi at high temperature, during summer in tropics

56	2017	Hamuro et al (2017)	*Pichia pastoris* KM71H	Thermostable mutant of the laccase isozyme Lac2a from the medicinal mushroom *Agaricus brasiliensis*	pPICZaB vector α-factor secretion signal *AOX1* promoter His tag	To improve stability by a phylogeny-based design method and to use the enzyme in polymerization of monomeric polyphenols
57	2017	Wang et al. (2017)	*Pichia pastoris* GS115	Alkaline thermostable *Thermobifida fusca* KW3 β-mannanase ManB	pAOhr vector α-factor secretion signal *AOX1* promoter Multiple copies of the gene	To enhance the expression with building of multi-copy vectors, and use of the enzyme for ramie (*Boehmeria nivea*) degumming with *Bacillus* sp. HG-28 expressing pectinase and xylanase
58	2017	Ma et al. (2017)	*Pichia pastoris* GS115	Laccase CueO from *Escherichia coli* K12	pHBM905BDM vector MF-4I leader sequence *AOX1* promoter His tag	Decolorization of synthetic dyes
59	2017	Ma et al. (2017)	*Pichia pastoris* GS115	Thermostable a-amylase from a deep-sea thermophile *Geobacillus* sp. 4j	pPIC9K, pGAPZ A and pPICZaB vectors α-factor secretion signal *AOX1* promoter	Overexpression of the unfolded protein response activator HAC1p to improve the expression level of the a-amylase
60	2017	Beneyton & Rossignol (2021)	Yarrowia lipolytica JMY2566	Thermostable variants of endo-β-1,4-xylanase C in error-prone PCR libraries	JMP62-type secretion vector Constitutive pTEF promoter	To construct a droplet-based microfluidics system for enzymatic high-throughput screening and its use to screen thermostable variants libraries
61	2017	Su et al. (2017)	*Picia pastoris* KM71	Phospholipase B from *Fusarium oxysporum*	pPIC9K vector α-factor secretion signal *AOX1* promoter	Oil degumming applications

(*Continued*)

Table 3.1 (Continued)

	Year	References	Yeast host	Thermozyme	Expression system	Aim—Application
62	2016	Sun et al. (2016)	*Pichia pastoris* GS115	Lipase LIP09 from the thermophilic fungus *Neosartorya fischeri* P1	pPIC9 vector α-factor secretion signal *AOX1* promoter	Characterization and high yield production of recombinant LIP9
63	2016	Kasprzak et al. (2016)	*Arxula adeninivorans* G1212, G1216, MS1006 and *Hansenula polymorpha* Kla8	Coexpression of *Carboxydothermus hydrogenoformans* alcohol dehydrogenase and *Bacillus megaterium* glucose dehydrogenase	pFPMT121 and Xplor®2 transformation/expression platforms Methanol induction *H. polymorpha* Constitutive expression *A. adeninivorans* His tag Intracellular expression	To synthesize ethyl (R)-mandelate, with yeast cells as catalysts
64	2016	Katsimpouras et al. (2016)	*Pichia pastoris* X33	*Endo-β-mannanase 26 A from Myceliophthora thermophila*	pPICZaC vector α-factor secretion signal *AOX1* promoter His tag	To enhance lignocellulose degradation by supplementing commercial enzyme mixtures
65	2016	Woon et al. (2016)	*Pichia pastoris* X33	Cellobiohydrolase 7B from the thermophilic fungus *Thielavia terrestris*	pPICZaC vector α-factor secretion signal *AOX1* promoter His tag	To enhance lignocellulose degradation by supplementing commercial enzyme mixtures

66	2016	Li et al. (2016)	*Yarrowia lipolytica* CLIB724	Trehalose synthase from the thermophilic archaea *Picrophilus torridus* DSM9790	pINA1313 vector Promoter hp4d Cell surface display expression	To develop a method to produce trehalose from maltose
67	2016	Xia et al. 2016)	*Pichia pastoris* GS115	Mutant (removing two potential O-glycosilation sites) of β-glucosidase 3 A from *Talaromyces leycettanus* JCM12802	pPIC9 vector α-factor secretion signal *AOX1* promoter	To develop a novel β-glucosidase with capacity of accelerating the conversion efficiency of cellulosic materials into glucose
68	2016	Ranjan & Satyanarayana (2016)	*Pichia pastoris* X33	Phytase of the thermophilic mold *Sporotrichum thermophile*	pPICZαA vector α-factor secretion signal *AOX1* promoter	To produce high levels of the enzyme and its use in dephytinization of chick feeds and as a haloperoxidase
69	2016	Harnpicharnchai et al. (2016)	*Pichia pastoris* KM71	*Aspergillus niger* BCC4525 β-mannanase	pPICZαA vector α-factor secretion signal *AOX1* promoter	Mannooligosaccharides production from low-cost biomass hydrolysis, utilization of the enzyme as a feed additive
70	2016	Chen et al. (2016)	*Yarrowia lipolytica* P01g	*Thermobifida fusca* BCRC19214 thermostable β-mannanase	P01g expression system Extracellular expression	Degradation of konjac glucomannan
71	2016	Gao et al. (2016)	*Pichia pastoris* GS115	Thermophilic fungal α-amylase from the psychrotolerant *Geomyces pannorum*	pPIC9K vector α-factor secretion signal *AOX1* promoter	A thermophilic amylase obtained from a psychrotolerant fungus used for starch hydrolysis. Potential application prospect in syrup industry
72	2016	Lan et al. (2016)	*Pichia pastoris* X-33	Thermostable lipase MAS1 from marine *Streptomyces* sp. strain	pPICZαA and pPIC9K vectors α-factor secretion signal *AOX1* promoter	To increase the production of the recombinant enzyme by chaperones coexpression

(*Continued*)

Table 3.1 (Continued)

	Year	References	Yeast host	Thermozyme	Expression system	Aim—Application
73	2016	Ma et al. (2016)	*Pichia pastoris* X-33	Thermostable exoinulinase from *Kluyveromyces cicerisporus* and mutants	pPICZαA vector α-factor secretion signal *AOX1* promoter His tag	To understand the catalytic mechanism of exoinulinase, and to convert inulin into high-fructose syrups and fructoligosaccharides to be used in pharmaceutical and food industries.
74	2015	Samahah et al. (2015)	*Pichia pastoris* X-33	Cellobiohydrolase Avi2 from the thermophilic fungus *Humicola insolens* ATCC16454	pPICZaC vector α-factor secretion signal *AOX1* promoter His tag	Characterization of the recombinant enzyme with potential to be exploited for the enzymatic hydrolysis of locally produced agricultural biomass
75	2015	Yu et al. (2015)	*Pichia pastoris* GS115	The thermostable endo-1,4-β-mannosidase from *Aspergillus niger* CBS 513.88	pHBM-905M vector Methanol induced expression Extracellular expression His tag	The high-level expression of a thermostable β-mannosidase for industrial applications
76	2015	Zhao et al. (2015)	*Pichia pastoris* GS115	β-glucosidase from *Myceliophthora thermophila* (mtbgl3b)	pPIC9 vector α-factor secretion signal *AOX1* promoter	Characterization of the recombinant enzyme with potential for industrial applications
77	2015	Zerva et al. (2015)	*Pichia pastoris* X-33	Class II thermophilic peroxidase from *Myceliophthora thermophila*	pPICZαC α-factor secretion signal *AOX1* promoter His tag	Polymerization of catechol
78	2015	Yang et al. (2015)	*Pichia pastoris* GS115	β-mannanase Man5P1from the thermophilic fungus *Neosartorya fischeri* P1	pPIC9 vector α-factor secretion signal *AOX1* promoter	Cost-effective recombinant enzyme production for commercialization and use in the food and feed industries

79	2015	Ali et al. (2015)	*Pichia pastoris* GS115	Laccase isoform Ppicc2 from the brown rot fungus *Postia placenta* MAD-698-R	pPIC9K vector α-factor secretion signal *AOX1* promoter His tag	To understand the roles played by laccases in brown rot fungi
80	2015	Lian et al. (2015)	*Kluyveromyces lactis* GKX31	Xylanase B from *Thermotoga maritima* MSB8	pKLAC1 vector α-factor secretion signal *LAC4* promoter	Construction of β-1,3-glucanosyltransferases KlGAS1-1 and KlGAS1–2 deletion strains to enhance heterologous xylanase secretion
81	2015	Wang et al. (2015)	*Pichia pastoris* GS115	*Thermococcus kodakaraensis* KOD1 DNA polymerase fused C-terminus to the dsDNA binding protein Sso7d from *Sulfolobus solfataricus*	pHBM905A vector α-factor secretion signal *AOX1* promoter	Cost-effective large-scale production of a high-performance thermostable DNA polymerase for use in PCR
82	2015	Oslan et al. (2015)	*Pichia guilliermondii* SO isolated from spoiled orange	Thermostable T1 lipase from *Geobacillus zalihae*	pPICZαB vector α-factor secretion signal *AOX1* promoter His tag	To develop a new host for expression of recombinant proteins
83	2015	Kumari & Gupta (2015)	*Pichia pastoris* X-33	Enantioselective lipase TALipA from *Trichosporon asahii* MSR54	pPICZαA vector α-factor secretion signal *AOX1* promoter His tag	Recombinant production and characterization of a lipase from a pathogen
84	2014	He et al. (2014)	*Pichia pastoris* KM71	Glucoamylase and α-amylase from thermophilic fungus *Rhizomucor pusillus* GX-3	pPIC9K and pPICZαA vectors α-factor secretion signal *AOX1* promoter	Coexpressing glucoamylase with α-amylase in one host to increase activity, to avoid the need to ferment repeatedly and to improve cost-effectiveness

(*Continued*)

Table 3.1 (Continued)

	Year	References	Yeast host	Thermozyme	Expression system	Aim—Application
85	2014	Huang et al. (2014)	*Pichia pastoris* X-33	Thermophilic β-mannanase from *Aspergillus niger* BK01 and mutants	pPICZαA vector α-factor secretion signal *AOX1* promoter His tag	Crystal structure and rational design to enhance enzymatic activity
86	2014	Fonseca-Maldonado et al. (2014)	*Pichia pastoris* GS115	A thermophilic variant of xylanase A (XynAG3) and the thermostable laccase (CotA), both from *Bacillus subtilis* (coproduced)	pPIC9K and pPICZαA vectors α-factor secretion signal *AOX1* promoter His tag	Hydrolysis of milled sugarcane bagasse
87	2014	Voutilainen et al. (2014)	*Saccharomyces cerevisiae* EGY48	Chimeric thermostable cellobiohydrolases	pSVEmpty_ENO episomal vector ENO1 promoter Extracellular expression	Lignocellulosic feedstock hydrolysis at high temperatures
88	2014	Haikarainen et al. (2014)	*Pichia pastoris* GS115	Manganese superoxide dismutase from the thermophilic fungus *Chaetomium thermophilum*	pPIC9K vector α-factor secretion signal *AOX1* promoter	High level expression, structure determination and its influence on thermostability
89	2014	Akbarzadeh et al. (2014)	*Pichia pastoris* GS115	Endoglucanase II from *Trichoderma reesei*	pPinkα-HC vector α-factor secretion signal *AOX1* promoter	Optimization of recombinant extracellular cellulase expression
90	2014	Karnaouri et al. (2014)	*Pichia pastoris* X-33	Thermostable endoglucanase from *Myceliophthora thermophila* ATCC 42464	pPICZαC vector α-factor secretion signal *AOX1* promoter His tag	Liquefaction of high-consistency pretreated wheat straw

91	2014	Morka et al. (2014)	*Saccharomyces cerevisiae* BY4741, *Kluyveromyces lactis* GG799, and *Hansenula polymorpha* KLA [illegible]-1	Lipase CAL-A from *Candida antarctica*	pSCL (episomal), pKLL and pHPL (integrative) vectors *GAL1*, *LAC4* and *FMD* inducible homologous promoters α-factor secretion signal His tag	To study the influence of the host on the production, thermostability and other characteristics of the recombinant enzyme
92	2014	Hu et al. (2014)	*Pichia pastoris* X-33	POXA1c laccase from *Pleurotus florida* (ACCC50153)	pGAPZA vector Native secretion signal Constitutive GAP promoter	To investigate the functions of the N and C termini of the enzyme using site-directed mutagenesis
93	2014	Liao et al. (2014)	*Pichia pastoris* GS115	Thermostable endo-1,4-β-mannanase from *Penicillium oxalicum* GZ-2	pPICZαA α-factor secretion signal *AOX1* promoter His tag	To hydrolyze various types of mannan polymers and release manno-oligosaccharides
94	2014	Hmida-Sayari et al. (2014)	*Pichia pastoris* SMD1168H	The thermostable *Bacillus subtilis* US417 phytase	pPICZαB vector α-factor secretion signal *AOX1* promoter His tag	To characterize the recombinant enzyme to be used as feed additive (resistant to the temperature used in the feed-pelleting process)
95	2014	Silva et al. (2014)	*Pichia pastoris*	Thermostable rhamnogalacturonan lyase mutants from *Bacillus licheniformis* (DSM 13/ ATCC14580)	pDP66K- RGL vector Extracellular expression His tag	Combinatorial protein engineering to increase thermostability and structure determination

(Honda et al., 1988; Pretorius et al., 1988). An endo-beta-glucanase, from an extremely thermophilic cellulolytic anaerobic bacterium isolated from a hot spring, was expressed in *S. cerevisiae* using an episomal plasmid without yeast promoter or secretion signal, merely cloning in the plasmid a native sequence containing the gene and its regulatory elements. The heat-stable enzyme specific activity was produced in the culture medium at levels 3–7 times higher than in *E. coli*, with a yield in the range of 29.4–73.7 and 9.2–14.2 U/g of protein, respectively (Honda et al., 1988). Also, a *Bacillus amyloliquefaciens* α-amylase gene was expressed from an episomal plasmid transformed into several *Saccharomyces* strains. In this case, the enzyme was successfully secreted, but using yeast promoter and secretory signals, not those of the native gene. The α-amylase optimal temperature ranged 50–65°C and production levels were similar in *E. coli* and *S. cerevisiae*, with a yield of 123 and 125.2 U/mL, respectively (Pretorius et al., 1988).

During the first years after 1988, the predominant yeast host for heterologous thermophilic enzymes expression was *S. cerevisiae*, with some incursions of *Kluyveromyces*, mainly *K. lactis* but also *K. marxianus*. However, *Pichia pastoris* (name still in use, but reassigned to *Komagataella phaffii*) soon started to emerge as a host producing high levels of recombinant proteins. Thus, in the period between 1989 and 2013, about 44% of recombinant thermophilic enzymes expression was from *S. cerevisiae*, 39% from *P. pastoris*, 15% from *K. lactis,* and 2% from *K. marxianus*. These percentages are estimated from a bibliographic search made during May 2024 using "thermophilic and yeast expression" as keywords. The use of *P. pastoris* was predominant in the second part of this 1989–2013 period, while *S. cerevisiae* was predominant in the first part. The use of the different yeast strains as hosts came hand in hand with the sequencing of their genomes and with the development of recombinant DNA tools for their genetic modification. No specific yeasts expression systems for heterologous thermophilic proteins were identified in the bibliographic search carried out. The same promoters, affinity tags, secretion signals, and terminator sequences used for mesophilic proteins resulted suitable also for thermophilic ones.

The plasmid YEpFlag1 with the *S. cerevisiae* strain BJ3505 is a pioneer highly used expression system that was commercialized by Eastman Kodak Company in the 90s under the name "Yeast N-terminal Flag Expression System" (Schuster et al., 2000). The YEpFlag1 vector bears, clockwise from position1: *ADH2* (alcohol dehydrogenase) promoter, mating factor α (MFα) leader sequence, FLAG peptide sequence, multiple cloning site (MCS), translational stop codon, *CYC1* bidirectional transcriptional terminator, 2-micron circle DNA, *TRP1* selectable marker. The FLAG peptide facilitates immunological detection and affinity purification of the expressed fusion protein. The *ADH2* promoter is regulated by glucose repression. BJ3505 cultures can be grown to high density in glucose, cell growth being essentially complete at 24 h of culture, possible toxic effects of the expressed heterologous protein are minimized, since protein expression does not occur until glucose is depleted (starting about 48 h of culture) and no other induction factor is required apart from glucose depletion. The BJ3505 strain contains a deficiency in a major vacuolar protease which greatly decreases the possibility of proteolytic degradation of the produced recombinant protein. Moreover, the high stability media YPHSM (1% dextrose, 3% glycerol, 1% yeast extract, 8% peptone, 20 mM $CaCl_2$) was designed to use with this expression system. The trp1-Δ101 marker deletion allows the selection of the BJ3505 cells transformed with the YEpFlag1 vector that carries de *TRP1* gene in a medium without tryptophan.

Kluyveromyces played an important role since it shows over *Saccharomyces* the advantages of a more respiratory metabolism and the capacity to use lactose as carbon source, lactose being responsible of the polluting power of the waste material cheese whey so that recombinant protein production from cheese whey by *Kluyveromyces* constituted an early example of circular bioeconomy (Rocha et al., 2011). Moreover, *K. marxianus* is thermotolerant and can grow at temperatures 10°C higher than *K. lactis* and *S. cerevisiae*, which is presumed to be positive for heterologous thermozymes folding (Gombert et al., 2016). The expression systems pSpGK1 and pKLAC1 are reported as being used in most of the cases of recombinant protein secretion by *Kluyveromyces*. The pSpGK1 vector is an episomal plasmid derived from pKD1 (Fleer et al., 1991) that features a clockwise arrangement of the constitutive promoter for the *S. cerevisiae* phosphoglycerate-kinase (PGK) gene, a secretion signal consisting of the 16-amino-acid presequence from the *K. lactis* killer toxin alpha-subunit, a MCS, and the PGK terminator sequence. It also contains the S11 replication origin, *URA3* as selection marker, and oriC and AmpR for propagation in *E. coli*. This plasmid was used successfully in *K. lactis* and *K. marxianus* (Rocha et al., 2011). The pKLAC1 expression system was developed later and commercialized as a Protein Expression Kit by New England Biolabs. The vector is capable of replication in *E. coli* and stable integration into the genome of *K. lactis*. *K. lactis* GG799 competent cells are transformed with the linearized pKLAC1 that integrates into the *K. lactis* genome at the *LAC4* locus. The high transformation efficiency makes the system suitable for methods such as, for example, expression cloning using cDNA libraries (Colussi & Taron, 2005). Yeast transformants are selected using the antibiotic-free acetamidase selection method, the amdS marker (source *Aspergillus nidulans*) being expressed under the control of the yeast *S. cerevisiae ADH1* promoter. Acetamidase expressed from pKLAC1 permits transformed cells to utilize acetamide as a sole nitrogen source on defined medium. Acetamide selection promotes formation of cells containing multiple integrations of pKLAC1 which results in higher yields of protein. Different to *K. lactis*, wild-type *K. marxianus* cells produce acetamidase, therefore this system is not advisable for *K. marxianus*. The MCS of pKLAC1 is positioned to allow translational fusion of the *K. lactis* MFα secretion leader sequence to the N-terminus of the recombinant target protein. Expression of the recombinant fusion protein is driven by the strong *K. lactisLAC4* (beta-galactosidase) promoter, repressed by glucose, and inducible by lactose and galactose. Production of the recombinant protein is achieved in a rich medium with galactose. This *LAC4* promoter has been modified to remove background expression in *E. coli*, so that genes toxic to *E. coli* can be cloned into pKLAC1 in bacteria prior to their expression in yeast. pKLAC2 is a subsequent variant with an alternative MCS.

3.3 The current situation and the success of *Pichia pastoris* as a host

In the last 10years, the yeast strain clearly predominant as a host for recombinant thermozymes production is *P. pastoris*. Table 3.1 provides that 80 out of 95 recombinant thermozymes were expressed from *P. pastoris* in the period 2014 to mid-2024. The other 15 recombinant thermozymes were expressed from *S. cerevisiae* (6 cases), *Yarrowia lipolytica* (3 cases), *K. lactis* and *Hansenula polymorpha* (2 cases each one), *Arxula adeninivorans* and *Pichia guilliermondii* (1 case each one) (Fig. 3.1).

P. pastoris, like all the other above-mentioned yeasts hosts, is a GRAS (safe) yeast. Characteristically, *P. pastoris* is a methylotrophic microorganism that uses methanol as sole carbon and energy source. It is reported to show high similarity of protein glycosylation to the mammalian cells, while the proteins expressed from *S. cerevisiae* are often N and O-hyperglycosylated, and to be very efficient in the production of membrane proteins (Karbalaei et al., 2020). The recombinant thermozymes expression is achieved in *P. pastoris* using integrative plasmids, mainly under the control of the exceptionally strong *AOX1* (alcohol oxidase) promoter that is inducible by methanol, even though methanol is toxic and flammable, which constitutes a safety risk (Vogl et al., 2018). Also, methanol must be added periodically to the culture medium to maintain the concentration above 0.5% (minimum inducing) and below 5% that is toxic for the cells (Karbalaei et al., 2020), this is a tedious task, but modern fermenters are equipped with automatic feed control. Despite these drawbacks, there are notably fewer cases given in Table 3.1 of methanol-free expression, in which the constitutive *GAP* (glyceraldehyde-3-phosphate dehydrogenase) or *PGK* promoters are used, than methanol-induced expression. The prevailing transformation method is electroporation. Integration into the genome can occur in several copies, but with a low frequency (1–10% of the selected transformants); the number of copies is then evidenced by the concentration of the selective antibiotic (Geneticin, Zeocin) that the transformed yeast can stand. Auxotrophic selectable markers are less used. As the number of copies of the gene increases to a certain extent, so generally does the

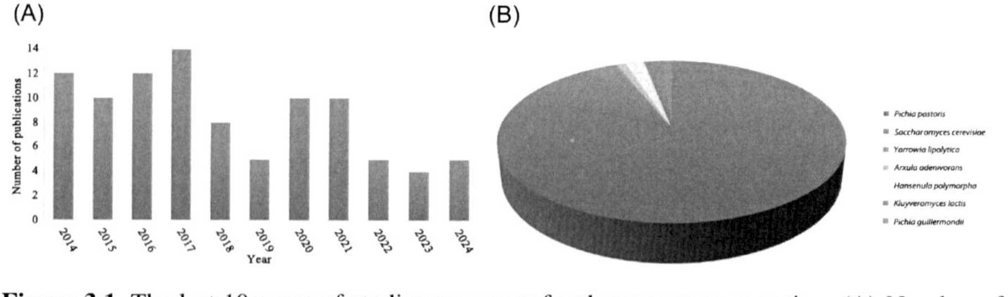

Figure 3.1 The last 10 years of studies on yeasts for thermozymes secretion. (A) Number of publications per year on heterologous production of thermozymes from yeasts. (B) Percentage of yeast strains used as host for thermozymes heterologous production in the last 10 years. More information in Table 3.1.

amount of protein produced. The thermozymes are efficiently secreted into the culture medium mainly using the *S. cerevisiae* MFα secretion signal and Kex2 signal peptidase. There are a few cases given in Table 3.1 that use the native secretion signal of the gene to be expressed. In general, MFα secretion signal gives optimal yields of secretion (Karim & Tasnim, 2022). The recombinant expression constructions also contain affinity tags to facilitate detection and purification of the protein. As can be seen in Table 3.1, the most used tag is oligoHis. The exceptionally low production of endogenous extracellular proteins by *P. pastoris* also contributes to facilitate the purification of recombinant proteins. However, *P. pastoris* produces proteases that can destroy the recombinant proteins, mainly considering that incubation time reaches about 100 hours. To avoid this drawback, protease-free strains have been constructed like PichiaPinkor SMD1168 (Karbalaei et al., 2020).

In the last 10 years, strains GS115 and X-33 with plasmid series pPIC, mostly, and pGAP, in second place, are the most popular expression systems used for thermozymes (Table 3.1). Recently, CRISPR-Cas9-mediated chromosomal integration of a thermozyme gene was reported for *P. pastoris* C1 (Dai et al., 2023). The generalization of these strains and vectors is supported by the commercialization of *P. pastoris* kits for protein expression, such as the most widespread EasySelect™ Pichia Expression Kit and Original Pichia Expression Kit, both of Invitrogen, the Pichia Secretory SUMOpro3 Expression Kit (LifeSensors), or the PichiaPink Yeast Expression Systems (ThermoFisher). There are also companies that offer customer-oriented services, for example, the UNLOCK PICHIA® protein expression technology platform of VALIDOGEN.

3.4 The activity of the recombinant thermozymes expressed from yeasts

Taken together, the largest group of thermozymes expressed from yeast in the last 10 years (Table 3.1) are those involved in lignocellulosic biomass degradation (cellulases, laccases, mannosidases, xylanases, pectinases, polygalacturonases, and many more). Lignocellulosic materials are very abundant in nature, in agricultural and forestry side streams, and in municipal wastes. Although extensively studied, the biotechnological valuation of these materials remains a challenge, mainly due to their recalcitrance to degradation (Escuder-Rodríguez et al., 2022). Thus, the development of suitable enzyme cocktails in this field offers great possibilities for improvement, thermophilic enzymes being specially appreciated, since high temperatures and extreme pH values of operation are often needed. The further applications of these enzymes given in Table 3.1 include prebiotic oligosaccharides production, monosaccharides fermentation, juice clarification, textile industry, and so on.

The other heterologous thermozymes expressed from yeasts and given in Table 3.1 are varied: amylases, phytases, chitinases, proteases, lipases, polymerases, oxidases, and many more. Most of them come from bacteria or fungi, strikingly only one from archaea, a trehalose synthase (Li et al., 2016). The scarcity of thermozymes from archaea draws attention because this microbial domain really likes thermophilic environments to live. The underexploitation of archaeal thermophilic xylanases has been recently reported (Saavedra-Bouza et al., 2023) and can be generalized to other enzymes as deduced from Table 3.1.

The main aim of thermozymes expression from yeasts, as inferred from the studies collected in Table 3.1, is to increase cost-effective production levels and facilitate purification so that extracellular secretion is preferred. Thermozymes expression from yeast is also used as a tool for protein engineering to improve thermostability or to change substrate affinity of enzymes.

Regarding methods for thermozymes purification, there is a predominance of affinity chromatography due to the presence of tag sequences in the commercially available vectors. Other traditional methods for protein purification have been also used like ammonium sulfate precipitation, exclusion or ionic exchange chromatography, and hydrophobic chromatography for lipases. Specifically, for thermozymes, as well as for thermotolerant proteins in general, protocols for purification by differential thermal denaturation have been developed (Olichon et al., 2007), in which samples are heated about 15 min at 70°C, time and temperature must be optimized for each protein, and thereafter cooled on ice before pelleting the denatured proteins of the host and recovering recombinant thermozymes from the supernatant. Moreover, Olichon et al., (2007) found that this simple and quick method provides considerable purification yields when compared with affinity chromatography (8.5 and 2.5 mg protein/L culture, respectively).

Table 3.1 provides a few examples of coexpression of several enzymes with synergic effect (Kasprzak et al., 2016), expression of chimeric enzymes bearing several activities in the same polypeptide (Abedi et al., 2021; Voutilainen et al., 2014), display of enzymes in the yeast surface (Li et al., 2016; Voutilainen et al., 2014; Xu et al., 2020), and production of metagenomics-origin enzymes (Bunterngsook et al., 2021; Escuder-Rodríguez et al., 2022; Han et al., 2024). Although in the last decade, there has been an explosion of metagenomic studies in extremophilic environments, including thermophilic ones, it is remarkable that Table 3.1 only contains three cases of metagenomics enzymes expression from yeasts. This is not proportional to the amount of new thermozymes that have been discovered by metagenomics bioprospection (DeCastro et al., 2016; Escuder-Rodríguez et al., 2018; Wohlgemuth et al., 2018). For this kind of enzymes, *E. coli* expression is mostly chosen in the first place, probably because it is a prokaryotic organism and therefore phylogenetically closer to the native producers of the enzymes. However, yeasts usually show better performance, in terms of amount of protein produced and ease of purification, than *E. coli* as hosts (Escuder-Rodríguez et al., 2022; López-López et al., 2015; Nisha & Satyanarayana, 2017; Simatupang et al., 2022) despite being eukaryotes.

Posttranslational modifications typical of eukaryotes have been reported to favor thermostability of proteins, which supports the advantages of yeasts over *E. coli* as hosts for recombinant proteins expression (Escuder-Rodríguez et al., 2018). Glycosylation is one of such modifications. In Table 3.1, several papers about influence of glycosylation on biochemical properties of thermozymes are found (Hu et al., 2019; Kashyap & Gupta, 2021; Wang et al., 2018). The general assumption that glycosylation increases thermostability is not fully confirmed from the reported results. Thus, the removal of some glycosylation points decreases (Kashyap & Gupta, 2021), but others increase (Kashyap & Gupta, 2021; Wang et al., 2018) or did not affect (Hu et al., 2019) thermostability. An indirect effect of glycosylation on thermostability through the changes induced in tertiary structure of the protein has been proposed to be the cause of these differences.

Besides glycosylation, several other biochemical characteristics of thermozymes have been previously proposed to be responsible of their thermostability (Escuder-Rodríguez et al., 2018), which have in common to provoke a reduced protein´s conformational flexibility. In Table 3.1, there are examples of increasing thermozymes stability by means of disulfide bond engineering (Navone et al., 2021; Yang et al., 2018), directed evolution (Schmidt et al., 2019), structural comparison of native and recombinant enzymes (Brito e Cunha et al., 2019), site-directed mutagenesis (Han et al., 2018), phylogeny-based design (Hamuro et al., 2017), or combinatorial protein engineering (Silva et al., 2014). These examples corroborate that increased structural rigidity is one basis of thermostability.

3.5 Conclusion and perspectives

Several yeast strains have proved to be suitable cell factories to produce recombinant enzymes from thermophilic origin in active form with their inherent biotechnological interest. Although the first yeast used for thermozymes heterologous expression was *S. cerevisiae*, the literature reviewed in this chapter reveals *P. pastoris* as the predominant host for this purpose within the last 10 years, with strains GS115 and X-33 being the most popular, in combination with plasmid series pPIC and pGAP. Advantages of yeasts over bacterial hosts are the same for thermophilic and mesophilic origin enzymes. No specific yeast expression system for thermozymes has been reported. Even though a considerable number of studies show that enzyme thermostability is maintained—or even improved—when expressed in yeast. From the reported results on thermozymes yeast expression, the effect of glycosilation on protein thermostability cannot be asserted. The type of thermozymes expressed in yeast hitherto comprises mostly those with activities related to vegetal biomass degradation and with bacterial or fungal origin. The relative low number of thermozymes from archaeal and metagenomic origin expressed in yeast highlights new research possibilities in these fields and reveals that there is still a lot of room for enhancement as well as in the development of additional or improved strong promoters for *P. pastoris* that do not require methanol induction or in the use of low-cost materials as culture broth components.

References

Abedi, E., Fatemi, F., Sefidbakht, Y., & Siadat, S. E. R. (2021). Development and characterization of a thermostable GH11/GH10 xylan degrading chimeric enzyme. *Enzyme and Microbial Technology, 149*. https://doi.org/10.1016/j.enzmictec.2021.109854, www.elsevier.com/locate/enzmictec.

Agrawal, D., Basotra, N., Balan, V., Tsang, A., & Chadha, B. S. (2020). Discovery and Expression of Thermostable LPMOs from Thermophilic Fungi for Producing Efficient Lignocellulolytic Enzyme Cocktails. *Applied Biochemistry and Biotechnology, 191*(2), 463–481. https://doi.org/10.1007/s12010-019-03198-5, http://www.springer.com/humana+press/journal/12010.

Agrawal, D., Tsang, A., & Chadha, B. S. (2021). Economizing the lignocellulosic hydrolysis process using heterologously expressed auxiliary enzymes feruloyl esterase D (CE1) and β-xylosidase (GH43) derived from thermophilic fungi Scytalidium thermophilum. *Bioresource Technology, 339*. https://doi.org/10.1016/j.biortech.2021.125603, www.elsevier.com/locate/biortech.

Ahmed, U., Stadelmann, T., Heid, D., Würtz, B., Pfannstiel, J., Ochsenreither, K., & Eisele, T. (2024). A novel, robust peptidyl-lys metalloendopeptidase from Trametes coccinea recombinantly expressed in Komagataella phaffii. *Applied Microbiology and Biotechnology, 108*(1), 1–12. https://doi.org/10.1007/s00253-023-12986-3, https://www.springer.com/journal/253.

Akbarzadeh, A., Ranaei Siadat, S. O., Motallebi, M., Zamani, M. R., Barshan Tashnizi, M., & Moshtaghi, S. (2014). Characterization and high level expression of acidic endoglucanase in Pichia pastoris. *Applied Biochemistry and Biotechnology, 172*(4), 2253–2265. https://doi.org/10.1007/s12010-013-0672-6.

An, H., Xiao, T., Fan, H., & Wei, D. (2015). Molecular characterization of a novel thermostable laccase PPLCC2 from the brown rot fungus Postia placenta MAD-698-R. *Electronic Journal of Biotechnology, 18*(6), 451–458. https://doi.org/10.1016/j.ejbt.2015.09.004.

Beklemishev, A. B., Pykhtina, M. B., Perminova, L. V., & Kovalenko, G. A. (2022). Recombinant strains producing thermomyces lanuginosus thermostable lipase and their use in heterogeneous biocatalysis, including processes of low-temperature synthesis of esters. *Federation Applied Biochemistry and Microbiology, 58*(8), 887–898. https://doi.org/10.1134/S0003683822080026, https://www.springer.com/journal/10438.

Beneyton, T., & Rossignol, T. (2021). Droplet-based microfluidic high-throughput screening of enzyme mutant libraries secreted by Yarrowia lipolytica. *Methods in Molecular Biology, 2307*, 205–219. https://doi.org/10.1007/978-1-0716-1414-3_14, http://www.springer.com/series/7651.

Berezina, O. V., Herlet, J., Rykov, S. V., Kornberger, P., Zavyalov, A., Kozlov, D., Sakhibgaraeva, L., Krestyanova, I., Schwarz, W. H., Zverlov, V. V., Liebl, W., & Yarotsky, S. V. (2017). Thermostable multifunctional GH74 xyloglucanase from Myceliophthora thermophila: high-level expression in Pichia pastoris and characterization of the recombinant protein. *Federation Applied Microbiology and Biotechnology, 101*(14), 5653–5666. https://doi.org/10.1007/s00253-017-8297-2, link.springer.de/link/service/journals/00253/index.htm.

Brito e Cunha, D. A., Bartkevihi, L., Robert, J. M., Cipolatti, E. P., Ferreira, A. T. S., Oliveira, D. M. P., Gomes-Neto, F., Almeida, R. V., Fernandez-Lafuente, R., Freire, D. M. G., & Anobom, C. D. (2019). Structural differences of commercial and recombinant lipase B from Candida antarctica: An important implication on enzymes thermostability. *International Journal of Biological Macromolecules, 140*, 761–770. https://doi.org/10.1016/j.ijbiomac.2019.08.148.

Bunterngsook, B., Mhuantong, W., Kanokratana, P., Iseki, Y., Watanabe, T., & Champreda, V. (2021). Identification and characterization of a novel AA9-type lytic polysaccharide monooxygenase from a bagasse metagenome. *Applied Microbiology and Biotechnology, 105*(1), 197–210. https://doi.org/10.1007/s00253-020-11002-2.

Burén, S., Jiang, X., López-Torrejón, G., Echavarri-Erasun, C., & Rubio, L. M. (2017). Purification and in vitro activity of mitochondria targeted nitrogenase cofactor maturase NifB. *Frontiers in Plant Science, 8*. https://doi.org/10.3389/fpls.2017.01567, http://journal.frontiersin.org/article/10.3389/fpls.2017.01567/full.

Cha, J. H., Yoon, J. J., & Cha, C. J. (2018). Functional characterization of a thermostable endoglucanase belonging to glycoside hydrolase family 45 from Fomitopsis palustris. *Applied Microbiology and Biotechnology, 102*(15), 6515–6523. https://doi.org/10.1007/s00253-018-9075-5, https://www.springer.com/journal/253.

Chen, C. Y., Huang, Y. C., Yang, T. Y., Jian, J. Y., Chen, W. L., & Yang, C. H. (2016). Degradation of konjac glucomannan by Thermobifida fusca thermostable β-mannanase from yeast transformant. *International Journal of Biological Macromolecules, 82*, 1–6. https://doi.org/10.1016/j.ijbiomac.2015.10.008, www.elsevier.com/locate/ijbiomac.

Chen, L., Wei, Y., Shi, M., Li, Z., & Zhang, S. H. (2020). Statistical optimization of a cellulase from Aspergillus glaucus CCHA for hydrolyzing corn and rice straw by RSM to enhance yield of reducing sugar. *Biotechnology Letters, 42*(4), 583–595. https://doi.org/10.1007/s10529-020-02804-5, www.wkap.nl/journalhome.htm/0141-5492.

Colussi, P. A., & Taron, C. H. (2005). Kluyveromyces lactis LAC4 promoter variants that lack function in bacteria but retain full function in K. lactis. *Applied and Environmental Microbiology, 71*(11), 7092–7098. https://doi.org/10.1128/AEM.71.11.7092-7098.2005.

Dai, W., Dong, H., Zhang, Z., Wu, X., Bao, T., Gao, L., & Chen, X. (2023). Enhancing the Heterologous expression of a thermophilic endoglucanase and its cost-effective production in pichia pastoris using multiple strategies. *International Journal of Molecular Sciences, 24*(19), 15017. https://doi.org/10.3390/ijms241915017.

DeCastro, M. E., Rodríguez-Belmonte, E., & González-Siso, M. I. (2016). Metagenomics of thermophiles with a focus on discovery of novel thermozymes. *Frontiers in Microbiology, 7*. https://doi.org/10.3389/fmicb.2016.01521, http://journal.frontiersin.org/article/10.3389/fmicb.2016.01521/full.

Dua, A., Joshi, S., & Satyanarayana, T. (2017). Recombinant exochitinase of the thermophilic mould Myceliopthora thermophila BJA: Characteristics and utility in generating N -acetyl glucosamine and in biocontrol of phytopathogenic fungi. *Biotechnology Progress, 33*(1), 70–80. https://doi.org/10.1002/btpr.2370.

Duan, X., Xiang, M., Wang, L., Yan, Q., Yang, S., & Jiang, Z. (2019). Biochemical characterization of a novel lipase from Malbranchea cinnamomea suitable for production of lipolyzed milkfat flavor and biodegradation of phthalate esters. *Food Chemistry, 297*, 124925. https://doi.org/10.1016/j.foodchem.2019.05.199.

Dwyer, K., Bentley, I. S., Tighe, E., McGleenan, E., Gaffney, D., & Walsh, G. (2021). Recombinant production and characterisation of two chitinases from Rasamsonia emersonii, and assessment of their potential industrial applicability. *Applied Microbiology and Biotechnology, 105*(20), 7769–7783. https://doi.org/10.1007/s00253-021-11578-3.

Elgharbi, F., Ben Hlima, H., Ben Mabrouk, S., & Hmida-Sayari, A. (2023). Expression of a copper activated xylanase in yeast: Location of the his-tag in the protein significantly affects the enzymatic properties. *Molecular Biotechnology, 65*(7), 1109–1118. https://doi.org/10.1007/s12033-022-00606-w.

Escuder-Rodríguez, J. J., Decastro, M. E., Cerdán, M. E., Rodríguez-Belmonte, E., Becerra, M., & González-Siso, M. I. (2018). Cellulases from thermophiles found by metagenomics. *Microorganisms, 6*(3). https://doi.org/10.3390/microorganisms6030066, https://www.mdpi.com/2076-2607/6/3/66/pdf.

Escuder-Rodríguez, J. J., González-Suarez, M., deCastro, M. E., Saavedra-Bouza, A., Becerra, M., & González-Siso, M. I. (2022). Characterization of a novel thermophilic metagenomic GH5 endoglucanase heterologously expressed in Escherichia coli and Saccharomyces cerevisiae. *Biotechnology for Biofuels and Bioproducts, 15*(1). https://doi.org/10.1186/s13068-022-02172-4, https://biotechnologyforbiofuels.biomedcentral.com/.

Fleer, R., Yeh, P., Amellal, N., Maury, I., Fournier, A., Bacchetta, F., Baduel, P., Jung, G., L'Hôte, H., Becquart, J., Fukuhara, H., & Mayaux, J. F. (1991). Stable multicopy vectors for high–level secretion of recombinant human serum albumin by kluyveromyces yeasts. *Bio/Technology, 9*(10), 968–975. https://doi.org/10.1038/nbt1091-968.

Fonseca-Maldonado, R., Ribeiro, L. F., Furtado, G. P., Arruda, L. M., Meleiro, L. P., Alponti, J. S., Botelho-Machado, C., Vieira, D. S., Bonneil, E., Furriel, R. D. P. M., Thibault, P., & Ward, R. J. (2014). Synergistic action of co-expressed xylanase/laccase mixtures against milled sugar cane bagasse. *Process Biochemistry, 49*(7), 1152–1161. https://doi.org/10.1016/j.procbio.2014.03.027, www.elsevier.com/inca/publications/store/4/2/2/8/5/7.

Gao, B., Mao, Y., Zhang, L., He, L., & Wei, D. (2016). A novel saccharifying α-amylase of Antarctic psychrotolerant fungi Geomyces pannorum: Gene cloning, functional expression, and characterization. *Starch/Staerke, 68*(1-2), 20–28. https://doi.org/10.1002/star.201500077, http://onlinelibrary.wiley.com/journal/10.1002/(ISSN)1521-379X.

Gao, J., Huang, J. W., Li, Q., Liu, W., Ko, T. P., Zheng, Y., Xiao, X., Kuo, C. J., Chen, C. C., & Guo, R. T. (2017). Characterization and crystal structure of a thermostable glycoside hydrolase family 45 1,4-β-endoglucanase from Thielavia terrestris. *Enzyme and Microbial Technology, 99*, 32–37. https://doi.org/10.1016/j.enzmictec.2017.01.005, www.elsevier.com/locate/enzmictec.

Gao, M. J., Yan, J. J., Zhao, Y., Zhu, L., Yang, G. S., & Zhan, X. B. (2021). Expression of a thermostable β-1,3-glucanase from Trichoderma harzianum in Pichia pastoris and use in oligoglucosides hydrolysis. *Process Biochemistry, 107*, 74–82. https://doi.org/10.1016/j.procbio.2021.05.010, www.elsevier.com/inca/publications/store/4/2/2/8/5/7.

Gombert, A. K., Madeira, J. V., Cerdán, M. E., & González-Siso, M. I. (2016). Kluyveromyces marxianus as a host for heterologous protein synthesis. *Applied Microbiology and Biotechnology, 100*(14), 6193–6208. https://doi.org/10.1007/s00253-016-7645-y, link.springer.de/link/service/journals/00253/index.htm.

Haikarainen, T., Frioux, C., Zhnag, L. Q., Li, D. C., & Papageorgiou, A. C. (2014). Crystal structure and biochemical characterization of a manganese superoxide dismutase from Chaetomium thermophilum. *Biochimica et Biophysica Acta - Proteins and Proteomics, 1844*(2), 422–429. https://doi.org/10.1016/j.bbapap.2013.11.014.

Hamuro, Y., Tajima, K., Matsumoto-Akanuma, A., Sakamoto, S., Furukawa, R., Yamagishi, A., Ohno, N., & Akanuma, S. (2017). Characterization of a thermostable mutant of Agaricus brasiliensis laccase created by phylogeny-based design. *Journal of Bioscience and Bioengineering, 124*(6), 623–629. https://doi.org/10.1016/j.jbiosc.2017.07.004.

Han, C., Li, W., Hua, C., Sun, F., Bi, P., & Wang, Q. (2018). Enhancement of catalytic activity and thermostability of a thermostable cellobiohydrolase from Chaetomium thermophilum by site-directed mutagenesis. *International Journal of Biological Macromolecules, 116*, 691–697. https://doi.org/10.1016/j.ijbiomac.2018.05.088.

Han, S., Zhang, W., Yan, Q., Jiang, Z., & Yang, S. (2024). High level production of a novel acidic and thermostable chitinase from Paecilomyces thermophila for the extraction of bioactive components from Ganoderma lucidum spores. *Process Biochemistry, 136*, 182–190. https://doi.org/10.1016/j.procbio.2023.11.023.

Harnpicharnchai, P., Pinngoen, W., Teanngam, W., Sornlake, W., Sae-Tang, K., Manitchotpisit, P., & Tanapongpipat, S. (2016). Production of high activity Aspergillus niger BCC4525 β-mannanase in Pichia pastoris and its application for mannooligosaccharides production from biomass hydrolysis. *Bioscience, Biotechnology, and Biochemistry, 80*(12), 2298–2305. https://doi.org/10.1080/09168451.2016.1230003.

He, Z., Zhang, L., Mao, Y., Gu, J., Pan, Q., Zhou, S., Gao, B., & Wei, D. (2014). Cloning of a novel thermostable glucoamylase from thermophilic fungus Rhizomucor pusillus and high-level co-expression with α-amylase in Pichia pastoris. *BMC Biotechnology, 14*(1). https://doi.org/10.1186/s12896-014-0114-8.

Hmida-Sayari, A., Elgharbi, F., Farhat, A., Rekik, H., Blondeau, K., & Bejar, S. (2014). Overexpression and biochemical characterization of a thermostable phytase from bacillus subtilis US417 in Pichia pastoris. *Molecular Biotechnology, 56*(9), 839–848. https://doi.org/10.1007/s12033-014-9764-y.

Honda, H., Iijima, S., & Kobayashi, T. (1988). Cloning and expression in Saccharomyces cerevisiae of an endo-β-glucanase gene from a thermophilic cellulolytic anaerobe. *Applied Microbiology and Biotechnology, 28*(1), 57–58. https://doi.org/10.1007/BF00250498.

Hou, Y. H., Yang, Z., Yin, Y., Meng, Z., Wang, J., Zhao, T., & Yang, Q. Z. (2021). Expression of an alkaline feruloyl esterases from thermophilic Chaetomium thermophilum and its boosting effect on delignification of pulp. *Elsevier Inc., China Enzyme and Microbial Technology, 150.* https://doi.org/10.1016/j.enzmictec.2021.109859, www.elsevier.com/locate/enzmictec.

Hu, M., Zhou, X., Shi, Y., Lin, J., Irfan, M., & Tao, Y. (2014). Essential role of the N- and C-terminals of laccase from pleurotus florida on the laccase activity and stability. *Applied Biochemistry and Biotechnology, 174*(5), 2007–2017. https://doi.org/10.1007/s12010-014-1147-0.

Hu, X., Yuan, X., He, N., Zhuang, T. Z., Wu, P., & Zhang, G. (2019). Expression of Bacillus licheniformis α-amylase in Pichia pastoris without antibiotics-resistant gene and effects of glycosylation on the enzymic thermostability. *3 Biotech, 9*(11). https://doi.org/10.1007/s13205-019-1943-x, http://www.springerlink.com/content/2190-572x/.

Huang, J. W., Chen, C. C., Huang, C. H., Huang, T. Y., Wu, T. H., Cheng, Y. S., Ko, T. P., Lin, C. Y., Liu, J. R., & Guo, R. T. (2014). Improving the specific activity of β-mannanase from Aspergillus niger BK01 by structure-based rational design. *Biochimica et Biophysica Acta - Proteins and Proteomics, 1844*(3), 663–669. https://doi.org/10.1016/j.bbapap.2014.01.011.

Ionata, E., Marcolongo, L., La Cara, F., Cetrangolo, G. P., & Febbraio, F. (2018). Improvement of functional properties of a thermostable β-glycosidase for milk lactose hydrolysis. *Biopolymers, 109*(10). https://doi.org/10.1002/bip.23118, http://onlinelibrary.wiley.com/journal/10.1002/(ISSN)1097-0282.

Jain, K. K., Kumar, S., Bhardwaj, K. N., & Kuhad, R. C. (2018). Functional expression of a thermostable endoglucanase from thermoascus aurantiacus RCKK in Pichia pastoris X-33 and its characterization. *Molecular Biotechnology, 60*(10), 736–748. https://doi.org/10.1007/s12033-018-0106-3, http://www.springer.com/humana+press/journal/12033.

Javanmard, A. S., Matin, M. M., & Bahrami, A. R. (2023). Polycistronic cellulase gene expression in Pichia pastoris. *Biomass Conversion and Biorefinery, 13*(8), 7151–7163. https://doi.org/10.1007/s13399-021-01765-7, https://www.springer.com/journal/13399.

Karbalaei, M., Rezaee, S. A., & Farsiani, H. (2020). Pichia pastoris: A highly successful expression system for optimal synthesis of heterologous proteins. *Journal of Cellular Physiology, 235*(9), 5867–5881. https://doi.org/10.1002/jcp.29583, http://onlinelibrary.wiley.com/journal/10.1002/(ISSN)1097-4652.

Karim, K. M. R., & Tasnim, T. (2022). Evaluation of recombinant glucoamylase expression by a native and α-mating factor secretion signal in pichia pastoris. *Journal of Microbiology,*

Biotechnology and Food Sciences, 11(5). https://doi.org/10.55251/jmbfs.3428, https://office2.jmbfs.org/index.php/JMBFS/index.

Karnaouri, A., Muraleedharan, M. N., Dimarogona, M., Topakas, E., Rova, U., Sandgren, M., & Christakopoulos, P. (2017). Recombinant expression of thermostable processive MtEG5 endoglucanase and its synergism with MtLPMO from Myceliophthora thermophila during the hydrolysis of lignocellulosic substrates. *Biotechnology for Biofuels, 10*(1). https://doi.org/10.1186/s13068-017-0813-1, http://www.biotechnologyforbiofuels.com/.

Karnaouri, A. C., Topakas, E., & Christakopoulos, P. (2014). Cloning, expression, and characterization of a thermostable GH7 endoglucanase from Myceliophthora thermophila capable of high-consistency enzymatic liquefaction. *Applied Microbiology and Biotechnology, 98*(1), 231–242. https://doi.org/10.1007/s00253-013-4895-9, link.springer.de/link/service/journals/00253/index.htm.

Kashyap, A., & Gupta, R. (2021). Disrupting putative N-glycosylation site N17 in lipase Lip11 of Yarrowia lipolytica yielded a catalytically efficient and thermostable variant accompanying conformational changes. *Enzyme and Microbial Technology, 151*, 109922. https://doi.org/10.1016/j.enzmictec.2021.109922.

Kasprzak, J., Rauter, M., Denter, S., Becker, K., Baronian, K., Bode, R., Schauer, F., Piontek, M., Vorbrodt, H.-M., & Kunze, G. (2016). Synthesis of ethyl (R)-mandelate using recombinant Carboxydothermus hydrogenoformans alcohol dehydrogenase produced by two yeast species. *Journal of Molecular Catalysis B: Enzymatic, 133*, 176–186. https://doi.org/10.1016/j.molcatb.2016.08.012.

Katsimpouras, C., Dimarogona, M., Petropoulos, P., Christakopoulos, P., & Topakas, E. (2016). A thermostable GH26 endo-β-mannanase from Myceliophthora thermophila capable of enhancing lignocellulose degradation. *Applied Microbiology and Biotechnology, 100*(19), 8385–8397. https://doi.org/10.1007/s00253-016-7609-2.

Kumari, A., & Gupta, R. (2015). Functional characterisation of novel enantioselective lipase TALipA from Trichosporon asahii MSR54: Sequence comparison revealed new signature sequence AXSXG among yeast lipases. *Applied Biochemistry and Biotechnology, 175*(1), 360–371. https://doi.org/10.1007/s12010-014-1268-5.

Lan, D., Qu, M., Yang, B., & Wang, Y. (2016). Enhancing production of lipase MAS1 from marine Streptomyces sp. strain in Pichia pastoris by chaperones co-expression. *Electronic Journal of Biotechnology, 22*, 62–67. https://doi.org/10.1016/j.ejbt.2016.06.003.

Li, N., Wang, H., Li, L., Cheng, H., Liu, D., Cheng, H., & Deng, Z. (2016). Integrated approach to producing high-purity trehalose from maltose by the yeast yarrowia lipolytica displaying trehalose synthase (TreS) on the cell surface. *Journal of Agricultural and Food Chemistry, 64*(31), 6179–6187. https://doi.org/10.1021/acs.jafc.6b02175.

Li, R. K., Hu, Y. J., He, Y. J., Ng, T. B., Zhou, Z. M., & Ye, X. Y. (2021). A thermophilic chitinase 1602 from the marine bacterium Microbulbifer sp. BN3 and its high-level expression in Pichia pastoris. *Biotechnology and Applied Biochemistry, 68*(5), 1076–1085. https://doi.org/10.1002/bab.2027, http://iubmb.onlinelibrary.wiley.com/hub/journal/10.1002/(ISSN)1470-8744/.

Li, Y. X., Wang, N. N., Yan, Q. J., Hua, X. H., Liu, Y., & Jiang, Z. Q. (2022). A novel neutral thermophilic β-mannanase from Malbranchea cinnamomea for controllable production of partially hydrolyzed konjac powder. *Applied Microbiology and Biotechnology, 106*(5-6), 1919–1932. https://doi.org/10.1007/s00253-022-11832-2, link.springer.de/link/service/journals/00253/index.htm.

Lian, Z., Jiang, J. B., Chi, S., Guan, G. H., Li, Y., & Li, J. L. (2015). Characterization of two GAS1 genes and their effects on expression and secretion of heterologous protein xylanase B in Kluyveromyces lactis. *Journal of Microbiology and Biotechnology,*

25(12), 1977–1988. https://doi.org/10.4014/jmb.1505.05058, http://www.jmb.or.kr/journal/download.php?Filedir=./submission/Journal/025/&num=7316.

Liao, H., Li, S., Zheng, H., Wei, Z., Liu, D., Raza, W., Shen, Q., & Xu, Y. (2014). A new acidophilic thermostable endo-1,4-β-mannanase from Penicillium oxalicum GZ-2: Cloning, characterization and functional expression in Pichia pastoris. *BMC Biotechnology, 14*(1). https://doi.org/10.1186/s12896-014-0090-z.

Liu, Z., Cao, L., Fu, X., Liang, Q., Sun, H., & Mou, H. (2021). A multi-functional genetic manipulation system and its use in high-level expression of a β-mannanase mutant with high specific activity in Pichia pastoris. *Microbial Biotechnology, 14*(4), 1525–1538. https://doi.org/10.1111/1751-7915.13812.

Liu, Z., Ning, C., Yuan, M., Yang, S., Wei, X., Xiao, M., Fu, X., Zhu, C., & Mou, H. (2020). High-level expression of a thermophilic and acidophilic β-mannanase from Aspergillus kawachii IFO 4308 with significant potential in mannooligosaccharide preparation. *Bioresource Technology, 295*, 122257. https://doi.org/10.1016/j.biortech.2019.122257.

López-López, O., Cerdán, M. E., & González-Siso, M. I. (2015). Thermus thermophilus as a source of thermostable lipolytic enzymes. *Microorganisms, 3*(4), 792–808. https://doi.org/10.3390/microorganisms3040792, https://www.mdpi.com/2076-2607/3/4/792/pdf.

Luo, Z., Miao, J., Li, G., Du, Y., & Yu, X. (2017). A Recombinant Highly Thermostable β-Mannanase (ReTMan26) from Thermophilic Bacillus subtilis (TBS2) Expressed in Pichia pastoris and Its pH and Temperature Stability. *Applied Biochemistry and Biotechnology, 182*(4), 1259–1275. https://doi.org/10.1007/s12010-017-2397-4.

Ma, J. Y., Cao, H. L., Tan, H. D., Hu, X. J., Liu, W. J., Du, Y. G., & Yin, H. (2016). Cloning, expression, characterization, and mutagenesis of a thermostable exoinulinase from kluyveromyces cicerisporus. *Applied Biochemistry and Biotechnology, 178*(1), 144–158. https://doi.org/10.1007/s12010-015-1864-z, http://www.springer.com/humana+press/journal/12010.

Ma, X., Liu, L., Li, Q., Liu, Y., Yi, L., Ma, L., & Zhai, C. (2017). High-level expression of a bacterial laccase, CueO from Escherichia coli K12 in Pichia pastoris GS115 and its application on the decolorization of synthetic dyes. *Enzyme and Microbial Technology, 103*, 34–41. https://doi.org/10.1016/j.enzmictec.2017.04.004.

Menéndez, C., Martínez, D., Pérez, E. R., Musacchio, A., Ramírez, R., López-Munguía, A., & Hernández, L. (2019). Engineered thermostable β–fructosidase from Thermotoga maritima with enhanced fructooligosaccharides synthesis. *Enzyme and Microbial Technology, 125*, 53–62. https://doi.org/10.1016/j.enzmictec.2019.02.002, www.elsevier.com/locate/enzmictec.

Morka, K., Pietruszka, J., & Meyer zu Berstenhorst, S. (2014). Comparative expression of lipase CAL-A in the yeasts Saccharomyces cerevisiae, Kluyveromyces lactis and Hansenula polymorpha to investigate a possible host influence. *Journal of Biotechnology, 191*, 176–186. https://doi.org/10.1016/j.jbiotec.2014.08.023.

Navone, L., Vogl, T., Luangthongkam, P., Blinco, J. A., Luna-Flores, C. H., Chen, X., von Hellens, J., Mahler, S., & Speight, R. (2021). Disulfide bond engineering of AppA phytase for increased thermostability requires co-expression of protein disulfide isomerase in Pichia pastoris. *Biotechnology for Biofuels, 14*(1). https://doi.org/10.1186/s13068-021-01936-8, http://www.biotechnologyforbiofuels.com/.

Niderhaus, C., Garrido, M., Insani, M., Campos, E., & Wirth, S. (2018). Heterologous production and characterization of a thermostable GH10 family endo-xylanase from Pycnoporus sanguineus BAFC 2126. *Process Biochemistry, 67*, 92–98. https://doi.org/10.1016/j.procbio.2018.01.017.

Nimpiboon, P., Tumhom, S., Nakapong, S., & Pongsawasdi, P. (2020). Amylomaltase from Thermus filiformis: Expression in Saccharomyces cerevisiae and its use in starch modification. *Journal of Applied Microbiology, 129*(5), 1287–1296. https://doi.org/10.1111/jam.14675.

Nisha, M., & Satyanarayana, T. (2017). Characteristics and applications of recombinant thermostable amylopullulanase of Geobacillus thermoleovorans secreted by Pichia pastoris. *Applied Microbiology and Biotechnology, 101*(6), 2357–2369. https://doi.org/10.1007/s00253-016-8025-3.

Olichon, A., Schweizer, D., Muyldermans, S., & de Marco, A. (2007). Heating as a rapid purification method for recovering correctly-folded thermotolerant VH and VHH domains. *BMC Biotechnology, 7*(1). https://doi.org/10.1186/1472-6750-7-7.

Oslan, S. N., Salleh, A. B., Raja Abd Rahman, R. N. Z., Leow, T. C., Sukamat, H., & Basri, M. (2015). A newly isolated yeast as an expression host for recombinant lipase. *Cellular and Molecular Biology Letters, 20*(2), 279–293. https://doi.org/10.1515/cmble-2015-0015, http://www.springerlink.com/content/120185/.

Pretorius, I. S., Laing, E., Pretorius, G. H. J., & Marmu, J. (1988). Expression of a Bacillus α-amylase gene in yeast. *Current Genetics, 14*(1), 1–8. https://doi.org/10.1007/BF00405846.

Qin, X., Tang, H., Cao, W., Dai, Z., Hu, L., Huang, Y., & Liu, T. (2020). An Orthogonal Tyrosyl-tRNA Synthetase/tRNA Pair from a Thermophilic Bacterium for an Expanded Eukaryotic Genetic Code. *Biochemistry, 59*(1), 90–99. https://doi.org/10.1021/acs.biochem.9b00757.

Ranjan, B., Pillai, S., Permaul, K., & Singh, S. (2017). Expression of a novel recombinant cyanate hydratase (rTl-Cyn) in Pichia pastoris, characteristics and applicability in the detoxification of cyanate. *Bioresource Technology, 238*, 582–588. https://doi.org/10.1016/j.biortech.2017.04.091.

Ranjan, B., & Satyanarayana, T. (2016). Recombinant HAP phytase of the thermophilic mold sporotrichum thermophile: Expression of the Codon-optimized phytase gene in pichia pastoris and applications. *Molecular Biotechnology, 58*(2), 137–147. https://doi.org/10.1007/s12033-015-9909-7.

Rocha, S. N., Abrahão-Neto, J., Cerdán, M. E., Gombert, A. K., & González-Siso, M. I. (2011). Heterologous expression of a thermophilic esterase in Kluyveromyces yeasts. *Applied Microbiology and Biotechnology, 89*(2), 375–385. https://doi.org/10.1007/s00253-010-2869-8.

Ronghua, Z., Xianqing, L., Fang, L., Qing, D., Wei, C., YaPing, W., & Ben, R. (2020). High-level Expression of an Acidic and Thermostable Chitosanase in Pichia pastoris Using Multi-copy Expression Strains and High-cell-density Cultivation. *Biotechnology and Bioprocess Engineering, 25*(4), 562–570. https://doi.org/10.1007/s12257-019-0445-3.

Saavedra-Bouza, A., Escuder-Rodríguez, J.-J., deCastro, M.-E., Becerra, M., & González-Siso, M.-I. (2023). Xylanases from thermophilic archaea: A hidden treasure. *Current Research in Biotechnology, 5*, 100116. https://doi.org/10.1016/j.crbiot.2022.11.003.

Samahah, A., Ali, M., Diba, F., Bakar, A., Md Illias, R., Hassan, O., Munir, A., & Murad, A. (2015). Cloning, heterologous expression and characterisation of a recombinant cellobiohydrolase from Humicola insolens ATCC16454 in Pichia pastoris.

Schmidt, A., Shvetsov, A., Soboleva, E., Kil, Y., Sergeev, V., Surzhik, M., & Anthony, W. (2019). Thermostability improvement of Aspergillus awamori glucoamylase via directed evolution of its gene located on episomal expression vector in Pichia pastoris cells. *Protein Engineering, Design and Selection, 32*(6), 251–259. https://doi.org/10.1093/protein/gzz048.

Schuster, M., Wasserbauer, E., Einhauer, A., Ortner, C., Hammerschmid, F., Werner, G., & Jungbauer, A. (2000). Protein expression strategies for identification of novel target proteins. *SLAS Discovery, 5*(2), 89–97. https://doi.org/10.1177/108705710000500205.

Serra, L. A., Mendes, T. D., Marco, J. L. D., & de Almeida, J. R. M. (2024). Application of Thermomyces lanuginosus polygalacturonase produced in Komagataella phaffii in biomass hydrolysis and textile bioscouring. *Enzyme and Microbial Technology, 177*. https://doi.org/10.1016/j.enzmictec.2024.110424, https://www.sciencedirect.com/science/journal/01410229.

Silva, I. R., Jers, C., Otten, H., Nyffenegger, C., Larsen, D. M., Derkx, P. M. F., Meyer, A. S., Mikkelsen, J. D., & Larsen, S. (2014). Design of thermostable rhamnogalacturonan lyase mutants from Bacillus licheniformis by combination of targeted single point mutations. *Applied Microbiology and Biotechnology, 98*(10), 4521–4531. https://doi.org/10.1007/s00253-013-5483-8, link.springer.de/link/service/journals/00253/index.htm.

Simatupang, D. F., Widhiastuty, M. P., Madayanti, F., & Akhmaloka, A. (2022). Lipolytic activity of Itb1.1 and Lk3 thermostable lipases expressed in Escherichia coli and Pichia pastoris. *Journal of Applied Pharmaceutical Science, 12*(9), 34–42. https://doi.org/10.7324/JAPS.2022.120905, www.japsonline.com/index.php.

Song, X., Shan, Y., Cao, L., Zhong, X., Wang, X., Gao, Y., Wang, K., Wang, W., & Zhu, T. (2024). Decolorization and detoxication of malachite green by engineered Saccharomyces cerevisiae expressing novel thermostable laccase from Trametes trogii. *Bioresource Technology, 399*, 130591. https://doi.org/10.1016/j.biortech.2024.130591.

Su, L., Ji, D., Tao, X., Yu, L., Wu, J., & Xia, Y. (2017). Recombinant expression, characterization, and application of a phospholipase B from Fusarium oxysporum. *Journal of Biotechnology, 242*, 92–100. https://doi.org/10.1016/j.jbiotec.2016.12.009.

Sun, Q., Chen, F., Geng, F., Luo, Y., Gong, S., & Jiang, Z. (2018). A novel aspartic protease from Rhizomucor miehei expressed in Pichia pastoris and its application on meat tenderization and preparation of turtle peptides. *Food Chemistry, 245*, 570–577. https://doi.org/10.1016/j.foodchem.2017.10.113.

Sun, Q., Wang, H., Zhang, H., Luo, H., Shi, P., Bai, Y., Lu, F., Yao, B., & Huang, H. (2016). Heterologous production of an acidic thermostable lipase with broad-range pH activity from thermophilic fungus Neosartorya fischeri P1. *Journal of Bioscience and Bioengineering, 122*(5), 539–544. https://doi.org/10.1016/j.jbiosc.2016.05.003.

Tan, H., Yang, G., Chen, W., Liu, Q., Li, K., & Yin, H. (2020). Identification and characterization of thermostable endo-polygalacturonase II B from Aspergillus luchuensis. *Journal of Food Biochemistry, 44*(3). https://doi.org/10.1111/jfbc.13133.

Vogl, T., Sturmberger, L., Fauland, P. C., Hyden, P., Fischer, J. E., Schmid, C., Thallinger, G. G., Geier, M., & Glieder, A. (2018). Methanol independent induction in Pichia pastoris by simple derepressed overexpression of single transcription factors. *Biotechnology and Bioengineering, 115*(4), 1037–1050. https://doi.org/10.1002/bit.26529, http://onlinelibrary.wiley.com/journal/10.1002/(ISSN)1097-0290.

Voutilainen, S. P., Nurmi-Rantala, S., Penttilä, M., & Koivula, A. (2014). Engineering chimeric thermostable GH7 cellobiohydrolases in Saccharomyces cerevisiae. *Applied Microbiology and Biotechnology, 98*(7), 2991–3001. https://doi.org/10.1007/s00253-013-5177-2, link.springer.de/link/service/journals/00253/index.htm.

Wang, F., Li, S., Zhao, H., Bian, L., Chen, L., Zhang, Z., Zhong, X., Ma, L., Yu, X., & Permyakov, E. A. (2015). Expression and characterization of the RKOD DNA polymerase in Pichia pastoris. *PLOS ONE, 10*(7), e0131757. https://doi.org/10.1371/journal.pone.0131757.

Wang, Y., Shu, T., Fan, P., Zhang, H., Turunen, O., Xiong, H., & Yu, L. (2017). Characterization of a recombinant alkaline thermostable β-mannanase and its application in eco-friendly ramie degumming. *Process Biochemistry, 61*, 73–79. https://doi.org/10.1016/j.procbio.2017.06.008.

Wang, Z., Guo, C., Liu, L., & Huang, H. (2018). Effects of N-glycosylation on the biochemical properties of recombinant bEKL expressed in Pichia pastoris. *Enzyme and Microbial Technology, 114*, 40–47. https://doi.org/10.1016/j.enzmictec.2018.03.004.

Wohlgemuth, R., Littlechild, J., Monti, D., Schnorr, K., van Rossum, T., Siebers, B., Menzel, P., Kublanov, I. V., Rike, A. G., Skretas, G., Szabo, Z., Peng, X., & Young, M. J. (2018). Discovering novel hydrolases from hot environments. *Biotechnology Advances, 36*(8), 2077–2100. https://doi.org/10.1016/j.biotechadv.2018.09.004.

Woon, J. S. K., Mackeen, M. M., Mahadi, N. M., Illias, R. M., Abdul Murad, A. M., & Abu Bakar, F. D. (2016). Expression and characterization of a cellobiohydrolase (CBH7B) from the thermophilic fungus Thielavia terrestris in Pichia pastoris. *Biotechnology and Applied Biochemistry, 63*(5), 690–698. https://doi.org/10.1002/bab.1431, http://onlinelibrary.wiley.com/journal/10.1002/(ISSN)1470-8744.

Xia, W., Xu, X., Qian, L., Shi, P., Bai, Y., Luo, H., Ma, R., & Yao, B. (2016). Engineering a highly active thermophilic β-glucosidase to enhance its pH stability and saccharification performance. *Biotechnology for Biofuels, 9*(1). https://doi.org/10.1186/s13068-016-0560-8.

Xu, L., Xiao, X., Wang, F., He, Y., Yang, X., Hu, J., Feng, Z., & Yan, Y. (2020). Surface-displayed thermostable candida rugosa lipase 1 for docosahexaenoic acid enrichment. *Applied Biochemistry and Biotechnology, 190*(1), 218–231. https://doi.org/10.1007/s12010-019-03077-z, http://www.springer.com/humana+press/journal/12010.

Yang, H., Shi, P., Lu, H., Wang, H., Luo, H., Huang, H., Yang, P., & Yao, B. (2015). A thermophilic β-mannanase from Neosartorya fischeri P1 with broad pH stability and significant hydrolysis ability of various mannan polymers. *Food Chemistry, 173*, 283–289. https://doi.org/10.1016/j.foodchem.2014.10.022.

Yang, H., Zhang, Y., Li, X., Bai, Y., Xia, W., Ma, R., Luo, H., Shi, P., & Yao, B. (2018). Impact of disulfide bonds on the folding and refolding capability of a novel thermostable GH45 cellulase. *Applied Microbiology and Biotechnology, 102*(21), 9183–9192. https://doi.org/10.1007/s00253-018-9256-2.

Yang, J., Ma, T., Shang-guan, F., & Han, Z. (2020). Improving the catalytic activity of thermostable xylanase from Thermotoga maritima via mutagenesis of non-catalytic residues at glycone subsites. *Enzyme and Microbial Technology, 139*, 109579. https://doi.org/10.1016/j.enzmictec.2020.109579.

Yu, S., Li, Z., Wang, Y., Chen, W., Fu, L., Tang, W., Chen, C., Liu, Y., Zhang, X., & Ma, L. (2015). High-level expression and characterization of a thermophilic β-mannanase from Aspergillus niger in Pichia pastoris. *Biotechnology Letters, 37*(9), 1853–1859. https://doi.org/10.1007/s10529-015-1848-7.

Zerva, A., Christakopoulos, P., & Topakas, E. (2015). Characterization and application of a novel class II thermophilic peroxidase from Myceliophthora thermophila in biosynthesis of polycatechol. *Enzyme and Microbial Technology, 75-76*, 49–56. https://doi.org/10.1016/j.enzmictec.2015.04.012.

Zhang, N., Wright, T., Wang, X., Savary, B. J., & Xu, J. (2020). Production of thermostable endo-1,5-α-l-arabinanase in Pichia pastoris for enzymatically releasing functional oligosaccharides from sugar beet pulp. *Applied Microbiology and Biotechnology, 104*(4), 1595–1607. https://doi.org/10.1007/s00253-019-10238-x.

Zhao, J., Guo, C., Tian, C., & Ma, Y. (2015). Heterologous Expression and Characterization of a GH3 β-Glucosidase from Thermophilic Fungi Myceliophthora thermophila in Pichia pastoris. *Applied Biochemistry and Biotechnology, 177*(2), 511–527. https://doi.org/10.1007/s12010-015-1759-z.

Zhou, W., Tong, S., Amin, F. R., Chen, W., Cai, J., & Li, D. (2023). Heterologous expression and biochemical characterization of a thermostable endoglucanase (MtEG5-1) from myceliophthora thermophila. *Fermentation, 9*(5). https://doi.org/10.3390/fermentation9050462.

Soil microalgae as biofertilizers

4

Timea I. Hajnal-Jafari, and Dragana Stamenov
Department of Field and Vegetable Crops, Faculty of Agriculture, University of Novi Sad, Novi Sad, Serbia

4.1 Microalgae in soil

Algae exhibit great diversity in shape, size, level of organization, life history, and metabolism (Lee, 2005). Biologists generally classify algae into four categories based on their size and morphological characteristics: unicellular, colonial, filamentous, and multicellular algae. This classification further divides algae into two main groups: macroalgae and microalgae. Macroalgae, macroscopic and multicellular, can grow up to 65 m in length. Commonly known as seaweeds, macroalgae include both freshwater and marine species (Ge et al., 2018; Lawton et al., 2013). In agriculture, seaweed extracts stimulate seed germination, plant growth, and defense mechanisms (Battacharyya et al., 2015). In contrast, microalgae are microscopic, existing as unicellular or filamentous organisms, with sizes ranging from approximately 1 to 900 μm (Nabti et al., 2017). Microalgae in phytoplankton inhabit nearly all aquatic, terrestrial, and subaerial surfaces (Tomaselli, 2004). Microalgae can be prokaryotic, such as cyanobacteria (divisions Cyanophyta and Prochlorophyta), which are gram-negative bacteria, or eukaryotic, such as green algae (division Chlorophyta), diatoms (class Bacillariophyceae), euglenoids (class Euglenophyceae), and dinoflagellates (division Dinophyta) (Barsanti & Gualtieri, 2014; Mata et al., 2010).

According to some scientists, a distinct group known as mesokaryotes is also recognized within algae (Lee, 2005). Mesokaryotes are considered intermediate forms that arose due to the loss of histones in eukaryotes but are still recognized as eukaryotes. According to these scientists, dinoflagellates are classified as mesokaryotes.

Although the term "microalgae" has long been used, it lacks a clear definition. It includes taxonomically unrelated unicells, colonial, and filamentous forms (Pulz & Gross, 2004).

Primarily aquatic, algae inhabit almost every habitat of our planet from deep oceans to hot and cold deserts, highly acidic postmining sites to Antarctic soils, air, and man-made subaerial surfaces (Sharma et al., 2007). Soil microalgae are terrestrial photoautotrophic prokaryotic and eukaryotic microbes, temporarily or permanently growing in soils. The primary role of soil microalgal communities encompasses the facilitation of soil formation and biochemical processes, such as dinitrogen fixation, aggregation stabilization, organic matter mineralization, enhancement of soil air and water retention, and promotion of soil microbial activity and structure (Alvarez et al., 2021). Furthermore, soil microalgae have an important role in soil and vegetation

Eukaryotic Microorganisms as Sources of Bioproducts. DOI: https://doi.org/10.1016/B978-0-443-30188-9.00004-4

restoration efforts in areas impacted by waste rock deposition and anthropogenically disturbed lands (Oliveira & MacIel-Silva, 2022).

Soil eukaryotic microalgae have been recognized as integral components of soil microflora since the late 19th century, gaining prominence in soil ecology during the second half of the 20th century (Starks et al., 2019). This diverse group of unicellular microorganisms varies in size from a few to hundreds of microns and inhabits nearly all terrestrial ecosystems on Earth. Apart from soil algae, subaerial algae, which are epiphytic (growing on different plant parts), epilithic (on rocks or pebbles), epizoic (on animals), or can grow on objects such as buildings (Lopez-Bautista et al., 2007), are also sometimes found in soils because to sustain themselves in their specific terrestrial habitats, they have to reside in soil at least at some stages in their life. Although the abundance of soil eukaryotic microalgae is much lower than that of soil microorganisms, they also exist widely in the soil of various habitats, such as woodland, grassland, farmland, tundra, and desert (Kabirov & Gaisina, 2009; Lin et al., 2013; Perera et al., 2018). They are considered to be important contributors to soil primary productivity and play key roles in soil nutrient cycling and plant growth (Abinandan et al., 2019; Alvarez et al., 2021).

Since there are various classifications of algae used by scientists, it is tough to determine which phyla of algae are dominant in the soil. The classification of a single algal species into a specific phylum can vary depending on the applied systematics and classification approach. The most widely accepted systematics classifies eukaryotic algae according to pigmentation, life cycle, and talus structure into the Chlorophyta (green algae), Rhodophyta (red algae), Xantophyta (yellow-green), Phaeophyta (brown algae), Bacillariophyta (diatoms), Pyrrophyta (yellowish-green to golden-brown), Euglenophyta, and Chrysophyta (golden-brown) phyla (Win et al., 2018). According to this classification, in soil algal species belonging to the phylum Chlorophyta, Bacilariophyta, and Chrysophyta dominate (Patova et al., 2023). According to algae within the Bacilariophyta belong to Chrysophyta and together form one phylum. The most reported algal species of soils included *Bracteacoccus minor, Chlorella vulgaris, Chlorococcum infusionum, Klebsormidium flaccidum, Stichococcus bacillaris* (Chlorophyta), *Hantzschia amphioxys, Pinnularia borealis* (Bacillariophyta). According to the other literature, soil eukaryotic algal species may be categorized into the Chlorophyta, Bacillariophyta, Streptophyta, Xanthophyceae, Eustigmatophyceae, and Euglenophyceae (Adesalu & Olugbemi, 2015; Jimel et al., 2021; Mikhailyuk et al., 2019; Vijayan & George Ray, 2015; Zhang et al., 2020). In general, the current literature analysis revealed that although microalgae of diverse groups exist in soil and other subaerial environments, the Chlorophyta are the most dominant group of microalgae in soil habitats (Patova & Novakovskaya, 2018).

Research efforts have revealed the composition of soil microalgae in various agroecosystems. Among the Chlorophyta, the common genera that have been observed in soil habitats include *Chlamydomonas, Chlorella*, and *Chlorococcum* (Novakovskaya et al., 2022). Zancan et al. (2006) investigated the composition in four distinct agroecosystems, revealing Chlorophyta (Chlamydophyceae and Chlorophyceae) as the main algal phylum. Similarly identified Chlorophyceae sequences were identified in sampled maize fields using PCR-denaturing gradient gel electrophoresis (PCR-DGGE). Bakieva et al. (2012) also concluded that microalgal species belonging to the Chlorophyta dominate in soil. Their study revealed that out of 80 species of isolated soil microalgae, 52 were classified as green algae. Lin and Wu (2014) delved into the diversity and abundance of Indigenous soil photomicrobes in arid subtropical Taiwan, pinpointing green algae such as *Bracteacoccus, Coccomyxa*, and *Koliella* as the predominant genus. Further analyses by Seppey et al. (2017) across forests, meadows, and croplands revealed Chlorophyceae as the dominant phototrophic microeukaryotes. In steppe ecosystems, Shcherbyna et al. (2021) successfully identified a range of soil microalgae species within the Chlorophyta division. Chlorophyta was the dominant algal group in studies of forest soils (Maltseva & Maltsev, 2021) and the biological soil crust in forests (Glaser et al., 2018). Notably, Chlorophyta's widespread distribution and high abundance in soil, as observed in various studies, underscore its significant diversity in soil ecosystems (Hodač et al., 2016).

Another significant group of microalgae in soils is Bacillariophyta, and the common genera include *Eunotia, Hantzchia, Luticola, Navicula, Nitzschia*, and *Pinnularia* (Flechtner et al., 2008; Zancan et al., 2006). In the crop fields of Nigeria, Adesalu and Olugbemi (2015) reported a high diversity of these algae. Among Bacillariophyta, *Pinullaria* is one of the most numerous genera. Pinnularia is often mentioned as the most diverse group in soil algocenoses, as many of its species are cosmopolitan and common (Barragán et al., 2018; Foets et al., 2021). Numerous species of *Pinnularia* thrive in soils of spruce forests of the middle and southern taiga (Novakovskaya & Patova, 2011) and volcanic soils of Kamchatka, Russia (Fazlutdinova et al., 2021). Several *Pinnularia* species (*P. borealis, P. obscura, P. schoenfelderi*, and *P. sinistra*) have been identified as dominating taxa in the soils of Poland (Stanek-Tarkowska et al., 2017; Stanek-Tarkowska et al., 2021). Pinnularia was also identified as the dominant diatom genus in paddy fields of Central Japan (Ohtsuka & Fujita, 2001).

Chrysophyta emerges as a dominant phylum in farmland and grassland soil (Fathi & Zaki, 2003). Furthermore, they reported that four eukaryotic algal groups, Bacillariophyta, Chlorophyta, Euglenophyta, and Chrysophyta, were recorded in soil cultivated with different plants. Lentendu et al. (2014) analyzed the eukaryotic microbial community in arable soil with differential fertilization via high-throughput sequencing. Although Chrysophyceae were detected, its abundance was lower than that of Chlorophyta. The reason for the lower diversity and abundance of algae belonging to the phylum Chrysophyta compared to algae of the division Chlorophyta varies. Some scientists suggest that the research method itself could contribute to this phenomenon. Traditional methods for surveying microalgal communities involve isolation, culture, and subsequent species identification based on morphological criteria (Hodač et al., 2016). However, compared to Chlorophyta, Chrysophyta species present challenges in isolation and cultivation due to inherent attributes (Xie & Fan, 2007). Consequently, while Chlorophyta's abundance and diversity in soil are often well-documented, limitations in analysis methods may bias the representation of soil Chrysophyta in the previous reports. Soil sources and microenvironments also likely influence the abundance and diversity of Chrysophyta (Zhang & Lv, 2022)

As primary producers, microalgae are particularly vulnerable to fluctuations in environmental conditions and natural disruptions. They demonstrate a heightened sensitivity to specific pollutants, which can precipitate quantitative and qualitative shifts within microalgal communities. The total biomass of microalgae serves as a crucial indicator of the overall health and suitability of aquatic and soil habitats. Certain pollutants can result in the mortality of microalgae and a subsequent decline in species diversity and mass reproduction (De Clerck et al., 2012).

The estimated number of microalgal species ranges in the hundreds of thousands, with population densities in topsoil varying between 3 and 100 million cells per gram (Chiaiese et al., 2018). According to some authors, soil algae exist in high numbers ($5.5\pm3.4\times10^6$) per gram of surface soil (Jassey et al., 2022). Soil microalgae, as photosynthetically active microorganisms, exhibit a dynamic community structure influenced by soil type, agricultural practices, and pesticide applications. While previous studies have emphasized the roles of water content and pH in the distribution and diversity of soil microalgae (Lin & Wu, 2014; Zancan et al., 2006), recent research highlights the strong correlation between soil eukaryotic microalgal community composition and edaphic factors such as pH, organic matter, and nitrogen (Rahman, 2022; Zhang & Lv, 2022).

4.2 Microalgae and plant's abiotic and biotic stress tolerance improvement

4.2.1 Microalgae and plant's abiotic stress tolerance improvement

Abiotic stress refers to any environmental factor that impedes plant growth and productivity. Due to their immobile nature, plants cannot relocate to evade the detrimental effects of such stressors. Common abiotic stresses include heavy metal (HM) contamination, cold, nutrient deficiency, soil salinity, high-temperature fluctuations, drought, and floods, all of which significantly reduce crop productivity (Ronga et al., 2018). Quantifying the precise impact of these stresses on agricultural lands remains challenging. However, Cramer et al. (2015), drawing on FAO reports, estimate that approximately 96.5% of global rural land is affected by these stressors. Their analysis also provides a detailed breakdown of the contributions from each specific stressor. The author highlights that soil salinity, drought, and floods are the most detrimental factors affecting agricultural land.

Abiotic stress affects plants in various ways, triggering various changes within the plant. The primary mechanism behind many of the negative effects of climate change-induced abiotic stress involves the production of reactive oxygen species (ROS). These ROS disrupt protein synthesis, cause lipid peroxidation, change membrane permeability, deactivate enzymes, and break down nucleic acids, ultimately resulting in cell death (Singh et al., 2019).

Recently, scientists have developed various strategies for managing plant stress to enhance crop productivity under challenging conditions. One such approach is the application of biofertilizers. Biofertilizers are substances that contain living microorganisms which, when applied to the soil, seed, or plant surface, colonize the rhizosphere and promote growth by increasing the supply or availability of nutrients to the host plant (Vessey, 2003). Microbial biofertilizers play a pivotal role in sustainable agriculture. Their main function is to enhance plant nutrient uptake and produce various physiologically active materials, which improve plant tolerance to stress, enhance the quality of farm products, and reduce the need for chemical fertilizers (Liu et al., 2001).

Eukaryotic and prokaryotic microalgae produce a wide range of biologically active compounds. These compounds include toxins, algaecides, plant growth regulators, and substances used in both the food and pharmaceutical industries (Bhattacharjee, 2016). In agriculture, the application of biofertilizers based on microalgal extracts to protect plants from abiotic stresses has shown promising results, with the intensity of the effect depending on factors such as the plant's developmental stage and the applied dose (Van Oosten et al., 2017; Vernier et al., 2005). Following the application of microalgae products, plants have demonstrated robust growth, increased yield, improved nutrient uptake, and enhanced resistance to abiotic stresses. Additionally, these treatments have led to increased quality and longer shelf life of fruits (Plaza et al., 2018; Righini et al., 2016; Singh et al., 2018). According to Kusvuran and Kusvuran (2019), the application of *Chlorella vulgaris*-based microbial fertilizer under drought conditions was favorable for the development, growth, and biochemical and physiological processes of guar plants. Therefore, microalgae application is a helpful strategy for enhancing the tolerance of guar plants when grown under drought conditions. Similarly, Duo et al. (2018) indicated that the application of microalgal extracts can help mitigate the detrimental effects of abiotic stresses such as drought and salinity. They found that microalgal extracts reduced salt and drought stress impacts on wheat, tomato, and guar. Kusvuran (2021) claimed that foliar application of the microalga *Chlorella vulgaris* significantly increased antioxidant levels, including total flavonoid and phenolic content, in broccoli plants subjected to drought stress. This antioxidant activity reduced ROS levels, thereby decreasing membrane damage and improving chlorophyll and carotenoid content as well as nutrient uptake. These changes facilitated the plants' adaptation to drought stress.

The application of microalgal extracts from *Dunaliella* spp. and *Phaeodactylum* spp. has been shown to mitigate the negative effects of salt stress during the germination of pepper seeds, significantly reducing the production of superoxide radicals and lipid peroxidation (Guzman-Murillo et al., 2013). In wheat plants irrigated with seawater, the application of microalgal extracts improved their tolerance to salinity. Specifically, *Chlorella* spp. extracts increased the antioxidant capacity and protein content of the kernels (Abd-El-Baky et al., 2010). Using microalgae to protect against various forms of abiotic stress activates plant antioxidant enzymes, including peroxidases, polyphenol oxidase, and phenylalanine ammonia-lyase (Arroussi et al., 2018; Renuka et al., 2018). Additionally, this protection is associated with increased stomatal closure frequency and enhanced water use efficiency (Li et al., 2014). In another study, it was observed that the application of combined microalgae-cyanobacteria extracts enhanced osmolyte accumulation, in this case, proline (Mutale-joan et al., 2020).

Arroussi et al. investigated the potential of sulfated exopolysaccharides extracted from *Dunaliella salina*, a halophilic green microalga, to enhance the salt stress adaptability of tomato plants. These sulfated exopolysaccharides significantly increased the length and dry weight of shoots and roots as well as the potassium (K+) content and K+/Na+ ratio in tomato plants under stress. Similarly, Abd-El-Baky et al. (2010) studied the potential of extracts from *Spirulina maxima* and *Chlorella ellipsoida* to enhance the salt stress adaptability of wheat plants. They found that the antioxidant compound content in whole grains of wheat plants irrigated with 10% and 20% seawater increased in response to these microalgal extracts. The results showed significantly higher levels of total carotenoids, tocopherol, phenolic compounds, and protein content. These increases attenuated the salinity stress in plants, primarily through the stimulation of antioxidant compounds and protein content. However, while numerous classes of microalgal metabolites with biostimulant and biofertilizer properties have been identified and documented (Kapoore et al., 2021), their precise mechanisms of action and effects on plant physiology are not yet fully understood.

Table 4.1 outlines the key mechanisms through which microalgae enhance plant resilience to abiotic stress.

Microalgae are rich in antioxidants such as carotenoids (alpha and beta carotene), tocopherols, flavonoids, and phenolic compounds. These antioxidants help in scavenging ROS generated during abiotic stress, thereby protecting plant cells from oxidative damage. Enhanced antioxidant activity reduces membrane damage and improves overall plant health under stress conditions. Commonly found phenolic compounds in various microalgae species include gallic, ferulic, caffeic, chlorogenic,

Table 4.1 Key mechanisms by which microalgae improve plant resilience to abiotic stress.

Mechanisms	References
Increased secondary metabolites	Del Mondo et al. (2021)
Production of polysaccharides	Gonçalves (2021)
Antioxidant production	Kusvuran (2021); Chanda et al. (2019)
Promotes osmolyte accumulation	Renuka et al. (2018)
Enhance nutrient availability and uptake	Plaza et al. (2018)
Phytohormones production	Gonçalves (2021); Lu and Xu (2015)
Stress signal modulation	Fahad et al. (2015)
Enhancement of photosynthetic efficiency	Mógor et al. (2018)
Induction of stomatal closure. Reduction in stomatal aperture through ROS mediated signaling pathway	Li et al. (2014)
Soil health improvement	Lee and Ryu (2021)
Bioactive compound production	Kapoore et al. (2021); Moreira et al. (2022)

synapic, and coumaric acids, along with hydroxybenzoates. Among flavonoids, rutin, quercetin, and kaempferol are the most frequently observed (Del Mondo et al., 2021).

The production of a wide range of extracellular polysaccharides has been documented in microalgae from various genera, including *Aphanothece, Arthrospira, Chlamydomonas, Chlorella, Cylindrotheca, Dunaliella, Navicula, Nostoc, Phaeodactylum, Porphyridium, Rhodella,* and *Scytonema.* In agriculture, these polysaccharides play a critical role in enhancing soil quality, serving as effective soil conditioners, nutrient carriers, and controlled-release systems for agrochemicals. Beyond their influence on soil properties, studies have also highlighted their potential to stimulate plant growth and bolster plant defenses against biotic and abiotic stresses (Gonçalves 2021). Polysaccharide extracts from various microalgae species, including *Spirulina platensis*, *Dunaliella salina*, *Porphyridium* spp., and *Phaeodactylum tricornutum*, have shown significant biostimulatory effects on tomato plants. These extracts enhance phenylalanine ammonia lyase and chitinase enzyme activity, increase total polyphenol content, and improve ROS scavenging capacity, while also stimulating the biosynthesis of very long-chain fatty acids essential for forming leaf cuticular wax (Rachidi et al., 2020). Furthermore, Mutale-joan et al. (2020) observed that injecting liquid extracts enriched with polysaccharides into tomato plants triggered de novo lipid biosynthesis. This resulted in elevated production of palmitic and stearic acids, vital for cuticular wax formation, as well as linoleic acid, which is integral to the jasmonate signaling pathway.

Through various mechanisms, microalgae enhance the activity of antioxidant enzymes in plants (Abd-El-Baky et al., 2010), including superoxide dismutase, catalase, ascorbate peroxidase, and glutathione reductase (Kusvuran, 2021). Microalgal polysaccharides stimulate plant responses via signaling pathways dependent on microbial-associated molecular patterns (Chanda et al., 2019). Soil enzymes like beta-glucanase and chitinase, secreted by microorganisms, hydrolyze these polysaccharides, releasing sugars that are recognized by plant receptors. This triggers signaling cascades involving Ca2+ influx, the activation of octadecanoid and phenylpropanoid pathways, salicylic acid and jasmonic acid signaling, and ROS-scavenging enzymes like catalase, peroxidase, and superoxide dismutase, along with the production of defense molecules (Chanda et al., 2019). Eukaryotic microalgae, such as *Chlorella* spp. and *Dunaliella salina*, produce polysaccharides with biostimulatory potential.

Microalgae produce osmoprotectants such as proline, glycine betaine, and other compatible solutes. When applied to plants, these compounds help in maintaining cellular osmotic balance, protecting cellular structures, and stabilizing proteins and membranes under osmotic stress conditions like drought and salinity (Renuka et al., 2018).

Microalgae can enhance nutrient availability and uptake by plants. They secrete organic acids and other compounds that solubilize essential nutrients in the soil, making them more accessible to plant roots. This improved nutrient availability supports better plant growth and stress resilience (Plaza et al., 2018).

Microalgae are known to produce both exogenous and endogenous phytohormones (Stirk et al., 2013). These include essential phytohormones such as auxins, cytokinins, gibberellins, and low-molecular-weight signaling molecules such as brassinosteroids. Brassinosteroids are steroidal phytohormones recognized as a new class of compounds that serve a dual purpose: they function as growth-promoting hormones and help protect against abiotic stresses, especially salt stress (Stirk et al., 2014). Phytohormones are small signaling molecules that coordinate cellular activities like division, elongation, and differentiation and are involved in processes such as pattern formation, organogenesis, reproduction, and responses to abiotic and biotic stress (Williams, 2011). Microalgal extracts rich in phytohormones offer promising potential for applications in crop production (Lu & Xu, 2015). These hormones can significantly influence plant stress responses by promoting root growth, enhancing shoot development, and improving overall stress tolerance (Gonçalves, 2021). Extracts and metabolites derived from microalgal species such as *Chlorella* spp., *Spirulina platensis*, *Acutodesmus* spp., *Scenedesmus* spp., *Dunaliella* spp., and *Calothrix elenkinii* are widely used as biostimulants (Colla & Rouphael, 2020; Ronga et al., 2019). Oancea et al. (2013) revealed that a biostimulant derived from *Nannochloris* sp. significantly enhanced the growth of tomato plants under nonwater-stressed conditions. Plaza et al. (2018) demonstrated that extracts from *Scenedesmus* spp. could enhance plant growth and development, increasing the number of flowers, shoots, and leaves in *Petunia hybrida*. Fahad et al. (2015) suggest that the mechanism of stress tolerance in plants, induced by microalgal cytokinins, may be due to the free radical scavenging properties of these cytokinins.

Table 4.2 provides an overview of the different types of hormones, along with specific microalgae species responsible for their production.

Microalgal extracts can influence the expression of stress-responsive genes in plants. They can activate or upregulate genes involved in stress tolerance pathways,

Table 4.2 The various hormone types and the corresponding microalgae species responsible for their production.

Phytohormones	Microalgal/Cyanobacterial sources	Role in agriculture	References
auxins; abscisic acid; cytokinins; ethylene; gibberellins; brassinosteroids	*Arthrospira*; *Protococcus*; *Scenedesmus (Scenedesmus armatus); Chlorella (Chlorella pyrenoidosa, Chlorella minutissima); Chlorococcum; Chlamydomonas (Chlamydomonas reinhardtii); Dunaliella (Dunaliella sp.); Draparnaldia (Draparnaldia mutabilis);*	Plant growth stimulation Regulation of cellular activities in crops Crop responses to stress conditions	Gonçalves (2021); Lu and Xu (2015)

leading to the production of stress-related proteins and enzymes that help plants cope with adverse conditions (Fahad et al., 2015).

Microalgae improve photosynthetic efficiency in plants by increasing chlorophyll content and enhancing the performance of the photosynthetic apparatus. This leads to better energy capture and utilization, which is crucial for plant survival and productivity under stress conditions (Mógor et al., 2018).

When the stomatal aperture decreases below a critical threshold, transpiration rates decline, resulting in enhanced instantaneous water use efficiency (WUE). This observation underpins the hypothesis that partial stomatal closure can significantly improve WUE in plants. To explore this, Li, et al. (2014) investigated whether *Chlorella* could be perceived by guard cells to trigger partial stomatal closure and enhance WUE when applied as a foliar spray. Their findings revealed that the exogenous application of optimal *Chlorella* suspension concentrations effectively improved the instantaneous WUE in *Vicia faba*, primarily by reducing transpiration rates without adversely affecting the net photosynthetic rate.

Microalgae contribute to soil health by improving soil structure, enhancing microbial activity, and increasing organic matter content. Healthy soils support robust plant growth and enhance the ability of plants to withstand stress (Lee & Ryu, 2021).

Microalgae produce a variety of bioactive compounds such as proteins and lipids that can have protective effects on plants. For example, sulfated exopolysaccharides from microalgae can enhance plant growth and stress tolerance by improving water retention and ion balance in plant tissues (Kapoore et al., 2021; Moreira et al., 2022).

4.2.2 Microalgae and plant's biotic stress tolerance improvement

Biotic stress in plants is caused by living organisms, such as viruses, bacteria, fungi, nematodes, insects, arachnids, and weeds (Gull et al., 2019). These biotic stress agents cause various types of diseases, infections, and damage to plants and ultimately affect plant productivity. Many biotic stresses affect photosynthesis, such as chewing insects, or illnesses affected by phytopathogenic microorganisms, reducing leaf area.

Biopesticides consist of naturally occurring organisms or compounds derived from biological sources that control the growth and development of plant pathogens, such as bacteria, fungi, insects, and nematodes (Fenibo et al., 2021). Biopesticides derived from microalgae are classified as both microbiological and biochemical agents (Parmar et al., 2023). As far as biotic stresses are concerned, microalgae can suppress fungal disease, inhibit fungal pathogen growth, activate plant immune responses, control pathogenic nematodes and insect pests by nematocidal or repellent activity, improve the control of bacterial pathogens, and promote prime innate plant immunity (Bileva, 2013; Choleva et al., 2005; Choleva et al., 2007; Hamouda & El-Ansary, 2017; Khan et al., 2007; Parmar et al., 2023).

Benderliev et al. (2003) proved the protective effect of microalgae biologically active compounds (siderophore) against pathogenic fungi. The attempts to find *Chlorella fusca* determinants allowed (Lee et al., 2020) to identify d-lactic acid secreted in the supernatant of *C. fusca* as a defense priming agent. This is the first report on the mechanism of innate immune activation by aquatic microalga *Chlorella* in higher plants. According to Ghasemi et al. (2007), the methanol and hexane extracts from *Chlamydomonas reinhardtii*, *Chlorella vulgaris*, *Scenedesmus obliquus*, and *Oocystis* sp. inhibited the growth of *Aspergillus niger*, *Candida kefyr*, and *Aspergillus fumigatus*.

In Hamouda and El-Ansary (2017), the effect of algal derivatives or their components on suppressing the root-knot nematode (*Meloidogyne incognita*) was examined. The application of the nematicidal potential of three microalgae, *Scenedesmus obliquus, Chlorella vulgaris*, and *Anabaena oryzae* was tested against *Meloidogyne incognita* on banana plants. All algal-tested materials significantly ($p < 0.05$) reduced the number of second-stage juveniles of *M. incognito* in the soil (52.6%).

Multiple studies have identified virus-inhibiting compounds in microalgae. For example, sulfated polysaccharides extracted from *Navicula directa* and *Chlorella autotrophica* have been shown to suppress the replication of viruses such as VHSV, ASFV, HSV-1, HSV-2, and the influenza A virus, as well as inhibit the enzyme hyaluronidase (Lee et al., 2006).

The bioactive compounds produced by microalgae can help enhance plant resistance and resilience against biotic stressors. Table 4.3 lists the primary mechanisms by which microalgae improve plant tolerance to biotic stress.

Microalgae can induce systemic resistance in plants, a state where the plant's immune system is primed to respond more effectively to pathogen attacks. This is achieved through the activation of defense-related genes and pathways. For example, compounds like polysaccharides and glycoproteins in microalgal extracts can activate plant defense mechanisms, leading to the production of pathogenesis-related proteins, enzymes, and other defensive compounds (Babu et al., 2015).

Microalgae produce a variety of antimicrobial compounds, including peptides, fatty acids, and secondary metabolites, that can directly inhibit the growth of plant pathogens. Microalgae synthesize peptides that have antifungal, antibacterial, and antiviral properties, which can help protect plants from a wide range of pathogens. Several microalgae were identified as possessing antimicrobial activity, with an

Table 4.3 Key mechanisms by which microalgae improve plant tolerance to biotic stress.

Mechanisms:	**References**
Induction of systemic resistance	Babu et al. (2015)
Production of antimicrobial compounds	Jokel et al. (2023)
Production of hydrolytic enzymes	Roberti et al. (2015)
Elicitation of phytoalexins	Poveda and Díez-Méndez (2023)
Improvement of soil microbiome	Arroussi et al. (2018); Li,et al. (2014)
Stimulation of plant growth and vitality	Arroussi et al. (2018); Van Oosten et al. (2017)

extract from *Chlorella sorokiniana* demonstrating the most potent growth inhibition against the plant pathogen *Phytophthora cactorum* (Jokel et al., 2023).

Microalgal and cyanobacterial extracts enhance plant defense mechanisms by increasing the activity of hydrolytic enzymes, such as endochitinases (including β-N-acetylhexosaminidase, chitin-1,4-β-chitobiosidase, and endoglucanase-β-1,3-glucanase), as well as enhancing polyphenol oxidase and peroxidase (POD) activity (Roberti et al., 2015).

Microalgae can stimulate the production of phytoalexins, which are antimicrobial substances synthesized by plants in response to pathogen attacks. The elicitors from microalgae trigger the biosynthesis of phytoalexins, enhancing the plant's capacity to inhibit pathogen growth and spread (Poveda & Díez-Méndez, 2023).

Microalgae contribute to a healthy soil microbiome, which is essential for plant health and resistance to biotic stress. The application of microalgae can promote the growth of beneficial soil microorganisms that compete with or inhibit harmful pathogens, improving overall soil health and plant resilience (Arroussi et al., 2018; Li et al., 2014).

Microalgae enhance overall plant growth and vitality, making plants more robust and capable of withstanding biotic stress. Microalgae produce phytohormones like auxins, cytokinins, and gibberellins, which promote root and shoot development, improving the plant's ability to absorb nutrients and water, and enhancing its overall health (Arroussi et al., 2018; Van Oosten et al., 2017).

The literature clearly shows that microalgae possess outstanding plant protection and plant growth promotion properties. However, large-scale commercial use remains limited. Key challenges hindering the widespread application of microalgae in the agrochemical industry include the high costs of cultivation, inconsistent biomass productivity, and the expensive, energy-intensive processes required to obtain purified metabolites with plant growth-promoting properties (Smetana et al., 2017).

4.3 Microalgae and soil quality improvement—soil conditioning

Soils are often, intentionally or unintentionally mismanaged or overused in modern agriculture, which consequently leads to losses of the organic matter quantity and quality, physico-chemical stability, reduction of soil fertility as well as overall productivity. Soil algae live in the topsoil, mainly up to 10 cm of the profile attached to and between soil particles. They are heavily exposed to and influenced by all seasonal weather changes and anthropogenic activities.

Soil is inhabited by different eukaryotic algae Chlorophytes such as *Scenedesmus, Chlorella*, *Chlorococcus*, *Cladophora*, *Klebsormidium*, and Chrysophytes (diatoms) including *Nitzschia*, *Hantzschia*, *Navicula*, and *Pinnularia* as the most frequent. This fascinating group of organisms contributes to organic matter enrichment in soil by performing photosynthesis. Algae are primary colonizers of soils, they use CO_2, water, and energy from the Sun to produce sugars. Microalgae also represent a major source of other primary metabolites; such are lipids and proteins. Jassey et al. (2022) stated that worldwide, soil algae take up carbon (C) in amounts that equal to about 6% of the net primary production of terrestrial vegetation. By releasing organic compounds into the soil, first of all, they contribute to total soil biomass, improve moisture and pH-holding capacity, increase aggregate stabilization, reduce the impact of wind erosion, and many more.

Microalgae are also very important in soil remediation (removal of different contaminants from soil).

4.3.1 Pollutants removal from the soil

Pollutants in agricultural soil are different substances and materials that enter the soil directly or indirectly. Pollutants that are directly linked to agriculture are pesticides, mineral and organic fertilizers, plastics (films for mulching and greenhouses, drip irrigation tubes and empty packaging), waste waters from irrigation and rural wastes. Indirect pollutants come from flooding and atmosphere by deposition of diffused molecules, or by other sources.

4.3.1.1 Benefits of microalgae on salt-affected soil remediation

Salt-affected soils contain soluble salts and/or exchangeable sodium, in amounts that impair plant growth and plant development. Halo tolerant or halophylic plants can thrive on these soils but traditional production or sustainable agriculture is only possible after remediation. Salt-affected soils, solonchaks, and solonetzes are usually formed and found in arid and semiarid regions but also in humid to subhumid climatic areas as well. Soil salinization is a natural process, characteristic of drylands. However, antropogenic influence is also present and related to agriculture, particularly intensive land use with irrigation which can speed up soil salinization. In Serbia, about 83000–95000 ha of salt affected soil can be found, mostly in northern part, in Vojvodina (Đorđević & Radmanović, 2016). These soils are not used in traditional or sustainable agriculture.

However, regardless of unfavorable conditions, some organisms inhabit these soils, among them are microalgae. Among the commonly recognized salt-tolerant eukaryotic microalgae generaare *Chlorella, Scenedesmus*, and *Dunaliella* (Chlorophyta). In order to survive and minimize the damages caused by salt stress, microalgae developed different defense mechanisms (Wang et al., 2018). These mechanisms include the accumulation of lipids and carbohydrates as storage molecules to maintain microalgae survival (Anand et al., 2019). Accumulation of ROS-scavenging enzymes such as catalase (CAT), peroxidase (POD), glutathione reductase (GR), ascorbate peroxidase (APX), superoxide dismutase (SOD), and molecules such as proline, betaine-glycine protect the cell from increased osmotic pressure and death.

This extraordinary capacity of microalgae to combat stress caused by elevated levels of salts is recognized by many researchers. The possibilities of using halophilic and halotolerant strains of microalgae, their consortia with other microbes and/or enzymes to ameliorate salt affected soil is investigated more intensely in the past 15 years.

Salt removals from saline environment are performed either by the mechanism of bioadsorption (physical adherence) or bioaccumulation or absorption (energy driven uptake). However, in soil only small amounts of Na+ can be sequestrated by sticky EPS, which is rich in polysaccharides and mucilage though once the Na+ is trapped, it will not threaten microbes and plants (Pei & Yu, 2023).

Al-Maliki & Ebreesum (2020) investigated the application of arbuscular mycorrhiza (AM) and green microalgae in arid saline soil on changes in soil carbon mineralization, and soil structure. They noticed the decline in soil pH from 7.7 (control) to 7.23 (treatment with *Chlamydomonas* sp.) and 6.96 (treatment with AM+ *Chlamydomonas* sp.), respectively. Also, significant improvement in carbon decomposition, soil aggregation, and bacterial biomass accumulation was recorded. In conclusion, it was stated that application of AM and green algae can be beneficial to soil ecology in arid saline soils.

In the study of Zhang et al. (2024), two indigenous algae, *Coelastrella* sp YJ-1 and *Scenedesmus* sp. YJ-2, were inoculated into saline soils with different salt levels with the aim to investigate their potential in enhancing soil health conducting a laboratory microcosm experiment. The results showed that introduction of two green algae in the saline soils led to formation of algal crusts, and the chlorophyll content gradually increased with the incubation time. The soil pH decreased significantly from the initial 8.15–9.45 to 6.97–7.56 after 60-day incubation. They also recorded increased secretion of EPSs as well as activities of catalase, sucrase, and urease. Algalization as a green and effective approach was recommended by the authors, for decreasing pH and increasing organic matter and enzyme activity of saline soils.

Chen et al. (2024) chose two indigenous green algae species, *Coelestrella* sp. SDEC-28 and *Dunaliella salina* SDEC-36, isolated from saline–alkali soil for the study. Both strains showed robust growth and reproduction capabilities under high-salinity conditions. After one growth period of these species, the electrical conductivities of the media decreased and the pH remained stable at 7.7 and 8.6, respectively. The authors verified the hypotheses that microalgae can remediate soil. The potential of using specific microalgal strains to improve soil conditions in eco-farms was demonstrated as well. In that way, the range of crops that can be cultivated in saline–alkali environments are extended.

The application of microalgae fertilizers in combination with phytoremediation, involving a complex selection of microalgae and plants and require large amounts of microalgal biomass (Nisha et al., 2018). However, this approach has its limitation due to the poor adaptability of applied microbes and other environmental factors such as lack of rainfall or irrigation. On the other hand, Singh (1950) proposed the stimulation of growth of soil indigenous microalgae population by irrigation or rainfall. It would enhance the production of extracellular algal polymers and eventually lead to salt adsorption.

Both approaches heavily rely on water and nutrients availability necessary for algae growth and functioning. Besides that, the complex interactions among algae, plant and environment are not sufficiently researched which limits a more widespread application of algal species for the purpose of saline soil remediation.

4.3.1.2 Benefits of microalgae on pesticide contaminated soil

Pesticides are by definition chemical substances that are used for pest control. Excessive and uncontrolled use of pesticides led to rapid accumulation in nontarget organisms and environmental systems. Soil and water are polluted by different pesticides regardless of being used in agriculture or not. Pesticides can be found in air, plants, animals, and ultimately in humans. Polluted soils must be remediated in order to prevent any negative and detrimental effects on living organisms.

Bioremediation using microalgae is one feasible alternative. The three mechanisms for organic pesticides removal by microalgae are bioaccumulation, biodegradation, and biosorption (Fig. 4.1).

Bioaccumulation or biouptake is defined as a metabolism-dependent, energy driven process which comprises of biosorption, and the active transportation of ions into cells. In bioaccumulation, pollutants are transferred into the cell. In the process of biodegradation, pesticides are broken down into smaller molecules which than act as nutrients for algal growth. This transformation is highly dependent on the microalgae enzymatic activity. Biosorption is a passive, metabolically independent process that is defined as the attachment of toxic pesticides to the surface of the microalgae cell. It is possible because of the specific cell wall structure made of carbohydrate.

Abdel-Razek et al. (2019) studied the potential of using a consortium of microalgae and cyanobacteria (*Chlorella vulgaris*, *Scenedesmus quadricuda,* and *Spirulina platensis*) to remove the organophosphate pesticide malathion from wastewater samples. The authors found that microalgae were able to remove malathion from the tested samples with up to 99% efficacy. Three hydrophobic pesticide active compounds (chlorpyrifos, cypermethrin, and oxadiazon) were selected to study the potential for their removal from aqueous media by a microalgae consortium of *Chlorella* sp. and *Scenedesmus* sp. (Avila et al., 2021). At the end of the study, biodegradation and photodegradation contributed to the removal of 55% of oxadiazon, 35% of chlorpyrifos, and 14% of cypermethrin. Furthermore, more than 60% of chlorpyrifos and cypermethrin were removed by sorption onto microalgae biomass.

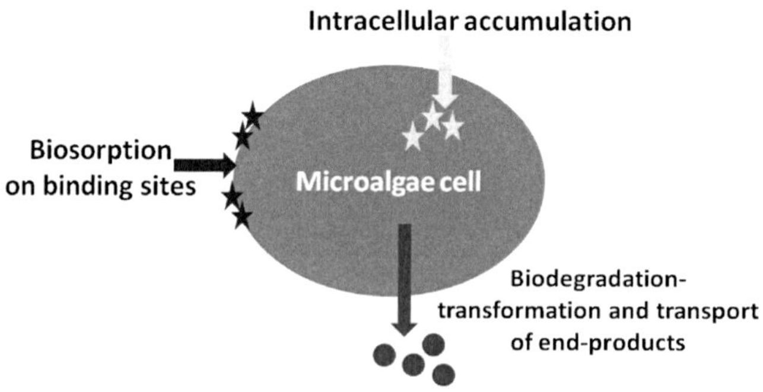

Figure 4.1 Mechanisms of pesticides bioremediation by microalgae. Remediation of pesticides by microalgae through biosorption, bioaccumulation, and biodegradation.

4.3.1.3 Benefits of microalgae on antibiotic contaminated soil

Antibiotics are substances for prevention and treatment of infectious bacterial, fungal, or parasitical diseases. Antibiotics are often overused and/or misused both in animals and humans. Contamination caused by antibiotics accumulation in the environment is becoming a serious issue and problem that needs to be addressed. Potential detrimental, harmful effect of antibiotics and its residues on living organisms and the niches they inhabit are well known. On the other hand, the subjects of antibiotics removal from contaminated sites, mechanisms, and techniques of soil bioremediation are still less researched, especially the potential usage of microalgae for that purpose.

Gojkovic et al. (2019) investigated the removal of 19 lipophilic compounds from the culture medium by the microalgae (*Coelastrella sp.* 3–4, *Coelastrum astroideum* RW10, *Chlorella vulgaris* 13-1, and *Chlorella saccharophila* RNY). They found that the potential of removal was highly efficient (>70% in average within 2 days). The most lipophilic compounds Biperiden, Trihexyphenidyl, Clomipramine, and Amitriptyline significantly accumulated in the biomass of most algal species, with a positive correlation between accumulation and their total biomass content. More persistent in the growth medium were hydrophilic compounds like Caffeine, Fluconazole, Trimetoprim, Codeine, Carbamazepin, Oxazepam, and Tramadol, which were detected in amounts of above 60% in average after algal treatment. *Coelastrella* sp. 3–4 and *Coelastrum astroideum* RW10 efficiently accumulated several compounds in their biomass, while *Chlorella vulgaris* 13-1 and *Chlorella saccharophila* RNY, were highly efficient in removing all 19 pharmaceuticals from the growth medium within 12 days. Hom-Diaz et al. (2022) investigated the ability of four selected microalgae (*Chlamydomonas reinhardtii*, *Chlorella sorokiniana*, *Dunaliella tertiolecta,* and *Pseudokirchneriella subcapitata*) to remove nine antibiotics and the antidepressant venlafaxine under different experimental conditions. They obtained good removal percentages for all tested antibiotics with the algae tested, with lower removals for trimethroprim (between 11 and 34%) in all cultures and higher removals for fluroquinolones in *D. tertiolecta* cultures (between 88 and 100%). Macrolides were mainly removed by biodegradation rather than photodegradation (which contributed less than 5%). The highest biodegradation removals were observed for erythromycin (23%), sulfapyridine (21%), and clarithromycin (20%). Based on the results, authors concluded that *C. reinhardtii* and *D. tertiolecta* cultures exhibited rapid growth and good removal percentages for some of the investigated antibiotics.

4.3.1.4 Benefits of microalgae on plastic and other contaminants removal from the soil

Plastics are synthetic organic polymers, which are hydrophobic, inert, high-molecular-weight long chains of molecules (monomers) that are joined together by covalent bonds. Algae, particularly microalgae, can degrade the plastic materials through the toxins systems or enzymes synthesized by microalgae itself while using the plastic polymers as carbon sources (Sanniyasi et al., 2021). Adhesion of algae on the surface will initiate the biodegradation and the production of ligninolytic and exopolysaccharide enzymes which is the key for plastic biodegradation (Fig. 4.2).

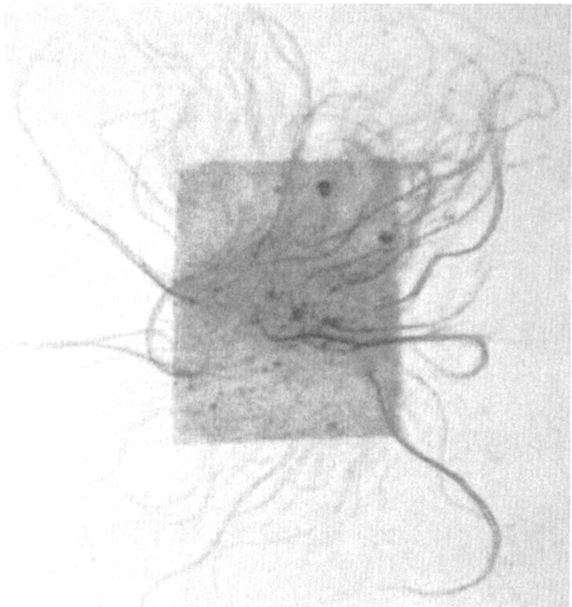

Figure 4.2 Microalga (*Uronema africanum* Borge) treated of low-density polyethylene (LDPE) sheet. The LDPE sheet (1 cm2) showing green hair-like structures protruding from its surface.

Source: Adapted from Sanniyasi, Elumalai, Rajesh Kanna Gopal, Dinesh Kumar Gunasekar, Preethy P. Raj (2021). Biodegradation of low-density polyethylene (LDPE) sheet by microalga, Uronema africanum Borge. *Scientific Reports 11*, 17233. https://doi.org/10.1038/s41598-021-96315-6.

Kumar et al. (2017) investigated the potential of polyethylene biodegradation and showed the high percentage of low-density polyethylene (LDPE) degradation by diatom *Navicula pupula* (4.44%) and green alga *Scenedesmus dimorphus* (3.74%). Gulnaz & Dincer (2009) investigated the biodegradation of bisphenol A (BPA) by a green algae *Chlorella vulgaris*. Authors found that degradation of BPA by algae was highly efficient. The results indicated that the BPA was easily degraded by algae and its concentration was 20 mgL-1 within 7 days. Degradation capacity of BPA by green alga *Chlorella fusca* var. *vacuolata* was also confirmed by Hirooka et al. (2005). This degradation capability is due to developed enzymatic system of microalgae. They are able to produce and secrete various plastic degrading enzymes. *Chlamydomonas reinhardtii* (*C. reinhardtii* CC-124 and CC-503), a green microalga was incubated with polyethylene terephthalate (PET) at 30 °C for up to 4 weeks (Kim et al., 2020). After the incubation period, dents and holes on the PET film surface were observed. Also, using high-performance liquid chromatography analysis terephthalic acid (TPA), the fully degraded form of PET was detected.

Bioremediation of heavy metals with microalgae is a process which can be divided two main stages. The first stage, named biosorption is a passive, metabolically independent process of heavy metal adsorption over the cell surface (Chugh et al., 2022). Heavy metals are then transported across the cell membrane into cytoplasm. This stage is called bioaccumulation, which is an active, metabolically dependent process that requires energy. Heavy metal ions can be accumulated inside vacuoles or thylakoids as well. After that, detoxification or biotransformation occurs in order to protect the cell from metal toxicity. This self-protection mechanism is actually a conversion process performed via chelating agents called chelators (organic compounds). This process is also known as complexation (complex formed with heavy metal ions and chelating molecules).

Scenedesmus sp., *Chlorococcum* sp., *Chlorella vulgaris* can be used for the bioremediation of lead (Pb), mercury (Hg), and cadmium (Cd). *Tetraselmis chuii* and *Scenedesmus bijuga* can remove copper (Cu). *Chlorella pyrenoidosa* can remove a wide range of heavy metals, Cu, zinc (Zn), Pb, Hg, arsenic (As), chromium (Cr), nickel (Ni), and Cd. *Chlorella minutissima* is capable of removing Cr (VI), and *Phaeodactylum tricornutum*, which has a high Cd 21 tolerance (CE50 5 22.3 mg/L), is also capable of removing Cr (VI) (Ammar et al., 2022). Hassan et al. (2017) listed several green algae namely *Chlorella vulgaris* which was capable of removing Cd (initial conc. 25–150 mg/L-final removal conc. 58.4 mg/g), as well as *Desmodes pleiomorphus* (initial conc. 0.5–5 mg/L-final removal conc. 61.2 mg/g), and *Chlamydomonas reinhardtii* (initial conc. 20–400 mg/L-final removal conc. 77.62 mg/g) while *Desmodes mupleiomorphus* removed Zn (initial conc. 1–30 mg/L-final removal conc. 360.2 mg/g), and *Chlorella vulgaris* removed Ni (initial conc. 100 mg/L-final removal conc. 28.6 immobilized).

4.3.2 Microalgae and soil structural stability

Soil aggregation is very important for moisture and nutrient retention, soil carbon sequestration but also in the prevention of water and wind erosion. Soil mineral particles (e.g., clays, silt, metal oxides, and aluminosilicates) and cation complexes binds together, and soil aggregates of different sizes are form (microaggregates <0.05mm and macroaggregates 0.05–0.25mm). This is possible due to exopolymer substances (EPSs) secreted by plant and/or microorganisms. Microalgal cells with EPSs act as glue in soil which contributes to micro- and macroaggregation provide suitable pores for plant growth and nutrients absorption. Well-structured soil is aerated with increased oxygen content and water holding capacity. Soil aggregates are mainly formed by algal and fungal filaments and excreted EPSs.

Redmile-Gordon et al. (2020) highlighted the importance of EPS-protein which was more closely linked to soil aggregate stability than EPS-polysaccharide (*p* values of 0.002 and 0.027, respectively). They confirmed that both EPS-protein and EPS polysaccharide were transient fractions: supporting the founding models for aggregate formation and suggested that the management of transient binding agents such as EPS—as opposed to simply increasing the total SOC content—may be a more feasible strategy to improve soil structural integrity. The effect of different microalgal biofertilizer on the stability of soil aggregates was studied and found that the retention of the available water and plant root growth could be improved by inoculation with

Chlorella sp. alone or in combination with vermiculite (Yilmaz & Sönmez, 2017). Also, microalgae as *Chlorella* sp., *Scenedesmus* sp., *Tribonema minus*, *Choricystis minor*, and *Klebsormidium subtile* can significantly contribute to improving soil stability (Song et al., 2022).

4.4 Microalgae and plant growth enhancement—plant conditioning

By using living cells of algae as well as algal extract, the effect of biofertilizer is achieved, which is suitable for improving the growth parameters of a large number of plant species (Chojnacka et al., 2012). In recent times, the application of live cultures and water extract of *Chlorella* sp. in the form of foliar nutrition of plants is becoming more common. In this way, the plant is provided with the necessary and accessible nutrients, amino acids, vitamins, and enzymes. Foliar application also ensures the formation of an algal biofilm on the above-ground part of the plant, which has a protective role, reduces evapotranspiration, provides a mechanical barrier, and protects the plant from attacks by phytopathogenic microorganisms and parasites.

Microalgae can secrete a variety of active substances, such as carotenoids, proteins, fatty acids, plant hormones, extracellular polysaccharides (EPSs), vitamins, and antibiotics (Fig. 4.3). All play a significant role in plant physiology, many act as chemical messengers to coordinate the activities of higher plant cells.

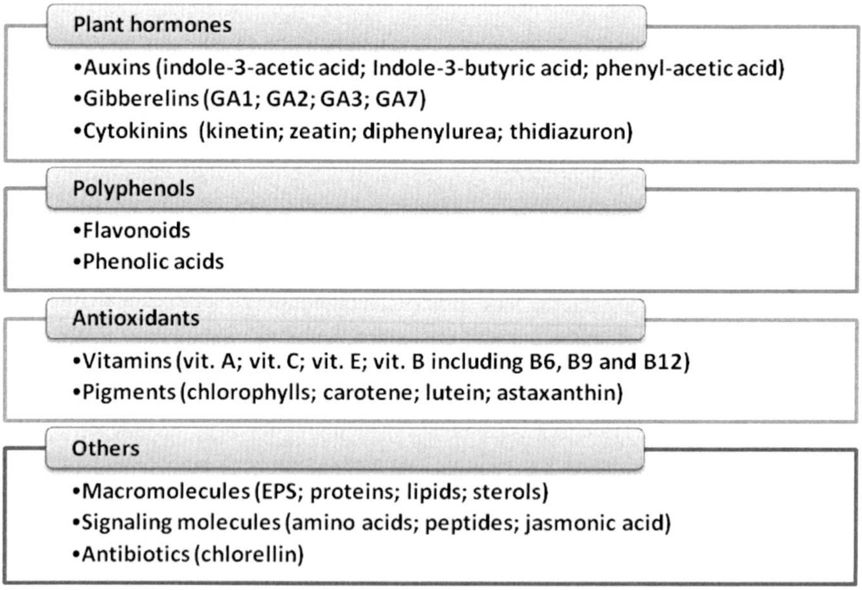

Figure 4.3 Bioactive compounds of microalgae. Compounds with plant growth promotion (PGP) properties present or produced by microalgae.

Physioactivators are groups of compounds found in algal extract and have a positive effect on plant growth and yield. The mechanisms of plant physiology activation take place simultaneously. It usually refers to plant response to stresses (Wahab et al., 2022). For example, the stimulation of the plant enzymatic activity that promotes mineral nutrition is one mechanism. Enzymes, especially nitrate reductase and phosphatases, are very important for better plant nutrition. If the uptake of nutrients from the soil is improved, primarily nitrogen, phosphorus, potassium, magnesium, manganese, and iron, the plant will have a slightly changed chemical composition. Another mechanism is the stimulation of the photosynthesis process by increasing the chlorophyll content in the leaves (Sherin et al., 2022) as well as greater chlorophyll activity. In this way, the synthesis of plant biomass increases. The formation of flowers and fruits is more pronounced due to the greater synthesis of polyamines, which stimulate cell division. A larger number of flowers also means that the percentage of pollination is higher as well as the process of fruit formation. Biostimulators obtained from algae are most often applied after the germination of plant seeds, in the form of foliar application, although it can also be applied directly to the soil.

4.4.1 Soil microalgae and plant growth enhancement-experience from Serbia

Soil algal diversity is also enormous, yet nothing or little is done to uncover their importance in the ecosystems they inhabit. Underestimating the algal diversity in these ecosystems is due to, first of all poor knowledge about their role in soil ecosystem maintenance as well as lack of significant scientific interest by researchers in algal taxonomy and phylogeny (Trbojevic & Predojević, 2021).

However, a handful of researchers from Serbia are interested and do mostly surveys and researches with freshwater algae and cyanobacteria. In aquatic ecosystems in Serbia, a total of 92 sites were surveyed where 17 species of Rhodophyta were recorded in the period between 1908 and 2017. *Bangia, Chantransia, Batrachospermum, Lemanea, Paralemanea, Thorea,* and *Hildenbrandia* were the most dominant genera (Simić & Krizmanić, 2018). Also, 1019 taxa of diatoms are recorded in Serbia including the widely distributed cosmopolitan *Achnanthidium minutissimum*, while *Geissleria gereckei* is more known for its specific area of distribution. A group of authors (Danilovic et al., 2017) isolated freshwater green algae species belonging to the genera *Chlorococcum*, *Chlorella*, *Desmodesmus*, and *Scenedesmus*. The primary goal was to evaluate the oil content in the algae isolates. It ranged from 15.8% (*Chlorococcum* sp.) to 33% (*Chlorella* sp). Since the microalgae isolates *Chlorella* sp. and *Scenedesmus* sp. had the highest oil productivity, high growth rate, and high oil content, these strains were subjected to further investigation in order to improve the oil yield and analyze the possibility of their use in biodiesel production. In the period between 2017 and 2022, freshwater red algae (Rhodophyta) were surveyed in Serbia by examining 411 localities of 196 springs, streams, and rivers. A total of 80 new records of algae were made: 18 species of the genera with *Batrachospermum*-like morphology as well as *Hildenbrandia rivularis* (30), *Lemanea* sp. (1), *L. fluviatilis* (2), *Paralemanea annu-lata* (15),

P. catenata (11) and 3 species of *P. torulosa* (Rakonjac & Simić, 2023). In the study of Khakimova et al., (2022), *Chlorella vulgaris* was cultivated in wastewater. During wastewater treatment, chemical oxygen demand (COD) and biological oxygen demand (BOD) removal efficiency was 93.7% and 98.1%, respectively. Total organic carbon (TOC) content decreased by 95.7%. Nitrites and nitrates decreased by 96%, while the biggest decrease in metal ions was achieved for Ca (82.7%) and Mn (97.6%). The authors concluded that coupling microalgae cultivation and wastewater treatment had a lot of potential for reducing contamination through biosorption, while also providing environmental advantages.

On the other hand, though the importance and application of specific soil microalgae are acknowledged by many researchers (Hajnal-Jafari et al., 2016; Žunić et al., 2022), there are no published articles about the diversity and taxonomy of green soil microalgae in Serbia till 2023 (Jafari et al., 2023). Also, no published research about the efficiency of soil microalgae and the possibility of application in plant production was done prior to 2021 (Seman et al., 2021).

Hajnal Jafari et al. (2022) isolated and determined five eukaryotic algal genera (Fig. 4.4) using light microscopy and confocal laser imaging tools. These isolates were also tested for antimicrobial activity (against four bacteria and six fungi isolates). In average, better inhibition was achieved against phytopathogenic fungi. *Chlorella* sp. 71 inhibited the growth of most tested fungi which can be attributed to production of secondary metabolites with antimicrobial properties.

Jafari et al. (2023) conducted research related to algae identification and taxonomy. The green microalgae were isolated from acidic arable soil of dystric cambisol type, using BG11 medium and the agar plate method. Firstly, the identification of the microalgae isolates was performed based on morphological characteristics using light microscopy (LM) and taxonomic monographs and reference books. Gene marker (ITS2) was used for molecular identification. Combination of morphology and DNA-based approach proved to be the most effective for obtaining a consistent species level identification. The first record of *Scenedesmus vacuolatus* 63 from soil in Vojvodina, Serbia was made.

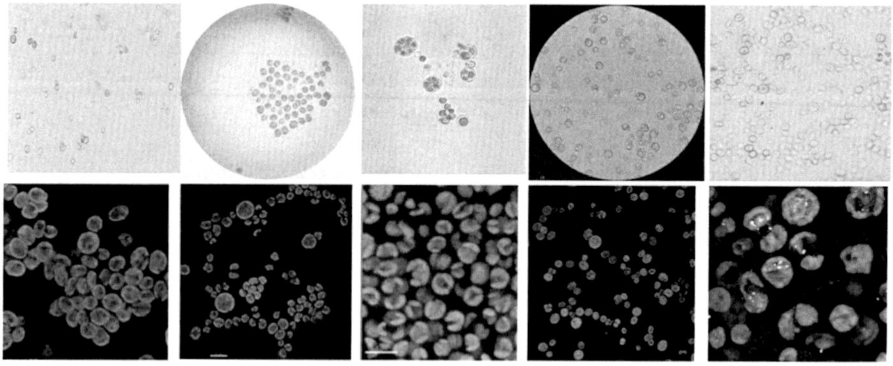

Figure 4.4 Soil microalgae from Serbia (magnification 1000 ×). *Chlorella* sp. (isolates 71 and 72) and *Dictyospherium* sp. (isolates 36, 37, and 69).

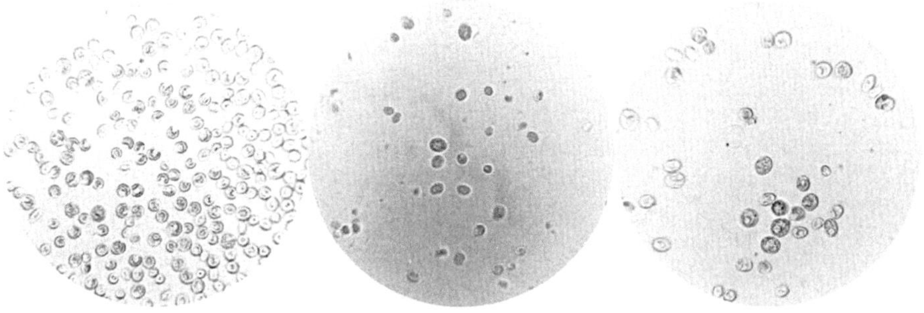

(A) (B) (C)

Figure 4.5 Soil microalgae. (A) *Chlorella vulgaris*, (B) *Scenedesmus vacuolatus*, (C) *Coelastrella* sp.

Žunić (2024) studied the effectiveness of foliar application of three microalgae species, *Chlorella vulgaris, Scenedesmus vacuolatus,* and *Coelastrella* sp. (Fig. 4.5) in two concentrations −5 and 10% on growth, productivity, and fruit quality of strawberry cultivar "Alba," in field conditions. Strawberry vegetative and generative potential (rosette height, number of leaves, flowers, and fruit per plant), physical fruit properties (weight, height, width, and shape index) was assessed as well as fruit quality parameters were measured (sugars, organic acids, and phenolic compounds) with high-performance liquid chromatography/mass spectrometry (HPLC/MS). Result showed that algae treatment (*Coelastrella* sp., 10%) applied by spraying positively affected the balance between vegetative and generative growth which consequently led to formation of 25% higher yields for the treated plants. Content of sugar and most of secondary metabolites were more than 10% higher compared to other tested treatments. Overall, the results of the study concluded that microalgae application may have positive outcomes in organic strawberry production. Furthermore, novel insights into influence of microalgae on important fruit quality components were made as well.

4.4.2 Final remarks

Some microalgae species are already commercially exploited in plant production such as *Chlorella vulgaris* (Natur Plasma™, Allfertis™), *Spirulina* (Spiragro™), *Chlorella* sp., and *Spirulina* sp. (Probio Ogród™). These products are all in liquid form. Despite their possibility to increase yield and concentration of functional compounds in food, microalgae-based plant biofertilizers are still poorly explored and implemented in agricultural practice (Žunić, 2024).

4.5 Production of commercial microalga-based biofertilizer

The production of a microalgae-based fertilizer is shown in Fig. 4.6. The following five main steps can be highlighted: 1. soil sampling; 2. isolation and characterization

Figure 4.6 Production of a microalgae-based biofertilizer. 1. soil sampling; 2. isolation and characterization of microalgae isolates; 3. trials with different crops in various conditions; 4. production of microalgae based biofertilizer; and 5. biofertilizer application.

of microalgae isolates; 3. trials with different crops in various conditions; 4. production of microalgae based biofertilizer; and 5. biofertilizer application. This is a simplified representation of a production process in which many activities within each step/substep are not described.

The microalgae used in the production of biofertilizer are most often isolated from the soil but also other substrates such as water or plants. Isolation and obtaining a pure culture of microalgae is a complex and time-consuming process that involves the use of certain methods and aseptic treatment, techniques under laboratory conditions (Hajnal Jafari et al., 2019). After obtaining a pure culture, further characterization based on a number of criteria is performed (morphological, biochemical, genetic properties, microalgae ecology, and cultivation conditions are examined) (Priyadarshani & Rath, 2012). Depending on the microalgae, specific criteria can also be defined (including toxicity, particulate metabolite production, etc.). Selected strains with desirable properties are used in trial with plants. First, they are tested in controlled, and then under semicontrolled conditions and eventually in field conditions. The tests are carried out on various types of plants, under different production conditions. This stage in the production is the longest and lasts for several years. Based on the results, the effectiveness of the applied microalgae on the plant is assessed by measuring a number of parameters (elements of yield and quality of the cultivated plant) as well as impact of the applied fertilizer on indigenous microbial populations, soil microbial activity.

Different formulations and dosage of biofertilizer are tested, along with methods, timing of administration in order to do application optimization and make a final recommendation for use. The last stage is the commercial production of the microalgae-based preparation. It involves large-scale production under strictly controlled conditions. The product is sold on the market under a trademarked name.

Liquid formulations are inoculums that use a nutrient medium that is prepared mainly in water but also in mineral and organic oils. The number of microalgae in these formulations is usually 2×10^9 CFU/mL living cells (Colony Forming Units per Milliliter). Additives such as sugars, glycerol, and cellulose can also be added to these formulations (carboxymethyl cellulose-CMC, arabic gum or synthetic additives). By adding them, the survival of microalgae is improved and prolonged. Liquid formulations have a long shelf life (up to 2 years). Application is simple and inexpensive, most often with the help of machines on large surfaces. Formulations are applied by spraying seedlings after planting, and during vegetation period spraying the aboveground part of the plant. In this way, there is direct contact with the plant and microalgae. One of the disadvantages is uneven distribution of microalgae on the plant material (when applied by foliar spraying). In addition, they require low storage temperatures, which make them more costly compared to solid formulations. Solid formulations of microalgae require suitable carriers. Wheat straw, manure, vermiculite, and peat are the most common ones. It is important to point out that in order to a product be successful, common interests of manufacturers and consumers must be satisfied. For the consumer, an important quality is practicality and simple usage that do not disturb the usual routine of applied techniques. For the manufacturer, an extended storage time (shelf life) more than one season is an additional bonus.

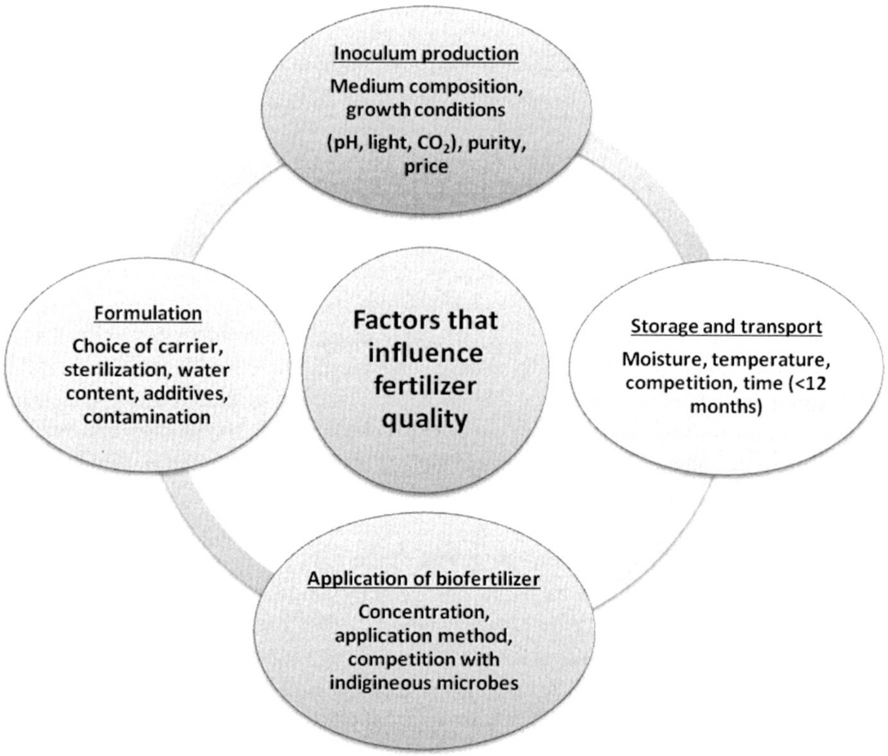

Figure 4.7 Factors that influence fertilizer quality. Factors related to inoculum production, formulation, storage, and transport as well as application.

Numerous factors affect the final quality and success of the microalgae-based fertilizer application, especially in field conditions (Fig. 4.7).

4.5.1 Production systems

The production of algae biomass is usually associated with wastewater treatment, that is, wastewater is used as a substrate for algae growth. The second specificity is that algae can use large amounts of CO_2, so there is no need for fossil fuels.

Open systems involve the production of algal biomass in pools of different volumes (Bibi et al., 2017). They are often used because of the relatively low cost of production. Cooling of the system is provided by evaporation of the growth medium, which increases water consumption, but eliminates the need for physical cooling of the culture, which is usually performed by pumping a large amount of water into the system. Since the pools are open, pollutants get into the system (soil particles, airborne particles, microorganisms, andbmany more), so the risk of contamination increases. Rapid detection, diagnosis, and adequate treatment are of most importance to maintain production at an optimal level.

Closed systems used for algal biomass production are called bioreactors or photobioreactors (Fig. 4.8). These systems are used for phototrophic algae cultivation where the light source is provided directly (solar radiation) or indirectly with artificial light sources (LED lamps). This production system is less exposed to abiotic factors, it is much easier to control, and the risk of contamination is significantly reduced. The circulation of algal biomass is such as to ensure uniform exposure of the algal culture to light. The energy consumption and cost are higher compared to open systems, because all liquids and gases in the reactor are carried in and out with the help of pumps. However, in this way, denser algal culture is obtained (higher concentration), which later makes the manipulation of the biomass much easier and reduces other costs. Maintaining the composition of water in the system is easier because it is not exposed to rain and evaporation. The exposure of algal culture to light is more uniform than in open systems. Closed systems can be different in size, shape, and are made from different materials.

The harvesting of algal biomass is carried out using various methods (Safi et al., 2014). The most commonly used methods are centrifugation, filtration, flocculation-aggregation of microalgae cells (with various organic and inorganic compounds, bioflocculation with microorganisms, electro-coagulation, and alkaline flocculation), sedimentation, and flotation. There is no universal method for harvesting algal biomass, each has certain advantages and disadvantages. However, in choosing an adequate method, the cost price must be taken into consideration, as it accounts for 20–30% of the total cost of algae production.

Partial or complete drying of algal biomass is carried out depending on the needs (Fig. 4.9). The resulting product is further processed (extraction of various metabolites) or used for a specific purpose, e.g., for fertilization when microalgae

Figure 4.8 Closed systems of microalgae production. Horizontal, flat-vertical panels and vertical tubular photo-bioreactors.

Figure 4.9 Algae biomass. Wet and dried algal biomass.

are used as biofertilizers and when introduced into the soil. This biomass is an excellent source of organic matter, which decomposes quickly and significantly improves the nutritional status of the soil, i.e., the absorption of macronutrients and micronutrients by plants.

4.5.2 Modes of application

Algae contain a variety of molecules such are starch, cellulose, proteins, oils, fats, polyunsaturated fatty acids (e.g., omega fatty acid), vitamins (B12, K, C, E, A, D), different minerals (K, Ca, Mg, Na, Zn, Cu, Co, I, B) as well as pigments (chlorophylls, carotenoids, and phycobilin proteins) (Sosa-Hernández et al., 2018). Live algae cells, algal metabolites/algal extracts or wet/dried biomass can be used as biostimulants, biofertilizers, and physioactivators, aquaculture feed and food colorant (Priyadarshani & Rath, 2012). The potential application of the extracts depends on the algae species, type of the BCs extracted, extraction technique. The application mode of a specific product depends on crop type and the specific need of the crop (enhanced nutrients uptake, increased phytohormone production, disease suppression, and many more). The methods of application and the various benefits are shown in Fig. 4.10.

Extraction is the recovery process of bioactive compounds (BCs) from algae. The traditional techniques for BCs extraction include maceration, hydro-distillation, mechanical pressing, and steam distillation. For the extraction procedures, the use of an appropriate solvent is crucial because of the efficiency of soluble fraction separation. Usually, very strong organic solvents, e.g., petroleum, chloroform, acetone, ether, toluene, benzene, and/or dichloromethane are used (Michalak & Chojnacka, 2014). Nowadays, novel extraction techniques include the usage of green solvents in BCs extraction from algae (Vo et al., 2024). These green solvents can be water and subcritical water, deep eutectic solvents (DES) including a subclass of natural deep eutectic solvents (NADES) and bio-based solvents (originated from agricultural biomass and food wastes).

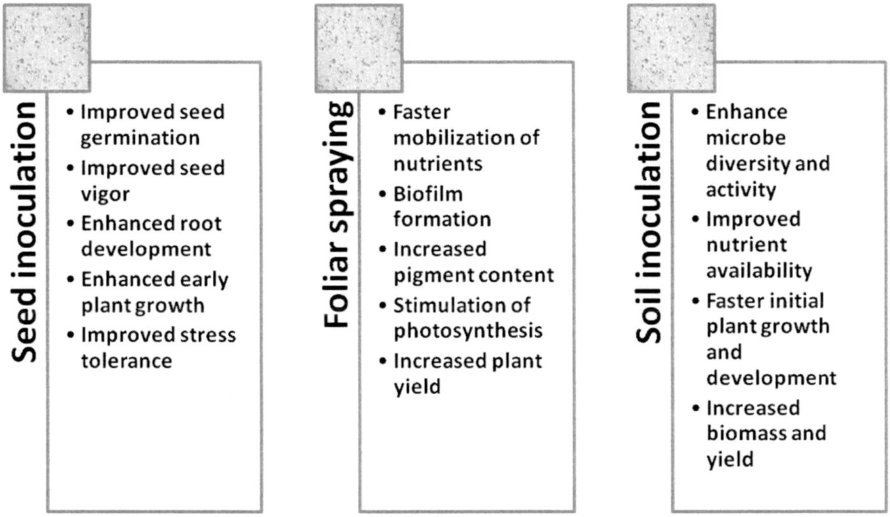

Figure 4.10 Application methods and benefits of algae-based fertilizer. Seed inoculation, foliar spraying, and soil inoculation with corresponding benefits.

Green solvents have many advantages in comparison to traditionally used organic solvents. Creation of less pollution, safer usage, better degradability, and higher efficiency are only few that make green solvents more desirable in extraction procedures. Supercritical fluid extraction, microwave-assisted extraction, ultrasound-assisted extraction, enzyme-assisted extraction, and pressurized liquid extraction are also novel techniques which can be used for BCs extraction, with or without green solvents.

Seed inoculation or seed biopriming is the most commonly used technique of seed coating with desirable microbes (live cells or extract). It is simple method and uses a small amount of inoculum. Plant seeds have to be surface sterilized before priming. Usually, the amount of microalgae cell that is applied should not be less than 10^6 CFU/mL. The priming itself can be done differently, soaking the prepared seeds for 2 hours minimum at room temperatures or during 2 days at 4°C in dark before sowing. In field conditions, the seeds can be treated with inoculum which is added to a suitable carrier, with or without added water and/or additives. Drying is carried out in situ. Each seed must be coated with inoculum, and since small amounts are used, this reduces the number of microorganisms that are applied to the seeds. This can reduce the desired effect of successful inoculation. If the inoculum is not applied well to the seed, it may be partially or completely removed during sowing. Also, when germinating seeds, they can rise to the surface of the soil and lead to drying and death of the applied microorganisms in the inoculum. However, seed priming can enhance seed germination and initial plant growth (mainly root development). Seman et al. (2021) examined the capacity of the microalgal cell suspension to improve germination and initial growth (root and shoot length) of white radish (*Raphanus sativus*) and kohlrabi (*Brassica oleracea* var. *gongylodes*). Compared to the control, the root length of white radish was increased by 64.24% and 41.32%, using *Chlorella* sp. and *Dictyosphaerium* sp. cell suspension, respectively. Kohlrabi root growth was stimulated by application of *Chlorella* sp. and *Dictyosphaerium* sp. suspension by 60.97% and 55.02%, respectively.

In recent times, the use of live cultures and aqueous extract of microorganisms, particularly algaehave become more and more common, in the form of foliar plant nutrition. By using live algae cells as well as algal extract, a biofertilizer effect is achieved, which is suitable for improving growth parameters in a large number of plant species. The plant is provided with the necessary and affordable nutrients, amino acids, vitamins, and enzymes. Foliar application also provides for the formation of an algal biofilm on the aboveground part of the plant, which has a protective role, reduces evapotranspiration, provides a mechanical barrier, and protects the plant from the attack of phytopathogenic microorganisms and parasites. Hajnal-Jafari et al. (2016) evaluated the effect of green algae *Chlorella vulgaris* on initial growth of wheat, maize, bean, and lettuce in in vitro conditions. The inoculum (Natur Plasma© – liquid fertilizer containing *Chlorella vulgaris*, $3x10^7$ CFU/ml, Hungary) were applied as foliar fertilizer by spraying. *Chlorella vulgaris* affected positively the length (28.5% increase) and fresh mass (17.9% increase) of maize root, stem length of wheat (24.2% increase), and stem mass of lettuce (56.34% increase).

Soil inoculation is used when a large population or amount of inoculum is introduced into the soil. Solid-state inoculums are usually used, in the form of granules (0.5–1.5 mm) that are placed below, next, to or above the seeds when sowing. The advantage of this inoculation technique is the ability to control the site of inoculum application as well as the dose administered. In this way, damage to the seed material is avoided, and the microorganisms in the inoculum are protected from the harmful effects of pesticides applied to the seeds. Small-seed cultures in particular benefit from this inoculation technique because they are exposed to a high concentration of inoculum. Algal biomass, left after metabolites extraction can be used as solid fertilizer. It is rich in nutrients for plants but can also positively influence the overall microbial population and activity in soil.

4.5.3 Future of microalgae-based fertilizers

For more successful and widespread application of microalgae-based fertilizers in sustainable agricultural production, it is certainly necessary to work on their improvement. Some of the areas of interest are as follows:

1. Assessment of existing and evaluation of future carriers.
2. Optimization of formulations that have shown good results in the field (conditions of application, amount of inoculum, ratio and selection of carriers and microalgae, and many more).
3. Improve the survival of microalgae in the inoculum.
4. Improve storage time.
5. Formulation of mixed inoculants containing different types of microorganisms with PGP properties including microalgae or combinations of PGP microorganisms with those used in the control of pathogens and pests.
6. Adding additives to formulations.

7. Creating an inoculum that is "farmer friendly," easy to use—does not involve numerous steps in their application.
8. The price of the product must be competitive in order to survive in the market.
9. Use of "local" microalgae. The concept of biofertilizer production is based on the isolation of microalgae from the soil or substrate to which they will return in much larger numbers. This is when the formulations are most effective. However, at the global level, in the world of developed trade, many "local" microalgae strains are often applied in different ecological conditions compared to those from which they originated, so their effectiveness varies.
10. Application of microalgae metabolites or extracts without living cells.

References

Abd-El-Baky, H. H., El-Baz, F. K., & El-Baroty, G. S. (2010). Enhancing antioxidant availability in grains of wheat plants irrigated with seawater using microalgae extracts. *Plant Physiology and Biochemistry, 48*(11), 963–969.

Abdel-Razek, M. A., Abozeid, A. M., Eltholth, M. M., Abouelenien, F. A., El-Midany, S. A., Moustafa, N. Y., & Mohamed, R. A. (2019). bioremediation of a pesticide and selected heavy metals in wastewater from various sources using a consortium of microalgae and cyanobacteria. *Slovenian Veterinary Research, 56*. https://doi.org/10.26873/svr-744-2019.

Abinandan, S., Subashchandrabose, S. R., Venkateswarlu, K., & Megharaj, M. (2019). Soil microalgae and cyanobacteria: the biotechnological potential in the maintenance of soil fertility and health. *Critical Reviews in Biotechnology, 39*(8), 981–998. https://doi.org/10.1080/07388551.2019.1654972, http://www.tandfonline.com/loi/ibty20.

Adesalu, A. T., & Olugbemi (2015). Soil algae: A case study of two vegetable farmlands in Lagos and Ogun States. *Ife Journal of Science, 17*.

Al-Maliki, S., & Ebreesum, H. (2020). Changes in soil carbon mineralization, soil microbes, roots density and soil structure following the application of the arbuscular mycorrhizal fungi and green algae in the arid saline soil. *Rhizosphere, 14*, 100203. https://doi.org/10.1016/j.rhisph.2020.100203.

Alvarez, A. L., Weyers, S. L., Goemann, H. M., Peyton, B. M., & Gardner, R. D. (2021). Microalgae, soil and plants: A critical review of microalgae as renewable resources for agriculture. *Algal Research, 54*. https://doi.org/10.1016/j.algal.2021.102200, http://www.sciencedirect.com/science/journal/aip/22119264.

Ammar, E. E., Aioub, A. A. A., Elesawy, A. E., Karkour, A. M., Mouhamed, M. S., Amer, A. A., & EL-Shershaby, N. A. (2022). Algae as Bio-fertilizers: Between current situation and future prospective. *Saudi Journal of Biological Sciences, 29*(5), 3083–3096. https://doi.org/10.1016/j.sjbs.2022.03.020.

Anand, V., Kashyap, M., Samadhiya, K., Ghosh, A., & Kiran, B. (2019). Salinity driven stress to enhance lipid production in Scenedesmus vacuolatus: A biodiesel trigger? *Biomass and Bioenergy, 127*, 105252. https://doi.org/10.1016/j.biombioe.2019.05.021.

Arroussi, H. E., Benhima, R., Mernissi, Mernissi, & Bennis, I. (2018). Microalgae polysaccharides: The green chemistry for sustainable agriculture. *African Journal of Biotechnology, 17*(31), 978–989.

Avila, R., Peris, A., Eljarrat, E., Vicent, T., & Blánquez, P. (2021). Biodegradation of hydrophobic pesticides by microalgae: Transformation products and impact on algae biochemical methane potential. *Science of The Total Environment, 754*, 142114. https://doi.org/10.1016/j.scitotenv.2020.142114.

Babu, S., Bidyarani, N., Chopra, P., Monga, D., Kumar, R., Prasanna, R., Kranthi, S., & Kumar Saxena, A. (2015). Evaluating microbe-plant interactions and varietal differences for enhancing biocontrol efficacy in root rot disease challenged cotton crop. *European Journal of Plant Pathology, 142*(2), 345–362. https://doi.org/10.1007/s10658-015-0619-6.

Bakieva, G. R., Khaibullina, L. S., Gaisina, L. A., & Kabirov, R. R. (2012). Ecological-floristic analysis of soil algae and cyanobacteria on the Tra-Tau and Yurak-Tau Mounts, Bashkiria. *Eurasian Soil Science, 45*(9), 873–881. https://doi.org/10.1134/s1064229312030027.

Barragán, C., Wetzel, C. E., & Ector, L. (2018). A standard method for the routine sampling of terrestrial diatom communities for soil quality assessment. *Journal of Applied Phycology, 30*(2), 1095–1113. https://doi.org/10.1007/s10811-017-1336-7.

Barsanti, L., & Gualtieri, P. (2014). *Algae: Anatomy, Biochemistry, And Biotechnology: Second Edition Algae: Anatomy, Biochemistry, and Biotechnology: Second Edition*. Italy: CRC Press, 1–327. http://www.tandfebooks.com/doi/book/9781439867334, https://doi.org/10.1201/b16544.

Battacharyya, D., Babgohari, M. Z., Rathor, P., & Prithiviraj, B. (2015). Seaweed extracts as biostimulants in horticulture. *Scientia Horticulturae, 196*, 39–48. https://doi.org/10.1016/j.scienta.2015.09.012, www.elsevier.com/inca/publications/store/5/0/3/3/1/6.

Benderliev, K. M., Ivanova, N. I., & Pilarski, P. S. (2003). Singlet oxygen and other reactive oxygen species are involved in regulation of release of iron-binding chelators from Scenedesmus cells. *Biologia Plantarum, 47*(4), 523–526. https://doi.org/10.1023/B:BIOP.0000041056.07819.df, http://www.kluweronline.com/issn/0006-3134.

Bhattacharjee, M. (2016). Pharmaceutically valuable bioactive compounds of algae. *Asian Journal of Pharmaceutical and Clinical Research, 9*(6), 43–47. https://doi.org/10.22159/ajpcr.2016.v9i6.14507, http://innovareacademics.in/journals/index.php/ajpcr/article/download/14507/8048.

Bibi, R., Ahmad, Z., Imran, M., Hussain, S., Ditta, A., Mahmood, S., & Khalid, A. (2017). Algal bioethanol production technology: A trend towards sustainable development. *Renewable and Sustainable Energy Reviews, 71*, 976–985. https://doi.org/10.1016/j.rser.2016.12.126.

Bileva, T. (2013). Influence of Green Algae Chlorella vulgaris on Infested with Xiphinema index Grape Seedlings. *Journal of Earth Science & Climatic Change, 04*(02). https://doi.org/10.4172/2157-7617.1000136.

Chanda, Mj, Merghoub, N., & Arroussi, H. E. L. (2019). Microalgae polysaccharides: the new sustainable bioactive products for the development of plant bio-stimulants? *Morocco World Journal of Microbiology and Biotechnology, 35*(11). https://doi.org/10.1007/s11274-019-2745-3, www.wkap.nl/journalhome.htm/0959-3993.

Chen, H., Yu, S., Yu, Z., Ma, M., Liu, M., & Pei, H. (2024). Phycoremediation Potential of Salt-Tolerant Microalgal Species: Motion, Metabolic Characteristics, and Their Application for Saline–Alkali Soil Improvement in Eco-Farms. *Microorganisms, 12*(4), 676. https://doi.org/10.3390/microorganisms12040676.

Chiaiese, P., Corrado, G., Colla, G., Kyriacou, M. C., & Rouphael, Y. (2018). Renewable Sources of Plant Biostimulation: Microalgae as a Sustainable Means to Improve Crop Performance. *Frontiers in Plant Science, 9*. https://doi.org/10.3389/fpls.2018.01782.

Chojnacka, K., Saeid, A., & Michalak, I. (2012). The possibilities of the application of algal biomass in the agriculture. *Chemik, 66*(11), 1235–1248. http://chemikinternational.com/images/pdf_files/2012_11/11_12_chemik%20_11.pdf.Poland.

Choleva, B., Bileva, T., Tzvetkov, Y., & Barakov, P. (2005). Preliminary study of the green algae chlorella (Chlorella vulgaris) for control on the root-knot nematode (Meloidogyne arenaria) in tomato plants and ectoparasite Xiphinema indexin grape seedlings. *Communications in Agricultural and Applied Biological Sciences, 70*(4), 915–926.

Choleva, B., Bileva, T., & Tsvetkov, J. (2007). Organo-biological means and methods for control of plant parasitic nematodes as alternative of agrochemicals. *Ecology and Future, 6*, 43–49.

Chugh, M., Kumar, L., Shah, M. P., & Bharadvaja, N. (2022). Algal Bioremediation of heavy metals: An insight into removal mechanisms, recovery of by-products, challenges, and future opportunities. *Energy Nexus, 7*, 100129. https://doi.org/10.1016/j.nexus.2022.100129.

Colla, G., & Rouphael, Y. (2020). Microalgae: New Source of Plant Biostimulants. *Agronomy, 10*(9), 1240. https://doi.org/10.3390/agronomy10091240.

Cramer, W., Yohe, G. W., Auffhammer, M., Huggel, C., Molau, U., Da Silva Dias, M. A. F., Solow, A., Stone, D. A., Tibig, L., Leemans, R., Seguin, B., Smith, N., & Hansen, G. (2015). Detection and attribution of observed impacts. In *Climate Change 2014 Impacts, Adaptation and Vulnerability: Part A: Global and Sectoral Aspects*. Germany: Cambridge University Press, 979–1038. http://doi.org/10.1017/CBO9781107415379, https://doi.org/10.1017/CBO9781107415379.023.

Danilovic, B., Cvetkovic-Rakic, J., Ciric, J., Simeunovic, J., Veljkovic, V., & Savic, D. (2017). The isolation and screening of microalgae for the production of oil. *Chemical Industry, 71*(1), 69–74. https://doi.org/10.2298/hemind151127019d.

De Clerck, O., Bogaert, K. A., & Leliaert, F. (2012). Diversity and Evolution of Algae. Primary Endosymbiosis. *Advances in Botanical Research, 64*, 55–86. https://doi.org/10.1016/B978-0-12-391499-6.00002-5, http://www.elsevier.com/wps/find/bookdescription.cws_home/675801/description#description.

Del Mondo, A., Smerilli, A., Ambrosino, L., Albini, A., Noonan, D. M., Sansone, C., & Brunet, C. (2021). Insights into phenolic compounds from microalgae: structural variety and complex beneficial activities from health to nutraceutics. *Critical Reviews in Biotechnology, 41*(2), 155–171. https://doi.org/10.1080/07388551.2021.1874284, http://www.tandfonline.com/loi/ibty20.

Duo, L. A., Liu, C. X., & Zhao, S. L. (2018). Alleviation of Drought Stress in Turfgrass by the Combined Application of Nano-compost and Microbes from Compost. *Russian Journal of Plant Physiology, 65*(3), 419–426. https://doi.org/10.1134/S102144371803010X, http://www.kluweronline.com/issn/1021-4437.

Đorđević, A., & Radmanović, S. (2016). Pedologija. Univerzitet u Beogradu, Poljoprivredni fakultet. Udžbenik, str.359.

Fahad, S., Hussain, S., Matloob, A., Khan, F. A., Khaliq, A., Saud, S., Hassan, S., Shan, D., Khan, F., Ullah, N., Faiq, M., Khan, M. R., Tareen, A. K., Khan, A., Ullah, A., Ullah, N., & Huang, J. (2015). Phytohormones and plant responses to salinity stress: a review. *Plant Growth Regulation, 75*(2), 391–404. https://doi.org/10.1007/s10725-014-0013-y, http://www.springeronline.com/sgw/cda/frontpage/0,11855,1-40109-70-35547532-0,00.html.

Fathi, A., & Zaki, F. (2003). Preliminary Survey Of Edaphic Algae In El-Minia Region, Nile Valley, Egypt. *Egyptian Journal of Phycology, 4*(2), 131–148. https://doi.org/10.21608/egyjs.2003.113478.

Fazlutdinova, A., Gabidullin, Y., Allaguvatova, R., & Gaysina, L. (2021). Diatoms in Volcanic Soils of Mutnovsky and Gorely Volcanoes (Kamchatka Peninsula, Russia). *Microorganisms, 9*(9), 1851. https://doi.org/10.3390/microorganisms9091851.

Fenibo, E. O., Ijoma, G. N., & Matambo, T. (2021). Biopesticides in Sustainable Agriculture: A Critical Sustainable Development Driver Governed by Green Chemistry Principles.

Frontiers in Sustainable Food Systems, 5. https://doi.org/10.3389/fsufs.2021.619058, www.frontiersin.org/journals/sustainable-food-systems#.

Flechtner, V. R., Johansen, J. R., & Belnap, J. (2008). The biological soil crusts of the San Nicolas Island: Enigmatic algae from a geographically isolated ecosystem. *Western North American Naturalist, 68*(4), 405–436. https://doi.org/10.3398/1527-0904-68.4.405.

Foets, J., Stanek-Tarkowska, J., Teuling, A. J., Van de Vijver, B., Wetzel, C. E., & Pfister, L. (2021). Autecology of terrestrial diatoms under anthropic disturbance and across climate zones. *Ecological Indicators, 122*, 107248. https://doi.org/10.1016/j.ecolind.2020.107248.

Ge, S., Madill, M., & Champagne, P. (2018). Use of freshwater macroalgae Spirogyra sp. for the treatment of municipal wastewaters and biomass production for biofuel applications. *Biomass and Bioenergy, 111*, 213–223. https://doi.org/10.1016/j.biombioe.2017.03.014, http://www.journals.elsevier.com/biomass-and-bioenergy.

Ghasemi, Y., Moradian, A., Mohagheghzadeh, A., Shokravi, S., & Morowvat, M. H. (2007). Antifungal and antibacterial activity of the microalgae collected from paddy fields of Iran: Characterization of antimicrobial activity of Chroococcus dispersus. *Journal of Biological Sciences, 7*(6), 904–910. https://doi.org/10.3923/jbs.2007.904.910, http://scialert.net/current.php?issn=1727-3048.

Glaser, K., Baumann, K., Leinweber, P., Mikhailyuk, T., & Karsten, U. (2018). Algal richness in BSCs in forests under different management intensity with some implications for P cycling. *Biogeosciences, 15*(13), 4181–4192. https://doi.org/10.5194/bg-15-4181-2018, http://www.biogeosciences.net/volumes_and_issues.html.

Gojkovic, Z., Lindberg, R. H., Tysklind, M., & Funk, C. (2019). Northern green algae have the capacity to remove active pharmaceutical ingredients. *Ecotoxicology and Environmental Safety, 170*, 644–656. https://doi.org/10.1016/j.ecoenv.2018.12.032.

Gonçalves, A. L. (2021). The use of microalgae and cyanobacteria in the improvement of agricultural practices: A review on their biofertilising, biostimulating and biopesticide roles. *Portugal Applied Sciences (Switzerland), 11*(2), 1–21. https://doi.org/10.3390/app11020871, https://www.mdpi.com/2076-3417/11/2/864/pdf.

Gull, A., Ahmad Lone, A., & Ul Islam Wani, N. (2019). *Biotic and Abiotic Stresses in Plants*. IntechOpen. https://doi.org/10.5772/intechopen.85832.

Gulnaz, O., & Dincer, S. (2009). Biodegradation of bisphenol a by Chlorella vulgaris and Aeromonas hydrophilia. *Journal of Applied Biological Science, 3*, 79–84.

Guzman-Murillo, M. A., Ascencio, F., & Bashan, Y. (2013). Potential of microalgae and cyanobacteria to enhance the salt stress tolerance of tomato plants by modulating their antioxidant system. *Plant Physiology and Biochemistry, 61*, 225–233.

Hajnal Jafari, T., D. Stamenov, D., & S. Đurić, S. (2019). Proizvodnja i primena biopreparata Edicija osnovni udžbenik. Univerzitet u Novom Sadu, Poljoprivredni fakultet, Novi Sad Univerzitet u Novom Sadu, Poljoprivredni fakultet, Novi Sad (2019), 978-86-7520-484-8.

Hajnal Jafari, T., Ž. Vladimira Đurić, Ž., & S. Stamenov Dragana., S. (2022). Simpozijum "Zemljište u doba precizne poljoprivrede i informacionih tehnologija" Novi Sad, Srbija. Univerzitet u Novom Sadu, Poljoprivredni fakultet. pp. 79–83. ISBN 978-86-7520-569-2. (In Serbian) Izolacija i karakterizacija autohtonih zemljišnih mikroalgi u Vojvodini, Srbija.

Hajnal-Jafari, T., Djuric, S., & Stamenov, D. (2016). Influence of green algae Chlorella vulgaris on initial growth of different agricultural crops. *Matica Srpska Journal for Natural Sciences,* (130), 29–33. https://doi.org/10.2298/zmspn1630029h.

Hajnal Jafari, T., Žunić, V., Djurić, S., & Stamenov, D. (2023). New Record of Scenedesmus Vacuolatus From Soil In Vojvodina, Serbia. *Bangladesh Journal of Plant Taxonomy, 30*(2), 255–261. https://doi.org/10.3329/bjpt.v30i2.70501.

Hamouda, R. A., & El-Ansary, M. S. M. (2017). Potential of plant-parasitic nematode control in banana plants by microalgae as a new approach towards resistance. *Egyptian Journal of Biological Pest Control, 27*(2), 165–172. www.ejbpc.com/index.html.

Hassan, Z.ul, Ali, S., Rizwan, M., Ibrahim, M., Nafees, M., & Waseem, M. (2017). *Role of Bioremediation Agents (Bacteria, Fungi, and Algae) in Alleviating Heavy Metal Toxicity*. Springer Science and Business Media LLC, 517–537. https://doi.org/10.1007/978-981-10-4059-7_27.

Hirooka, T., Nagase, H., Uchida, K., Hiroshige, Y., Ehara, Y., Nishikawa, J. I., Nishihara, T., Miyamoto, K., & Hirata, Z. (2005). Biodegradation of bisphenol A and disappearance of its estrogenic activity by the green alga Chlorella fusca var. vacuolata. *Environmental Toxicology and Chemistry, 24*(8), 1896–1901. https://doi.org/10.1897/04-259R.1.

Hodač, L., Hallmann, C., Spitzer, K., Elster, J., Faßhauer, F., Brinkmann, N., Lepka, D., Diwan, V., Friedl, T., & Margesin, R. (2016). Widespread green algae Chlorella and Stichococcus exhibit polar-temperate and tropical-temperate biogeography. *FEMS Microbiology Ecology, 92*(8), fiw122. https://doi.org/10.1093/femsec/fiw122.

Hom-Diaz, A., Jaén-Gil, A., Rodríguez-Mozaz, S., Barceló, D., Vicent, T., & Blánquez, P. (2022). Insights into removal of antibiotics by selected microalgae (Chlamydomonas reinhardtii, Chlorella sorokiniana, Dunaliella tertiolecta and Pseudokirchneriella subcapitata). *Algal Research, 61*, 102560. https://doi.org/10.1016/j.algal.2021.102560.

Jassey, V. E. J., Walcker, R., Kardol, P., Geisen, S., Heger, T., Lamentowicz, M., Hamard, S., & Lara, E. (2022). Contribution of soil algae to the global carbon cycle. *New Phytologist, 234*(1), 64–76. https://doi.org/10.1111/nph.17950.

Jimel, M., Kvíderová, J., & Elster, J. (2021). Annual Cycle of Mat-Forming Filamentous Alga Tribonema cf. minus (Stramenopiles, Xanthophyceae) in Hydro-Terrestrial Habitats in the High Arctic Revealed By Multiparameter Fluorescent Staining. *Republic Journal of Phycology, 57*(3), 780–796. https://doi.org/10.1111/jpy.13109, http://onlinelibrary.wiley.com/journal/10.1111/(ISSN)1529-8817.

Jokel, M., Salazar, J., Chovancek, E., Sirin, S., & Allahverdiyeva, Y. (2023). Screening of several microalgae revealed biopesticide properties of Chlorella sorokiniana against the strawberry pathogen Phytophthora cactorum. *Journal of Applied Phycology, 35*(6), 2675–2687. https://doi.org/10.1007/s10811-023-03015-x.

Khan, Z., Kim, Y., Kim, S., & Kim, H. (2007). Observations on the suppression of root-knot nematode (Meloidogyne arenaria) on tomato by incorporation of cyanobacterial powder (Oscillatoria chlorina) into potting field soil. *Bioresource Technology, 98*(1), 69–73. https://doi.org/10.1016/j.biortech.2005.11.029.

Kabirov, R. R., & Gaisina, L. A. (2009). Parameters of the productivity of soil algae in terrestrial ecosystems. *Eurasian Soil Science, 42*(12), 1374–1379. https://doi.org/10.1134/s1064229309120072

Kapoore, R. V., Wood, E. E., & Llewellyn, C. A. (2021). Algae biostimulants: A critical look at microalgal biostimulants for sustainable agricultural practices. *Biotechnology Advances, 49*. https://doi.org/10.1016/j.biotechadv.2021.107754, www.elsevier.com/inca/publications/store/5/2/5/4/5/5/index.htt.

Khakimova, N., Maravić, N., Davidović, P., Blagojević, D., Bečelić-Tomin, M., Simeunović, J., Pešić, V., Šereš, Z., Mandić, A., Pojić, M., & Mišan, A. (2022). Sugar Beet Processing Wastewater Treatment by Microalgae through Biosorption. *Water, 14*(6), 860. https://doi.org/10.3390/w14060860.

Kim, J. W., Park, S.-B., Tran, Q.-G., Cho, D.-H., Choi, D.-Y., Lee, Y. J., & Kim, H.-S. (2020). Functional expression of polyethylene terephthalate-degrading enzyme (PETase) in green microalgae. *Microbial Cell Factories, 19*(1). https://doi.org/10.1186/s12934-020-01355-8.

Kumar, R. V., Kanna, G., & Elumalai, S. (2017). Biodegradation of Polyethylene by Green Photosynthetic Microalgae. *Journal of Bioremediation and Biodegradation, 8*(381). https://doi.org/10.4172/2155-6199.1000381.

Kusvuran, S., & Kusvuran, Z. (2019). The effect of Chlorella vulgaris-based microbial fertilizer on guar plants under drought stress. *International Journal of Agriculture and Biology, 21*(4), 769–778.

Lawton, R. J., de Nys, R., & Paul, N. A. (2013). Selecting Reliable and Robust Freshwater Macroalgae for Biomass Applications. *PLoS One, 8*(5). https://doi.org/10.1371/journal.pone.0064168Australia, http://www.plosone.org/article/fetchObject.action?uri=info%3Adoi%2F10.1371%2Fjournal.pone.0064168&representation=PDF.

Kusvuran, S. (2021). Microalgae (Chlorella vulgaris Beijerinck) alleviates drought stress of broccoli plants by improving nutrient uptake, secondary metabolites, and antioxidative defense system. *Horticultural Plant Journal, 7*(3), 221–231. https://doi.org/10.1016/j.hpj.2021.03.007.

Lee, J. B., Hayashi, K., Hirata, M., Kuroda, E., Suzuki, E., Kubo, Y., & Hayashi, T. (2006). Antiviral sulfated polysaccharide from Navicula directa, a diatom collected from deep-sea water in Toyama Bay. *Biological and Pharmaceutical Bulletin, 29*(10), 2135–2139. https://doi.org/10.1248/bpb.29.2135Japan, http://www.jstage.jst.go.jp/article/bpb/29/10/2135/_pdf.

Lee, R. E. (2005). Phycology. Cambrigde University Press. UK. pp. 614.

Lee, S. M., & Ryu, C. M. (2021). Algae as New Kids in the Beneficial Plant Microbiome. *Frontiers in Plant Science, 12*. https://doi.org/10.3389/fpls.2021.599742, https://www.frontiersin.org/journals/plant-science.

Lee, S. M., Kim, S. K., Lee, N., Ahn, C. Y., & Ryu, C. M. (2020). d-Lactic acid secreted by Chlorella fusca primes pattern-triggered immunity against Pseudomonas syringae in Arabidopsis. *Plant Journal, 102*(4), 761–778. https://doi.org/10.1111/tpj.14661, http://onlinelibrary.wiley.com/journal/10.1111/(ISSN)1365-313X.

Lentendu, G. (2014). Data from: effects of long term differential fertilization on eukaryotic microbial communities in an arable soil: a multiple barcoding approach. *Dryad Digital Repository, 10*.

Li, Q., Chen, L., Chen, X., & Li, N. (2014). Microalgae as a potential high-value bioresource for wastewater treatment. *Science of The Total Environment, 652*, 180–198.

Li, Y., Xu, S. S., Gao, J., Pan, S., & Wang, G. X. (2014). Chlorella induces stomatal closure via NADPH oxidase-dependent ROS production and its effects on instantaneous water use efficiency in Vicia faba. *PLoS One, 9*(3). https://doi.org/10.1371/journal.pone.0093290, http://www.plosone.org/article/fetchObject.action?uri=info%3Adoi%2F10.1371%2Fjournal.pone.0093290&representation=PDF.

Lin, C. S., & Wu, J. T. (2014). Tolerance of soil algae and cyanobacteria to drought stress. *Journal of Phycology, 50*(1), 131–139. https://doi.org/10.1111/jpy.12141.

Lin, H., Fang, L., Low, C. S., Chow, Y., & Lee, Y. K. (2013). Occurrence of glycerol uptake in Dunaliella tertiolecta under hyperosmotic stress. *FEBS Journal, 280*(4), 1064–1072. https://doi.org/10.1111/febs.12100.

Liu, J., Li, J., & Ge, C. (2001). Advance in role mechanism of microbial fertilizer. *Journal of Microbiology, 21*(3), 33–46.

Lopez-Bautista, J. M., Rindi, F., & Casamatta, D. (2007). *The Systematics of Subaerial Algae*. Springer Science and Business Media LLC, 599–617. https://doi.org/10.1007/978-1-4020-6112-7_33.

Lu, Y., & Xu, J. (2015). Phytohormones in microalgae: a new opportunity for microalgal biotechnology? *Trends in Plant Science, 20*(5), 273–282. https://doi.org/10.1016/j.tplants.2015.01.006.

Maltseva, I. A., & Maltsev, Y. I. (2021). Diversity of cyanobacteria and algae in dependence to forest-forming tree species and properties rocks of dump. *International Journal of Environmental Science and Technology, 18*(3), 545–560. https://doi.org/10.1007/s13762-020-02868-w.

Mata, T. M., Martins, A. A., & Caetano, N. S. (2010). Microalgae for biodiesel production and other applications: A review. *Renewable and Sustainable Energy Reviews, 14*(1), 217–232. https://doi.org/10.1016/j.rser.2009.07.020.

Michalak, I., & Chojnacka, K. (2014). Algal extracts: Technology and advances. *Engineering in Life Sciences, 14*(6), 581–591. https://doi.org/10.1002/elsc.201400139.

Mikhailyuk, T., Glaser, K., Tsarenko, P., Demchenko, E., & Karsten, U. (2019). Composition of biological soil crusts from sand dunes of the Baltic Sea coast in the context of an integrative approach to the taxonomy of microalgae and cyanobacteria. *European Journal of Phycology, 54*(3), 263–290. https://doi.org/10.1080/09670262.2018.1557257, https://www.tandfonline.com/loi/tejp20.

Moreira, J. B., Vaz, Bd. S., Cardias, B. B., Cruz, C. G., Almeida, A. C. Ad, Costa, J. A. V., & Morais, M. Gd (2022). Microalgae Polysaccharides: An Alternative Source for Food Production and Sustainable Agriculture. *Polysaccharides, 3*(2), 441–457. https://doi.org/10.3390/polysaccharides3020027, https://www.mdpi.com/journal/polysaccharides.

Rahman, M. M. (2022). Dynamics of epibenthic copepods in relation to environmental factors and phytoplankton abundance in tropical river, estuarine and coastal environments. *Estuarine, Coastal and Shelf Science, 266*, 107748. https://doi.org/10.1016/j.ecss.2022.107748.

Mutale-joan, C., Redouane, B., Najib, E., Yassine, K., Lyamlouli, K., Laila, S., Zeroual, Y., & Hicham, E. A. (2020). Screening of microalgae liquid extracts for their bio stimulant properties on plant growth, nutrient uptake and metabolite profile of Solanum lycopersicum L. *Scientific Reports, 10*(1). https://doi.org/10.1038/s41598-020-59840-4.

Mógor, Á. F., Ördög, V., Lima, G. P. P., Molnár, Z., & Mógor, G. (2018). Biostimulant properties of cyanobacterial hydrolysate related to polyamines. *Journal of Applied Phycology, 30*(1), 453–460. https://doi.org/10.1007/s10811-017-1242-z, www.wkap.nl/journalhome.htm/0921-8971.

Nabti, E., Jha, B., & Hartmann, A. (2017). Impact of seaweeds on agricultural crop production as biofertilizer. *International Journal of Environmental Science and Technology, 14*(5), 1119–1134. https://doi.org/10.1007/s13762-016-1202-1.

Nisha, R., Kiran, B., Kaushik, A., & Kaushik, C. P. (2018). Bioremediation of salt affected soils using cyanobacteria in terms of physical structure, nutrient status and microbial activity. *International Journal of Environmental Science and Technology, 15*(3), 571–580. https://doi.org/10.1007/s13762-017-1419-7.

Novakovskaya, I. V., & Patova, E. N. (2011). Soil Algae of Spruce Forests and Their Change under Conditions of Aero-Technogenic Pollution. Biologia: Syktyvkar, Russian.

Novakovskaya, I. V., Patova, E. N., Dubrovskiy, Y. A., Novakovskiy, A. B., & Kulyugina, E. E. (2022). Distribution of algae and cyanobacteria of biological soil crusts along the elevation gradient in mountain plant communities at the Northern Urals (Russian European Northeast). *Journal of Mountain Science, 19*(3), 637–646. https://doi.org/10.1007/s11629-021-6952-7.

Oancea, F., Velea, S., Fătu, V., Mincea, C., & Ilie, L. (2013). Micro-algae based plant biostimulant and its effect on water stressed tomato plants. *Rom. J. Plant Prot. 6*, 104–117.

Ohtsuka, T., & Fujita, Y. (2001). The diatom flora and its seasonal changes in a paddy field in Central Japan. *Nova Hedwigia, 73*(1-2), 97–128. https://doi.org/10.1127/nova.hedwigia/73/2001/97.

Oliveira, M. F., & MacIel-Silva, A. S. (2022). Biological soil crusts and how they might colonize other worlds: insights from these Brazilian ecosystem engineers. *Journal of Experimental Botany, 73*(13), 4362–4379. https://doi.org/10.1093/jxb/erac162, http://jxb.oxfordjournals.org.

Parmar, P., Kumar, R., Neha, Y., & Srivatsan, V. (2023). Microalgae as next generation plant growth additives: Functions, applications, challenges and circular bioeconomy based solutions. *Frontiers in Plant Science, 14*. https://doi.org/10.3389/fpls.2023.1073546.

Patova, E. N., & Novakovskaya, I. V. (2018). Soil algae of the Northeastern European Russia. *Novosti sistematiki nizshikh rastenii, 52*(2), 311–353. https://doi.org/10.31111/nsnr/2018.52.2.311.

Patova, E., Novakovskaya, I., Gusev, E., & Martynenko, N. (2023). Diversity of Cyanobacteria and Algae in Biological Soil Crusts of the Northern Ural Mountain Region Assessed through Morphological and Metabarcoding Approaches. *Diversity, 15*(10), 1080. https://doi.org/10.3390/d15101080.

Pei, H., & Yu, Z. (2023). Microalgae: a revolution for salt-affected soil remediation. *Trends in Biotechnology, 42*(2), 147–149. https://doi.org/10.1016/j.tibtech.2022.08.009.

Perera, I., Subashchandrabose, S. R., Venkateswarlu, K., Naidu, R., & Megharaj, M. (2018). Consortia of cyanobacteria/microalgae and bacteria in desert soils: an underexplored microbiota. *Applied Microbiology and Biotechnology, 102*(17), 7351–7363. https://doi.org/10.1007/s00253-018-9192-1, http:/link.springer.de/link/service/journals/00253/index.htm.

Plaza, B. M., Gómez-Serrano, C., Acién-Fernández, F. G., & Jimenez-Becker, S. (2018). Effect of microalgae hydrolysate foliar application (Arthrospira platensis and Scenedesmus sp.) on Petunia x hybrida growth. *Journal of Applied Phycology, 30*(4), 2359–2365. https://doi.org/10.1007/s10811-018-1427-0, www.wkap.nl/journalhome.htm/0921-8971.

Poveda, J., & Díez-Méndez, A. (2023). Use of elicitors from macroalgae and microalgae in the management of pests and diseases in agriculture. *Phytoparasitica, 51*(4), 667–701. https://doi.org/10.1007/s12600-022-01009-y, https://www.springer.com/journal/12600.

Priyadarshani, I., & Rath, B. (2012). Commercial and industrial applications of micro algae – A review. *Journal of. Algal Biomass Utilization. 3*(4), 2229.

Pulz, O., & Gross, W. (2004). Valuable products from biotechnology of microalgae. *Applied Microbiology and Biotechnology, 65*(6), 635–648. https://doi.org/10.1007/s00253-004-1647-x.

Rachidi, F., Benhima, R., Sbabou, L., & El Arroussi, H. (2020). Microalgae polysaccharides bio-stimulating effect on tomato plants: Growth and metabolic distribution. *Biotechnology Reports, 25*, e00426. https://doi.org/10.1016/j.btre.2020.e00426.

Rakonjac, A., & Simić, S. (2023). Updating knowledge on the distribution of red algae (Rhodophyta) in Serbia (Southeast Europe). *Kragujevac Journal of Science,* (45), 239–244. https://doi.org/10.5937/kgjsci2345239r.

Redmile-Gordon, M., Gregory, A. S., White, R. P., & Watts, C. W. (2020). Soil organic carbon, extracellular polymeric substances (EPS), and soil structural stability as affected by previous and current land-use. *Geoderma, 363*, 114143. https://doi.org/10.1016/j.geoderma.2019.114143.

Renuka, N., Guldhe, A., Prasanna, R., Singh, P., & Bux, F. (2018). Microalgae as multi-functional options in modern agriculture: current trends, prospects and challenges. *Biotechnology Advances, 36*(4), 1255–1273. https://doi.org/10.1016/j.biotechadv.2018.04.004.

Righini, S., Parati, K., & Marini, G. (2016). Bio-fertilizers based on microalgae for the improvement of crop production. *Frontiers in Plant Science, 7*.

Roberti, R., Galletti, S., Burzi, P. L., Righini, H., Cetrullo, S., & Perez, C. (2015). Induction of defence responses in zucchini (Cucurbita pepo) by Anabaena sp. water extract. *Biological Control, 82*, 61–68. https://doi.org/10.1016/j.biocontrol.2014.12.006.

Ronga, D., Orzes, G., Catellani, M., Francia, E., Milc, J., & Pecchioni, N. (2018). Tackling heavy metal contamination in crops: The effects of beneficial microorganisms and biostimulants on phytoremediation. *Applied Soil Ecology, 123*, 642–659.

Ronga, D., Caradonia, F., Setti, L., Hagassou, D., Giaretta Azevedo, C. V., Milc, J., Pedrazzi, S., Allesina, G., Arru, L., & Francia, E. (2019). Effects of innovative biofertilizers on yield of processing tomato cultivated in organic cropping systems in northern Italy. *Acta Horticulturae,* (1233), 129–136. https://doi.org/10.17660/actahortic.2019.1233.19.

Simić, S., & Krizmanić, J. (2018). Istraživanja bentosnih crvenih i silikatnih algi u Srbiji – prošlost, sadašnjost, perspektiva. II Kongres biologa Srbije, Knjiga sažetaka: 200 str. (2018).

Safi, C., Zebib, B., Merah, O., Pontalier, P.-Y., & Vaca-Garcia, C. (2014). Morphology, composition, production, processing and applications of Chlorella vulgaris: A review. *Renewable and Sustainable Energy Reviews, 35*, 265–278. https://doi.org/10.1016/j.rser.2014.04.007.

Sanniyasi, E., Gopal, R. K., Gunasekar, D. K., & Raj, P. P. (2021). Biodegradation of low-density polyethylene (LDPE) sheet by microalga, Uronema africanum Borge. *Scientific Reports, 11*(1). https://doi.org/10.1038/s41598-021-96315-6.

Seman, V., Hajnal-Jafari, T., Đurić, S., & Stamenov, D. (2021). The Capacity of Soil Microalgae to Improve Germination and Initial Growth of White Radish and Kohlrabi. *Contemporary Agriculture, 70*(1-2), 36–40. https://doi.org/10.2478/contagri-2021-0007.

Seppey, C. V. W., Singer, D., Dumack, K., Fournier, B., Belbahri, L., Mitchell, E.A.D. & Lara, E. (2017). Distribution patterns of siol microbial eukaryotes suggest widespread algivory by phagotrophic protists as an alternative pathway for nutrient cycling. *Soil Biology and Biochemistry, 112*, 68–76. https://doi.org/10.1016/j.soilbio.2017.05.002.

Sharma, N. K., Rai, A. K., Singh, S., & Brown, R. M. (2007). Airborne algae: Their present status and relevance. *Journal of Phycology, 43*(4), 615–627. https://doi.org/10.1111/j.1529-8817.2007.00373.x.

Shcherbyna, R., Gontcharov, A., & Tikhonenkov, D. (2021). Soil microalgae diversity in steppe ecosystems. *European Journal of Phycology, 56*(4), 504–515. https://doi.org/10.1080/09670262.2021.1915701.

Sherin, G., Aswathi, K. P. R., & Puthur, J. T. (2022). Photosynthetic functions in plants subjected to stresses are positively influenced by priming. *Plant Stress, 4*, 100079. https://doi.org/10.1016/j.stress.2022.100079.

Singh, A., Kumar, A., Yadav, S., & Singh, I. K. (2019). Reactive oxygen species mediated signaling during abiotic stress. *Plant Gene, 18*. https://doi.org/10.1016/j.plgene.2019.100173.

Singh, R., Upadhyay, A. K., Chandra, P., & Singh, D. P. (2018). Sodium chloride incites reactive oxygen species in green algae Chlorococcum humicola and Chlorella vulgaris: Implication on lipid synthesis, mineral nutrients and antioxidant system. *Bioresource Technology, 270*, 489–497. https://doi.org/10.1016/j.biortech.2018.09.065.

Singh, R. N. (1950). Reclamation of 'Usar' Lands in India through Blue-green Algæ *Nature, 165*(4191), 325–326. https://doi.org/10.1038/165325b0.

Smetana, S., Sandmann, M., Rohn, S., Pleissner, D., & Heinz, V. (2017). Autotrophic and heterotrophic microalgae and cyanobacteria cultivation for food and feed: life cycle assessment. *Bioresource Technology, 245*, 162–170. https://doi.org/10.1016/j.biortech.2017.08.113, www.elsevier.com/locate/biortech.

Song, X., Bo, Y., Feng, Y., Tan, Y., Zhou, C., Yan, X., Ruan, R., Xu, Q., & Cheng, P. (2022). Potential applications for multifunctional microalgae in soil improvement. *Frontiers in*

Environmental Science, 10. https://doi.org/10.3389/fenvs.2022.1035332, http://journal.frontiersin.org/journal/environmental-science.

Sosa-Hernández, J. E., Escobedo-Avellaneda, Z., Iqbal, H. M. N., & Welti-Chanes, J. (2018). State-of-the-Art Extraction Methodologies for Bioactive Compounds from Algal Biome to Meet Bio-Economy Challenges and Opportunities. *Molecules (Basel, Switzerland), 23*(11), 2953. https://doi.org/10.3390/molecules23112953.

Stanek-Tarkowska, J., Czyz, E. A., Kaniuczak, J., & Poradowska, A. (2017). Physicochemical properties of silt loamy soil and diversity of diatom species under winter wheat and oats. *Journal of Ecological Engineering, 18*(6), 142–151. https://doi.org/10.12911/22998993/76828, http://www.jeeng.net.

Stanek-Tarkowska, J., Szostek, M., & Rybak, M. (2021). Effect of Different Doses of Ash from Biomass Combustion on the Development of Diatom Assemblages on Podzolic Soil under Oilseed Rape Cultivation. *Agronomy, 11*(12), 2422. https://doi.org/10.3390/agronomy11122422.

Starks, T. L., Elliot Shubert, L., & Trainor, F. R. (2019). Ecology of soil algae: a review. *Phycologia, 20*(1), 65–80. https://doi.org/10.2216/i0031-8884-20-1-65.1.

Stirk, W. A., Bálint, P., Tarkowská, D., Novák, O., Strnad, M., Ördög, V., & van Staden, J. (2013). Hormone profiles in microalgae: Gibberellins and brassinosteroids. *Plant Physiology and Biochemistry, 70*, 348–353. https://doi.org/10.1016/j.plaphy.2013.05.037.

Stirk, W. A., Bálint, P., Tarkowská, D., Novák, O., Maróti, G., Ljung, K., Ture˘cková, V., Strnad, VÖrdög, V., & Staden. 2014). Efect of light on growth and endogenous hormones in Chlorella minutissima (Trebouxiophyceae).

Tomaselli, L. (2004). The microbial cell, In: A. Richmond (Ed.), Handbook of Microalgal Culture: Biotechnology and Applied Phycology. Blackwell Science Ltd., Oxford, UK, pp. 3-19.

Trbojevic, I., & redojević, D. (2021). Small Water Bodies of the Western BalkansAlgae in Shallow and Small Water Bodies of Serbia: A Frame for Species and Habitat Protection. https://doi.org/10.1007/978-3-030-86478-1_8.

Van Oosten, M. J., Pepe, O., De Pascale, S., Silletti, S., & Maggio, A. (2017). The role of biostimulants and bioeffectors as alleviators of abiotic stress in crop plants. *Chemical and Biological Technologies in Agriculture, 4*(1). https://doi.org/10.1186/s40538-017-0089-5, http://chembioagro.springeropen.com/.

Vernier, P., Decker, A., Kulaeva, O., & McDonald, I. (2005). The role of microalgal extracts in plant growth and stress tolerance. *Journal of Applied Phycology, 17*(5), 395–402.

Vessey, J. K. (2003). Plant growth promoting rhizobacteria as biofertilizers. *Plant and Soil, 255*(2), 571–586. https://doi.org/10.1023/A:1026037216893.

Vijayan, D., & George Ray, J. (2015). Green algae of a unique tropical wetland, Kuttanadu, Kerala, India, in relation to soil regions, seasons, and paddy growth stages. *International Journal of Science, Environment and Technology, 4*, 770–803.

Vo, T. P., Nguyen, D. Q., Thi Ho, T. A., Nguyen, T. M., Nguyen, Huy Ha, M., & Vo, P. H. N. (2024). Novel extraction of bioactive compounds from algae using green solvent: Principles, Applications, and Future perspectives. *Journal of Agriculture and Food Research, 18*, 101535. https://doi.org/10.1016/j.jafr.2024.101535.

Wahab, A., Abdi, G., Saleem, M. H., Ali, B., Ullah, S., Shah, W., Mumtaz, S., Yasin, G., Muresan, C. C., & Marc, R. A. (2022). Plants' Physio-Biochemical and Phyto-Hormonal Responses to Alleviate the Adverse Effects of Drought Stress: A Comprehensive Review. *Plants, 11*(13). https://doi.org/10.3390/plants11131620, https://www.mdpi.com/2223-7747/11/13/1620/pdf?version=1655777519.

Wang, X., Lin, L., Lu, H., Liu, Z., Duan, N., Dong, T., Xiao, H., Li, B., & Xu, P. (2018). Microalgae cultivation and culture medium recycling by a two-stage cultivation system.

Frontiers of Environmental Science & Engineering, 12(6). https://doi.org/10.1007/s11783-018-1078-z.

Williams, M. E. (2011). Introduction to Phytohormones. *Teaching tools in plant biology: lecture notes.*

Win, T. T., Barone, G. D., Secundo, F., & Fu, P. (2018). Algal Biofertilizers and Plant Growth Stimulants for Sustainable Agriculture. *Industrial Biotechnology, 14*(4), 203–211. https://doi.org/10.1089/ind.2018.0010, www.liebertonline.com/ind.

Xie, C., & Fan, J. (2007). Assessment of second-order velocity-slip boundary conditions of the navier-stokes equations. *Lixue Xuebao/Chinese Journal of Theoretical and Applied Mechanics, 39*(1), 1–6.

Yilmaz, E., & Sönmez, M. (2017). The role of organic/bio–fertilizer amendment on aggregate stability and organic carbon content in different aggregate scales. *Soil and Tillage Research, 168*, 118–124. https://doi.org/10.1016/j.still.2017.01.003, www.elsevier.com/inca/publications/store/5/0/3/3/1/8.

Zancan, S., Trevisan, R., & Maurizio, G. (2006). Paoletti, Soil algae composition under different agro-ecosystems in North-Eastern Italy. *Agriculture, Ecosystems & Environment, 112*(1), 1–12. https://doi.org/10.1016/j.agee.2005.06.018.

Zhang, C., Yu, X., Laipan, M., Wei, T., & Guo, J. (2024). Soil health improvement by inoculation of indigenous microalgae in saline soil. *Environmental Geochemistry and Health, 46*(1). https://doi.org/10.1007/s10653-023-01790-7.

Zhang, H., Gong, W., Zeng, W., Yan, Z., Jia, B., Li, G., & Liang, H. (2020). Organic carbon promotes algae proliferation in membrane-aeration based bacteria-algae symbiosis system (MA-BA). *Water Research, 176*, 115736. https://doi.org/10.1016/j.watres.2020.115736.

Zhang, L., & Lv, J. (2022). Linkages between soil respiration and soil eukaryotic microalgae following vegetation restoration in the Loess Plateau, China. *Land Degradation & Development, 33*(6), 881–891. https://doi.org/10.1002/ldr.4189.

Žunić, V., Jafari, T. H., Grabić, J., Đurić, S., & Stamenov, D. (2022). Hydroponic systems: exploring the balance between co-cultivation of Chlorella vulgaris and Swiss chard (Beta vulgaris L. subsp. cicla). *Journal of Applied Phycology, 34*(2), 903–913. https://doi.org/10.1007/s10811-021-02673-z, www.wkap.nl/journalhome.htm/0921-8971.

Žunić, V. (2024). Izolacija i karakterizacija zemljišnih mikroalgi za primenu u biljnoj proizvodnji. Doktorska disertacija, Poljoprivredni fakultet, Univerzitet u Novom Sadu. UDK: 561.26/27(043.3).

Food-grade oil from microalgae: a natural and sustainable source of nutraceuticals

5

Aysun Yurdunuseven Yildiz
Department of Food Engineering, Pamukkale University, Denizli, Turkey

5.1 Introduction

Edible oils are complex lipid mixtures, primarily composed of triglycerides, with smaller quantities of other lipids, vitamins, and antioxidants. These oils play a crucial role in energy storage and nutrient transport within the body. Traditional edible oils are primarily derived from plant and animal fats, which are influenced by seasonal and environmental factors and require large areas of land for their production (Li et al., 2021). As agricultural practices are central to edible oil production, they also play a significant role in climate change, contributing approximately 30% of global greenhouse gas emissions and consuming around 90% of the world's freshwater resources through irrigation. Beyond these impacts, the extensive use of fertilizers exacerbates environmental issues such as eutrophication (Fawcett et al., 2022). Furthermore, the large area of arable land dedicated to oilseed crops raises concerns about the sustainability of land use, particularly as climate change imposes additional constraints on land and water resources (Sarma et al., 2021).

Given the environmental challenges and resource limitations associated with traditional edible oil production, there is a growing need to explore alternative sources that are both sustainable and nutritionally valuable. In this context, the quest for more sustainable food production methods has sparked increased interest in microalgae as a promising alternative for producing food-grade oils. To address the increasing demand for sustainable oil sources, it is essential to explore alternative options, such as microalgae, which hold promise not only for food applications but also for a wide range of industrial uses. Microalgae can thrive without arable land, utilize waste as nutrients, and can be cultivated year-round in various climates, offering environmental benefits and increased production flexibility (Fawcett et al., 2022). Moreover, by using wastewater as a nutrient source, microalgae not only support their own growth but also contribute to environmental sustainability by helping to treat and recycle waste. Microalgae are a sustainable source of edible oil, requiring only sunlight and minimal nutrients for large-scale cultivation, while thriving in nonarable environments such as seawater and wastewater. Their fatty acid profiles are very similar to those of plant oils; some species are particularly rich in ω-3 polyunsaturated fatty acids (PUFAs), making microalgae oil a promising functional ingredient in the food industry (Xue et al., 2020). Some microalgae oils can replace

Eukaryotic Microorganisms as Sources of Bioproducts. DOI: https://doi.org/10.1016/B978-0-443-30188-9.00024-X

traditional vegetable oils, while others rich in long-chain PUFAs (LC-PUFAs) can boost cardiovascular health in food products (Draaisma et al., 2013).

Algae are classified into macroalgae, which are multicellular plants visible to the naked eye, and microalgae, which are microscopic, photosynthetic organisms that can thrive in diverse environments, with over 57,000 species identified, of which 40,000 have been explored for commercial applications (Rehman et al., 2022). Microalgae can convert sunlight, CO_2, and nutrients into biomass, making them an efficient and eco-friendly source of lipids. These organisms, ranging from 0.0002 to 0.01 mm in size, are classified within the *Protista* kingdom, which includes various types such as red, green, brown, and blue-green algae (Mane et al., 2024). As eukaryotes, their cellular structure includes organelles such as chloroplasts, endoplasmic reticulum, and mitochondria, mirroring those found in higher plants (Xue et al., 2020). Microalgae can be cultivated using a variety of systems, such as open ponds, raceway configurations, and photobioreactors (Gharib et al., 2024). To meet the growing demand for microalgae biomass, researchers are focused on developing efficient technologies for large-scale microalgae production, often utilizing open pond systems despite their limitations in controlling operational conditions. However, improving the design of photobioreactors has become a key focus, as it offers significant potential for enhancing productivity. The main objectives in this area include increasing biomass yield, ensuring cost-effectiveness, maintaining product purity, and improving the system's adaptability to varying environmental conditions (Rehman et al., 2022). Achieving these goals is crucial for establishing a sustainable and high-output microalgae production system.

With increasing issues related to food security, environmental degradation, and the depletion of traditional oil sources, exploring alternative sources such as microalgae oils is both timely and necessary. Microalgae are rich in proteins, lipids, and carbohydrates, and contain various beneficial enzymes, pigments, sterols, and vitamins, making them nutritionally valuable (Su et al., 2023). Cultivation conditions significantly affect the nutrient and bioactive profiles of microalgae, with lipid content reaching up to 77%, protein up to 70%, and carbohydrates around 50% of dry weight (Su et al., 2023). The biomass composition of microalgae varies by species, with *Arthrospira platensis* being high in protein, while *Phaeodactylum tricornutum* and *Tisochrysis lutea* are rich in PUFAs (Morais et al., 2021). Currently, several microalgae species, including *Spirulina* sp., *Chlorella* sp., *Dunaliella* sp., *Haematococcus* sp., *Schizochytrium* sp., *Porphyridium cruentum*, and *Crypthecodinium cohnii*, hold "Generally Recognized as Safe (GRAS)" status, along with specific oil and protein products derived from these organisms (Su et al., 2023). In the EU, *A. platensis* and *Chlorella vulgaris* are well established for human consumption, while *Odontella aurita* and *Tetraselmis chui* have also been recently approved as novel foods (Morais et al., 2021). The annual commercial production of microalgae is around 5000 tons of dry matter, with approximately 110 producers in the Asia–Pacific region, predominantly in China, Taiwan, and India (Sathasivam et al., 2019). Microalgae metabolites, such as those derived from *Spirulina* and *Chlorella*, serve as valuable dietary supplements with recognized health benefits and have been widely commercialized across various countries, including Taiwan, Japan, and Mexico (Katiyar & Arora, 2020).

The lipid production by microalgae is 15–300 times higher compared to oil-bearing crops such as corn, soybean, sunflower, and oil palm (Poh et al., 2020). Optimal growth conditions for microalgae, including light intensity, temperature, and pH, vary by species, with factors such as light wavelength, temperature range (25°C–50°C), and pH range (6.5–8.5) playing crucial roles in maximizing biomass and lipid productivity (Rehman et al., 2022). Lipid productivity is the key metric for assessing the oil-producing potential of microalgae, with oil yields reaching 6000–15,000 gallons per acre, significantly higher than traditional vegetable oils (Sajjadi et al., 2018). Additionally, microalgae contain 20%–60% PUFAs, with notable species like *P. tricornutum*, *Botryococcus braunii*, and *Dunaliella salina* (Xue et al., 2020). Algal oil is viewed for its significant commercial potential, as microalgae can yield not only oil but also protein and carbohydrates, allowing for the creation of valuable coproducts from algal biomass (Huang et al., 2016). The lipid content of microalgae can vary significantly from 1.5% to 75% of dry biomass, influenced by strain, culture medium, and growth conditions, leading to diverse fatty acid profiles across species (Cvejic et al., 2019). Microalgae strains known for their high PUFA content include *Schizochytrium*, *P. cruentum*, *P. tricornutum*, and *Nannochloropsis oculata* (Xue et al., 2021). Cultivation conditions significantly influence the nutrient composition of biomass, and optimizing these factors can enhance yields of valuable compounds like proteins and lipids (Fawcett et al., 2022).

This chapter offers a comprehensive overview of microalgae oil production, including its chemical composition, extraction methods, and safety and regulatory considerations. Additionally, it examines the use of microalgae oils in food products and nutraceuticals, while also addressing the economic and environmental aspects of large-scale production. Furthermore, this chapter aims to provide insights into the current status of microalgae oils, their nutritional benefits, and the challenges involved in their commercial application within the food industry.

5.2 Microalgae oils: fatty acid composition, quality, and health benefits

Microalgae have the potential to produce high-quality oils rich in essential fatty acids, antioxidants, and other bioactive compounds. The composition of microalgae can vary significantly among species, but certain trends are commonly observed across many genera. Microalgae lipids are primarily classified into polar and nonpolar types, with polar lipids like glycerophospholipids contributing to cell structure, while nonpolar lipids, such as triacylglycerols (TAGs), serve as energy storage (Kwan et al., 2021). Microalgae play a crucial role in producing essential fatty acids, particularly LC-PUFAs like γ-linolenic acid, arachidonic acid, eicosapentaenoic acid (EPA), and docosahexaenoic acid (DHA). As humans and animals cannot synthesize these fatty acids, they must be sourced from the diet, predominantly through fish and fish oil. The high PUFA content in fish is largely attributed to their consumption of microalgae, highlighting the potential of microalgae as a sustainable alternative for PUFA production (Sathasivam et al., 2019). However, the high PUFA content also makes these oils prone to oxidation; research indicates that incorporating natural antioxidants can significantly improve oxidative stability (Bañares et al., 2022). Table 5.1 summarizes the fatty acid profiles of

Table 5.1 Fatty acid composition (%) of some microalgae species.

Microalgae species	Miristic acid (C14:0)	Palmitic acid (C16:0)	Palmitoleic acid (C16:1)	Stearic acid (C18:0)	Oleic acid (C18:1)	Linoleic acid (C18:2)	Alpha-linolenic acid (C18:3)	EPA (C20:5)	DHA (C22:6)	References
Crypthecodinium cohnii	28.23%	22.28%	0.56%	1.32%	6.73%	0.18%	-	0.02%	13.93%	Kose (2023)
Isochrysis galbana	9.5%	5.2%	10.7%	8.7%	0.2%	0.4%	9.5%	10.3%	33.6%	Suh et al. (2015)
Schizochytrium sp.	5.5%	19.1%	-	1.2%	16.2%	4.11%	-	-	38.61%	Wang et al. (2015)
Nannochloropsis oculata	0.6%	6.6%	9.0%	35.5%	0.6%	0.2%	29.1%	11.5%	3.6%	Suh et al. (2015)
Phaeodactylum tricornutum	5.5%	1.4%	18.6%	8.3%	0.8%	2.8%	-	52.6%	5.5%	Suh et al. (2015)
Tisochrysis lutea	16.8%	11.4%	8.3%	0.3%	7.7%	4.4%	5.5%	0.6%	11.1%	Mai et al. (2021)
Botryococcus braunii	-	11.32%	2.73%	2.15%	52.25%	6.69%	15.81%	2.72%	-	Cheng et al. (2013)
Chlorella vulgaris	0.3%	-	2.4	6.5%	4.1%	17.6%	26.3%	-	-	Conde et al. (2021)
Spirulina sp.	0.1%	38.6%	5.1%	5.4%	2.6%	21.4%	23.3%	-	-	Conde et al. (2021)
Haematococcus pluvialis	3.1%	4.38%	0.33%	2.16%	16.11%	7.04%	21.2%	3.54%	-	Scodelaro Bilbao et al. (2016)
Dunaliella salina	1.29%	12.16%	2.88%	3.64%	32.81%	12.14%	1.32%	-	4.09%	Almutairi (2020)
Porphyridium cruentum	0.22%	33.1%	4.21%	2.8%	1.71%	17.8%	-	13.15%	-	Russo et al. (2024)
Arthrospira platensis	3.26%	51.33%	4.83%	1.1%	5.23%	18.94%	4.19%	-	-	Aouir et al. (2017)

several microalgae species, highlighting the diversity of lipid compositions. This variation in lipid profiles further emphasizes the versatility of microalgae in various biotechnological and nutritional applications.

Microalgae, particularly species such as *Isochrysis galbana* and *P. tricornutum*, are known for their ability to produce high levels of omega-3 fatty acids, including EPA and DHA (Lu et al., 2024). This makes them a more sustainable source of these essential nutrients compared to animal-derived oils like cod liver oil, while also providing a direct source of EPA and DHA, unlike terrestrial plants which lack these crucial fatty acids (Katiyar & Arora, 2020). The presence of these essential fatty acids makes microalgae oils a desirable alternative to fish oils in dietary supplements and functional foods. A balanced intake of omega-3 and omega-6 is crucial, with imbalances potentially leading to health risks such as cardiovascular diseases, diabetes, and cancer. Deep-sea fish oil is a rich source of omega-3 fatty acids like EPA and DHA, but its use is limited by concerns over contamination with toxins such as mercury, environmental issues like overfishing, and consumer dislike for its strong odor (Fernandes & Cordeiro, 2021). Omega-3 fatty acids are vital for human health, as mammals cannot produce them and must obtain them from food sources. The main omega-3s include alpha-linolenic acid (ALA), EPA, and DHA, all of which provide significant health benefits, such as reducing inflammation and supporting heart and brain health (Mane et al., 2024). Furthermore, the antioxidant properties of carotenoids found in microalgae, such as astaxanthin and β-carotene, contribute additional health benefits, including antiinflammatory effects and potential cancer prevention (Ávila-Román et al., 2021). Their crucial roles in various physiological functions have made them a focus of interest within the food industry. EPA and DHA are often lacking in human diets, primarily sourced from marine algae and fish (Fawcett et al., 2022). Microalgae serve as a valuable source of marine lipids rich in unsaturated fatty acids, offering potential as a substitute for traditional fish oils. For example, *I. galbana* yields nearly double the amount of EPA compared to cod liver oil, while *C. cohnii* contains DHA at levels almost six times greater (Xue et al., 2021). Microalgae oils offer a higher content of PUFAs and a more appealing taste, making them an attractive option for vegetarians and expanding their potential use in the edible oil market (Xue et al., 2021). Vegetarians often have lower dietary levels of n-3 LC-PUFAs, while vegans typically have none. Therefore, alternative sources of these essential fatty acids are crucial for those following plant-based diets. Certain microalgae, including *B. braunii* and *Nannochloropsis salina*, are rich in saturated and monounsaturated fatty acids, while others like *Chlamydomonas mexicana* and *Pavlova lutheri* contain higher levels of PUFAs (Jadhav, 2024). This suggests that including microalgae oils in the diet may be a health-beneficial strategy, such as managing cardiovascular risk factors. Notably, species such as *P. tricornutum* and *Skeletonema* sp. are particularly promising for food applications due to their significant accumulation of EPA and DHA (Morais et al., 2021). In contrast, plant-based omega-3 sources like flaxseed, canola, and walnut oils contain ALA, which has a very low conversion rate to the more beneficial EPA and DHA in the human body (Katiyar & Arora, 2020).

The production of essential LC-PUFAs in microalgae is influenced by genetic factors and environmental conditions and integrating this knowledge can enhance the growth and lipid accumulation of specific strains (Morais et al., 2021). Microalgae can significantly increase their lipid content when exposed to various environmental stresses, such as strong light, extreme temperatures, and nutrient limitations, particularly nitrogen stress (Xue et al., 2020). Microalgae can increase lipid production under stress conditions because these factors stimulate cellular processes that promote lipid accumulation (Bibi et al., 2022). Additionally, genetic modifications can further improve lipid synthesis by enhancing the metabolic pathways that favor fatty acid and triglyceride production while reducing competing processes (Xue et al., 2020). Implementing techniques such as nitrogen deprivation and sodium chloride addition has shown effectiveness in increasing lipid concentrations in key microalgae species like *Scenedesmus* and *Chlorella* (Morais et al., 2021). However, it is difficult to achieve both high biomass productivity and high lipid content at the same time, as conditions that promote growth often lead to lower lipid levels, while strategies like nutrient starvation or reducing phosphorus can increase lipid content but significantly hinder growth due to reduced photosynthesis (Poh et al., 2020). A two-stage culture method is often employed, where microalgae first grow in nutrient-rich conditions to build biomass, followed by a second stage that promotes lipid accumulation under stress. A promising two-stage cultivation approach begins with ideal growth conditions followed by stress induction to boost lipid reserves. In contrast, to conventional stress-inducing methods during microalgae cultivation, postharvest stress conditions, such as nutrient starvation and potassium chloride addition under dark conditions, significantly enhanced lipid accumulation in *C. vulgaris*, achieving lipid yields as high as 40.28% (dry weight), with fatty acid profiles similar to those of traditional vegetable oils like corn, soybean, and sunflower (Poh et al., 2020). Increasing salinity in culture media, whether through seawater or added salt, significantly enhances lipid accumulation, particularly PUFAs, in microalgae. For instance, adding 50g/L NaCl to *C. vulgaris* cultures raised the lipid content to 28% (w/w), while higher concentrations of NaCl (60 to 400 mmol L) further increased lipid levels from 31.5% to 49.5% (w/w) (Morais et al., 2021).

5.3 Extraction methods for food-grade oils

The production of microalgae oils involves cultivating microalgae under controlled conditions and extracting the lipids from the algal cells. The extraction of food-grade oils from microalgae is a critical step in the production process, influencing both yield and quality. Microalgae biomass presents significant potential for oil extraction, yet various challenges hinder its technological development. To isolate valuable compounds like lipids, cell lysis is essential, which can be achieved through mechanical, chemical, or biological methods (Valdovinos-García et al., 2022). The extraction of food-grade oils from microalgae involves several steps, including cell disruption, oil recovery, and purification. The choice of extraction method depends on factors such as the species of microalgae, the desired oil yield, and the desired quality of the final product. The first step in extracting microalgae oils is to disrupt the cell walls to release the intracellular lipids. Once the cells have been disrupted, the oil can be recovered from the resulting slurry. The extracted oil may contain impurities, such as pigments, proteins, and other cellular components. Purification is necessary to remove these impurities and obtain a high-quality food-grade oil. The microalgae biorefinery sector encompasses cultivation, dewatering, lipid extraction and purification, and upgrading of fractions (Cvejic et al., 2019). Fig. 5.1 outlines the key stages involved in microalgae oil production, from the initial cultivation of microalgae to the final oil extraction and refining processes. These steps are essential for optimizing the yield and quality of oils that can be used in a variety of applications, including nutraceuticals, biofuels, and other industrial uses. Lipid extraction from microalgae typically follows two main approaches: the "dry route," which involves a dewatering process with centrifugation, mechanical dehydration, and thermal drying, and the "wet route," which bypasses drying and extracts lipids directly from the aqueous phase after cell disruption using methods like stirred ball mills and solvent extraction (Russell et al., 2022).

Several methods have been developed to efficiently extract lipids from microalgae biomass, each with its advantages and limitations. The most commonly used techniques include solvent extraction, supercritical fluid extraction, and mechanical pressing. The most common technique is organic solvent extraction, where solvents such as hexane and ethanol are utilized to dissolve and extract oils from disrupted algal cells. This method is straightforward and effective, but concerns regarding the toxicity and volatility of organic solvents necessitate the exploration of greener alternatives. The extraction of food-grade oils from microalgae has been the focus of

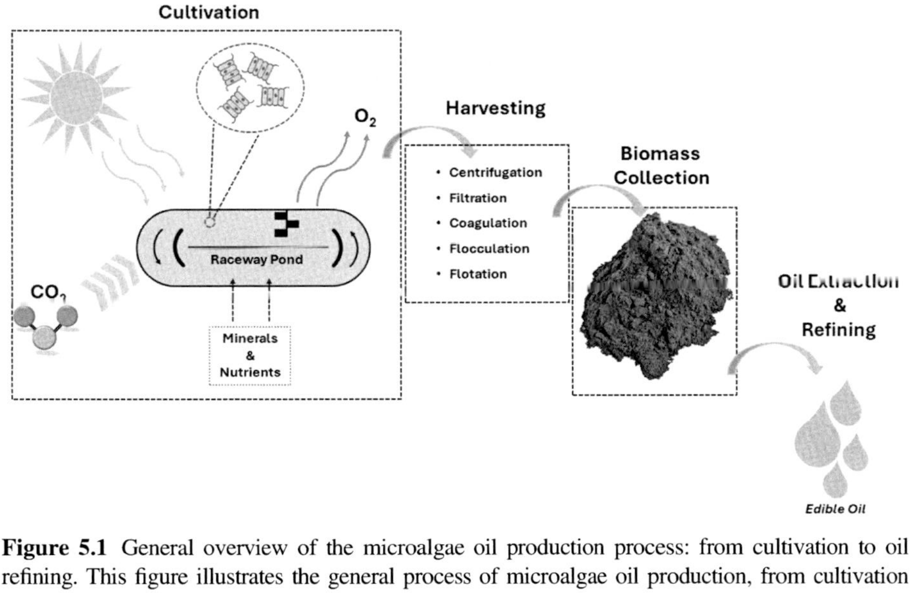

Figure 5.1 General overview of the microalgae oil production process: from cultivation to oil refining. This figure illustrates the general process of microalgae oil production, from cultivation through harvesting and oil extraction, to the final stage of oil refining.

various studies, exploring different methodologies to maximize oil yield while ensuring the quality of the final product. Conventional methods such as solvent extraction and mechanical pressing have proven effective, but innovative techniques such as supercritical fluid extraction and ultrasonic-assisted extraction are being developed to enhance oil recovery and preserve the integrity of sensitive compounds. These advancements are crucial in addressing the economic feasibility of microalgae oil production, which remains a barrier to widespread commercialization. The extraction of lipids from microalgae involves multiple processes with varying operating conditions, making it challenging to find a one-size-fits-all solution, as each species requires specific optimization for cultivation, pretreatment, extraction, and quantification (Russell et al., 2022).

Microalgae extracts contain a diverse range of lipid molecules, including long-chain hydrocarbons, wax esters, and free fatty acids, while vegetable oils are predominantly made up of TAGs. Typically, microalgae have lower TAG content, with their lipid profile consisting of about 53% TAGs, 29% phospholipids, and 16% glycolipids (Morais et al., 2021). Exploring effective extraction techniques is crucial for enhancing the yield and reducing costs of microalgae lipids while maintaining their bioactive properties, facilitating broader industrial applications. Chloroform/methanol offers the highest oil extraction efficiency; however, due to chloroform's carcinogenic properties, it is unsuitable for edible oil production. Low-toxicity solvents like ethanol, ethyl acetate, and n-hexane, including their combinations, are preferable for extracting oils from microalgae (Li et al., 2021). According to a study found that polar and nonpolar solvent mixtures enhance oil extraction from wet algal mud, while single solvents yield better results with dry algal powder, indicating the need for tailored solvent systems based on the algal form (Li et al., 2021). For effective edible oil production, a high level of neutral lipids is essential, as polar lipids can create insoluble contaminants that compromise oil quality and stability (Huang et al., 2016). Due to the presence of impurities such as chlorophyll and trace metals, crude microalgae extracts often require refining to enhance their quality for human consumption. This refining process involves several steps, including neutralization, degumming, bleaching, and dewaxing, to remove unwanted substances and improve the oil's characteristics (Morais et al., 2021).

Traditional techniques like Soxhlet extraction and Folch extraction, although cost-effective and straightforward, have significant drawbacks, including high organic reagent consumption, environmental pollution, and lengthy extraction times (Zhou et al., 2022). Supercritical fluid extraction, particularly using supercritical CO_2, has gained popularity due to its efficiency and environmental safety. Supercritical fluid extraction can selectively extract high-value lipids while minimizing degradation, making it suitable for food-grade applications (Molino et al., 2020). Supercritical fluid extraction, particularly using CO_2, has gained traction for its ability to extract oils without residual solvents, which is particularly important for food-grade applications (Jiménez Callejón et al., 2022). Additionally, enzyme-assisted extraction methods are being investigated, where enzymes such as cellulase are employed to facilitate cell wall disruption, enhancing lipid recovery without the need for harsh chemicals (Mat Husin et al., 2024). Emerging extraction techniques, such as enzymatic extraction and microwave-assisted extraction, are also being investigated for their potential to enhance oil yield and quality while minimizing environmental impact. Overall, while traditional methods like solvent extraction remain prevalent, advancements in green technologies and enzymatic approaches are paving the way for more sustainable and efficient extraction processes for food-grade oils from microalgae. The extraction method is an important factor affecting both the quality of microalgae oils and their potential application in food products, and further research is needed to optimize these methods for commercial feasibility. The choice of extraction method for microalgae oils depends on various factors, including the desired oil yield, quality, and cost. By optimizing the extraction process, it is possible to obtain high-quality food-grade oils from microalgae.

5.4 Applications of microalgae oils in nutraceuticals and food products

Microalgae oils have emerged as a significant source of nutraceuticals, particularly due to their rich content of omega-3 LC-PUFAs, such as EPA and DHA. These oils are increasingly utilized as alternatives to fish oils in various health products, including dietary supplements and infant formula (Liu et al., 2022). The polar lipid fraction of microalgae oils, which includes essential fatty acids, has been shown to enhance bioavailability and stability, making these oils particularly valuable in functional food applications (Jiménez Callejón et al., 2020). The incorporation of microalgae oils into food products such as beverages, snacks, and fortified foods is becoming more common, with studies indicating that these oils can significantly improve the nutritional profile of conventional products (Sidari & Tofalo, 2019). As shown in Fig. 5.2, microalgae oils are utilized across a wide range of industries due to their rich content of omega-3 fatty acids and other bioactive compounds. These oils are applied in sectors such as nutraceuticals, food, cosmetics, pharmaceuticals, biofuels, aquaculture, agriculture, and industrial applications, offering sustainable solutions for these diverse fields.

Microalgae, often referred to as "green gold," offer a wide range of sustainable applications, including as a source of plant protein, animal feed, health supplements, and valuable compounds like carotenoids and omega-3 fatty acids (Kwan et al., 2021). Microalgae are a promising source of bioactive secondary metabolites, including PUFAs, proteins, vitamins, and polysaccharides, which have significant applications in pharmaceuticals, public health, and nutrition, particularly in the prevention of noncommunicable diseases (Katiyar & Arora, 2020). The variety of food-grade microalgae is expanding as new strains are evaluated each year. In Europe, foods not significantly consumed before May 15, 1997 are classified as Novel Food under regulation (Reg. 2015/2283), which took effect on January 1, 2018 (Cruz & Vasconcelos, 2024). Prior to this date, only a few microalgae, including *Arthrospira* and *Chlorella*, were considered food grade, while products like PUFA-rich oils from *Schizochytrium* and *Ulkenia*, as well as astaxanthin-rich oleoresin from *Haematococcus*, were later recognized in the EU novel food catalog (Cvejic et al.,

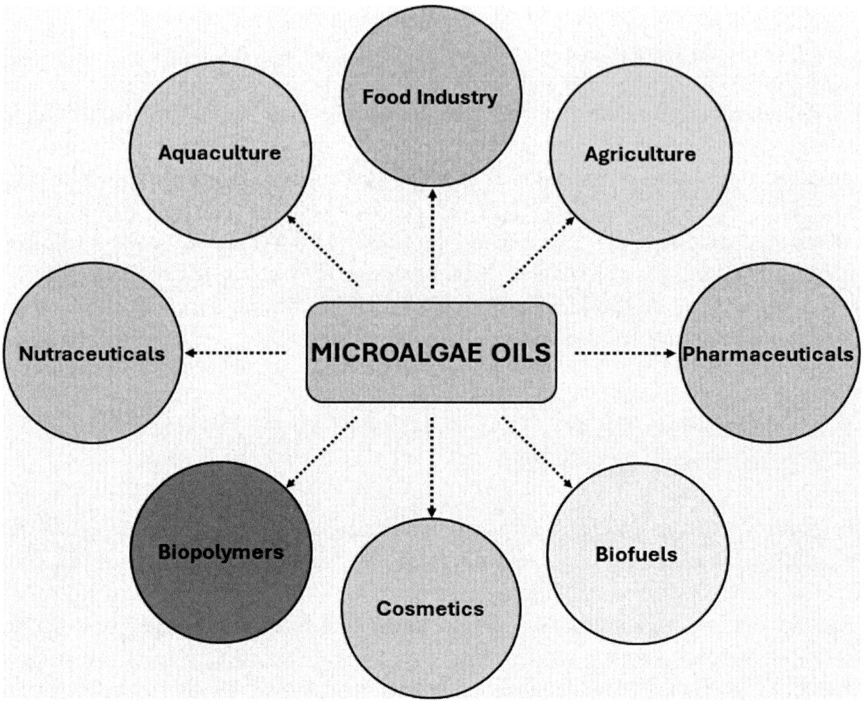

Figure 5.2 Overview of key sectors utilizing microalgae oils in various applications. This figure provides an overview of the key sectors utilizing microalgae oils in various applications, highlighting their potential use in areas such as agriculture, food, and biofuels.

2019). Microalgae such as *Spirulina* and *Chlorella* are produced in large quantities worldwide for use as nutrient-rich supplements, while specific strains like *Crypthecodinium* and *Schizochytrium* are increasingly utilized in the production of omega-3 enriched products, including infant milk formulas and "Omega" eggs, as well as in the aquaculture and feed industries (Katiyar & Arora, 2020). Several companies, including Oceans Alive (USA), Blue Biotech (Germany), and Flora Health (USA), are already producing omega-3 PUFAs from microalgae for use as dietary supplements or food ingredients (Fernandes & Cordeiro, 2021). The influence of microalgae oils on flavor and the potential negative effects of rare fatty acids in their profiles must be thoroughly investigated prior to implementation (Draaisma et al., 2013). Although most EPA and DHA are derived from fish, the share of microalgae oils is increasing, driven by consumer demand for options that are less fishy, free from harmful contaminants, and suitable for vegetarian and vegan labeling. Major companies in the microalgae EPA/DHA market include DSM, Lonza, SourceOmega, Qualitas Health, and Fermentalg, with algae-based DHA primarily used in infant formulas and dietary supplements (Cvejic et al., 2019). Martek, USA, is a leading producer of DHA derived from *C. cohnii*, while *Nannochloropsis* shows potential for

EPA production, suitable for use in functional foods and animal feeds (Udayan et al., 2021).

Microalgae are increasingly used in the food industry, with various value-added products such as capsules, powders, and liquids being marketed to provide essential functional lipids and proteins, while also being incorporated into a wide range of food items like noodles, snacks, beverages, pasta, and confectioneries to enhance their nutritional value (Katiyar et al., 2021). Algal oils are increasingly being incorporated into foodstuffs, providing omega-3 PUFAs without fish, making them attractive for vegan and organic markets. Additionally, key companies have developed algae-based oil concentrates omega-3 content, catering to both the nutraceutical and pharmaceutical markets, with products available in infant formulas (Liu et al., 2022). However, despite their rich nutritional profile, some microalgae oils face challenges in food applications due to color and flavor changes, which can affect consumer acceptance (Katiyar & Arora, 2020). In particular, the unpleasant sensory properties of microalgae products pose a significant challenge to their commercialization and further research is needed to explore these properties and their impact on food quality (Hosseinkhani et al., 2022).

5.5 Regulatory and safety considerations

The use of microalgae oils in food products and dietary supplements presents promising opportunities; however, concerns regarding their safety, bioavailability, and regulatory compliance remain contentious (Udayan et al., 2021). These concerns primarily revolve around potential toxicity, contamination risks, and the need for rigorous safety assessments to ensure these products are safe for human consumption. As microalgae-based oils continue to gain attention in the food industry, a comprehensive understanding of their safety profile is crucial. Safety evaluations of microalgae oils should thoroughly address several forms of toxicity, including mutagenicity, systemic toxicity, and reproductive toxicity (Xue et al., 2020). Additionally, risks associated with the presence of heavy metals and pathogenic microorganisms due to production processes must be carefully assessed. Such evaluations are essential for ensuring that these oils do not pose health risks. For instance, certain microalgae species, such as *Spirulina*, *Chlorella*, *Dunaliella*, *Haematococcus*, *Schizochytrium*, *P. cruentum*, *Prototheca moriformis*, *C. cohnii*, *Chlamydomonas reinhardtii*, and *Euglena gracilis*, have been classified as GRAS by the U.S. Food and Drug Administration (FDA), indicating their safety for human consumption (Yang et al., 2024). Despite this recognition, more extensive toxicological studies are needed to establish a broader safety profile for microalgae oils. The European Union's "Novel Foods" regulation serves as a critical framework to guarantee that microalgae oils meet safety and quality standards prior to commercialization (Cruz & Vasconcelos, 2024). In Europe, the novel foods regulation mandates that any new food source undergo thorough safety assessments before being introduced to the market. These evaluations must identify any undesirable substances that may arise during production, ensuring that the final product is free from contaminants that could harm consumers. Regulatory frameworks play a pivotal role in ensuring that these products are safe and effective for human consumption. A well-defined regulatory process ensures that microalgae oils meet both safety and quality standards, ultimately fostering consumer trust and encouraging the growth of the microalgae-based food sector. Although food safety assessments of some microalgae products and strains show promise, further data and in-depth studies are required to strengthen the safety profile of these oils.

5.6 Economic and environmental considerations

The production of food-grade oils from microalgae presents significant economic and environmental opportunities, positioning them as sustainable alternatives to traditional vegetable oils. Economically, microalgae oils can be advantageous due to their high lipid content and rapid growth rates, which can lead to efficient oil yields. Studies have shown that integrating microalgae cultivation with wastewater treatment can provide a dual benefit: cleaning polluted water while simultaneously producing valuable oils (Udayan et al., 2023). Furthermore, microalgae require less land and freshwater than traditional crops, making them an attractive option for sustainable food production, particularly in regions facing water scarcity (Fawcett et al., 2022). However, the economic feasibility of microalgae oil production is often hindered by high cultivation and extraction costs. Microalgae oils are a promising alternative to fish oil for consumers looking for natural food options, but their production cost of about €8.30 per kg is much higher than the €1.00 per kg for vegetable oils (Xue et al., 2021). By comparison, omega-3 soft-gel capsules derived from algae are priced around €1212 per kg, significantly higher than fish oil at €294 per kg, though lower than krill oil, which is priced at €3523 per kg (Cvejic et al., 2019). Advances in biotechnological methods and economies of scale are essential to reduce these costs and make microalgae oil more competitive with conventional oils. The industrial-scale production of microalgae relies on optimizing biomass productivity, with strategies including adjustments to cultivation conditions, strain selection, and genetic modification to enhance output and reduce costs (Fernandes & Cordeiro, 2021)

Microalgae cultivation for oil production offers several environmental benefits. Through photosynthesis, microalgae absorb CO_2, which not only supports their growth but also reduces atmospheric CO_2 levels. This characteristic makes them an environmentally sustainable source for producing edible oils. Microalgae are considered "sustainable biofactories" because of their ability to reduce atmospheric CO_2 levels while producing a wide range of valuable compounds, though their effective industrial use requires selecting the appropriate species and optimizing growth conditions (Fernandes & Cordeiro, 2021). Their ability to absorb CO_2 during photosynthesis makes them a promising solution for reducing greenhouse gases, while their biomass can be utilized for biofuels, biogas, and other eco-friendly products, enhancing the economic viability of algae-based industries (Kwan et al., 2021). Microalgae, with their high CO_2 fixation efficiency, play a crucial role in addressing climate change, as they can fix significant amounts of CO_2, with estimates suggesting that 280 tons of microalgae dry biomass per hectare per year can capture approximately 513 tons of CO_2 (Fernandes & Cordeiro, 2021). Additionally, microalgae-based wastewater treatment offers a cost-effective and environmentally friendly alternative to conventional methods, capable of efficiently removing nutrients like nitrogen and phosphorus (Han et al., 2021). Beyond this, the process also holds potential for producing valuable lipids that can be used for biofuel production. Several studies have investigated the use of wastewater from various sources—such as municipal, industrial, and agricultural effluents—for microalgae cultivation focused on lipid production. However, optimizing this process remains challenging due to factors like fluctuating wastewater composition and the need for pretreatment to enhance microalgae growth (Fernandes & Cordeiro, 2021).

5.7 Challenges and future perspectives

As the global demand for sustainable, plant-based oils continues to rise, microalgae oils are gaining recognition as a viable alternative to fish oils, particularly for providing plant-based sources of omega-3 fatty acids. Microalgae offer a vegan-friendly solution to the challenges posed by overfishing and the depletion of marine ecosystems. Furthermore, microalgae-derived omega-3s hold potential for combating malnutrition and improving public health outcomes globally (Katiyar & Arora, 2020). These oils not only offer a vegan-friendly alternative to fish-derived omega-3 sources but also have the potential to reduce the environmental impact associated with overfishing and the depletion of marine ecosystems.

Despite their potential, there are significant hurdles to overcome before microalgae oils can become a commercially viable alternative to conventional oil sources. One of the key challenges is the economic viability of large-scale microalgae oil production. The cultivation process for microalgae requires controlled environments, which can be energy-intensive and costly. Photobioreactors, often used to cultivate algae, demand substantial investments in infrastructure, and scaling these systems for industrial use remains a significant hurdle (Katiyar & Arora, 2020). Additionally, harvesting and lipid extraction processes are relatively inefficient and more expensive compared to traditional oil crops, such as soybeans or palm oil. These high production costs, combined with relatively low biomass productivity for many microalgae strains, hinder the economic feasibility of mass production (Torres-Tiji et al., 2020). In addition, to economic challenges, scaling up production presents significant technical hurdles. Transitioning from laboratory-scale experiments to industrial-scale production introduces complexities related to maintaining optimal growth conditions, ensuring consistent oil quality, and optimizing yield (Hossain et al., 2021). Large-scale production requires substantial infrastructure, such as photobioreactors or large open-pond systems, as well as highly trained personnel to manage the process. Regulatory hurdles also add to the complexity, as microalgae oils must meet strict food safety and quality standards before they can be used in consumer products. Moreover, the limited market demand for microalgae oils, combined with the high costs associated with large-scale production, makes it difficult to justify the infrastructure investments needed to scale up production.

Another obstacle is the chemical instability of microalgae oils, particularly due to their high levels of PUFAs. While these PUFAs are valuable from a nutritional standpoint, they are highly susceptible to oxidation, which leads to reduced shelf life and diminished oil quality. This oxidation not only impacts the nutritional profile of the oil but also results in the formation of off-flavors and odors, which can be off-putting in food applications (Siddiqui et al., 2024). As a result, enhancing the oxidative stability of microalgae oils is a critical area of ongoing research. Strategies such as modifying the lipid composition of algae strains or developing new preservation techniques could help mitigate this issue. Additionally, sensory properties, including the strong color and fishy odor that are often associated with certain microalgae oils, present further challenges for their incorporation into food products (Morais et al., 2021). These sensory attributes may limit the oils' appeal to consumers and make it more difficult for manufacturers to integrate them into a variety of food formulations. Overcoming these sensory challenges will be key to expanding the market for microalgae oils in food products, particularly in mainstream consumer markets.

Despite these challenges, the future potential of microalgae oils remains strong. Researchers are exploring various solutions to improve the efficiency and cost-effectiveness of production. Advances in biotechnology, including genetic modification and selective breeding of high-yielding algae strains, are helping to increase lipid content and improve oil quality. Moreover, ongoing developments in cultivation methods, such as optimizing light, nutrient, and temperature conditions, are expected to improve biomass productivity and, consequently, oil yield. Additionally, innovations in extraction technologies, such as enzymatic and solvent-free methods, hold promise for reducing production costs while maintaining high oil quality (Jeevan Kumar et al., 2017). In conclusion, while significant challenges remain in terms of cost, scalability, and stability, ongoing advancements in production technologies and strain optimization offer promising solutions. With continued research and innovation, microalgae oils have the potential to become a sustainable and economically viable alternative to traditional oils, making them a valuable component of future food systems.

5.8 Conclusion

Microalgae oils represent a promising alternative to traditional oil sources, offering a wealth of benefits for both health and sustainability. Their rich composition of essential fatty acids, vitamins, and bioactive compounds positions them as an ideal candidate for use in nutraceuticals, functional foods, and dietary supplements. However, despite their potential, the economic and technical barriers to large-scale production remain significant. High cultivation costs, challenges in scaling up production, and issues related to oil stability and sensory properties are key obstacles that must be overcome. Ongoing research into strain optimization, cultivation methods, and more efficient extraction techniques is essential to improving the viability of microalgae oils as a mainstream food ingredient. As the global demand for sustainable, plant-based omega-3 sources grows, microalgae-derived oils could play a critical role in meeting these needs, especially for individuals seeking vegan or environmentally friendly alternatives to fish oils. If these challenges can be addressed, microalgae oils have the potential to revolutionize the food industry, offering a sustainable, nutritious, and economically feasible source of oils for future generations.

References

Almutairi, A. W. (2020). Effects of nitrogen and phosphorus limitations on fatty acid methyl esters and fuel properties of *Dunaliella salina*. *Environmental Science and Pollution Research, 27*(26), 32296–32303. https://doi.org/10.1007/s11356-020-08531-8.

Aouir, A., Amiali, M., Bitam, A., Benchabane, A., & Raghavan, V. G. (2017). Comparison of the biochemical composition of different *Arthrospira platensis* strains from Algeria, Chad and the USA. *Journal of Food Measurement and Characterization, 11*(2), 913–923. https://doi.org/10.1007/s11694-016-9463-4.

Ávila-Román, J., García-Gil, S., Rodríguez-Luna, A., Motilva, V., & Talero, E. (2021). Antiinflammatory and anticancer effects of microalgal carotenoids. *Marine Drugs, 19*(10), 531. https://doi.org/10.3390/md19100531.

Bañares, C., Chabni, A., Reglero, G., & Torres, C. F. (2022). Oxidative stability of microalgae oil and its acylglycerol mixture obtained by enzymatic glycerolysis and the antioxidant effect of supercritical rosemary extract. *LWT, 171*, 114150. https://doi.org/10.1016/j.lwt.2022.114150.

Bibi, F., Jamal, A., Huang, Z., Urynowicz, M., & Ishtiaq Ali, M. (2022). Advancement and role of abiotic stresses in microalgae biorefinery with a focus on lipid production. *Fuel, 316*, 123192. https://doi.org/10.1016/j.fuel.2022.123192.

Cheng, P., Ji, B., Gao, L., Zhang, W., Wang, J., & Liu, T. (2013). The growth, lipid and hydrocarbon production of *Botryococcus braunii* with attached cultivation. *Bioresource Technology, 138*, 95–100. https://doi.org/10.1016/j.biortech.2013.03.150.

Conde, T. A., Neves, B. F., Couto, D., Melo, T., Neves, B., Costa, M., Silva, J., Domingues, P., & Domingues, M. R. (2021). Microalgae as sustainable bio-factories of healthy lipids: Evaluating fatty acid content and antioxidant activity. *Marine Drugs, 19*(7), 357. https://doi.org/10.3390/md19070357.

Cruz, J. D., & Vasconcelos, V. (2024). Legal aspects of microalgae in the European food sector. *Foods, 13*(1), 124. https://doi.org/10.3390/foods13010124.

Cvejic, J. H., Langellotti, A. L., Bonnefond, H., Verardo, V., & Bernard, O. (2019). Microalgae as a source of edible oils. In *Lipids and edible oils: Properties, processing and applications*. Elsevier, 175–211. 10.1016/B978-0-12-817105-9.00005-7.

Draaisma, R. B., Wijffels, R. H., Slegers, P. M., Brentner, L. B., Roy, A., & Barbosa, M. J. (2013). Food commodities from microalgae. *Current Opinion in Biotechnology, 24*(2), 169–177. https://doi.org/10.1016/j.copbio.2012.09.012.

Fawcett, C. A., Senhorinho, G. N. A., Laamanen, C. A., & Scott, J. A. (2022). Microalgae as an alternative to oil crops for edible oils and animal feed. *Algal Research, 64*, 102663. https://doi.org/10.1016/j.algal.2022.102663.

Fernandes, T., & Cordeiro, N. (2021). Microalgae as sustainable biofactories to produce high-value lipids: Biodiversity, exploitation, and biotechnological applications. *Marine Drugs, 19*(10), 573. https://doi.org/10.3390/md19100573.

Gharib, A., Djema, W., Moñino Fernández, P., Chin-On, R., Janssen, M., & Bernard, O. (2024). Validation of an adaptive temperature model for closed microalgae cultivation systems. *Algal Research, 86*, 103838. https://doi.org/10.1016/j.algal.2024.103838.

Han, W., Jin, W., Li, Z., Wei, Y., He, Z., Chen, C., Qin, C., Chen, Y., Tu, R., & Zhou, X. (2021). Cultivation of microalgae for lipid production using municipal wastewater. *Process Safety and Environmental Protection, 155*, 155–165. https://doi.org/10.1016/j.psep.2021.09.014.

Hossain, N., Chowdhury, T., Chowdhury, H., Ahmed, A., Nizamuddin, S., Griffin, G., Mahlia, T. M. I., & Park, Y.-K. (2021). *Edible bio-oil production from microalgae and application of nano-technology*. Elsevier BV, 91–116. 10.1016/b978-0-12-821218-9.00004-9.

Hosseinkhani, N., McCauley, J. I., & Ralph, P. J. (2022). Key challenges for the commercial expansion of ingredients from algae into human food products. *Algal Research, 64*, 102696. https://doi.org/10.1016/j.algal.2022.102696.

Huang, Y., Zhang, D., Xue, S., Wang, M., & Cong, W. (2016). The potential of microalgae lipids for edible oil production. *Applied Biochemistry and Biotechnology, 180*(3), 438–451. https://doi.org/10.1007/s12010-016-2108-6.

Jadhav, H. B. (2024). Functional triacylglycerols from microalgae and their use in the formulation of functional foods—Review. *Food Chemistry Advances, 4*, 100695. https://doi.org/10.1016/j.focha.2024.100695.

Jeevan Kumar, S. P., Vijay Kumar, G., Dash, A., Scholz, P., & Banerjee, R. (2017). Sustainable green solvents and techniques for lipid extraction from microalgae: A review. *Algal Research, 21*, 138–147. https://doi.org/10.1016/j.algal.2016.11.014.

Jiménez Callejón, M. J., Robles Medina, A., González Moreno, P. A., Esteban Cerdán, L., Orta Guillén, S., & Molina Grima, E. (2020). Simultaneous extraction and fractionation of lipids from the microalga Nannochloropsis sp. for the production of EPA-rich polar lipid concentrates. *Journal of Applied Phycology, 32*(2), 1117–1128. https://doi.org/10.1007/s10811-020-02037-z.

Jiménez Callejón, M. J., Robles Medina, A., Macías Sánchez, M. D., González Moreno, P. A., Navarro López, E., Esteban Cerdán, L., & Molina Grima, E. (2022). Supercritical fluid extraction and pressurized liquid extraction processes applied to eicosapentaenoic acid-rich polar lipid recovery from the microalga Nannochloropsis sp. *Algal Research, 61*, 102586. https://doi.org/10.1016/j.algal.2021.102586.

Katiyar, R., & Arora, A. (2020). Health promoting functional lipids from microalgae pool: A review. *Algal Research, 46*, 101800. https://doi.org/10.1016/j.algal.2020.101800.

Katiyar, R., Banerjee, S., & Arora, A. (2021). Recent advances in the integrated biorefinery concept for the valorization of algal biomass through sustainable routes. *Biofuels, Bioproducts and Biorefining, 15*(3), 879–898. https://doi.org/10.1002/bbb.2187.

Kose, A. (2023). Chemical composition and tyrosinase inhibitory activities of fatty acids obtained from heterotrophic microalgae, *S. limacinum* and *C. cohnii*. *Applied Biochemistry and Biotechnology, 195*(1), 369–385. https://doi.org/10.1007/s12010-022-04143-9.

Kwan, P. P., Banerjee, S., Shariff, M., & Yusoff, F. Md (2021). Influence of light on biomass and lipid production in microalgae cultivation. *Aquaculture Research, 52*(4), 1337–1347. https://doi.org/10.1111/are.15023.

Li, C. J., Xin, M. R., & Sun, Z. L. (2021). Selection of extraction solvents for edible oils from microalgae and improvement of the oxidative stability. *Journal of Bioscience and Bioengineering, 132*(4), 365–371. https://doi.org/10.1016/j.jbiosc.2021.06.008.

Liu, Y., Ren, X., Fan, C., Wu, W., Zhang, W., & Wang, Y. (2022). Health benefits, food applications, and sustainability of microalgae-derived N-3 PUFA. *Foods, 11*(13), 1883. https://doi.org/10.3390/foods11131883.

Lu, X., Yang, S., He, Y., Zhao, W., Nie, M., & Sun, H. (2024). Nutritional value and productivity potential of the marine microalgae nitzschia laevis, *Phaeodactylum tricornutum* and Isochrysis galbana. *Marine Drugs, 22*(9), 386. https://doi.org/10.3390/md22090386.

Mai, T. D., Lee-Chang, K. J., Jameson, I. D., Hoang, T., Cai, N. B. A., & Pham, H. Q. (2021). Fatty acid profiles of selected microalgae used as live feeds for shrimp postlarvae in Vietnam. *Aquaculture Journal, 1*(1), 26–38. https://doi.org/10.3390/aquacj1010004.

Mane, S., Kumari, P., Singh, A., Taneja, N. K., & Chopra, R. (2024). Amelioration for oxidative stability and bioavailability of N-3 PUFA enriched microalgae oil: An overview. *Critical Reviews in Food Science and Nutrition, 64*(9), 2579–2600. https://doi.org/10.1080/10408398.2022.2124505.

Mat Husin, M. A., Mohd Yasin, N. H., Takriff, M. S., & Jamar, N. H. (2024). A review on pretreatment methods for lipid extraction from microalgae biomass. *Preparative Biochemistry and Biotechnology, 54*(2), 159–174. https://doi.org/10.1080/10826068.2023.2214923.

Molino, A., Mehariya, S., Di Sanzo, G., Larocca, V., Martino, M., Leone, G. P., Marino, T., Chianese, S., Balducchi, R., & Musmarra, D. (2020). Recent developments in supercritical fluid extraction of bioactive compounds from microalgae: Role of key parameters, technological achievements and challenges. *Journal of CO2 Utilization, 36*, 196–209. https://doi.org/10.1016/j.jcou.2019.11.014.

Morais, E. G., Schüler, L., Pereira, H., Maia, I., Gangadhar, K. N., Costa, J. A. V., Morais, M. G., Varela, J., & Barreira, L. (2021). Microalgae as source of edible lipids. In *Cultured microalgae for the food industry: Current and potential applications*. Elsevier, 147–175. 10.1016/B978-0-12-821080-2.00015-0.

Poh, Z. L., Amalina Kadir, W. N., Lam, M. K., Uemura, Y., Suparmaniam, U., Lim, J. W., Show, P. L., & Lee, K. T. (2020). The effect of stress environment towards lipid accumulation in microalgae after harvesting. *Renewable Energy, 154*, 1083–1091. https://doi.org/10.1016/j.renene.2020.03.081.

Rehman, M., Kesharvani, S., Dwivedi, G., & Gidwani Suneja, K. (2022). Impact of cultivation conditions on microalgae biomass productivity and lipid content. *Materials Today: Proceedings, 56*, 282–290. https://doi.org/10.1016/j.matpr.2022.01.152.

Russell, C., Rodriguez, C., & Yaseen, M. (2022). Microalgae for lipid production: Cultivation, extraction & detection. *Algal Research, 66*, 102765. https://doi.org/10.1016/j.algal.2022.102765.

Russo, G. L., Langellotti, A. L., Verardo, V., Garcia, B. M., Oliviero, M., & Masi, P. (2024). Sustainable cultivation of *Porphyridium cruentum* via agro-industrial by-products: A study on biomass and lipid enhancement. *Biocatalysis and Agricultural Biotechnology, 60*, 103341. https://doi.org/10.1016/j.bcab.2024.103341.

Sajjadi, B., Chen, W. Y., Raman, A. A. A., & Ibrahim, S. (2018). Microalgae lipid and biomass for biofuel production: A comprehensive review on lipid enhancement strategies and their effects on fatty acid composition. *Renewable and Sustainable Energy Reviews, 97*, 200–232. https://doi.org/10.1016/j.rser.2018.07.050.

Sarma, S., Sharma, S., Rudakiya, D., Upadhyay, J., Rathod, V., Patel, A., & Narra, M. (2021). Valorization of microalgae biomass into bioproducts promoting circular bioeconomy: A holistic approach of bioremediation and biorefinery. *3 Biotech, 11*(8). https://doi.org/10.1007/s13205-021-02911-8.

Sathasivam, R., Radhakrishnan, R., Hashem, A., & Abd_Allah, E. F. (2019). Microalgae metabolites: A rich source for food and medicine. *Saudi Journal of Biological Sciences, 26*(4), 709–722. https://doi.org/10.1016/j.sjbs.2017.11.003.

Scodelaro Bilbao, P. G., Damiani, C., Salvador, G. A., & Leonardi, P. (2016). Haematococcus pluvialis as a source of fatty acids and phytosterols: Potential nutritional and biological implications. *Journal of Applied Phycology, 28*(6), 3283–3294. https://doi.org/10.1007/s10811-016-0899-z.

Sidari, R., & Tofalo, R. (2019). A comprehensive overview on microalgal-fortified/based food and beverages. *Food Reviews International, 35*(8), 778–805. https://doi.org/10.1080/87559129.2019.1608557.

Siddiqui, S. A., Ucak, İ., Afreen, M., Sasidharan, A., Yunusa, B. M., Bhowmik, S., Pandiselvam, R., Ambartsumov, T. G., & Shah, M. A. (2024). Microalgae as a potential raw material for plant-based seafood alternatives: A comprehensive review. *Food Science and Nutrition, 12*(11), 8559–8593. https://doi.org/10.1002/fsn3.4313.

Su, M., Bastiaens, L., Verspreet, J., & Hayes, M. (2023). Applications of microalgae in foods, pharma and feeds and their use as fertilizers and biostimulants: Legislation and regulatory aspects for consideration. *Foods, 12*(20), 3878. https://doi.org/10.3390/foods12203878.

Suh, S. S., Kim, S. J., Hwang, J., Park, M., Lee, T. K., Kil, E. J., & Lee, S. (2015). Fatty acid methyl ester profiles and nutritive values of 20 marine microalgae in Korea. *Asian Pacific Journal of Tropical Medicine, 8*(3), 191–196. https://doi.org/10.1016/S1995-7645(14)60313-8.

Torres-Tiji, Y., Fields, F. J., & Mayfield, S. P. (2020). Microalgae as a future food source. *Biotechnology Advances, 41*. https://doi.org/10.1016/j.biotechadv.2020.107536.

Udayan, A., Pandey, A. K., Sharma, P., Sreekumar, N., & Kumar, S. (2021). Emerging industrial applications of microalgae: Challenges and future perspectives. *Systems Microbiology and Biomanufacturing, 1*(4), 411–431. https://doi.org/10.1007/s43393-021-00038-8.

Udayan, A., Pandey, A. K., Sirohi, R., Sreekumar, N., Sang, B. I., Sim, S. J., Kim, S. H., & Pandey, A. (2023). Production of microalgae with high lipid content and their potential as sources of nutraceuticals. *Phytochemistry Reviews, 22*(4), 833–860. https://doi.org/10.1007/s11101-021-09784-y.

Valdovinos-García, E. M., Bravo-Sánchez, M. G., Olán-Acosta, MdlÁ., Barajas-Fernández, J., Guzmán-López, A., & Petriz-Prieto, M. A. (2022). Technoeconomic evaluation of microalgae oil production: Effect of cell disruption method. *Fermentation, 8*(7). https://doi.org/10.3390/fermentation8070301.

Wang, J., Wang, X. D., Zhao, X. Y., Liu, X., Dong, T., & Wu, F. A. (2015). From microalgae oil to produce novel structured triacylglycerols enriched with unsaturated fatty acids. *Bioresource Technology, 184*, 405–414. https://doi.org/10.1016/j.biortech.2014.09.133.

Xue, Z., Li, S., Yu, W., Gao, X., Zheng, X., Yu, Y., & Kou, X. (2021). Research advancement and commercialization of microalgae edible oil: A review. *Journal of the Science of Food and Agriculture, 101*(14), 5763–5774. https://doi.org/10.1002/jsfa.11390.

Xue, Z., Yu, Y., Yu, W., Gao, X., Zhang, Y., & Kou, X. (2020). Development prospect and preparation technology of edible oil from microalgae. *Frontiers in Marine Science, 7*. https://doi.org/10.3389/fmars.2020.00402.

Yang, S., Xu, H., Chen, J. H., Liu, B., & Cheng, K. W. (2024). Microalgae as a sustainable protein source: Key issues related to their production, application, and the way forward. *Food and Bioprocess Technology, 17*(8), 1–33. https://doi.org/10.1007/s11947-023-03194-y.

Zhou, J., Wang, M., Saraiva, J. A., Martins, A. P., Pinto, C. A., Prieto, M. A., Simal-Gandara, J., Cao, H., Xiao, J., & Barba, F. J. (2022). Extraction of lipids from microalgae using classical and innovative approaches. *Food Chemistry, 384*. https://doi.org/10.1016/j.foodchem.2022.132236.

Downstream processes for yeast microbial oils extraction

6

Marta de Vicente[1], *Cristina González-Fernández*[1,2,3], *and Elia Tomás-Pejó*[1]
[1]Biotechnological Processes Unit, IMDEA Energy, Móstoles (Madrid), Spain, [2]Department of Chemical Engineering and Environmental Technology School of Industrial Engineering, University of Valladolid, Valladolid, Spain, [3]Institute of Sustainable Processes, Valladolid, Spain

6.1 Introduction

Considering the need to find cost-effective bio-based alternatives to fossil fuels and petrochemicals, microbial oils (MO) have emerged as promising sources of bioproducts for many industrial sectors (Morales-Palomo et al., 2022). Up to now, plants and animal fats have been the primary source of oils but their competition with agriculture and livestock has recently raised the interest in alternative sources of lipids. MO, also known as single cell oils (SCO), are accumulated inside the microorganisms and are mainly formed by fatty acids (FAs), including acylglycerols [mono-, di-, and triacylglycerols (TAG)], free FAs, sterols, and pigments (Ghazani & Marangoni, 2022). In the microorganism cell, lipids play an important role as energy store biomolecules, intervene in signaling, and act as structural components of cell membranes (Saini et al., 2021).

MO have some advantages when compared to oils derived from plants or animals. Microorganisms growth is less dependent on environmental conditions, they can grow on unsuitable lands for plants, have shorter life cycle and can achieve higher lipid yields (Wang et al., 2021). MO also display similar FA profile than plant oils, which makes them a feasible substitute of oils derived from plants (Morales-Palomo et al., 2022; Tomás-Pejó et al., 2021).

MO can be produced by many microorganisms such as bacteria, microalgae, and fungi (Zainuddin et al., 2021). Those that accumulate more than 20% of their dry weight in the form of SCO are considered oleaginous (Khot et al., 2020; Salvador López et al., 2022). Some microorganisms can even accumulate up to 70–80% w/w of lipids in their cells (Ma et al., 2018). In the particular case of yeast, some strains are able to achieve high lipid yields from waste-derived carbon sources, reaching more than 0.26 g/g when grown sugar-based feedstock, equivalent to 80 % of the theoretical value (Qiao et al., 2015). Up to 0.36 g/g have been accumulated when using short-chain fatty acids (SCFAs) as carbon source (Tomás-Pejó et al., 2023).

When comparing all oleaginous microorganisms, yeast offer significant advantages due to their fast growth rate and their ability to convert waste-derived carbon sources (glycerol, short chain fatty acids (SCFAs) and lignocellulosic sugars) into lipids at high yields. The recovery of MO is also eased by the thinner yeast cell wall when

Eukaryotic Microorganisms as Sources of Bioproducts. DOI: https://doi.org/10.1016/B978-0-443-30188-9.00008-1

compared to microalgae (Saini et al., 2021). Additionally, lipid extraction in yeast is generally easier than in bacteria, where lipids are tightly integrated into the bacterial membrane (Bao et al., 2021). Besides, yeast-based processes are easier to scale-up when compared to other oils producing microorganisms such as those accumulated in microalgae (Robles-Iglesias et al., 2023; Tomás-Pejó et al., 2021; Zainuddin et al., 2021).

When pure glucose is used for MO production, carbon source can account for up to 94% of the total process cost (Gallego-García et al., 2022). Thus, there is an urgent need to find cost-effective carbon sources for yeast fermentation. SCFAs derived from the anaerobic fermentations (AF) of wastes and other industrial by-products are promising alternative carbon sources for microbial FA production (Llamas, et al., 2020; Greses et al., 2020). In this sense, tons of wastes globally produced by the food industry can be diverted toward MO production with appropriated microorganisms (FAO, 2022; Tomás-Pejó et al., 2021). For example, fish industries capture over 140 million tons of food, with 25% considered waste. Fabiszewska et al., (2021) used fish oil as a carbon source to produce lipids rich in unsaturated FA like docosahexaenoic acid (DHA) and eicosapentaenoic acid (EPA). Li et al., (2021) produced terpenoids from waste cooking oil by converting it into acetyl-CoA, a precursor for their biosynthesis in *Yarrowia lipolytica*. Together with the previous ones, olive oil mill wastewater and food waste leachate can also be used as carbon source (Dias et al., 2021).

MO downstream is the primary bottleneck when aiming at an efficient lipid recovery process. Life cycle sustainability assessments emphasized the need to find alternative disruption methods, both mechanical and non-mechanical, to further enhance lipid extraction (Abelleira et al., 2024). This would improve process efficiency, as downstream can account up to 60 % of the overall cost. These disruption methods include acid hydrolysis, high-pressure homogenization, microwave -assisted extraction, enzymatic pretreatment, bead milling beating, etc. Their disruption effectiveness and the use of different extraction methods, both conventional and green solvents, are evaluated. Green solvents are a sustainable environmental, and ecological alternative, to the use of chloroform, methanol, sodium hydroxide, and sulfuric acid. The disruption and green methods should substitute acidic thermolysis found responsible for 80% of total impacts in the climate change category (Abelleira et al., 2024). Also, the feasibility of using these methods at an industrial scale, the economic impact of their implementation, and the cost of the feedstock haves to be considered, as electricity, heat consumption, and the selected feedstock are a primary concerns (Gallego-García et al., 2022; Abelleira et., 2024.

This chapter explores the main challenges and recent advances in MO production and applications. It also addresses the oil-based products achievable through yeast fermentation together with their current and potential applications. The existing technologies for yeast cell disruption and lipids downstream processing are discussed. Therefore, this review highlights new strategies to stablish MO as sustainable bioproduct alternatives to those conventionally produced via petrochemical routes.

6.2 Oleaginous yeast

More than 160 species of yeasts have been described as oleaginous so far. Nevertheless, due to the relevance of the topic and the research efforts to isolate new candidates from different environments such as sea water or Antarctic soil, the number of oleaginous yeasts species is increasing every year (Adel et al., 2021). Among oleaginous yeast, the most relevant species, due to their high intracellular lipid content, carbon source versatility, growth performance, and/or genetic accessibility, are *Candida tropicalis*, *Rhodosporidium toruloides*, *Cutaneotrichosporon oleaginosus*, *Lipomyces starkeyi*, *Rhodotorula glutinis*, and undoubtedly, *Y. lipolytica* (Abeln & Chuck, 2021; Salvador López et al., 2022).

Y. lipolytica stores approximately 95% of intracellular lipids as neutral TAGs in lipid bodies. TAGs are synthesized via de novo (from hydrophilic carbon sources including sugars, alcohols, or organic acids) or ex novo pathways (from hydrophobic carbon sources such as free FAs, TAGs, or alkanes) (Llamas, et al., 2020).

Recent efforts have been focused on producing high lipid content using *Y. lipolytica* and utilizing carbon sources derived from residual sources (agricultural and food waste, oils, and lignocellulosic materials). Tomás-Pejó et al., (2023) achieved 41.9% w/w lipid content when growing *Y. lipolytica* ACA DC 50109 in media containing 11.5 g/L of food waste derived SCFA after struvite precipitation. Limitation of nitrogen, which is essential for nucleic acid and protein synthesis, has been identified as a crucial factor to promote lipid accumulation in microorganisms (Ghazani & Marangoni, 2022; Morales-Palomo et al., 2022). In this sense, the high nitrogen content present in real digestates hinders lipid accumulation unless nitrogen removal is carried out. Pereira et al., (2023) carried out a two-step batch process, where *Y. lipolytica* was first grown on SCFAs followed by a glycerol fermentation to boost the lipogenic phase. This approach allowed a lipid content increase from 7.2% w/w to 22% w/w. Limitation of nitrogen, which is essential for nucleic acid and protein synthesis, has been identified as a crucial factor to promote lipid accumulation in microorganisms (Ghazani & Marangoni, 2022; Morales-Palomo et al., 2022).

Traditional metabolic engineering strategies have been applied for decades to improve the lipid accumulation in *Y. lipolytica* (Friedlander et al., 2016) or to inhibit the intracellular lipid degradation (Beopoulos et al., 2008; Dulermo & Nicaud, 2011). Lipid metabolism has been also modified at regulatory level through transcriptional factors activity (Leplat et al., 2018). The reduction of competing by-products such as glycogen (Bhutada et al., 2017) or citric acid (Sagnak et al., 2018) has been also accomplished. Improved yeast fitness and robustness has also been targeted by engineering redox metabolism to increase the NADPH pool (Qiao et al., 2017), designing oxidative stress defense pathways (Zhang, et al., 2020; Xu et al., 2017) and adaptative laboratory evolution (ALE) strategies to boost lipid accumulation (Daskalaki et al., 2019). In the particular case of SCFAs utilization, ALE has been successfully applied not only to increase *Y. lipolytica* lipid production capacity but to favor its caproic acid utilization capacity (Morales-Palomo et al., 2024).

In order to produce lipids in an economically feasible manner, *Y. lipolytica* metabolism has been expanded to use other carbon sources such as xylose (Ledesma-Amaro et al., 2016), to avoid lipids extraction by targeting at lipids secretion (Ledesma-Amaro, Dulermo, et al., 2016) and to unlink lipogenesis from nitrogen starvation (Xu et al., 2017).

6.3 Microbial oils applications

MO have a broad application range that goes from the food industry to the health, cosmetic, and energy sectors (Figure 6.1. MO are an alternative source for oleochemicals and a biotechnological conversion option for the production of biofuels and bioproducts (Llamas, et al., 2020; Morales-Palomo et al., 2022; Tomás-Pejó et al., 2023). Carbon source origin for MO influences future applications. Thus, food industry and medical applications have strict regulations regarding using SCO derived from organic wastes.

6.3.1 Food

Within the food sector, FAs are essential components of diet. FAs affect some metabolic functions of the body, offering many health benefits (Szczepańska et al., 2022). Some FAs can help with brain development while others can reduce cancer and heart diseases (Xiong et al., 2022). Based on their interest as food additives, the production of

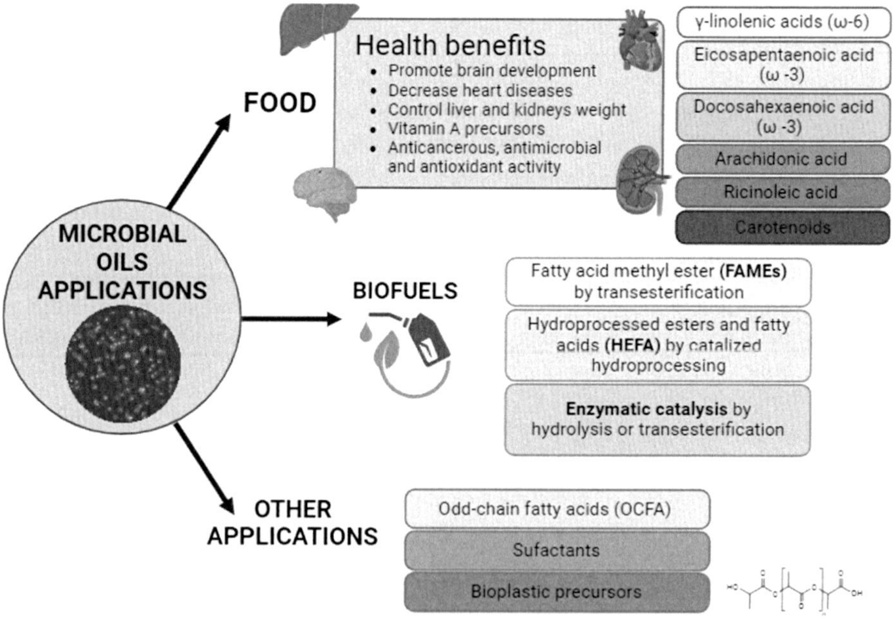

Figure 6.1 Microbial oil applications derived from yeast.

polyunsaturated fatty acids (PUFAs), such as linoleic and γ-linolenic acids (ω-6) or EPA (ω-3), take almost all the attention (Ochsenreither et al., 2016).

EPA is a PUFA mainly obtained from fish with high health benefits such as cholesterol regulation and improvement of liver and kidney weight. Genetically modified *Sacharomyces cerevisiae* only reached 1% w/w of EPA of total FA (Xue et al., 2013). However, the introduction of 30 copies of 8 different genes, together with the suppression of the peroxisomal biogenesis factor (*PEX10*), lipase 1 (LIP1), and sterol carrier protein (SCP2), led to a *Y. lipolytica* strain accumulating up to 56.6% w/w of EPA and 17.2% w/w of linoleic acid of the total FAs. This strain was commercially used for human supplements and for farmed salmon feed (Cao et al., 2022; Xue et al., 2013).

Gamma linoleic acid, an essential FA, is usually extracted from plant seeds, specially evening primrose and black currant (Farag et al., 2023). However, the insufficient natural production has increased the interest of microbial production. Recent efforts have been carried out to produce γ- linoleic acid-rich *Y. lipolytica* (Gu et al., 2023). *Y. lipolytica* also has the inherent capacity of producing α-linoleic acid, a precursor of DHA and EPA, reaching up to 87.96% of PUFAs, out of which 9.25% corresponded to α-linoleic acid (Mattanna et al., 2014).

DHA and arachidonic acid are essential FAs present in fish oils and have associated beneficial health effects when used during pregnancy and when given to infants. These acids are mainly microbiologically obtained from microalgae, through the anaerobic polyketide synthases pathway (Sun et al., 2019). The higher yeast growing capability at industrial scale and the available genetic tools for the modification of the FAs pathways, make yeast a suitable alternative for the production of both DHA and arachidonic acid (Xue et al., 2013). Gemperlein et al., (2019) developed a *Y. lipolytica* strain to produce high DHA yields by integrating a codon-optimized *pfa* biosynthetic gene clusters encoding PUFA synthase. In this study, by the control of culture conditions and media composition, 16.8% of DHA was achieved without limiting nitrogen presence and reducing the amount of phosphate buffer.

Ricinoleic acid derivatives can act as pigments, surfactants, emulsifiers, inks, adhesives, paints, and many more. Castor oil is the principal source of this acid but its exploitation is really limited. Plants for castor oil production only grow under specific climate regions, their seeds need to be harvested by hand present toxic components (Ledesma-Amaro & Nicaud, 2016). For these reasons, several yeasts species (*S. cerevisiae, S. pombe,* and *Y. lipolytica*) have been engineered to produce high concentrations of this acid (Beopoulos et al., 2014; Park & Hahn, 2024). Biotransformation of ricinoleic acid is used to produce γ-decalactone, a peach aroma used in the food industry. Other aromas than can be derived from microorganisms include hexanal, derived from linoleic acid, and limonene, which production reached 23.56 mg/L in *Y. lipolytica* (Cao et al., 2016).

Due to the current high demand of cacao and the limited availability of cacao beans, some authors have engineered FA profile in *Y. lipolytica* to resemble cacao butter through modifying FA desaturase (Ole1p and Fad2p) expression while also reducing linoleic and oleic acid to 40% w/w (Konzock et al., 2022).

Carotenoids cannot be naturally biosynthesized by the human body, needing to be consumed as additives or food supplements (Paul et al., 2023). These microbially-produced pigments can be obtained at higher yields and lower production costs than those chemically produced (Mata-Gómez et al., 2014). More specifically, some yeasts like *Rhodotorula, Sporidiobolus,* and *Sporobolomyces* are capable of synthesizing carotenoids (β-carotene, torulene, and torularhodin) and xanthophylls (astaxanthin and canthaxanthin) that can be precursors of vitamin A and have anticancerous, antimicrobial, and antioxidant activity or reduce the risk of cardiovascular disease (Rapoport et al., 2021; Rodrigues et al., 2022; Sen et al., 2019). The highest concentrations of torulene and torularhodin were achieved with *R. glutinis*, reaching 3.2×10^{-4} g/g_{dw} and 7.2×10^{-4} g/g_{dw}, respectively, under different fermentation conditions (Buzzini & Martini, 2000; Sakaki et al., 2001).

6.3.2 Biofuels

MO are considered relevant sources for biofuels production. Different biofuels can be obtained from MO including biodiesel, sustainable aviation fuels, or biogasoline.

Nowadays, there are two widely commercialized diesel alternatives, fatty acid methyl ester (FAMEs) produced through transesterification, and hydroprocessed esters and fatty acids (HEFA) obtained by catalyzed hydroprocessing of FA (Dimian & Rothenberg, 2016).

Biodiesel produced through the transesterification or esterification of TAG or free FA has been claimed to be a sustainable alternative to diesel. The high content of C16 and C18 acids in MO makes them a valuable source for biofuels production. Alkyl esters are formed by the reaction between TAG and short chain alcohols (e.g., methanol). This process allows a high-octane number with low biofuel viscosity, which helps maintaining the liquid state of the fuel under cold conditions (Kruger et al., 2021; Patel et al., 2018). Transesterification reaction can be catalyzed or noncatalyzed. For the catalyzed reaction, heterogeneous catalysts (zirconia and zeolite-based catalyst) provide a high range of selectivity. However, they are not as stable as homogeneous catalysts and do not provide sufficient conversion efficiency for biodiesel production. Conversely, at the expenses of a high energy consumption, homologous catalysts (sodium hydroxide, potassium hydroxide, or sulfuric acid) allow high bioconversion yields in a short time (Changmai et al., 2020; Rizwanul Fattah et al., 2020).

Besides transesterification, MO can be chemically transformed into "drop in" hydrocarbons fuels by catalyzed hydroprocessing. This transformation is based on deoxygenation and saturation of MO using high pressure H_2 and a metallic catalyst for product hydroisomerization (Monteiro et al., 2022). Unsaturated carbon chains and oxygen are removed to increase biofuel stability and calorific power while reducing water reactivity as well. HEFA focuses on the conversion of esters and FA into hydrocarbons, free of aromatics and sulfur, with carbon chain lengths typically between C8 and C16, making useful for sustainable aviation fuels. HEFA jet fuels have already been included by some airlines in commercial flights, making MO utilization a valuable alternative for fossil fuels replacements (Monteiro et al., 2022).

However, the reliance on fossil-derived H_2 for HEFA still presents a bottleneck for jet fuels (kerosene) production (Su et al., 2022).

Catalyzed hydroprocessing has been mainly carried out by using microalgae-derived lipids as feedstock, even though a limited amount of studies on yeast have been also reported. Sànchez i Nogué et al., (2018) obtained 68% liquid hydrocarbon yield, with 70% C12–C20 hydrocarbon derived from *R. toruloides*. Conversely, the diverse MO composition, which includes terpenes and carotenoids, increases hydrogen consumption due to the high degree of unsaturation of these compounds. Additionally, the polar lipids within MO can deactivate metal or acid catalysts, affecting the hydroprocessing efficiency (Kruger et al., 2021).

In order to avoid chemical routes, enzymatic catalysis appear as an alternative for the breakdown of MO (mainly TAGs) into free FAs and glycerol through hydrolysis or transesterification. Enzymes are more stable and work at milder temperatures exhibiting a high substrate specificity and selectivity. These benefits help reduce by-products formation while avoiding/reducing sequential purification steps. However, enzymes reutilization and short-term storage difficult the scale up process (Monteiro et al., 2022).

6.3.3 Other applications

Metabolic engineering of *Y. lipolytica* is also being carried out to obtain unusual lipids with long length, odd-chain FAs (OCFAs) or FA with different functional groups (Park & Ledesma-Amaro, & Nicaud, 2020; Park & Nicaud, 2020; Park et al., 2021). Some MO, such as OCFAs (i.e., C15:0, C17:0 and C17:1), are high added-value lipids that are not conventionally available through traditional petrochemical processes (Park et al., 2021). OCFAs have unique pharmacological properties while also being used in the pesticides, lubricants, cosmetics, food, and plastic industries. OCFAs are also commercially important due to their antifungal and anti-inflammatory properties (Park et al., 2018; Park & Nicaud, 2020; Zhang, et al., 2020). For example, cis-9-heptadecenoic acid has antiinflammatory effects and can help in psoriasis treatment, allergies, and autoimmune diseases (Park et al., 2018).

Despite their numerous applications, only small amounts of OCFAs are being detected in the yeast FA profile, as they are produced by the incorporation of propionyl-CoA instead of acetyl-CoA, with the activation of propionate being the limiting step in OCFAs synthesis (Park et al., 2018). Targeting at propionic acid consumption and finding the optimum balance between the precursors of acetyl CoA and propionyl-CoA is necessary to reach high OCFAs concentrations. Llamas, et al., (2020) reached 6.4% w/w of C17:0 when *Y. lipolytica* was grown on 5 g /L of SCFAs (Park et al., 2021) developed a strain capable of accumulating up to 50% w/w of OCFAs by enhancing the propionic acid precursors pool.

Surfactants are compounds able to reduce surface tension, with applications in the oil industry (Adetunji & Olaniran, 2021). Natural surfactants offer high biodegradability and foaming capacity while not being negatively influenced by high temperatures. In this sense, extracellular lipids are produced by *R. glutinis* and *Candida ingens*. *Y. lipolytica* NCIM 3589 was reported to produce an extracellular emulsifier under nitrogen-limiting conditions, in the presence of NaCl and alkanes or crude oil, which increased hydrophobicity and facilitated hydrocarbon degradation (Zinjarde & Pant, 2002). Other compounds that can be produced from MOs are biolubricants, waxes, and emollients (Papadaki et al., 2018; Uğur et al., 2024),

Yeast can be utilized as cell factories for biodegradable plastics production like polylactic acid, polyhydroxyalkanoate (PHA), and polybutylenesuccinate (PBS) (Zhang et al., 2023). These plastics can be synthesized from several compounds derived from yeast fermentation such as lactic, succinic, and muconic acids, which act as essential building blocks in the polymerization process. Engineered yeast strains of *Y. lipolytica* and *S. cerevisiae* can also directly produce poly D-lactic acid (Lajus et al., 2020; Ylinen et al., 2021). In the case of PHAs, *Y. lipolytica* produced 0.00205 $g/g_{d.w.}$ of PHA by expressing the *Pha*C1 gene from *P. aeruginosa* PAO1 with a PTS1 peroxisomal signal (Gao et al., 2015). Within PHAs, polyhydroxybutyrate (PHB) can be accumulated also at higher levels [10.2% w/w of the total dry cell weight (DCW)] (Li et al., 2017). Halophilic yeast strain, *Pichia kudriavzevii* TSLS24 has also been described as PHB producer, reaching 43.4% w/w and 1.8 g/L (Thu et al., 2023).

6.4 Downstream processing of microbial oils: cell disruption and lipid extraction

Microorganisms do not naturally secrete lipids extracellularly. Because of that, cell disruption methods and pretreatments have to be carried out to break up the cell wall prior to lipid extraction (Ochsenreither et al., 2016). This downstream processing of biomass can entail high costs. As an example, it has been estimated that cell disruption, lipid extraction, and transesterification can involve more than 40% of the total biofuels cost when derived from MO (Huang et al., 2018).

Selection of the pretreatment method depends on the yeast cell wall composition, which concomitantly affects the efficiency of many solvents. The thickness and rigidness conferred by cell wall proteins, polysaccharides (β1,6-glucan, β1,3-glucan and mannan), and chitin make difficult the extraction of intracellular components (Zainuddin et al., 2021). Cell wall makes up for 30% w/w of the DCW (Nguyen et al., 1998) and their composition is influenced by environmental conditions, such pH, temperature, aeration, and many more (Khot et al., 2020; Zainuddin et al., 2021). Pretreatments damage mainly cell wall glucans, affecting cell wall strength and elasticity of the inner wall.

As shown in Figure 6.2, yeast cells disruption methods can be classified in two different groups: (1) mechanical (homogenization, ultrasonication, bead mill, freeze-thaw, microwave-assisted extraction, autoclaving) and (2) nonmechanical, which can be physical, chemical or enzymatic (Gautério et al., 2023; Kot et al., 2020).

6.4.1 Mechanical disruption methods prior MO extraction

Mechanical methods allow a nonselective microbial cell wall disruption for intracellular product release and are more widely used at industrial scale (Gautério et al., 2023). Mechanical methods include sonication, high-pressure homogenization, bead milling, and microwave-assisted extraction (Figure 6.2). Reviewing these methods, brief mechanisms of action include:

- **Sonication** implies the use of sound energy at a frequency higher than 20 kHz, creating gas-filled bubbles during cavitation phenomena (Liu et al., 2016; Zainuddin et al., 2021). These microscopic bubbles grow during the rarefaction phase and then compress. Bubbles explode causing an increase in temperature and pressure of several atmospheres, which ultimately leads to the disruption of cells (Zhang et al., 2014). Both protein and lipids can be recovered with this process (Liu et al., 2016). Output power, duty cycle, and temperature are some factors that can affect the effectiveness of this method (Liu et al., 2016). Two different routes are available: (1) high frequency and low power ultrasonication (ex, 520 kHz, 40 W) or (2) low frequency and high power ultrasonication (e.g., 50 Hz, 2800 W). Zhang et al., (2014) reported that out of the two available sonication frequencies, the breakage of the cells was higher at low frequency. To avoid potential lipid oxidation, sonication has been also performed with the sample in an ice bath (Narisetty et al., 2022). The high efficiency of sonication is, however, compromised by the high cost, as it requires high amount of energy, making it not viable at a high scale. Alternatives to increase the efficiency include increasing the sonic amplitudes and the exposure time (Zainuddin et al., 2021). Likewise, optimal yeast biomass concentration has to be established for maximum efficiency as at high cell concentration, the recovery yield decreased due to a lower energy transfer. However, a low cell concentration reduces the collision chances between them and also reduces the lipid extraction efficiency (Dong et al., 2016; Zhang et al., 2014).

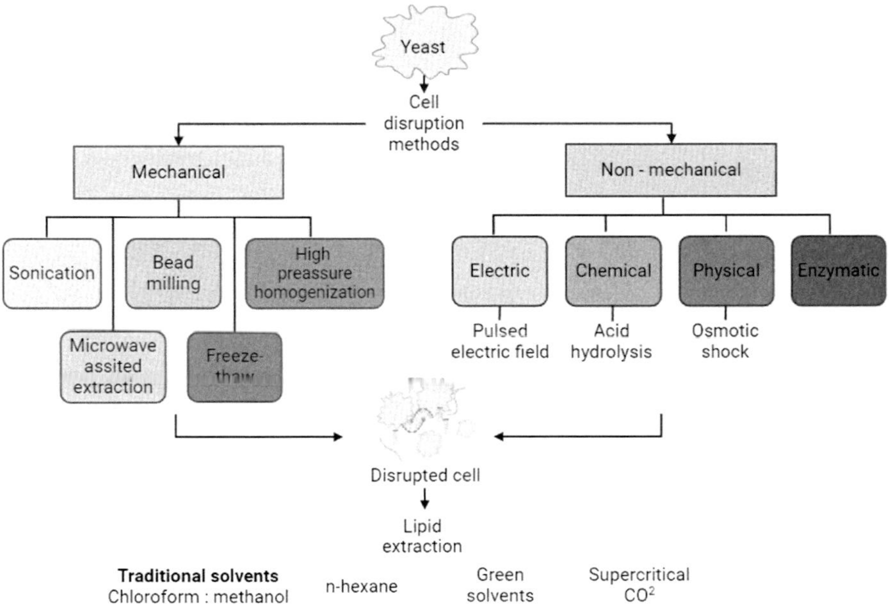

Figure 6.2 Microbial oils extraction downstream process. Disruption methods and lipid extraction in yeast.

- **High-pressure homogenization**. This method, where cells are disrupted by applying high pressure, is one of the preferred methods for cell disruption by the industry (Drévillon et al., 2018). In this case, cells go through a small gap in the equipment that provokes cavitation phenomena thanks to an increase in speed. This causes mechanical stress after a pressure release (Koubaa et al., 2020; Patel et al., 2018). This method is well stablished in the industry for the recovery of different compounds (cell-free proteins, recombinant proteins, and oil extraction from yeast) (Zainuddin et al., 2021). When working with yeast, 1500 bar and 5 cycles were found optimal for maximum disruption and lipid recovery when dried cells of *Y. lipolytica* were used (Drévillon et al., 2018).
- **Bead milling** pretreatment is based on the impact of different material beads against the cells with the presence of an impeller for better mixing. In this case, grinding is responsible of cell disruption. The effectiveness of this method depends on the yeast strain, the cell concentration and composition (Koubaa et al., 2020). Bead milling can be considered a fast and efficient way to disrupt both wet and dry yeast cells. This method has a high disruption efficiency and can be used at different operation scales, explaining its wide implementation in industry (Liu et al., 2016; Saini et al., 2021).
- **Microwaved-assisted reaction** shares similar bases with sonication as both require high electromagnetic energy to heat up the sample. Microwave reaction works by heating up the intracellular water through high-frequency waves leading to cell disruption, unlike sonication which induces cavitation phenomena (Zhou et al., 2019). Some advantages of this process derive from its reproducibility, lower energy consumption, reduced processing time, and no need for water removal (Meullemiestre et al., 2016; Patel et al., 2019). In microwave-assisted reaction, the high temperatures may transform diacylglycerol into free-FAs, changing the final lipid profile when targeting at intracellular lipids recovery (Meullemiestre et al., 2016).
- **Freeze -thaw** – This pretreatment implies cells freezing for more than 24 h followed by defrosting. This disruption methods is only conventionally applied at laboratory scale. It should be highlighted that one of the main factors when choosing a disruption method is the effect that it might exert on FA composition. In this sense, PUFAs, although easily degraded during the disruption process, remained stable when using this pretreatment.

6.4.2 Non-mechanical disruption methods prior MO extraction

Non-mechanical methods allow a more selective cell disruption while causing less damage on the targeted compound; however, the scalability of some of these processes is more complicated than mechanicals ones. These methods are less scalable due to the complexity of the downstream processes, the cost, and the possible degradation of the target compounds (Gautério et al., 2023). Nonmechanical methods can be divided into four groups; electrical, enzymatic, chemical, and physical.

- **Electrical method—Pulsed electric field (PEF)**—This method consists on several short electrical pulses sent to the cells to increase the transmembrane voltage and membrane permeability with the ultimate goal of releasing the compound of interest. This disruption method can be carried out through two routes: washed and unwashed (by directly using the harvested yeasts). When aiming for high lipid yields, the washed method allows a higher recovery due to the weakened cell wall after washing. However, scalability at an industrial scale is more limited due to the higher water and sample volume usage.
- **Enzymatic—Enzyme-assisted disruption** can reduce lipid degradation as no temperature increase is linked to this disruption method. This method focuses on targeting both the

mannoprotein complex and glucan structure through the utilization various enzymes, namely proteases (hydrolyze the protein structure), glucanases (for inner wall disruption), mannases, and chitinases (Liu et al., 2016). Enzyme-assisted disruption/hydrolysis is yet not economically feasible at industrial scale, due to the high enzyme cost. Their cell wall degradation ability should also be improved to achieve a cost-effective process (Giovannoni et al., 2020).

- **Chemical—Acid hydrolysis** pretreatment is the most widely used method for cell disruption and provides the highest efficiency (Kot et al., 2020; Yu et al., 2015). Acids act on the lipid-starch and lipid-proteins bonds to release the lipids while also acting on the cell wall polymers (Khot et al., 2020). This method involves the use of acids followed by high temperature > 60 ºC. Acid hydrolysis also has associated some disadvantages since other cellular components may be degraded and the production of harmful side streams might also take place (Jin et al., 2012; Kot et al., 2020).
- **Chemical- Detergent**—Cell wall solubilization by detergents depends on their ability to solubilize proteins, phospholipids, and many more (Schuck et al., 2003). As an example, amino acid derived detergent, like N-lauroyl sarcosine, can be used for wet cell-disruption of *Y. lipolytica.* 23 mg/g dry biomass of N-lauroyl sarcosine, 30°C and 12.5 min were determined to be optimal conditions for lipid recovery (0.99 g/g), followed by 1: 1 chloroform: methanol extraction (Yellapu et al., 2016).
- **Physical—Osmotic shock**. This method implies the use of salts to cause osmotic pressure difference between the inside of the cell and the outside, favoring water entry into the cell from the medium, which thereby increases the intracellular pressure (Zainuddin et al., 2021). Cell does not resist the pressure increase and ends up breaking and releasing lipids for further wet lipid extraction (Yoo et al., 2012).

6.4.3 Cell disruption efficiency and lipid extraction applied to yeasts

The vast majority of authors assess cell disruption efficiency based on lipid extraction yield. In this sense, the previous discussed cell disruption methods have been compared to determine the most efficient strategy in terms of lipid extraction but also related to cost-effectiveness and scalability.

Kot et al., (2020) compared both mechanical and nonmechanical methods (acid hydrolysis, sonication, osmotic shock, freeze thaw, pasteurization, and homogenization). Out of all the disruption methods, acid hydrolysis was considered the most effective, achieving a lipid yield of 0.20 and 0.21 $g/g_{d.w.}$ for two *Cyberlidnera jadinni* and *R. glutinis*, respectively (Table 6.1). The second highest yields were reached with sonication, recovering 0.20 $g/g_{d.w.}$ and 0.17 $g/g_{d.w.}$, respectively. However, it was concluded that sonication should be carefully implemented since it can result in the degradation of unsaturated FA in some strains. Based on their results, Kot et al., (2020) indicated that acid hydrolysis was the only disruption method capable of breaking yeast cell wall while avoiding unsaturated FA degradation. Regarding acid hydrolysis, Yu et al., (2015) also found this pretreatment as the most effective, allowing the highest lipid recovery from *Cryptococcus curvatus* when compared to sonication, bead milling, and Soxhlet method (0.47 $g/g_{d.w}$ vs 0.43, 0.45, and 0.41 $g/g_{d.w}$, respectively). The lowest lipid recovery for Yu et al., (2015) was attained with microwaved disruption, achieving less than 0.25 $g/g_{d.w.}$ (Table 6.1).

Table 6.1 Overview of the disruption and extraction methods used for efficient lipid recovery from various yeast strains.

Strain	Disruption method	Extraction method	Lipid yield (g/g d.w.)	% recovery - disruption efficiency	Biomass	References
Cyberlidnera jadinni	Acid hydrolysis	Bligh & Dyer. Chloroform: methanol (1:2 v/v)	0.2	–	Dry	Kot et al. (2020)
	Sonication		0.2	–		
	Osmotic shock		0.12	–		
	Freeze thaw		0.14	–		
	Pasteurization		0.13	-		
	Homogenization		0.15	–		
	No disruption—control sample		0.12	–		
Rhodotorula glutinis	Acid hydrolysis	Bligh and Dyer. Chloroform: methanol (1:2 v/v)	0.21	–	Dry	Kot et al. (2020)
	Sonication		0.17	–		
	Osmotic shock		0.14	–		
	Freeze thaw		0.16	-		
	Pasteurization		0.13	–		
	Homogenization		0.13	–		
	No disruption—control sample		0.14	–		
Cryptococcus curvatus	Acid hydrolysis	Chloroform: methanol (1:1 v/v)	0.47	–	Wet	Yu et al. (2015)
	Sonication		0.43	–		
	Bead milling		0.45	–		
	Microwaved assisted		0.25	–		
	Autoclaving		0.1	–		
	Bead milling	Chloroform: methanol (1:1 v/v)	0.3	–	Dry	Huang et al. (2018)
	Sonication	Chloroform: methanol (1:1 v/v)	0.28	–	Dry	Chatterjee and Mohan (2018)
	No pretreatment	Automated pressurized liquid extraction. Chloroform-methanol (1:1, v/v)	–	89.5	–	Li et al. (2020)
Trichosporon oleaginosus	Ultrasonication	Chloroform: methanol (2:1 v/v)	–	100	–	Zhang et al. (2014)

Yarrowia lipolytica	Freeze thaw	Bligh & Dyer. Chloroform: Methanol (1:2 v/v)				
	0.06	–	Wet	Meullemiestre et al. (2016)		
	Cold drying		0.14	–		
	Microwaved assisted		0.08	–		
	Bead milling		0.13	–		
	No pretreatment	Chloroform: methanol	0.06	–		
		Bead milling	0.13	–		
		Ultrasonication	0.08	–		
	Detergent (N-lauroyl sarcosine)	Chloroform: methanol (1:1 v/v)	-	95.3	–	Yellapu et al. (2016)
	No pretreatment	Dimethylcyclohexylamine	0.58	–	Wet	Yook et al. (2019)
		Ethylbutylamine	0.57	–		
		Dipropylamine	0.32	–		
		Bligh & Dyer	0.56	–		
	Acid-catalyzed hot-water	Chloroform-methanol (2:1v/v)	0.49	–	Dry	Johnravindar et al. (2018)
	No pretreatment	Hexane	0.14	–	Freeze-dried	Breil et al. (2016)
		Isopropanol	0.15	–		
		Ethanol	0.15	–		
		Ethyl acetate	0.13	–		
		Ethyl lactate	0.16	–		
		DMC	0.15	–		
		p-cymene	0.16	–		
		α-pinene	0.14	–		
		d-Limonene	0.14	–		
		MeTHF	0.16	–		
		CPME	0.15	–		
	Homogenization Triton solution	Folch Chloroform:methanol (2:1 v/v)				
	0.11	–	Dried prior extraction	Fabiszewska et al. (2023)		
	Homogenization Triton+ glass beads		0.06	–	Dry	
	Pulsed electric field		0.28	–		
	High pressure processing		0.06			
	Ultrasound		0.09			

(Continued)

Table 6.1 (Continued)

Strain	Disruption method	Extraction method	Lipid yield (g/g d.w.)	% recovery - disruption efficiency	Biomass	References
Saitozyma podzolica	High pressure homogenization	Bligh and Dyer (1:1 v/v)	0.43	95	Dry	Gorte et al. (2020a)
		Ethanol-hexane extraction (2,3: 1 v/v)	0.47	95		
		Folch chloroform:methanol (2:1 v/v)	0.38	95		
	Bead milling	Bligh and Dyer (1:1 v/v)	0.32	93		
		Ethanol-hexane extraction (2,3: 1 v/v)	0.37	–		
		Folch chloroform:methanol (2:1 v/v)	0.28	–		
	Ultrasonic	Bligh and Dyer (1:1 v/v)	0.21	27		
		Folch chloroform:methanol (2:1 v/v)	0.17	27		
	Pulsed electric field	Ethanol-hexane extraction				
	-	54	Unwashed	Gorte et al. (2020b)		
	Pulsed electric field		–	99	Washed	
	No pretreatment		–	7	Wet	
Rhodotorula toruloides	High pressure homogenization	–	0.46	–	–	Dias et al. (2022)
	No pretreatment	n-hexane	–	6.9	Wet	
	Jin et al. (2012)					
	Enzyme assisted extraction (MAN5C)	n-hexane, chloroform or ethyl acetate	–	11.9		
	Microwaved radiation	–	–	1.5		
	Enzyme assisted extraction + microwave radiation	n-hexane, chloroform Ethyl acetate	–	9.574.786.9		
	No pretreatment	Automated pressurized liquid extraction +Folch (1:1 chloroform:methanol)	–	89 (at 105°C) 95 % (with addition of diatomite)	–	Li et al. (2020)
	Acid hydrolysis	Bligh and Dyer + water (2:2:1.8)	0.25	–	Dry	Bonturi et al. (2015)
	No pretreatment		0.23	–		
	Acid hydrolysis	Folch chloroform:methanol (2:1 v/v)	0.34	–		
	Enzymatic		0.31	–		
	No pretreatment		0.42	–		
	Acid hydrolysis	Hexane	0.42	–		
Rhodosporidium kratochvilovae	Microwave + sonication	Chloroform: methanol (2:1 v/v)	0.7	–	Wet	Patel et al. (2019)
	Acid catalyzed hot water	Hexane	0.62	–		
	Microwaved assisted	Chloroform: methanol (2:1 v/v)	0.67	–		
	Ultrasonication	Chloroform: methanol (2:1 v/v)	0.6	–	Dry	

Classical methods based on organic solvents utilization are still predominant for lipid extraction, being Folch et al., (1957) and Bligh & Dyer, (1959) the most predominant methods for yeast. These extraction methods are mainly used at laboratory scale, due to the safety, environmental, and recovery concerns associated with the use of chloroform and methanol (solvents used for the lipid extraction). The difference of both methods relies on the ratio of both solvents. Folch et al., (1957) developed a solvent mixture of chloroform/methanol (2:1). Lipids dissolve in the chloroform, whereas all salts dissolve in the methanol, leaving a multilayer separation. Cell debris remains between both phases. Bligh and Dyer method is preferred when working with high biomass content and when quantitative lipid composition is not needed. In this case, some FA (like triglycerides) are more efficiently extracted than others (e.g., phospholipids) (Khot et al., 2020).

Meullemiestre et al., (2016) also compared different cell disruption methods (freeze thaw, microwave-assisted extraction, sonication, and bead milling), followed by chloroform:methanol (1:2, v: v) extraction. Lipid recovery through freeze thaw allowed the highest lipid recovery, reaching 0.14 $g/g_{d.w.}$ followed by bead milling (0.13 $g/g_{d.w}$) (Table 6.1). Bead milling and freeze thaw doubled the lipid recovery yield when compared to only directly using the Bligh and Dyer method. Bead milling, microwave, and ultrasound, previously mentioned as disruption methods, can also be used simultaneously with solvents (chloroform: methanol) for direct lipid extraction.Meullemiestre et al., (2016) highlighted the use of bead-milling and solvents for lipid yield recovery (0.13 $g/g_{d.w.}$), reaching a higher value than with conventional maceration (0.06 $g/g_{d.w}$) (Table 6.1). The biggest difference was observed in lipid composition, as diacylglycerols were degraded to FFA in ultrasound and microwave pretreatments. Likewise, the combination of ultrasonication was followed by chloroform: methanol evidenced showed 100% lipid recovery when using *Trichosporon* sp. (Zhang et al., 2014).

Another consideration for lipid extraction is the utilization of dry or wet biomass. Dried biomass is preferred as the presence of water reduces the efficiency of lipid extraction when a solvent-based process is carried out (Yoo et al., 2012). Yeast drying is conventionally carried out by centrifugation, sedimentation, micro or ultrafiltration, followed by a heating or lyophilization step (Khot et al., 2020). When wet biomass is used, lower lipid recovery can be attributed to a reduced lipid accessibility, reduced mass transfer, and emulsions caused by the water presence (de Jesus & Filho, 2020; Dong et al., 2016, Zainuddin et al., 2021). Due to the higher cost that drying biomass implies, dry biomass is mainly used on research at a smaller scale or for analysis while the use of wet biomass would be preferable at large scale. At this point, it should be highlighted that when using wet biomass, cell wall disruption is commonly performed by mechanical pretreatments. When evaluating wet cell biomass for lipids extraction (Patel et al., 2019) established an efficient method based on microwave-sonication. This disruption method allowed for the highest lipid recoveries (0.70 $g/g_{d.w}$) when compared to acid-catalyzed hot water and microwave-assisted disruption (0.62 $g/g_{d.w.}$ and 0.67 $g/g_{d.w.}$, respectively). A higher recovery was also allowed when compared to the use of dry biomass (0.60 $g/g_{d.w.}$) (Table 6.1).

Not only cell wall moisture should be considered when selecting the pretreatment and extraction methods but also the cell wall composition. For instance, high-pressure homogenization showed different efficiencies as yeast cell wall greatly varies among strains (Zainuddin et al., 2021). This method allowed higher disruption efficiency than bead milling for *S. podzolica* (Gorte, Hollenbach, et al., 2020). Similarly, Dias et al., (2022) also reported the highest recovery yield (0.46 $g/g_{d.w}$) when high-pressure homogenization was applied on *R. toruloides*. High-pressure homogenization is of high consideration when scaling up given its high efficiency when working with high biomass contents (Gomes et al., 2020).

Regarding nonmechanical methods, Gorte et al., (2020b) tested PEF on *Saitozyma podzolica*. Due to the high conductivity of the yeast biomass, the washed route allowed an energy-cost reduction. The best PEF condition was attained with 1-μs pulses and an electric field of 40 kV/cm. The unwashed route increased lipid extraction efficiency from 7% to 54%, after PEF disruption of the yeast cells. Higher recoveries were reached in the "washed route," as 99% recovery efficiency were attained after PEF (Table 6.1).

Few studies have dealt with enzyme-assisted disruption. Recombinant β-1,3-glucomannanase, able of breaking glucomannan present in the cell wall, combined with microwave irradiation resulted in a 5.2-fold increase in the lipid recovery yield, when compared with only the enzyme hydrolysis for *R. toruloides* (Table 6.1) (Jin et al., 2012).

Out of all these pretreatments, mechanical methods are mostly applied at industrial scale due to their higher efficiency for disruption at higher biomass concentrations. The use of dry or wet biomass has to be considered depending on the selected disruption and extraction methods, together with process cost and potential scalability.

6.4.4 Innovative lipid extraction methods: green solvents

As mentioned earlier, Folch et al., (1957) and Bligh & Dyer (1959) methods are the most commonly applied for yeast extraction at laboratory level. However, toxicity associated with their utilization along with their potential harmful environmental effect, hinder their utilization at industrial scale. Several alternatives for lipid extraction like the use of green solvents, derived from organic/inorganic (ionic liquids) sources are being investigated (de Jesus & Filho, 2020) (Figure 6.3). Most extraction methods combine both solvent/green solvent with a pretreatment, as they reduce the extraction time while allowing higher lipid recoveries (higher than 90%) (Table 6.1). Yellapu et al., (2016) achieved 95.3 % recovery when working with N-lauroyl sarcosine. According to the lipids to be extracted, solvent selection determines the extraction efficiency. Nonpolar solvents are used to solubilize neutral lipids covalently bonded and polar solvents are used for polar lipids like glycerophospholipids.

Another solvent is n-hexane, a less toxic and a cheap option that also enables a phase separation as with chloroform. A 98% extraction yield was reached by Imatoukene et al., (2020) when n-hexane was used as solvent for lipid extraction after bead milling and HPH disruption. However, the extraction yield was still slightly lower than when using chloroform/methanol, which reached up to 99.6%.

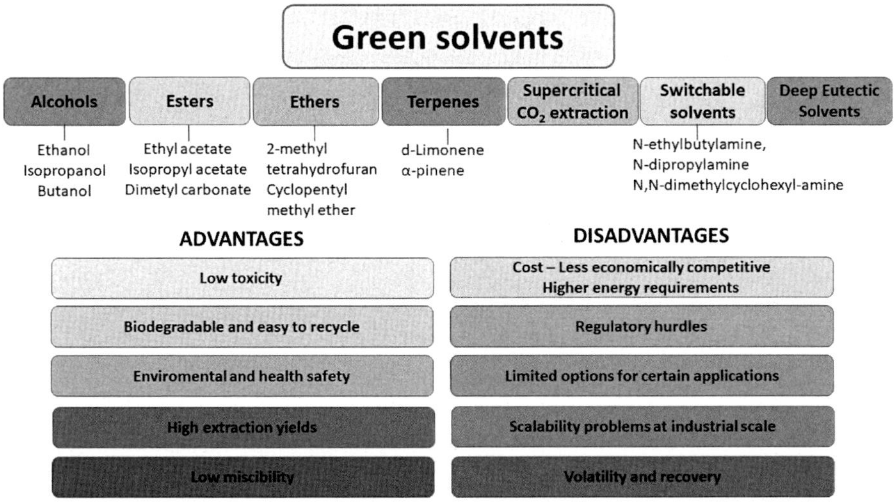

Figure 6.3 Overview of different types of green solvents and their main advantages and disadvantages.

Isopropanol was found to be another promising alternative due to their reduced cost, higher recovery, and lower toxicity on *Y. lipolytica* (Breil et al., 2016). Ethanol:-hexane combination has also been used for lipid extraction. Gorte et al., (2020a) and Imatoukene et al., (2020) identified isoamyl acetate as the best green solvent for lipid extraction when using wet biomass.

Other organic green solvents are cyclopentyl methyl ether and 2-methyltetrahydrofuran (de Jesus & Filho, 2020). Nonorganic solvents have also been used for yeast lipid extraction allowing a higher recovery when compared with chloroform/methanol. These solvents are expensive but require a lower energy consumption and can be used on wet biomass. Some of the solvents tried out by Yook et al., (2019) were switchable hydrophilicity solvents like N-ethylbutylamine, N-dipropylamine, and N,N-dimethylcyclohexyl-amine on a 1:1 ratio. These solvents change their hydrophobic/hydrophilic trait in the presence of CO_2, facilitating lipid recovery and solvent reutilization. N-ethylbutylamine and N,N-dimethylcyclohexyl amine allowed high recovery yields (0.61 $g/g_{d.w}$ and 0.64 $g/g_{d.w.}$). This implied a 13% increase when compared to traditional solvents (Bligh and Dyer method), which allowed 0.56 $g/g_{d.w}$ in *Y. lipolytica* (Yook et al., 2019).

Lastly, an alternative to using traditional solvent-based extraction techniques is supercritical CO_2 extraction. This method has appeared as an unconventional approach for bioactive compound recovery from yeast, offering the selective extraction of the saponificable lipid fraction without chemicals use, making it more environmentally safe. This technique allows an inexpensive process, that could be coupled with a disruption method reaching high lipid yields. Furthermore, supercritical CO_2 extraction can be scalable to industrial use.

6.5 Waste-derived MO production and limitation on their use

Tons of food residues are thrown away (FAO., 2022), leading to economic and environmental problems and finding economically-feasible alternatives for the revalorization of these wastes is of outmost importance. In this context, the European Commission is putting in place not only regulations for waste reduction but also initiatives for waste valorization into bioproducts. More specifically, the European Waste Framework Directive 2008/98, in Article 3(15), states that when "transforming waste into a product, the product escapes the legal regime of waste from the very moment when the application of such a legal regime is no longer warranted." However, despite the recent novel technologies for wastes valorization, the origin of wastes should be carefully considered. In this line, the Regulation (EC) No 1069/2009 of the European parliament states that the revalorization of animal wastes in the energy sector should be minimized due to the potential health risks that it might imply. Thus, wastes origin and application of the targeted product should jointly be considered to attain efficient marketability of bioproducts derived from wastes.

When waste valorization involves microorganisms-based processes, the GRAS (generally recognized as safe) status of the microorganisms may be also required depending on the further application of the product. This fact draws particular attention when aiming and products derived from wastes for pharmaceutical and food sectors.

As previously mentioned, yeast are microbial cell factories that can efficiently convert some hydrophobic (animals fats, vegetable and fish oils, and soap-derived products) and hydrophilic wastes (vegetables, fruits, lignocellulosic materials, and dairy) (Gottardi et al., 2021; Tomás-Pejó et al., 2021). Indeed, different new startups food companies are taking advantage of this fact to produce waste-derived proteins and oils with yeast (ÄIO., 2024).

More concretely, the yeast specie *Y. lipolytica* has arisen as one of the most promising MO producers from different waste streams. *Y. lipolytica* is also capable of utilizing SCFAs derived from the AF of wastes (agroindustrial, lignocellulosic, or household wastes) as carbon sources for bioproduction. As a prove of this promising technology for wastes valorization, Morales-Palomo et al., (2023) and Tomás-Pejó et al., (2023) reported up to 37.7% w/w and 41.9 ± 2.6% (w/w) lipids DW in *Y. lipolytica* from waste-derived SCFAs under PO_4^{3-} and N limitation conditions, respectively. Leiva-Candia et al., (2014) also showed the capacity of *Y. lipolytica* to use waste oil derived from fish frying as carbon source for MO production, reaching a lipid content as high as 45.5% w/w.

Conventionally, *Y. lipolytica* accumulates 30–40% w/w of proteins per DW containing all the essential amino acids (Gottardi et al., 2021). Thus, apart from MO production, some yeast strains can also be used for animal feed due to its high protein content. This fact opens new perspectives and suggests *Y. lipolytica* as a versatile yeast specie not only applied for MO production but for single-cell protein.

6.6 Concluding remarks

MO are promising sources for relevant bioproducts (biofuels, food and mgoedical supplements, surfactants, and many more) in different industrial sectors. The utilization of carbon sources derived from wastes might be essential for reducing the associated production costs compared to petrochemical routes. A key step for MO is the downstream processing of yeast biomass. In this sense, yeast cell wall composition determines the most appropriated MO extraction method. High-pressure homogenization, sonication, and acid hydrolysis are the most effective disruption methods for subsequent lipid recovery in the extraction step. However, scalability and economic feasibility of the selected disruption process should be carefully considered for its industrial implementation. Lastly, the utilization of green solvents that replace the traditional chloroform: methanol solvents also needs more research to achieve more sustainable and scalable processes.

Acknowledgments

This work has been supported by the Spanish Ministry of Economy and Competitiveness through the grand RYC2019–027773-I, BIOMIO (PID2020–119403RB-C21) funded by MCIN/AEI/10.13039/501100011033 and the Ministry of Science and innovation for the predoctoral contract for the formation of future doctors (FPI 2021) (PID2020–119403RB-C21). The project LILIS (EUR2022–134030) funded by MICINN/AEI_NEXT GENERATION-UE also supported this work.

References

Abeln, F., & Chuck, C. J. (2021). The history, state of the art and future prospects for oleaginous yeast research. *Microbial Cell Factories, 20*(1). https://doi.org/10.1186/s12934-021-01712-1.

Adel, A., El-Baz, A. , Shetaia, Y. , & Sorour, N.M. Biosynthesis of polyunsaturated fatty acids by two newly cold-adapted Egyptian marine yeast. 3 Biotech, 11(11), 461. https://doi.org/10.1007/s13205-021-03010-4.

Abelleira, S., Cruz, P. L., & Iribarren, D. (2024). Life cycle sustainability assessment of microbial oil from organic waste. *Cleaner Environmental Systems, 15*, 100236. https://doi.org/10.1016/j.cesys.2024.100236

Adetunji, A. I., & Olaniran, A. O. (2021). Production and potential biotechnological applications of microbial surfactants: An overview. *Journal of Biological Sciences, 28*(1), 669–679. https://doi.org/10.1016/j.sjbs.2020.10.058.

ÄIO • Innovative food technologies to replace palm oil, coconut oil and animal fats with sustainable and healthier alternatives. (2024).

Bao, W., Li, Z., Wang, X., Gao, R., Zhou, X., Cheng, S., Men, Y., & Zheng, L. (2021). Approaches to improve the lipid synthesis of oleaginous yeast *Yarrowia lipolytica:* A review. *Renewable and Sustainable Energy Reviews, 149*, 111386. https://doi.org/10.1016/j.rser.2021.111386.

Beopoulos, A., Mrozova, Z., Thevenieau, F., Le Dall, M. T., Hapala, I., Papanikolaou, S., Chardot, T., & Nicaud, J. M. (2008). Control of lipid accumulation in the yeast *Yarrowia lipolytica. Applied and Environmental Microbiology, 74*(24), 7779–7789. https://doi.org/10.1128/AEM.01412-08.

Beopoulos, A., Verbeke, J., Bordes, F., Guicherd, M., Bressy, M., Marty, A., & Nicaud, J.-M. (2014). Metabolic engineering for ricinoleic acid production in the oleaginous yeast *Yarrowia lipolytica. Applied Microbiology and Biotechnology, 98*(1), 251–262. https://doi.org/10.1007/s00253-013-5295-x.

Bhutada, G., Kavšček, M., Ledesma-Amaro, R., Thomas, S., Rechberger, G. N., Nicaud, J. M., & Natter, K. (2017). Sugar versus fat: Elimination of glycogen storage improves lipid accumulation in *Yarrowia lipolytica. FEMS Yeast Research, 17*(3). https://doi.org/10.1093/femsyr/fox020.

Bligh, E. G., & Dyer, W. J. (1959). A rapid method of total lipid extraction and purification. *Canadian Journal of Biochemistry and Physiology, 37*(8), 911–917. https://doi.org/10.1139/o59-099.

Breil, C., Meullemiestre, A., Vian, M., & Chemat, F. (2016). Bio-based solvents for green extraction of lipids from oleaginous yeast biomass for sustainable aviation biofuel. *Molecules, 21*(2), 196. https://doi.org/10.3390/molecules21020196.

Bonturi, Nemailla, Matsakas, Leonidas, Nilsson, Robert, Christakopoulos, Paul, Miranda, Everson, Berglund, Kris, & Rova, Ulrika (2015). Single Cell Oil Producing Yeasts *Lipomyces starkeyi* and *Rhodosporidium toruloides*: Selection of Extraction Strategies and Biodiesel Property Prediction. *Energies, 8*(6), 5040–5052. 10.3390/en8065040.

Buzzini, P., & Martini, A. (2000). Production of carotenoids by strains of *Rhodotorula glutinis* cultured in raw materials of agro-industrial origin. *Bioresource Technology, 71*(1), 41–44. https://doi.org/10.1016/s0960-8524(99)00056-5.

Cao, L., Yin, M., Shi, T. Q., Lin, L., Ledesma-Amaro, R., & Ji, X. J. (2022). Engineering *Yarrowia lipolytica* to produce nutritional fatty acids: Current status and future perspectives. *Synthetic and Systems Biotechnology, 7*(4), 1024–1033. https://doi.org/10.1016/j.synbio.2022.06.002.

Cao, X., Lv, Y. B., Chen, J., Imanaka, T., Wei, L. J., & Hua, Q. (2016). Metabolic engineering of oleaginous yeast *Yarrowia lipolytica* for limonene overproduction. *Biotechnology for Biofuels, 9*(1). https://doi.org/10.1186/s13068-016-0626-7.

Changmai, B., Vanlalveni, C., Ingle, A. P., Bhagat, R., & Rokhum, L. (2020). Widely used catalysts in biodiesel production: A review. *RSC Advances, 10*(68), 41625–41679. https://doi.org/10.1039/d0ra07931f.

Chatterjee, Sulogna, & Mohan, S. Venkata (2018). Microbial lipid production by *Cryptococcus curvatus* from vegetable waste hydrolysate. *Bioresource Technology, 254*, 284–289. 10.1016/j.biortech.2018.01.079.

Daskalaki, A., Perdikouli, N., Aggeli, D., & Aggelis, G. (2019). Laboratory evolution strategies for improving lipid accumulation in *Yarrowia lipolytica. Applied Microbiology and Biotechnology, 103*(20), 8585–8596. https://doi.org/10.1007/s00253-019-10088-7.

Dias, B., Lopes, M., Ramôa, R., Pereira, A. S., & Belo, I. (2021). *Candida tropicalis* as a promising oleaginous yeast for olive mill waste water bioconversion. *Energies, 14*(3). https://doi.org/10.3390/en14030640.

Dias, C., Nobre, B. P., Santos, J. A. L., Lopes da Silva, T., & Reis, A. (2022). Direct lipid and carotenoid extraction from *Rhodosporidium toruloides* broth culture after high pressure homogenization cell disruption: Strategies, methodologies, and yields. *Biochemical Engineering Journal, 189*. 10.1016/j.bej.2022.108712.

Dimian, A. C., & Rothenberg, G. (2016). An effective modular process for biodiesel manufacturing using heterogeneous catalysis. *Catalysis Science and Technology, 6*(15), 6097–6108. https://doi.org/10.1039/c6cy00426a.

Dong, T., Knoshaug, E. P., Pienkos, P. T., & Laurens, L. M. L. (2016). Lipid recovery from wet oleaginous microbial biomass for biofuel production: A critical review. *Applied Energy, 177*, 879–895. https://doi.org/10.1016/j.apenergy.2016.06.002.

Drévillon, L., Koubaa, M., & Vorobiev, E. (2018). Lipid extraction from *Yarrowia lipolytica* biomass using high-pressure homogenization. *Biomass and Bioenergy, 115*, 143–150. https://doi.org/10.1016/j.biombioe.2018.04.014.

Dulermo, T., & Nicaud, J. M. (2011). Involvement of the G3P shuttle and B-oxidation pathway in the control of TAG synthesis and lipid accumulation in *Yarrowia lipolytica*. *Metabolic Engineering, 13*(5), 482–491. https://doi.org/10.1016/j.ymben.2011.05.002.

Fabiszewska, A. U., Zieniuk, B., Kozłowska, M., Mazurczak-Zieniuk, P. M., Wołoszynowska, M., Misiukiewicz-Stępień, P., & Nowak, D. (2021). Studies on upgradation of waste fish oil to lipid-rich yeast biomass in *Yarrowia lipolytica* batch cultures. *Foods, 10*(2), 1–15. https://doi.org/10.3390/foods10020436.

Fabiszewska, Agata, Pakulska, Anna, Zieniuk, Bartłomiej, Wierzchowska, Katarzyna, Jasińska, Karina, Małajowicz, Jolanta, & Nowak, Dorota (2023). Unconventional Extraction Methods of Oleaginous Yeast Cell Pretreatment and Disruption. *Applied Sciences, 13*(24), 13135. https://doi.org/10.3390/app132413135.

FAO (2022). *The state of food security and nutrition in the world 2022*. FAO. https://doi.org/10.4060/cc0639en.

Farag, M. A., Reda, A., Nabil, M., Elimam, D. M., & Zayed, A. (2023). Evening primrose oil: a comprehensive review of its bioactives, extraction, analysis, oil quality, therapeutic merits, and safety. *Food and Function, 14*(18), 8049–8070. https://doi.org/10.1039/d3fo01949g.

Folch, J., Lees, M., & Stanley, G. H. S. (1957). A simple method for the isolation and purification of total lipides from animal tissues. *Journal of Biological Chemistry, 226*(1), 497–509. https://doi.org/10.1016/s0021-9258(18)64849-5.

Friedlander, J., Tsakraklides, V., Kamineni, A., Greenhagen, E. H., Consiglio, A. L., MacEwen, K., Crabtree, D. V., Afshar, J., Nugent, R. L., Hamilton, M. A., Shaw, A. J., South, C. R., Stephanopoulos, G., & Brevnova, E. E. (2016). Engineering of a high lipid producing *Yarrowia lipolytica* strain. *Biotechnology for Biofuels, 9*(1). https://doi.org/10.1186/s13068-016-0492-3.

Gallego-García, M., Susmozas, A., Moreno, A. D., & Negro, M. J. (2022). Evaluation and identification of key economic bottlenecks for cost-effective microbial oil production from fruit and vegetable residues. *Fermentation, 8*(7), 23115637. https://doi.org/10.3390/fermentation8070334.

Gao, C., Qi, Q., Madzak, C., & Lin, C. S. K. (2015). Exploring medium-chain-length polyhydroxyalkanoates production in the engineered yeast *Yarrowia lipolytica*. *Journal of Industrial Microbiology and Biotechnology, 42*(9), 1255–1262. https://doi.org/10.1007/s10295-015-1649-y.

Gautério, G. V., da Silva, R. M., Karraz, F. C., Coelho, M. A. Z., Ribeiro, B. D., & Lemes, A. C. (2023). Cell disruption and permeabilization methods for obtaining yeast bioproducts. *Cleaner Chemical Engineering, 6*, 100112. https://doi.org/10.1016/j.clce.2023.100112.

Gemperlein, K., Dietrich, D., Kohlstedt, M., Zipf, G., Bernauer, H. S., Wittmann, C., Wenzel, S. C., & Müller, R. (2019). Polyunsaturated fatty acid production by *Yarrowia lipolytica* employing designed myxobacterial PUFA synthases. *Nature Communications, 10*(1). https://doi.org/10.1038/s41467-019-12025-8.

Ghazani, S. M., & Marangoni, A. G. (2022). Microbial lipids for foods. *Trends in Food Science and Technology, 119*, 593–607. https://doi.org/10.1016/j.tifs.2021.10.014.

Giovannoni, M., Gramegna, G., Benedetti, M., & Mattei, B. (2020). Industrial use of cell wall degrading enzymes: The fine line between production strategy and economic feasibility. *Frontiers in Bioengineering and Biotechnology, 8*. https://doi.org/10.3389/fbioe.2020.00356.

Gomes, T. A., Zanette, C. M., & Spier, M. R. (2020). An overview of cell disruption methods for intracellular biomolecules recovery. *Preparative Biochemistry and Biotechnology, 50*(7), 635–654. https://doi.org/10.1080/10826068.2020.1728696 .

Gorte, O., Hollenbach, R., Papachristou, I., Steinweg, C., Silve, A., Frey, W., Syldatk, C., & Ochsenreither, K. (2020a). Evaluation of downstream processing, extraction, and quantification strategies for single cell oil produced by the oleaginous yeasts *Saitozyma podzolica* DSM 27192 and *Apiotrichum porosum* DSM 27194. *Frontiers in Bioengineering and Biotechnology, 8*. https://doi.org/10.3389/fbioe.2020.00355.

Gorte, O., Nazarova, N., Papachristou, I., Wüstner, R., Leber, K., Syldatk, C., Ochsenreither, K., Frey, W., & Silve, A. (2020b). Pulsed electric field treatment promotes lipid extraction on fresh oleaginous yeast *Saitozyma podzolica* DSM 27192. *Frontiers in Bioengineering and Biotechnology, 8*. https://doi.org/10.3389/fbioe.2020.575379.

Gottardi, D., Siroli, L., Vannini, L., Patrignani, F., & Lanciotti, R. (2021). Recovery and valorization of agri-food wastes and by-products using the non-conventional yeast *Yarrowia lipolytica*. *Trends in Food Science & Technology, 115*, 74–86. https://doi.org/10.1016/j.tifs.2021.06.025.

Greses, S., Tomás-Pejó, E., & Gónzalez-Fernández, C. (2020). Agroindustrial waste as a resource for volatile fatty acids production via anaerobic fermentation. *Bioresource Technology, 297*, 122486. https://doi.org/10.1016/j.biortech.2019.122486.

Grubišić, M., Mihajlovski, K., Gruičić, A. M., Beluhan, S., Šantek, B., & Šantek, M. I. (2021). Strategies for improvement of lipid production by yeast *Trichosporon oleaginosus* from lignocellulosic biomass. *Journal of Fungi, 7*(11). https://doi.org/10.3390/jof7110934.

Gu, X., Huang, L., & Lian, J. (2023). Biomanufacturing of γ-linolenic acid-enriched galactosyldiacylglycerols: Challenges in microalgae and potential in oleaginous yeasts. *Synthetic and Systems Biotechnology, 8*(3), 469–478. https://doi.org/10.1016/j.synbio.2023.06.007.

Huang, Xf, Shen, Y., Luo, Hj, Liu, Jn, & Liu, J. (2018). Enhancement of extracellular lipid production by oleaginous yeast through preculture and sequencing batch culture strategy with acetic acid. *Bioresource Technology, 247*, 395–401. https://doi.org/10.1016/j.biortech.2017.09.096.

Imatoukene, N., Koubaa, M., Perdrix, E., Benali, M., & Vorobiev, E. (2020). Combination of cell disruption technologies for lipid recovery from dry and wet biomass of *Yarrowia lipolytica* and using green solvents. *Process Biochemistry, 90*, 139–147. https://doi.org/10.1016/j.procbio.2019.11.011.

de Jesus, S. S., & Filho, R. M. (2020). Recent advances in lipid extraction using green solvents. *Renewable and Sustainable Energy Reviews, 133*. https://doi.org/10.1016/j.rser.2020.110289.

Jin, G., Yang, F., Hu, C., Shen, H., & Zhao, Z. K. (2012). Enzyme-assisted extraction of lipids directly from the culture of the oleaginous yeast *Rhodosporidium toruloides*. *Bioresource Technology, 111*, 378–382. https://doi.org/10.1016/j.biortech.2012.01.152.

Johnravindar, Davidraj, Karthikeyan, Obulisamy Parthiba, Selvam, Ammaiyappan, Murugesan, Kumarasamy, & Wong, Jonathan W. C. (2018). Lipid accumulation potential of oleaginous yeasts: A comparative evaluation using food waste leachate as a substrate. *Bioresource Technology, 248*, 221–228. https://doi.org/10.1016/j.biortech.2017.06.151.

Khot, M., Raut, G., Ghosh, D., Alarcón-Vivero, M., Contreras, D., & Ravikumar, A. (2020). Lipid recovery from oleaginous yeasts: Perspectives and challenges for industrial applications. *Fuel, 259*. https://doi.org/10.1016/j.fuel.2019.116292.

Konzock, O., Matsushita, Y., Zaghen, S., Sako, A., & Norbeck, J. (2022). Altering the fatty acid profile of *Yarrowia lipolytica* to mimic cocoa butter by genetic engineering of desaturases. *Microbial Cell Factories, 21*(1). https://doi.org/10.1186/s12934-022-01748-x.

Kot, A. M., Gientka, I., Bzducha-Wróbel, A., Błażejak, S., & Kurcz, A. (2020). Comparison of simple and rapid cell wall disruption methods for improving lipid extraction from yeast cells. *Journal of Microbiological Methods, 176*. https://doi.org/10.1016/j.mimet.2020.105999.

Koubaa, M., Imatoukene, N., Drévillon, L., & Vorobiev, E. (2020). Current insights in yeast cell disruption technologies for oil recovery: A review. *Chemical Engineering and Processing - Process Intensification, 150*. https://doi.org/10.1016/j.cep.2020.107868 107868.

Kruger, J. S., Knoshaug, E. P., Dong, T., Hull, T. C., & Pienkos, P. T. (2021). Catalytic hydroprocessing of single-cell oils to hydrocarbon fuels: Converting microbial lipids to fuels is a promising approach to replace fossil fuels. *Johnson Matthey Technology Review, 65*(2), 227–246. https://doi.org/10.1595/205651321x16013966874707.

Lajus, S., Dusséaux, S., Verbeke, J., Rigouin, C., Guo, Z., Fatarova, M., Bellvert, F., Borsenberger, V., Bressy, M., Nicaud, J. M., Marty, A., & Bordes, F. (2020). Engineering the yeast *Yarrowia lipolytica* for production of polylactic acid homopolymer. *Frontiers in Bioengineering and Biotechnology, 8*. https://doi.org/10.3389/fbioe.2020.00954.

Ledesma-Amaro, R., & Nicaud, J. M. (2016). Metabolic engineering for expanding the substrate range of *Yarrowia lipolytica*. *Trends in Biotechnology, 34*(10), 798–809. https://doi.org/10.1016/j.tibtech.2016.04.010.

Ledesma-Amaro, R., Lazar, Z., Rakicka, M., Guo, Z., Fouchard, F., Coq, A. M. C. L., & Nicaud, J. M. (2016). Metabolic engineering of *Yarrowia lipolytica* to produce chemicals and fuels from xylose. *Metabolic Engineering, 38*, 115–124. https://doi.org/10.1016/j.ymben.2016.07.001.

Ledesma-Amaro, R., Dulermo, R., Niehus, X., & Nicaud, J. M. (2016). Combining metabolic engineering and process optimization to improve production and secretion of fatty acids. *Metabolic Engineering, 38*, 38–46. https://doi.org/10.1016/j.ymben.2016.06.004.

Leiva-Candia, D. E., Pinzi, S., Redel-Macías, M. D., Koutinas, A., Webb, C., & Dorado, M. P. (2014). The potential for agro-industrial waste utilization using oleaginous yeast for the production of biodiesel. *Fuel, 123*, 33–42. https://doi.org/10.1016/j.fuel.2014.01.054.

Leplat, C., Nicaud, J. M., & Rossignol, T. (2018). Overexpression screen reveals transcription factors involved in lipid accumulation in *Yarrowia lipolytica*. *FEMS Yeast Research, 18*(5). https://doi.org/10.1093/femsyr/foy037.

Li, Z.-J., Wang, Y.-Z., Wang, L.-R., Shi, T.-Q., Sun, X.-M., & Huang, H. (2021). Advanced strategies for the synthesis of terpenoids in *Yarrowia lipolytica*. *Journal of Agricultural and Food Chemistry, 69*(8), 2367–2381. https://doi.org/10.1021/ACS.JAFC.1C00350.

Li, Qiang, Kamal, Rasool, Chu, Yadong, Wang, Qian, Yu, Xue, & Huang, Qitian (2020). Automated Pressurized Liquid Extraction of Microbial Lipids from Oleaginous Yeasts. *Applied Biochemistry and Biotechnology, 192*(1), 283–295. https://doi.org/10.1007/s12010-020-03331-9.

Li, Z. J., Qiao, K., Liu, N., & Stephanopoulos, G. (2017). Engineering *Yarrowia lipolytica* for poly-3-hydroxybutyrate production. *Journal of Industrial Microbiology and Biotechnology, 44*(4-5), 605–612. https://doi.org/10.1007/s10295-016-1864-1.

Liu, D., Ding, L., Sun, J., Boussetta, N., & Vorobiev, E. (2016). Yeast cell disruption strategies for recovery of intracellular bio-active compounds—A review. *Innovative Food Science and Emerging Technologies, 36*, 181–192. https://doi.org/10.1016/j.ifset.2016.06.017.

Llamas, M., Magdalena, J. A., González-Fernández, C., & Tomás-Pejó, E. (2020). Volatile fatty acids as novel building blocks for oil-based chemistry via oleaginous yeast fermentation. *Biotechnology and Bioengineering, 117*(1), 238–250. https://doi.org/10.1002/bit.27180.

Llamas, M., Dourou, M., González-Fernández, C., Aggelis, G., & Tomás-Pejó, E. (2020). Screening of oleaginous yeasts for lipid production using volatile fatty acids as substrate. *Biomass and Bioenergy, 138*, 105553. https://doi.org/10.1016/j.biombioe.2020.105553.

Ma, Y., Gao, Z., Wang, Q., & Liu, Y. (2018). Biodiesels from microbial oils: Opportunity and challenges. *Bioresource Technology, 263*, 631–641. https://doi.org/10.1016/j.biortech.2018.05.028.

Mata-Gómez, L. C., Montañez, J. C., Méndez-Zavala, A., & Aguilar, C. N. (2014). Biotechnological production of carotenoids by yeasts: An overview. *Microbial Cell Factories, 13*(1). https://doi.org/10.1186/1475-2859-13-12.

Mattanna, P., Rosa, P. D., da, Poli, J. S., Richards, N. S. P., Daboit, T. C., Scroferneker, M. L., Corção, G., Bertoldi, F. C., Deschamps, F. C., & Silva, P. V. da (2014). Lipid profile and antimicrobial activity of microbial oils from 16 oleaginous yeasts isolated from artisanal cheese. *Revista Brasileira de Biociências : Brazilian Journal of Biosciences, 12*(2), 121–126.

Meullemiestre, A., Breil, C., bert-Vian, M., & Chemat, F. (2016). Microwave, ultrasound, thermal treatments, and bead milling as intensification techniques for extraction of lipids from oleaginous *Yarrowia lipolytica* yeast for a biojetfuel application. *Bioresource Technology, 211*, 190–199. https://doi.org/10.1016/j.biortech.2016.03.040.

Monteiro, R. R. C., Dos Santos, I. A., Arcanjo, M. R. A., Cavalcante, C. L., de Luna, F. M. T., Fernandez-Lafuente, R., & Vieira, R. S. (2022). Production of jet biofuels by catalytic hydroprocessing of esters and fatty acids: A review. *Catalysts, 12*(2). https://doi.org/10.3390/catal12020237.

Morales-Palomo, S., González-Fernández, C., & Tomás-Pejó, E. (2022). Prevailing acid determines the efficiency of oleaginous fermentation from volatile fatty acids. *Journal of Environmental Chemical Engineering, 10*(2). https://doi.org/10.1016/j.jece.2022.107354 107354.

Morales-Palomo, S., Navarrete, C., Martínez, J. L., González-Fernández, C., & Tomás-Pejó, E. (2024). Transcriptomic profiling of an evolved *Yarrowia lipolytica* strain: tackling hexanoic acid fermentation to increase lipid production from short-chain fatty acids. *Microbial Cell Factories, 23*(1), 101. https://doi.org/10.1186/s12934-024-02367-4.

Morales-Palomo, S., Tomás-Pejó, E., & González-Fernández, C. (2023). Phosphate limitation as crucial factor to enhance yeast lipid production from short-chain fatty acids. *Microbial Biotechnology, 16*(2), 372–380. https://doi.org/10.1111/1751-7915.14197.

Narisetty, V., Prabhu, A. A., Bommareddy, R. R., Cox, R., Agrawal, D., Misra, A., Haider, M. A., Bhatnagar, A., Pandey, A., & Kumar, V. (2022). Development of hypertolerant strain of *Yarrowia lipolytica* accumulating succinic acid using high levels of acetate. *ACS Sustainable Chemistry and Engineering, 10*(33), 10858–10869. https://doi.org/10.1021/acssuschemeng.2c02408.

Nguyen, T. H., Fleet, G. H., & Rogers, P. L. (1998). Composition of the cell walls of several yeast species. *Applied Microbiology and Biotechnology, 50*(2), 206–212. https://doi.org/10.1007/s002530051278.

Ochsenreither, K., Glück, C., Stressler, T., Fischer, L., & Syldatk, C. (2016). Production strategies and applications of microbial single cell oils. *Frontiers in Microbiology, 7*. https://doi.org/10.3389/fmicb.2016.01539.

Papadaki, A., Fernandes, K. V., Chatzifragkou, A., Aguieiras, E. C. G., da Silva, J. A. C., Fernandez-Lafuente, R., Papanikolaou, S., Koutinas, A., & Freire, D. M. G. (2018). Bioprocess development for biolubricant production using microbial oil derived via fermentation from confectionery industry wastes. *Bioresource Technology, 267*, 311–318. https://doi.org/10.1016/j.biortech.2018.07.016.

Park, K., & Hahn, J. S. (2024). Engineering *Yarrowia lipolytica* for sustainable ricinoleic acid production: A pathway to free fatty acid synthesis. *Metabolic Engineering, 81*, 197–209. 10.1016/j.ymben.2023.12.002.

Park, Y. K., & Nicaud, J. M. (2020). Metabolic Engineering for Unusual Lipid Production in *Yarrowia lipolytica. Microorganisms, 8*(12). https://doi.org/10.3390/microorganisms8121937.

Park, Y. K., Dulermo, T., Ledesma-Amaro, R., & Nicaud, J. M. (2018). Optimization of odd chain fatty acid production by *Yarrowia lipolytica. Biotechnology for Biofuels, 11*(1). https://doi.org/10.1186/s13068-018-1154-4.

Park, Y. K., Ledesma-Amaro, R., & Nicaud, J. M. (2020). De novo biosynthesis of odd-chain fatty acids in *Yarrowia lipolytica* enabled by modular pathway engineering. *Frontiers in Bioengineering and Biotechnology, 7*. https://doi.org/10.3389/fbioe.2019.00484.

Park, Y. K., Bordes, F., Letisse, F., & Nicaud, J. M. (2021). Engineering precursor pools for increasing production of odd-chain fatty acids in *Yarrowia lipolytica. Metabolic Engineering Communications, 12*. https://doi.org/10.1016/j.mec.2020.e00158.

Patel, A., Mikes, F., & Matsakas, L. (2018). An overview of current pretreatment methods used to improve lipid extraction from oleaginous microorganisms. *Molecules, 23*(7), 1562. https://doi.org/10.3390/molecules23071562.

Patel, A., Arora, N., Pruthi, V., & Pruthi, P. A. (2019). A novel rapid ultrasonication-microwave treatment for total lipid extraction from wet oleaginous yeast biomass for sustainable biodiesel production. *Ultrasonics Sonochemistry, 51*, 504–516. https://doi.org/10.1016/j.ultsonch.2018.05.002.

Paul, D., Kumari, P. K., & Siddiqui, N. (2023). Yeast carotenoids: Cost-effective fermentation strategies for health care applications. *Fermentation, 9*(2). https://doi.org/10.3390/fermentation9020147.

Pereira, A. S., Lopes, M., Duarte, M. S., Alves, M. M., & Belo, I. (2023). Integrated bioprocess of microbial lipids production in *Yarrowia lipolytica* using food-waste derived volatile fatty acids. *Renewable Energy, 202*, 1470–1478. https://doi.org/10.1016/j.renene.2022.12.012.

Qiao, K., Imam Abidi, S. H., Liu, H., Zhang, H., Chakraborty, S., Watson, N., Kumaran Ajikumar, P., & Stephanopoulos, G. (2015). Engineering lipid overproduction in the oleaginous yeast *Yarrowia lipolytica*. *Metabolic Engineering, 29*, 56–65. https://doi.org/10.1016/j.ymben.2015.02.005.

Qiao, K., Wasylenko, T. M., Zhou, K., Xu, P., & Stephanopoulos, G. (2017). Lipid production in *Yarrowia lipolytica* is maximized by engineering cytosolic redox metabolism. *Nature Biotechnology, 35*(2), 173–177. https://doi.org/10.1038/nbt.3763.

Rapoport, A., Guzhova, I., Bernetti, L., Buzzini, P., Kieliszek, M., & Kot, A. M. (2021). Carotenoids and some other pigments from fungi and yeasts. *Metabolites, 11*(2), 92. https://doi.org/10.3390/metabo11020092.

Rizwanul Fattah, I. M., Ong, H. C., Mahlia, T. M. I., Mofijur, M., Silitonga, A. S., Rahman, S. M. A., & Ahmad, A. (2020). State of the art of catalysts for biodiesel production. *Frontiers in Energy Research, 8*. https://doi.org/10.3389/fenrg.2020.00101.

Robles-Iglesias, R., Naveira-Pazos, C., Fernández-Blanco, C., Veiga, M. C., & Kennes, C. (2023). Factors affecting the optimisation and scale-up of lipid accumulation in oleaginous yeasts for sustainable biofuels production. *Renewable and Sustainable Energy Reviews, 171*, 113043. https://doi.org/10.1016/j.rser.2022.113043.

Rodrigues, T. V. D., Teixeira, E. C., Macedo, L. P., dos Santos, G. M., Burkert, C. A. V., & de Medeiros Burkert, J. F. (2022). Agroindustrial byproduct-based media in the production of microbial oil rich in oleic acid and carotenoids. *Bioprocess and biosystems engineering, 45*(4), 721–732. https://doi.org/10.1007/s00449-022-02692-1.

Sagnak, R., Cochot, S., Molina-Jouve, C., Nicaud, J. M., & Guillouet, S. E. (2018). Modulation of the Glycerol Phosphate availability led to concomitant reduction in the citric acid excretion and increase in lipid content and yield in *Yarrowia lipolytica. Journal of Biotechnology, 265*, 40–45 http://10.1016/j.jbiotec.2017.11.001tec.2017.11.001.

Saini, R. K., Prasad, P., Shang, X., & Keum, Y. S. (2021). Advances in lipid extraction methods—A review. *International Journal of Molecular Sciences, 22*(24). https://doi.org/10.3390/ijms222413643.

Sakaki, H., Nakanishi, T., Tada, A., Miki, W., & Komemushi, S. (2001). Activation of torularhodin production by *Rhodotorula glutinis* using weak white light irradiation. *Journal of Bioscience and Bioengineering, 92*(3), 294–297. https://doi.org/10.1016/s1389-1723(01)80265-6.

Salvador López, J. M., Vandeputte, M., & Van Bogaert, I. N. A. (2022). Oleaginous yeasts: Time to rethink the definition? *Belgium Yeast, 39*(11-12), 553–606. https://doi.org/10.1002/yea.3827.

Schuck, S., Honsho, M., Ekroos, K., Shevchenko, A., & Simons, K. (2003). Resistance of cell membranes to different detergents. *Proceedings of the National Academy of Sciences of the United States of America, 100*(10), 5795–5800. https://doi.org/10.1073/pnas.0631579100.

Sen, T., Barrow, C. J., & Deshmukh, S. K. (2019). Microbial pigments in the food industry—challenges and the way forward. *Frontiers in Nutrition, 6*. https://doi.org/10.3389/fnut.2019.00007.

Su, J., van Dyk, S., & Saddler, J. (2022). Repurposing oil refineries to "stand-alone units" that refine lipids/oleochemicals to produce low-carbon intensive, drop-in biofuels. *Journal of Cleaner Production, 376*, 134335. https://doi.org/10.1016/j.jclepro.2022.134335.

Sun, X. M., Ren, L. J., Zhao, Q. Y., Ji, X. J., & Huang, H. (2019). Enhancement of lipid accumulation in microalgae by metabolic engineering. *Biochimica et Biophysica Acta - Molecular and Cell Biology of Lipids, 1864*(4), 552–566. https://doi.org/10.1016/j.bbalip.2018.10.004.

Szczepańska, P., Hapeta, P., & Lazar, Z. (2022). Advances in production of high-value lipids by oleaginous yeasts. *Critical Reviews in Biotechnology, 42*(1), 1–22. https://doi.org/10.1080/07388551.2021.1922353.

Sànchez i Nogué, V., Black, B. A., Kruger, J. S., Singer, C. A., Ramirez, K. J., Reed, M. L., Cleveland, N. S., Singer, E. R., Yi, X., Yeap, R. Y., Linger, J. G., & Beckham, G. T. (2018). Integrated diesel production from lignocellulosic sugars: Via oleaginous yeast. *Green Chemistry, 20*(18), 4349–4365 10.1039/c8gc01905c.

Thu, N. T. T., Hoang, L. H., Cuong, P. K., Viet-Linh, N., Nga, T. T. H., Kim, D. D., Leong, Y. K., & Nhi-Cong, L. T. (2023). Evaluation of polyhydroxyalkanoate (PHA) synthesis by *Pichia* sp. TSLS24 yeast isolated in Vietnam. *Scientific Reports, 13*(1). https://doi.org/10.1038/s41598-023-28220-z.

Tomás-Pejó, E., Morales-Palomo, S., & González-Fernández, C. (2021). Microbial lipids from organic wastes: Outlook and challenges. *Bioresource Technology, 323*, 124612. https://doi.org/10.1016/j.biortech.2020.124612.

Tomás-Pejó, E., Morales-Palomo, S., & González-Fernández, C. (2023). *Cutaneotrichosporon curvatum* and *Yarrowia lipolytica* as key players for green chemistry: Efficient oil producers from food waste via the carboxylate platform. *Bioengineered, 14*(1). https://doi.org/10.1080/21655979.2023.2286723.

Uğur, Ş., Zieniuk, B., & Fabiszewska, A. (2024). Nutritional and medicinal properties of microbial oil. *Applied Sciences, 14*(10), 4232. https://doi.org/10.3390/app14104232.

Wang, H., Peng, X., Zhang, H., Yang, S., & Li, H. (2021). Microorganisms-promoted biodiesel production from biomass: A review. *Energy Conversion and Management: X, 12*, 100137. https://doi.org/10.1016/j.ecmx.2021.100137.

Xiong, R. G., Zhou, D. D., Wu, S. X., Huang, S. Y., Saimaiti, A., Yang, Z. J., Shang, A., Zhao, C. N., Gan, R. Y., & Li, H. B. (2022). Health benefits and side effects of short-chain fatty acids. *Foods, 11*(18), 2863. https://doi.org/10.3390/foods11182863.

Xu, P., Qiao, K., & Stephanopoulos, G. (2017). Engineering oxidative stress defense pathways to build a robust lipid production platform in *Yarrowia lipolytica. Biotechnology and Bioengineering, 114*(7), 1521–1530. https://doi.org/10.1002/bit.26285.

Xue, Z., Sharpe, P. L., Hong, S. P., Yadav, N. S., Xie, D., Short, D. R., Damude, H. G., Rupert, R. A., Seip, J. E., Wang, J., Pollak, D. W., Bostick, M. W., Bosak, M. D., Macool, D. J., Hollerbach, D. H., Zhang, H., Arcilla, D. M., Bledsoe, S. A., Croker, K., ... Zhu, Q. (2013). Production of omega-3 eicosapentaenoic acid by metabolic engineering of *Yarrowia lipolytica. Nature Biotechnology, 31*(8), 734–740. https://doi.org/10.1038/nbt.2622.

Yellapu, S. K., Bezawada, J., Kaur, R., Kuttiraja, M., & Tyagi, R. D. (2016). Detergent assisted lipid extraction from wet yeast biomass for biodiesel: A response surface methodology approach. *Bioresource Technology, 218*, 667–673. https://doi.org/10.1016/j.biortech.2016.07.011.

Ylinen, A., Maaheimo, H., Anghelescu-Hakala, A., Penttilä, M., Salusjärvi, L., & Toivari, M. (2021). Production of D-lactic acid containing polyhydroxyalkanoate polymers in yeast *Saccharomyces cerevisiae. Journal of Industrial Microbiology and Biotechnology, 48*(5-6). https://doi.org/10.1093/jimb/kuab028.

Yoo, G., Park, W. K., Kim, C. W., Choi, Y. E., & Yang, J. W. (2012). Direct lipid extraction from wet *Chlamydomonas reinhardtii* biomass using osmotic shock. *Bioresource Technology, 123*, 717–722,. https://doi.org/10.1016/j.biortech.2012.07.102.

Yook, S. D., Kim, J., Woo, H. M., Um, Y., & Lee, S. M. (2019). Efficient lipid extraction from the oleaginous yeast *Yarrowia lipolytica* using switchable solvents. *Renewable Energy, 132*, 61–67. 10.1016/j.renene.2018.07.129.

Yu, X., Dong, T., Zheng, Y., Miao, C., & Chen, S. (2015). Investigations on cell disruption of oleaginous microorganisms: Hydrochloric acid digestion is an effective method for lipid extraction. *European Journal of Lipid Science and Technology, 117*(5), 730–737. https://doi.org/10.1002/ejlt.201400195.

Zainuddin, M. F., Fai, C. K., Ariff, A. B., Rios-Solis, L., & Halim, M. (2021). Current pretreatment/cell disruption and extraction methods used to improve intracellular lipid recovery from oleaginous yeasts. *Microorganisms, 9*(2), 251. https://doi.org/10.3390/microorganisms9020251.

Zhang, F. L., Zhang, L., Zeng, D. W., Liao, S., Fan, Y., Champreda, V., Runguphan, W., & Zhao, X. Q. (2023). Engineering yeast cell factories to produce biodegradable plastics and their monomers: Current status and prospects. *Biotechnology Advances, 68*. https://doi.org/10.1016/j.biotechadv.2023.108222.

Zhang, L. S., Liang, S., Zong, M. H., Yang, J. G., & Lou, W. Y. (2020). Microbial synthesis of functional odd-chain fatty acids: A review. *World Journal of Microbiology and Biotechnology, 36*(3). https://doi.org/10.1007/s11274-020-02814-5.

Zhang, S., He, Y., Sen, B., & Wang, G. (2020). Reactive oxygen species and their applications toward enhanced lipid accumulation in oleaginous microorganisms. *Bioresource Technology, 307*, 123234. https://doi.org/10.1016/j.biortech.2020.123234.

Zhang, X., Yan, S., Tyagi, R. D., Drogui, P., & Surampalli, R. Y. (2014). Ultrasonication assisted lipid extraction from oleaginous microorganisms. *Bioresource Technology, 158*, 253–261. https://doi.org/10.1016/j.biortech.2014.01.132.

Zhou, X., Jin, W., Tu, R., Guo, Q., Han, Sf, Chen, C., Wang, Q., Liu, W., Jensen, P. D., & Wang, Q. (2019). Optimization of microwave assisted lipid extraction from microalga *Scenedesmus obliquus* grown on municipal wastewater. *Journal of Cleaner Production, 221*, 502–508. https://doi.org/10.1016/j.jclepro.2019.02.260.

Zinjarde, S. S., & Pant, A. (2002). Emulsifier from a tropical marine yeast, *Yarrowia lipolytica* NCIM 3589. *Journal of Basic Microbiology, 42*(1), 67–73 https://doi.org/10.1002/1521-4028(200203)42:1<67::AID-JOBM67>3.0.CO;2-M.

Industrial enzyme production by yeast

7

Fakhra Liaqat[1], *Bahar Aslanbay Guler*[2], *and Esra Imamoglu*[2]
[1]Center of Environmental and Biological Studies, Arabian Gulf University, College of Education, Administrative and Technical Sciences, Manama, Bahrain, [2]Bioengineering Department, Ege University, Izmir, Turkey

7.1 Introduction

Enzymes play a pivotal role in a myriad of industrial processes, offering a sustainable and efficient means of catalysing biochemical reactions. Röhm and Haas, a German company, commercialized an enzyme formulation for the first time in 1907. Over a century after this groundbreaking achievement, enzymes are emerging as an attractive alternative to an expanding range of industrial chemical processes. The global enzymes market is expected to increase at a compounds annual growth rate (CAGR) of 7.8% from 2023 to 2029 reaching $20.4 billion by 2029 (Global Enzyme Mark, 2023). Novozymes, a Danish company currently known as Novensis, occupies nearly half of the global market. Dutch multinational corporation (DSM) in the Netherlands and Badische Anilinund Sodafabrik (BASF) in Germany are prominent European companies, while DuPont, a part of International Flavors & Fragrences (IFF) Inc., has gained a significant position in the enzyme industry in the United States (Talens-Perales et al., 2015). Currently, enzyme industries face several significant challenges, including limited enzyme stability under demanding industrial processing conditions, issues with microbial contamination, and the low recyclability of biocatalysts (Mesbah, 2022).

In the production of industrial enzymes, yeast has garnered significant attention due to the ease of genetic manipulation, rapid growth rates, and adaptability to diverse environmental conditions (Walker, 1998; Yang et al., 2024). Yeasts such as *Saccharomyces* and *Pichia* have been extensively researched for their potential in producing several important enzymes that find applications across various sectors, including textiles, paper and pulp, pharmaceuticals, agriculture, detergents, waste management, biorefineries, and food industries (Chi et al., 2009; Parapouli et al., 2020; Shrivastava et al., 2023). Major industrially important yeast enzymes include, amylase, cellulase, gelatinase, β-glucosidase, inulinase, lipase, pectinase, protease, urease, and tannase. This book chapter focuses on exploring the production of these key enzymes by yeast, addressing the recent advances in their production and the challenges that must be overcome to optimize their production. This chapter will also explore the application of these enzymes across different industries. For instance, amylases are used in the food industry for starch liquefaction and brewing, cellulases in biofuel production, lipases in detergent formulations, and proteases in the

Eukaryotic Microorganisms as Sources of Bioproducts. DOI: https://doi.org/10.1016/B978-0-443-30188-9.00013-5

pharmaceutical and food processing industries. Each enzyme has unique applications and associated challenges in production, such as maintaining stability, optimizing yield, and controlling contamination. Addressing these challenges is crucial for the successful industrial production of enzymes by yeast.

7.2 Industrial enzymes

Industrially important yeast enzymes, their culture conditions, enzyme activities, and applications are summarized in Table 7.1.

7.2.1 Amylase

Amylases are enzymes that hydrolyze starch into sugars. These enzymes can be categorized into four types: α-amylase (EC 3.2.1.1), which breaks down starch by breaking the 1,4-α-glycosidic bonds; β-amylase (EC 3.2.1.2), which converts starch into maltose by hydrolyzing the α-1,4-glucan linkages; glucoamylase (EC 3.2.1.3), which releases glucose from the non-reducing ends of starch molecules; and isoamylase (EC 3.2.1.68), which plays a role in breaking down the α-1,6-glucosidic linkage in the side chain of starch polysaccharide (Mondal et al., 2022; Shad et al., 2023). It is worth mentioning that the enzyme classification system, which includes EC numbers developed by the Enzyme Nomenclature Committee of the International Union of Biochemistry and Molecular Biology (IUBMB), has been widely used to describe enzymes and their reactions. However, a current trend in enzyme classification—the Carbohydrate-Active EnZymes (CAZy) classification system—has gained significant attention for categorizing and annotating enzymes involved in the synthesis, modification, and breakdown of carbohydrates (Drula et al., 2022). According to CAZy system, amylases are classified primarily under the glycoside hydrolase (GH) families. These enzymes belong to GH families GH13, GH57, GH119, and GH126 (Janeček & Svensson, 2022). Amylases account for almost 30% of the worldwide industrial enzyme market. The amylases industry has made a total contribution of $7.5 billion, with a CAGR of 8.2% between 2015 and 2020. The global need for enzymes is met by the main manufacturers: Novensis, IFF Inc., and Roche based in Switzerland (Shad et al., 2023). Amylases have diverse applications in various industries, including starch liquefaction and saccharification, bread and baking, textile desizing, paper production, detergents, medical and clinical chemistry analysis, as well as the food and pharmaceutical sectors (Far et al., 2020; Mondal et al., 2022; Salah et al., 2021).

Yeast, particularly strains of *S. cerevisiae*, has been extensively studied for its ability to produce amylase efficiently (Augustín et al., 1978). During the last 30 years, there has been substantial advancement in the expression of several amylase genes in *S. cerevisiae* for the purpose of starch conversion (van Zyl, Bloom, & Viktor, 2012). The production of amylase through fermentation in yeast can be achieved using either submerged fermentation (SmF) or solid-state fermentation (SSF) methods, with the selection of the method based on the particular requirements of the process. The production of industrially significant enzymes has typically depended on SmF

Table 7.1 Production and applications of industrially important enzymes by yeast (*NA=Not available).

Enzyme	Strain	Culture conditions	Enzyme activity	Application	References
Amylase					
	Cryptococcus flavus	pH 5.5 and 50°C	0.25 U/ml	Hydrolysis of starch, amylose, and amylopectin	Wanderley et al. (2004)
	Apiotrichum mycotoxinivorans	0.5 % soluble starch, at 28°C for 96 h	NA	Hydrolysis of starch	Carvalho et al. (2021)
	Pichia membranifaciens	pH 7.0 and 35°C	32.5 U/mL	NA	Salah et al. (2021)
	Candida sorboxylosa and *Cyberlindnera fabianii*	SSF for 48 h at pH 4 and 25°C	17.1 U/mL and 31.2 U/mL	NA	Sukmawati et al. (2021)
	Clavispora lusitanice ABS7	pH 9.0 and 75°C	352.41 U/mg	Laundry detergent	Dakhmouche Djekrif et al. (2021)
	Candida wangnamkhiaoensis	Single-stage steady-state continuous cultures in an airlift bioreactor	14000 U/L/h	Production of biofuels	Chávez-Camarillo et al. (2022)
Cellulase and β-glucosidase					
	Candida tropicalis MK-118	SmF	17.39 and 12.32 IU/ mL	Bioconversion of lignocellulosic biomass	Shariq and Sohail (2020)
	Trichosporon laibachii	SmF	0.3 U/mL	NA	Giese et al. (2017)
	Trichosporon asahii	pH 6.5 and 30°C	35.70 U	Saccharification of biomass	Valliammai et al. (2021)

(*Continued*)

Table 7.1 (Continued)

Enzyme	Strain	Culture conditions	Enzyme activity	Application	References
	Apiotrichum mycotoxinivorans	1% carboxymethylcellulose at 28°C for 96 h	NA	NA	Carvalho et al. (2021)
	Rhodotorula oryzicola	pH 6.02 and 65°C	1.44 U/mg	Bioethanol production	de Araujo Ribeiro and Assis (2023)
	Cyberlindnera rhodanensis	48 h cultivation with xylan substrate at 37°C	9.45 U/mL	NA	Kham et al. (2024)
Gelatinase					
	Metschnikowia sp., *Leucosporidium fragarium*, and *Mrakia* sp.,	pH 2.6–7.6 and 4°C–37°C	NA	NA	Yuivar et al. (2019)
	Candida kefyr 41 PSB	pH 7.0 and 105°C	360 U/mg	Gelatin hydrolysis	Yavuz et al. (2017)
Inulinase					
	Rhodotorula mucilaginosa	SmF, pH 8.0 and 75°C	3.54 U	Fructooligosaccharides production	de Araujo Ribeiro et al. (2021)
	Saccharomyces sp.	SSF, pH 5.5 and 35°C	91.0 ± 0.60 U/g of dry substrate	NA	Onilude et al. (2012)

Lipase					
	Yarrowia lipolytica	SSF, pH ~6.0, 27°C, and sunflower/ olive cakes as substrates	85 U/g of dry mass substrates	Fatty acids esterification during biodiesel production	Costa et al. (2023)
	Yarrowia lipolytica	SSF, pH 8.0, 28°C, and soybean meal	106.7±3.7 U/g	Synthesis of esters for food industry	de Souza et al. (2019)
	Yarrowia lipolytica	SSF, 28°C, 14 h and soybean bran	16 U/g (μmoles of acid min^{-1} g^{-1})	Biocatalytic esterification of fatty acids	Silva et al. (2019)
	Yarrowia lipolytica	SSF, 28°C, soybean bran supplemented with watermelon peels	75.22 U/g	Biocatalysts in poly (ethylene terephthalate) depolymerization reactions	Sales et al. (2020)
	Candida rugosa Geotrichum candidum	SmF, 27.1°C, pH 3.5, and soybean molasses	12.3 U/mL and 11.48 U/mL	NA	De Morais et al. (2016)
	Metschnikowia pulcherrima Zygosaccharomyces mrakii Candida boidinii Saccharomyces cerevisiae	SmF, 30°C, and onion skins	25.63 ± 0.38 U/mL 45.45 ± 2.53 U/mL 123.23 ± 2.78 U/mL 68.65 ± 4.42 U/mL	NA	Ramires et al. (2023)
	Magnusiomyces capitatus	30°C, pH 6.1, 100 rpm mixing, and 1 vvm aeration	3.96 U/mL	NA	Salgado et al. (2020)
	Rhodotorula sp.	SmF, pH 8.0, and 15°C	4.77 U/mg	Potential candidate in detergent formulation	Maharana and Singh (2018)
	Meyerozyma guilliermondii	SmF, pH 4.0, and cheese whey	285.82 U/mL	Potential use as feed enzymes	Knob et al. (2020)

(Continued)

Table 7.1 (Continued)

Enzyme	Strain	Culture conditions	Enzyme activity	Application	References
Pectinase					
	Yarrowia phangngaensis	SmF, pH 7.5, and banana peel	1.601 U/mL	Apple juice clarification	Ezeh et al. (2023)
	Rhodosporidiobolus, Cystofilobasidium, and *Yamadazyma*	SmF, 15°C, and 150 rpm mixing	0.76 U/mL - 1.73 U/mL	NA	Daskaya-Dikmen et al. (2018)
	Geotrichum candidum	pH 5.3 and 25°C	0.25 IU/mL	NA	Ahmed et al. (2019)
	Meyerozyma sp.	pH 7.0, 25°C, and 180 rpm mixing	~14 U/mL min	Fruit juice clarification	Meena et al. (2021)
Protease					
	Hanseniaspora uvarum *Candida californica* *Cryptococcus albidus* *Pichia membranifaciens* *Rhodotorula mucilaginosa*	30°C and 120 rpm mixing	187 U/mL at pH 6.0 185 U/mL at pH 7.0 180 U/mL at pH 7.0 152 U/mL at pH 6.0 175 U/mL at pH 6.0	NA	Hesham et al. (2017)
	Yarrowia lipolytica	30°C, 210 rpm mixing, and chrome-tanned leather shavings as carbon source	800 U/mg of protein	NA	Wieczorek et al. (2023)
	Rhodosporidiobolus, Cystofilobasidium, and *Yamadazyma*	SmF, 15°C, and 150 rpm mixing	2.11–10.53 U/mL	NA	Daskaya-Dikmen et al. (2018)

Tannase					
	Rhodosporidium diobovatum	25°C and 15 g/L tannic acid	26.4 U/mL	Tannin biodegradation and gallic acid production	Pan et al. (2020)
	Geotrichum cucujoidarum	SSF, 30°C, pH 5.0, Jamun leaves and lemon peel powder	26.43 U/g dry solids	NA	Thangavelu et al. (2024)
	Sporidiobolus ruineniae	30°C and 150 rpm mixing	31.1 mU/mL	Feed industry	Kanpiengjai et al. (2020)
	Saccharomyces cerevisiae	SmF, 30°C, and pH 6.0	NA	Beverage manufacturing and gallic acid production	Lopes et al. (2018)
	Kluyveromyces marxianus	SSF, 45°C, olive pomace waste, 48 h	42.37 U/mg	NA	Mahmoud et al. (2018)
	Rhodotorula glutinis	pH 6.0, 25°C, 3 days, 150 rpm mixing	3.33 U/mg	NA	Ong et al. (2024)

because of its simplicity of controlling a variety of parameters, including pH, temperature, aeration, oxygen transfer, and moisture (de Souza & de Oliveira e Magalhães, 2010). In SmF, mass transfer is the main challenge. To address the limitations of mass transfer in SmF and to facilitate the scaling up of solid-state and liquid-surface cultures, an alternating liquid phase–air phase bioreactor was developed. The setup included of a bioreactor that was fitted with a siphon system and a reservoir. The yeast *S. cerevisiae* IR2 was immobilized in the reservoir and used for the concurrent production of amylase and ethanol from untreated cassava starch. The method exhibited exceptional stability during seven consecutive batches, with an outstanding ethanol yield (0.46 g-ethanol per g-starch) and productivity (1.73 g-ethanol/L/h) (Roble et al., 2020). The substrate for amylase production by yeast typically includes starch-rich materials such as wheat, potato, corn, and cassava starch (Chen et al., 2010; Prado et al., 2016). A recent study explored the production kinetics of α-amylase from *Candida wangnamkhiaoensis* in single-stage steady-state continuous cultures within an airlift bioreactor, using a range of carbon and energy sources as fermentation substrates. In *C. wangnamkhiaoensis*, α-amylase production was stimulated by maltose and starch, with starch being the more effective inducer. Using starch as the carbon source, the maximum α-amylase specific activity (16900 ± 00 U/g biomass) was obtained at a dilution rate of 0.24/h (Chávez-Camarillo et al., 2022). The adaptability and versatility make yeast an invaluable organism for biotechnological applications involving amylase production.

7.2.2 Cellulase

Cellulase enzymes, which are crucial for breaking down cellulose into glucose units, play an important role in industries such as biofuel production, textiles, and paper and pulp. The enzymes are categorized into three primary groups: endoglucanase (endo-1,4-β-D-glucanase, EG, EC 3.2.1.4), exoglucanase (exo-1,4-β-D-glucanase, CBH, EC 3.2.1.9), and β-glucosidase (1,4-β-D-glucanase, BG, EC 3.2.1) (Korsa et al., 2023). Cellulases in the CAZy database are distributed among several GH families including GH1, GH3, GH5, GH6, GH7, GH8, GH9, GH12, GH45, and GH48 as well as auxiliary activities (AA) Families AA9, and AA10 (Co & Hug, 2020). Cellulases, together with pectinase and hemicellulases, collectively account for around 20% of the worldwide enzyme market. Glucose and other types of simple sugars, which are produced when cellulase breaks down cellulose, may be used as an inexpensive substance for fermentation. This process can then be used to generate ethanol and other highly valuable products (Sohail et al., 2022). Cellulase enzymes hold significant potential in several key industries, including biofuel production, textile processing and finishing, paper and pulp manufacturing, brewery and wine industry, animal feed industry, agriculture and waste management (Sutaoney et al., 2024). The market for cellulases is projected to reach over $9 billion by 2030. IFF Inc. and Novensis are the dominant global manufacturers of cellulase enzyme, providing cellulase blends for various applications (Areeshi, 2022). Other international companies manufacturing and marketing cellulases are DMS, AB Enzymes (Darmstadt, Germany), and Sunson Industry Group (Beijing, China) (Ilić et al., 2023).

While fungi and bacteria are known producers of cellulase enzymes, yeast cellulases have recently garnered attention owing to their activity across a wide range of pH and at elevated temperatures (Choudhary et al., 2016; Poomani et al., 2024; Touijer et al., 2019). For instance, Shariq and Shoail Shariq & Sohail (2020) documented the concurrent production of cellulase and xylanase enzymes using a native *C. tropicalis* MK-118 strain (Table 7.1). Beyond terrestrial yeasts, marine yeasts have also been extensively studied for their cellulose-degrading capabilities. Rong et al., (2015) examined the yeast populations in surface seawater from the Yellow Sea in China, specifically targeting yeast strains with potential for cellulase production. The heterologous expression of cellulase genes from fungi or bacteria in *S. cerevisiae* has been achieved, allowing efficient cellulose fermentation (Oh & Jin, 2020). Additionally, various yeast strains, including *Candida* sp., *Dendrobium spathilingue*, and *Moesziomyces* sp., have been reported to effectively hydrolyze crude lignocellulosic substrates (Sohail et al., 2022).

Current advances in cellulase production include the use of more affordable substrates such as readily accessible fruit wastes as well as the deployment of a cost-effective fermentation technique like SSF. These advancements have the potential to reduce the overall production cost of cellulases. In addition, SSF closely replicates the natural environment for fungus, resulting in lower energy and capital requirements while yielding a higher amount of enzymes compared to SmF (Abduh et al., 2022; Areeshi, 2022). SSF is particularly suitable for the bioconversion of agricultural residues and other lignocellulosic materials (Behera et al., 2023). SSF has the potential to yield higher enzyme titers and activities due to better substrate utilization (Leite et al., 2021). Optimization of fermentation conditions, genetic engineering techniques to enhance enzyme expression, and the development of more efficient and cost-effective purification methods are ongoing areas of research. Addressing these challenges is critical for the large scale and cost-effective synthesis of cellulase enzymes using yeast, which holds promise for sustainable industrial applications.

7.2.2.1 β-Glucosidase

β-Glucosidases (EC 3.2.1.21) are glycoside hydrolase enzymes that break β-glucosidic bonds in disaccharides and oligosaccharides, releasing glucose from diverse substrates (Sirilun et al., 2016). While β Glucosidase can be found in animals and plants, the majority of commercially used β-Glucosidase enzymes are of microbial origin, highlighting the significance of microbial production for this enzyme (Tiwari et al., 2013). The cellulolytic process of bacteria and fungi is facilitated by β-glucosidases in yeasts, such as *Debaryomyces hansenii*, these enzymes are responsible for the release of flavoring compounds including nerol, geraniol, linalool, benzyl, and phenyl ethyl alcohols (Sirilun et al., 2016). In biorefineries, β-glucosidases are the rate-limiting enzymes for cellulose breakdown, resulting in glucose release from cellobiose. β-glucosidases are used in medicines, cosmetics, brewing, dairy, and food processing industries (Kannan et al., 2023). β-glucosidase is often used in the fruit processing and wine manufacturing industries to improve desirable flavor characteristics. Yeast β-glucosidases have been examined extensively for their potential to enhance the taste of alcoholic beverages (Godse et al., 2021). The purified β-glucosidase derived from *Issatchenkia terricola* induced the liberation of volatile chemicals from their glucosides, hence augmenting the desirable fragrance profile of Cabernet Sauvignon wine. In addition, the enzyme was incapable of hydrolyzing anthocyanin glucosides, resulting in the preservation of the red color in wines (de Ovalle et al., 2018). The potential enological use of β-glucosidase from *Trichosporon asahii* has been investigated, revealing notable relative activity and stability in conditions of 10% ethanol, 20% sugar, and pH 3.0 (Wang et al., 2015). Several other wine-related yeasts including *Pichia anomala* C84, *Hanseniaspora guilliermondii* A27-3-4, *Metschnikowia pulcherrima* F-1–6, and *Clavispora lusitaniae* C117 had shown remarkable β-glucosidase activity (Huang et al., 2021), depicting their potential applications in enhancing wine aroma.

The applications of β-glucosidase extend to bioethanol production sector. β-glucosidase has been studied for its ability to effectively convert cellulosic biomass into fermentable sugars, making it a promising candidate for bioethanol production. Recently, de Araujo Ribeiro & Assis (2023) used β-glucosidase from yeast *Rhodotorula oryzicola* to hydrolyze delignified sugarcane bagasse with conversion rate of $72.46 \pm 1.60\%$. *Cyberlindnera rhodanensis*, an ethanol-producing yeast, secretes a thermostable extracellular β-glucosidase when cultivated with 0.5% (w/v) carboxy-methyl cellulose as carbon source. At 35, 45, and 55°C, this crude β-glucosidase demonstrated a remarkable stability of 100%, 88%, and 75% for a period of 3 hours, respectively (Kham et al., 2024). Companies like IFF Inc. and Novensis are actively involved in developing enzyme cocktails containing β-glucosidase for bioethanol applications, showcasing the growing interest in utilizing this enzyme for sustainable energy production.

7.2.3 Gelatinase

Gelatinase is an extracellular metalloproteinase which hydrolyzes gelatin into polypeptides, peptides, and amino acids. Gelatinases can also act on pheromone, collagen, casein, proteoglycan, elastin, laminin, fibronectin, and fibrinogen (Deepa BK & Nanjundaiah, 2023). Gelatinase activity is not directly classified in the CAZy database because CAZy focuses on carbohydrate-active enzymes. However, gelatinases are proteases and are classified in a protease-focused database MEROPS. Common families for gelatinases in the MEROPS database include Matrix Metalloproteinases (MMPs), such as M10 (MEROPS, 2024). Gelatinases find applications in medicine, food industry, pharmaceuticals, and biorefineries. Studies have shown that gelatinases participate in numerous neurodegenerative and autoimmune diseases (Malemud, 2006). In pharmaceuticals, gelatinases are employed in drug delivery systems. In the cosmetics industry, they are used in antiaging products to enhance skin texture and appearance. In the food, they are used to clarify the beverages, in meat processing and manufacturing of gel-based products. Furthermore, gelatinases participate in the bioremediation of gelatinous waste, reducing environmental pollution. Additionally, gelatinases have applications in molecular biology in the extraction and purification of nucleic acids and proteins (Pathade, 2024). Major commercial players in gelatinase manufacturing are Sigma-Aldrich (now Merck) and Creative Enzymes. Microbial sources of gelatinase are mainly bacteria; however, yeasts can also produce these enzymes (Yuivar et al., 2019). Microbial gelatinases offer several advantages over plant and animal sources, including simple production method, scalability, and a lower risk of contamination (Pathade, 2024). Researchers have isolated various psychrotolerant yeast species and confirmed the synthesis of extracellular gelatinase in *Leucosporidium fragarium*, *Mrakia* sp., and *Metschnikowia* sp. Among these yeasts, *Mrakia* sp. has been shown to have the greatest level of gelatinase activity (Yuivar et al., 2019). A thermostable metalloprotease was isolated from *C. kefyr* 41 PSB in earlier research. The enzyme was found to have optimal activity at a pH of 7.0 and a temperature of 105°C. When casein was used as a substrate, the enzyme exhibited a Km of 3.5 mg/mL and a V_{max} of 4.4 μmol/mL/min, whereas with gelatin as the substrate, the Km and V_{max} values were 1.65 mg/mL and 6.1 μmol/mL/min, respectively (Yavuz et al., 2017). Challenges in gelatinase production by yeast may include optimizing fermentation conditions for maximum enzyme yield, improving substrate specificity, and addressing issues related to enzyme stability. Overcoming these challenges is crucial to ensure the efficient production of gelatinase enzymes and meet the diverse industrial demands for this versatile enzyme.

7.2.4 Inulinase

Inulinase, also referred to as β-2,1-D fructan fructanohydrolase (EC 3.2.1.7), is an enzyme that breaks down inulin into fructose. Inulinases are categorized into endo and exoinulinase based on their mode of action. Endoinulinase, also known as 2,1-β-D fructan fructanohydrolase (EC 3.2.1.7), cleaves the bonds between fructose units that are situated away from the end of the inulin polymer. Conversely, exoinulinase (β-D-fructohydrolase; EC 3.2.1.80) cleaves the end units of fructose in inulin, raffinose, and sucrose, resulting in fructose as the primary product (Othman et al., 2020). According to CAZy database, primary family for inulinases is GH32, which also includes other fructosyltransferases and invertases (CAZY database, 2024). Inulinase has garnered significant interest due to its extensive applicability in the production of fructose syrup and ethanol from inulin (Vijayaraghavan et al., 2009). They are extensively used in the production of fructooligosaccharides, food products, pharmaceuticals, and beverages. They also play a key role in the production of ethanol, sorbitol, citric acid, lactic acid, and other chemicals as well as in energy generation and the manufacturing of bioplastics (de Carvalho Cardoso et al., 2025). The key players in the global inulinase market are Novensis, Creative Enzymes, IFF Inc., DMS, Amano Enzyme Inc., Megazyme, BASF, AB enzymes, and Sunson Industry Group (Inulinase Market Res, 2024). Inulinase is synthesized by several microorganisms including yeasts, such as *Kluyveromyces* sp., *C. kefyr*, *D. cantarelli* and *Cryptococcus aureus* (Sheng et al., 2007). *Kluyveromyces* are the most frequently employed yeast species for producing inulinase (Kalil et al., 2005). The variables that influence the production of inulinase by *K. marxianus* include aeration, agitation, and shear stress. It has been shown that the optimal conditions for fermentation include aeration at a rate of 1 vvm, agitation at 450 rpm using a pitched blade up impeller in a stirred reactor. Under these conditions, a production of 176 IU/ml of inulinase was achieved (Silva-Santisteban et al., 2006). Previously, Paixao et al. isolated a novel yeast strain (*Zygosaccharomyces bailii* Talf1) for the bioprocesses of inulin-rich materials, attaining an inulinase activity of 8.67 U/ml with *Jerusalem artichoke* juice as the inducer substrate (Paixão et al., 2013). Inulinase may be used to produce fructooligosaccharides (FOS) from inulin. For instance, de Araujo Ribeiro et al. (2021) used SmF to produce inulinase from *Rhodotorula mucilaginosa*. The inulinase was then immobilized on chicken eggshell. Subsequently, fructooligosaccharides were synthesized utilizing the immobilized inulinase (at a concentration of 5% w/v) and inulin as the substrate (de Araujo Ribeiro et al., 2021). These investigations demonstrated the potential role of inulinases in the processing of agricultural residues to produce high-value oligosaccharides.

7.2.5 Lipase

Lipases (EC 3.1.1.3) are an important class of hydrolytic enzymes that facilitate breakdown of triglycerides into glycerol and free fatty acids through hydrolysis. They are typically not classified within the CAZy classification unless they specifically target carbohydrate esters. However, the Carbohydrate Esterase 5 family in the CAZy database includes cutinases, which share structural and functional similarities with lipases, as both types of enzymes act on ester bonds (Nakamura et al., 2017). The catalytic characteristics of lipases are crucial affecting the application of lipase in various sectors, including biofuels, medicine, cosmetics, food, and feed (Raoufi & Mousavi Gargari, 2018). It is extensively utilized in these sectors for a wide variety of purposes, such as enhancing flavors, improving texture, extending the shelf life of various products, removing lipid-based stains, improving nutrient absorption and digestion, and promoting better health and growth in livestock. Currently, particular attention has been given to biodiesel production because lipases exhibit strong catalytic activity, high conversion rate, stereo selectivity, and regioselectivity under mild conditions (Cavalcante et al., 2021; Kumar et al., 2023). Beyond the energy sector, lipases have interesting practices in the leather industry as tanning agents, in pulp and paper production to reduce pitch problems, in biosensor development, in fabrication of biodegradable compounds, and in wastewater bioremediation (Kumar et al., 2023). However, its commercial use faces several challenges, including low stability, difficulties in separation, and limited reusability. To address these issues, the immobilization of lipase has been widely employed using various physical and chemical methods. Recently, metal-organic framework (MOF)-mediated lipase immobilization has been proposed as a novel and efficient strategy to enhance the enzyme's efficiency, stability, and recyclability, thereby expanding its application fields. One promising area where MOF-mediated lipase can be utilized is enantio-separation, which plays a critical role in the pharmaceutical and chemical industries for producing single enantiomers, especially with the advent of chiral drugs (Patil et al., 2023). Given the current and potential applications of lipase, its global industrial market has been experiencing significant growth and is projected to reach approximately USD 645.18 million by 2029 (Lipase Market Res, 2024). This growth reflects the enzyme's critical role in driving innovation across above-mentioned industries. Several major companies including Novensis, DSM, Associated British Foods, and IFF Inc. dominate this market (Guerrand, 2017).

While lipases can be easily obtained from plants and animals, microbial sources are more preferable for commercial applications due to their high production rate, ease of cultivation, susceptibility to genetic modification, and broad substrate specificity (Filho et al., 2019). Among these microorganisms, yeasts and filamentous fungi have been widely studied because they are capable of producing extracellular lipases, which allows the extraction step to be skipped. Most common lipase-producing yeasts are *Y. lipolytica, S. cerevisiae, Candida* spp., *Rhodotorula* spp., and *Pichia* spp. (Kuncharoen et al., 2020; Sakpuntoon et al., 2020; Salgado et al., 2020). Currently, studies on lipase production by yeasts has mainly focused on determining the lipase activity of various strains, optimizing process conditions, identifying or constructing novel species, and applying the produced lipases in different fields (Table 7.1). Costa, Salgado, and Belo compared *Y. divulgata* and *Y. lipolytica* to produce lipase by using sunflower and olive cakes as substrates in a SSF system. *Y. lipolytica*, the most extensively employed yeast for lipase production, possessed the maximal lipase activity of 85 U/g in tray bioreactors (Costa et al., 2023). Within the scope of lipase production from *Y. lipolytica*, de Souza, Ribeiro, and Coelho stated the highest lipase activity of 106.7 U/g using soybean meal in SSF (de Souza et al., 2019), while Silva et al. (2019) assessed a lipase activity of 16 U/g through fermentation of soybean bran as a low-cost biocatalyst for biodiesel production. Sales et al. (2020) reported a 31% rise in lipase activity of *Y. lipolytica* with the supplementation of soybean bran with peels of watermelon in a SSF process.

Considering different yeast species, De Morais et al. (2016) examined the optimum fermentation conditions of *C. rugose* and *G. candidum* for lipase production in a SmF containing soybean molasses. They found that using 200 g/L of soybean molasses at 27 ± 1°C and pH 3.5 resulted in 12.3 U/mL and 11.48 U/mL for *C. rugosa* and *G. candidum*, respectively (De Morais et al., 2016). Ramires et al. (2023) assayed 45 different yeast strains for synthesis of polyphenol and lipase using red and yellow onion skins in liquid SmF. *Metschnikowia pulcherrima* and *Zygosaccharomyces mrakii* showed the highest lipase activity (25.63 U/mL and 45.45 U/mL, respectively) with red onion skin, while the highest activities with yellow skin were exhibited by *C. boidinii* (123.23 U/mL) and *S. cerevisiae* (68.65 U/mL) (Ramires et al., 2023).

Although many yeast species have been reported as lipase producers, there is a need for novel strains to meet the increasing industrial demand. Thus, some novel yeasts have been isolated from various environmental sources. *M. capitatus*, isolated from olive mill wastewaters and cultivated in shake flasks, produced extracellular lipase with a 7-fold higher activity than the control strain *Y. lipolytica* under the same conditions (Salgado et al., 2020). A psychotolerant yeast, *Rhodotorula* sp., isolated from East Antarctica, produced lipase having a maximal activity of 4.77 U/mg at 15°C and pH 8.0, using 5% (v/v) palmolein oil and 2.5% inoculum size. Purification of the enzyme via dialysis led to a 5.63-fold increase in its activity (Maharana & Singh, 2018).

A novel isolate from oil mill effluent, *M. guilliermondii*, was reported as a highly efficient lipase producer. It demonstrated a 6.7-fold increase in lipase production under optimal conditions, significantly surpassing the lipase activity reported in other studies (Knob et al., 2020). These findings indicate a promising approach for discovering new strains for lipase production, which could be suitable for various industrial applications.

7.2.6 Pectinase

Pectinases are structural polysaccharides that catalyze the degradation of pectins, which are natural heteropolysaccharides found in the higher plant cells (Alimardani-Theuil et al., 2011). These enzymes are categorized into four classes based on how they act on the galacturonan backbone of pectins: pectin esterase (EC 3.1.1.11), polygalacturonases (EC 3.2.1.15, Glycoside hydrolases in CAZy database), proto-pectinase (EC 3.2.1.99), and pectin lyase (EC 4.2. 2.10, Polysaccharide lyase in CAZy database). Among them, pectin esterase removes acetyl and methoxyl groups from pectin, whereas polygalacturonases are responsible for hydrolyzing the glycosidic α-1,4 bonds (Arya et al., 2022; Haile et al., 2022). The global demand for pectinase, similar to other industrial enzymes, is steadily increasing because of its wide range of applications in many commercial areas including paper and pulp sector, oil extraction, textile industry, functional foods, wastewater treatment, and food and beverage sector, especially juice clarification and fruit processing. Among different types of pectinase, protopectinase segment dominated the market due to its growing use in the food processing and pharmaceutical industries. The food and beverage segment also took a considerable proportion, owing mostly to the rapid development in the consumption of healthier and more natural food products. However, the industrial pectinase market still faces critical challenges due to high production costs at large scale, requirement of advanced microbial fermentation and purification processes, and strict food safety standards (Haile et al., 2022). Despite the challenges, companies like AB Enzymes, Amanos Enzyme, Novensis, and IFF Inc., which lead the enzyme sector, hold a significant market share in pectinase production (AB Enzymes, 2024; Pectinase Market Res, 2024).

Pectinases are widely found in many higher plants, but microbial sources such as filamentous fungi, bacteria, and yeasts are more preferred for industrial pectinase production. Until now, pectinolytic activity has been reported in several yeasts including *S. cerevisia*, *Wickerhamomyces anomalus*, *Saccharomycopsis fibuligera*, *Pichia kudriavzevii*, *Torulopsis kefyr*, *Candida pseudotropicalis*, *Aureobasidium pullulans*, and many more (Haile et al., 2022; Shrestha et al., 2021). Several researchers focused on determining the optimum conditions for producing pectinase from different yeast genera (Table 7.1). Ezeh et al. (2023) reported the optimum process conditions for *Y. phangngaensis* as pH 7.5, inoculum size of 1.4×10^4 cells/ml, 5% banana peel substrate, and incubation time of 36 hours. According to Daskaya-Dikmen and co-workers, cold-adapted yeasts belonging to the *Rhodosporidiobolus*, *Cystofilobasidium*, and *Yamadazyma* had pectinase activity ranging from 0.76 U/mL to 1.73 U/mL after 3 days of incubation at 15°C with agitation at 150 rpm (Daskaya-Dikmen et al., 2018). A similar study reported that *G. candidum*, which is known as a yeast-like fungus but has been classified as a yeast for more than 25 years, produced 0.25 IU/mL pectinase under optimum conditions of pH 5.3, inoculum size of 3%, and substrate concentration of 3% (Ahmed et al., 2019). The same research group have focused on the immobilization of *G. candidum* with different agents to increase pectinase production. The immobilization of yeast with corncob, an agro industrial residue from corn leaves and husks, resulted in an approximately 2-fold enhancement of pectinase activity (Ahmed et al., 2020). Immobilization with sugarcane bagasse increased enzyme activity from 0.047 IU/mL min to 0.26 IU/mL min compared to free cells (Ejaz & Sohail, 2021), while sodium alginate immobilization improved pectinase activity 2.5-fold. These results suggested that immobilization of *G. candidum* with different agents is a promising approach for producing pectinase for commercial applications, but more studies are needed for a wider range of yeast species (Ejaz et al., 2020). Another effective technique for enhancing pectinase production and efficiency is the use of genetic engineering strategies, which focus on improving existing sources or developing novel strains. Abd El-Aziz et al. applied physical and chemical mutagenesis to *R. mucilaginosa* PY18 and reported a 3.3-fold increase in pectinase-specific activity under optimized conditions, attributed to enhanced expression of endo-polygalacturonase (Abd El-Aziz et al., 2023). In another study, the impact of codon optimization and promoter selection on pectinase production in *P. pastoris* was examined, and the optimal combination of codon and promoter was identified. The results showed that the optimized codon-promoter combination achieved the highest pectinase activity of 42.33 U/mL, which was four times higher than that reported in similar studies (Karaoğlan & Erden-Karaoğlan, 2020).

Cell wall degrading enzymes, especially pectinase, have important applications in the fruit juice industry by enhancing yield, softening and clarifying the juice, and reducing sugar content (Patel et al., 2022). Pectinase produced from *Y. phangngaensis XB3* showed considerable clarification of apple juice, reducing turbidity by 71.95% and color by 50.12% after treatment with 7.03 IU/mL of the enzyme (Ezeh et al., 2023). Meena et al., (2021) reported that pectinase obtained from *Meyerozyma sp.* caused a significant decrease in the viscosity of fruit juice. These results confirm the use of pectinase in clarifying fruit juice; however, more studies should be conducted to reveal the toxicity and coproducts of enzymatic treatment. The wine industry uses pectinase to clarify wine, improve the filtration process, maximize yield, enhance wine quality by releasing color and flavor components from grapes, and alter the alcohol content (Haile et al., 2022). The addition of pectinase-producing yeasts, including *Issatchenkia orientalis, Pichia kluyveri*, and *Cryptococcus adeliensis*, to the fermentation process led to considerable variation in the chemical content of both white and red wines and minor changes in their tastes (Paup et al., 2022). Similarly, inoculation of *Kluyveromyces marxianus* yeast, a well-known pectinase producer, into the alcoholic fermentation increased methanol concentration, altered volatile compounds, and improved free-run wine (Rollero et al., 2018). Pectinases are crucial in the coffee and tea fermentation processes, serving as another key application in the beverage industry. They degrade pectins, enhance drink quality, reduce foams in instant tea and coffee powders, speed up the fermentation process, and prevent the mucilage formation on coffee beans (Haile et al., 2022; Shrestha et al., 2021). Haile and Kang highlighted the potential use of several yeast species, *S. cerevisiae, Pichia kudriavzevii, Wickerhamomyces anomalus, Papiliotrema flavescens*, and *Saccharomycopsis fibuligera*, as starter culture for coffee fermentation. *S. cerevisiae* showed great potential in producing pectin methylesterase, while *S. fibuligera* and *W. anomalus* produced higher amounts of pectin lyase and polygalacturonase. After pH and Brix analyzes, all tested yeasts were found to be promising sources for initiating coffee fermentation. According to the results, each yeast species may contribute differently to the fermentation process of coffee, potentially influencing flavor profiles and overall quality through their enzymatic activities (Haile & Kang, 2019).

7.2.7 Protease

Proteases (EC 3.421.40) are another common enzyme group that hydrolyze proteins by catalyzing the hydrolysis of peptide bonds. They possess many functional roles in living systems, including cell proliferation, differentiation, signal transduction, apoptosis, DNA replication, and transcription (Bond, 2019; Naveed et al., 2021). Proteases are divided into different categories depending on their active site (aspartic, cysteine, serine, and metalloproteases), optimal pH values (acid, neutral, and alkaline proteases), and type of reaction catalyzed (endopeptidases, exopeptidases) (dos Santos Aguilar & Sato, 2018). From an industrial perspective, proteases are highly valuable across numerous sectors, accounting for about 60% of the overall enzyme market and 40% of global enzyme sales. The global protease market is expected to grow at a compound annual growth rate of 6% to 8% across various regions, reaching USD 4.6 billion by 2033 (Protease Market Res, 2023). A key factor driving this growth is the increasing use of proteases in the food, beverage, and pharmaceutical industries, particularly in protein hydrolysis, meat tenderization, cheese making, clarification of beer haze, baking, cereal meshing, and reduction of food protein allergy (Naveed et al., 2021; Tavano et al., 2018). To meet the protease demand across these industries, various enzyme products marketed by Novensis, Savinase, Alcalase, Esperase, Neutrase, Protamex, BioFeed Pro, and many more are commercially available (Sujitha & Shanthi, 2023). In the pharmaceutical sector, they have significant potential to fight against several diseases, including cancer, cystic fibrosis, inflammation, digestive issues, and cardiovascular problems (dos Santos Aguilar & Sato, 2018; Naveed et al., 2021). Additionally, these enzymes are utilized in detergent sector to remove tough stains by hydrolyzing proteins. In leather industry, proteases are employed for soaking, dehairing, and degreasing operation. In cosmetic industry, they are employed for skin softening and peeling. Proteases are also used in silk degumming and in waste management for degrading organic waste and removing feathers (Naveed et al., 2021). Currently, nano-structured immobilized protease enzyme attracts great attention that shows important advantages over conventional enzyme in terms of stability, reusability, and biocatalytic efficiency. Integration of this approach to the commercial protease market will pave the way for applications of immobilized enzymes in bioremediation and biomedical sectors (Bilal et al., 2024).

Among various sources of proteases, animals are insufficient to meet global demand and raise ethical concerns. Plant-based protease production is time-consuming and sensitive to climatic conditions. Therefore, microbial sources, bacteria, fungi, and yeasts, have gained attention as protease producer (Gurumallesh et al., 2019; Naveed et al., 2021; Razzaq et al., 2019). *Bacillus* species are extensively used for large-scale protease production because of their high capacity for protein secretion (Contesini et al., 2018). *Trichoderma* and *Aspergillus* are some of the prominent fungi species for industrial applications (Gurumallesh et al., 2019). However, there are only a few reports on the production of proteases from yeast species at a commercial level (Table 7.1). Hesham et al. (2017) reported the best protease producers from 23 yeast isolates from Kingdom of Saudi Arabia. The top five yeast species identified were *Hanseniaspora uvarum*, *Cryptococcus albidus*, *Pichia membranifaciens*, *Rhodotorula mucilaginosa*, and *Candida californica*, which showed the highest acidic and neutral protease activity. In another research, *Y. lipolytica* exhibited higher proteolytic activity of 800 U/mg of protein in the presence of chrome-tanned leather shavings as additional carbon source (Wieczorek et al., 2023). Daskaya-Dikmen and coworkers evaluated microbial composition of environmental samples from some regions of Turkey and found that protease activities were in the range of 2.11–10.53 U/mL for the yeast isolates, among which *Yamadazyma spp*. demonstrated the highest activity (Daskaya-Dikmen et al., 2018). Overall, the significant potential of yeasts for protease production should be thoroughly investigated and implemented in commercial-scale applications.

7.2.8 Urease

Urease (EC 3.5.1.5) is a nickel-containing metalloenzyme that converts urea into ammonia and carbon dioxide. Urea is the primary substrate of urease, but it can also hydrolyze other substrates such as formamide, hydroxyurea, acetamide, and semicarbazide (Follmer, 2008). The main commercial application of urease involves the removal of urea from aqueous solutions in several industries including, medicine, agriculture, and environmental management. Although urea is a low-toxicity compound, its prolonged presence can produce lethal substrates, highlighting the importance of its removal. In patients with renal failure, uremic toxins accumulate in the blood due to the kidneys' impaired ability to eliminate them, leading to elevated levels that the body cannot tolerate. Incorporating urease into dialysis treatment enhances the removal of these toxins by enzymatically hydrolyzing urea, thereby improving the efficiency of blood detoxification. Additionally, the unique properties of urease in detecting urea have expanded its application in biosensors, which play a significant role in the global urease market (Krajewska, 2009; Zhang et al., 2020). In agricultural applications, urease is utilized to convert urea in the soil into ammonia. This process not only helps to remove excess urea but also provides ammonia as a valuable nitrogen source for plants, enhancing soil fertility and supporting plant growth (Modolo et al., 2018). The extensive use of urease in agriculture is a major driver of its increasing market demand. With the global focus on sustainable agricultural practices, urease plays a critical role in reducing nitrogen loss from fertilizers, thereby improving crop yields and promoting soil health (Urease Market Res, 2023). From an environmental perspective, urease-producing microorganisms can be used to treat urea-contaminated wastewaters. By converting urea into ammonia and carbon dioxide, these microorganisms help reduce the concentration of urea in the water, thereby preventing pollution and protecting aquatic ecosystems (Weerakoon et al., 2023). Urease is increasingly being utilized beyond its conventional roles in agriculture and environmental management. A notable emerging application is in the biomedical field, where urease-powered micro- and nanoscale devices demonstrate significant potential for diagnosing and treating various conditions, such as gastric disorders, cardiovascular diseases, and cancer. Additionally, these enzyme-driven devices offer promising uses in medical imaging and innovative drug delivery systems, thanks to their biocompatibility and high payload capacities (Li et al., 2024).

Urease is widely distributed in plants, bacteria, fungi, algae, and yeasts but is not found in animals. Compared to plant and bacterial sources, urease-producing yeasts are much rarer. Until now, *Cryptococcus neoformans* (Toplis et al., 2020), *Xanthophyllomyces dendrorhous* (Persike et al., 2002), and *Cryptococcus gattii* (Feder et al., 2015) are some of the yeasts known for their urease-producing capability. However, industrially available urease is mainly derived from the *Canavalia ensiformis* due to its high urease content and the extensive research conducted on this plant (Hanane et al., 2022). To fully explore the potential of yeasts and expand their industrial applications, further studies are needed. They should focus on optimizing cultivation conditions, improving enzyme purification techniques, and exploring various applications.

7.2.9 Tannase

Tannase (E.C 3.1. 1.20) is a biotechnologically critical enzyme involved in the hydrolysis of ester bonds in gallotannin, tannic acid, and gallic acid esters, resulting in the release of glucose and gallic acid (Yao et al., 2014). Tannins are polyphenolic compounds commonly found in the vascular tissues of plants. They are utilized as stabilizers, antioxidants, therapeutics, and tanning agents in many industries, including leather, pharmaceutical, food, beverage, and paper. The uncontrolled discharge of tannin from factories can lead to severe issues for both environmental and public health (Govindarajan et al., 2016). Therefore, the bioremediation of tannin waste using tannase enzymes plays a crucial role in environmental protection (Sutaoney et al., 2024). The food and beverage industry are the largest sector in the global tannase market, driven by the enzyme's extensive application in enhancing the quality of products such as tea, wine, beer, and fruit juices. Tannase improves flavor by breaking down tannins, reduces bitterness, and stabilizes beverages. It plays a vital role in clarifying tea and refining the taste profiles of wine and beer. This sector continues to drive growth in the tannase market. However, expanding uses in pharmaceuticals, cosmetics, and environmental solutions also contribute to its rise (Lekshmi et al., 2021). According to market research, the Global Tannase Market is expected to increase from USD 363.2 Million in 2022 to USD 608.2 Million by 2032 (Tannase Market Res, 2022).

The biochemical and functional properties of tannase vary based on culture conditions and the producer organisms. Its main sources include bacteria, yeasts, and fungi, while fewer plant species produce the enzyme through their metabolic pathways. Among microbial organisms, yeasts and fungi produce tannases as glycoproteins that undergo post-translational modification, whereas bacteria do not possess this capability (Lekshmi et al., 2021). To date, fungal species, especially *Penicillium* and *Aspergillus*, are the major sources of industrial tannase, while numerous efforts have been made to explore novel yeast strains. *P. pastoris*, *Candida* sp., and *Arxula adeninivorans* have been reported as the most frequent sources (Lopes et al., 2018). Pan et al. (2020) identified a yeast strain, *R. diobovatum*, as a tannase producer with potential applications in tannin biodegradation and gallic acid production. Thangavelu et al. (2024) reported that a newly-isolated yeast, *G. cucujoidarum*, produced tannase at a high titer in SSF using agro-industrial wastes (Jamun leaves and lemon peel). Kanpiengjai et al. (2020) studied the coproduction of gallic acid and tannase from *S. ruineniae*, a pigment-producing yeast. The enzyme was obtained in a cell-associated form with a high degree of efficiency for gallic acid production and was resistant to high temperatures. It can be applied as a feed additive to enhance feed absorption and digestibility. Lopes et al. (2018) discovered tannase from *S. cerevisiae* in a SmF system, reporting it is a promising enzyme for the beverage industry and for gallic acid production. The enzyme yield was considerably affected by culture conditions, including temperature, tannic acid concentration, and agitation rate. Additionally, its activity was maximal at pH 6.0 and a temperature 30°C in the presence of salt. Mahmoud et al. (2018) reported that *K. marxianus* produced tannase in a SSF system utilizing olive pomace as support material. The enzyme, purified using aliminium sulfate precipitation, showed maximal activity at 35°C and at both pH 4.5 and 8.5. The research group suggested that the produced tannase had strong potential for use in the pharmaceutical industry due to its high activitiy for tannin conversion. In a similar study, purification of tannase produced by *R. glutinis* yeast was successfully performed in a step-wise process involving ultrafiltration, gel filtration chromatography, and anion exchange chromatography. The maximum activity of the purified enzyme was at pH 7.0 and 40°C (Ong et al., 2024). These results prove that the biochemical properties of tannase, and therefore its activity, vary greatly depending on the species from which it is produced, and the cultivation conditions. However, these studies have generally been limited to the laboratory scale. There is a need for research aimed at industrial applications, as is the case with most other types of enzymes.

7.3 Limitations

Despite the significant advancements in the production of yeast-derived enzymes, several limitations persist. The optimization of fermentation conditions for different enzymes, such as amylases and cellulases, often requires extensive experimentation and fine-tuning, which can be resource-intensive. The scalability of enzyme production from laboratory to industrial scale remains a challenge, particularly for enzymes like proteases and tannases, where commercial production is still limited. It is well recognized that achieving high enzyme yields in small-scale laboratory settings does not always translate to large-scale industrial production. During this transition, it is important to maintain product quality and consistency. With increasing production volume, a wide range of chemical, physical, and design factors can negatively impact the yeast growth and enzyme yield. The most critical issues are oxygen transfer, mixing characteristics, managing shear stress, mass and heat transfer limitations, and process control that may become more complex in larger systems. Economic concerns, including the high cost of media and energy demands, also play a significant role in industrial scale productions. Additionally, maintaining enzyme stability and activity under industrial conditions can be problematic, as enzymes may degrade or lose efficacy over time. Enzyme efficiency in yeast-based production is affected by several critical factors that can limit its industrial viability. One significant challenge is the yield per unit of biomass, which may decrease at high cell densities due to oxygen and nutrient limitations. Additionally, stress responses triggered by industrial fermentation conditions, including shear forces, temperature fluctuations, or pH changes, can lead to enzyme degradation or reduced activity. Furthermore, bottlenecks in enzyme secretion can result in intracellular accumulation, lowering the overall yield of active enzymes. Issues related to contamination and yield optimization further complicate industrial production processes. For enzymes like gelatinase and urease, substrate specificity and environmental conditions play crucial roles in determining enzyme efficiency, yet these factors are often difficult to control consistently. Furthermore, while genetic engineering holds promise for enhancing enzyme production, the practical application of these techniques is often constrained by cost and technical difficulties.

7.4 Conclusion and future perspectives

Yeast-derived enzymes are invaluable across a range of industries due to their diverse catalytic functions and applications. This chapter highlights the advancements in the production of key yeast enzymes, including amylase, cellulase, gelatinase, β-glucosidase, inulinase, lipase, pectinase, protease, urease, and tannase. These enzymes play critical roles in food processing, biofuel production, environmental management, and other sectors. However, the production of these enzymes faces challenges related to optimizing yield, ensuring stability, and scaling up from laboratory to industrial levels. Addressing these challenges is essential for harnessing the full potential of yeast-derived enzymes and meeting industrial demands efficiently.

Research focused on improving enzyme stability under industrial conditions and developing cost-effective, scalable purification methods is crucial. Advances in enzyme immobilization techniques and stabilization strategies could address these issues. Exploring novel fermentation technologies and reactor designs could also be beneficial. Conducting comprehensive studies on the economic feasibility and environmental impact of yeast-derived enzyme production will provide valuable insights into the sustainability and cost-effectiveness of these processes. At this point, life cycle analysis (LCA) provides a comprehensive assessment of the environmental impacts associated with the process, from raw material acquisition to the final product. This analysis encompasses the cultivation of yeast, downstream processes, waste management, enzyme delivery, and their utilization by consumers. A comprehensive LCA identifies hotspots in the production chain, such as high energy demand during the fermentation stage or the use of toxic chemicals in downstream processing, and proposes strategies to improve the system in terms of environmental impact, overall yield, and economic feasibility. Conducting an LCA in industrial enzyme production offers numerous advantages, including the optimization of enzyme production, minimization of resource use, reduction of greenhouse gas emissions, and enhancement of the overall sustainability of industrial biotechnology. By addressing these areas, future research can overcome current limitations and enhance the efficiency, cost-effectiveness, and applicability of yeast-derived enzymes across various industrial sectors.

References

AB Enzymes. (2024). https://www.abenzymes.com/en/.

Abd El-Aziz, N. M., Moharam, M. E., El-Gamal, N. N., & Khalil, B. E. (2023). Enhancement of novel Endo-polygalacturonase expression in Rhodotorula mucilaginosa PY18: Insights from mutagenesis and molecular docking. *Microbial Cell Factories, 22*(1). https://doi.org/10.1186/s12934-023-02253-5, https://microbialcellfactories.biomedcentral.com/.

Abduh, M. Y., Ramadhan, C. R., Fadhlilah, A. P., Abdul, S. D. N., & Burhan, K. H. (2022). Solid-state fermentation of groundnut (Arachis hypogaea) shell using Trichoderma sp., tape yeast, and tempeh yeast to produce cellulase. *Journal of Applied Biology and Biotechnology, 10*(4), 153–160. https://doi.org/10.7324/JABB.2022.100421, jabonline.in/index.php.

Ahmed, A., Ejaz, U., & Sohail, M. (2020). Pectinase production from immobilized and free cells of Geotrichum candidum AA15 in galacturonic acid and sugars containing medium. *Journal of King Saud University - Science, 32*(1), 952–954. https://doi.org/10.1016/j.jksus.2019.07.003.

Ahmed, A., Khan, M. N., Ahmad, A., Khan, S. A., & Sohail, M. (2019). Optimization of pectinase production from geotrichum candidum AA15 using response surface methodology. *Pakistan Journal of Botany, 51*(2), 743–750. https://doi.org/10.30848/PJB2019-2(41), http://www.pakbs.org/pjbot/papers/1550951489.pdf.

Alimardani Thouil, P., Gainvors Claisse, A., & Duchiron, F. (2011). Yeasts: An attractive source of pectinases—From gene expression to potential applications: A review. *Process Biochemistry, 46*(8), 1525–1537. https://doi.org/10.1016/j.procbio.2011.05.010.

de Araujo Ribeiro, G. C., & Assis, S. A. D. (2023). Production of β-glucosidase by Rhodotorula oryzicola and use of enzyme for hydrolysis of sugarcane bagasse delignified. *Journal of Food Science and Technology, 60*(11), 2761–2771. https://doi.org/10.1007/s13197-023-05783-3, https://www.springer.com/journal/13197.

de Araujo Ribeiro, G. C., Fernandes, P., Silva, D. A. A., Brandão, H. N., & de Assis, S. A. (2021). Inulinase from Rhodotorula mucilaginosa: immobilization and application in the production of fructooligosaccharides. *Food Science and Biotechnology, 30*(7), 959–969. https://doi.org/10.1007/s10068-021-00931-x, http://www.springerlink.com/content/n3031612t262/.

Areeshi, M. Y. (2022). Microbial cellulase production using fruit wastes and its applications in biofuels production. *International Journal of Food Microbiology, 378*, 109814. https://doi.org/10.1016/j.ijfoodmicro.2022.109814.

Arya, P. S., Yagnik, S. M., Rajput, K. N., Panchal, R. R., & Raval, V. H. (2022). Valorization of agro-food wastes: Ease of concomitant-enzymes production with application in food and biofuel industries. *Bioresource Technology, 361*. https://doi.org/10.1016/j.biortech.2022.127738, www.elsevier.com/locate/biortech.

Augustín, J., Zemek, J., Kocková-Kratochvílová, A., & Kuniak, L.' (1978). Production of α-amylase by yeasts and yeast-like organisms. *Folia Microbiologica, 23*(5), 353–361. https://doi.org/10.1007/bf02876434.

Behera, S. S., Kerketta, A., & Ray, R. C. (2023). Solid-state fermentation for the production of microbial cellulases. In *Biotechnology of microbial enzymes: Production, biocatalysis, and industrial applications, Second Edition*. India: Elsevier, 59–88. 10.1016/B978-0-443-19059-9.00012-8, https://www.sciencedirect.com/book/9780443190599.

Bilal, M., Qamar, S. A., Carballares, D., Berenguer-Murcia, Á., & Fernandez-Lafuente, R. (2024). Proteases immobilized on nanomaterials for biocatalytic, environmental and biomedical applications: Advantages and drawbacks. *Biotechnology Advances, 70*, 108304. https://doi.org/10.1016/j.biotechadv.2023.108304.

Bond, J. S. (2019). Proteases: History, discovery, and roles in health and disease. *Journal of Biological Chemistry, 294*(5), 1643–1651. https://doi.org/10.1074/jbc.TM118.004156, http://www.jbc.org/content/294/5/1643.full.pdf.

Carvalho, J. K., Panatta, A. A. S., Silveira, M. A. D., Tav, C., Johann, S., Rodrigues, M. L. F., & Martins, C. V. B. (2021). Yeasts isolated from a lotic continental environment in Brazil show potential to produce amylase, cellulase and protease. *Biotechnology Reports, 30*. https://doi.org/10.1016/j.btre.2021.e00630, http://www.journals.elsevier.com/biotechnology-reports/.

de Carvalho Cardoso, E. S., Martins, N. Q., Azevedo, R. A., Palmeira, L. S., Quintanilha-Peixoto, G., Andrade, B., Santos, M. P. F., Uetanabaro, A. P. T., da Silva, E. G. P., Góes-Neto, A., & da Costa, A. M. (2025). Production and application of inulinase by new isolates of Aspergillus welwitschiae from fermented peach-palm waste for the production of fructooligosaccharides. *Food Chemistry, 465*. https://doi.org/10.1016/j.foodchem.2024.141978, https://www.sciencedirect.com/science/journal/03088146.

Cavalcante, F. T. T., Neto, F. S., Rafael de Aguiar Falcão, I., Erick da Silva Souza, J., de Moura Junior, L. S., da Silva Sousa, P., Rocha, T. G., de Sousa, I. G., de Lima Gomes, P. H., de Souza, M. C. M., & dos Santos, J. C. S. (2021). Opportunities for improving biodiesel production via lipase catalysis. *Fuel, 288*. https://doi.org/10.1016/j.fuel.2020.119577, http://www.journals.elsevier.com/fuel/.

CAZY database. (2024). https://www.cazy.org/GH32.html.

Chen, L., Chi, Z. M., Chi, Z., & Li, M. (2010). Amylase production by saccharomycopsis fibuligera A11 in solid-state fermentation for hydrolysis of cassava starch. *Applied Biochemistry and Biotechnology, 162*(1), 252–263. https://doi.org/10.1007/s12010-009-8744-3.

Chi, Z., Chi, Z., Liu, G., Wang, F., Ju, L., & Zhang, T. (2009). Saccharomycopsis fibuligera and its applications in biotechnology. *Biotechnology Advances, 27*(4), 423–431. https://doi.org/10.1016/j.biotechadv.2009.03.003.

Choudhary, J., Singh, S., & Nain, L. (2016). Thermotolerant fermenting yeasts for simultaneous saccharification fermentation of lignocellulosic biomass. *Electronic Journal of Biotechnology, 21*, 82–92. https://doi.org/10.1016/j.ejbt.2016.02.007.

Chávez-Camarillo, G. M., Lopez-Nuñez, P. V., Jiménez-Nava, R. A., Aranda-García, E., & Cristiani-Urbina, E. (2022). Production of extracellular α-amylase by single-stage steady-state continuous cultures of Candida wangnamkhiaoensis in an airlift bioreactor. *PLoS One, 17*(3). https://doi.org/10.1371/journal.pone.0264734, https://journals.plos.org/plosone/article/file?id=10.1371/journal.pone.0264734&type=printable.

Co, R., & Hug, L. A. (2020). A Need for Improved Cellulase Identification from Metagenomic Sequence Data. *Applied and Environmental Microbiology, 87*(1), 1–8. https://doi.org/10.1128/AEM.01928-20, http://aem.asm.org/.

Contesini, F. J., Melo, R. Rd, & Sato, H. H. (2018). An overview of Bacillus proteases: from production to application. *Critical Reviews in Biotechnology, 38*(3), 321–334. https://doi.org/10.1080/07388551.2017.1354354.

Costa, A. R., Salgado, J. M., & Belo, I. (2023). Olive and sunflower cakes as suitable substrates for lipase production by Yarrowia spp.: From flasks to bioreactor. *Biocatalysis and Agricultural Biotechnology, 51*, 102783. https://doi.org/10.1016/j.bcab.2023.102783.

Dakhmouche Djekrif, S., Bennamoun, L., Labbani, F. Z. K., Ait Kaki, A., Nouadri, T., Pauss, A., Meraihi, Z., & Gillmann, L. (2021). An Alkalothermophilic Amylopullulanase from the Yeast Clavispora lusitaniae ABS7: Purification, Characterization and Potential Application in Laundry Detergent. *Catalysts, 11*(12), 1438. https://doi.org/10.3390/catal11121438.

Daskaya-Dikmen, C., Karbancioglu-Guler, F., & Ozcelik, B. (2018). Cold active pectinase, amylase and protease production by yeast isolates obtained from environmental samples. *Extremophiles: Life Under Extreme Conditions, 22*(4), 599–606. https://doi.org/10.1007/s00792-018-1020-0.

Deepa BK, P., & Nanjundaiah, S. M. (2023). Extremotolerant gelatinase enzyme production by UCM4 bacterial strain isolated from the rhizospheric soil of rhizophora mangle. *Asian Journal of Biological and Life Sciences, 11*(3), 769–775. https://doi.org/10.5530/ajbls.2022.11.103.

Drula, E., Garron, M. L., Dogan, S., Lombard, V., Henrissat, B., & Terrapon, N. (2022). The carbohydrate-active enzyme database: Functions and literature. *Nucleic Acids Research, 50*(1), D571. https://doi.org/10.1093/nar/gkab1045, https://academic.oup.com/nar/issue.

Ejaz, U., Hanif, H., & Sohail, M. (2020). Two layered strategy for cost effective production of pectinase: immobilization of yeast and utilization of crude substrate. *Heliyon, 6*(11), e05456. https://doi.org/10.1016/j.heliyon.2020.e05456.

Ejaz, U., & Sohail, M. (2021). Supporting role of lignin in immobilization of yeast on sugarcane bagasse for continuous pectinase production. *Journal of the Science of Food and Agriculture, 101*(4), 1709–1714. https://doi.org/10.1002/jsfa.10764.

Ezeh, N. O., Nwagu, T. N. T., Amadi, O. C., Okolo, B., Moneke, A., & Agu, R. (2023). Extracellular pectinase production from a novel Yarrowia phangngaensis XB3 grown on banana waste and its application in fruit juice clarification. *Biocatalysis and Agricultural Biotechnology, 47*. https://doi.org/10.1016/j.bcab.2023.102614, http://www.journals.elsevier.com/biocatalysis-and-agricultural-biotechnology/.

Far, B. E., Ahmadi, Y., Khosroushahi, A. Y., & Dilmaghani, A. (2020). Microbial alpha-amylase production: Progress, challenges and perspectives. *Advanced Pharmaceutical Bulletin, 10*(3), 350–358. https://doi.org/10.34172/apb.2020.043, http://apb.tbzmed.ac.ir/.

Feder, V., Kmetzsch, L., Staats, C. C., Vidal-Figueiredo, N., Ligabue-Braun, R., Carlini, C. R., & Vainstein, M. H. (2015). Cryptococcus gattii urease as a virulence factor and the relevance of enzymatic activity in cryptococcosis pathogenesis. *FEBS Journal, 282*(8), 1406–1418. https://doi.org/10.1111/febs.13229, http://onlinelibrary.wiley.com/journal/10.1111/(ISSN)1742-4658.

Filho, D. G., Silva, A. G., & Guidini, C. Z. (2019). Lipases: Sources, immobilization methods, and industrial applications. *Applied Microbiology and Biotechnology, 103*(18), 7399–7423. https://doi.org/10.1007/s00253-019-10027-6, link.springer.de/link/service/journals/00253/index.htm.

Follmer, C. (2008). Insights into the role and structure of plant ureases. *Phytochemistry, 69*(1), 18–28. https://doi.org/10.1016/j.phytochem.2007.06.034.

Giese, E. C., Dussán, K. J., Pierozzi, M., Chandel, A. K., Pagnocca, F. C., & Da Silva, S. S. (2017). Cellulase production by trichosporon laibachii. *Orbital, 9*(4), 271–278. https://doi.org/10.17807/orbital.v9i4.1024, http://www.orbital.ufms.br/index.php/Chemistry/article/download/1024/pdf.

Global Enzyme Mark. (2023). http://www.marketsandmarkets.com.

Godse, R., Bawane, H., Tripathi, J., & Kulkarni, R. (2021). Unconventional β-Glucosidases: A Promising Biocatalyst for Industrial Biotechnology. *Applied Biochemistry and Biotechnology, 193*(9), 2993–3016. https://doi.org/10.1007/s12010-021-03568-y.

Govindarajan, R. K., Revathi, S., Rameshkumar, N., Krishnan, M., & Kayalvizhi, N. (2016). Microbial tannase: Current perspectives and biotechnological advances. *Biocatalysis and Agricultural Biotechnology, 6*, 168–175. https://doi.org/10.1016/j.bcab.2016.03.011.

Guerrand, D. (2017). Lipases industrial applications: Focus on food and agroindustries. *OCL - Oilseeds and fats, Crops and Lipids, 24*(4). https://doi.org/10.1051/ocl/2017031, http://www.ocl-journal.org/.

Gurumallesh, P., Alagu, K., Ramakrishnan, B., & Muthusamy, S. (2019). A systematic reconsideration on proteases. *International Journal of Biological Macromolecules, 128*, 254–267. https://doi.org/10.1016/j.ijbiomac.2019.01.081.

Haile, M., & Kang, W. H. (2019). Isolation, identification, and characterization of pectinolytic yeasts for starter culture in coffee fermentation. *Microorganisms, 7*(10), 401. https://doi.org/10.3390/microorganisms7100401.

Haile, S., Ayele, A., & Rodriguez-Couto, S. (2022). Pectinase from microorganisms and its industrial applications. *The Scientific World Journal, 2022*, 1–15. https://doi.org/10.1155/2022/1881305.

Hanane, T., Najoua, B., Salsabil, H., Abdellatif, J. I., Dalila, B., Ahmad, I., BukharI, S. A. R., Irfan, M., Chen, L., & Hicham, B. (2022). Qualitative Screening of Yeast Biodiversity for Hydrolytic Enzymes Isolated from the Gastrointestinal Tract of a Coprophage "Gymnopleurus sturmi" and Dung of Ruminants. *Fermentation, 8*(12). 10.3390/fermentation8120692, www.mdpi.com/journal/fermentation.

Hesham, A. E. L., Alrumman, S. A., Al-Dayel, M. A., & Salah, H. A. (2017). Screening and genetic identification of acidic and neutral protease-producing yeasts strains by 26S rRNA gene sequencing. *Cytology and Genetics, 51*(3), 221–229. https://doi.org/10.3103/s0095452717030033.

Huang, R., Zhang, F., Yan, X., Qin, Y., Jiang, J., Liu, Y., & Song, Y. (2021). Characterization of the β -Glucosidase activity in indigenous yeast isolated from wine regions in China. *Journal of Food Science, 86*(6), 2327–2345. https://doi.org/10.1111/1750-3841.15741.

Ilić, N., Milić, M., Beluhan, S., & Dimitrijević-Branković, S. (2023). Cellulases: From lignocellulosic biomass to improved production. *Energies, 16*(8), 3598. https://doi.org/10.3390/en16083598.

Inulinase Market Res, (2024). https://markwideresearch.com/press-release/inulinase-market-expands-its-reach-expected-to-exceed-280-million-by-2035/.

Janeček, Š., & Svensson, B. (2022). How many α-amylase GH families are there in the CAZy database? *Amylase, 6*(1), 1–10. https://doi.org/10.1515/amylase-2022-0001.

Kalil, S. J., Maugeri-Filho, F., & Rodrigues, M. I. (2005). Ion exchange expanded bed chromatography for the purification of an extracelular inulinase from Kluyveromyces marxianus. *Process Biochemistry, 40*(2), 581–586. https://doi.org/10.1016/j.procbio.2004.01.034.

Kannan, P., Shafreen, M., Achudhan, A. B., Gupta, A., & Saleena, L. M. (2023). A review on applications of β-glucosidase in food, brewery, pharmaceutical and cosmetic industries. *Carbohydrate Research, 530.* https://doi.org/10.1016/j.carres.2023.108855, www.elsevier.com/locate/carres.

Kanpiengjai, A., Khanongnuch, C., Lumyong, S., Haltrich, D., Nguyen, T. H., & Kittibunchakul, S. (2020). Co-production of gallic acid and a novel cell-associated tannase by a pigment-producing yeast, Sporidiobolus ruineniae A45.2. *Microbial Cell Factories, 19*(1). https://doi.org/10.1186/s12934-020-01353-w, http://www.microbialcellfactories.com/home/.

Karaoğlan, M., & Erden-Karaoğlan, F. (2020). Effect of codon optimization and promoter choice on recombinant endo-polygalacturonase production in Pichia pastoris. *Enzyme and Microbial Technology, 139*, 109589. https://doi.org/10.1016/j.enzmictec.2020.109589.

Kham, N. N. N., Phovisay, S., Unban, K., Kanpiengjai, A., Saenjum, C., Lumyong, S., Shetty, K., & Khanongnuch, C. (2024). A thermotolerant yeast cyberlindnera rhodanensis DK Isolated from laphet-so capable of extracellular thermostable β-glucosidase production. *Journal of Fungi, 10*(4). https://doi.org/10.3390/jof10040243, www.mdpi.com/journal/jof.

Knob, A., Izidoro, S. C., Lacerda, L. T., Rodrigues, A., & de Lima, V. A. (2020). A novel lipolytic yeast Meyerozyma guilliermondii: Efficient and low-cost production of acid and promising feed lipase using cheese whey. *Biocatalysis and Agricultural Biotechnology, 24.* https://doi.org/10.1016/j.bcab.2020.101565, http://www.journals.elsevier.com/biocatalysis-and-agricultural-biotechnology/.

Korsa, G., Konwarh, R., Masi, C., Ayele, A., & Haile, S. (2023). Microbial cellulase production and its potential application for textile industries. *Annals of Microbiology, 73*(1). https://doi.org/10.1186/s13213-023-01715-w.

Krajewska, B. (2009). Ureases. II. Properties and their customizing by enzyme immobilizations: A review. *Journal of Molecular Catalysis B: Enzymatic, 59*(1-3), 22–40. https://doi.org/10.1016/j.molcatb.2009.01.004.

Kumar, A., Verma, V., Dubey, V. K., Srivastava, A., Garg, S. K., Singh, V. P., & Arora, P. K. (2023). Industrial applications of fungal lipases: a review. *Frontiers in Microbiology, 14.* https://doi.org/10.3389/fmicb.2023.1142536, https://www.frontiersin.org/journals/microbiology#.

Kuncharoen, N., Techo, S., Savarajara, A., & Tanasupawat, S. (2020). Identification and lipolytic activity of yeasts isolated from foods and wastes. *Mycology, 11*(4), 279–286. https://doi.org/10.1080/21501203.2020.1745922.

Leite, P., Sousa, D., Fernandes, H., Ferreira, M., Costa, A. R., Filipe, D., Gonçalves, M., Peres, H., Belo, I., & Salgado, J. M. (2021). Recent advances in production of lignocellulolytic enzymes by solid-state fermentation of agro-industrial wastes. *Current Opinion in Green and Sustainable Chemistry, 27.* https://doi.org/10.1016/j.cogsc.2020.100407, http://www.journals.elsevier.com/current-opinion-in-green-and-sustainable-chemistry.

Lekshmi, R., Arif Nisha, S., Thirumalai Vasan, P., & Kaleeswaran, B. (2021). A comprehensive review on tannase: Microbes associated production of tannase exploiting tannin rich agro-industrial wastes with special reference to its potential environmental and industrial applications. *Environmental Research, 201*, 111625. https://doi.org/10.1016/j.envres.2021.111625.

Li, W.-W., Yu, Z.-L., & Jia, J. (2024). Urease-powered micro/nanomotors: Current progress and challenges. *Journal of Pharmaceutical Analysis*, 101095. https://doi.org/10.1016/j.jpha.2024.101095.

Lipase Market Res. (2024). https://www.mordorintelligence.com/industry-reports/lipase-market.

Lopes, L. M. D. M., Costa Batista, L. H., Gouveia, M. J., Leite, T. C. C., de Mello, M. R. F., de Assis, S. A., & de Sena, A. R. (2018). Kinetic and thermodynamic parameters, and partial characterization of the crude extract of tannase produced by Saccharomyces cerevisiae CCMB 520. *Natural Product Research, 32*(9), 1068–1075. https://doi.org/10.1080/14786419.2017.1380010, www.tandf.co.uk/journals/titles/14786419.asp.

Maharana, A. K., & Singh, S. M. (2018). A cold and organic solvent tolerant lipase produced by Antarctic strain Rhodotorula sp. Y-23. *Journal of Basic Microbiology, 58*(4), 331–342. https://doi.org/10.1002/jobm.201700638, http://onlinelibrary.wiley.com/journal/10.1002/(ISSN)1521-4028.

Mahmoud, A. E., Fathy, S. A., Rashad, M. M., Ezz, M. K., & Mohammed, A. T. (2018). Purification and characterization of a novel tannase produced by Kluyveromyces marxianus using olive pomace as solid support, and its promising role in gallic acid production. *International Journal of Biological Macromolecules, 107*, 2342–2350. https://doi.org/10.1016/j.ijbiomac.2017.10.117, https://www.sciencedirect.com/journal/international-journal-of-biological-macromolecules.

Malemud, C. J. (2006). Matrix metalloproteinases (MMPs) in health and disease: An overview. *Frontiers in Bioscience, 11*(2), 1696–1701. https://doi.org/10.2741/1915, https://www.fbscience.com/archives.

Meena, B., Sowmeya, V. G., Praveen, A. B., Swetha, A., Sarath Chandra, D. N., & Kavitha, M. (2021). Pectin Degradation in Fruit Juices by Pectinase from Meyerozyma sp. VITPCT75 Isolated from Phyllanthus emblica. *Journal of Pure and Applied Microbiology, 15*(2), 926–935. https://doi.org/10.22207/jpam.15.2.51.

MEROPS. (2024). http://www.marketsandmarkets.com.

Mesbah, N. M. (2022). Industrial biotechnology based on enzymes from extreme environments. *Frontiers in Bioengineering and Biotechnology, 10*. https://doi.org/10.3389/fbioe.2022.870083, http://journal.frontiersin.org/journal/bioengineering-and-biotechnology#archive.

Modolo, L. V., da-Silva, C. J., Brandão, D. S., & Chaves, I. S. (2018). A minireview on what we have learned about urease inhibitors of agricultural interest since mid-2000s. *Journal of Advanced Research, 13*, 29–37. https://doi.org/10.1016/j.jare.2018.04.001, http://www.elsevier.com/wps/find/journaldescription.cws_home/722881/description#description.

Mondal, S., Mondal, K., Halder, S. K., Thakur, N., & Mondal, K. C. (2022). Microbial Amylase: Old but still at the forefront of all major industrial enzymes. *Biocatalysis and Agricultural Biotechnology, 45*. https://doi.org/10.1016/j.bcab.2022.102509, http://www.journals.elsevier.com/biocatalysis-and-agricultural-biotechnology/.

De Morais, W. G., Kamimura, E. S., Ribeiro, E. J., Pessela, B. C., Cardoso, V. L., & De Resende, M. M. (2016). Optimization of the production and characterization of lipase from Candida rugosa and Geotrichum candidum in soybean molasses by submerged fermentation. *Protein Expression and Purification, 123*, 26–34. https://doi.org/10.1016/j.pep.2016.04.001, http://www.elsevier.com/inca/publications/store/6/2/2/9/3/5/index.htt.

Nakamura, A. M., Nascimento, A. S., & Polikarpov, I. (2017). Structural diversity of carbohydrate esterases. *Biotechnology Research and Innovation, 1*(1), 35–51. https://doi.org/10.1016/j.biori.2017.02.001.

Naveed, M., Nadeem, F., Mehmood, T., Bilal, M., Anwar, Z., & Amjad, F. (2021). Protease—a versatile and ecofriendly biocatalyst with multi-industrial applications: An updated review. *Catalysis Letters, 151*(2), 307–323. https://doi.org/10.1007/s10562-020-03316-7.

Oh, E. J., & Jin, Y. S. (2020). Engineering of Saccharomyces cerevisiae for efficient fermentation of cellulose. *FEMS Yeast Research, 20*(1). https://doi.org/10.1093/femsyr/foz089, http://femsyr.oxfordjournals.org/.

Ong, C. B., Ibrahim, D., & Mohd Kassim, M. J. N. (2024). The Tannase from red yeast Rhodotorula glutinis: purification and characterization. *Biocatalysis and Biotransformation, 42*(2), 110–117. https://doi.org/10.1080/10242422.2022.2136523, http://www.tandfonline.com/loi/ibab20.

Onilude, A. A., Fadaunsi, I. F., & Garuba, E. O. (2012). Inulinase production by Saccharomyces sp. in solid state fermentation using wheat bran as substrate. *Annals of Microbiology, 62*(2), 843–848. https://doi.org/10.1007/s13213-011-0325-3.

Othman, Z., Selim, A., Bayoumy, S., & Saber, W. (2020). Inulinase Production from Plant Materials by some Local Yeast Strains. *Journal of Agricultural Chemistry and Biotechnology, 11*(3), 71–77. https://doi.org/10.21608/jacb.2020.87000.

de Ovalle, S., Cavello, I., Brena, B. M., Cavalitto, S., & González-Pombo, P. (2018). Production and characterization of a β-glucosidase from Issatchenkia terricola and its use for hydrolysis of aromatic precursors in Cabernet Sauvignon wine. *LWT, 87*, 515–522. https://doi.org/10.1016/j.lwt.2017.09.026, https://www.journals.elsevier.com/lwt.

Paixão, S. M., Teixeira, P. D., Silva, T. P., Teixeira, A. V., & Alves, L. (2013). Screening of novel yeast inulinases and further application to bioprocesses. *New Biotechnology, 30*(6), 598–606. https://doi.org/10.1016/j.nbt.2013.02.002.

Pan, J., Wang, N. N., Yin, X. J., Liang, X. L., & Wang, Z. P. (2020). Characterization of a robust and pH-stable tannase from mangrove-derived yeast rhodosporidium diobovatum Q95. *Marine Drugs, 18*(11). https://doi.org/10.3390/md18110546, https://www.mdpi.com/1660-3397/18/11/546.

Parapouli, M., Vasileiadis, A., Afendra, A. S., & Hatziloukas, E. (2020). Saccharomyces cerevisiae and its industrial applications. *AIMS Microbiology, 6*(1), 1–31. https://doi.org/10.3934/microbiol.2020001, https://www.aimspress.com/fileOther/PDF/microbiology/microbiol-06-01-001.pdf.

Patel, V. B., Chatterjee, S., & Dhoble, A. S. (2022). A review on pectinase properties, application in juice clarification, and membranes as immobilization support. *Journal of Food Science, 87*(8), 3338–3354. https://doi.org/10.1111/1750-3841.16233, http://onlinelibrary.wiley.com/journal/10.1111/(ISSN)1750-3841.

Pathade, A. (2024). Isolation and Characterization of Gelatinase Enzyme-Producing Microorganisms from Soil for Industrial Applications. *Bulletin of Pure and Applied Sciences-Zoology*, 461–474.

Patil, P. J., Dong, X., Usman, M., Bhambore, N. R., Shah, H., Zhang, C., & Li, X. (2023). Recent trends in metal-organic frameworks mediated lipase immobilization: A state-of-the-art review. *Food Safety and Health, 1*(2), 139–169. https://doi.org/10.1002/fsh3.12021, https://onlinelibrary.wiley.com/journal/28351096.

Paup, V. D., Barton, T. L., Edwards, C. G., Lange, I., Lange, B. M., Lee, J., & Ross, C. F. (2022). Improving the chemical and sensory characteristics of red and white wines with pectinase-producing non-Saccharomyces yeasts. *Journal of Food Science, 87*(12), 5402–5417. https://doi.org/10.1111/1750-3841.16371, http://onlinelibrary.wiley.com/journal/10.1111/(ISSN)1750-3841.

Pectinase Market Res. (2024). https://www.alliedmarketresearch.com/pectinase-market.

Persike, D. S., Bonfim, T. M. B., Santos, M. H. R., Lyng, S. M. O., Chiarello, M. D., & Fontana, J. D. (2002). Invertase and urease activities in the carotenogenic yeast Xanthophyllomyces dendrorhous (formerly Phaffia rhodozyma). *Bioresource Technology, 82*(1), 79–85. https://doi.org/10.1016/S0960-8524(01)00121-3.

Poomani, M. S., James, S. B., Muthan, K., & Subramanian, V. (2024). Unravelling yeast cellulase potential: A computational approach to structural study, cellulolytic activity, and docking. *Journal of Molecular Structure, 1300*. https://doi.org/10.1016/j.molstruc.2023.137254, https://www.sciencedirect.com/science/journal/00222860.

Prado, H. F. A., Reis, A. A., & Oliveira (2016). Fungal amylase: Production system. *Fungal Enzymes*.

Protease Market Res. (2023). https://www.futuremarketinsights.com/reports/protease-market.

Ramires, F. A., Bavaro, A. R., D'Antuono, I., Linsalata, V., D'Amico, L., Baruzzi, F., Pinto, L., Tarantini, A., Garbetta, A., Cardinali, A., & Bleve, G. (2023). Liquid submerged fermentation by selected microbial strains for onion skins valorization and its effects on polyphenols. *World Journal of Microbiology and Biotechnology, 39*(10). https://doi.org/10.1007/s11274-023-03708-y, https://www.springer.com/journal/11274.

Raoufi, Z., & Mousavi Gargari, S. L. (2018). Biodiesel production from microalgae oil by lipase from Pseudomonas aeruginosa displayed on yeast cell surface. *Biochemical Engineering Journal, 140*, 1–8. https://doi.org/10.1016/j.bej.2018.09.008, www.elsevier.com/locate/bej.

Razzaq, A., Shamsi, S., Ali, A., Ali, Q., Sajjad, M., Malik, A., & Ashraf, M. (2019). Microbial proteases applications. *Frontiers in bioengineering and biotechnology, 7*.

Roble, N. D., Ogbonna, J., & Tanaka, H. (2020). Simultaneous amylase production, raw cassava starch hydrolysis and ethanol production by immobilized Aspergillus awamori and Saccharomyces cerevisiae in a novel alternating liquid phase–air phase system. *Process Biochemistry, 95*, 115–121. https://doi.org/10.1016/j.procbio.2020.05.020, www.elsevier.com/inca/publications/store/4/2/2/8/5/7.

Rollero, S., Zietsman, A. J. J., Buffetto, F., Schückel, J., Ortiz-Julien, A., & Divol, B. (2018). Kluyveromyces marxianus secretes a pectinase in shiraz grape must that impacts technological properties and aroma profile of wine. *Journal of Agricultural and Food Chemistry, 66*(44), 11739–11747. https://doi.org/10.1021/acs.jafc.8b03977, http://pubs.acs.org/journal/jafcau.

Rong, Y., Zhang, L., Chi, Z., & Wang, X. (2015). A carboxymethyl cellulase from a marine yeast (Aureobasidium pullulans 98): Its purification, characterization, gene cloning and carboxymethyl cellulose digestion. *Journal of Ocean University of China, 14*(5), 913–921. https://doi.org/10.1007/s11802-015-2574-4.

Sakpuntoon, V., Angchuan, J., Boontham, W., Khunnamwong, P., Boonmak, C., & Srisuk, N. (2020). Grease waste as a reservoir of lipase-producing yeast and description of limtongella siamensis gen. nov., sp. nov. *Microorganisms, 8*(1), 27. https://doi.org/10.3390/microorganisms8010027.

Salah, H. A., Temerk, H. A., Salah, N. A., Alshehri, S. R. Z., Al-Harbi, J. A., Mawad, A. M. M., Khaled, K. A. M., Hesham, A. E. L., & Amein, K. A. (2021). Production and optimization of xylanase and α-amylase from non-saccharomyces yeasts (Pichia membranifaciens). *Journal of Pure and Applied Microbiology, 15*(1). https://doi.org/10.22207/JPAM.15.1.43, https://microbiologyjournal.org/production-and-optimization-of-xylanase-and-%ce%b1-amylase-from-non-saccharomyces-yeasts-pichia-membranifaciens/.

Sales, J. C. S., de Castro, A. M., & Ribeiro, B. D. (2020). M.A. Maria, Supplementation of watermelon peels as an enhancer of lipase and esterase production by Yarrowia lipolytica in solid-state fermentation and their potential use as biocatalysts in poly (ethylene terephthalate) (PET) depolymerization reactions. *Biocatalysis and Biotransformation, 38*(6), 457–468. https://doi.org/10.1080/10242422.2020.1782387, http://www.tandfonline.com/loi/ibab20.

Salgado, V., Fonseca, C., Lopes da Silva, T., Roseiro, J. C., & Eusébio, A. (2020). Isolation and identification of magnusiomyces capitatus as a lipase-producing yeast from olive mill wastewater. *Waste and Biomass Valorization, 11*(7), 3207–3221. https://doi.org/10.1007/s12649-019-00725-7, http://www.springer.com/engineering/journal/12649.

dos Santos Aguilar, J. G., & Sato, H. H. (2018). Microbial proteases: Production and application in obtaining protein hydrolysates. *Food Research International, 103*, 253–262. https://doi.org/10.1016/j.foodres.2017.10.044, www.elsevier.com/inca/publications/store/4/2/2/9/7/0.

Shad, M., Hussain, N., Usman, M., Akhtar, M. W., & Sajjad, M. (2023). Exploration of computational approaches to predict the structural features and recent trends in α-amylase production for industrial applications. *Biotechnology and Bioengineering, 120*(8), 2092–2116. https://doi.org/10.1002/bit.28504, http://onlinelibrary.wiley.com/journal/10.1002/(ISSN)1097-0290.

Shariq, M., & Sohail, M. (2020). Production of cellulase and xylanase from candida tropicalis (Mk-118) on purified and crude substrates. *Pakistan Journal of Botany, 52*(1), 323–328. https://doi.org/10.30848/PJB2020-1(14), http://www.pakbs.org/pjbot/papers/1580204220.pdf.

Sheng, J., Chi, Z., Li, J., Gao, L., & Gong, F. (2007). Inulinase production by the marine yeast Cryptococcus aureus G7a and inulin hydrolysis by the crude inulinase. *Process Biochemistry, 42*(5), 805–811. https://doi.org/10.1016/j.procbio.2007.01.016.

Shrestha, S., Rahman, M. S., & Qin, W. (2021). New insights in pectinase production development and industrial applications. *Applied Microbiology and Biotechnology, 105*(24), 9069–9087. https://doi.org/10.1007/s00253-021-11705-0, https://www.springer.com/journal/253.

Shrivastava, A., Pal, M., & Sharma, R. K. (2023). Pichia as yeast cell factory for production of industrially important bio-products: Current trends, challenges, and future prospects. *Journal of Bioresources and Bioproducts, 8*(2), 108–124. https://doi.org/10.1016/j.jobab.2023.01.007, https://www.sciencedirect.com/journal/journal-of-bioresources-and-bioproducts.

Silva, Souza, Valoni, E., Castro, Coelho, M. A. Z., Ribeiro, B. D., Henriques, C. A., & Langone, M. A. P. (2019). Biocatalytic esterification of fatty acids using a low-cost fermented solid from solid-state fermentation with Yarrowia lipolytica. *Biotech, 9*.

Silva-Santisteban, B. O. Y., Converti, A., & Filho, F. M. (2006). Intrinsic activity of inulinase From Kluyveromyces marxianus ATCC 16045 and carbon and nitrogen balances. *Food Technology and Biotechnology, 44*(4), 479–483. http://public.carnet.hr/ftbrfd/44-479.pdf Brazil.

Sirilun, S., Chaiyasut, C., Pengkumsri, N., Peerajan, S., Chaiyasut, K., Suwannalert, P., & Sivamaruthi, B. S. (2016). Screening and characterization of β-glucosidase production by Saccharomyces cerevisiae. *Journal of Applied Pharmaceutical Science, 6*(5), 29–035.

Sohail, M., Barzkar, N., Michaud, P., Tamadoni Jahromi, S., Babich, O., Sukhikh, S., Das, R., & Nahavandi, R. (2022). Cellulolytic and xylanolytic enzymes from yeasts: properties and industrial applications. *Molecules (Basel, Switzerland), 27*(12), 3783. https://doi.org/10.3390/molecules27123783.

de Souza, C. E. C., Ribeiro, B. D., & Coelho, M. A. Z. (2019). Characterization and application of yarrowia lipolytica lipase obtained by solid-state fermentation in the synthesis of different esters used in the food industry. *Applied Biochemistry and Biotechnology, 189*(3), 933–959. https://doi.org/10.1007/s12010-019-03047-5, http://www.springer.com/humana+press/journal/12010.

de Souza, P. M., & de Oliveira e Magalhães, P. (2010). Application of microbial α-amylase in industry - A review. *Brazilian Journal of Microbiology, 41*(4), 850–861. https://doi.org/10.1590/S1517-83822010000400004.

Sujitha, P., & Shanthi, C. (2023). Importance of enzyme specificity and stability for the application of proteases in greener industrial processing- a review. *Journal of Cleaner Production, 425*, 138915. https://doi.org/10.1016/j.jclepro.2023.138915.

Sukmawati, D., Nurkhasanah, S., Afifah, Z. N., Husna, S. N. A., Widowati, R., Enshasy, H. E., & Dailin, D. J. (2021). Metagenomic-based approach for the analysis of yeast diversity associated with amylase production in Lai (Durio kutejensis). *Journal of Pure and Applied Microbiology, 15*(1), 75–90. https://doi.org/10.22207/JPAM.15.1.02, https://microbiologyjournal.org/metagenomic-based-approach-for-the-analysis-of-yeast-diversity-associated-with-amylase-production-in-lai-durio-kutejensis/.

Sutaoney, P., Rai, S. N., Sinha, S., Choudhary, R., Gupta, A. K., Singh, S. K., & Banerjee, P. (2024). Current perspective in research and industrial applications of microbial cellulases. *International Journal of Biological Macromolecules, 264*. https://doi.org/10.1016/j.ijbiomac.2024.130639, https://www.sciencedirect.com/science/journal/01418130.

Sutaoney, P., Akhand, A., Meshram, M., Sinha, S., Joshi, V., & Shahadat, M. (2024). Tannase production using green biotechnology and its applications: A review. *Biochemical Engineering Journal, 202*, 109163. https://doi.org/10.1016/j.bej.2023.109163.

Talens-Perales, D., Marín-Navarro, J., & Polaina, J. (2015). Enzymes: Functions and Characteristics. In *Encyclopedia of Food and Health*. Spain: Elsevier Inc, 532–538. 10.1016/B978-0-12-384947-2.00256-7, http://www.sciencedirect.com/science/book/9780123849533.

Tannase Market Res. (2022). https://www.futuremarketinsights.com/reports/tannase-market.

Tavano, O. L., Berenguer-Murcia, A., Secundo, F., & Fernandez-Lafuente, R. (2018). Biotechnological applications of proteases in food technology. *Comprehensive Reviews in Food Science and Food Safety, 17*(2), 412–436. https://doi.org/10.1111/1541-4337.12326, http://www.interscience.wiley.com/jpages/1541-4337.

Thangavelu, N., Jeyabalan, J., Veluchamy, A., & Belur, P. D. (2024). Production of tannase from a newly isolated yeast, Geotrichum cucujoidarum using agro-residues. *Preparative Biochemistry and Biotechnology, 54*(4), 564–572. https://doi.org/10.1080/10826068.2023.2256011, www.tandf.co.uk/journals/titles/10826068.asp.

Tiwari, P., Misra, B., & Sangwan, N. S. (2013). β-glucosidases from the fungus Trichoderma: an efficient cellulase machinery in biotechnological applications. *BioMed Research International*, (1).

Toplis, B., Bosch, C., Schwartz, I. S., Kenyon, C., Boekhout, T., Perfect, J. R., & Botha, A. (2020). The virulence factor urease and its unexplored role in the metabolism of Cryptococcus neoformans. *FEMS Yeast Research, 20*(4). https://doi.org/10.1093/FEMSYR/FOAA031, http://femsyr.oxfordjournals.org/.

Touijer, H., Benchemsi, N., Ettayebi, M., Idrissi, A. J., Chaouni, B., & Bekkari, H. (2019). Thermostable cellulases from the yeast trichosporon sp. *Enzyme Research, 2019*, 1–6. https://doi.org/10.1155/2019/2790414.

Urease Market Res. https://www.researchandmarkets.com/reports/5898049/urease-market-report-forecast.

Valliammai, M. G., Gopal, N. O., & Anandham, R. (2021). Probing cellulolytic yeast from forest ecosystem for the saccharification of Napier fodder biomass. *Journal of Environmental Biology, 42*(4(SI)), 1152–1161. https://doi.org/10.22438/jeb/42/4(si)/mrn-1538a.

Vijayaraghavan, K., Yamini, D., Ambika, V., & Sravya Sowdamini, N. (2009). Trends in inulinase production – a review. *Critical Reviews in Biotechnology, 29*(1), 67–77. https://doi.org/10.1080/07388550802685389.

Walker, G. M. (1998). *Yeast physiology and biotechnology*. John Wiley & Sons.

Wanderley, K. J., Torres, F. A. G., Moraes, L. M. P., & Ulhoa, C. J. (2004). Biochemical characterization of α-amylase from the yeast Cryptococcus flavus. *FEMS Microbiology Letters, 231*(2), 165–169. https://doi.org/10.1016/S0378-1097(03)00955-8, http://femsle.oxfordjournals.org/.

Wang, Y., Zhang, C., Xu, Y., & Li, J. (2015). Evaluating potential applications of indigenous yeasts and their β -glucosidases. *Journal of the Institute of Brewing, 121*(4), 642–648. https://doi.org/10.1002/jib.256.

Weerakoon, D., Bansal, B., Padhye, L. P., Rachmani, A., James Wright, L., Silyn Roberts, G., & Baroutian, S. (2023). A critical review on current urea removal technologies from water: An approach for pollution prevention and resource recovery. *Separation and Purification Technology, 314*. https://doi.org/10.1016/j.seppur.2023.123652, http://www.journals.elsevier.com/separation-and-purification-technology/.

Wieczorek, D., Miśkiewicz, K., Gendaszewska, D., Pipiak, P., Lasoń-Rydel, M., Sieczyńska, K., & Ławińska, K. (2023). Extracellular activity of proteases from Yarrowia lipolytica IPS21 as a function of the carbon and nitrogen source. *Fibres & Textiles in Eastern Europe, 31*(5), 66–74. https://doi.org/10.2478/ftee-2023-0046.

Yang, S., Song, L., Wang, J., Zhao, J., Tang, H., & Bao, X. (2024). Engineering Saccharomyces cerevisiae for efficient production of recombinant proteins. *Engineering Microbiology, 4*(1), 100122. https://doi.org/10.1016/j.engmic.2023.100122.

Yao, J., Guo, G. S., Ren, G. H., & Liu, Y. H. (2014). Production, characterization and applications of tannase. *Journal of Molecular Catalysis B: Enzymatic, 101*, 137–147. https://doi.org/10.1016/j.molcatb.2013.11.018.

Yavuz, S., Kocabay, S., Çetinkaya, S., Akkaya, B., Akkaya, R., Yenidunya, A. F., & Bakıcı, M. Z. (2017). Production, purification, and characterization of metalloprotease from Candida kefyr 41 PSB. *International Journal of Biological Macromolecules, 94*, 106–113. https://doi.org/10.1016/j.ijbiomac.2016.10.006, www.elsevier.com/locate/ijbiomac.

Yuivar, Y., Alcaino, J., Cifuentes, V., & Baeza, M. (2019). Characterization of gelatinase produced by Antarctic Mrakia sp. *Journal of Basic Microbiology, 59*(8), 846–852. https://doi.org/10.1002/jobm.201900126.

Zhang, J., Yan, B., He, C., Hao, Y., Sun, S., Zhao, W., & Zhao, C. (2020). Urease-immobilized magnetic graphene oxide as a safe and effective urea removal recyclable nanocatalyst for blood purification. *Industrial & Engineering Chemistry Research, 59*(19), 8955–8964. https://doi.org/10.1021/acs.iecr.0c00302.

van Zyl, W. H., Bloom, M., & Viktor, M. J. (2012). Engineering yeasts for raw starch conversion. *Applied Microbiology and Biotechnology, 95*(6), 1377–1388. https://doi.org/10.1007/s00253-012-4248-0.

Microalgae biomass as feedstock in anaerobic technologies

8

Rudolphus Antonius Timmers[1], *Octavio García-Depraect*[2,3], *Rebecca Serna-García*[4], *Alejandra Martorell-Múgica*[5], *and Silvia Greses*[4]
[1]Circular Economy Area, CARTIF Technology Centre, Valladolid, Spain, [2]Institute of Sustainable Processes, University of Valladolid, Valladolid, Spain, [3]Department of Chemical Engineering and Environmental Technology, University of Valladolid, Valladolid, Spain, [4]Department of Chemical Engineering, University of Valencia, Valencia, Spain, [5]Biotechnological Processes Unit, IMDEA Energia, Móstoles, Spain

8.1 Introduction

The depletion of fossil fuels and the negative environmental impacts derived from their massive use have created an urgent need to find alternative renewable feedstocks and sustainable technologies to mitigate energy and chemicals supply risks. In this context, anaerobic technologies using open mixed cultures represent a promising approach due to their low energy demand, low technological complexity, and reduced environmental impact. Traditionally, anaerobic digestion (AD) has been at the forefront of anaerobic technologies, which are being implemented for the treatment of residues with the concomitant production of an energy stream (biogas). Beyond biogas, AD has recently gained attention for producing an alternative source of energy and biocompounds of high industrial interest, such as hydrogen (H_2) and carboxylates. While H_2 has been recognized as an excellent, safe, and environmentally friendly alternative to fossil fuels that does not contribute to climate change and is crucial for the transition to clean energy, carboxylates are important building blocks for the chemical industry and also serve as a sustainable carbon source for a wide variety of biological processes (e.g., yeast fermentation, polyhydroxyalkanoate bioaccumulation). The advantages of anaerobic technologies over traditional petrochemical processes to produce energy and green chemicals have increased the number of feedstocks explored to maximize the profitability of these biotechnologies and expand their implementation. In this regard, microalgae have emerged as an interesting feedstock due to the capability to fix carbon dioxide (CO_2), recover nutrients from residual streams and their rapid growth rates thereby producing a high amount of biomass to be biotechnologically valorized, boosting the environmental impact mitigation while continuously producing a renewable feedstock (microalgal biomass). This chapter describes the anaerobic technologies to valorized microalgae biomass into bioenergy and biocompounds, presenting the most recent developments as well as the challenges and opportunities found in the use of microalgal biomass as feedstock.

Eukaryotic Microorganisms as Sources of Bioproducts. DOI: https://doi.org/10.1016/B978-0-443-30188-9.00020-2

8.2 Microalgae biomass as a feedstock for anaerobic bioprocesses

The idea of using algae as a fuel source was first introduced by Meier (1955), who proposed methane production from carbohydrate-rich algal cells. This concept was further explored by Golueke et al. (1957), who conducted laboratory-scale studies on the AD of *Scenedesmus* and *Chlorella*, comparing their performance to that of raw sewage sludge. Since then, interest in microalgae for bioenergy production has steadily increased, primarily due to their high biomass yield and sustainable cultivation systems. Unlike terrestrial crops, microalgae present several advantages since algae do not compete for arable land, freshwater, or food resources, making them a promising candidate for third-generation biofuels (Table 8.1). Another benefit of using microalgae biomass is the absence of lignin, which facilitates the conversion via biological processes. Moreover, the environmental benefits of microalgae also involve the ability to grow in waste streams, contributing to waste treatment while simultaneously sequestering CO_2 through photosynthesis (Leong et al., 2024; Satheesh et al., 2023). This dual functionality enhances their sustainability as a

Table 8.1 Advantages and disadvantages of using algal biomass as feedstock.

Advantages	Disadvantages
Algae have a rapid photosynthesis rate and grow 10–50 times faster than plants.	Maintaining optimal temperatures for algal cultivation is difficult, especially on a large scale.
They can use CO_2 and water effectively for biomass production.	Algae are susceptible to microbial contamination.
Algae do not require fertile land and can grow in areas with poor soil.	Algal cultivation demands significant nutrients and water.
Most algae species can withstand high levels of CO_2, sulfur oxides, and nitrogen oxides.	Their high-water content makes the drying process energy intensive.
Algal biomass contains a high level of carbohydrates, promoting fermentation and boostıng hydrogen production.	Algae have low biomass density, leading to expensive harvesting processes.
They do not contain lignin, meaning less intensive pretreatment is needed to enable bioaccessibility.	Resistant nature of the microalgal cell wall to biological degradation
Algae can be grown without the need for herbicides or pesticides.	
Algal biomass can be utilized to create a variety of commercially valuable products.	

No Permission Required.

renewable feedstock for bioenergy, biofuels, and green chemicals production, positioning them as a key player in the future of clean transition.

8.2.1 Microalgae cultivation

Microalgae are highly versatile microorganisms with flexible metabolic pathways. Although they are generally photoautotrophic organisms, their metabolism can also be mixotrophic, photoheterotrophic, or heterotrophic. Apart from light and a carbon source, microalgae also require nutrients for growth, such as ammonium and phosphate, which can be sourced from nutrient-rich waste streams. Their adaptability allows them to be cultivated in diverse environments, whether saline or freshwater, including wastewater streams, concentrate brine streams, agro-industrial wastes, and AD effluents.

Algae-based wastewater bioremediation appears as a promising option for next-generation wastewater resource recovery facilities, in which water, energy, and nutrients are recovered from waste streams. In these systems, microalgae not only utilize nutrients present in wastewater but also have the potential to remove a range of other pollutants such as organic pollutants, heavy metals, emerging contaminants, and other drugs (Vo et al., 2019). Microalgae have been reported to effectively grow by assimilating nutrients from various types of wastewater, including urban (Nordio et al., 2024), agro-industrial (de Carvalho et al., 2022), agricultural (Khalid et al., 2019), and industrial (Zabochnicka, 2022) wastewaters.

Among the residual streams investigated, the nutrient-rich supernatant resulting from AD represents an interesting, low-cost substrate for microalgae cultivation (Chong et al., 2022). Although numerous studies have demonstrated the capacity of microalgae to grow on digestate (Fernandes et al., 2022; Marazzi et al., 2019), relatively few have focused on optimizing this process, specifically, enhancing the uptake of digestate by microalgae. AD supernatant is characterized by high turbidity, ammonium, and alkaline pH, which favors the free-ammonia form of nitrogen and can substantially inhibit microalgae growth. Various methods have been investigated to overcome these challenges, such as the dilution of the stream (Barzee et al., 2022) or the pretreatment using filtration or combining with other biological processes. Recent studies have demonstrated that *Chlorella* can adapt to high-strength AD effluent by imposing long exposure times, resulting in high microalgal biomass production while removing 100% of the nitrogen in continuous mode (Greses et al., 2022).

Moreover, with freshwater shortage becoming an increasingly pressing problem, microalgae grown in saline effluents have attracted considerable interest in recent decades. The use of desalination concentrate, mainly from desalination facilities, as a medium for microalgae cultivation can boost process feasibility. Several microalgae species as *Nannochloropsis oculate, Chlorella vulgaris, Scendesmus quadricauda, S. platensis*, and *Dunaliella tertiolecta* can be cultivated on desalination brine (Al Bazedi et al., 2023).

8.2.2 Chemical composition of microalgal biomass

The chemical composition of microalgae (namely carbohydrates, lipids, and proteins) varies considerably depending on species, class, and environmental conditions, which ultimately determines the anaerobic technology selected and the process efficiency achieved according to the biomass biodegradability. For example, total lipid content ranges from 20% to 60% of dry cell weight depending on strain characteristics and growth conditions (Bibi et al., 2022), which makes microalgae a promising alternative for biofuel production in biorefineries.

Lipids and carbohydrate content in microalgal species can be boosted under different stress conditions, such as temperature, pH, light intensity, salinity, and nutrient availability. Nevertheless, applying these stressors must be economically feasible and should have a negligible impact on the growth and survival of the algae. Although the scientific community has focused on optimizing lipids content for biofuel production, it is now widely recognized that fully utilizing the entire microalgal biomass is the most feasible strategy to increase the viability of microalgae-based processes. For instance, Tibbetts et al. (2015) performed a complete characterization of different microalgal species and found that microalgae can contain low to moderate ash content (5%–17%), moderate to high carbohydrates (18%–46%) and crude protein (18%–46%), high crude lipid (12%–48%), and an energy content of 19–27 MJ kg^{-1}. Table 8.2 provides further details on the chemical composition of various microalgae species.

Lage and Gentili (2023) analyzed the biochemical composition of microalgae when using wastewater as culture media and flue gases as CO_2 source, finding carbohydrates and lipids as main compounds in a consortium dominated by *Scenedesmus dimorphus*, *S. quadricauda*, *Desmodesmus armatus*, and *Chlorella* sp., while the content of lipids remarkable increased when these conditions were replicated at greenhouse conditions.

8.2.3 Microalgae harvesting

The production process of microalgal biomass contains three major steps: cultivation, harvesting or collection, and processing. Of these, microalgal biomass harvesting appears as one of the main bottlenecks for the full-scale implementation of microalgae-based technologies. Small cell size (<30 μm), low concentration, dilute growth medium, electronegative cell membrane surface, and low settling velocity are the main obstacles in microalgae harvesting and dewatering processes. Harvesting costs alone can account for 20%–30% of the total production cost, which presents a challenge for the industrialization of microalgal biomass production. Innovative, cost-effective cultivation techniques and harvesting methods are crucial for improving the viability of bioenergy production. Different harvesting techniques are currently being employed to concentrate microalgae, including electrical, biological, chemical, and mechanical technologies. Among the most commercially utilized techniques are filtration, centrifugation, electrocoagulation, sonication, and chemical-based flocculation, which typically achieve concentration ranges from 0.02–0.25 to 1–5% (w/w) (de Morais et al., 2023).

Centrifugation is a fast and efficient method used for microalgae harvesting but is a time-consuming and energy-intensive method. More cost-effective and suitable alternatives for microalgae harvesting include sedimentation, which is a natural but slow process, and coagulation-flocculation, which can be very effective as a

Table 8.2 Chemical composition of various microalgae % dry weight (DW).

Species	Nitrogen (%)	Proteins (%)	Lipids (%)	Carbohydrate (%)	Ash (%)	Total phenolic content (mg GAE gDW^{-1})[a]	References
Spirulina[b]	11.67	55.8	14.2	22.2	7.8	10.7	Tibbetts et al., (2015)
Chlorella[c]	11.16	53.3	15.7	25.2	5.8	7.7	Tibbetts et al., (2015)
Phaeodactylum tricornutum (CCMP-1327)	8.28	39.6	18.2	25.2	17	9.9	Tibbetts et al., (2015)
Nannochloropsis granulata (CCMP-535)	3.74–7.01	17.9–33.5	23.6–47.8	27.4–36.2	6.7–6.9	6.0–8.0	Tibbetts et al., (2015)
Botryococcus braunii (UTEX-572)	8.17–8.35	39.1–39.9	24.9–34.4	18.5–30.6	5.4–7.2	7.1–13.2	Tibbetts et al., (2015)
Euglena sp.	-	39–61	14–20	14–18	-	-	Roy and Pal (2015)
Spirulina platensis	-	50–65	4–9	8–14	-	-	Roy and Pal (2015)
Porphyridum sp.	-	28–39	-	50–57	-	-	Roy and Pal (2015)
Scenedesmus sp.	-	50–56	12–14	10–52			Roy and Pal (2015)
Nannochloropsis gaditana	6.94[d]	33.17[d]	27.89[d]	15.90[d]	11.52[d]	-	Di Lena et al., (2020)
Isochrysis galbana	4.09[d]	28.98[d]	31.09[d]	17.67[d]	15.16[d]	-	Di Lena et al., (2020)
Tetraselmis suecica	5.45[d]	26.05[d]	14.68[d]	24.01[d]	17.99[d]	-	Di Lena et al., (2020)

[a] mg Gallic Acid Equivalents · g dry $weight^{-1}$.
[b] Earthrise Nutritionals LCC (Irvine, CA, USA), species *Arthrospira platensis*.
[c] Sequel Naturals Ltd. (Port Coquitlam, BC, Canada), species *Chlorella pyrenoidosa* and *Chlorella vulgaris*.
[d] Value refers to a gram per 100 g dry mass.

Adapted from Bhatia, S.K., Rajesh Banu, J., Singh, V., Kumar, G., & Yang, Y.H. (2023). Algal biomass to biohydrogen: Pretreatment, influencing factors, and conversion strategies. *Bioresource Technology*, *368*, 128332. https://doi.org/10.1016/j.biortech.2022.128332. No Permission Required.

pre-concentration step. In flocculation, microalgae cells are aggregated into larger particles, or flocs, by adding chemical or bio-flocculants to neutralize the surface charge of the cells. However, using chemicals in the coagulation-flocculation process can limit the reusability of treated water and the harvested biomass. In line with this, organic biodegradable flocculants as chitosan, starch, tannins, and other commercially available products have emerged as an environmentally friendly alternative. Filtration works by physically separating microalgae through a semipermeable membrane, which retains the algae while allowing the liquid medium to pass through. While this method offers several advantages as simplicity of operation and high biomass recovery rates, its major limitation is related to membrane fouling and clogging, leading to frequent membrane replacements, and increased maintenance costs. Different types of permeable membranes and filtration techniques such as microfiltration, ultrafiltration, vacuum filtration, and pressure filtration, have been tested, though large-scale application remains challenging.

Recently, innovative methods such as magnetic and nanotechnologies have been investigated (de Lima Barizão et al., 2021), offering fast, automated, and efficient microalgae separation, but their economic viability on a large scale still requires assessment before they can be deemed practical alternatives. Selecting the optimal harvesting method depends on the nature of the microalgal species, the biomass cell density and size, the end-product conditions, and the reutilization of the growth medium.

8.2.4 Biomass pretreatments: disruption and hydrolysis

Microalgae culture in residual streams generally has resilient cell walls primarily composed of organic compounds with slow biodegradability, which hampers their degradation by microbial activity. This structural barrier retains organic matter within the cytoplasm, rendering it inaccessible to anaerobic microorganisms. Eukaryotic microalgae such as *Chlorella*, *Scenedesmus*, and *Nannochloropsis* typically present a polysaccharide-based cell wall, mainly composed of cellulose and hemicellulose layers, along with recalcitrant compounds like sporopollenin and polyterpenes, which significantly hinder their biological degradation (D'Hondt et al., 2017). To overcome this drawback, several pretreatment strategies have been applied to microalgae biomass to enhance the anaerobic biodegradability (Fig. 8.1). A summary of recently used pretreatments for microalgae biodegradation improvement is givenin Table 8.3.

It is important to highlight that the anaerobic biodegradability of feedstocks is assessed by performing a biochemical methane potential test (BMP). BMPs allow to obtain the maximum methane volume production per gram of organic matter present in the feedstock (also known as methane yield), which can be expressed as a gram of chemical oxygen demand (COD) or a gram of volatile solids (VS) (Gonzalez-Fernandez et al., 2018). Thus, the anaerobic biodegradability efficiency is determined by comparing the BMP result with the maximum theoretical value, which is 350 mL per gram of COD at 0°C and 1 atm. The biodegradability can be obtained as follows:

$$\text{Biodegradability}(\%) = \frac{(\text{NmLCH}_4/\text{gCOD})_{\text{experimental}}}{(\text{NmLCH}_4/\text{gCOD})_{\text{theoretical}}} \times 100 \tag{8.1}$$

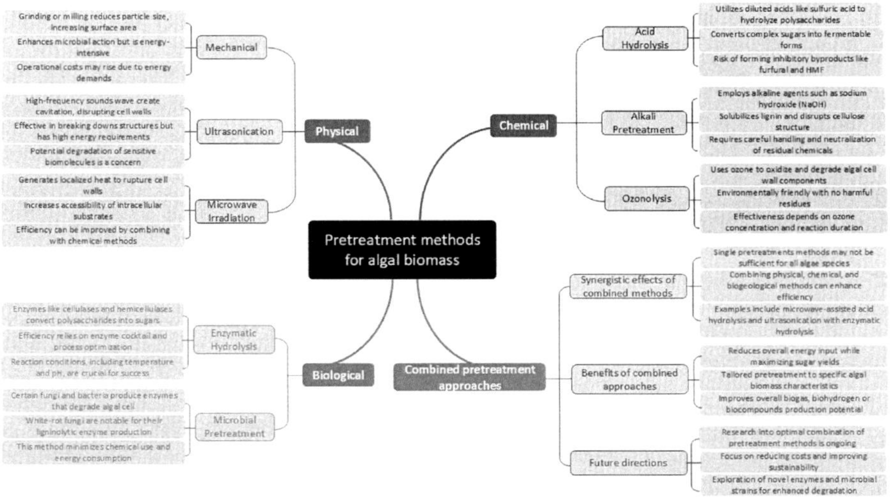

Figure 8.1 Algal biomass pretreatments for anaerobic degradation increase.
No Permission Required.

Each pretreatment efficiency relies on microalgal characteristics, such as the cell wall structure and the biochemical composition of the cells. Among the pretreatments studied (Fig. 8.1), thermal, mechanical, chemical, and biological pretreatments are the most commonly used.

Thermal pretreatment involves heating biomass to solubilize its components, within a temperature range of 50°C–270°C and varying reaction times from minutes to hours (Magdalena, et al., 2018). Studies summarized in Table 8.3 show that, although thermal pretreatment often leads to increased methane yields, the optimal temperature and exposure time depend on the microalgae species. However, the methane yield increase does not offset the energy consumed during pretreatment, resulting in an overall negative energy balance. Moreover, these methods present other drawbacks such as the potential formation of recalcitrant compounds, particularly under hydrothermal conditions, which may reduce process performance.

Mechanical pretreatment methods, such as milling, ultrasounds, and microwave treatments, focus on disrupting microalgal cells using physical force. Bead milling is a complex process where effectiveness depends on variables like bead size, density, filling, and mixing speed. Ultrasound pretreatment promotes cell wall disruption and organic matter solubilization, though its success depends on factors like the applied specific energy and substrate concentration. Similarly, microwave pretreatment studies report improved biomass solubilization and methane production with the applied specific energy (Table 8.3).

Chemical pretreatments are frequently combined with thermal pretreatments (Table 8.3). Acid and alkaline reagents are commonly used to solubilize polymers (hemicellulose and lignin) favoring the enzymatic attack. Nevertheless, some chemicals must be removed before the anaerobic process, and combining chemical

Table 8.3 Biochemical methane potential of microalgal biomass subjected to various pretreatments.

Pretreatment		Conditions	Microalgae	Methane yield (raw microalgae) ($mLCH_4 \cdot gVS^{-1}$)	Methane yield (treated microalgae) ($mLCH_4 \cdot gVS^{-1}$)	Methane yield improve (%)	References
Thermal		75°C–10 h	*Chlorella Stigeoclonium*	90	146	62	Solé-Bundó et al., (2018)
		70°C–0.5 h	*Chlorella*	155	215	39	Wang et al., (2017)
		90°C–0.5 h		155	228	47	
		121°C–0.3 h		155	322	108	
		60°C–3 h	Microalgal biomass[a]	37	173	368	Estevam et al., (2024)
		78C–7 h			291	687	
		95°C–3 h			236	538	
		121°C–0.1 h	Microalgal biomass[a]	295	420	42	Bohutskyi et al., (2023)
		121°C–0.5 h			440	49	
		121°C–1 h			450	52	
		75°C–10 h	Microalgal biomass[a]	140	126	61	Solé-Bundó et al., (2020)
Mechanical	Bead milling (glass beads)	1 mm, 4 h	*Scenedesmus*	54	97	80	Inglesby et al., (2015)
		4 mm, 1 h		133	166	20	
		0.35 mm, 0.4 h	*Acutodesmus*	191	289	51	Gruber-Brunhumer et al., (2015)
		1.25 mm, 0.08 h	Microalgal biomass[b]	162	170	5	Martín et al., (2018)
		2.5 mm, 1 h		162	180	11	

Table 8.3 (Continued)

Pretreatment		Conditions	Microalgae	Methane yield (raw microalgae) ($mLCH_4 \cdot gVS^{-1}$)	Methane yield (treated microalgae) ($mLCH_4 \cdot gVS^{-1}$)	Methane yield improve (%)	References
	Ultrasound	57 $kJ \cdot gTS^{-1}$	*T. suecica*	169	356	-2	Paul et al., (2023)
		57 $kJ \cdot gTS^{-1}$	*N. oceanica*	123	165	-22	
		20 $kJ \cdot gTS^{-1}$	*Acutodesmus*	191	101	53	Gruber-Brunhumer et al., (2015)
		200 W, 30 min	*Chlorella sorokiniana*	317	458	44	Córdova et al., (2018)
		400 W, 20 min			414	30	
		400 W, 30 min			425	34	
		400 W, 40 min			422	33	
		26.7 $kJ \cdot gTS^{-1}$	Microalgal biomass[c]	106	114	8	Passos et al., (2015)
		77 $kJ \cdot gTS^{-1}$	*Scenedesmus*	82[d]	93[d]	13	González-Fernández et al., (2012)
		130 $kJ \cdot gTS^{-1}$			154[d]	88	
	Microwave	22 $kJ \cdot gTS^{-1}$	*Scenedesmus*	117	150	28	Passos et al., (2014)
		44 $kJ \cdot gTS^{-1}$	*Chlorella*		167	43	
		65 $kJ \cdot gTS^{-1}$			210	80	
		34.3 $J \cdot gTS^{-1}$	Microalgal biomass[c]	106	128	21	Passos et al., (2015)

(Continued)

Table 8.3 (Continued)

Pretreatment		Conditions	Microalgae	Methane yield (raw microalgae) ($mLCH_4 \cdot gVS^{-1}$)	Methane yield (treated microalgae) ($mLCH_4 \cdot gVS^{-1}$)	Methane yield improve (%)	References
Chemical	Alkali	0.5 M NaOH 2 M NaOH 2 M NaOH + H_2O_2 0.5% w/w	Microalgal biomass[b]	162	173 377 279	7 133 72	Martín et al., (2018)
	Thermo-alkali	2% NaOH + 50°C, 48 h	*Chlorella*	137[d]	157[d]	17	Mahdy, et al., (2014)
		5% NaOH + 50°C, 24 h	*Scenedesmus*	141[d]	170[d]	20	Mahdy, et al., (2014)
		0.5 N NaOH+ 121°C, 0.5 h	*Chlorella* *Nannochloropsis*	340 355	481 432	41 22	Bohutskyi et al., (2014)
		3 N NaOH, 90°C, 3 h, pH 11	Microalgal biomass[a]	37	193	429	Estevam et al., (2024)
	Thermo-acid	3 N HCl, 15 h: +85°C, pH 3 +90°C, pH 5	Microalgal biomass[a]	37	 31 148	 -14 306	Estevam et al., (2024)

Table 8.3 (Continued)

Pretreatment	Conditions	Microalgae	Methane yield (raw microalgae) ($mLCH_4 \cdot gVS^{-1}$)	Methane yield (treated microalgae) ($mLCH_4 \cdot gVS^{-1}$)	Methane yield improve (%)	References
Biological	Cellulase	*Oocystis and diatoms*	189	203	8	Passos et al., (2016)
	Enzyme mix[e]			217	15	
	Alcalase	*Chlorella*	191[d]	287[d]	50	Mahdy et al., (2014)
	+ 75°C, 0.5 h			270[d]	41	
	Viscozyme			218[d]	14	
	+ 75°C, 0.5 h			220[d]	15	
	Alcalase and viscozyme			270[d]	41	
	+ 75°C, 0.5 h'			299[d]	57	
	Cellulase + 55°C	*Porphyridium cruentum*	130	153	18	Kendir Çakmak and Ugurlu (2020)
	Protease + 35°C			230	77	
	Protease + 55°C			260	100	
	Viscozyme + 35°C			172	32	
	Viscozyme + 55°C			230	77	
	Viscozyme + 50°C, 5 h	*Scenedesmus*	206	358	73	Greses (2017)

[a] Microalgae biomass grown in high-rate algal ponds treating wastewater.
[b] Microalgae biomass grown in high-rate algal ponds mainly composed of Stigeoclonium, Monoraphidium, Nitzschia, and Navicula.
[c] Microalgae biomass grown in photobioreactor: Tetradesmus (55%), Desmodesmus (16%), Aphanothce (11%), Chlorella (5%), Scenedesmus (4%), Parachlorella kessleri (3%).
[d] Expressed as $mLCH_4 \cdot gCOD_{inf}^{-1}$.
[e] Enzyme mix (cellulase, glucohydrolase, and xylanase).
No Permission Required.

treatments with thermal processes is often required to significantly improve methane production and biomass solubilization.

An alternative to those high energy-intensive pretreatments is the use of biological reactions, which offer high selectivity and avoid introducing external inhibitory compounds. Enhancing biodegradability via enzymatic hydrolysis of the microalgal cell wall and other polymers has been proven to be effective (Table 8.3), with the combination of different enzymes yielding the best outcomes.

8.3 Microalgae biomass in anaerobic digestion for methane production

8.3.1 Description of the AD process

AD is a well-known process in which a complex and synergic microbial network is able to transform the organic matter into a source of energy (biogas), regardless of the macromolecular composition (proteins, carbohydrates, and lipids). During AD, the organic matter is biologically degraded through four main steps: Hydrolysis, acidogenesis, acetogenesis, and methanogenesis (Fig. 8.2). In the hydrolysis step, the complex organic matter is decomposed into monomers, namely sugars, amino acids, and long-chain fatty acids, which are enabled by the extracellular enzymes excreted by hydrolytic bacteria. Thereafter, the acidogenic bacteria ferment those soluble monomers in the acidogenesis step, giving rise to volatile fatty acids (VFAs) also known as short-chain fatty acids (SCFAs), lactic acid, alcohols, and direct precursors of methane formation, such as acetic acid and hydrogen (H_2) and CO_2. During the acetogenesis step, acetic acid, CO_2, and H_2 are the main metabolic products released from the degradation of the compounds resulting from the previous steps, either hydrolysis or acidogenesis. Unlike previous steps, the metabolisms involved in acetogenesis generate most of the H_2 in the system. In the last step, methanogenesis, methanogenic archaea are responsible for the production of biogas (mainly CH_4 and CO_2) through two main pathways depending on the metabolized substrate. Whereas the acetoclastic pathway involves the consumption of acetic acid to the ultimate product CH_4 and CO_2, the hydrogenotrophic pathway releases CH_4 as a product from the metabolization of CO_2 and H_2. Acetogenesis and methanogenesis exhibit high synergy, being the acetogenic acetate-H_2 ratio crucial to determine the methanogenic pathway followed in the last step of AD.

AD is one of the most exploited bioprocesses to valorize organic residues into a source of energy since it is a low complex and low energy demand technology that can be applied to a wide variety of residues, provided that biodegradable matter content exceeds 60% on dry matter basis to ensure process profitability. Among the organic residues assessed to produce biogas via AD, the microalgal biomass has been pointed out as a promising feedstock for AD due to the several advantages over other biomass sources (Table 8.1). More specifically, microalgae culture for the treatment of brine or waste streams represents the most feasible approach since the use of low-cost sources of nutrients considerably increases the technological cost-effectiveness. Nevertheless, the

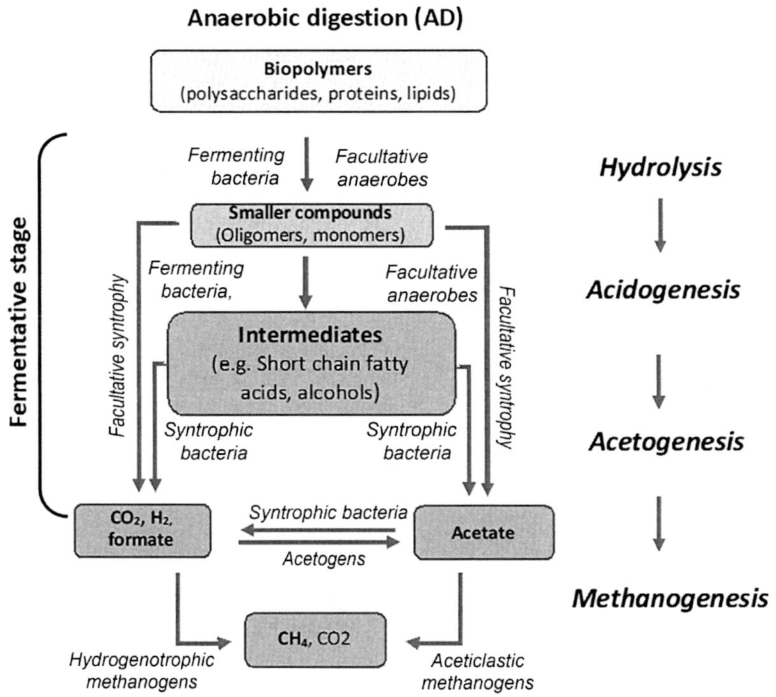

Figure 8.2 Schemes of anaerobic digestion including the main stages considered.
Source: Adapted from Tomás-Pejó, E., González-Fernández, C., Greses, S., Kennes, C., Otero-Logilde, N., Veiga, M.C., Bolzonella, D., Müller, B., & Passoth, V. (2023). Production of short-chain fatty acids (SCFAs) as chemicals or substrates for microbes to obtain biochemicals. *Biotechnology for Biofuels and Bioproducts*, *16* (1). https://doi.org/10.1186/s13068-023-02349-5.

use of residual streams as culture media promotes the growth of specific microalgae strains, such as *Chlorella vulgaris* and *Scendesmus quadricauda* in wastewater (Greses et al., 2022) and *Nannochloropsis oculata* in desalination brine, that exhibit a highly resistant cell wall in their structure that hampers the microbial anaerobic degradation by limiting the bacterial bioaccessibility. To increase the anaerobic biodegradability of microalgal biomass, costly pretreatments are normally required (Fig. 8.1), which also involves the production of inhibitors (e.g., furfural) that can be detrimental to biogas production. As an alternative, innovative reactor configurations can play a significant role in microalgae degradation. Technologies able to decouple the hydraulic retention time (HRT) and the solid retention time (SRT) can mediate an increase in the intrinsic hydrolytic activity of the microbiome by subjecting the microbiome to long contact periods with the feedstock (long SRTs) (Greses et al., 2017). Likewise, these technologies can promote the adaptation and evolution of the microbiome to deal with the osmotic pressure stress provoked by the use of microalgal biomass with high salinity content (De Vrieze et al., 2017). Moreover, the codigestion of saline microalgae with other substrates, such as sewage sludge, can also mitigate the osmotic stress and enhance biogas production (Parsy et al., 2024).

8.3.2 Yields and main insights

Although AD of microalgal biomass has been studied since 1957 (Golueke et al., 1957), the implementation at the industrial scale still requires overcoming some technical barriers mainly related to biodegradation efficiency and process stability. The development of new technologies, pretreatments, and the increase in microbial insights have boosted the AD of microalgal biomass in the last decade.

Microalgae species lacking cell wall (e.g., *Chlamydomonas reinhardtii*) have been claimed as optimal biomass to be transformed into biogas via AD, since high methane yields varying from 306 to 400 mL CH_4/gVS have been reported in the absence of pretreatment (Frigon et al., 2013). However, the use of waste streams as low-cost source of nutrients and culture media promote the natural growth of highly resistant microalgae strains, such as *Chlorella* sp., *Scenedesmus* sp., *Dunaliella salina*, *Nannochloropsis salina*, *Botryococcus braunii*, *Micractinium* sp., or *Selenastrum capricornutum*., due to their capability to grow in the presence of contaminants or high salinity. These microalgae not only exhibit high resistance to biological degradation but also can release toxics and inhibitors that significantly hamper the biogas production process. In reactors operated in continuous mode, the AD of those raw microalgae normally results in biodegradabilities between 12% and 23%, corresponding to methane yields of 35.7–73 mL $CH_4/gCOD_{fed}$, (Greses et al., 2018; Lunprom et al., 2019). Those values represent remarkably low AD efficiencies when compared with common organic residues used as feedstock for AD. For instance, AD of the organic fraction of municipal solid waste normally results in a bioconversion efficiency ranging between 40% and 50% (Jiang et al., 2020).

As detailed in the previous section, pretreatments are required to increase the microalgae bioaccessibility and the organic matter bioavailability by breaking down the microalgae cell wall. Among the traditional pretreatments applied to microalgal biomass (Table 8.3), enzymatic pretreatments demonstrated the highest efficiency for biogas production. As an example, Mahdy et al., (2015) found a methane yield of 128 mL CH_4/g COD_{fed} (36% biodegradability) in continuous mode when *Chlorella* was subjected to an enzymatic pretreatment using a proteolytic enzymatic cocktail (Alcalase, Novozyme). Likewise, Greses et al. (2024) reached 171.3 mLCH_4/g COD_{fed} (48% biodegradability) when a microalgae-bacteria consortium dominated by *Chlorella* was subject to similar enzymatic pretreatment. However, although the pretreatments are effective in improving AD efficiency, the biogas production increase cannot offset the high economic cost related to the addition of reagents and the energy demand. Moreover, microalgae culture in waste streams normally exhibit a high protein content, thereby showing a low carbon-to-nitrogen ratio (C/N). This feature can create a double limitation for anaerobic microorganisms: (1) the C/N unbalance (<20) cannot meet the nutritional requirements of the microorganisms and (2) the anaerobic degradation of proteins releases NH_4^+ as the main end soluble product, which acts as an inhibitor at high concentrations. This limitation is exacerbated when microalgae cultures in saline media are used as feedstock because the high salinity content represents an additional inhibitor for the anaerobic microbiome.

Currently, potential solutions to AD limitations (low efficiency and inhibitors) are under intense investigation. The low C/N ratio can be easily solved by adding carbon-rich co-substrates, which is a common approach in AD. The main strategies are focused on improving the metabolic capabilities of the anaerobic microbial network instead of using costly pretreatments to overcome the hydrolysis limitation. In this regard, the use of a nonconventional reactor configuration can be beneficial to promote the hydrolytic potential of the anaerobic microbiome, thereby increasing the microalgal biodegradability without applying pretreatments. For instance, Greses et al. (2017) reported a microalgae biodegradability of 70% (242.1 $mLCH_4/g$ COD_{fed}) using an anaerobic membrane bioreactor (AnMBR). AnMBR can decouple HRT and SRT, allowing to work at extremely high SRTs (i.e., 100 d), while keeping a high flow rate of microalgal biomass (low HRTs). This configuration enables a hydrolytic bacteria enrichment and increases the biomass-bacteria contact time, ultimately resulting in striking high AD efficiencies. In this line, the use of two-stage AD also represents an interesting alternative since this configuration allows for the separation of the fermentative stage and the methanogenic stage (Fig. 8.2), taking advantage of the different growth rates and optimal conditions of the microorganisms to reach microalgae biodegradability of 51.2% (Zabed et al., 2020). As an alternative, the microbiome can be also bioaugmented with highly active hydrolytic bacteria, such as *Serratia marcescens* EGD-HP20, which can boost the hydrolysis of microalgae cell wall and increase the biomethanation potential (Poddar & Khardenavis, 2024).

Likewise, the microbiome can also be acclimated to the presence of inhibitors, such as salinity or high NH_4^+ content, reaching high AD efficiencies by evolving the conventional anaerobic microbiome through an adaptation period. Hence, the use of salt-adapted inoculum in AD enabled the stable degradation of *D. salina* reaching a high biodegradability efficiency of 48% (Roberts et al., 2019). Similarly, acclimation of the microbiome to extremely high ammonium levels (>7 g NH_4^+-N/L) is possible, reaching 327 mL CH_4/g VS (48% biodegradation efficiency). The exposure of the microbiome to the inhibitor for long periods following a stepwise-increase strategy promotes the adaptation of the microbial network, specially of those exhibiting high sensitivity, such as methanogenic archaea (Tian et al., 2018). To overcome AD inhibitions, the microbiome can be also redesigned by applying a bioaugmentation strategy. Recent studies demonstrated the process bioaugmentation with bacteria isolated from highly toxic environments (e.g., *Sphingopyxis* sp. YF1) can effectively degrade some toxins resulting from microalgae degradation, such as microcystin (Massey et al., 2022).

The latest advances in AD of microalgae have pointed out the use of nanoparticles, nanotubes, and biochar as promising approaches to increase AD efficiency. Sendra et al. (2018) found that the structure of *Chlorella* can be altered by applying nanoparticles of silver prior AD, which modifies the bioaccessibility of the biomass and increases the microalgal biodegradability. Moreover, the addition of nanoparticles of iron, graphene, carbon nanotubes, or biochar can improve AD by increasing the microbiome concentration since these materials can act as structural supports to retain microorganisms in the system. The addition of those nanomaterials also facilitates electron transfer throughout the microbial network, increasing AD efficiency by 25%–50% (Saravanakumar et al., 2023).

8.3.3 Challenges in the AD technology

The fundamental challenge in AD of microalgal biomass still relies on the biodegradability of the microalgae strain culture in waste streams. Although the use of residual streams is crucial to ensure the economic feasibility of the process and enable the recovery of resources (i.e., water and nutrients), the high resistance of the microalgae able to grow in harsh conditions and the presence of inhibitors still hamper the process. Nevertheless, the technological advances, such as the use of AnMBR or nanoparticles, have considerably improved the process performance, forwarding the scientific efforts to decrease technological complexity and reduce the price of carbon-based nanomaterials. Thus, the next step to implement AD of microalgae at full scale is related to ensuring process stability considering both processes, microalgae culture, and AD, by developing a warning system to prevent process failure against unforeseen events (e.g., presence of inhibitors, overloading).

8.4 Microalgae biomass in anaerobic digestion for hydrogen production

8.4.1 Description of the process

Dark fermentation (DF) is a complex biological process that operates in the absence of light under anaerobic conditions, facilitated by hydrolytic-acidogenic bacteria that metabolize organic compounds, primarily carbohydrates, through enzymatic reactions (Martínez-Fraile et al., 2024). DF comprises the first steps of AD (Fig. 8.2, fermentative stage) and involves several metabolic pathways for fermentative hydrogen production, including acetic-type fermentation, butyric-type fermentation, ethanol-type fermentation, and lactate-based fermentation. The main by-products of DF include hydrogen, carbon dioxide (CO_2) in the gaseous phase, and other by-products such as VFAs (e.g., acetate and butyrate) in the fermentation broth (Martínez-Fraile et al., 2024). Furthermore, the scalability of DF makes it adaptable to different setups, whether small-scale decentralized units or larger industrial operations, making it a flexible solution for various regions and capacities (Das & Veziroğlu, 2001). Lastly, DF can be integrated with various other bioprocesses, such as photofermentation (Rai & Singh, 2016), AD (Sarkar et al., 2021), bioplastics (polyhydroxyalkanoates) synthesis (Kumar et al., 2019), and biooil synthesis (Pereira et al., 2022), among others, to increase overall energy yield and/or enhance the profitability of the overall process.

Several factors influence the efficiency and performance of the DF process for biohydrogen production. Substrate composition plays a significant role, with carbohydrate-rich wastes being the most favorable for hydrogen production. Additionally, the microbial community composition also affects the process. Key operational factors

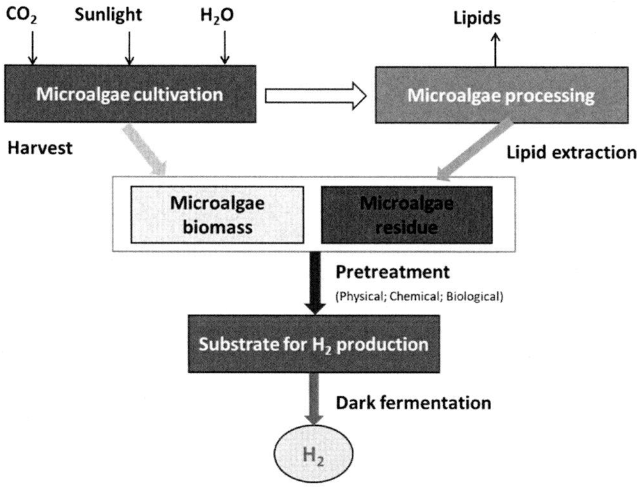

Figure 8.3 A biotechnological approach to cultivating microalgae as feedstock for biohydrogen production via DF.
From Wang, J., & Yin, Y. (2018). Fermentative hydrogen production using pretreated microalgal biomass as feedstock. *Microbial Cell Factories*, *17*(1). https://doi.org/10.1186/s12934-018-0871-5.

such as pH and temperature are critical, with pH values between 5.5 and 6.5 being optimal for hydrogen production, while mesophilic temperatures often tend to favor higher hydrogen yields compared to thermophilic processes. HRT and organic loading rate (OLR) are operating parameters that also impact microbial activity, with shorter HRTs and higher OLRs generally promoting hydrogen production under controlled conditions. Reactor design and configuration, along with the use of appropriate inoculum sources and pretreatment methods, further shape the microbial dynamics and overall efficiency of the DF process. Finally, macro- and micronutrient availability, such as iron, calcium, and nitrogen, also play a crucial role in ensuring proper microbial metabolism and optimizing hydrogen production rates (García-Depraect et al., 2021).

As shown in Fig. 8.3, algal biomass is widely recognized as a promising substrate for biohydrogen production due to its high carbohydrate content (e.g., starch, glycogen, and cellulose), which can be efficiently converted into hydrogen through fermentation processes (Pal et al., 2024). In the following sections, the latest advances in the field of DF of algal biomass, along with its associated barriers and future directions for improvement, will be discussed in detail.

8.4.2 Latest advances in DF of algal biomass

In recent years, numerous efforts have been made to advance the field of DF using algal biomass as feedstock. Sim et al. (2021) investigated the continuous H_2 production from red algal biomass (*Eucheuma spinosum*) using a dynamic membrane bioreactor (DMBR) over 109 days. The maximum H_2 production rate (HPR) reached 21.58 L/L-d at an HRT of 3 h, with a peak H_2 yield of 1.28 mol H_2/mol hexose at 4.5 h HRT. The main challenges included persistent H_2 consumption by the propionate pathway, even after heat treatment, and managing biomass retention and membrane performance over time. Despite these issues, the system maintained a low filtration pressure of 3.41 kPa, eliminating the need for backwashing, and exhibited high stability and efficiency in H_2 production. Later, Singh et al. (2022) used *Scenedesmus obliquus* UTEX 393 and achieved a maximum H_2 yield of 97.6 mL/g VS_{added} through acid-thermal pretreatment with 0.5 N H_2SO_4 at 121°C. This pretreatment significantly enhanced H_2 production, yielding nearly 10 times more than the untreated biomass, which produced only 9.8 mL/g VS_{added}. Sriyod et al. (2021) reported the development and optimization of a one-step multienzyme (cocktail) pretreatment for the DF of *Chlorella* sp., yielding 43.16 mL H_2/g VS_{added} under optimal conditions, representing an increase of 39% compared to the yield from untreated biomass. Kumar et al. (2019) utilized a two-step pretreatment process, combining ultrasonication with enzymatic saccharification (Viscozyme L, Novozyme) to enhance H_2 production via DF from mixed microalgal biomass cultivated in diluted swine manure. Findings reported by Stanislaus et al. (2018) also endorse the fact that optimization of pretreatment techniques and the ratio of substrate to inoculum can boost the DF production of H_2 from microalgal biomass (*C. vulgaris*), achieving a peak H_2 yield of 190 mL H_2/g VS_{added}.

In another study, the low C/N ratio commonly found in microalgal biomass was addressed by Sun et al. (2018), who investigated the cofermentation of protein-rich microalgae (*Chlorella pyrenoidosa*) and carbohydrate-rich rice residue to improve the production of H_2 through DF. The co-fermentation of pretreated rice residue and microalgae at a 5:1 mixing ratio resulted in a peak H_2 yield of 201.8 mL/g VS_{added}, representing a 10.7-fold enhancement compared to the mono-fermentation of pretreated microalgae. Khalekuzzaman et al. (2023) also evaluated a dark co-fermentation system using fecal sludge as a co-substrate, resulting the 50:50 microalgae-fecal sludge mixture in the highest H_2 yield of 24.5 mL/g VS_{added}.

More recently, Ramprakash et al. (2024) investigated the use of iron oxide nanoparticles (IONPs) as a chemical additive to enhance both microalgae biomass growth and lipid accumulation in *Chlorella* sp. The hydrolysate obtained from this pretreated biomass was used as a substrate for fermentative H_2 production using *Enterobacter aerogenes*, yielding 2.6 mol H_2/mol sugar, which represented a 44% increase in H_2 yield compared to the control. This dual approach, combining enhanced biomass and lipid accumulation with H_2 production, demonstrated that IONPs were a promising catalyst for biorefinery applications.

Microalgae cultivated in various types of wastewater have also been documented for their potential in biohydrogen production and wastewater treatment. The production of dark fermentative hydrogen using as feedstock different microalgae, i.e., *Chlorella pyrenoidosa*, *Scenedesmus obliquus*, and *Chlorella sorokiniana* cultivated in domestic wastewater was reported by Satheesh et al. (2023). The maximum H_2 production yield (mL H_2/g VS_{added}) was 45.5 for *C. pyrenoidosa*, 38.4 for *S. obliquus*, and 34.8 for *C. sorokiniana*. It is noteworthy that H_2 production as high as 116.3 mL H_2/g VS_{added} has been reported from *S. obliquus* biomass through DF driven by *Clostridium butyricum* (Ortigueira et al., 2015).

8.4.3 Challenges and opportunities in the DF technology

The use of microalgae for DF presents several challenges that hinder its widespread adoption and commercial viability (Fig. 8.4). Similar to AD, one of the primary obstacles is the recalcitrant nature of the microalgal cell wall, which makes the biomass resistant to microbial degradation, limiting its fermentability (Aslam et al., 2023). This issue necessitates pretreatment methods to effectively break down the cell walls and release fermentable substrates like carbohydrates (Parakh et al., 2023). However, no single pretreatment method works universally due to the diverse nature of microalgae (Bhatia et al., 2023). Thus, the development of more cost-effective, energy-efficient, and scalable pretreatment methods that minimize the formation of inhibitors, such as acids, furfurals, and its derivates, is critical for enhancing H_2 yields (Rady et al., 2024). Moreover, the high protein content in microalgal biomass can lead to a low carbon-to-nitrogen ratio, which is not favorable for efficient fermentation, and may require nutrient adjustments to optimize the process. In this regard, future research must explore the use of co-substrates such as food waste or agricultural residues (Khalekuzzaman et al., 2023).

Scaling up DF remains a key challenge due to several technical and operational factors. One of the primary difficulties is maintaining stable hydrogen production at a larger scale while ensuring the optimal retention of active microbial communities

Figure 8.4 Barriers and future directions for improving fermentative H_2 production from algal biomass.
No Permission Required.

(García-Depraect et al., 2021). The overall low efficiency of biohydrogen production during DF is another significant challenge. To improve the yield, researchers often combine DF with photo-fermentation, although this integration comes with its own set of technical complexities and increased costs due to the need for specialized photobioreactors. Hybrid systems of DF and microbial electrolysis cells have also been proposed as a strategy to increase overall hydrogen production. Another strategy is genetic engineering, which plays a critical role in modifying hydrogenase enzymes and pathways using approaches such as random mutagenesis, gene overexpression, and inducible genes to improve H_2 yields (Zhang et al., 2023). Nutrient deprivation, particularly of sulfur, nitrogen, or phosphorus, further enhances biohydrogen production by triggering anaerobic conditions that activate hydrogenase. The design of continuous DF systems is also essential for maximizing H_2 production efficiency. A well-designed reactor can improve the retention of hydrogen-producing microorganisms, reduce the risk of wash-out at shorter HRT, and enhance substrate conversion efficiency being DMBRs, a promising reactor configuration to traditional designs.

Furthermore, the separation of biohydrogen from other gases produced during DF, such as CO_2, remains an issue, requiring further innovations in gas separation technologies to ensure cost-effective scalability. Thus, future research should also investigate large-scale, low-cost cultivation systems for microalgae, such as open pond systems combined with advances in photobioreactor designs, to reduce operational costs and increase biomass productivity. In this context, the cultivation of microalgae on DF effluents, an approach aligned with the principles of circular economy, remains in the early stages of development (Lacroux et al., 2023). Finally, in the future, it will be important for studies to carry out thorough techno-economic analyzes and life cycle assessments to assess the environmental and economic impacts of large-scale biohydrogen production from algal biomass.

8.5 Microalgae biomass in anaerobic fermentation for carboxylates production

8.5.1 Description of the process

Beyond biogas, nowadays significant research efforts are focused on the production of intermediates metabolites of AD within the so-called carboxylates platform. The carboxylate platform is a promising technology for replacing petrochemicals to produce a wide range of chemicals and liquid fuels. In this manner, carboxylates are set to be an environmentally friendly option to replace petroderivatives. Consequently, their market demand keeps on increasing (Table 8.4).

VFAs, also known as carboxylates, are intermediate products released after the hydrolytic and the acidogenic stages of AD (Fig. 8.2). These VFAs are organic molecules that range from 2-carbon atoms (acetic acid) to 6-carbon atoms (caproic acid). Acetic, propionic (C3), iso-butyric and butyric (C4), iso-valeric and valeric (C5) and caproic acids are the VFAs normally encountered in AD. In conventional

Table 8.4 Market projection of volatile fatty acids.[a]

Compound	Market	Projected market
Acetic acid	6.9 billion USD in 2020	15 billion USD in 2027
Propionic acid	1.1 billion USD in 2020	1.4 billion USD in 2028
Butyric acid	450.9 million USD in 2020	1330.71 million USD in 2030
Caproic acid	226.5 million USD in 2023	358.8 million USD in 2030

[a] References: Market research (2023) and Research and market (2023).
No Permission Required.

AD, their concentration is normally low since methanogenic archaea used them as substrates to ultimately produce biogas. Similar to DF, the solution to produce VFAs is to shorten the traditional AD bioprocess through the inhibition of the methanogenesis stage and promote the fermentative one in the so-called anaerobic fermentation (AF). The inhibition of archaea activity or their washout should be thereby targeted when operating reactors for VFAs production. These two strategies might be conducted by setting harsh operational conditions or different reactors configuration.

With regard to operational conditions, conventional AD is conducted at neutral pH, mesophilic, and thermophilic temperatures (35°C–55°C), long HRTs and SRTs, and conservative OLRs. On the opposite, when aiming at decreasing archaea activity to avoid VFAs consumption, acidic pH, low process temperatures (25°C), and high OLRs have been shown to be beneficial (Gonçalves et al., 2024). Moreover, as explained in the previous sections, an important issue of microalgae biomass is their recalcitrant nature conferred by their resistant cell wall. Similar to what has been observed for methane production via conventional AD, the hydrolysis stage of AF is limited by the cell wall of microalgae (Magdalena, et al., 2018). Because of that, it is also necessary to pretreat microalgae biomass prior to AF to attain optimum VFAs production yields. In principle, as AD, DF, and AF shared the first steps of degradation, similar pretreatments can be employed to disrupt and hydrolyze microalgae biomass (Fig. 8.1).

8.5.2 Yields and main insights

It should be highlighted that the most appropriate way of providing results from the attained data is the comparison of the organic matter employed as feedstock (in terms of COD) and the VFAs attained in the culture broth of anaerobic fermenters (also in terms of COD). More specifically, the formula to be employed responds to equation (8.2):

$$\text{Bioconversion efficiency}(\%) = \frac{(\text{VFAs}_{\text{COD}})_{\text{effluent}}}{\text{totalCOD}_{\text{influent}}} \times 100 \tag{8.2}$$

In this case, the COD equivalents of each VFAs to be employed for the calculation of (8.2) can be seen in Table 8.5:

Table 8.5 Conversion factors for each VFA.

VFA	Carbon (%)	COD equivalent (g COD/g VFA)
Acetic acid	40.00	1.066
Propionic acid	48.64	1.512
Butyric acid	54.53	1.816
Valeric acid	58.80	2.036
Caproic acid	62.04	2.204

No Permission Required.

Another important calculation in this process is linked to methane production. In this case, COD removal can be calculated to estimate the amount of COD that has been converted to gas when the process reached the steady state, by employing the following equation (8.3):

$$\text{COD removal (\%)} = \frac{(\text{COD}_{\text{incluent}} - \text{COD}_{\text{effluent}})}{\text{COD}_{\text{influent}}} \times 100 \quad (8.3)$$

The greater the value determined for COD removal, the worse would be for VFAs production as this would mean that VFAs are mainly being consumed for gas production purposes either CH_4 or H_2.

8.5.2.1 Batch operation mode

Batch assays are an appropriate tool for decision-making when targeting at discriminating for the use of pretreatments or operational conditions (e.g., pH or temperature). Similar to BMPs, biochemical carboxylate potential (BCPs) assays have been used for such a purpose. The main difference between BMP and BCP is the feedstock-microorganisms ratio (in terms of $COD_{feedstock}/VS_{seeding\ inoculum}$) set in the assay. In the case of BCPs, aiming at the accumulation of VFAs, an organic overload is employed for concomitant inhibition of archaea.

Results conducted in batch assays have been used to test the efficiency of pretreatments to the anaerobic inoculum aiming at decreasing archaea activity, defining operational conditions such as temperature and pH. For instance, BCPs employing *Chlorella* sp. as feedstock were set at 25°C, 35°C, and 50°C and different initial pH values (5.5 and 7.5) (Magdalena, et al., 2018). Results showed conversion yields ranging 40%–48% at 25°C for both pH values (5.5 and 7.5) and at 35°C for pH 5.5. In line with these results, Cerdán et al. (2021) also stated that neutral pH were more suitable for VFAs production when using microalgae biomass as feedstock in AF at thermophilic conditions (ca. 50°C). When operating at neutral pH (6.5), Kumar et al., 2019 concluded that fermenting temperature increase resulted in a higher VFAs production when using *Chlorella vulgaris*. The same conclusion was attained by Cho et al. (2015) when using a mixture of microalgae including *Desmodesmus* sp., *Scenedesmus* sp., and *Chlamydomonas* sp. These authors also reported that not only acidogens activity was promoted at increasing temperature but the hydrolytic bacteria also showed a positive correlation with temperature.

Most of the times, microalgae biomass is rich in proteins (Greses et al., 2024), needing alternative co-substrates to balance the C/N ratio. Given the commonly low carbon-to-nitrogen ratio (C/N) of microalgae, it is also frequent to find the cofermentation of this substrate with alternative ones that enhance the carbon content. Similarly to conventional AD and DF, AF also relies on a high feedstock C/N. Cerdán et al. (2021) reported a C/N for microalgae biomass (59.5% *Mychonastes homospherea* and 39.5% *Pseudoanabaena* sp.) of around 7%. Those authors reported a VFA yield of 0.24 g VFA/g $COD_{influent}$ when microalgae biomass was subjected to AF, while the addition of a co-substrate (sewage sludge) increased the C/N, which ultimately raised the VFAs yield to 0.31 g VFA/g CODin. The beneficial effect of co-AF has been also shown when using microalgae biomass with vinasses (Pinheiro et al., 2024). Cofermentation is not only regarded as a tool to increase the C/N ratio but also relevant as the addition of another substrate might bring additional micronutrients and dilute other components that might cause a negative effect in AF.

The fact that microalgae biomass is rich in proteins does not only affect the C/N but also the ammonium released during AF from complex proteins upon degradation. In fact, it has been identified that around 3 g N/L inhibits the acidogenic step (Magdalena, et al., 2018), thus limiting the VFAs production. This is of particular importance when tuning the pH of AF to inhibit methanogens as this operation also reflects in nitrogen mineralization. Under alkaline conditions, a higher amount of ammonium is released and thereby, acidogenic bacteria metabolism is hampered. The key role of microalgae proteins in AF has been extensively reviewed by Magdalena et al. (2018). As it can be seen in Table 8.6, BCPs revealed that the carboxylates yields might be quite broad as it heavily depends on microalgae strain and composition as well as on the operational conditions imposed (Magdalena, 2021).

At this point, it should be highlighted that operation in BCPs can provide a limited amount of information, as carbon and nutrients availability decline along the assay. Because they are not replenished, microbial growth rates and community structure might evolve differently when operating reactor in semicontinuous feeding mode.

8.5.2.2 Semicontinuous operation mode

When operating reactors in semicontinuous mode, additional parameters should be considered. Most frequently, these parameters include OLR fed to the fermenter and the HRT, in addition to pH or temperature (as it is normally the case in BCPs). With regard to temperature, the general conclusion in batch mode was that high temperatures were more beneficial while in semicontinuous operation this tendency was not repeated. Most likely this can be attributed to the accumulation of not only the targeted products (VFAs) but also to other chemicals that might have a negative effect along fermentation time. In this sense, low temperature operation (25°C) gave the best results in terms of organic matter bioconversion into VFAs (35%) with respect to mesophilic (25%) when using *Chlorella* sp. in continuously stirred tank reactors (CSTRs; Magdalena, et al., 2019). It was reported by Llamas Llamas, (2021)

Table 8.6 VFAs distribution profile and conversion yields of different microalgae strain attained in BCPs.

Strain	Temperature (°C)	pH	Pretreatment	Substrate composition (% VS)			VFAs profile (%)							$VFAsCOD/COD_{influent}$ (%)
				Carbohydrates	Proteins	Lipids	Acetic acid	Propionic acid	Butyric acid	isoButyric acid	Valeric acid	isoValeric acid	Caproic acid	
Chlorella vulgaris	15	6.4	NO	26	62	12	70	30	–	–	–	–	–	17.4
	35						70	20	10	–	–	–	–	38.2
	55						70	10	20	–	–	–	–	40.5
Chlorella vulgaris	15	6.4	After lipid extraction	33	66	1	70	30	–	–	–	–	–	9.4
	35						50	40	10	–	–	–	–	33.4
	55						60	10	30	–	–	–	–	42.1
Chlorella sp.	25	5.5	Enzymatic (protease)	25	65	10	41	28	7	9	7	–	–	47.7
	35						26	35	9	12	8	–	–	39.1
	50						33	11	15	14	–	–	–	34.5
Chlorella sp.	25	7.5	Enzymatic (protease)	25	65	10	54	21	6	6	6	7	–	45.1
	35						57	21	6	8	1	7	–	48.3
	50						46	17	12	15	–	9	–	37.1
Microcystis	25	10	NO	6	63	4	38	14	–	36	–	12	–	13.1
			0.5 g/L Act. carbon				50	13	–	21.1	–	15.9	–	31.5
Chlorella Pyrenoidosa	35	6.5	NO	34	48	18	52	11	35	–	–	–	3	4.3
			140°C - 10 min - 1% H_2SO_4				41	14	37	–	–	–	8	9.1
Arthrospira Plantesis	37	6	2.5% H_2SO_4 – 135°C - 15 min	19	76	5	40	5	28	5	5	9	8	NA
Desmodesmus, Scenedesmus and Chlamydomona	35	6.9	NO	NA	NA	NA	86	2	2	–	–	10	–	20
	45						74	8	5	2	–	10	–	33
	55						66	15	5	1	2	11	–	50
Ettlia	35	7.2	1% NaOH - ultrasounds	53	41	6	64	25	11	–	–	–	–	25.3

Adapted from Magdalena, J. A. (2021). Anaerobic digestion of microalgae biomass for volatile fatty acids production.
NA, Not available.

that thermophilic conditions might mediate a higher methanogenic activity when compared to 25°C, as explained by the higher COD removals attained in the acidogenic fermentation of *Chlorella vulgaris*. Thereby, it can be concluded that the low COD removal reported at 25°C is responsible for a reduced methanogenic activity, which was beneficial for VFAs accumulation. Moreover, not only the methanogenesis stage might be affected by temperature but the hydrolysis stage is also highly dependent on this parameter. In principle, it is assumed that hydrolysis evolves more favorable at thermophilic range, as temperatures around 50°C are the most optimal temperature for enzymes to conduct hydrolysis. Gruhn et al. (2016), who used *Scenedesmus* sp. for AF at mesophilic and thermophilic conditions, found out that both reactors resulted similarly in hydrolytic activity (approx. 0.4 g SCOD/g VS) but VFAs accumulation was more favored at the mesophilic rage. Once again favouring the use of low temperatures when targeting at carboxylates production. Also related to the hydrolytic stage, Llamas (2021) did not report any effect upon the different operating temperatures as the microalgae biomass was pretreated prior to being used as feedstock. In this regard, it should be highlighted that it is highly relevant to consider whether microalgae biomass has been pretreated or not to fully comprehend the anaerobic process.

Regarding the OLR, high OLRs are targeted for VFAs production as their accumulation also helps in the inhibition of archaea. Because of that, investigations have been directed toward the study of this parameter. Magdalena et al. (2019) increased the OLR from 1.5 to 12 g COD/Ld in CSTRs employing *Chlorella* sp. as feedstock. Their results evidenced that the bioconversion efficiency remained at 37% regardless of the OLR. However, going further to 15 g COD/Ld was shown to be detrimental to the bioconversion. More specifically, as displayed in the acidogenic efficiency ratio, authors identified that the acidogenic stage was inhibited (0.6 against 0.8–0.9 $VFAs_{COD}$/sCODout).

HRT is a parameter that establishes the flow rate daily fed into the reactor and thus, it determines the available time for organic matter degradation, microbial growth, and intermediate compounds accumulation in the reactor. Considering the lower growth rate of methanogenic archaea in comparison with that of acidogenic bacteria (Henze et al., 2008), the HRT control can adjust the microbial activity of the different bacterial groups and archaea. Controlling the HRT is essential not only for adjusting microbial activity but also in a full-scale perspective. HRTs employed for VFAs production (8–15 days) in CSTRs are substantially lower than those found optimal for methane production using microalgae as feedstock (15–20 days). Magdalena et al. (2019) found that the bioconversion into VFAs was maintained at values close to 37% at HRT of 8 days when using an anaerobic microbiome adapted to work within the acidogenic conditions. However, when the same HRT (8 days) was assessed in a CSTR inoculated with non-adapted sludge (i.e., inoculum collected from an AD of sewage sludge), the bioconversion dropped to 25%. Llamas et al. (2022) evaluated even lower HRT and evidenced the benefits of using a step-wise strategy to decrease this value and thereby, bioconverting more microalgae biomass into the targeted carboxylates.

As explained at the beginning of this section, VFAs distribution profile is also of paramount importance as their price in the market differ. VFAs profile was affected by the different temperatures, OLR, and reactor configurations assessed (e.g., CSTR and UASB). For instance, butyric acid prevailed in the VFAs profile in CSTR configuration at high OLRs (3 COD/Ld) and 25°C, whereas acetic acid was the most abundant when the OLR was decreased to 1.5 g COD/Ld (Magdalena et al., 2020). Llamas, et al. (2022) also observed different VFAs profile depending on the temperature employed in the AF. In this study, when thermophilic range was employed in the fermentation, acetic acid was consumed for methane production purposes and thereby, propionic acid was the main contributor of the VFAs pool. Butyric acid was also the prevailing acid when using *Scenedesmus* sp. in fermenters operated at meso and thermophilic range (Gruhn et al., 2016). Alternatively, propionic acid outstood as the most abundant carboxylate in UASB reactor configuration at 25°C and high OLRs (>8 gCOD L^{-1} d^{-1}) employing *Chlorella* sp. as feedstock (Magdalena et al., 2020). The prevalence of propionic acid is not unexpected as odd-chain acids may be related to the high protein content of the microalgae biomass (Regueira et al., 2020). Overall, it can be stated that not only operational conditions in the fermenter but also the characteristics of the microalgae biomass exhibit an effect on VFAs production selectivity. Elucidating this interplay remains key for determining the prevalence of one or other VFA in the cultivation broth, as this would determine as well their market application.

8.5.3 Challenges in the technology

When compared to other substrates, microalgae biomass has provided an average of 0.33 g VFA_{COD}/g CODin, which is far below the values obtained for other wastes (0.5–0.6 g VFA_{COD}/g CODin) (Gonçalves et al., 2024; Lago et al., 2023). The most obvious difference between microalgae biomass and alternative residues used for anaerobic technologies lies in their macromolecular distribution. In this regard, carbohydrate-rich biomass commonly provides higher process yields than that of microalgae biomass, which is commonly rich in proteins. One option to increase their content in carbohydrates would be to grow microalgae in nitrogen poor effluents or the use of cyanobacteria. In this latter case, while being still photosynthetic microorganisms, their ability to accumulate glycogen might be important when dealing with their macromolecular distribution. The use of alternative reactor configuration has been poorly studied, leaving room to further improve yield and production kinetics by employing anaerobic systems that allow to decoupling HRT from SRT. The downstream processing should not be disregarded either, as once produced, carboxylates need to be extracted for further use and even in some cases, purified. One alternative to avoid those costs would be to directly use the VFAs-rich effluents for different purposes. An innovative example is the replacement of the sugar platform by these carboxylates. Recently, the potential of carboxylates has been proven as an innovative carbon source for microbial oils (Tomás-Pejó et al., 2023) and polyhydroxyalkanoates production (Park et al., 2023). While the VFA platform is still an emerging technology when employing microalgae biomass as a feedstock, further development should be targeted at attaining higher process yield (g VFA/g substrate), productivity (g VFA/L-day), cost-effective, energy-efficient separation methods to recover VFA from the fermentation broth and direct usage of VFAs-rich effluents for the production of alternative microbial chemical platforms. The cost-effectiveness of the anaerobic processes using microalgae biomass as feedstock also depends on the bioprocess stability since it determines the persistence and quality of the resulting product. Microalgae culture is a highly sensitive process that can be influenced by various factors, including light, temperature, variability in the influent, and the presence of predators. Thus, the development of early warning systems to mitigate microalgae culture oscillations or failure are crucial to ensure the stability of the anaerobic technology implemented and the related cost-efficiency.

8.6 Conclusions

Microalgal biomass is a sustainable feedstock for energy and green chemicals production using anaerobic technologies, presenting promising results at the laboratory scale for the production of biogas, H_2, and VFAs. Notwithstanding, the large-scale implementation of anaerobic technologies using microalgal biomass as feedstock faces several constraints and challenges. The low biodegradability of microalgae culture in waste streams, the presence of inhibitors, maintaining the process stability, and recovery of the product outcome need to find economic solutions to ensure process feasibility and profitability. The intense investigation into this topic is creating significant progress in optimizing these processes, lowering costs, and improving scalability. Therefore, the integration of microalgal biomass with anaerobic technologies is expected to be feasible in the near future, contributing to the transition to more renewable production systems.

References

Al Bazedi, G., Ismail, M. M., Mugwanya, M., & Sewilam, H. (2023). Desalination concentrate microalgae cultivation: Biomass production and applications. *Sustainable Water Resources Management, 9*(4). https://doi.org/10.1007/s40899-023-00887-2.

Aslam, A., Bahadar, A., Liaquat, R., & Muddasar, M. (2023). Recent advances in biological hydrogen production from algal biomass: A comprehensive review. *Fuel, 350*, 128816. https://doi.org/10.1016/j.fuel.2023.128816.

Barzee, T. J., Yothers, C., Edalati, A., Rude, K., Chio, A., El Mashad, H. M., Franz, A., & Zhang, R. (2022). Pilot microalgae cultivation using food waste digestate with minimal resource inputs. *Bioresource Technology Reports, 19*, 101200. https://doi.org/10.1016/j.biteb.2022.101200.

Bhatia, S. K., Rajesh Banu, J., Singh, V., Kumar, G., & Yang, Y. H. (2023). Algal biomass to biohydrogen: Pretreatment, influencing factors, and conversion strategies. *Bioresource Technology, 368*. https://doi.org/10.1016/j.biortech.2022.128332.

Bibi, F., Ishtiaq Ali, M., Ahmad, M., Bokhari, A., Shiong Khoo, K., Zafar, M., Asif, S., Mubashir, M., Han, N., & Loke Show, P. (2022). Production of lipids biosynthesis from Tetradesmus nygaardii microalgae as a feedstock for biodiesel production. *Fuel, 326*. https://doi.org/10.1016/j.fuel.2022.124985.

Bohutskyi, P., Betenbaugh, M. J., & Bouwer, E. J. (2014). The effects of alternative pretreatment strategies on anaerobic digestion and methane production from different algal strains. *Bioresource Technology, 155*, 366–372. https://doi.org/10.1016/j.biortech.2013.12.095.

Bohutskyi, P., Phan, D., Spierling, R. E., & Lundquist, T. J. (2023). Hydrothermal but not mechanical pretreatment of wastewater algae enhanced anaerobic digestion energy balance due to improved biomass disintegration and methane production kinetics. *Energies, 16*(20), 7146. https://doi.org/10.3390/en16207146.

de Carvalho, J. C., Molina-Aulestia, D. T., Martinez-Burgos, W. J., Karp, S. G., Manzoki, M. C., Medeiros, A. B. P., Rodrigues, C., Scapini, T., Vandenberghe, L. Pd. S., Vieira, S., Woiciechowski, A. L., Soccol, V. T., & Soccol, C. R. (2022). Agro-industrial wastewaters for algal biomass production, bio-based products, and biofuels in a circular bioeconomy. *Fermentation, 8*(12), 728. https://doi.org/10.3390/fermentation8120728.

Cerdán, J. M. A., Tejido-Nuñez, Y., Aymerich, E., de GoñiGoñi, J. G. M., & Garcia-Aguirre, J. (2021). A comprehensive comparison of methane and bio-based volatile fatty acids production from urban and agro-industrial sources. *Waste and Biomass Valorization, 12*(3), 1357–1369. https://doi.org/10.1007/s12649-020-01093-3.

Cho, H. U., Kim, Y. M., Choi, Y. N., Kim, H. G., & Park, J. M. (2015). Influence of temperature on volatile fatty acid production and microbial community structure during anaerobic fermentation of microalgae. *Bioresource Technology, 191*, 475–480. https://doi.org/10.1016/j.biortech.2015.03.009.

Chong, C. C., Cheng, Y. W., Ishak, S., Lam, M. K., Lim, J. W., Tan, I. S., Show, P. L., & Lee, K. T. (2022). Anaerobic digestate as a low-cost nutrient source for sustainable microalgae cultivation: A way forward through waste valorization approach. *Science of The Total Environment, 803*(1), 47–55. https://doi.org/10.1016/j.scitotenv.2021.150070.

Córdova, O., Passos, F., & Chamy, R. (2018). Physical pretreatment methods for improving microalgae anaerobic biodegradability. *Applied Biochemistry and Biotechnology, 185*(1), 114–126. https://doi.org/10.1007/s12010-017-2646-6.

Das, D., & Veziroğlu, T. N. (2001). Hydrogen production by biological processes: A survey of literature. *International Journal of Hydrogen Energy, 26*(1), 13–28. https://doi.org/10.1016/S0360-3199(00)00058-6.

D'Hondt, E., Martín-Juárez, J., Bolado, S., Kasperoviciene, J., Koreiviene, J., Sulcius, S., Elst, K., & Bastiaens, L. (2017). Cell disruption technologies. In *Microalgae-based biofuels and bioproducts: From feedstock cultivation to end-products*. Elsevier Inc., 133–154. https://doi.org/10.1016/B978-0-08-101023-5.00006-6.

Di Lena, G., Casini, I., Lucarini, M., Sanchez del Pulgar, J., Aguzzi, A., Caproni, R., Gabrielli, P., & Lombardi-Boccia, G. (2020). Chemical characterization and nutritional evaluation of microalgal biomass from large-scale production: A comparative study of five species. *European Food Research and Technology, 246*(2), 323–332. https://doi.org/10.1007/s00217-019-03346-5.

De Vrieze, J., Christiaens, M. E. R., Walraedt, D., Devooght, A., Ijaz, U. Z., & Boon, N. (2017). Microbial community redundancy in anaerobic digestion drives process recovery after salinity exposure. *Water Research, 111*, 109–117. https://doi.org/10.1016/j.watres.2016.12.042.

Estevam, R., Franci Gonçalves, R., Oss, R. N., Sampaio, I. C. F., & Cassini, S. T. (2024). Effects of different thermal and thermochemical pretreatments on the anaerobic digestion of algal biomass cultivated in urban wastewater and collected with and without chemical coagulants. *Algal Research, 78*. https://doi.org/10.1016/j.algal.2024.103417.

Fernandes, F., Silkina, A., Gayo-Peláez, J. I., Kapoore, R. V., de la Broise, D., & Llewellyn, C. A. (2022). Microalgae cultivation on nutrient rich digestate: The importance of strain and digestate tailoring under PH control. *Applied Sciences (Switzerland), 12*(11). https://doi.org/10.3390/app12115429.

Frigon, J.-C., Matteau-Lebrun, F., Hamani Abdou, R., McGinn, P. J., O'Leary, S. J. B., & Guiot, S. R. (2013). Screening microalgae strains for their productivity in methane following anaerobic digestion. *Applied Energy, 108*, 100–107. https://doi.org/10.1016/j.apenergy.2013.02.051.

García-Depraect, O., Castro-Muñoz, R., Muñoz, R., Rene, E. R., León-Becerril, E., Valdez-Vazquez, I., Kumar, G., Reyes-Alvarado, L. C., Martínez-Mendoza, L. J., Carrillo-Reyes, J., & Buitrón, G. (2021). A review on the factors influencing biohydrogen production from lactate: The key to unlocking enhanced dark fermentative processes. *Bioresource Technology, 324*, 124595. https://doi.org/10.1016/j.biortech.2020.124595.

Golueke, C. G., Oswald, W. J., & Gotaas, H. B. (1957). Anaerobic digestion of algae. *Applied Microbiology, 5*(1), 47–55. https://doi.org/10.1128/aem.5.1.47-55.1957.

Gonçalves, M. J., González-Fernández, C., & Greses, S. (2024). Long hydraulic retention time mediates stable volatile fatty acids production against slight pH oscillations. *Waste Management, 176*, 140–148. https://doi.org/10.1016/j.wasman.2024.01.012.

Gonzalez-Fernandez, C., Barreiro-Vescovo, S., de Godos, I., Fernandez, M., Zouhayr, A., & Ballesteros, M. (2018). Biochemical methane potential of microalgae biomass using different microbial inocula. *Biotechnology for Biofuels, 11*(1). https://doi.org/10.1186/s13068-018-1188-7.

González-Fernández, C., Sialve, B., Bernet, N., & Steyer, J. P. (2012). Comparison of ultrasound and thermal pretreatment of Scenedesmus biomass on methane production. *Bioresource Technology, 110*, 610–616. https://doi.org/10.1016/j.biortech.2012.01.043.

Greses, S. (2017). Anaerobic degradation of microalgae grown in the effluent from an anaerobic membrane bioreactor (AnMBR) treating urban wastewater.

Greses, S., Zamorano-López, N., Borrás, L., Ferrer, J., Seco, A., & Aguado, D. (2018). Effect of long residence time and high temperature over anaerobic biodegradation of Scenedesmus microalgae grown in wastewater. *Journal of Environmental Management, 218*, 425–434. https://doi.org/10.1016/j.jenvman.2018.04.086.

Greses, S., Gaby, J. C., Aguado, D., Ferrer, J., Seco, A., & Horn, S. J. (2017). Microbial community characterization during anaerobic digestion of Scenedesmus spp. under mesophilic and thermophilic conditions. *Algal Research, 27*, 121–130. https://doi.org/10.1016/j.algal.2017.09.002.

Greses, S., Jimenez, J., González-Fernández, C., & Steyer, J.-P. (2024). Modelling of anaerobic digestion of microalgae biomass: Effect of overloading perturbation. *Bioresource Technology, 399*, 130625. https://doi.org/10.1016/j.biortech.2024.130625.

Greses, S., Tomás-Pejó, E., Markou, G., & González-Fernández, C. (2022). Microalgae production for nitrogen recovery of high-strength dry anaerobic digestion effluent. *Waste Management, 139*, 321–329. https://doi.org/10.1016/j.wasman.2021.12.043.

Gruber-Brunhumer, M. R., Jerney, J., Zohar, E., Nussbaumer, M., Hieger, C., Bochmann, G., Schagerl, M., Obbard, J. P., Fuchs, W., & Drosg, B. (2015). Acutodesmus obliquus as a benchmark strain for evaluating methane production from microalgae: Influence of different storage and pretreatment methods on biogas yield. *Algal Research, 12*, 230–238. https://doi.org/10.1016/j.algal.2015.08.022.

Gruhn, M., Frigon, J. C., & Guiot, S. R. (2016). Acidogenic fermentation of Scenedesmus sp.-AMDD: Comparison of volatile fatty acids yields between mesophilic and thermophilic conditions. *Bioresource Technology, 200*, 624–630. https://doi.org/10.1016/j.biortech.2015.10.087.

Henze, M., van Loosdrecht, M. C. M., Ekama, G. A., & Brdjanovic, D. (2008). *Biological wastewater treatment: Principles, modelling and design*. IWA Publishing. 10.2166/9781780401867.

Inglesby, A. E., Griffiths, M. J., Harrison, S. T. L., & van Hille, R. P. (2015). Anaerobic digestion of Spirulina sp. and Scenedesmus sp.: A comparison and investigation of the impact of mechanical pre-treatment. *Journal of Applied Phycology, 27*(5), 1891–1900. https://doi.org/10.1007/s10811-015-0669-3.

Jiang, J., He, S., Kang, X., Sun, Y., Yuan, Z., Xing, T., Guo, Y., & Li, L. (2020). Effect of organic loading rate and temperature on the anaerobic digestion of municipal solid waste: Process performance and energy recovery. *Frontiers in Energy Research, 8*. https://doi.org/10.3389/fenrg.2020.00089.

Kendir Çakmak, E., & Ugurlu, A. (2020). Enhanced biogas production of red microalgae via enzymatic pretreatment and preliminary economic assessment. *Algal Research, 50*, 101979. https://doi.org/10.1016/j.algal.2020.101979.

Khalekuzzaman, M., Jahan, N., Kabir, S. B., Hasan, M., Fayshal, M. A., & Roy Chowdhury, D. (2023). Substituting microalgae with fecal sludge for biohythane production enhancement and cost saving through two-stage anaerobic digestion. *Journal of Cleaner Production, 427*. https://doi.org/10.1016/j.jclepro.2023.139352.

Khalid, A. A. H., Yaakob, Z., Abdullah, S. R. S., & Takriff, M. S. (2019). Assessing the feasibility of microalgae cultivation in agricultural wastewater: The nutrient characteristics. *Environmental Technology and Innovation, 15*. https://doi.org/10.1016/j.eti.2019.100402.

Kumar, G., Ponnusamy, V. K., Bhosale, R. R., Shobana, S., Yoon, J. J., Bhatia, S. K., Rajesh Banu, J., & Kim, S. H. (2019). A review on the conversion of volatile fatty acids to polyhydroxyalkanoates using dark fermentative effluents from hydrogen production. *Bioresource Technology, 287*. https://doi.org/10.1016/j.biortech.2019.121427.

Lacroux, J., Llamas, M., Dauptain, K., Avila, R., Steyer, J.-P., van Lis, R., & Trably, E. (2023). Dark fermentation and microalgae cultivation coupled systems: Outlook and challenges. *Science of The Total Environment, 865*, 161136. https://doi.org/10.1016/j.scitotenv.2022.161136.

Lage, S., & Gentili, F. G. (2023). Chemical composition and species identification of microalgal biomass grown at pilot-scale with municipal wastewater and CO2 from flue gases. *Chemosphere, 313*, 137344. https://doi.org/10.1016/j.chemosphere.2022.137344.

Lago, A., Greses, S., Aboudi, K., Moreno, I., & González-Fernández, C. (2023). Effect of decoupling hydraulic and solid retention times on carbohydrate-rich residue valorization into carboxylic acids. *Scientific Reports, 13*(1). https://doi.org/10.1038/s41598-023-48097-2.

Leong, Y. K., Su, R.-H., Varjani, S., Lee, D.-J., & Chang, J.-S. (2024). Integrated biohydrogen production and dairy manure wastewater treatment via a microalgae platform. *International Journal of Hydrogen Energy, 52*, 404–417. https://doi.org/10.1016/j.ijhydene.2023.06.308.

de Lima Barizão, A. C., de Oliveira, J. P., Gonçalves, R. F., & Cassini, S. T. (2021). Nanomagnetic approach applied to microalgae biomass harvesting: Advances, gaps, and perspectives. *Environmental Science and Pollution Research, 28*(33), 44795–44811. https://doi.org/10.1007/s11356-021-15260-z.

Llamas, M. (2021). Microalgae biomass valorization for volatile fatty acids, microbial oils and biogas production in a biorefinery context.

Llamas, M., Greses, S., Tomás-Pejó, E., & González-Fernández, C. (2022). Carboxylic acids production via anaerobic fermentation: Microbial communities' responses to stepwise and direct hydraulic retention time decrease. *Bioresource Technology, 344*, 126282. https://doi.org/10.1016/j.biortech.2021.126282.

Lunprom, S., Phanduang, O., Salakkam, A., Liao, Q., Imai, T., & Reungsang, A. (2019). Biohythane production from residual biomass of Chlorella sp. biomass through a two-stage anaerobic digestion. *International Journal of Hydrogen Energy, 44*(6), 3339–3346. https://doi.org/10.1016/j.ijhydene.2018.09.064.

Magdalena, J. A. (2021). Anaerobic digestion of microalgae biomass for volatile fatty acids production.

Magdalena, J. A., Greses, S., & González-Fernández, C. (2019). Impact of organic loading rate in volatile fatty acids production and population dynamics using microalgae biomass as substrate. *Scientific Reports, 9*(1). https://doi.org/10.1038/s41598-019-54914-4.

Magdalena, J. A., Llamas, M., Tomás-Pejó, E., & González-Fernández, C. (2019). Semicontinuous anaerobic digestion of protease pretreated Chlorella biomass for volatile fatty acids production. *Journal of Chemical Technology and Biotechnology, 94*(6), 1861–1869. https://doi.org/10.1002/jctb.5960.

Magdalena, J. A., Tomás-Pejó, E., Ballesteros, M., & González-Fernandez, C. (2018). Volatile fatty acids production from protease pretreated Chlorella biomass via anaerobic digestion. *Biotechnology Progress, 34*(6), 1363–1369. https://doi.org/10.1002/btpr.2696.

Magdalena, J. A., Ballesteros, M., & González-Fernandez, C. (2018). Efficient Anaerobic Digestion of Microalgae Biomass: Proteins as a Key Macromolecule. *Molecules, 23*(5), 1098. https://doi.org/10.3390/molecules23051098.

Magdalena, J. A., Greses, S., & González-Fernández, C. (2020). Anaerobic degradation of protein-rich biomass in an UASB reactor: Organic loading rate effect on product output and microbial communities dynamics. *Journal of Environmental Management, 274*, 111201. https://doi.org/10.1016/j.jenvman.2020.111201.

Mahdy, A., Mendez, L., Ballesteros, M., & González-Fernández, C. (2014). Enhanced methane production of Chlorella vulgaris and Chlamydomonas reinhardtii by hydrolytic enzymes addition. *Energy Conversion and Management, 85*, 551–557. https://doi.org/10.1016/j.enconman.2014.04.097.

Mahdy, A., Mendez, L., Ballesteros, M., & González-Fernández, C. (2014). Autohydrolysis and alkaline pretreatment effect on Chlorella vulgaris and Scenedesmus sp. methane production. *Energy, 78*, 48–52. https://doi.org/10.1016/j.energy.2014.05.052.

Mahdy, A., Mendez, L., Ballesteros, M., & González-Fernández, C. (2015). Protease pretreated Chlorella vulgaris biomass bioconversion to methane via semi-continuous anaerobic digestion. *Fuel, 158*, 35–41. https://doi.org/10.1016/j.fuel.2015.04.052.

Marazzi, F., Bellucci, M., Rossi, S., Fornaroli, R., Ficara, E., & Mezzanotte, V. (2019). Outdoor pilot trial integrating a sidestream microalgae process for the treatment of centrate under non optimal climate conditions. *Algal Research, 39*, 101430. https://doi.org/10.1016/j.algal.2019.101430.

Market research. (2023). https://www.verifiedmarketresearch.com/product.

Martín, J., Riol, E., Fernández, J. M., Muñoz, R., García-Encina, P. A., & Bolado, S. (2018). Effect of pretreatments on biogas production from microalgae biomass grown in pig manure treatment plants. *Bioresource Technology, 257*, 30–38. https://doi.org/10.1016/j.biortech.2018.02.063.

Martínez-Fraile, C., Muñoz, R., Teresa Simorte, M., Sanz, I., & García-Depraect, O. (2024). Biohydrogen production by lactate-driven dark fermentation of real organic wastes derived from solid waste treatment plants. *Bioresource Technology, 403*, 130846. https://doi.org/10.1016/j.biortech.2024.130846.

Massey, I. Y., Peng, T., Danping, C., & Yang, F. (2022). Optimization of Biodegradation Characteristics of Sphingopyxis sp. YF1 against Crude Microcystin-LR Using Response Surface Methodology. *Toxins, 14*(4), 240. https://doi.org/10.3390/toxins14040240.

Meier, R. L. (1955). Biological cycles in the transformation of solar energy into useful fuels. In F. Daniels, & J. A. Duffie (Eds.). *Solar energy research*. Madison, WI, USA: University of Wisconsin Press.

de Morais, E. G., Sampaio, I. C. F., Gonzalez-Flo, E., Ferrer, I., Uggetti, E., & García, J. (2023). Microalgae harvesting for wastewater treatment and resources recovery: A review. *New Biotechnology, 78*, 84–94. https://doi.org/10.1016/j.nbt.2023.10.002.

Nordio, R., Belachqer-El Attar, S., Clagnan, E., Sánchez-Zurano, A., Pichel, N., Viviano, E., Adani, F., Luis Guzmán, J., & Acién, G. (2024). Exploring microbial growth dynamics in a pilot-scale microalgae raceway fed with urban wastewater: Insights into the effect of operational variables. *Journal of Environmental Management, 369*, 122385. https://doi.org/10.1016/j.jenvman.2024.122385.

Ortigueira, J., Alves, L., Gouveia, L., & Moura, P. (2015). Third generation biohydrogen production by Clostridium butyricum and adapted mixed cultures from Scenedesmus obliquus microalga biomass. *Fuel, 153*, 128–134. https://doi.org/10.1016/j.fuel.2015.02.093.

Pal, D., Banerjee, S., Chandra, S., Das, D., Pandit, S., Roy, A., Hasan, M., Khandaker, M. U., & Alreshidi, M. A. (2024). The generation of biohydrogen from pretreated algal biomass in batch fermentation mode. *International Journal of Hydrogen Energy*. https://doi.org/10.1016/j.ijhydene.2024.01.078.

Parakh, S. K., Tian, Z., Wong, J. Z. E., & Tong, Y. W. (2023). From microalgae to bioenergy: Recent advances in biochemical conversion processes. *Fermentation, 9*(6). https://doi.org/10.3390/fermentation9060529.

Park, Y., Jeon, J. M., Park, J. K., Yang, Y. H., Choi, S. S., & Yoon, J. J. (2023). Optimization of polyhydroxyalkanoate production in Halomonas sp. YLGW01 using mixed volatile fatty acids: A study on mixture analysis and fed-batch strategy. *Microbial Cell Factories, 22*(1). https://doi.org/10.1186/s12934-023-02188-x.

Parsy, A., Ficara, E., Mezzanotte, V., Guerreschi, A., Guyoneaud, R., Monlau, F., & Sambusiti, C. (2024). Incorporating saline microalgae biomass in anaerobic digester treating sewage sludge: Impact on performance and microbial populations. *Bioresource Technology, 397*, 130444. https://doi.org/10.1016/j.biortech.2024.130444.

Passos, F., Hom-Diaz, A., Blanquez, P., Vicent, T., & Ferrer, I. (2016). Improving biogas production from microalgae by enzymatic pretreatment. *Bioresource Technology, 199*, 347–351. https://doi.org/10.1016/j.biortech.2015.08.084.

Passos, F., Carretero, J., & Ferrer, I. (2015). Comparing pretreatment methods for improving microalgae anaerobic digestion: Thermal, hydrothermal, microwave and ultrasound. *Chemical Engineering Journal, 279*, 667–672. https://doi.org/10.1016/j.cej.2015.05.065.

Passos, F., Hernández-Mariné, M., García, J., & Ferrer, I. (2014). Long-term anaerobic digestion of microalgae grown in HRAP for wastewater treatment. Effect of microwave pretreatment. *Water Research, 49*, 351–359. https://doi.org/10.1016/j.watres.2013.10.013.

Paul, R., Silkina, A., Melville, L., Suhartini, S., & Sulu, M. (2023). Optimisation of ultrasound pretreatment of microalgal biomass for effective biogas production through anaerobic digestion process. *Energies, 16*(1), 553. https://doi.org/10.3390/en16010553.

Pereira, A. S., Lopes, M., Miranda, S. M., & Belo, I. (2022). Bio-oil production for biodiesel industry by Yarrowia lipolytica from volatile fatty acids in two-stage batch culture.

Applied Microbiology and Biotechnology, 106(8), 2869–2881. https://doi.org/10.1007/s00253-022-11900-7.

Pinheiro, A. A. D., da Silva, E. M., de Oliveira, D. C. P., Magnus, B. S., Motteran, F., Florencio, L., & Leite, W. R. M. (2024). Volatile fatty acid and methane production from vinasse and microalgae using two-stage anaerobic co-digestion. *Environmental Science and Pollution Research*. https://doi.org/10.1007/s11356-024-34089-w.

Poddar, B. J., & Khardenavis, A. A. (2024). Genomic characterization of bacteria reveals their bioaugmentation and pre-treatment potential for improved hydrolysis and biomethanation of protein-rich substrates. *Biomass Conversion and Biorefinery*. https://doi.org/10.1007/s13399-024-06113-z.

Rady, H. A., Ali, S. S., & El-Sheekh, M. M. (2024). Strategies to enhance biohydrogen production from microalgae: A comprehensive review. *Journal of Environmental Management, 356*. https://doi.org/10.1016/j.jenvman.2024.120611.

Rai, P. K., & Singh, S. P. (2016). Integrated dark- and photo-fermentation: Recent advances and provisions for improvement. *International Journal of Hydrogen Energy, 41*(44), 19957–19971. https://doi.org/10.1016/j.ijhydene.2016.08.084.

Ramprakash, B., Sivaramakrishnan, R., Subramani, K., & Incharoensakdi, A. (2024). Iron oxide treated Chlorella sp. for enhanced biomass and lipid content coupled to fermentative hydrogen production by Enterobacter aerogenes using hydrolyzate from pretreated biomass. *International Journal of Hydrogen Energy, 73*, 43–53. https://doi.org/10.1016/j.ijhydene.2024.06.003.

Regueira, A., Bevilacqua, R., Lema, J. M., Carballa, M., & Mauricio-Iglesias, M. (2020). A metabolic model for targeted volatile fatty acids production by cofermentation of carbohydrates and proteins. *Bioresource Technology, 298*, 122535. https://doi.org/10.1016/j.biortech.2019.122535.

Research and market. (2023). https://www.researchandmarkets.com/reports/5302114/caproic-acid-global-market-trajectory.

Roberts, K. P., Heaven, S., & Banks, C. J. (2019). Semi-continuous anaerobic digestion of the marine micro-algal species *I. galbana* and *D. salina* grown under low and high sulphate conditions. *Algal Research, 41*, 101564. https://doi.org/10.1016/j.algal.2019.101564.

Roy, S. S., & Pal, R. (2015). Microalgae in aquaculture: A review with special references to nutritional value and fish dietetics. *Proceedings of the Zoological Society, 68*(1), 1–8. https://doi.org/10.1007/s12595-013-0089-9.

Saravanakumar, A., Sudha, M. R., Chen, W.-H., Pradeshwaran, V., Ashokkumar, V., & Selvarajoo, A. (2023). Biomethane production as a green energy source from anaerobic digestion of municipal solid waste: A state-of-the-art review. *Biocatalysis and Agricultural Biotechnology, 53*, 102866. https://doi.org/10.1016/j.bcab.2023.102866.

Sarkar, O., Santhosh, J., Dhar, A., & Venkata Mohan, S. (2021). Green hythane production from food waste: Integration of dark-fermentation and methanogenic process towards biogas up-gradation. *International Journal of Hydrogen Energy, 46*(36), 18832–18843. https://doi.org/10.1016/j.ijhydene.2021.03.053.

Satheesh, S., Pugazhendi, A., Al-Mur, B. A., & Balasubramani, R. (2023). Biohydrogen production coupled with wastewater treatment using selected microalgae. *Chemosphere, 334*. https://doi.org/10.1016/j.chemosphere.2023.138932.

Sendra, M., Blasco, J., & Araújo, C. V. M. (2018). Is the cell wall of marine phytoplankton a protective barrier or a nanoparticle interaction site? Toxicological responses of Chlorella autotrophica and Dunaliella salina to Ag and CeO_2 nanoparticles. *Ecological Indicators, 95*, 1053–1067. https://doi.org/10.1016/j.ecolind.2017.08.050.

Sim, Y. B., Jung, J. H., Baik, J. H., Park, J. H., Kumar, G., Rajesh Banu, J., & Kim, S. H. (2021). Dynamic membrane bioreactor for high rate continuous biohydrogen production from algal biomass. *Bioresource Technology, 340*. https://doi.org/10.1016/j.biortech.2021.125562.

Singh, H., Rout, S., & Das, D. (2022). Dark fermentative biohydrogen production using pretreated Scenedesmus obliquus biomass under an integrated paradigm of biorefinery. *International Journal of Hydrogen Energy, 47*(1), 102–116. https://doi.org/10.1016/j.ijhydene.2021.10.018.

Solé-Bundó, M., Garfí, M., & Ferrer, I. (2020). Pretreatment and co-digestion of microalgae, sludge and fat oil and grease (FOG) from microalgae-based wastewater treatment plants. *Bioresource Technology, 298*, 122563. https://doi.org/10.1016/j.biortech.2019.122563.

Solé-Bundó, M., Salvadó, H., Passos, F., Garfí, M., & Ferrer, I. (2018). Strategies to optimize microalgae conversion to biogas: Co-digestion, pretreatment and hydraulic retention time. *Molecules, 23*(9), 2096. https://doi.org/10.3390/molecules23092096.

Sriyod, K., Reungsang, A., & Plangklang, P. (2021). Single and combined enzymatic saccharification and biohydrogen production from Chlorella sp. biomass. *BioEnergy Research, 14*(3), 940–953. https://doi.org/10.1007/s12155-020-10191-y.

Stanislaus, M. S., Zhang, N., Yuan, Y., Zheng, H., Zhao, C., Hu, X., Zhu, Q., & Yang, Y. (2018). Improvement of biohydrogen production by optimization of pretreatment method and substrate to inoculum ratio from microalgal biomass and digested sludge. *Renewable Energy, 127*, 670–677. https://doi.org/10.1016/j.renene.2018.05.022.

Sun, C., Xia, A., Liao, Q., Fu, Q., Huang, Y., Zhu, X., Wei, P., Lin, R., & Murphy, J. D. (2018). Improving production of volatile fatty acids and hydrogen from microalgae and rice residue: Effects of physicochemical characteristics and mix ratios. *Applied Energy, 230*, 1082–1092. https://doi.org/10.1016/j.apenergy.2018.09.066.

Tian, H., Fotidis, I. A., Mancini, E., Treu, L., Mahdy, A., Ballesteros, M., González-Fernández, C., & Angelidaki, I. (2018). Acclimation to extremely high ammonia levels in continuous biomethanation process and the associated microbial community dynamics. *Bioresource Technology, 247*, 616–623. https://doi.org/10.1016/j.biortech.2017.09.148.

Tibbetts, S. M., Milley, J. E., & Lall, S. P. (2015). Chemical composition and nutritional properties of freshwater and marine microalgal biomass cultured in photobioreactors. *Journal of Applied Phycology, 27*(3), 1109–1119. https://doi.org/10.1007/s10811-014-0428-x.

Tomás-Pejó, E., González-Fernández, C., Greses, S., Kennes, C., Otero-Logilde, N., Veiga, M. C., Bolzonella, D., Müller, B., & Passoth, V. (2023). Production of short-chain fatty acids (SCFAs) as chemicals or substrates for microbes to obtain biochemicals. *Biotechnology for Biofuels and Bioproducts, 16*(1). https://doi.org/10.1186/s13068-023-02349-5.

Vo, H. N. P., Ngo, H. H., Guo, W., Nguyen, T. M. H., Liu, Y., Liu, Y., Nguyen, D. D., & Chang, S. W. (2019). A critical review on designs and applications of microalgae-based photobioreactors for pollutants treatment. *Science of the Total Environment, 651*, 1549–1568. https://doi.org/10.1016/j.scitotenv.2018.09.282.

Wang, M., Lee, E., Dilbeck, M. P., Liebelt, M., Zhang, Q., & Ergas, S. J. (2017). Thermal pretreatment of microalgae for biomethane production: Experimental studies, kinetics and energy analysis. *Journal of Chemical Technology & Biotechnology, 92*(2), 399–407. https://doi.org/10.1002/jctb.5018.

Zabed, H. M., Akter, S., Yun, J., Zhang, G., Zhang, Y., & Qi, X. (2020). Biogas from microalgae: Technologies, challenges and opportunities. *Renewable and Sustainable Energy Reviews, 117*, 109503. https://doi.org/10.1016/j.rser.2019.109503.

Zabochnicka, M. (2022). Industrial wastewater as a growth medium for microalgal biomass for a sustainable circular bioeconomy. *Applied Sciences, 12*(20), 10299. https://doi.org/10.3390/app122010299.

Zhang, J., Xue, D., Wang, C., Fang, D., Cao, L., & Gong, C. (2023). Genetic engineering for biohydrogen production from microalgae. *iScience, 26*(8), 107255. https://doi.org/10.1016/j.isci.2023.107255.

9 From volatile fatty acids to carotenoids: bioconversion of alternative carbon sources for pigment production in yeasts

Cecilia Naveira-Pazos[1], Volkmar Passoth[2], Bettina Müller[2], María C. Veiga[1], and Christian Kennes[1]
[1]Chemical Engineering Laboratory, Faculty of Sciences and Interdisciplinary Centre of Chemistry and Biology, Centro Interdisciplinar de Química y Biología (CICA), BIOENGIN Group, University of La Coruña, A Coruña, Spain, [2]Department of Molecular Sciences, Swedish University of Agricultural Sciences, Uppsala, Sweden

9.1 Introduction

In the early 19th century, the expression "We are what we eat" emerged to reflect the fact that human health is intimately linked to diet. Indeed, today, concerns about environmental pollution go hand in hand with increased demand for safer, healthier, and more functional foods. In addition, recent studies on the negative effects of synthetic food additives have prompted the limitation and substitution of these compounds by natural products, which continues to expand (Papapostolou et al., 2023).

Carotenoids are no longer considered essential only for their function as natural coloring agents, but these food additives also stand out for their antioxidant capacity and their role as vitamin A precursors. Vitamin A is crucial for the proper functioning of the eyes, brain, and for reproduction, among others; hence the importance of carotenoids in human nutrition is an indisputable fact (Papapostolou et al., 2023). The global carotenoids market was estimated to be USD 2.5 billion in 2024 while it is expected to reach USD 3.4 billion by 2029, at a compound annual growth rate (CAGR) of 6.3% during this 5-year period (Research and Markets, 2024).

Although synthetic pigments currently dominate the market, there is growing interest in natural pigments due to safety and environmental concerns. Although plant extraction remains a source for some natural pigments, it is limited by regional and seasonal factors. However, the production of natural pigments by microorganisms is a promising alternative that is not constrained by these limitations and can be more easily scaled. In addition to being able to synthesize a wide variety of microbial pigments, microorganisms can also be genetically modified to produce plant and animal pigments (Qin et al., 2023). Among the carotenoid-producing microorganisms, yeasts have proven to be efficient, with *Rhodotorula* spp. and *Phaffia rhodozyma* strains being the focus of the recently published research (Papapostolou et al., 2023). Generally, red yeasts allow the production of mixtures of carotenoids,

Eukaryotic Microorganisms as Sources of Bioproducts. DOI: https://doi.org/10.1016/B978-0-443-30188-9.00019-6

mainly β-carotene, γ-carotene, lycopene, torulene, and torularhodin (Mannazzu et al., 2015; Sereti et al., 2023). Furthermore, another advantage of yeasts is their ability to assimilate low-cost substrates to produce high value-added metabolites, which not only significantly reduces the cost of the process but also contributes to environmental sustainability. Among low-cost substrates, VFAs stand out because they can be produced from various wastes, effluents, and even polluting gases, and they have the potential to be used to produce carotenoids using yeasts.

With this in mind, this chapter will focus on the microbial production of carotenoids, specifically using yeasts. It will first explore the microbial production of natural pigments, which are important not only for food coloring but also for human health. This chapter will then provide detailed information on the chemistry of carotenoids produced by yeasts, including their classification and nomenclature. Influential parameters in the synthesis of these compounds will be discussed, as well as the potential of VFAs as alternative substrates for their biosynthesis, moving away from less economically viable substrates.

9.2 Microbial production of natural pigments (carotenoids)

Pigments, used as dyes since prehistoric times (Sharma et al., 2024), provide color diversity to humans and are utilized in a wide range of manufactured products. Their application spans various sectors such as medicine, cosmetics, food industry, and textile production. Although synthetic pigments dominate the industry today due to their low prices and stability, their use has been considerably reduced due to hyperallergenicity, carcinogenicity, and other toxicological problems associated with some synthetic dyes. These adverse effects have motivated the scientific community to favor the use of natural pigments derived from plants, animals, and microorganisms. This shift aims to promote a healthier and more environmentally-friendly lifestyle, considering also the biodegradability of these compounds (Qin et al., 2023; Sharma et al., 2024).

According to their structure, natural pigments may be classified into flavonoids (i.e., anthocyanins), isoprenoids (i.e., carotenoids), pyrroles (chlorophylls), and nitrogen-heterocyclic derivatives (i.e., betalains). Another classification is based on their hue, which is particularly relevant in the food industry. For example, warm colors including yellow and red hues attract consumers and can stimulate the appetite (Sigurdson et al., 2017).

Natural pigments produced by microorganisms (bacteria, fungi, or microalgae) (Barreto et al., 2023) have significant advantages over those obtained from plants and animals. Microbial production is not limited by regional and seasonal restrictions and is easily scalable. Furthermore, microbial fermentation offers higher yields, more cost-effective production, and ease in improvement of microbial strains (Aman Mohammadi et al., 2022). Among natural pigments, carotenoids stand out as fundamental components of the human diet. In addition, some of them are precursors of vitamin A and are recognized as bioactive phytochemicals that help to reduce the risk of various degenerative and cardiovascular diseases, among others. In addition to their nutritional importance, carotenoids are valued in various industries, including chemical, pharmaceutical, poultry, and cosmetics. However, their production by chemical methods or through extraction from plants involves significant costs considering their low yields compared to microbial production.

Microbes exhibit greater efficiency in this process, offering the possibility of further reducing costs through the use of low-cost substrates, such as agroindustrial wastes. Microbial metabolism efficiently utilizes the sources of carbon, nitrogen, and other nutrients present in the waste, resulting in a considerable decrease in the price of the process while favouring environmental friendliness (Mata-Gómez et al., 2014).

9.3 The potential of oleaginous yeasts for carotenoid biosynthesis

9.3.1 Classification and nomenclature of yeast-produced carotenoids

Carotenoids represent a class of terpenoid pigments with 40 carbon atoms, which are synthesized from two geranylgeranyl pyrophosphate units. Mainly soluble in nonpolar solvents, these types of pigments are divided into two groups: carotenes and xanthophylls. One difference between the two groups is that some carotenes, such as β-carotene and torulene, are composed only of carbon and hydrogen, whereas xanthophylls, e.g., astaxanthin and canthaxanthin, have a chemical structure containing also oxygen (Mata-Gómez et al., 2014). It has been demonstrated that these four carotenoids (Fig. 9.1) can be produced by yeasts (Mata-Gómez et al., 2014). Oleaginous yeasts of the genus *Rhodotorula* can synthesize β-carotene, torulene, and torularhodin in varying amounts. The genus exhibits variable carotenoid yields influenced by growth conditions, ranging from low (less than 100 μg/g) to very high (more than 500 μg/g) levels. Astaxanthin is produced mainly by the yeast *Xanthophyllomyces dendrorhous* (*Phaffia rhodozyma*) (Mata-Gómez et al., 2014). β-carotene, $C_{40}H_{56}$, is the most common and stable pigment found in nature. In addition to being used as a food colorant, it can be converted into vitamin A in the human body, exerting oxidation resistance, anticancer activity, and cardiovascular properties (Wang et al., 2021), making it suitable for use as a dietary supplement as well. Torulene, $C_{40}H_{54}$ (3′,4′-didehydro-β,ψ-carotene), exhibits an exceptional color due to its molecular structure containing 13 double conjugated bonds. Related to this, β-carotene presents less antioxidant efficacy than torulene due to the presence of two less conjugated bonds in its chemical structure (Mata-Gómez et al., 2014). Astaxanthin, $C_{40}H_{52}O_4$, whose biosynthesis is limited exclusively to microorganisms, is an important carotenoid used mainly as an additive in animal feed due to its organoleptic properties, being especially relevant in aquaculture. Moreover, this carotenoid is noted for its potent antioxidant capabilities, having an oxygen radical absorbance capacity (ORAC) 100–500 times higher than other antioxidants and a free radical inhibitory activity 10 times higher than related carotenoids such as α-carotene, β-carotene, lutein, and lycopene. It also serves

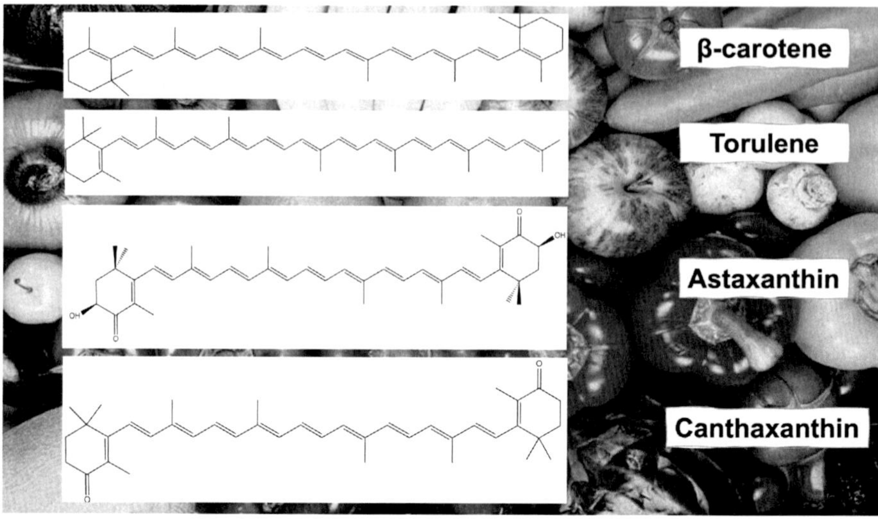

Figure 9.1 Molecular structures of carotenoids: β-carotene, torulene, astaxanthin, and canthaxanthin.

as a nutraceutical and medicinal ingredient in the treatment of degenerative diseases. Finally, canthaxanthin ($C_{40}H_{52}O_2$), lacks provitamin A activity but presents strong antioxidant activity, having therefore also multiple medicinal applications (e.g., protection against skin cancers). In addition, it is also used as a coloring additive in fish feed to enhance the color of fish flesh thereby attracting consumers (Bhosale & Bernstein, 2005).

9.3.2 Key factors influencing the production of *β*-carotene and other carotenoids in yeasts

Recent interest in biotechnological methods to generate valuable metabolic compounds, such as carotenoids, is growing due to their advantages across various industries.

While carotenoids can also be obtained from bacteria, filamentous fungi, and algae, their production using yeasts becomes particularly interesting due to certain advantages of these microorganisms over others. For instance, yeasts are not limited by physicochemical factors such as light or temperature, unlike microalgae (Naveira-Pazos et al., 2023). However, it is important to note that while most microalgae species are grown phototrophically, some have the ability to grow heterotrophically, using only organic substances as carbon and energy sources. With heterotrophic cultivation of microalgae, the need for light would be eliminated and, in addition, the volumetric productivities of biomass and the productivities of cell mass, proteins, and lipids are considerably increased compared to photoautotrophic growth (Morales-Sánchez & Martinez-Rodriguez, 2017). In addition, while light is not essential for yeast growth, it can still influence certain metabolic processes, particularly in carotenoid biosynthesis, as will be discussed in the following sections.

Microbial carotenoid production must prioritize cost-efficiency, productivity, and environmental sustainability to achieve efficient large-scale mass production in the near future. Cell growth and carotenoid synthesis is influenced by nutritional factors and environmental conditions and, in fact, the optimal conditions for production may not be the same as those required for cell growth (Wang et al., 2021). Some of the most important factors to consider in the synthesis of carotenoids by yeast are discussed in the following sections (Table 9.1).

9.3.2.1 Light

Carotenogenesis in various organisms is regulated by light, although the intensity and protocol of illumination varies depending on the microorganism. To combat potential cell damage from light exposure, microorganisms possess protective mechanisms. Carotenoids act as photoprotective compounds against the harmful effects of light and are produced during carotenogenesis (Mata-Gómez et al., 2014), a process of biosynthesis that can occur under various conditions, including light. To understand the role of white light illumination as a stimulant, it is essential to determine the levels of biosynthetic enzymes, which link illumination to carotenoid production (Frengova & Beshkova 2009). Pham et al. (2020) studied the effect of light on carotenoid (and lipid) production in the oleaginous yeast *Rhodotorula toruloides* (formerly *Rhodosporidium* sp.), a red yeast known for its ability to produce carotenoids, mainly β-carotene, γ-carotene, torulene, and torularhodin (Buzzini et al., 2007; J.J.L. Lee et al., 2014; Nagaraj et al., 2022). In this study, of Pham et al, *R. toruloides* was cultured under light and dark conditions in 15% glucose medium. Under dark conditions, the amounts of β-carotene and torulene were similar. However, under light conditions, a significant increase in β-carotene production was observed. Analysis of carotenoid composition revealed a higher proportion of β-carotene under light, whereas the proportion of torulene was higher in darkness. These findings suggest that light irradiation specifically affects the metabolic steps involved in β-carotene biosynthesis, resulting in more intense pigmentation and higher total carotenoid production in the presence of light. Furthermore, they confirmed that light affects not only the pigment concentration obtained but also the carotenoid profile. Stachowiak & Czarnecki (2007) also aimed to study the influence of light on carotenogenesis. In this case, they worked with the yeast *P. rhodozyma* (CBS 5626), which is characterized by a carotenogenic composition in which astaxanthin can represent between 83% and 87% of the total pigments. In their research, the highest carotenoid concentrations were obtained using an illuminance of 400 (8.83 mg/L) and 600 lux (8.49 mg/L), while the lowest were obtained in cultures carried out under dark conditions (1.65 mg/L). Furthermore, the constant light stress which the yeast cells are exposed to proved to be key in enhancing carotenogenesis to a greater extent than a single light treatment. These results are also in agreement with those of Yen & Zhang (2011), who found 67% increase in β-carotene content in batch cultures with *Rhodotorula glutinis*, using two LED lamps, with respect to the

Table 9.1 Parameters that affect yeast growth and carotenoid production, according to different studies.

Yeast	Carbon source (g/L)	Carbon source origin	T (°C)	pH	Nitrogen source	C/N ratio	Aeration rate (vvm)	Carotenoid content (mg/g)	Carotenoid concentration (mg/L)	References
Yarrowia lipolytica JMY6862	Glucose	Synthetic[a]	28	5.5	Yeast extract Peptone	-	2	90 (β-carotene)	6500	Larroude et al. (2018)
Rhodotorula glutinis TISTR 5159	Crude glycerol	Biodiesel plant	30	6	$(NH_4)_2SO_4$	85	2	-	135.25	Saenge et al. (2011)
R. glutinis	Molasses sucrose	Industry	30	6	$(NH_4)_2SO_4$	-	2.4	28.4	125	Aksu and Eren (2007)
R. glutinis	Whey lactose	Industry	30	6	$(NH_4)_2SO_4$	-	2.4	35.5	56.8	Aksu and Eren (2007)
Y. lipolytica JMY6862	Acetic and butyric acid	Anaerobic fermentation with syngas + synthetic[a]	30	6	Yeast extract NH_4Cl	25	2	88.4	759	Robles-Iglesias et al. (2023a)
Y. lipolytica JMY6862	Acetic acid	Anaerobic fermentation with syngas + synthetic[a]	32	6	Yeast extract NH_4Cl	∞	0.1	44	164	Robles-Iglesias et al. (2023b)
Y. lipolytica JMY6862	Acetic acid	Anaerobic fermentation with syngas + synthetic	30	6	Yeast extract NH_4Cl	10	0.5-2	62.3	382.7	Robles-Iglesias et al. (2024)

[a] Synthetic: refers to VFAs that have been industrially produced through controlled and purified chemical processes and are commercially purchased, as opposed to acids extracted from more complex natural sources, which may contain impurities or additional compounds.

nonirradiated batch, reaching levels of 24.60 and 14.69 mg/g, respectively. However, it was also observed that a higher light intensity, provided with three LED lamps, caused damage to the yeast cells, resulting in a lower β-carotene content of 17.38 mg/g.

9.3.2.2 Nitrogen source and C/N ratio

The nitrogen source is essential for proper growth of microorganisms. Some of the most common sources include yeast extract, ammonium sulfate, and peptone, among others (Naveira-Pazos et al., 2023). Nitrogen concentration also affects carotenoid production. On the one hand, nitrogen concentrations that exceed the optimal requirements for cell growth may disfavor the production of this type of metabolites. This phenomenon was observed in the study of Mata-Gómez et al. (2023), in which β-carotene production and cell growth of *R. glutinis* were indirectly proportional to the urea concentration used. Specifically, urea concentrations higher than 3 g/L (equivalent to 0.3% w/v) resulted in a decrease in carotenoid production. On the other hand, variation of nitrogen sources in culture media affects carotenoid production and carotenoid ratio. This was corroborated by the studies of El-Banna et al. (2012) in which the medium containing ammonium sulfate resulted in the production of higher cellular and volumetric carotenoids using *R. glutinis var. glutinis*. Furthermore, while ammonium nitrate yielded a higher percentage of β-carotene, peptone gave the highest percentage of torulene. In line with these findings are the results obtained by Bhosale & Gadre (2002), who concluded in their study with *R. glutinis* mutant 32 that complex organic nitrogen sources, e.g., peptone, yeast extract, urea, resulted in a higher accumulation of carotenoids compared to inorganic sources such as ammonium sulfate, ammonium chloride, and ammonium nitrate. These results are partially consistent with those obtained by El-Salam et al. (2014), which also support that although ammonium sulfate favors the growth of oleaginous yeasts, it tends to decrease β-carotene production compared to complex nitrogen sources such as yeast extract.

Information on the influence of C/N ratios used in the culture media on carotenoid production by yeasts is ambiguous, with results showing either a positive or an inverse correlation. Braunwald et al. (2013) reported that lipid and carotenoid production by *R. glutinis* increased at high C/N ratios. However, they observed that both low and too high ammonium concentrations had a detrimental effect on total carotenoid production, as at C/N=70 more carotenoids were produced than at low and high ammonium concentrations. Furthermore, the authors also reported that variation in C/N ratio affects the carotenoid profile; although low yields were obtained working at a low C/N ratio of 20, it presented the highest proportion of β-carotene. Conversely, other authors claim that C/N ratios higher than 50 trigger the diversion of acetyl-CoA towards fatty acid synthesis and not towards carotenoid production through mevalonate synthesis (Paul et al., 2023). Consistent with this, another study with *R. glutinis* showed that its growth in media with C/N ratios of 70:1 and 100:1 disfavored carotenoid biosynthesis compared to lower ratios of 20:1 and 50:1 (Tkáčová et al., 2017).

9.3.2.3 Temperature

Another critical factor in carotenoid production by yeasts is temperature, as it impacts on cell growth and metabolite synthesis. Temperature regulates biosynthetic pathways, including carotenogenesis. The effect of temperature on microorganisms depends on the species and results in changes in the amount of carotenoids synthesized (Frengova & Beshkova, 2009; Mata-Gómez et al., 2014). Temperatures between 20 and 30°C were reported to be adequate for the growth of *R. glutinis* and growth was considerably slower at 20°C. Carotenoid synthesis had its maximum at 20°C and decreased above 30°C (Bhosale & Gadre, 2002). The reason for the higher accumulation of β-carotene at 20°C may be explained by the results of Simpson et al. (1964), who showed that γ-carotene acts as a branch point for pigment synthesis. When using lower temperatures, the dehydrogenation and decarboxylation activities leading to torulene synthesis are less pronounced compared to the activity of β-carotene synthase (Bhosale & Gadre, 2002). For *R. mucilaginosa*, Cheng & Yang (2016) reported 25°C as the optimum temperature for carotenoid production. A temperature of 25°C allowed reaching 2506 μg/L carotenoids, while the lowest carotenoid concentration recorded, 1630 μg/L, was obtained at 31°C. In terms of carotenoid content, the maximum value of 322.0 μg/g corresponded to growth at 25°C, and the lowest value of 257.3 μg/g, was obtained at a temperature of 22°C. These and other studies suggest that an optimum temperature for carotenoid production cannot be universally established for all yeasts, as it depends on the specific microorganism and other fermentation conditions.

9.3.2.4 Aeration

Carotenogenesis is a predominantly aerobic process, although it has also been observed in some anaerobic organisms. In the case of carotenoid-producing yeasts, which are mainly aerobic, aeration rate is another important parameter considering that it strongly influences substrate assimilation, growth, cell mass, and carotenoid synthesis. As with the other parameters, the effect of aeration also depends on the microorganism (Mata-Gómez et al., 2014). In the literature review of Frengova & Beshkova (2009), the authors concluded that the optimal air flow rate and agitation for carotenogenesis with *Rhodotorula* and *Phaffia* are in the range of 0.5–1.9 vvm and 180–900 rpm, respectively. In addition, the effect of aeration usually results in the variation of the amount of carotenoids synthesized and even in changes in their composition. Ribeiro et al. (2023) examined the impact of aeration and agitation on carotenoid production by *R. toruloides*. Their results showed that the highest carotenoid yields were obtained by increasing the agitation speed from 300 to 700 rpm and the aeration rate from 0.5 to 1.5 vvm, reaching carotenoid levels of almost 120 μg/mL. Besides, variation in the carotenoid profile as a function of these two parameters was also confirmed. The main carotenoid produced at 0.5 vvm and 300 rpm was β-carotene, representing approximately 70% of the total carotenoid content. Torularhodin and torulene followed in smaller proportions. It was demonstrated that, under low aeration conditions, the conversion of β-carotene to other carotenoids, such as torularhodin or torulene, does not occur efficiently. For example, the highest torularhodin content, approximately 35%, was obtained at a high aeration rate (1.5 vvm) and at 300 rpm. In addition, it is important to clarify that aeration does not only provide oxygen for growth and production of carotenoids. For example, it has been reported that decreased levels of oxygen available to yeast influences the synthesis of carotenes and xanthophylls, due to the oxidation of carotenes into astaxanthin, canthaxanthin, and others (Mata-Gómez et al., 2014).

9.3.2.5 Metal ions and trace elements

The presence of metal ions and salts in the culture medium is another significant factor that can greatly influence carotenoid synthesis by yeasts. Some metals such as Ba, Fe, Mg, Ca, Zn, and Co can stimulate carotenoid production in certain strains. However, this may depend on the concentration, since, for example, it is known that Mn can act favorably on carotenoid accumulation at certain concentrations but can also have an inhibitory effect on this process. Additionally, trace elements can selectively influence the carotenoid profile via activating specific carotenogenic enzymes. For example, while Zn^{2+} and Mn^{2+} can sometimes inhibit the production of torulene and torularhodin, Al^{3+} and Zn^{2+} can increase the production of β-carotene and γ-carotene (Bhosale, 2004; Frengova & Beshkova, 2009; Mata-Gómez et al., 2014).

9.3.2.6 Solvents and other chemical agents

Solvents such as ethanol, methanol, isopropanol, and ethylene glycol can stimulate the production of carotenoids in microorganisms and even the formation of certain specific carotenoids, which can modify the profile obtained. Specifically, ethanol has been shown to induce key enzymes in the biosynthetic pathway such as HMG-CoA reductase, which in turn increases carotenoid production (Bhosale, 2004). The effects of other solvents on carotenoid biosynthesis may involve different mechanisms. Terpenes, amines, alkaloids, and antibiotics can also be involved in the synthesis of caroitenoids by microorganisms. The cyclase inhibitor 2-(4-chlorophenylthio) triethylamine (CPTA) can stimulate the accumulation of lycopene and γ-carotene, while, for example, the addition of amines can increase the formation of lycopene and the decrease of amine concentration can result in the increase of β-carotene (Bhosale, 2004).

9.3.2.7 Substrates

The late logarithmic growth phase marks the onset of carotenoid production in yeasts, which continues into the stationary phase. The presence of an appropriate carbon source is essential for effective carotenoid synthesis during this non-growth period (Frengova & Beshkova, 2009). Yeasts can use a variety of conventional carbon sources for carotenoid synthesis, such as glucose (Aksu & Eren, 2007), xylose (Parajó et al., 1998), glycerol (Saenge et al., 2011), and sucrose (Aksu & Eren, 2007), among others. Depending on the carbon source used, the effects caused on different yeast strains are variable (Paul et al., 2023). A comprehensive review of the literature reveals that carbohydrates such as glucose and sucrose are among the preferred carbon sources for the bioproduction of carotenoids by yeasts. Generally, without focusing exclusively on carotenoid production, bioprocesses mainly use feedstocks based on sugars as a source of carbon and energy for the production of value-added compounds.

However, these types of carbon sources are costly. To reduce costs and, in some cases, enhance carotenoid (and other bioproducts) yields, several studies have focused on indentifying low-cost and even more environmentally sustainable substrates (Soong et al., 2023). Consequently, there has been a growing interest in utilizing by-products from industrial processes that both pollute the environment and cause high treatment costs. Examples of sustainable and viable substrates for carotenoid production by yeasts are carbohydrate-rich agroindustrial by-products such as grape must (Buzzini & Martini, 2000), sugar cane juice (Squina et al., 2002), milk whey (Frengova et al., 1994), as well as pretreated lignocellulosic residues, such as wheat straw and wood pulp (Nagaraj et al., 2022; Sinha et al., 2021), wastewater (Hladnik et al., 2024) and, more recently, even polluting gases, such as carbon dioxide and also syngas (Robles-Iglesias et al., 2023a), among others. Using these carbon sources, yeasts accumulate and store hydrocarbon-rich fats as primary metabolites during the early logarithmic phase of their growth, and carotenoids as secondary metabolites in the late stationary phase (Nagaraj et al., 2022; Singh et al., 2016).

Another increasingly recognized feedstock is volatile fatty acids (VFAs), which are considered low-cost compounds as they can be sourced from renewable and potentially polluting sources. Considering the variety of substrates that can be assimilated by some yeasts, it has been reported that the assimilation of VFAs or a combination of sugars and VFAs (Bao et al., 2021; Morales-Palomo et al., 2022; Wu et al., 2023) also accompanies the accumulation of metabolites, such as carotenoids, among others.

9.4 VFAs as alternative carbon source for carotenoid production

VFAs are carboxylic acids, with six and fewer carbon atoms (C2: acetic acid, C3: propionic acid, C4: butyric acid, C5: valeric acid, C6: caproic acid), that can be produced from various waste materials, wastewater, and C1 gases, such as carbon dioxide. The rapid growth of the world's population is generating a significant increase in organic waste and industrial by-products, many of which, if not properly managed, can have a negative environmental impact. However, these materials present considerable potential as feedstocks for microbial VFAs production. The valorization of these materials through bioconversion into metabolites of great industrial interest, such as carotenoids, is emerging as a highly promising strategy. A significant portion of them comes from the agroindustrial sector, including sugar cane molasses, by-products from paper mills, residues from palm oil mills and even sewage sludge, among others (Tomás-Pejó et al., 2023). Given the yeast's potential in carotenoid biosynthesis and their ability to utilize VFAs as carbon source, the subsequent section will delve into the valorization of organic wastes through anaerobic fermentation (AF) into VFAs and their yeast-based bioconversion into carotenoids.

9.4.1 VFAs production through anaerobic fermentation

The conversion of various organic waste streams to VFAs primarily occurs through AF. During AF, biopolymers are hydrolyzed into soluble monomers, which are then fermented into VFAs such as acetic, propionic, and butyric acids, as well as lactate, alcohols, and hydrogen. This process is also known as acidogenic or dark fermentation (W. S. Lee et al., 2014). VFAs can be produced from diverse substrates, including sludge, food waste, the organic fraction of municipal solid waste, and paper mill effluents. Basically, VFAs are intermediate compounds synthetized during anaerobic digestion (AD) of organic biomass. AD involves a complex and coordinated interaction among diverse microorganisms, ultimately resulting in the production of methane and carbon dioxide. Initially, polymeric substrates such as polysaccharides, proteins, and lipids undergo enzymatic hydrolysis, breaking down into monomers. During acidogenesis, these monomers are fermented into VFAs, longer-chain fatty acids, lactate, alcohols, CO_2, formate, and H_2. In the acetogenesis stage, intermediates are converted into acetate, H_2, and formate. Finally, during methanogenesis, these substrates are further converted into methane and CO_2 (Robles et al., 2018). To accumulate VFAs, it is necessary to inhibit the methanogenic step resulting in a shortened process (AF). This can be achieved by adjusting the fermentation conditions such as temperature, pH, hydraulic and solid retention time (HRT and SRT), organic loading rate (OLR), and feedstock composition. Precise control of these parameters enables the optimization of VFA yield, production rate, and VFA profile (W. S. Lee et al., 2014; Moestedt et al., 2020). Valorizing complex organic waste trough AF yields VFAs production with diverse acid profiles ranging from C2 to C6 (Magdalena et al., 2020). Consequently, various strategies have been developed to regulate specific VFAs production, taking into account both the macromolecular composition of the organic waste and the operational conditions to favor AF and VFAs production.

9.4.2 Strategies for maximizing VFAs production in anaerobic fermentation

The operational conditions in the bioreactor are pivotal in shaping microbial composition and structure, thereby influencing metabolic pathway dominance, and the end products produced. Proper selection of AF conditions is essential to minimize acetogenic and methanogenic activity while enhancing hydrolytic and acidogenic activities. High organic loading rates (OLR) and short hydraulic residence times (HRT) foster hydrolytic and acidogenic bacteria growth, while slower methanogens are washed out, thereby minimizing VFAs degradation. Additionally, process pH and temperature variably influence the physiology and activities of different metabolic groups. For instance, methanogenic activity is favored at neutral pH, while acidogenesis peaks at pH 5.5–6.0. Slightly acidic pH in processes fermenting carbohydrate-rich residues limits methanogenesis, boosting the VFAs accumulation with high acidification yields (60–80% of soluble chemical oxygen demand) and bioconversion efficiency (30–60%) (Carvalheira & Duque, 2021; Cavinato et al.,

2017; Greses et al., 2020). Natural acidification of food or agricultural residues can lower the process pH beneficially, but pH below 5.5 inhibits both methanogenic and acidogenic activities, favoring lactic acid or ethanol accumulation (Greses et al., 2022a). A slightly alkaline pH (pH 8-9) can be beneficially for the bioconversion of protein-rich biomass into VFAs (Bermúdez-Penabad et al., 2017). Further increase in pH levels up to 10–12 have shown both negative and positive effects on the acidogenic activity (Bermúdez-Penabad et al., 2017; Huang & Xiang, 2024). The process pH impacts both feedstock degradability and the VFA profile. For example, Carvalheira and Duque, 2021 reported increased butyric acid concentrations at alkaline pH, while acetic and propionic acids dominated during AF at acidic pH (5–6) (Carvalheira & Duque, 2021). In contrast, another study found butyric acid to be the dominating VFA at pH 5–6 (Wang et al., 2014), suggesting that additional factors, such as feed stock composition, play a significant role.

Hydraulic retention time (HRT) and OLR are interconnected parameters that regulate microbial growth and organic matter intake in the reactor. Extending HRT promotes complex feedstock degradation by fostering bacterial growth and enabling the microbial community to hydrolyze polymeric compounds, thereby increasing soluble organic matter for the acidogenic process. However, short HRTs are preferred because longer HRTs allow materials to remain in the reactor for an extended period, promoting slow-growing microorganisms such as methanogenic archaea (Schnürer, 2016). Additionally, short HRTs are linked to high OLR, leading to an organic matter overload that lowers the pH. This occurs because the methanogens are slow growers and cannot adapt, leading to acetate accumulation. This pH drop, in turn, promotes further VFAs production and reduces methanogenic activity. Selecting the appropriate operational conditions depends strongly on the macromolecular composition of the feedstock and the desired VFA profile. For instance, HRT shorter than 6 days enhances the VFAs production (Cavinato et al., 2017), while a higher HRT of 19–41 days may be needed to increase the VFAs production when digesting more complex organic matter (Ribau Teixeira et al., 2020). Moreover, HRT influences the VFA profile itself: short HRTs typically produce shorter VFAs like acetic and propionic acids (Carvalheira & Duque, 2021), whereas longer HRTs applied on carbohydrate-rich feedstock and an acidic pH result in the accumulation of longer VFAs such as butyric and caproic acids (Greses et al., 2021). The configuration of the reactor is crucial due to the impact of HRT and OLR on process performance and VFAs profiles. Continuous stirred tank reactors (CSTR), sequencing batch reactors (SBR), upflow anaerobic sludge blanket (UASB), and anaerobic membrane bioreactors (AnMBR), are being applied and studied. Among these, SBR, UASB, and AnMBR allow decoupling the HRT from solid retention time (SRT), enabling high flow rates without increasing tank volume.

The process temperature significantly impacts microbial metabolic rates and influences the microbiota composition based on the composition and complexity of the feedstock. Mesophilic processes, which typically run at 35–37°C, range from 20 to 43°C. Thermophilic processes operate at 50–60°C (Nie et al., 2021). Additionally, psychrophilic processes, running below 20°C, may also be applied (Greses et al., 2022b; Nie et al., 2021). Mesophilic processes are optimal for feedstocks like cow manure, maize silage, and the organic fraction of municipal solid waste, as they enhance both hydrolytic and acidogenic activities (Cavinato et al., 2017; Garcia-Aguirre et al., 2017). Thermophilic processes are commonly used for recalcitrant feedstocks such as protein-rich substrates (Battista et al., 2020). Despite their high energy demand, thermophilic processes can achieve high bioconversion efficiencies and produce a highly specific VFA profile (Greses et al., 2022b). Elevated temperatures promote the growth of methanogens, so other process parameters as mentioned above must be carefully managed to enhance VFAs production and prevent methane production. In contrast, low temperatures significantly limit methanogen growth rates and promote VFAs accumulation (Magdalena et al., 2018). However, efficient VFAs production in psychrophilic processes can be best achieved when the feedstock has a high content of readily biodegradable carbohydrates (Magdalena et al., 2018).

9.4.3 The role of substrate composition in shaping VFA profiles

The substrate composition in terms of the macromolecules proteins, carbohydrates, and lipids ultimately determines the VFA profile, as the metabolic pathways required for degradation vary between macromolecules, leading to different end products or VFA profiles. The initial substrate composition determines the establishment of specific microbial communities adapted to the substrate, specializing in the hydrolysis and acidogenesis of its specific compounds. Within these communities, representatives from the phylum *Bacillota* (formerly *Firmicutes*) play a crucial role in maximizing VFAs production. However, variations within *Bacillota* genera due to substrate composition led to significant variations in the resulting VFA profiles (Greses et al., 2021; Llamas et al., 2021a).

To illustrate the interrelationship between substrate composition, microbiota, and VFA profile, the differences in the fermentation of protein-rich and carbohydrate-rich substrates are discussed in the following paragraph:

The hydrolysis of protein-rich substrates like sewage sludge or microalgae releases amino acids. These amino acids are degraded to VFAs through two primary pathways: (i) The Stickland fermentation, where one amino acid act as electron acceptors, yielding VFAs of equivalent carbon length to the original amino acid, and another amino acid act as electron donor, resulting in a product shorter by one carbon; (ii) The fermentation of single amino acids that necessitates the presence of hydrogen-utilizing bacteria or archaea (Ramsay & Pullammanappallil, 2001). Acetic acid and propionic acid are the primary VFAs produced from the fermentation of proteinaceous materials. When microalgae biomass undergoes fermentation, the VFA profile is typically dominated by these VFAs, between 50% and 65% of the total (Llamas et al., 2021a; Magdalena et al., 2019). When sewage sludge is used as a substrate, the presence of acetic acid and propionic acid can increase to 70% to 80% of the VFAs produced (Iglesias-Iglesias et al., 2019). Butyric and valeric acids also occur in smaller and varying amounts produced from both substrates. The phylum *Bacillota* typically dominates this protein-degrading microbial community, comprising 65% to 83% of relative abundance. The amino acids resulting from protein degradation promotes the growth of amino acid degraders within this phylum, including *Peptostreptococcus*, *Sporoanaerobacter*, and members of the Clostridiales order (especially *Clostridium*) (Shen et al., 2017). *Peptostreptococcus* and *Sporoanaerobacter* produce propionic and acetic acid, respectively. Representatives of the genus *Clostridium* are known for their ability to perform Stickland fermentation under anaerobic conditions (Fonknechten et al., 2010). This diversity in amino acid degraders suggests that protein conversion into VFAs typically involves both metabolic pathways (see above) (Llamas et al., 2021b; Shen et al., 2017).

VFA profiles from carbohydrate-rich substrates such as food wastes differ significantly, with acetic acid and butyric acid as the most prevalent VFAs. The hydrolysis of carbohydrates generates various sugars (fructose, glucose, and galactose) that are subsequently fermented to VFAs.

The carbohydrate content of food waste varies with location and seasonal factors, influencing the efficiency of VFAs production and the concentrations of acetic and butyric acids. The fermentation of different food wastes results in different VFAs profiles: for instance, acetic acid (40%) dominated the VFA profile when fermenting potato peels (Zhang et al., 2020), while butyric acid prevailed (41.1%–43.5%) when cucumber and tomato residues were used (Greses et al., 2020). Generally, a higher carbohydrate content in the substrates appears to correlate with increased concentrations of butyric acid (Greses et al., 2022b). In AF of carbohydrate-rich substrates, acetic and butyric fermentations are commonly observed, though other metabolic pathways may also play roles. Genera correlating with carbohydrate degradation include those from the *Lactobacillales* and *Clostridiales* orders within the *Bacillota* phylum. *Lactobacillales* encompass a diverse group that includes lactic acid bacteria and homoacetogenic bacteria. Lactic acid can be converted into VFAs, and homoacetogens transform H_2 and CO_2 into acetic acid, thereby increasing its abundance (Zhou et al., 2018). Prominent bacteria include *Clostridium tyrobutyricum*, *Ruminococcus* spp, and the homoacetogen *Acetobacterium woodii*. *C. tyrobutyricum* is known for converting lactic acid into butyric acid through butyrate-type fermentation (Liu et al., 2013), while members of the genus *Ruminococcus* are involved in breaking down carbohydrates to produce both butyric and acetic acid (Greses et al., 2022a).

Moreover, the fermentation of organic matter into VFAs depends on its intrinsic characteristics, typically converting 10% to 50% of the chemical oxygen demand (COD) in the feedstock (Jimenez et al., 2020). While operational conditions can enhance microbial activity for substrate degradation, some wastes contain recalcitrant compounds like lignin, hindering microbial breakdown. Lignocellulosic biomass (e.g., forest residues, herbaceous plants, wheat straw, rice straw, corn residues, sugarcane bagasse) requires pretreatments (e.g., chemical catalysts, enzymes, steam explosion, physical disruption) to increase the bioavailability of organic matter, given its complex structure, before fermentation (Chaudhary et al., 2024; Greses et al., 2023; Passoth & Sandgren, 2019). Table 9.2 presents various substrates primarily derived from by-products and waste materials that have been successfully utilized for VFAs production via AF. It details the predominant VFAs obtained from each substrate, along with relevant process parameters as mentioned above.

Table 9.2 Substrates and operational parameters used in anaerobic fermentation for VFAs production, including the predominant acids in the VFA pool. Key operational parameters such as temperature, hydraulic retention time, organic loading rate, (soluble) chemical oxygen demand ((S)COD), volatile solids, and duration in days (d) are indicated. AD refers to anaerobic digestion (Tomás-Pejó et al., 2023).

Feedstock	Reactor type	Inoculum source	Operational conditions	Prevailing VFAs (sorted by order of magnitude)	References
Cheese whey	AnSBR*	From AD of cheese whey	30°C, pH=5, OLR=6 g COD/Ld, HRT=2 d	Acetic, propionic, butyric, valeric acids	Lagoa-Costa et al. (2020)
Sewage sludge	CSTR*,*	Sewage sludge	37°C, pH=5.6, OLR=0.9 g COD/Ld, HRT=10 d	Acetic, propionic, iso-butyric, butyric, iso-valeric, valeric acids	Iglesias-Iglesias et al. (2019)
Tuna waste	CSTR*,*	From AD of tuna waste	37°C, pH=5-10, OLR=4 g COD/Ld, HRT=10 d	Acetic, propionic, butyric, iso-valeric acids	Bermúdez-Penabad et al. (2017)
Brewery wastewater	CSTR*,*	From AD of brewery wastewater	30°C, pH=5, OLR=6 g COD/Ld, HRT=2 d	Acetic, propionic, butyric, valeric acids	Ben et al. (2016)
Cheese whey + brewery wastewater (different ratios)	UASB*,*,*	From AD of cheese whey	30°C, pH=5, OLR=6 g COD/Ld, HRT=2 d	Acetic, propionic, butyric, valeric acids	Lagoa-Costa et al. (2022)
Cheese whey + sewage sludge (different ratios)	CSTR*,*	From AD of brewery wastewater	37°C, pH=5.5, OLR=1–2.4 g COD/Ld, HRT=15 d	Caproic, acetic, butyric, iso-butyric acids	Iglesias-Iglesias et al. (2021b)
Agroindustrial wastes: potato solid waste, grape marc distilled, grape marc, brewery spent grain	Batch	From AD of brewery wastewater	30°C	Acetic, butyric, propionic acids	Iglesias-Iglesias et al. (2021a)
Agroindustrial waste: Cucumber	CSTR*,*	From a conventional AD of sewage sludge at 35°C.	25°C, pH=3.8-4.6 (NO pH control), OLR=3 g VS/Ld	Butyric, caproic acids	Greses et al. (2022a)
Tomato				Acetic, butyric acids	
Lettuce				Acetic, butyric acids	
Agroindustrial waste: Cucumber	CSTR*,*			Acetic, butyric, caproic acids	Greses et al. (2020)

(*Continued*)

Table 9.2 (Continued)

Feedstock	Reactor type	Inoculum source	Operational conditions	Prevailing VFAs (sorted by order of magnitude)	References
Tomato		From a conventional AD of sewage sludge at 35°C.	25°C, pH=5.5-6 (pH control), OLR=3 g VS/Ld, HRT=9.3-10.2 d	Acetic, butyric, caproic acids	
Lettuce				Acetic, butyric acids	
Agroindustrial waste: melon and watermelon	CSTR*,*	From a conventional AD of sewage sludge at 35°C.	25°C, pH=5.5–6 (pH control), OLR=3 g VS/Ld, HRT=27 d and 20d	Iso-butyric, caproic acids	Greses et al. (2021)
Microalgae biomass (pre-treated)	CSTR*,*	From a conventional AD of sewage sludge at 35°C.	25°C, pH=5.7–6.3 (NO pH control), OLR=1.5 g COD/Ld, HRT=8 d	Acetic, propionic acids	Lamas et al. (2021b)
Food waste	Fed-batch	From a conventional AD of sewage (pre-treated with acide shock)	28°C, pH=10 OLR=15 g COD/Ld	Acetate, butyrate, propionate	Sarkar et al. (2016)
Brewer's spent grain	CSTR*,*	Sludge from waste water treatment AD	30°C, pH=4.7 (pH control), HRT=16d–41d, OLR=4.3–16 g TS/Ld	Propionate, acetate, butyrate, valerate	Ribau Teixeira et al. (2020)
Potato peel waste	CSTR*,*	Sludge from mesophilic AD feed with cuttle dung	37°C, pH=5 and 7 (pH control), 20 g SCOD/L (pH5) 47.4 g SCOD/L (pH7)	Acetate, propionate, butyrate (pH7) Butyrate, acetate, propionate (pH5)	Lu et al. (2020)

* Anaerobic sequencing batch reactor.
** Continuous stirred tank reactor.
*** Up-flow anaerobic sludge blanket.

Modified from Tomás-Pejó, E., González-Fernández, C., Greses, S., Kennes, C., Otero-Logilde, N., Veiga, M.C., Bolzonella, D., Müller, B., & Passoth, V. (2023). Production of short-chain fatty acids (SCFAs) as chemicals or substrates for microbes to obtain biochemicals. BioMed Central Ltd., Spain Biotechnology for Biofuels and Bioproducts, 16(1). https://doi.org/10.1186/s13068-023-02349-5, https://biotechnologyforbiofuels.biomedcentral.com/.

9.4.4 Carotenoids accumulation from VFAs in yeasts

Various oleaginous yeast strains exhibit a notable ability to metabolize VFAs, effectively directing them toward lipid production when grown on these substrates (Park et al., 2021; Wu et al., 2023). However, carotenoid production by yeasts using VFAs is a relatively recent and still limited field of research and more studies are needed to understand and optimize this process. This scarce literature information is summarized hereafter, where VFAs are present either as co-substrates or, otherwise, as single carbon sources. Regardless of the use of glucose as a substrate, being one of the most efficient carbon sources for cell growth, it has been shown that the addition of acetic acid to the sugar-rich culture medium can significantly influence yeast carotenogenesis. Fig. 9.2 summarizes the metabolic processes involved in carotenoid production in yeasts using acetic acid as an example of VFAs as carbon sources.

Kim et al. (2003) stimulated astaxanthin production by the addition of ethanol (10 g/L) or acetic acid (5 g/L) to a batch-fed culture of *P. rhodozyma*. This resulted in an astaxanthin concentration of 45.62 g/L with ethanol and 43.87 mg/L with acetic acid compared to 34.77 mg/L obtained without the addition of these compounds. Nagaraj et al. (2022) investigated the production of carotenoids and lipids by *R. toruloides* CBS 14 using wheat straw hydrolysate as carbon source, obtaining a β-carotene content of 1.48 mg/100 g dry weight, and being able to identify also, in smaller amounts, γ-carotene, torularhodin, and torulene. However, the carbon source in this study contained mainly glucose, followed by the sugar xylose and acetic acid in only very low concentration, with xylose present at approximately 17–19 g/L, which is about 40% of the glucose concentration, and acetic acid present at around 3–5 g/L, which is approximately 7–11% of the glucose concentration.

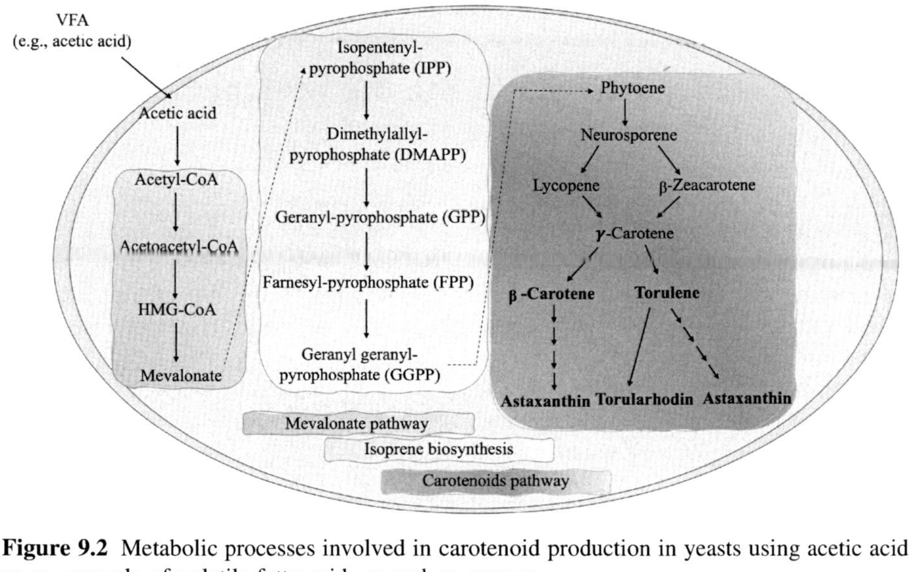

Figure 9.2 Metabolic processes involved in carotenoid production in yeasts using acetic acid as an example of volatile fatty acids as carbon sources.

VFAs obtained from AF in pure cultures with *Clostridium carboxidivorans* or *Clostridium aceticum* using C1 gases as the sole substrate also have significant potential for subsequent carotenoid biosynthesis by yeasts. Acetogenic bacteria can grow on C1 compounds and produce VFAs, primarily acetic acid, through the Wood-Ljungdahl pathway (Fernández-Naveira et al., 2017). This has been shown recently by Robles-Iglesias et al. (2023a). In this study, the authors investigated the simultaneous accumulation of lipids and β-carotene in an engineered strain of *Yarrowia lipolytica* from acetic and butyric acids produced by *C. carboxidivorans* from CO and CO_2 as carbon sources. Since *Y. lipolytica* is not able to produce carotenoids naturally, the synthesis genes were introduced by genetic engineering (Gottardi et al., 2021). Robles-Iglesias et al. (2023a) reached a concentration and β-carotene content of 759 mg/L and 88.4 mg/g cell, respectively, for 79 g/L VFAs consumed. Another study using the same engineered yeast strain, but grown on acetic acid obtained from syngas fermentation by the acetogen *C. aceticum*, reported somewhat lower yields of β-carotene. Here, optimal concentration and β-carotene content were found at pH 6.0, reaching 164 mg/L and 44 mg/g cell, respectively (Robles-Iglesias et al., 2023b). In contrast, another study found significantly higher β-carotene concentration (382.7 mg/L) and β-carotene content (62.3 mg/g) in *Y. lipolytica* after the consumption of 41 g/L of fermentation-derived acetate from *C. aceticum* (Robles-Iglesias et al., 2024). Moreover, these three 2-stage fermentation studies reveal a significant relationship between culture conditions and the final carotenoid yields. Despite using the same yeast strain, variations in parameters such as pH, aeration, temperature, acetogenic medium composition, acetogenic species, and acid type and concentration all substantially impact the final carotenoid content and concentration. Thus, combining an initial AF process to convert C1 gases into VFAs with a subsequent aerobic fermentation using oleaginous yeasts to bioconvert the VFAs into carotenoids is both innovative and promising and could address various energy and environmental challenges.

9.5 Legislative barriers to the use of genetically modified organisms and residues

Genetic modification of microorganisms has been key to advances in research and the development of agricultural, industrial, and biomedical applications. However, regulatory frameworks and new genome editing technologies have not evolved at the same pace, creating challenges regarding their safety and monitoring. In addition, legislative barriers to the use of genetically modified organisms (GMOs) differ from country to country and region to region. In the case of Europe, the European Union (EU) is considered to be restrictive in establishing high biosafety standards for human and animal consumption, environmental impact and consumer interests. Applicants have the responsibility to demonstrate the safety of transgenic products and their use, in accordance with EU Regulation 1829/2003 (Rozas et al., 2022). Thus, this regulation requires exhaustive prior safety studies before a GMO can be marketed, which can result in long and costly processes that can sometimes discourage innovation. In addition, Regulation 1830/2003 establishes rules to ensure that products containing GMOs and food and feed obtained from them follow a traceability that requires labeling and monitoring of environmental and health risks, which allows consumers to make informed decisions, but can also generate distrust toward these products (Reglamento (CE) n° 1830/2003 del Parlamento Europeo y del Consejo, de septiembre de 2003, relativo a la trazabilidad y al etiquetado de organismos modificados genéticamente y a la trazabilidad de los alimentos y piensos producidos a partir de éstos, y por el que se modifica la Directiva, 2001/18/CE, 268 OJ L, 2003). Despite all these restrictions, the EU is one of the main importers of GMOs (Rozas et al., 2022). However, the lack of harmonization in regulations among Member States may also create uncertainty when seeking to use GMOs in industry. As mentioned in the previous sections, the genetic modification of microorganisms for the production of carotenoids has great potential for improving the efficiency of the process. A relevant example mentioned above is the case of the yeast *Y. lipolytica*, which does not naturally produce this type of metabolite.

On the other hand, the interest in using certain agroindustrial wastes as raw materials to obtain carotenoids was also discussed throughout this chapter in order to reduce waste and even reduce the negative environmental impact attributed to them. In this case, food safety and quality regulations require that all materials used meet specific purity and safety standards, which could also be a challenge if residues are used due to restrictive legislation. It is worth mentioning Regulation (EC) No. 178/2002, which lays down the general principles and requirements of food law in the EU with the aim of ensuring that food is safe and therefore protecting the health of consumers. Regarding the use of waste specifically, reference can be made to law 7/2022, of April 8, which establishes guidelines on waste management, and promotes the reduction and recovery of these by-products or directive 2008/98/EC on waste (Directiva 2008/98/CE del Parlamento Europeo y del Consejo, de 19 de noviembre de 2008, sobre los residuos y por la que se derogan determinadas Directivas, 2008). Even so, each type of waste presents unique characteristics and may be subject to different regulations. Moreover, the final application of the product obtained entails additional legal requirements. It is therefore essential to carry out a detailed and customized study of the applicable legislation in each particular case.

9.6 Conclusions

The importance of carotenoids is well recognized, playing an increasingly important role in human health. Yeasts have been shown to be promising microorganisms for the natural production of carotenoids, providing a safer alternative to synthetic pigments which may pose health risks. Utilizing yeasts for this purpose with low-cost substrates such as VFAs offers a sustainable strategy that may significantly reduce production costs and facilitate the scalability of the process in the future. Although carotenoid production, based on the valorization of VFAs from various pollutants, is still under study, this chapter demonstrates that yeasts have the potential to become a key pillar in transforming the carotenoid industry. This approach is not only more sustainable and innovative but also offers significant health benefits.

Related titles

Bioenergy: Biomass to Biofuels (ISBN: 978-0-12-407909-0).
Biofuels from Algae (ISBN: 978-0-444-59558-4)
Advances in Biorefineries: Biomass and Waste Supply Chain Exploitation (ISBN: 978-0-85709-521-3)
Handbook of Biofuels Production (ISBN: 978-1-84569-679-5)
Biomass Gasification, Pyrolysis and Torrefaction: Practical Design and Theory (ISBN: 978-0-12-396488-5)

Acknowledgments

This chapter is based upon work from COST Action Yeast4Bio (CA18229), supported by COST (European Cooperation in Science and Technology). Research undertaken at BIOENGIN Group, Spain, on this topic is mainly financed by the Spanish Ministry of Science and Innovation (PID2020-117805RB-I00 and TED2021-130055B-I00). The BIOENGIN Group at UDC thanks Xunta de Galicia for financial support to Competitive Reference Research Groups (ED431C 2021/55). VP and BM were supported by the Swedish Research Council for Environment, Agricultural Sciences and Spatial Planning (Formas) [grant number 2018-01877].

References

Aksu, Z., & Eren, A. T. (2007). Production of carotenoids by the isolated yeast of *Rhodotorula glutinis*. *Biochemical Engineering Journal, 35*(2), 107–113. https://doi.org/10.1016/j.bej.2007.01.004.

Aman Mohammadi, M., Ahangari, H., Mousazadeh, S., Seyede, Hosseini, M., & Dufossé, L. (2022). Microbial pigments as an alternative to synthetic dyes and food additives: a brief review of recent studies. *Bioprocess and Biosystems Engineering, 45*(1), 1–12. https://doi.org/10.1007/s00449-021-02621-8.

Bao, W., Li, Z., Wang, X., Gao, R., Zhou, X., Cheng, S., Men, Y., & Zheng, L. (2021). Approaches to improve the lipid synthesis of oleaginous yeast *Yarrowia lipolytica*: A review. *Renewable and Sustainable Energy Reviews, 149*, 111–386. https://doi.org/10.1016/J.RSER.2021.111386.

Barreto, J. V.de O., Casanova, L. M., Junior, A. N., Reis-Mansur, M. C. P. P., & Vermelho, A. B. (2023). Microbial pigments: Major groups and industrial applications. *Microorganisms, 11*, 2920. https://doi.org/10.3390/microorganisms11122920.

Battista, F., Frison, N., Pavan, P., Cavinato, C., Gottardo, M., Fatone, F., Eusebi, A. L., Majone, M., Zeppilli, M., Valentino, F., Fino, D., Tommasi, T., & Bolzonella, D. (2020). Food wastes and sewage sludge as feedstock for an urban biorefinery producing biofuels and added-value bioproducts. *Journal of Chemical Technology and Biotechnology, 95*(2), 328–338. https://doi.org/10.1002/jctb.6096.

Ben, M., Kennes, C., & Veiga, M. C. (2016). Optimization of polyhydroxyalkanoate storage using mixed cultures and brewery wastewater. *Journal of Chemical Technology and Biotechnology, 91*(11), 2817–2826. https://doi.org/10.1002/jctb.4891.

Bermúdez-Penabad, N., Kennes, C., & Veiga, M. C. (2017). Anaerobic digestion of tuna waste for the production of volatile fatty acids. *Waste Management, 68*, 96–102. https://doi.org/10.1016/J.WASMAN.2017.06.010.

Bhosale, P. (2004). Environmental and cultural stimulants in the production of carotenoids from microorganisms. *Applied Microbiology and Biotechnology, 63*(4), 351–361. https://doi.org/10.1007/s00253-003-1441-1.

Bhosale, P., & Bernstein, P. S. (2005). Microbial xanthophylls. *Applied Microbiology and Biotechnology, 68*(4), 445–455. https://doi.org/10.1007/s00253-005-0032-8.

Bhosale, P., & Gadre, R. V. (2002). Manipulation of temperature and illumination conditions for enhanced beta-carotene production by mutant 32 of *Rhodotorula glutinis. Letters in Applied Microbiology, 34*(5), 349–353. https://doi.org/10.1046/j.1472-765x.2002.01095.x.

Braunwald, T., Schwemmlein, L., Graeff-Hönninger, S., French, W. T., Hernandez, R., Holmes, W. E., & Claupein, W. (2013). Effect of different C/N ratios on carotenoid and lipid production by *Rhodotorula glutinis. Applied Microbiology and Biotechnology, 97*(14), 6581–6588. https://doi.org/10.1007/s00253-013-5005-8.

Buzzini, P., & Martini, A. (2000). Production of carotenoids by strains of *Rhodotorula glutinis* cultured in raw materials of agro-industrial origin. *Bioresource Technology, 71*(1), 41–44. https://doi.org/10.1016/s0960-8524(99)00056-5.

Buzzini, P., Innocenti, M., Turchetti, B., Libkind, D., van Broock, M., & Mulinacci, N. (2007). Carotenoid profiles of yeasts belonging to the genera *Rhodotorula, Rhodosporidium, Sporobolomyces,* and *Sporidiobolus. Canadian Journal of Microbiology, 53*(8), 1024–1031. https://doi.org/10.1139/w07-068.

Carvalheira, M., & Duque, A. F. (2021). Chapter 9: From food waste to volatile fatty acids towards a circular economy, in: Fermentation: Processes, benefits and risks. BoD – Books on Demand, London.

Cavinato, C., Da Ros, C., Pavan, P., & Bolzonella, D. (2017). Influence of temperature and hydraulic retention on the production of volatile fatty acids during anaerobic fermentation of cow manure and maize silage. *Bioresource Technology, 223*, 59–64. https://doi.org/10.1016/j.biortech.2016.10.041.

Chaudhary, G., Chaudhary, N., Saini, S., Gupta, Y., Vivekanand, V., & Panghal, A. (2024). Assessment of pretreatment strategies for valorization of lignocellulosic biomass: Path forwarding towards lignocellulosic biorefinery. *Waste and Biomass Valorization, 15*(1), 1–36. https://doi.org/10.1007/s12649-023-02219-z.

Cheng, Y. T., & Yang, C. F. (2016). Using strain *Rhodotorula mucilaginosa* to produce carotenoids using food wastes. *Taiwan Institute of Chemical Engineers, 61*, 270–275. https://doi.org/10.1016/j.jtice.2015.12.027.

Directiva 2008/98/CE del Parlamento Europeo y del Consejo, de 19 de noviembre de 2008, sobre los residuos y por la que se derogan determinadas Directivas. (2008).

El-Banna, A. A., Abd El-Razek, A. M., & El-Mahdy, A. R. (2012). Some factors affecting the production of carotenoids by *Rhodotorula glutinis var. Glutinis. Food and Nutrition Sciences, 03*(01), 64–71. https://doi.org/10.4236/fns.2012.31011.

El-Salam, B. A. A., Amer, A. E. A., & Amer, M. E. (2014). Using whey for production of carotenoids by *Rhodotorula glutinis. Middle East Journal of Applied Sciences, 4*, 385–391.

Fernández-Naveira, Á., Veiga, M. C., & Kennes, C. (2017). Effect of pH control on the anaerobic H-B-E fermentation of syngas in bioreactors. *Journal of Chemical Technology and Biotechnology, 92*(6), 1178–1185. https://doi.org/10.1002/JCTB.5232.

Fonknechten, N., Chaussonnerie, S., Tricot, S., Lajus, A., Andreesen, J. R., Perchat, N., Pelletier, E., Gouyvenoux, M., Barbe, V., Salanoubat, M., Le Paslier, D., Weissenbach, J., Cohen, G. N., & Kreimeyer, A. (2010). *Clostridium sticklandii,* a specialist in amino acid degradation: Revisiting its metabolism through its genome sequence. *BMC Genomics, 11*, 555. https://doi.org/10.1186/1471-2164-11-555.

Frengova, G. I., & Beshkova, D. M. (2009). Carotenoids from *Rhodotorula* and *Phaffia*: Yeasts of biotechnological importance. *Journal of Industrial Microbiology and Biotechnology, 36*(2), 163–180. https://doi.org/10.1007/s10295-008-0492-9.

Frengova, G., Simova, E., Pavlova, K., Beshkova, D., & Grigorova, D. (1994). Formation of carotenoids by *Rhodotorula glutinis* in whey ultrafiltrate. *Biotechnology and Bioengineering, 44*(8), 888–894. https://doi.org/10.1002/bit.260440804.

Garcia-Aguirre, J., Aymerich, E., de Goñi, J. G.-M., & Esteban-Gutiérrez, M. (2017). Selective VFA production potential from organic waste streams: Assessing temperature and pH influence. *Bioresource Technology, 244*, 1081–1088. https://doi.org/10.1016/j.biortech.2017.07.187.

Gottardi, D., Siroli, L., Vannini, L., Patrignani, F., & Lanciotti, R. (2021). Recovery and valorization of agri-food wastes and by-products using the non-conventional yeast *Yarrowia lipolytica. Trends in Food Science & Technology, 115*, 74–86. https://doi.org/10.1016/j.tifs.2021.06.025.

Greses, S., Llamas, M., Morales-Palomo, S., González-Fernández, C., & Tomás-Pejó, E. (2022). *Microbial production of oleochemicals. in: Bisaria, V. (Ed.), Handbook of Biorefinery Research and Technology*. Dordrecht, Netherlands: Springer, 1–23. https://doi.org/10.1007/978-94-007-6724-9_17-1.

Greses, S., Tomás-Pejó, E., & Gónzalez-Fernández, C. (2020). Agroindustrial waste as a resource for volatile fatty acids production via anaerobic fermentation. *Bioresource Technology, 297*, 122486. https://doi.org/10.1016/j.biortech.2019.122486.

Greses, S., Tomás-Pejó, E., & González-Fernández, C. (2021). Short-chain fatty acids and hydrogen production in one single anaerobic fermentation stage using carbohydrate-rich food waste. *Journal of Cleaner Production, 284*, 124727. https://doi.org/10.1016/j.jclepro.2020.124727.

Greses, S., Tomás-Pejó, E., & González-Fernández, C. (2022a). Assessing the relevance of acidic pH on primary intermediate compounds when targeting at carboxylate accumulation. *Biomass Conversion and Biorefinery, 12*(10), 4519–4529. https://doi.org/10.1007/s13399-022-02416-1.

Greses, S., Tomás-Pejó, E., & González-Fernández, C. (2022b). Statistical correlation between waste macromolecular composition and anaerobic fermentation temperature for specific short-chain fatty acid production. *Environmental Research, 206*. https://doi.org/10.1016/j.envres.2021.112288.

Hladnik, L., Vicente, F. A., Grilc, M., & Likozar, B. (2024). β-Carotene production and extraction: A case study of olive mill wastewater bioremediation by *Rhodotorula glutinis* with simultaneous carotenoid production. *Biomass Conversion and Biorefinery, 14*(7), 8459–8467. https://doi.org/10.1007/s13399-022-03081-0.

Huang, X., & Xiang, Z. (2024). Enhanced volatile fatty acids production in thermal hydrolysis pretreatment with constant pH regulation: Effects evaluation and mechanism elucidation. *Journal of Environmental Chemical Engineering, 12*(6), 114830. https://doi.org/10.1016/j.jece.2024.114830.

Iglesias-Iglesias, R., Campanaro, S., Treu, L., Kennes, C., & Veiga, M. C. (2019). Valorization of sewage sludge for volatile fatty acids production and role of microbiome on acidogenic fermentation. *Bioresource Technology, 291*, 121817. https://doi.org/10.1016/J.BIORTECH.2019.121817.

Iglesias-Iglesias, R., Fernandez-Feal, M. M.del C., Kennes, C., & Veiga, M. C. (2021a). Valorization of agro-industrial wastes to produce volatile fatty acids: Combined effect of substrate/inoculum ratio and initial alkalinity. *Environmental Technology, 42*(25), 3889–3899. https://doi.org/10.1080/09593330.2020.1743370.

Iglesias-Iglesias, R., Portela-Grandío, A., Treu, L., Campanaro, S., Kennes, C., & Veiga, M. C. (2021b). Co-digestion of cheese whey with sewage sludge for caproic acid production: Role of microbiome and polyhydroxyalkanoates potential production. *Bioresource Technology, 337*, 125388. https://doi.org/10.1016/j.biortech.2021.125388.

Jimenez, J., Charnier, C., Kouas, M., Latrille, E., Torrijos, M., Harmand, J., Patureau, D., Spérandio, M., Morgenroth, E., Béline, F., Ekama, G., Vanrolleghem, P. A., Robles, A., Seco, A., Batstone, D. J., & Steyer, J. P. (2020). Modelling hydrolysis: Simultaneous versus sequential biodegradation of the hydrolysable fractions. *Waste Management, 101*, 150–160. https://doi.org/10.1016/j.wasman.2019.10.004.

Kim, S. J., Kim, G. J., Park, D. H., & Ryu, Y. W. (2003). High-level production of astaxanthin by fed-batch culture of mutant strain *Phaffia rhodozyma* AJ-6-1. *Journal of Microbiology and Biotechnology, 13*(2), 175–181.

Lagoa-Costa, B., Kennes, C., & Veiga, M. C. (2020). Cheese whey fermentation into volatile fatty acids in an anaerobic sequencing batch reactor. *Bioresource Technology, 308*, 123226. https://doi.org/10.1016/j.biortech.2020.123226.

Lagoa-Costa, B., Kennes, C., & Veiga, M. C. (2022). Influence of feedstock mix ratio on microbial dynamics during acidogenic fermentation for polyhydroxyalkanoates production. *Journal of Environmental Management, 303*, 114132. https://doi.org/10.1016/j.jenvman.2021.114132.

Larroude, M., Celinska, E., Back, A., Thomas, S., Nicaud, J. M., & Ledesma-Amaro, R. (2018). A synthetic biology approach to transform *Yarrowia lipolytica* into a competitive biotechnological producer of β-carotene. *Biotechnology and Bioengineering, 115*(2), 464–472. https://doi.org/10.1002/bit.26473.

Lee, J. J. L., Chen, L., Shi, J., Trzcinski, A., & Chen, W. N. (2014). Metabolomic profiling of *Rhodosporidium toruloides* grown on glycerol for carotenoid production during different growth phases. *Journal of Agricultural and Food Chemistry, 62*(41), 10203–10209. https://doi.org/10.1021/jf502987q.

Lee, W. S., Chua, A. S. M., Yeoh, H. K., & Ngoh, G. C. (2014). A review of the production and applications of waste-derived volatile fatty acids. *Chemical Engineering Journal, 235*, 83–99. https://doi.org/10.1016/j.cej.2013.09.002.

Liu, S., Bischoff, K. M., Leathers, T. D., Qureshi, N., Rich, J. O., & Hughes, S. R. (2013). Butyric acid from anaerobic fermentation of lignocellulosic biomass hydrolysates by *Clostridium tyrobutyricum* strain RPT-4213. *Bioresource Technology, 143*, 322–329. https://doi.org/10.1016/j.biortech.2013.06.015.

Llamas, M., Greses, S., Tomás-Pejó, E., & González-Fernández, C. (2021a). Tuning microbial community in non-conventional two-stage anaerobic bioprocess for microalgae biomass valorization into targeted bioproducts. *Bioresource Technology, 337*, 125387. https://doi.org/10.1016/j.biortech.2021.125387.

Llamas, M., Magdalena, J. A., Greses, S., Tomás-Pejó, E., & González-Fernández, C. (2021b). Insights on the microbial communities developed during the anaerobic fermentation of raw and pretreated microalgae biomass. *Chemosphere, 263*, 127942. https://doi.org/10.1016/j.chemosphere.2020.127942.

Lu, Y., Zhang, Q., Wang, X., Zhou, X., & Zhu, J. (2020). Effect of pH on volatile fatty acid production from anaerobic digestion of potato peel waste. *Bioresource Technology, 316*, 123851. https://doi.org/10.1016/j.biortech.2020.123851.

Magdalena, J. A., Tomás-Pejó, E., Ballesteros, M., & González-Fernandez, C. (2018). Volatile fatty acids production from protease pretreated *Chlorella* biomass via anaerobic digestion. *Biotechnology Progress, 34*(6), 1363–1369. https://doi.org/10.1002/btpr.2696.

Magdalena, J. A., Llamas, M., Tomás-Pejó, E., & González-Fernández, C. (2019). Semicontinuous anaerobic digestion of protease pretreated *Chlorella* biomass for volatile fatty acids production. *Journal of Chemical Technology and Biotechnology, 94*(6), 1861–1869. https://doi.org/10.1002/jctb.5960.

Magdalena, J. A., Ballesteros, M., & González-Fernández, C. (2020). Acidogenesis and chain elongation for bioproduct development. *Wastewater Treatment Residues as Resources for Biorefinery Products and Biofuels,* 391–414. https://doi.org/10.1016/B978-0-12-816204-0.00017-5.

Mannazzu, I., Landolfo, S., da Silva, T. L., & Buzzini, P. (2015). Red yeasts and carotenoid production: Outlining a future for non-conventional yeasts of biotechnological interest. *Journal of Microbiology and Biotechnology, 31*(11), 1665–1673. https://doi.org/10.1007/s11274-015-1927-x.

Mata-Gómez, L. C., Montañez, J. C., Méndez-Zavala, A., & Aguilar, C. N. (2014). Biotechnological production of carotenoids by yeasts: An overview. *Microbial cell factories, 13*(1), 1–11. https://doi.org/10.1186/1475-2859-13-12.

Mata-Gómez, L. C., Mapelli-Brahm, P., Meléndez-Martínez, A. J., Méndez-Zavala, A., Morales-Oyervides, L., & Montañez, J. (2023). Microbial carotenoid synthesis optimization in goat cheese whey using the Robust Taguchi Method: A sustainable approach to help tackle vitamin A deficiency. *Foods, 12*(3), 658. https://doi.org/10.3390/foods12030658.

Moestedt, J., Müller, B., Nagaraj, Y. N., & Schnürer, A. (2020). Acetate and lactate production during two-stage anaerobic digestion of food waste driven by *Lactobacillus* and *Aeriscardovia*. *Frontiers in Energy Research, 8,* 105. https://doi.org/10.3389/fenrg.2020.00105.

Morales-Palomo, S., González-Fernández, C., & Tomás-Pejó, E. (2022). Prevailing acid determines the efficiency of oleaginous fermentation from volatile fatty acids. *Journal of Environmental Chemical Engineering, 10*(2), 107354. https://doi.org/10.1016/j.jece.2022.107354.

Morales-Sánchez, D., & Martinez-Rodriguez, O. A. (2017). Heterotrophic cultivation of microalgae: Production of metabolites of commercial interest. *Journal of Chemical Technology & Biotechnology, 92*(5), 925–936. https://doi.org/10.1002/jctb.5115.

Nagaraj, Y. N., Burkina, V., Okmane, L., Blomqvist, J., Rapoport, A., Sandgren, M., Pickova, J., Sampels, S., & Passoth, V. (2022). Identification, quantification and kinetic study of carotenoids and lipids in *Rhodotorula toruloides* CBS 14 cultivated on wheat straw hydrolysate. *Fermentation, 8*(7), 300. https://doi.org/10.3390/fermentation8070300.

Naveira-Pazos, C., Robles-Iglesias, R., Fernández-Blanco, C., Veiga, M. C., & Kennes, C. (2023). State-of-the-art in the accumulation of lipids and other bioproducts from sustainable sources by *Yarrowia lipolytica*. *Reviews in Environmental Science and Biotechnology, 22*(4), 1131–1158. https://doi.org/10.1007/s11157-023-09670-3.

Nie, E., He, P., Zhang, H., Hao, L., Shao, L., & Lü, F. (2021). How does temperature regulate anaerobic digestion? *Renewable and Sustainable Energy Reviews, 150*, 111453. https://doi.org/10.1016/j.rser.2021.111453.

Papapostolou, H., Kachrimanidou, V., Alexandri, M., Plessas, S., Papadaki, A., & Kopsahelis, N. (2023). Natural carotenoids: Recent advances on separation from microbial biomass and methods of analysis. *Antioxidants, 12*(5), 1030. https://doi.org/10.3390/antiox12051030.

Parajó, J. C., Santos, V., & Vázquez, M. (1998). Optimization of carotenoid production by *Phaffia rhodozyma* cells grown on xylose. *Process Biochemistry, 33*(2), 181–187. https://doi.org/10.1016/s0032-9592(97)00045-9.

Park, Y. K., González-Fernández, C., Robles-Iglesias, R., Vidal, L., Fontanille, P., Kennes, C., Tomás Pejó, E., Nicaud, J. M., & Fickers, P. (2021). Bioproducts generation from carboxylate platforms by the non-conventional yeast *Yarrowia lipolytica. FEMS Yeast Research, 21*(6), foab047. https://doi.org/10.1093/femsyr/foab047.

Passoth, V., & Sandgren, M. (2019). Biofuel production from straw hydrolysates: Current achievements and perspectives. *Applied Microbiology and Biotechnology, 103*(13), 5105–5116. https://doi.org/10.1007/s00253-019-09863-3.

Paul, D., Kumari, P. K., & Siddiqui, N. (2023). Yeast carotenoids: Cost-effective fermentation strategies for health care applications. *Fermentation, 9*(2), 147. https://doi.org/10.3390/fermentation9020147.

Pham, K. D., Shida, Y., Miyata, A., Takamizawa, T., Suzuki, Y., Ara, S., Yamazaki, H., Masaki, K., Mori, K., Aburatani, S., Hirakawa, H., Tashiro, K., Kuhara, S., Takaku, H., & Ogasawara, W. (2020). Effect of light on carotenoid and lipid production in the oleaginous yeast *Rhodosporidium toruloides. Bioscience, Biotechnology and Biochemistry, 84*(7), 1501–1512. https://doi.org/10.1080/09168451.2020.1740581.

Qin, Z., Wang, X., Gao, S., Li, D., & Zhou, J. (2023). Production of natural pigments using microorganisms. *Journal of Agricultural and Food Chemistry, 71*(24), 9243–9254. https://doi.org/10.1021/acs.jafc.3c02222.

Ramsay, I. R., & Pullammanappallil, P. C. (2001). Protein degradation during anaerobic wastewater treatment: Derivation of stoichiometry. *Biodegradation, 12*(4), 247–256. https://doi.org/10.1023/A:1013116728817.

Reglamento (CE) n° 1830/2003 del Parlamento Europeo y del Consejo, de 22 de septiembre de 2003, relativo a la trazabilidad y al etiquetado de organismos modificados genéticamente y a la trazabilidad de los alimentos y piensos producidos a partir de éstos, y por el que se modifica la Directiva 2001/18/CE, 268 OJ L (2003). http://data.europa.eu/eli/reg/2003/1830/oj/spa.

Research and Markets. (2024). Global Carotenoids Market by Type (Astaxanthin, Beta-Carotene, Lutein, Lycopene, Canthaxanthin, Zeaxanthin), Formulations (Oil Suspension, Powder, Beadlet, and Emulsion), Source (Natural And Synthetic), Application and Region—Forecast to 2029. https://www.researchandmarkets.com/reports/4987693/global-carotenoids-market-by-type-astaxanthin.

Ribau Teixeira, M., Guarda, E. C., Freitas, E. B., Galinha, C. F., Duque, A. F., & Reis, M. A. M. (2020). Valorization of raw brewers' spent grain through the production of volatile fatty acids. *New Biotechnology, 57*, 4–10. https://doi.org/10.1016/j.nbt.2020.01.007.

Ribeiro, R. M. M. G. P., Pião, B. W., Gonçalves, D. O., Scontri, M., Mazziero, V. T., Mussagy, C. U., Raghavan, V., Astudillo-Castro, C., Córdova, A., Cerri, M. O., & Tambourgi, E. B. (2023). Synergistic effects of stirring and aeration rate on carotenoid production in yeast *Rhodotorula toruloides* CCT 7815 envisioning their application as soap additives. *Fermentation, 9*(9), 828. https://doi.org/10.3390/fermentation9090828.

Robles, G., Nair, R.B., Kleinsteuber, S., Nikolausz, M., Sárvári Horváth, I. (2018). Biogas production: Microbiological aspects, In: Tabatabaei, M., Ghanavati, H. (Eds.), Biogas: Fundamentals, Process, and Operation. Springer International Publishing, Cham, pp. 163–198. https://doi.org/10.1007/978-3-319-77335-3_7.

Robles-Iglesias, R., Naveira-Pazos, C., Nicaud, J.-M., Veiga, M. C., & Kennes, C. (2023a). Two-stage syngas fermentation into microbial oils and β-carotene with *Clostridium carboxidivorans* and engineered *Yarrowia lipolytica*. *Journal of CO_2 Utilization, 76*, 102593. https://doi.org/10.1016/j.jcou.2023.102593.

Robles-Iglesias, R., Nicaud, J.-M., Veiga, M. C., & Kennes, C. (2023b). Integrated fermentative process for lipid and β-carotene production from acetogenic syngas fermentation using an engineered oleaginous *Yarrowia lipolytica* yeast. *Bioresource Technology, 389*, 129815. https://doi.org/10.1016/j.biortech.2023.129815.

Robles-Iglesias, R., Fernández-Blanco, C., Nicaud, J.-M., Veiga, M. C., & Kennes, C. (2024). Unlocking the potential of one-carbon gases (CO_2, CO) for concomitant bioproduction of β-carotene and lipids. *Ecotoxicology and Environmental Safety, 271*, 115950. https://doi.org/10.1016/j.ecoenv.2024.115950.

Rozas, P., Kessi-Pérez, E. I., & Martínez, C. (2022). Genetically modified organisms: Adapting regulatory frameworks for evolving genome editing technologies. *Biological Research, 55*(1). https://doi.org/10.1186/s40659-022-00399-x.

Saenge, C., Cheirsilp, B., Suksaroge, T. T., & Bourtoom, T. (2011). Potential use of oleaginous red yeast *Rhodotorula glutinis* for the bioconversion of crude glycerol from biodiesel plant to lipids and carotenoids. *Process Biochemistry, 46*(1), 210–218. https://doi.org/10.1016/j.procbio.2010.08.009.

Sarkar, O., Kumar, A. N., Dahiya, S., Krishna, K. V., Yeruva, D. K., & Mohan, S. V. (2016). Regulation of acidogenic metabolism towards enhanced short chain fatty acid biosynthesis from waste: Metagenomic profiling. *RSC Advances, 6*(22), 18641–18653. https://doi.org/10.1039/c5ra24254a.

Schnürer, A. (2016). Biogas production: Microbiology and technology. *Anaerobes in biotechnology*, 195–234. https://doi.org/10.1007/10_2016_5.

Sereti, F., Papadaki, A., Alexandri, M., Kachrimanidou, V., & Kopsahelis, N. (2023). Exploring the potential of novel *R. kratochvilovae* red yeasts towards the sustainable synthesis of natural carotenoids. *Sustainable Chemistry and Pharmacy, 31*, 100927. https://doi.org/10.1016/j.scp.2022.100927.

Sharma, N., Shekhar, P., Kumar, V., Kaur, H., & Jayasena, V. (2024). Microbial pigments: Sources, current status, future challenges in cosmetics and therapeutic applications. *Journal of Basic Microbiology, 64*(1), 4–21. https://doi.org/10.1002/jobm.202300214.

Shen, D., Yin, J., Yu, X., Wang, M., Long, Y., Shentu, J., & Chen, T. (2017). Acidogenic fermentation characteristics of different types of protein-rich substrates in food waste to produce volatile fatty acids. *Bioresource Technology, 227*, 125–132. https://doi.org/10.1016/j.biortech.2016.12.048.

Sigurdson, G. T., Tang, P., & Giusti, M. M. (2017). Natural colorants: Food colorants from natural sources. *Annual Review of Food Science and Technology, 8*, 261–280. https://doi.org/10.1146/annurev-food-030216-025923.

Simpson, K. L., Nakayama, T. O., & Chichester, C. O. (1964). Biosynthesis of yeast carotenoids. *Journal of Bacteriology, 88*(6), 1688–1694. https://doi.org/10.1128/jb.88.6.1688-1694.1964.

Singh, G., Jawed, A., Paul, D., Bandyopadhyay, K. K., Kumari, A., & Haque, S. (2016). Concomitant production of lipids and carotenoids in *Rhodosporidium toruloides* under osmotic stress using response surface methodology. *Frontiers in Microbiology, 7*, 1686. https://doi.org/10.3389/fmicb.2016.01686.

Sinha, S., Singh, G., Paul, D., & Amity Institute of Biotechnology, Amity University, Noida, India. (2021). Lipid and carotenoid production by *Rhodosporodium toruloides* ATCC 204091 using C5 and C6 sugars obtained from lignocellulosic hydrolysate. *Journal of Environmental Biology, 42*(4), 938–944. https://doi.org/10.22438/jeb/42/4/MRN-1583.

Soong, Y. H. V., Coleman, S. M., Liu, N., Qin, J., Lawton, C., Alper, H. S., & Xie, D. (2023). Using oils and fats to replace sugars as feedstocks for biomanufacturing: Challenges and opportunities for the yeast *Yarrowia lipolytica*. *Biotechnology Advances, 65*, 108–128. https://doi.org/10.1016/j.biotechadv.2023.108128.

Squina, F. M., Yamashita, F., Pereira, J. L., & Mercadante, A. Z. (2002). Production of carotenoids by *Rhodotorula rubra* and *R. glutinis* in culture medium supplemented with sugar cane juice. *Food Biotechnology, 16*(3), 227–235. https://doi.org/10.1081/FBT-120016776.

Stachowiak, B., & Czarnecki, Z. (2007). Effect of light on carotenoid yield in fed cultures of *Phaffia rhodozyma* CBS 5626. *Polish Journal of Food and Nutrition Sciences, 57*(3A), 129–131.

Tkáčová, J., Čaplová, J., Klempová, T., & Čertík, M. (2017). Correlation between lipid and carotenoid synthesis in torularhodin-producing *Rhodotorula glutinis*. *Annals of Microbiology, 67*(8), 541–551. https://doi.org/10.1007/s13213-017-1284-0.

Tomás-Pejó, E., González-Fernández, C., Greses, S., Kennes, C., Otero-Logilde, N., Veiga, M. C., Bolzonella, D., Müller, B., & Passoth, V. (2023). Production of short-chain fatty acids (SCFAs) as chemicals or substrates for microbes to obtain biochemicals. *Biotechnology for Biofuels and Bioproducts, 16*(1), 96. https://doi.org/10.1186/s13068-023-02349-5.

Wang, K., Yin, J., Shen, D., & Li, N. (2014). Anaerobic digestion of food waste for volatile fatty acids (VFAs) production with different types of inoculum: Effect of pH. *Bioresource Technology, 161*, 395–401. https://doi.org/10.1016/j.biortech.2014.03.088.

Wang, L., Liu, Z., Jiang, H., & Mao, X. (2021). Biotechnology advances in β-carotene production by microorganisms. *Trends in Food Science & Technology, 111*, 322–332. https://doi.org/10.1016/j.tifs.2021.02.077.

Wu, C.-C., Honda, K., & Kazuhito, F. (2023). Current advances in alteration of fatty acid profile in *Rhodotorula toruloides*: A mini-review. *World Journal of Microbiology and Biotechnology, 39*(9), 234. https://doi.org/10.1007/s11274-023-03595-3.

Yen, H. W., & Zhang, Z. (2011). Enhancement of cell growth rate by light irradiation in the cultivation of *Rhodotorula glutinis*. *Bioresource Technology, 102*(19), 9279–9281. https://doi.org/10.1016/j.biortech.2011.06.062.

Zhang, Q., Lu, Y., Zhou, X., Wang, X., & Zhu, J. (2020). Effect of different vegetable wastes on the performance of volatile fatty acids production by anaerobic fermentation. *Science of the Total Environment, 748*, 142390. https://doi.org/10.1016/j.scitotenv.2020.142390.

Zhou, M., Yan, B., Wong, J. W. C., & Zhang, Y. (2018). Enhanced volatile fatty acids production from anaerobic fermentation of food waste: A mini-review focusing on acidogenic metabolic pathways. *Bioresource Technology, 248*, 68–78. https://doi.org/10.1016/j.biortech.2017.06.121.

The potential of photosynthetic biogas upgrading combined with biopolymers production by anoxygenic phototrophic bacteria

10

Roxana Ángeles[1,2,3,4], *André Freches*[3,4], *Yadira Rodríguez*[1,2,3,4], *Joana C. Fradinho*[3,4], *Maria A.M. Reis*[3,4], *Raúl Muñoz*[1,2], *and Raquel Lebrero*[1,2]
[1]Institute of Sustainable Processes, University of Valladolid, Valladolid, Spain, [2]Department of Chemical Engineering and Environmental Technology, University of Valladolid, Valladolid, Spain, [3]Associate Laboratory i4HB—Institute for Health and Bioeconomy, NOVA School of Science and Technology, NOVA University Lisbon, Caparica, Portugal, [4]UCIBIO—Applied Molecular Biosciences Unit, Department of Chemistry, NOVA School of Science and Technology, NOVA University Lisbon, Caparica, Portugal

10.1 Introduction: biogas and biomethane

Anaerobic digestion represents an effective platform for recovering carbon and nutrients contained in organic waste and wastewater, capable of producing biofertilizers, industrial heat, and electricity, biomethane for injection into natural gas networks or use as biofuel, or biogas to produce commodities and fine chemical products (Ángeles Torres et al., 2020). The number of biogas plants in Europe has increased rapidly in the last 15 years, from 7634 plants in 2009 to 19491 in operation at the end of 2022 (EBA, 2023). Primary energy production from biogas has also grown exponentially in the last decade in the EU27 (reaching an annual output of 15.8 million tons of oil equivalent, Mtoe, at the end of 2019). However, in recent years, this production has stabilized due to limitations on using energy crops in Germany and Austria, and the lower tax incentives and bonuses to produce electricity from biogas. Of these 15.8 Mtoe, 1.14 Mtoe were generated in wastewater treatment plants (WWTPs), 1.20 Mtoe in landfills, and 13.29 Mtoe in the treatment of urban, agricultural, livestock, and energy crop waste (EurObserv, 2023). By 2030, the European Biogas Association and the European Commission have estimated that biogas production will reach 40 Mtoe.

The number of biomethane production plants in the European Union has also increased, from 182 at the end of 2011 to 1548 in June 2024, which has led to an increase in biomethane production from 0.5 bcm in 2011 up to 6.4 bcm in June 2024 (EBA European Biogas Association, 2024; EBA, 2023). Most biogas (67%) and biomethane (64%) today in Europe are produced from agricultural waste or energy crops (EBA, 2023). The price of biomethane is largely determined by the treatment plant's capacity, varying from 12 € cents to produce 80 m^3 biomethane h^{-1} to 8 €

Eukaryotic Microorganisms as Sources of Bioproducts. DOI: https://doi.org/10.1016/B978-0-443-30188-9.00017-2

cents to produce 500 m^3 biomethane h^{-1}. A comprehensive purification of biogas to biomethane requires the elimination of CO_2, H_2S, siloxanes, and volatile organic compounds, among others, and represents approximately 30% of the cost of biomethane.

Today, a wide range of CO_2 removal technologies are available commercially: pressurized water scrubbers, chemical scrubbers, organic solvent scrubbers, pressure swing adsorption (PSA) systems, and membrane separation systems. In 2021, membrane separation units dominated the EU biomethane market (with a 51% share), followed by pressurized water scrubbers (14%), chemical scrubbers (12%), and PSA systems (10%). The electricity consumption of physicochemical technologies ranges between 0.2 and 0.3 kWh Nm^{-3}, except for chemical scrubbers which require an extra input of thermal energy of ~ 0.5 kWh Nm^{-3} for solvent regeneration and electricity consumption of 0.1–0.15 kWh Nm^{-3}. On the other hand, in recent years, several biological processes have been developed to eliminate CO_2 from biogas, which entail lower operating costs, lower environmental impacts, and the valorization of the CO_2 present in the biogas. The most relevant biotechnologies for upgrading biogas are photosynthetic purification, hydrogenotrophic, and electromethanogenesis processes (Severi et al., 2023).

In this context, photosynthetic biogas upgrading biotechnology by microalgae/cyanobacteria is based on CO_2 fixation via oxygenic photosynthesis, where the oxygen photosynthetically produced is utilized by sulfur-oxidizing bacteria to oxidize H_2S to sulfate (SO_4^{2-}). Photosynthetic biogas upgrading has rapidly gained ground due to its cost-effectiveness and as a sustainable alternative with large-scale potential. Nevertheless, this biotechnology presents challenges such as contamination of the upgraded biomethane with N_2 and O_2, the toxic effect of H_2S over microalgae when present at high concentrations in the raw biogas, the presence of microorganisms that may affect the production of microalgae, the high dependence on light, the footprint of this technology and the limitations in valorizing the algal biomass (Rodero et al., 2023) (Table 10.1).

As an alternative, anoxygenic phototrophic bacteria (APB), including phototrophic purple bacteria (PPB), purple sulfur bacteria (PSB), purple nonsulfur bacteria (PNSB), green sulfur bacteria (GSB), and green nonsulfur bacteria (GNSB), have been explored due to their unique metabolism capable of adapting to a variety of environments. In this sence, PPB have been repoted to grow without light using organic carbon as carbon source (Fradinho et al., 2021) (Table 10.1). APB are known for their ability to perform anoxygenic photosynthesis, which allows them to use light as an energy source without producing oxygen. They can fix CO_2 via the Calvin-Benson-Bassham (CBB) cycle or other carbon fixation pathways such as the reverse tricarboxylic acid (rTCA) cycle and the 3-Hydroxypropionate bicycle, incorporating it into organic compounds. This process can be linked to the synthesis of biopolymers, where fixed carbon is used as a building block. Additionally, APB can oxidize H_2S as an electron donor in their metabolic processes (Egger et al., 2020). Thus, using APB for biogas upgrading and subsequent biomass valorization into marketable compounds constitutes an interesting approach.

Table 10.1 Advantages and disadvantages of using microalgae and APB for photosynthetic biogas upgrading.

	Advantages	Disadvantages
Microalgae-bacteria	• Simultaneous removal of CO_2 **and H_2S using a microalgae-bacteria consortium**.	• Light dependency of microalgae cultivation. • Photosynthetic oxygen desorbed from the cultivation broth to the biomethane. • Toxic effects of high H_2S **concentration on microalgae**. • Limitations in valorizing algal biomass.
APB[a]	• Simultaneous removal of CO_2 **and H_2S by a single strain**. • Tolerance and adaptability to high H_2S **concentration**. • Enhanced photosynthetic efficiencies compared to microalgae. • Ability to grow without light using organic compounds as energy and carbon sources through chemotrophy under aerobic and anoxic/anaerobic conditions (strain-dependent). • Potential production of hydrogen, single-cell proteins, fertilizers, pigments, polyhydroxyalkanoates (PHA), co-enzyme Q10, and 5-aminolevulinic acid.	• Limited knowledge of their application to biogas upgrading. • Additional cost of precursor substrates. • Limited substrate yield efficiency. • Limited volumetric productivity. • Costly downstream processing.

[a] Including *PPB*, phototrophic purple bacteria; *PSB*, purple sulfur bacteria; *PNSB*, purple nonsulfur bacteria; *GSB*, green sulfur bacteria; *GNSB*, green nonsulfur bacteria.

This chapter explores the potential of APB as a promising biotechnology for upgrading biogas to biomethane, emphasizing the critical factors involved in synthesizing valuable bioproducts, including biofertilizers, single-cell proteins (SCPs), pigments, and biopolymers like PHA. Additionally, it covers the fundamentals of photosynthetic biogas upgrading and the essential elements for advancing this biotechnology in conjunction with biopolymer production with APB, which offers a sustainable alternative to conventional plastics and enhances the overall cost-effectiveness of biogas upgrading technologies.

10.2 Emerging biotechnologies for biogas upgrading to biomethane

Photosynthetic biogas upgrading is based on the symbiotic action of microalgae and bacteria. The CO_2 and H_2S from biogas are selectively absorbed into the photobioreactor (PBR) culture broth due to the high operational pH and buffer capacity. CO_2 is photosynthetically fixed in the form of algal biomass in the presence of light, releasing oxygen that is used by lithoautotrophic bacteria to carry out the biological oxidation of H_2S to sulfate (Severi et al., 2023). The nutrients necessary to support the growth of microalgae and bacteria in the PBR (typically raceways) can be supplied by the liquid fraction of the digestate, with the subsequent treatment of this effluent from the anaerobic digestion. This technology has been optimized on an outdoors pilot scale at the University of Valladolid by members of the Institute of Sustainable Processes within the H2020 INCOVER project, obtaining biomethane concentrations close to 95% *v/v* in spring, summer, and autumn, and 92% *v/v* in winter. Subsequently, the process was validated at a demo scale at El Torno WWTP in Chiclana de la Frontera, where methane concentrations of 95% *v/v* in the treated biogas were consistently reached. Within the BBI-JU-H2020 URBIOFIN project, the technology was demonstrated in the context of the anaerobic treatment of urban waste in a 272 m^2 PBR in Zaragoza (Spain). A comparative analysis of this photosynthetic upgrading technology versus a conventional process based on an activated carbon filter for the removal of H_2S followed by a pressurized water scrubber for the removal of CO_2 revealed that the photosynthetic process can reduce electricity consumption from 0.2 kWh Nm^{-3} for the physical-chemical technology to 0.08 kWh Nm^{-3} for the case of photosynthetic upgrading. This led to a decrease in biogas purification costs from 0.2 € Nm^{-3} for the conventional system to 0.03 € Nm^{-3} for the photosynthetic system (Toledo-Cervantes et al., 2017).

Hydrogenotrophic biogas purification is carried out by the action of hydrogenotrophic archaea capable of enzymatically reducing CO_2 to CH_4 using H_2 as an electron donor. In this process, 1 mol of CO_2 (used as an electron acceptor and carbon source by archaea) is reduced with four moles of H_2, which ideally should be produced from renewable energy. This technology can be carried out in situ, by injecting H_2 into the anaerobic digester, or ex situ, in an external reactor enriched with hydrogenotrophic archaea. The in situ pressurized biogas upgrading can generate biomethane with a methane content of 90% *v/v* at 3 bars. In situ upgrading can be also carried out supported by an internal biogas recirculation, using diffusion membranes to achieve gas-liquid H_2 transfer coefficients that allow an efficient use of the H_2 supplied (Fachal-Suárez et al., 2024). The fact that anaerobic digesters are not designed to facilitate gas–liquid transport, together with the increase in the pH of the anaerobic broth inherent to the consumption of CO_2, has made ex situ upgrading the most popular configuration. This upgrading can be carried out in bubble columns with fine bubble diffusers (and internal biogas recirculation) or in trickling biofilters. The latter allows obtaining CH_4 concentrations of 98% *v/v* with different concentrations of starting biogas and methanation capacities of 10–12 $Nm^3\ m^{-3}\ d^{-1}$ (Ghofrani-Isfahani et al., 2021).

Finally, **electromethanogenesis** represents an innovative technology with high potential, which requires significant improvements to reduce its current operating costs (0.8 € Nm^{-3}). This technology uses electricity produced from renewable sources to generate H_2 at the cathode of an electrochemical cell supplied with raw biogas. The high operating pH of the cathode entails a partial purification of the biogas, which leaves the cathode enriched in H_2 and CH_4, and with a low CO_2 content (ideally the $H_2{:}CO_2$ ratio of this biogas leaving the cathode is 4). The partially purified biogas is taken to a methanization reactor with hydrogenotrophic archaea similar to the one described in the previous paragraph, which allows CH_4 concentrations greater than 98% *v/v*. The electrochemical cell where the biogas purification is carried out is built with an anionic membrane, which allows the diffusion of $HCO_3^-/CO_3^{2-}/HS^-/S^{2-}$ ions to maintain the charge balance between anode and cathode during electrolysis. This transfer of carbonate and bicarbonate anions from the cathode to the anode keeps the pH of the cathode high and facilitates the absorption of CO_2 (Verbeeck et al., 2019).

10.2.1 Principles of microalgal photosynthetic biogas upgrading

Photosynthetic biogas upgrading is supported by the symbiotic action of microalgae and bacteria in PBRs (Fig. 10.1). Photosynthetic microorganisms such as microalgae and cyanobacteria use solar energy to fix CO_2 present in raw biogas via the CBB cycle during oxygenic photosynthesis, producing O_2, which is used by sulfur-oxidizing bacteria to oxidize H_2S to SO_4^{2-} (Ángeles Torres et al., 2020). The CBB cycle occurs in the chloroplasts of microalgae and the plasmalemma membrane of cyanobacteria. This pathway consists of 13 reactions catalyzed by 11 enzymes, with ribulose-1,5-bisphosphate carboxylase/oxygenase (RuBisCO) serving as the primary enzyme responsible for CO_2 fixation (Severo et al., 2019).

Before biological conversion, CO_2 and H_2S gases contained in the raw biogas need to be transferred to the liquid broth. To this aim, the raw biogas is fed to a scrubbing column where both gases are absorbed into the liquid broth that is recirculated from the PBR. In addition, a settler is placed between the PBR and the scrubbing column to control biomass concentration and productivity. Photosynthetic biogas upgrading technology has been commonly implemented in open PBR configurations, such as raceway ponds, due to their simple design, ease of construction and operation, as well as their low energy requirements (Bahr et al., 2014). These factors contribute to reduced overall costs Several authors have demonstrated the effectiveness of this

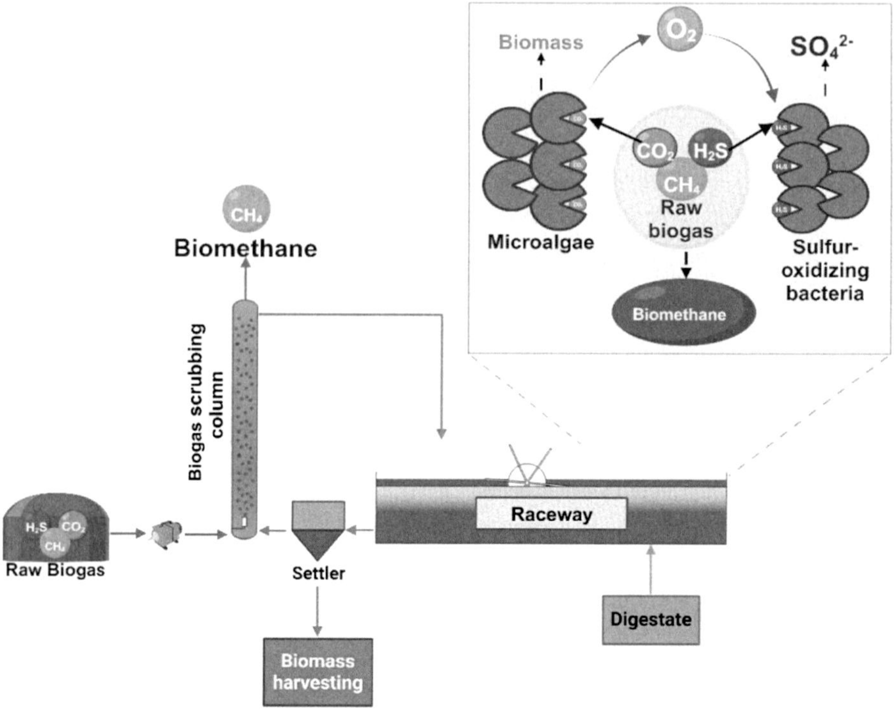

Figure 10.1 Schematic representation of a photosynthetic biogas upgrading system consisting of a raceway pond interconnected to a biogas scrubbing column.
This figure is a new imagen. Figure was created with BioRender.com.

configuration in removing CO_2 and H_2S simultaneously, both indoors and outdoors, in pilot and demo plants, accomplishing biomethane composition of CH_4 > 90% *v/v*, CO_2 ≈ 0.6–12% *v/v*, H_2S=0% *v/v*, N_2 ≈ 0.7–9% *v/v*, and O_2 ≈ 0–2% *v/v* (Franco-Morgado et al., 2017; Marín et al., 2018; Rodero et al., 2020). The operating parameters of the scrubbing column and the cultivation broth, as well as environmental factors and the microalgae population, can significantly influence the quality of thc biomethane.

10.2.1.1 Operating parameters of the scrubbing column

The external biogas scrubbing column facilitates the transfer of CO_2 and H_2S from the gas phase into the cultivation broth. Operating the system at a liquid-to-biogas flow rate (L/G) ratio of less than 1 is crucial for enhancing the removal efficiency of CO_2 and H_2S due to their higher aqueous solubility according to Henry's Law constant (H_{CO_2} ≈ 0.83 and H_{H_2S} ≈ 2.45 at 25°C) compared to CH_4 (H_{CH_4} ≈ 0.03 at 25°C) (Toledo-Cervantes et al., 2017). To limit N_2 and O_2 desorption, previous studies have optimized the process using a hollow fiber polydimethylsiloxane

(PDMS) gas–liquid membrane to degasification the cultivation broth, where a degassing performance at 100–300 of vacuum pressure was demonstrated with dissolved oxygen removal in the cultivation broth of ~62% resulting a decrease of N_2 and O_2 concentration into the biomethane to ~0.8 and ~0.1% *v/v*, respectively (Ángeles et al., 2020).

10.2.1.2 Operating parameters of the cultivation broth

The absorption of both acidic gases, CO_2 and H_2S, is enhanced at pH greater than 9 and high alkalinity (> 1000 mg IC L^{-1}) in the cultivation broth, which promotes higher mass transfer rates. H_2S is absorbed and oxidized in the biogas scrubbing column using the photosynthetic O_2 dissolved in the cultivation broth recirculated from the PBR (10.1). In contrast, CO_2 is photosynthetically fixed in the PBR (10.2) (Muñoz et al., 2015). In this sense, the CO_2 absorbed in the highly alkaline liquid broth of the biogas scrubbing column is converted to bicarbonate (HCO_3^-, (10.3)). The bicarbonate is then incorporated into the cells as CO_2 by the action of the enzyme carbonic anhydrase (10.4), raising the pH and regenerating carbonate (CO_3^{-2}) according to bicarbonate-carbonate equilibrium (10.5) (Bose et al., 2019).

$$H_2S + 2O_2 \rightarrow SO_4^{2-} + H^+ \tag{10.1}$$

$$CO_2 + \text{solar energy} \rightarrow O_2 + \text{biomass} \tag{10.2}$$

$$CO_2 + H_2O + CO_3^{2-} \rightarrow 2HCO_3^- \tag{10.3}$$

$$HCO_3^- \xrightarrow{\text{enzyme carbonic anhydrase}} CO_2 + OH_3^- \tag{10.4}$$

$$HCO_3^- + OH^- \rightarrow CO_3^{2-} + H_2O \tag{10.5}$$

10.2.1.3 Environmental Parameters

Microalgae and cyanobacteria are photosynthetic microorganisms that rely on light for growth. While light positively influences photosynthesis, this effect plateaus at a saturation irradiance of ~200 μmol m^{-2} s^{-1} for most microalgal species. Beyond this point, photosynthesis is no longer promoted with light intensity and might be even inhibited at higher values (Carvalho et al., 2011). Moreover, during biogas upgrading with microalgae, photosynthesis ceases in the absence of light (dark period), leading to CO_2 accumulation that decreases the pH of the cultivation broth. This drop in pH, along with the build-up of H_2S, inhibits microalgae growth (Ángeles et al., 2021).

For most mesophilic microalgae species, the optimal growth temperature ranges between 15 to 30 °C, though this depends on each species. Nevertheless, in the context of photosynthetic biogas upgrading at IC concentration > 500 mg IC L^{-1}, it has been demonstrated that the temperature has a minimal impact on the biomethane quality (Rodero et al., 2018).

10.2.1.4 Microalgae population

The most common microalgae and cyanobacteria species found during photosynthetic biogas upgrading are *Pseudanabaena* sp., *Chlorellaminutissima*, *Chlorella vulgaris*, *Chlorellahomosphaera*, *Spirullina*, *Nannochloropsis oleoabundans*, *Scenedesmus* sp., and *Scendesmus obliquus* (Ouyang et al., 2015; Prandini et al., 2016; Zhao et al., 2015). These microorganisms exhibit high tolerance to high alkalinity and pH values and can grow in environments with CO_2 contents ranging from 0.03 up to 30% *v/v*.

The valorization of microalgal biomass during photosynthetic biogas upgrading has been wildely demonstrated. For instance, its use as a biofertilizer, due to its high content of biostimulants and biopesticides, has been proven by Toledo-Cercantes et al. (2017). Kikuchi et al. (2024) also highlighted the potential applications in the nutraceutical, food, livestock, and aquaculture industries of a microalgal consortium of *Micractinium*, *Chlorella*, *Scenedesmus*, and *Desmodesmus* sp. during biogas upgrading. Furthermore, a microalgal-cyanobacteria consortium, dominated by *Pseudanabaena* sp. and characterized by a high carbohydrate content, was reported by Ángeles et al. (2020) as a potential feedstock for bioethanol production when nitrogen feast famine cycles were applied during biogas upgrading. Nevertheless, despite its potential use, microalgal biomass productivity and composition are highly dependent on several factors such as pH, environmental temperature, light availability, dissolved oxygen in the cultivation broth, herbivorous predators and pathogens, hydraulic retention time (HRT), nutrient availability, and the composition of the microalgal population (Narayanan, 2024).

10.3 The potential of anoxygenic phototrophic bacteria as a platform for biogas upgrading to biomethane

10.3.1 CO_2 and H_2S removal by APB

In the context of photosynthetic biogas upgrading, APB can utilize different metabolic pathways to consume compounds like CO_2 and H_2S. These pathways are integral to processes like anoxygenic photosynthesis and biogas purification and include the CBB cycle, the rTCA cycle, the 3-Hydroxypropionate bicycle, and the anoxygenic photoautotrophic sulfide oxidation (Fig. 10.2).

The CBB cycle was a groundbreaking discovery, marking the first reported mechanism for carbon fixation (Li et al., 2020). The CBB cycle is prevalent in various organisms, including algae, plants, cyanobacteria, and purple phototrophic bacteria, playing a vital role in CO_2 fixation within microorganisms (Jensen et al., 2017). Key enzymes in the CBB pathway are ribulose-1,5-bisphosphate carboxylase/oxygenase (RuBisCO), phosphoribulokinase (PRK), and glyceraldehyde-3-phosphate dehydrogenase (GAPDH) (Yu et al., 2020). In the CBB cycle, three molecules of CO_2 are assimilated by RuBisCO, ultimately resulting in the formation of one molecule of glyceraldehyde 3-phosphate. This entire process involves a series of 29

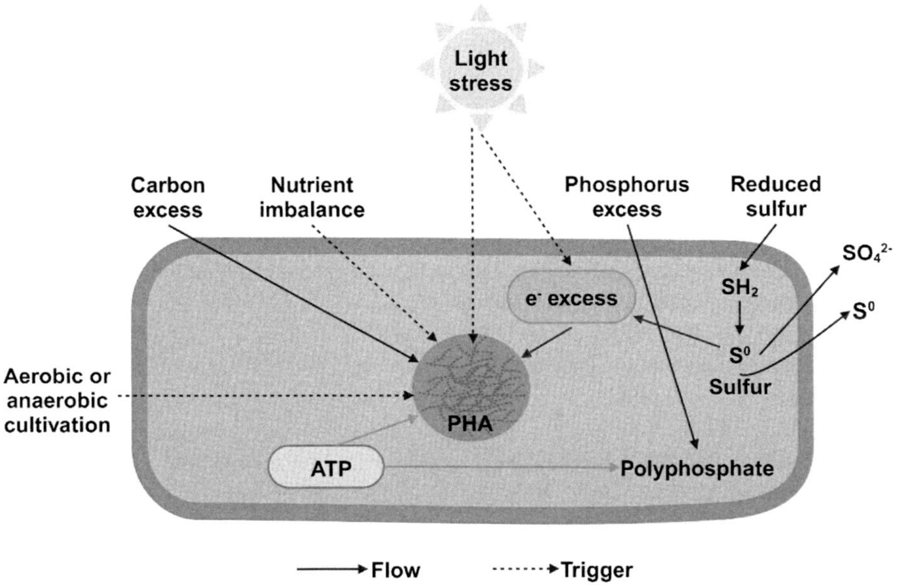

Figure 10.2 Schematic representation of APB metabolism that supports biopolymer accumulation.
Source: Modified from, Fradinho, J., Allegue, L. D., Ventura, M., Melero, J. A., Reis, M. A. M. & Puyol, D. (2021). Up-scale challenges on biopolymer production from waste streams by Purple Phototrophic Bacteria mixed cultures: A critical review. *Bioresource Technology, 327*. https://doi.org/10.1016/j.biortech.2021.124820. Figure was created with BioRender.com.

reactions catalyzed by 13 enzymes. RuBisCO plays a pivotal role in driving the conversion of atmospheric CO_2 into carbohydrates, a crucial step that supplies a source of carbohydrates to a wide range of organisms (Sharkey, 2019).

The discovery of the second carbon-fixation pathway occurred in 1966, attributed to the work of Evans and their colleagues (1966). This pathway was unraveled in *Chlorobium limicola*, a green sulfur bacterium. Interestingly, experiments with fermentative bacteria revealed that ferredoxin could serve as a direct electron donor for the synthesis of pyruvate and alpha-ketoglutarate from CO_2 and reduced ferredoxin, alongside acetyl-CoA or succinyl-CoA, through an irreversible pathway (Lo et al., 2021). This pathway became known as the rTCA or the Arnon–Buchanan cycle, and it is considered a highly plausible candidate for the earliest autotrophic metabolism (Rubin-Blum et al., 2019). In the rTCA cycle, the primary product of carbon fixation is acetyl-CoA. This acetyl-CoA is a crucial building block that is further transformed into various central intermediates in carbon metabolism. These intermediates include pyruvate/phosphoenolpyruvate (PEP), oxaloacetate, and 2-oxoglutarate (Kitadai et al., 2017). The conversion of acetyl-CoA into pyruvate is facilitated by ferredoxin-dependent pyruvate synthase, and pyruvate itself can also be converted into PEP. Additionally, oxaloacetate is synthesized through the actions of pyruvate or PEP carboxylase (Evans et al., 1966). Notably, the enzymes involved in this cycle are sensitive to oxygen. Consequently, the rTCA cycle is predominantly observed in anaerobic organisms or facultative anaerobes that can tolerate lower oxygen concentrations than atmospheric levels (Steffens et al., 2021).

The 3-hydroxypropionate bicycle, also known as the Fuchs–Holo route, was first identified in *Chloroflexus aurantiacus*, a thermophilic green nonsulfur bacterium (Holo, 1989). This pathway is distinctive in that it utilizes three molecules of bicarbonate (HCO_3^-) to synthesize one molecule of pyruvate, through a series of reactions divided into two cycles. In the first cycle, glyoxylate is generated by carboxylating acetyl-CoA, resulting in the production of malonyl-CoA. This malonyl-CoA is then reduced to proprionyl-CoA through a hydroxypropionate intermediate. Subsequent carboxylation of proprionyl-CoA leads to the formation of succinyl-CoA. Succinyl-CoA is further converted to (S)-malyl-CoA, which regenerates acetyl-CoA, the initial starting molecule, while releasing glyoxylate as the first carbon fixation product. The second cycle begins with the assimilation of glyoxylate, transforming it into methylmalyl-CoA. This compound is then converted to citramalyl-CoA through mesaconyl-CoA. Finally, citramalyl-CoA is further converted into pyruvate and acetyl-CoA, effectively regenerating the precursors of the cycle (Mattozzi et al., 2013). Notably, this carbon-fixation pathway appears to be exclusive to a specific clade of photosynthetic *Chloroflexus* bacteria (Garritano et al., 2022). *Chloroflexus aurantiacus* utilizes the 3-hydroxypropionate bicycle in conjunction with the glyoxylate cycle to metabolize various organic substrates like glycolate, acetate, propionate, 3-hydroxypropionate, lactate, butyrate, or succinate for its central carbon metabolism, both during autotrophic and heterotrophic growth conditions (Klatt et al., 2013).

Under the same perspective, PSB and GSB are responsible for the photoautotrophic oxidation of reduced sulfur compounds (Pokorna & Zabranska, 2015). These organisms not only fix CO_2 but also utilize light energy to oxidize reduced sulfur compounds (Pfennig, 1975). The reaction mechanisms for the biological desulfurization process, performed by phototropic bacteria to remove H_2S, can be described by (10.6), which also involves the formation of elemental sulfur. However, in the presence of low sulfide concentrations and high constant light intensity, the reaction may shift according to (10.7), ultimately leading to the production of sulfates (Findlay et al., 2015; Hamilton et al., 2014).

$$2H_2S + CO_2 + \xrightarrow{\text{light energy}} 2S^0 + CH_2O + H_2O \quad (10.6)$$

$$H_2S + 2CO_2 + 3H_2O \xrightarrow{\text{light energy}} SO_4^{2-} + 2H^+ + 2CH_2O \quad (10.7)$$

Overall, APB presents several metabolic pathways suitable for biogas upgrading, with mechanisms for CO_2 and H_2S uptake. These processes typically derive energy from light (in the case of phototrophic bacteria) or from the oxidation of inorganic compounds. In anoxygenic photosynthesis, light energy drives the electron transport chain, generating a proton motive force for ATP synthesis, which is then used for various metabolic processes. Although the research on APB's ability to assimilate

Table 10.2 Investigations on CO_2 and H_2S removal using APB.

APB strain	Photobioreactor configuration	Working volume (L)	Gas composition (%*v*/*v*)	Removal capacity	References
Chromatiaceae sp. (PSB)dominant PPB mixed culture	Bubble column	2	2000/5000 ppm_V H_2S 30% CO_2 CH_4 balance	243.12 mg-S L^{-1} d^{-1}	Egger et al. (2023).
Chromatiaceae sp. (PSB)dominant PPB mixed culture	Serum bottles	0.5	[a]150 mg-S L^{-1}	42.96 mg-S L^{-1} d^{-1}	Egger et al. (2020).
Allochromatium vinosum (GSB)	Gastight glass bottles	0.27	1% H_2S 1.5% H_2S 2% H_2S	2.4 mg-S L^{-1} d^{-1} 3.5 mg-S L^{-1} d^{-1} 4.6 mg-S L^{-1} d^{-1}	Struk et al. (2023).
Chlorobium limicola (PSB)	Enclosed glass cylinder	2.3	0.5% H_2S 29.5% CO_2 70% CH_4	30.6 mg-S L^{-1} d^{-1}	Struk et al. (2023).
Rhodopseudomonas sp. (PNSB) dominant species	Cylindrical column	3.6	2.5% $CO_2$97.5% N_2	0.22 mg-CO_2 L^{-1} d^{-1}	Zhang et al. (2024)
PPB mixed culture	Gastight glass bottles	0.4	0.5% H_2S 29.5% CO_2 70% CH_4	89 % CO_2 removal efficiency 71 % H_2S removal efficiency	Marín et al. (2019)
Chloroflexus aurantiacus Y-400-fl (GNSB)	Stirred tank	1	[a]45 mg-S L^{-1}	18.35 mg-S L^{-1} d^{-1}	Ángeles et al. (2024)

[a] Sulfide was added in the medium as Na_2SNo Permission Required

CO_2 and H_2S is less extensive compared to that on microalgae and cyanobacteria, there is growing interest in this process, as summarized in Table 10.2.

10.3.2 Bioproducts derived from APB cultivation

10.3.2.1 Biofertilizers

APB, particularly PPB, has recently gained attention as a potential alternative to conventional fertilizers due to its useful traits and long-term benefits for plant growth (Sakarika et al., 2020). APB-derived biofertilizers can be classified based on their form and functionality: (1) dead cells—biomass powder as a direct source of nutrients, (2) living microorganisms—active cells converting nutrients into plant-available forms, and (3) cell-derived products—growth promoters/biostimulants present in the supernatant (Capson-Tojo et al., 2020).

Direct application of APB dried biomass (i.e., direct fertilization) possesses great potential due to its nitrogen/phosphorous/potassium (N/P/K) content and its slow-release pattern (nutrients are rendered to the plant by the microbiome present in the soil) (Sakarika et al., 2020b). Particularly, phosphorous content in *Rhodobacter* sp. [8.5/2.4/0.5% in dried weight (DW)] was found to be substantially higher than that present in microalgae (8.6/0.3/0.7% in DW) (Spanoghe et al., 2020), which is attributed to the polyphosphate accumulating capacity of PNSB (Lai et al., 2017). Despite limited application and a lack of systematic studies (e.g., reference fertilizer choice, growth medium, individual or microbial blends, and many more), only a few studies have demonstrated the beneficial use of APB as a microbial fertilizer. For instance, the use of autoclaved *Rhodopseudomonas* sp. BL6 and KL6 supported an increase of shoot and root dry weight as well as tomato fruit yield (Lee et al., 2008). However, no positive effects were observed in Chinese cabbage using *Rhodopseudomonaspalustris* PS3, YSC3, and YSC4 (Wong et al., 2014). Recent studies suggest that PNSB performs comparably to reference organic fertilizers in parsley or tomato growth (Wambacq et al., 2022). Nevertheless, these studies only examined PNSB blended with microalgae and aerobic heterotrophic bacteria, with PNSB contents ranging from 5% to 15%.

The use of APB in its living form enables indirect fertilization, as it enhances nutrient availability through biological nitrogen fixation and phosphate solubilization (Sakarika et al., 2020). The N_2-fixing capability belongs to most of APB. It can be mediated by Mo-Fe, V-Fe, and Fe-Fe nitrogenase enzymes, which convert atmospheric N_2 into its reduced form (NH_3) under light anaerobic conditions (Olivares et al., 2013). During the last decade, the use of PNSB for nitrogen uptake has been particularly investigated in paddy soils for rice production (Nookongbut et al., 2018). For instance, an increase of 50%–65% of the nitrogen content in the roots of rice plants was observed by inoculating hydroponic solutions with *Rhodobacter capsulatus* Kb1. Additionally, seed coating using the same strain increased by 9.2% the nitrogen content of rice straw (Gamal-Eldin & Elbanna, 2011). Other crops, such as pak choi and lettuce, demonstrated enhanced nitrogen efficiencies when inoculated with *R. palustris* PS3 (Hsu, 2015). Similarly, APB offers a bio-based solution for the solubilization of inorganic phosphate complexes. Phosphorus availability in soils is usually poor as it is immobilized in soils as aluminum, calcium, or iron phosphates (Rawat et al., 2022). APB achieves solubilization through mechanisms such as the secretion of siderophores (chelating agents), organic acids, and protons; or the excretion of extracellular enzymes, improving soil fertility and further increasing plant nutrient uptake (Khuong et al., 2017; Sakarika et al., 2020) found that a PNSB consortium of *R. palustris* strains dissolved strengite ($FePO_4 \cdot 2H_2O$) and variscite ($AlPO_4 \cdot 2H_2O$) in rice trials, increasing grain yield by 34% at a P_2O_5 concentration of 45 kg ha^{-1}, with the highest P uptake and lowest levels of Al and Fe.

Besides their fertilizing functionality, APB produces active compounds that are involved in several regulatory mechanisms in plants, commonly referred to as plant growth-promoting substances (PGPSs): 5-aminolevulinic acid (ALA), indole-3-acetic acid (IAA), or lipopolysaccharide (LPS) (Iwai et al., 2022). ALA is an intermediate involved in the biosynthesis of tetrapyrrole compounds, serving as the precursor for the production of chlorophylls, vitamin B_{12}, and bacteriochlorophylls (BChls) through the C_4 pathway in APB (Sakpirom et al., 2017) and increasing their resistance to abiotic stressors such as salinity, drought, UV, low-light or extreme temperatures (Sun et al., 2009). It also plays an essential role as a natural pesticide, herbicide, or fungicide (>800 mg L^{-1}) (Jiang et al., 2022; Su et al., 2017). *R. palustris* TN114 and PP803 were identified as the excellent ALA producers (25.6 and 21.1 μM, respectively). The former tolerated up to 6% NaCl, with optimal ALA production of 3.4 mg L^{-1} when cultivated at 0.25% NaCl, indicating its potential for crop cultivation in saline conditions. IAA, an auxin hormone crucial for plant growth and root activation, can inhibit plant growth at high concentrations (Sakarika et al., 2020; Sakpirom et al., 2017).

10.3.2.2 Single-cell protein

The term SCP, also known as microbial protein, refers to high-quality (or food-grade) protein derived from microbial biomass (i.e., bacteria, algae, yeast, or fungi) (Matassa et al., 2016). The production of "cultured" meat offers a sustainable solution to alleviate the rising global pressure for protein demand, which is predicted to double from the current 202 million tons by 2050. Indeed, a socio-economic analysis revealed that SCP could substitute 10%–19% of animal-based and plant-based protein by 2050 (Henchion et al., 2017). Conventional SCP production relies on the use of natural gas or food grade substrates (classified as first generation SCP despite being land-independent), with commercial products available in the market for animal feed such as FeedKind® and UniProtein® (Ritala et al., 2017). Current research has been directed to the development of second generation processes, in which SCP production can be integrated into resource recovery and carbon-fixation technologies, including biogas upgrading (Khoshnevisan et al., 2022).

In this context, APB is already well-known for being an SCP mediator (with protein contents up to 80%) and a promising nexus of resource recovery and food security (Hülsen et al., 2018). Besides being regarded as proteinaceous biomass, APB are also characterized by their balanced essential amino acid profile (particularly for their sulfur-containing methionine and cysteine content) and other functional compounds such as phospholipids, vitamins, fatty acids, ubiquinone-10 (CoQ10), and photopigments, being commonly tested as low-inclusion nutritional supplement in aquaculture (Saejung & Ampornpat, 2019). A recent study demonstrated that cultures grown in real effluents with *Rhodoseudomonas* and *Rhodobacter* as dominant species had comparable or even higher crude protein contents compared to pure PPB cultures axenically cultured on synthetic media (Capson-Tojo et al., 2020).

10.3.2.3 Pigments

BChls and carotenoids are natural pigments present in APB, both serving as active components of the light-harvesting antenna complexes for CO_2 fixation in APB (Kuo et al., 2012). BChls absorb sunlight in the near-infrared region (>700 nm) while carotenoids absorb light energy between 400 and 550 nm (Krayer et al., 2011; Mandal et al., 2020). BChl *a* is found in most APB, while BChl *b* is characteristic of PNSB, and BChl *c*, *d*, and *e* are present in GSB and GNSB. The light-harvesting complexes in APB are the light-harvesting ring 1, light-harvesting 1-like structures, light-harvesting ring 2, and chlorosomes (Bryant & Frigaard, 2006; Saer & Blankenship, 2017). However, the specific light-harvesting antennas depend on the type of microorganism, for example, GNSB and GSB have chlorosomes, whereas PPB poses light-harvesting ring 1.

Carotenoids play an essential role in photosynthesis by broadening the range of light absorption and protecting the photosynthetic machinery from excess light (Novoveská et al., 2019). They are classified into carotenes and xanthophylls based on their chemical composition. The synthesis of carotenoids by various APBs is influenced by their specific biochemical pathways and cultivation conditions. For example, β-carotene is present in *Chloroflexus*, *Oscillochloris*, and *Chloronema*, whereas *Roseiflexus castenholzii* and *Heliothrix oregonensis* contain solely BChl *a*(Freches & Fradinho, 2024).

BChls and carotenoids have many benefits such as antioxidants, immune system stimulants, and antiinflammatory (Pucelik et al., 2020). Thus, strategies have been developed to enhance pigment production in APB, for example, adjusting the light intensity, wavelength, and photoperiod or the supplementation of Mg^{2+} and salt (Liu et al., 2019; Spanoghe et al., 2021). For example, the PNSB *R. palustris* 42OL exhibited high pigment accumulation as carotenoids and BChl *a* under low light intensities of 250 μmol photons m^{-2} s^{-1} compared to high light intensities of 1500 μmol photons m^{-2} s^{-1} (Muzziotti et al., 2017). A recent investigation was carried out on the pigment production capabilities of *Chloroflexus aurantiacus* Y-400-fl in a stirred-tank photobioreactor under continuous sulfide supplementation, resulting in an accumulation of total carotenoids and BChl *c* up to 2.2 and 2.9% dcw, respectively, at a volumetric light intensity of ~3.6 W L^{-1} (Ángeles et al., 2024).

10.3.2.4 Biopolymers

APB constitutes a promising platform for producing biopolymers. One of the primary biopolymers synthesized by APB is PHA, a family of biodegradable plastics (Higuchi-Takeuchi et al., 2016). APB can accumulate PHA under certain conditions, such as nutrient limitation (e.g., nitrogen or phosphorus) coupled with an excess of carbon sources (Foong et al., 2019) (Table 10.3). Besides PHA, APB can also synthesize glycogen and other storage polymers (polyphosphate), serving as energy reserves (Fradinho et al., 2021). While pure cultures are often employed for PHA production, they come with the added cost and complexity of maintaining sterility. An alternative approach is the use of mixed cultures, specifically PPB-enriched cultures, which offer high adaptability and the ability to efficiently degrade waste substrates. Despite these mixed cultures can be easily obtained in wastewater treatment plants and can accumulate similar amounts of PHA as pure cultures, there is still limited information on their performance (Almeida et al., 2024).

The production of biopolymers such as PHA by PPB and phototrophic mixed cultures can be coupled with the utilization of waste gases like CO_2 and H_2S. This approach not only helps in waste mitigation but also contributes to the circular economy and carbon neutrality goals. The field of biopolymer synthesis using PPB and phototrophic mixed cultures is still evolving, with ongoing research focused on optimizing conditions for higher yield, understanding the metabolic pathways involved, and scaling up the process for industrial applications (Martin-Gamboa et al., 2023; Srisawat et al., 2022). This approach holds promise for sustainable production of biopolymers, leveraging the unique metabolic abilities of PPB and the synergistic potential of mixed microbial cultures.

10.3.3 Key factors for the development of biogas upgrading bioprocess combined with biopolymer production by APB

As is well known, APB has metabolic versatility, however, light intensity, nutrient concentration, pH, temperature, and aerobic and anaerobic cultivation can affect biopolymer production (Fig. 10.3).

10.3.3.1 Light intensity

Light intensity and quality are crucial for APB's growth and metabolic activities, as they rely on light for photosynthesis. The intensity and wavelength of light significantly influence these processes, affecting energy provision and the types of synthesized biopolymers. High light intensities have been found to stimulate PHA production by APB, by helping dissipate excess reducing power and protecting cells from photooxidative stress, with PHA synthesis serving as an electron sink in this context (Higuchi-Takeuchi & Numata, 2019). For instance, high PHB accumulations of up to 30% dcw have been reported under high light intensities using low acetate concentration (<30 CmM) in phototrophic mixed culture enriched in PPB (Fradinho et al., 2019). Therefore, increasing light intensity can significantly enhance substrate consumption and PHB production rates, potentially offering economic advantages for outdoor biopolymer production processes.

10.3.3.2 Nutrients

Nutrient limitation, particularly nitrogen, can trigger the accumulation of biopolymers like PHA as a storage compound (Fig. 10.3). Optimizing bacterial growth and biopolymer synthesis requires finding the right balance of macronutrients and

Table 10.3 PHA production by APB.

APB strain	Photobioreactor configuration	Working volume (L)	PHA content (%)	PHA productivity (mg $PHAL^{-1}$ d^{-1})	References
PNSB mixed culture	Stirred tank	0.7	20% wt*	7	Luongo et al. (2017)
PPB mixed culture	Sequential biological photobioreactor	1.4	30% VSS	490	Fradinho et al. (2019)
PNSB mixed culture	Serum bottle	0.4	14% wt[a]	36	Policastro et al. (2020)
Consortium of microalgae and PPB	Stirred tank	0.5	30.8% VSS	2670	Almeida et al. (2021)
PPB mixed culture	Serum bottle	0.16	21% wt	-	Allegue et al. (2021)
PPB mixed culture	Membrane photobioreactor,	2	42% wt	-	Allegue et al. (2022)
PPB mixed culture	High-rate algal pond	20000	36% VSS	140	Almeida et al. (2024)
PPB mixed culture	Stirred tank	10	33% wt	-	Srivastava et al. (2024)

[a] Polyhydroxybutyrate (PHB) content

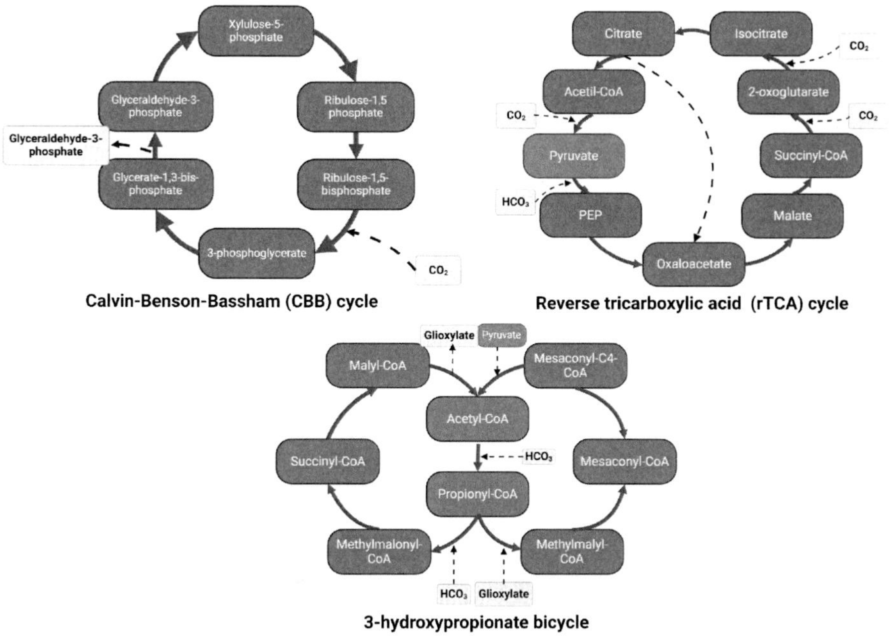

Figure 10.3 Metabolic CO_2 fixation pathways via anoxygenic photosynthesis.
Source: Modified from, Santos Correa, Sulamita, Schultz, Junia, Lauersen, Kyle J. & Soares Rosado, Alexandre. (2023). Natural carbon fixation and advances in synthetic engineering for redesigning and creating new fixation pathways. *Journal of Advanced Research*, *47*, 75–92. https://doi.org/10.1016/j.jare.2022.07.011. Figure was created with https://www.BioRender.com.

micronutrients such as carbon, phosphorous, sulfur, and magnesium have been found to lead to higher PHA production (Monroy & Buitrón, 2020). The typical approach to induce PHA accumulation involves a feast/famine regime; however, this operating mode is not always practical or economically convenient (Koller, 2018). However, the effect of these nutrients on PHB synthesis can vary depending on the specific growing conditions and the strain used. For example, PHB production by marine PPB under aerobic conditions was favored in a growth medium rich in phosphorus and nitrogen (Higuchi-Takeuchi & Numata, 2019). This illustrates the importance of considering different bacterial strains' unique requirements and responses when optimizing PHA production processes.

10.3.3.3 pH and temperature

Another important parameter is the pH of the growth medium, which affects the enzyme activities and overall metabolism of APB, with each strain of APB presenting an optimal pH range. An optimal temperature range is also necessary for efficient biopolymer synthesis and to promote bacterial growth. The impact of temperature, pH, and light irradiance on biomass production by enriched PPB cultures was evaluated, obtaining the highest biomass productivity at >30 W m^{-2}, pH7, and 25°C, compared to 15°C and ≥38°C, where the biomass productivity decreased (Rodero et al., 2024). To improve their tolerance to adverse environmental conditions, APB might be genetically manipulated, also enhancing their capability to synthesize specific biopolymers (Vivek et al., 2021).

10.3.3.4 Aerobic and anaerobic conditions

APB can accumulate PHA under aerobic and anaerobic conditions, specifically, PNSB can synthesize PHAs under both conditions (Montiel-Corona & Buitrón, 2021). The PHB accumulation was stimulated under the aerobic chemoheterotrophic growth of *Rhodospirillum rubrum* (Narancic et al., 2016). *Rhodobacter sphaeroides* 14 F showed high PHA content of up to 60% dcw under aerobic and dark conditions compared to anaerobic and light conditions, concluded that anaerobic and light conditions were not feasible for PHA production (Lorrungruang et al., 2006). However, the anaerobic and anaerobic cultivation conditions impact PHA production and significantly affect the co-production of other valuable by-products in APB. For example, under microaerobic-light conditions, PPB tends to release more 5-aminolevulinic acid (5-ALA) into the environment compared to aerobic-dark conditions (Nunkaew et al., 2014). Moreover, Coenzyme Q10 (CoQ10) is typically generated under aerobic-dark cultivation conditions (Lu et al., 2019).

10.3.4 Modeling the metabolism to biopolymer production

Modeling is essential for developing, optimizing, and scaling up biotechnological processes involving APB for biopolymer production (Fradinho et al., 2021). However, modeling APB metabolism is challenging due to their complex pathways and the need for practical yet accurate models. Key considerations include mapping APB metabolic pathways for CO_2 and H_2S conversion into biopolymers, understanding enzyme kinetics, and choosing the appropriate model type.

Models can be stoichiometric, such as Constraint-Based Reconstruction and Analysis (COBRA) to describe the biochemical reactions in the metabolic network and predict growth and product formation rates; dynamic for simulating metabolic processes, providing a more realistic view of the bacterial growth and biopolymer production process over time; or kinetic models, focusing on reaction rates to predict how changes in conditions such as substrate concentration or light intensity could impact the rate of biopolymer synthesis in APB. A systems biology approach integrates genomics, transcriptomics, proteomics, and metabolomics for a comprehensive view of metabolism (Rawat et al., 2023). Optimization and sensitivity analysis are crucial for identifying optimal production conditions and assessing process sensitivity to parameter changes (Renardy et al., 2019). Environmental and operational factors must be incorporated and models should be validated with experimental data (Finore et al., 2014).

Overall, modeling APB metabolism is a multidisciplinary task involving microbiology, biochemistry, chemical engineering, and computational biology.

Effective models guide experimental work, enhance understanding of metabolic processes, and improve biopolymer production efficiency and yield.

10.4 Conclusions

APB possess a unique metabolism that allows them to adapt to diverse environments, utilize light as an energy source, remove sulfur compounds, and capture CO_2 through alternative pathways to the CBB cycle, such as the rTCA cycle and the 3-hydroxypropionate bicycle. Achieving optimal conditions for APB to thrive and efficiently produce biopolymers, however, requires a delicate balance. The production of PHAs by APB is influenced by a complex interplay of factors, which also affect the synthesis of other valuable bioproducts, including pigments, biofertilizers, SCP, CoQ10, and 5-ALA, among others. By carefully managing these factors, the efficiency of value-added product synthesis by APB using CO_2 and H_2S can be significantly enhanced, improving the viability of this process for industrial applications such as photosynthetic biogas upgrading. This, in turn, could greatly increase the economic sustainability of biogas upgrading technology.

References

Allegue, L. D., Ventura, M., Melero, J. A., & Puyol, D. (2021). Integrated sustainable process for polyhydroxyalkanoates production from lignocellulosic waste by purple phototrophic bacteria. *GCB Bioenergy, 13*(5), 862–875. https://doi.org/10.1111/gcbb.12807.

Allegue, L. D., Ventura, M., Melero, J. A., & Puyol, D. (2022). Unraveling PHA production from urban organic waste with purple phototrophic bacteria via organic overload. *Renewable and Sustainable Energy Reviews, 166*, 112687. https://doi.org/10.1016/j.rser.2022.112687.

Almeida, J. R., Serrano, E., Fernandez, M., Fradinho, J. C., Oehmen, A., & Reis, M. A. M. (2021). Polyhydroxyalkanoates production from fermented domestic wastewater using phototrophic mixed cultures. *Water Research, 197*, 117101. https://doi.org/10.1016/j.watres.2021.117101.

Almeida, J. R., Serrano León, E., Rogalla, F., Fradinho, J. C., Oehmen, A., & Reis, M. A. M. (2024). Polyhydroxyalkanoates production in purple phototrophic bacteria ponds: A breakthrough in outdoor pilot-scale operation. *Science of The Total Environment, 912*, 168899. https://doi.org/10.1016/j.scitotenv.2023.168899.

Ángeles, R., Raquel,L., Maria, R., Fradinho, A. M., Joana, C. (2024). 1st International Conference on Novel Photorefineries for Resource Recovery 195 198 Enhanced Photosynthetic Bacterial Pigment Production During Biological Sulfide Removal by Anoxygenic Photosynthetic Bacteria. https://doi.org/10.5281/zenodo.13476761.

Ángeles, R., Arnaiz, E., Gutiérrez, J., Sepúlveda-Muñoz, C. A., Fernández-Ramos, O., Muñoz, R., & Lebrero, R. (2020). Optimization of photosynthetic biogas upgrading in closed photobioreactors combined with algal biomass production. *Journal of Water Process Engineering, 38*, 101554. https://doi.org/10.1016/j.jwpe.2020.101554.

Ángeles, R., Rodríguez, Á., Domínguez, C., García, J., Prádanos, P., Muñoz, R., & Lebrero, R. (2020). Strategies for N_2 and O_2 removal during biogas upgrading in a pilot algal-bacterial photobioreactor. *Algal Research, 48*, 101920. https://doi.org/10.1016/j.algal.2020.101920.

Ángeles, R., Vega-Quiel, M. J., Batista, A., Fernández-Ramos, O., Lebrero, R., & Muñoz, R. (2021). Influence of biogas supply regime on photosynthetic biogas upgrading performance in an enclosed algal-bacterial photobioreactor. *Algal Research, 57*, 102350. https://doi.org/10.1016/j.algal.2021.102350.

Ángeles Torres, R., Marín, D., Marín, D., Rodero, M.del R., Rodero, M.del R., Pascual, C., González-Sanchez, A., de Godos Crespo, I., de Godos Crespo, I., Lebrero, R., Lebrero, R., & Torre, R. M. (2020). Biogas treatment for H_2S, CO_2, and other contaminants removal. In *From biofiltration to promising options in gaseous fluxes biotreatment: Recent developments, new trends, advances, and opportunities*. Elsevier, 153–176. https://doi.org/10.1016/B978-0-12–819064-7.00008-X.

Bahr, M., Díaz, I., Dominguez, A., González Sánchez, A., & Muñoz, R. (2014). Microalgal-biotechnology as a platform for an integral biogas upgrading and nutrient removal from anaerobic effluents. *Environmental Science & Technology, 48*(1), 573–581. https://doi.org/10.1021/es403596m.

Bose, A., Lin, R., Rajendran, K., O'Shea, R., Xia, A., & Murphy, J. D. (2019). How to optimise photosynthetic biogas upgrading: A perspective on system design and microalgae selection. *Biotechnology Advances, 37*(8), 107444. https://doi.org/10.1016/j.biotechadv.2019.107444.

Bryant, D. A., & Frigaard, N. U. (2006). Prokaryotic photosynthesis and phototrophy illuminated. *Trends in Microbiology, 14*(11), 488–496. https://doi.org/10.1016/j.tim.2006.09.001.

Capson-Tojo, G., Batstone, D. J., Grassino, M., Vlaeminck, S. E., Puyol, D., Verstraete, W., Kleerebezem, R., Oehmen, A., Ghimire, A., Pikaar, I., Lema, J. M., & Hülsen, T. (2020). Purple phototrophic bacteria for resource recovery: Challenges and opportunities. *Biotechnology Advances, 43*, 107567. https://doi.org/10.1016/j.biotechadv.2020.107567.

Carvalho, A. P., Silva, S. O., Baptista, J. M., & Malcata, F. X. (2011). Light requirements in microalgal photobioreactors: An overview of biophotonic aspects. *Applied Microbiology and Biotechnology, 89*(5), 1275–1288. https://doi.org/10.1007/s00253-010-3047-8.

EBA, Statistical Report 2023: Tracking biogas and biomethane deployment across Europe. (2023), https://www.europeanbiogas.eu/eba-statistical-report-2023/.

EBA European Biogas Association. (2024). European Biomethane Map 2024 https://www.europeanbiogas.eu/european-biomethane-map-2024/.

Egger, F., Hülsen, T., Tait, S., & Batstone, D. J. (2020). Autotrophic sulfide removal by mixed culture purple phototrophic bacteria. *Water Research, 182*, 115896. https://doi.org/10.1016/j.watres.2020.115896.

Egger, F., Hülsen, T., & Batstone, D. J. (2023). Continuous H_2S removal from biogas using purple phototrophic bacteria. *Chemical Engineering Journal, 454*, 140449. https://doi.org/10.1016/j.cej.2022.140449.

EurObserv, E. R. (2023). Biogas barometer 2023 https://www.eurobserv-er.org/biogas-barometer-2023/.

Evans, Buchanan, & Arnon (1966). A new ferredoxin-dependent carbon reduction cycle in a photosynthetic bacterium. *Proceedings of the National Academy of Sciences of the United States of America, 55*(4), 928–934. https://doi.org/10.1073/pnas.55.4.928.

Fachal-Suárez, M., Krishnan, S., Chaiprapat, S., González, D., & Gabriel, D. (2024). An overview of biomethanation and the use of membrane technologies as a candidate to overcome H2 mass transfer limitations. *Biotechnology Advances, 77*, 108465. https://doi.org/10.1016/j.biotechadv.2024.108465.

Findlay, A. J., Bennett, A. J., Hanson, T. E., & Luther, G. W. (2015). Light-dependent sulfide oxidation in the anoxic zone of the chesapeake bay can be explained by small populations of phototrophic bacteria. *Applied and Environmental Microbiology, 81*(21), 7560–7569. https://doi.org/10.1128/AEM.02062–15.

Foong, C. P., Higuchi-Takeuchi, M., Numata, K., & Koller, M. (2019). Optimal iron concentrations for growth-associated polyhydroxyalkanoate biosynthesis in the marine photosynthetic purple bacterium Rhodovulum sulfidophilum under photoheterotrophic condition. *PLoS One, 14*(4), e0212654. https://doi.org/10.1371/journal.pone.0212654.

Finore, I., Di Donato, P., Mastascusa, V., Nicolaus, B., & Poli, A. (2014). Fermentation technologies for the optimization of marine microbial exopolysaccharide production. *Marine Drugs, 12*(5), 3005–3024. https://doi.org/10.3390/md12053005.

Fradinho, J., Allegue, L. D., Ventura, M., Melero, J. A., Reis, M. A. M., & Puyol, D. (2021). Up-scale challenges on biopolymer production from waste streams by Purple Phototrophic Bacteria mixed cultures: A critical review. *Bioresource Technology, 327*, 124820. https://doi.org/10.1016/j.biortech.2021.124820.

Fradinho, J. C., Oehmen, A., & Reis, M. A. M. (2019). Improving polyhydroxyalkanoates production in phototrophic mixed cultures by optimizing accumulator reactor operating conditions. *International Journal of Biological Macromolecules, 126*, 1085–1092. https://doi.org/10.1016/j.ijbiomac.2018.12.270.

Franco-Morgado, M., Alcántara, C., Noyola, A., Muñoz, R., & González-Sánchez, A. (2017). A study of photosynthetic biogas upgrading based on a high rate algal pond under alkaline conditions: Influence of the illumination regime. *Science of The Total Environment, 592*, 419–425. https://doi.org/10.1016/j.scitotenv.2017.03.077.

Freches, A., & Fradinho, J. C. (2024). The biotechnological potential of the Chloroflexota phylum. *Applied and Environmental Microbiology, 90*(6), e01756-23. https://doi.org/10.1128/aem.01756-23.

Gamal-Eldin, H., & Elbanna, K. (2011). Field evidence for the potential of Rhodobacter capsulatus as biofertilizer for flooded rice. *Current Microbiology, 62*(2), 391–395. https://doi.org/10.1007/s00284-010-9719-x.

Garritano, A. N., Song, W., & Thomas, T. (2022). Carbon fixation pathways across the bacterial and archaeal tree of life. *PNAS Nexus, 1*(5), pgac226. https://doi.org/10.1093/pnasnexus/pgac226.

Ghofrani-Isfahani, P., Tsapekos, P., Peprah, M., Kougias, P., Zhu, X., Kovalovszki, A., Zervas, A., Zha, X., Jacobsen, C. S., & Angelidaki, I. (2021). Ex-situ biogas upgrading in thermophilic up-flow reactors: The effect of different gas diffusers and gas retention times. *Bioresource Technology, 340*, 125694. https://doi.org/10.1016/j.biortech.2021.125694.

Hamilton, T. L., Bovee, R. J., Thiel, V., Sattin, S. R., Mohr, W., Schaperdoth, I., Vogl, K., Gilhooly, W. P., Lyons, T. W., Tomsho, L. P., Schuster, S. C., Overmann, J., Bryant, D. A., Pearson, A., & Macalady, J. L. (2014). Coupled reductive and oxidative sulfur cycling in the phototrophic plate of a meromictic lake. *Geobiology, 12*(5), 451–468. https://doi.org/10.1111/gbi.12092.

Henchion, M., Hayes, M., Mullen, A., Fenelon, M., & Tiwari, B. (2017). Future protein supply and demand: Strategies and factors influencing a sustainable equilibrium. *Foods, 6*(7), 53. https://doi.org/10.3390/foods6070053.

Higuchi-Takeuchi, M., & Numata, K. (2019). Acetate-inducing metabolic states enhance polyhydroxyalkanoate production in marine purple non-sulfur bacteria under aerobic conditions. *Frontiers in Bioengineering and Biotechnology, 7*, 118. https://doi.org/10.3389/fbioe.2019.00118.

Higuchi-Takeuchi, M., Morisaki, K., Toyooka, K., Numata, K., & Chen, G.-Q. (2016). Synthesis of high-molecular-weight polyhydroxyalkanoates by marine photosynthetic purple bacteria. *PLoS One, 11*(8), e0160981. https://doi.org/10.1371/journal.pone.0160981.

Holo, H. (1989). Chloroflexus aurantiacus secretes 3-hydroxypropionate, a possible intermediate in the assimilation of CO_2 and acetate. *Archives of Microbiology, 151*(3), 252–256. https://doi.org/10.1007/bf00413138.

Hsu, S. (2015). Application of phototrophic bacterial inoculant to reduce nitrate content in hydroponic leafy vegetables. *Crop Environment Bioinformatics, 12*, 30–41. https://doi.org/10.30061/CEB.201503_12(1).0003.

Hülsen, T., Hsieh, K., Lu, Y., Tait, S., & Batstone, D. J. (2018). Simultaneous treatment and single cell protein production from agri-industrial wastewaters using purple phototrophic bacteria or microalgae – A comparison. *Bioresource Technology, 254*, 214–223. https://doi.org/10.1016/j.biortech.2018.01.032.

Iwai, R., Uchida, S., Yamaguchi, S., Sonoda, F., Tsunoda, K., Nagata, H., Nagata, D., Koga, A., Goto, M., Maki, T.-aki, Hayashi, S., Yamamoto, S., & Miyasaka, H. (2022). Effects of seed bio-priming by purple non-sulfur bacteria (PNSB) on the root development of rice. *Microorganisms, 10*(11), 2197. https://doi.org/10.3390/microorganisms10112197.

Jensen, E., Clément, R., Maberly, S. C., & Gontero, B. (2017). Regulation of the Calvin–Benson–Bassham cycle in the enigmatic diatoms: biochemical and evolutionary variations on an original theme. *Philosophical Transactions of the Royal Society B: Biological Sciences, 372*(1728), 20160401. https://doi.org/10.1098/rstb.2016.0401.

Jiang, M., Hong, K., Mao, Y., Ma, H., Chen, T., & Wang, Z. (2022). Natural 5-aminolevulinic acid: Sources, biosynthesis, detection and applications. *Frontiers in Bioengineering and Biotechnology, 10*, 841443. https://doi.org/10.3389/fbioe.2022.841443.

Khoshnevisan, B., He, L., Xu, M., Valverde-Pérez, B., Sillman, J., Mitraka, G.-C., Kougias, P. G., Zhang, Y., Yan, S., Ji, L., Carbajales-Dale, M., Elyasi, S. N., Marami, H., Tsapekos, P., Liu, H., & Angelidaki, I. (2022). From renewable energy to sustainable protein sources: Advancement, challenges, and future roadmaps. *Renewable and Sustainable Energy Reviews, 157*, 112041. https://doi.org/10.1016/j.rser.2021.112041.

Khuong, N. Q., Kantachote, D., Onthong, J., & Sukhoom, A. (2017). The potential of acid-resistant purple nonsulfur bacteria isolated from acid sulfate soils for reducing toxicity of Al^{3+} and Fe^{2+} using biosorption for agricultural application. *Biocatalysis and Agricultural Biotechnology, 12*, 329–340. https://doi.org/10.1016/j.bcab.2017.10.022.

Kikuchi, Y., Kanai, D., Sugiyama, K., & Fujii, K. (2024). Biogas upgrading by wild alkaliphilic microalgae and the application potential of their biomass in the carbon capture and utilization technology. *Fermentation, 10*(3), 134. https://doi.org/10.3390/fermentation10030134.

Kitadai, N., Kameya, M., & Fujishima, K. (2017). Origin of the reductive tricarboxylic acid (rTCA) cycle-type CO_2 fixation: A perspective. *Life (Chicago, Ill. 1978), 7*(4), 39. https://doi.org/10.3390/life7040039.

Klatt, C. G., Liu, Z., Ludwig, M., Kühl, M., Jensen, S. I., Bryant, D. A., & Ward, D. M. (2013). Temporal metatranscriptomic patterning in phototrophic Chloroflexi inhabiting a microbial mat in a geothermal spring. *The ISME Journal, 7*(9), 1775–1789. https://doi.org/10.1038/ismej.2013.52.

Koller, M. (2018). A review on established and emerging fermentation schemes for microbial production of polyhydroxyalkanoate (PHA) biopolyesters. *Fermentation, 4*(2), 30. https://doi.org/10.3390/fermentation4020030.

Krayer, M., Yang, E., Diers, J. R., Bocian, D. F., Holten, D., & Lindsey, J. S. (2011). De novo synthesis and photophysical characterization of annulated bacteriochlorins. Mimicking and extending the properties of bacteriochlorophylls. *New Journal of Chemistry, 35*(3), 587. https://doi.org/10.1039/c0nj00771d.

Kuo, F. S., Chien, Y. H., & Chen, C. J. (2012). Effects of light sources on growth and carotenoid content of photosynthetic bacteria Rhodopseudomonas palustris. *Bioresource Technology, 113*, 315–318. https://doi.org/10.1016/j.biortech.2012.01.087.

Lai, Y. C., Liang, C. M., Hsu, S. C., Hsieh, P. H., & Hung, C. H. (2017). Polyphosphate metabolism by purple non-sulfur bacteria and its possible application on photo-microbial fuel cell. *Journal of Bioscience and Bioengineering, 123*(6), 722–730. https://doi.org/10.1016/j.jbiosc.2017.01.012.

Lee, K. H., Koh, R. H., & Song, H. G. (2008). Enhancement of growth and yield of tomato by Rhodopseudomonas sp. under greenhouse conditions. *Journal of Microbiology, 46*(6), 641–646. https://doi.org/10.1007/s12275-008-0159-2.

Li, Z., Xin, X., Xiong, B., Zhao, D., Zhang, X., & Bi, C. (2020). Engineering the Calvin–Benson–Bassham cycle and hydrogen utilization pathway of Ralstonia eutropha for improved autotrophic growth and polyhydroxybutyrate production. *Microbial Cell Factories, 19*(1), 228. https://doi.org/10.1186/s12934-020-01494-y.

Liu, S., Daigger, G. T., Kang, J., & Zhang, G. (2019). Effects of light intensity and photoperiod on pigments production and corresponding key gene expression of Rhodopseudomonas palustris in a photobioreactor system. *Bioresource Technology, 294*, 122172. https://doi.org/10.1016/j.biortech.2019.122172.

Lo, S.-C., Chiang, E.-P. I., Yang, Y.-T., Li, S.-Y., Peng, J.-H., Tsai, S.-Y., Wu, D.-Y., Yu, C.-H., Huang, C.-H., Su, T.-T., Tsuge, K., & Huang, C.-C. (2021). Growth enhancement facilitated by gaseous CO_2 through heterologous expression of reductive tricarboxylic acid cycle genes in Escherichia coli. *Fermentation, 7*(2), 98. https://doi.org/10.3390/fermentation7020098.

Lorrungruang, C., Martthong, J., Sasaki, K., & Noparatnaraporn, N. (2006). Selection of photosynthetic bacterium Rhodobacter sphaeroides 14 F for polyhydroxyalkanoate production with two-stage aerobic dark cultivation. *Journal of Bioscience and Bioengineering, 102*(2), 128–131. https://doi.org/10.1263/jbb.102.128.

Lu, H., Peng, M., Zhang, G., Li, B., & Li, Y. (2019). Brewery wastewater treatment and resource recovery through long term continuous-mode operation in pilot photosynthetic bacteria-membrane bioreactor. *Science of The Total Environment, 646*, 196–205. https://doi.org/10.1016/j.scitotenv.2018.07.268.

Luongo, V., Ghimire, A., Frunzo, L., Fabbricino, M., d'Antonio, G., Pirozzi, F., & Esposito, G. (2017). Photofermentative production of hydrogen and poly-β-hydroxybutyrate from dark fermentation products. *Bioresource Technology, 228*, 171–175. https://doi.org/10.1016/j.biortech.2016.12.079.

Mandal, M. K., Chanu, N. K., & Chaurasia, N. (2020). Cyanobacterial pigments and their fluorescence characteristics: Applications in research and industry. In *Advances in cyanobacterial biology*. Elsevier, 55–72. 10.1016/B978-0-12–819311-2.00005-X.

Martin-Gamboa, M., Allegue, L. D., Puyol, D., Melero, J. A., & Dufour, J. (2023). Environmental life cycle assessment of polyhydroxyalkanoates production by purple phototrophic bacteria mixed cultures. *Journal of Cleaner Production, 428*, 139421. https://doi.org/10.1016/j.jclepro.2023.139421.

Marín, D., Posadas, E., Cano, P., Pérez, V., Blanco, S., Lebrero, R., & Muñoz, R. (2018). Seasonal variation of biogas upgrading coupled with digestate treatment in an outdoors pilot scale algal-bacterial photobioreactor. *Bioresource Technology, 263*, 58–66. https://doi.org/10.1016/j.biortech.2018.04.117.

Marín, D., Posadas, E., García, D., Puyol, D., Lebrero, R., & Muñoz, R. (2019). Assessing the potential of purple phototrophic bacteria for the simultaneous treatment of piggery wastewater and upgrading of biogas. *Bioresource Technology, 281*, 10–17. https://doi.org/10.1016/j.biortech.2019.02.073.

Matassa, S., Boon, N., Pikaar, I., & Verstraete, W. (2016). Microbial protein: future sustainable food supply route with low environmental footprint. *Microbial Biotechnology, 9*(5), 568–575. https://doi.org/10.1111/1751-7915.12369.

Mattozzi, Md, Ziesack, M., Voges, M. J., Silver, P. A., & Way, J. C. (2013). Expression of the sub-pathways of the Chloroflexus aurantiacus 3-hydroxypropionate carbon fixation bicycle in *E. coli*: Toward horizontal transfer of autotrophic growth. *Metabolic Engineering, 16*(1), 130–139. https://doi.org/10.1016/j.ymben.2013.01.005.

Monroy, I., & Buitrón, G. (2020). Production of polyhydroxybutyrate by pure and mixed cultures of purple non-sulfur bacteria: A review. *Journal of Biotechnology, 317*, 39–47. https://doi.org/10.1016/j.jbiotec.2020.04.012.

Montiel-Corona, V., & Buitrón, G. (2021). Polyhydroxyalkanoates from organic waste streams using purple non-sulfur bacteria. *Bioresource Technology, 323*, 124610. https://doi.org/10.1016/j.biortech.2020.124610.

Muzziotti, D., Adessi, A., Faraloni, C., Torzillo, G., & De Philippis, R. (2017). Acclimation strategy of Rhodopseudomonas palustris to high light irradiance. *Microbiological Research, 197*, 49–55. https://doi.org/10.1016/j.micres.2017.01.007.

Muñoz, R., Meier, L., Diaz, I., & Jeison, D. (2015). A review on the state-of-the-art of physical/chemical and biological technologies for biogas upgrading. *Reviews in Environmental Science and Bio/Technology, 14*(4), 727–759. https://doi.org/10.1007/s11157-015-9379-1.

Narancic, T., Scollica, E., Kenny, S. T., Gibbons, H., Carr, E., Brennan, L., Cagney, G., Wynne, K., Murphy, C., Raberg, M., Heinrich, D., Steinbüchel, A., & O'Connor, K. E. (2016). Understanding the physiological roles of polyhydroxybutyrate (PHB) in Rhodospirillum rubrum S1 under aerobic chemoheterotrophic conditions. *Applied Microbiology and Biotechnology, 100*(20), 8901–8912. https://doi.org/10.1007/s00253-016-7711-5.

Narayanan, M. (2024). Promising biorefinery products from marine macro and microalgal biomass: A review. *Renewable and Sustainable Energy Reviews, 190*, 114081. https://doi.org/10.1016/j.rser.2023.114081.

Nookongbut, P., Kantachote, D., Megharaj, M., & Naidu, R. (2018). Reduction in arsenic toxicity and uptake in rice (*Oryza sativa* L.) by As-resistant purple nonsulfur bacteria. *Environmental Science and Pollution Research, 25*(36), 36530–36544. https://doi.org/10.1007/s11356-018-3568-8.

Novovceská, L., Ross, M. E., Stanley, M. S., Pradelles, R., Wasiolek, V., & Sassi, J.-F. (2019). Microalgal carotenoids: A review of production, current markets, regulations, and future direction. *Marine Drugs, 17*(11), 640. https://doi.org/10.3390/md17110640.

Nunkaew, T., Kantachote, D., Kanzaki, H., Nitoda, T., & Ritchie, R. J. (2014). Effects of 5-aminolevulinic acid (ALA)-containing supernatants from selected Rhodopseudomonas palustris strains on rice growth under NaCl stress, with mediating effects on chlorophyll, photosynthetic electron transport and antioxidative enzymes. *Electronic Journal of Biotechnology, 17*(1), 19–26. https://doi.org/10.1016/j.ejbt.2013.12.004.

Olivares, J., Bedmar, E. J., & Sanjuán, J. (2013). Biological nitrogen fixation in the context of global change. *Molecular Plant-Microbe Interactions, 26*(5), 486–494. https://doi.org/10.1094/MPMI-12-12-0293-CR.

Ouyang, Y., Zhao, Y., Sun, S., Hu, C., & Ping, L. (2015). Effect of light intensity on the capability of different microalgae species for simultaneous biogas upgrading and biogas slurry nutrient reduction. *International Biodeterioration & Biodegradation, 104*, 157–163. https://doi.org/10.1016/j.ibiod.2015.05.027.

Pfennig, N. (1975). The phototrophic bacteria and their role in the sulfur cycle. *Plant and Soil, 43*(1-3), 1–16. https://doi.org/10.1007/bf01928472.

Pokorna, D., & Zabranska, J. (2015). Sulfur-oxidizing bacteria in environmental technology. *Biotechnology Advances, 33*(6), 1246–1259. https://doi.org/10.1016/j.biotechadv.2015.02.007.

Policastro, G., Luongo, V., & Fabbricino, M. (2020). Biohydrogen and poly-β-hydroxybutyrate production by winery wastewater photofermentation: Effect of substrate concentration and nitrogen source. *Journal of Environmental Management, 271*, 111006. https://doi.org/10.1016/j.jenvman.2020.111006.

Prandini, J. M., da Silva, M. L. B., Mezzari, M. P., Pirolli, M., Michelon, W., & Soares, H. M. (2016). Enhancement of nutrient removal from swine wastewater digestate coupled to biogas purification by microalgae Scenedesmus spp. *Bioresource Technology, 202*, 67–75. https://doi.org/10.1016/j.biortech.2015.11.082.

Pucelik, B., Sułek, A., & Dąbrowski, J. M. (2020). Bacteriochlorins and their metal complexes as NIR-absorbing photosensitizers: Properties, mechanisms, and applications. *Coordination Chemistry Reviews, 416*, 213340. https://doi.org/10.1016/j.ccr.2020.213340.

Rawat, C. D., Phian, S., Gupta, R., Verma, H., Kumar, M., Kaur, J., & Rawat, V. S. (2023). Microbial bioprocesses in remediation of contaminated environments and resource recovery. In *Microbial bioprocesses: Applications and perspectives*. Elsevier, 225–274. 10.1016/B978-0-323-95332-0.00005-3.

Rawat, P., Sharma, A., Shankhdhar, D., & Shankhdhar, S. C. (2022). Improvement of phosphorus uptake, phosphorus use efficiency, and grain yield of upland rice (Oryza sativa L.) in response to phosphate-solubilizing bacteria blended with phosphorus fertilizer. *Pedosphere, 32*(5), 752–763. https://doi.org/10.1016/j.pedsph.2022.06.005.

Renardy, M., Hult, C., Evans, S., Linderman, J. J., & Kirschner, D. E. (2019). Global sensitivity analysis of biological multiscale models. *Current Opinion in Biomedical Engineering, 11*, 109–116. https://doi.org/10.1016/j.cobme.2019.09.012.

Ritala, A., Häkkinen, S. T., Toivari, M., & Wiebe, M. G. (2017). Single cell protein—state-of-the-art, industrial landscape and patents 2001–2016. *Frontiers in Microbiology, 8*, 2009. https://doi.org/10.3389/fmicb.2017.02009.

Rodero, M.del R., Posadas, E., Toledo-Cervantes, A., Lebrero, R., & Muñoz, R. (2018). Influence of alkalinity and temperature on photosynthetic biogas upgrading efficiency in high rate algal ponds. *Algal Research, 33*, 284–290. https://doi.org/10.1016/j.algal.2018.06.001.

Rodero, M.del R., Severi, C. A., Rocher-Rivas, R., Quijano, G., & Muñoz, R. (2020). Long-term influence of high alkalinity on the performance of photosynthetic biogas upgrading. *Fuel, 281*, 118804. https://doi.org/10.1016/j.fuel.2020.118804.

Rodero, M.del R., Magdalena, J. A., Steyer, J.-P., Escudié, R., & Capson-Tojo, G. (2024). Potential of enriched phototrophic purple bacteria for H2 bioconversion into single cell protein. *Science of The Total Environment, 908*, 168471. https://doi.org/10.1016/j.scitotenv.2023.168471.

Rodero, Md. R., Ángeles, R., García-Depraect, O., Lebrero, R., & Muñoz, R. (2023). Recent advances on photosynthetic biogas upgrading to biomethane. In *Biogas to biomethane:*

Engineering, production, sustainability. Elsevier, 117–140. 10.1016/B978-0-443-18479-6.00010-7.

Rubin-Blum, M., Dubilier, N., & Kleiner, M. (2019). Genetic evidence for two carbon fixation pathways (the Calvin-Benson-Bassham cycle and the reverse tricarboxylic acid cycle) in symbiotic and free-living bacteria. *mSphere, 4*(1). https://doi.org/10.1128/mSphere.00394-18.

Saejung, C., & Ampornpat, W. (2019). Production and nutritional performance of carotenoid-producing photosynthetic bacterium Rhodopseudomonas faecalis PA2 grown in domestic wastewater intended for animal feed production. *Waste and Biomass Valorization, 10*(2), 299–310. https://doi.org/10.1007/s12649-017-0070-3.

Saer, R. G., & Blankenship, R. E. (2017). Light harvesting in phototrophic bacteria: Structure and function. *Biochemical Journal, 474*(13), 2107–2131. https://doi.org/10.1042/BCJ20160753.

Sakarika, M., Spanoghe, J., Sui, Y., Wambacq, E., Grunert, O., Haesaert, G., Spiller, M., & Vlaeminck, S. E. (2020). Purple non-sulphur bacteria and plant production: Benefits for fertilization, stress resistance and the environment. *Microbial Biotechnology, 13*(5), 1336–1365. https://doi.org/10.1111/1751-7915.13474.

Sakpirom, J., Kantachote, D., Nunkaew, T., & Khan, E. (2017). Characterizations of purple non-sulfur bacteria isolated from paddy fields, and identification of strains with potential for plant growth-promotion, greenhouse gas mitigation and heavy metal bioremediation. *Research in Microbiology, 168*(3), 266–275. https://doi.org/10.1016/j.resmic.2016.12.001.

Severi, C. A., Lebrero, R., & Muñoz, R. (2023). Biogas upgrading using microalgae. In *Sustainable industrial processes based on microalgae*. Elsevier, 157–184. 10.1016/B978-0-443-19213-5.00008-X.

Severo, I. A., Deprá, M. C., Zepka, L. Q., & Jacob-Lopes, E. (2019). Carbon dioxide capture and use by microalgae in photobioreactors. In *Bioenergy with carbon capture and storage: Using natural resources for sustainable development*. Elsevier, 151–171. 10.1016/B978-0-12-816229-3.00008-9.

Sharkey, T. D. (2019). Discovery of the canonical Calvin–Benson cycle. *Photosynthesis Research, 140*(2), 235–252. https://doi.org/10.1007/s11120-018-0600-2.

Spanoghe, J., Grunert, O., Wambacq, E., Sakarika, M., Papini, G., Alloul, A., Spiller, M., Derycke, V., Stragier, L., Verstraete, H., Fauconnier, K., Verstraete, W., Haesaert, G., & Vlaeminck, S. E. (2020). Storage, fertilization and cost properties highlight the potential of dried microbial biomass as organic fertilizer. *Microbial Biotechnology, 13*(5), 1377–1389. https://doi.org/10.1111/1751-7915.13554.

Spanoghe, J., Vermeir, P., & Vlaeminck, S. E. (2021). Microbial food from light, carbon dioxide and hydrogen gas: Kinetic, stoichiometric and nutritional potential of three purple bacteria. *Bioresource Technology, 337*, 125364. https://doi.org/10.1016/j.biortech.2021.125364.

Srisawat, P., Higuchi-Takeuchi, M., & Numata, K. (2022). Microbial autotrophic biorefineries: Perspectives for biopolymer production. *Polymer Journal, 54*(10), 1139–1151. https://doi.org/10.1038/s41428-022-00675-3.

Srivastava, P., Villamil, J. A., Melero, J. A., Martínez, F., & Puyol, D. (2024). Using inorganic acids to stop purple phototrophic bacteria metabolism improves PHA recovery at a large scale. *Biomass Conversion and Biorefinery, 14*(15), 17693–17703. https://doi.org/10.1007/s13399-023-03810-z.

Steffens, L., Pettinato, E., Steiner, T. M., Mall, A., König, S., Eisenreich, W., & Berg, I. A. (2021). High CO_2 levels drive the TCA cycle backwards towards autotrophy. *Nature, 592*(7856), 784–788. https://doi.org/10.1038/s41586-021-03456-9.

Struk, M., Sepúlveda-Muñoz, C. A., Kushkevych, I., & Muñoz, R. (2023). Photoautotrophic removal of hydrogen sulfide from biogas using purple and green sulfur bacteria. *Journal of Hazardous Materials, 443*, 130337. https://doi.org/10.1016/j.jhazmat.2022.130337.

Su, P., Tan, X., Li, C., Zhang, D., Cheng, J.'e, Zhang, S., Zhou, X., Yan, Q., Peng, J., Zhang, Z., Liu, Y., & Lu, X. (2017). Photosynthetic bacterium R hodopseudomonas palustris GJ -22 induces systemic resistance against viruses. *Microbial Biotechnology, 10*(3), 612–624. https://doi.org/10.1111/1751-7915.12704.

Sun, Y. P., Zhang, Z. P., & Wang, L. J. (2009). Promotion of 5-aminolevulinic acid treatment on leaf photosynthesis is related with increase of antioxidant enzyme activity in watermelon seedlings grown under shade condition. *Photosynthetica, 47*(3). https://doi.org/10.1007/s11099-009-0055-y.

Toledo-Cervantes, A., Estrada, J. M., Lebrero, R., & Muñoz, R. (2017). A comparative analysis of biogas upgrading technologies: Photosynthetic vs physical/chemical processes. *Algal Research, 25*, 237–243. https://doi.org/10.1016/j.algal.2017.05.006.

Toledo-Cervantes, A., Madrid-Chirinos, C., Cantera, S., Lebrero, R., & Muñoz, R. (2017). Influence of the gas-liquid flow configuration in the absorption column on photosynthetic biogas upgrading in algal-bacterial photobioreactors. *Bioresource Technology, 225*, 336–342. https://doi.org/10.1016/j.biortech.2016.11.087.

Verbeeck, K., De Vrieze, J., Biesemans, M., & Rabaey, K. (2019). Membrane electrolysis-assisted CO2 and H2S extraction as innovative pretreatment method for biological biogas upgrading. *Chemical Engineering Journal, 361*, 1479–1486. https://doi.org/10.1016/j.cej.2018.09.120.

Vivek, N., Hazeena, S. H., Alphy, M. P., Kumar, V., Magdouli, S., Sindhu, R., Pandey, A., & Binod, P. (2021). Recent advances in microbial biosynthesis of C3 – C5 diols: Genetics and process engineering approaches. *Bioresource Technology, 322*, 124527. https://doi.org/10.1016/j.biortech.2020.124527.

Wambacq, E., Alloul, A., Grunert, O., Carrette, J., Vermeir, P., Spanoghe, J., Sakarika, M., Vlaeminck, S. E., Haesaert, G., & Aschonitis, V. G. (2022). Aerobes and phototrophs as microbial organic fertilizers: Exploring mineralization, fertilization and plant protection features. *PLoS One, 17*(2), e0262497. https://doi.org/10.1371/journal.pone.0262497.

Wong, W. T., Tseng, C. H., Hsu, S. H., Lur, H. S., Mo, C. W., Huang, C. N., Hsu, S. C., Lee, K. T., & Liu, C. T. (2014). Promoting effects of a single Rhodopseudomonas palustris inoculant on plant growth by Brassica rapa chinensis under low fertilizer input. *Microbes and Environments, 29*(3), 303–313. https://doi.org/10.1264/jsme2.ME14056.

Yu, A., Xie, Y., Pan, X., Zhang, H., Cao, P., Su, X., Chang, W., & Li, M. (2020). Photosynthetic phosphoribulokinase structures: Enzymatic mechanisms and the redox regulation of the Calvin-Benson-Bassham cycle. *The Plant Cell, 32*(5), 1556–1573. https://doi.org/10.1105/tpc.19.00642.

Zhang, T., Quijano, G., & Wang, M. (2024). Fast development of purple nonsulfur bacteria (PNSB) in a bubble column photobioreactor: Influence of carbon source and dissolved O2 availability on wastewater treatment performance. *Journal of Water Process Engineering, 65*, 105885. https://doi.org/10.1016/j.jwpe.2024.105885.

Zhao, Y., Sun, S., Hu, C., Zhang, H., Xu, J., & Ping, L. (2015). Performance of three microalgal strains in biogas slurry purification and biogas upgrade in response to various mixed light-emitting diode light wavelengths. *Bioresource Technology, 187*, 338–345. https://doi.org/10.1016/j.biortech.2015.03.130.

The role of yeasts in the production of alternative food protein

11

Anna Bzducha-Wróbel[1], *and Elisabete Coelho*[2]
[1]Department of Food Biotechnology and Microbiology, Institute of Food Sciences, Warsaw University of Life Sciences, Warsaw, Poland, [2]LAQV-REQUIMTE, "GlycoFoodChem" Group, Department of Chemistry, University of Aveiro, Aveiro, Portugal

11.1 Introduction

The latest study of the European Parliamentary Research Service—Panel for the Future of Science and Technology underlines the increasing interest in alternative proteins in terms of their potential to improve the global food security and reduce the negative impacts of food and feed production on environment (Scientific Foresight Unit (STOA) EPRS, 2024).

The current global model of food production is based on the efficiency of the photosynthesis process, which determines both the yield of plant biomass and animal protein. It is estimated that the production of 1 kg of animal protein for human consumption requires about 6–7 kg of plant biomass, which is food also for humans. The availability of the main categories of raw materials in food production therefore depends primarily on the efficiency of photosynthetic conversion of carbon dioxide into organic biomass. The projected increase in the world's population to about 9.8 billion in 2050 forces the search for new solutions to increase the efficiency of agricultural production and distribution of high-quality proteins for the entire world population. The demand for food in the forecast period will amount to about 11.10 billion tons, with the current food production at the level of about 9 billion tons (Bell et al., 2022; https://ourworldindata.org/food-supply).

Undoubtedly, the possibilities of increasing the efficiency of agricultural production based on photosynthesis processes will be limited by the deepening problems related to climate change.

Conventional agriculture and food production alone are responsible for the emission of approx. 26% of greenhouse gases, which negatively affect the balance in the natural environment, causing, among other things, the growth of areas with water shortages, but also plagued by natural disasters such as droughts or floods. Another problem is the limited availability of arable land, the area of which is decreasing as a result of soil degradation. Excessive release of biogenic elements, such as nitrogen or phosphorus, discharged into aquatic ecosystems together with industrial and municipal sewage and as a result of the intensification of agriculture, leads to the eutrophication, which is harmful to aquatic organisms. The loss of

Eukaryotic Microorganisms as Sources of Bioproducts. DOI: https://doi.org/10.1016/B978-0-443-30188-9.00006-8

biodiversity also affects terrestrial ecosystems as a result of deforestation and basing agricultural production only on specific species of plants and animals. At the same time, food processing requires energy inputs, so the increase in energy demand will be problematic in the face of limited fossil fuel resources. Considering above, innovative solutions are needed that will make the food and food ingredient production system resistant, at least to some extent, to climate change and the insufficiently efficient photosynthesis process. One of the directions should be solutions based on the biotechnological use of microorganisms, including yeast, for the production of food and its ingredients, while simultaneously managing the streams of waste raw materials from agrifood production. Such solutions would implement the assumptions of a circular economy, rationalizing the use of natural resources, currently largely wasted.

Many countries of the world are currently striving to expand the raw material base with nonconventional food sources in the form of microbiological biomass or its purified components. Microbial cells, including yeast, have been consumed by humans for centuries as a component of fermented foods of both plant and animal origin. Microbial proteins known as single-cell protein (SCP) are the dried cells of unicellular microorganisms or are protein concentrates extracted from the biomass, mainly fungi like, yeasts and mold, microalgae, and bacteria. Application of microbial protein in human diets and animal feeding has long history dating back to World War I, when the Torula yeasts (current taxonomic name *Cyberlindnera jadinii*) have been supplemented to meet protein requirements in Germany (Barnett, 2004; Sousa-Silva et al., 2021). SCPs can be used as whole cells after inactivation by heating (inactive dry yeast), as autolysates and hydrolysates, or as purified extracts of valuable intracellular compounds (Ganeva et al., 2020).

The industrial production of *Saccharomyces cerevisiae* yeast has been established for over 100 years, and currently there are approx. 2,86 mln tons produced per year, approx. 35% in Europe to obtain bread, beer, and wine for human consumption as well as direct feed supplement in agriculture and aquaculture (Bell et al., 2022; Ganeva et al., 2020). Additionally, the biomass of *Saccharomyces cerevisiae* var. *boulardii* is used nowadays as probiotic dietary supplements in human nutrition (Choi et al., 2022), while the inactivated biomass of nonconventional yeast *Yarrowia lipolytica* has been recently authorized by European Food Safety Authority (EFSA) as novel food with the possibility of its use as a protein dietary supplement (Turck et al., 2019). Different nonconventional yeast strains of species, e.g., *Debaryomyces hansenii, Kazachstania unispora, Cyberlindnera jadinii*, Kluyveromyces lactis, *Pichia membranifaciens*, *Hanseniaspora uvarum*, *Kluyveromyces thermotolerans*, *Kluyveromyces marxianus*, *Pichia kudriavzevii,* or *Wickerhamomyces anomalus* are also promising considering probiotic traits and development of pro-healthy food matrices (Comitini et al., 2023; Hugot et al., 2023).

To produce a unit amount of microbial biomass less time, water and land is requires and yeast cultivation process is responsible for lower carbon footprint (e.g., carbon dioxide) comparing with plant and animal farming (Choi et al., 2022). Yeast biomass could be obtained rapidly considering short time of cell doubling, ex. approx. 90 min for *S. cerevisiae* under optimal growth conditions. According to a report by the Confederation of European Yeast Producers, obtaining 1 kg of yeast biomass corresponds to emissions of between 363 eq. of CO2 in the case of liquid yeast (18% of dry matter) and 3204 eq. of CO2 in the case of dry yeast (95% dry matter content) while production of animal proteins like: eggs generates 4500 g CO2 eq./ kg, pork—12300 g CO2 eq./ kg, beef—99500 kg CO2 eq./ kg when in case of plant origin proteins it is approx. 3200 CO2 eq./ kg for soya bean, 1600 CO2 eq./ kg for wheat (Epsztein, 2022; Carbon Foot Print Cofalec; Choi et al., 2022). Industrial yeast production is a very good example of a circular economy, which takes care of sustainable management of natural resources and environmental protection. In large-scale yeast production, coproducts from beet and cane sugar production, such as molasses and sugar syrup, are currently used. In turn, the components remaining in the culture medium after yeast multiplication are transformed into cattle feed, biofertilizers rich in potassium (potash extract) and vinasse for spraying on field, which contributes to reducing the emission of biogenic elements to sewage water, but also as a consequence to reducing the amount of mineral fertilizers used. Currently, the largest share in the carbon footprint of the yeast industry is the level of energy consumption (39%) and the emission intensity of processes related to sugar-based products (approx. 24%) (Carbon Foot Print Cofalec). The inclusion of renewable energy sources (solar energy, wind energy) in the yeast production process, e.g., for powering bioreactors, as well as the recovery of biogas produced in the yeast cultivation process and its use in a single-unit yeast production processes requiring energy input, will contribute to increasing the level of circularity in the yeast production.

Considering the forecasted dynamic development of the yeast protein market, which will amount to approximately USD 274.20 million by 2030, with approximately USD 143.95 million in 2022 (CAGR of 8.39% in the forecast period 2022–2030) (Brainy Insights, Report ID: TBI-13114, 2022), it is very reasonable to look for cheap carbon sources, other than molasses and sugar syrup, because they are becoming a very scarce raw material. Thus, new bioprocess technologies face various challenges not encountered in conventional fermentation. One such challenge is the narrow spectrum of assimilated carbon sources typical of conventional *Saccharomyces* yeasts and its limitation in tolerance to environmental stresses, requirement of supplementary vitamins and other complex organic compounds to grow efficiently. Many non-*Saccharomyces* yeasts, called "nonconventional," have been identified with metabolic potential that can overcome the limitations typical of the former. At the same time, nonconventional species are characterized by rapid growth rates and high efficiency of cellular biomass production on waste carbon and nitrogen sources and much less sensitivity to various unfavorable environmental factors (like extreme pH, high temperature, hyperosmotic stress, and fermentation inhibitors) (Geijer et al., 2022). The introduction of new type of raw materials for yeast cultivation, e.g., lignocellulosic biomass, will force the use of non-*Saccharomyces* yeast in these processes since pentoses (xylose, arabinose) are commonly accumulated in inedible parts of plant biomass. *Saccharomyces cerevisiae* yeast are capable of efficient growth in the presence of balanced quantities of glucose and sucrose only, which means growth directed at biomass production without losses of substrate used for ethanol production. Application of non-Saccharomyces yeast in the food technology is accelerating direction nowadays (Comitini et al., 2023).

Only microorganisms that do not pose a threat to consumer health may be used in food and feed production. In the European Union, new species of microorganisms intended to be used in food chain production are subjected to regulation and safety assessment by the European Food Safety Authority. The assessment is proceeded at the request of an entity interested in introducing new species of bacteria or yeast to qualified presumption of safety (QPS) list of food grade microorganisms and intended for feed production. The status of qualified presumed safety (QPS) is granted based on the results of taxonomic affiliation, body of knowledge, and evaluation of potential safety risk, related to, for example, antibiotic resistance genes or pathogenicity genes or history of infections in humans and animals (Herman et al., 2019). There are currently 17 species of yeast on the QPS list, including three species of *Saccharomyces* sensu stricto (*Saccharomyces cerevisiae, Saccharomyces bayanus*, and *Saccharomyces pastorianus*) and 14 nonconventional yeast species including: *Schizosaccharomyces pombe*, *Cyberlindnera jadinii*, *Komagatella phaffi*, *Komagataella pastoris*, *Yarrowia lypolitica*, *Wickerhamomyces anomalus*, *Kluyveromyces lactis*, *Kluyveromyces marxianus*, *Candida cylindracea*, *Debaryomyces hansenii*, *Hanseniaspora uvarum*, *Ogataea angusta*, *Xanthophyllomyces dendrorhous* (current name *Phaffiarhodozyma*), and *Zygosaccharomyces rouxii*. In most cases of nonconventional yeast, the status given is QPS "for production purposes only," which means ensuring the absence of living cells of the production organism in the final product, which may be, for example, biomass (https://www.efsa.europa.eu/en/topics/topic/qualified-presumption-safety-qps).

Another challenge to the possibility of increasing global production of yeast proteins is the limited availability of technical infrastructure. State of Global Fermentation Capacity Report (Capacitor) published in 2023 presents data on the global availability of bioreactors. The data collected indicate that the global bioreactors' capacity is about 50.3 million L, and approximately 38% of the products generated by fermentation are proteins, with approximately 36% of the products being cellular biomass (total wet and dried) (https://www.capacitor.bio/facilities/). It should therefore be noted that microbiological protein, alongside proteins with specific activity, i.e., enzymes, is one of the main products of biotechnological processes in the world. Given the increasing demand for alternative, healthy and sustainable food grade protein sources, scientific and industrial efforts should be made to improve technical and technological performance of sustainable processing operations of microbial protein production, to build enough industrial scale infrastructure and to power the cost-effective aspects of single cell protein production of high nutritional value (Fasolin et al., 2019).

A lot of information available today about protein secretion in eukarya has been obtained from studies of yeast (Huang et al., 2017). This contributes to the possibility of using yeast in precision fermentation solutions aimed at the secretion of specific proteins, e.g., of animal and plant origin. There are some examples of precision yeast fermentation products available, like sweet proteins (Oobli, USA), whey proteins (Perfect Day Inc., USA), casein (Nutropy, France), milk proteins (Remilk, Izrael) or leghemoglobine (Impossible Foods Inc., USA) (Graham & Ledesma-Amaro, 2023). Considering naturally sweet proteins they can be used as a high-intensity, low-calorie sweetener in food and beverages with no glycemic effect. Mentioned protein was expressed in *Komagataella phaffii* (formerly *Pichia pastoris*) cells and produced *via* precision fermentation. The results of a safety and risk assessment (in silico allergenicity, in vitro genotoxicity, and 90-day dietary toxicity tests) evidenced that the protein, named brazzein, derived from *K. phaffii* is considered safe for use as a sweetener in food technology (Lynch et al., 2023). In a world that will be faced with protein shortages, however, the white possibility of obtaining animal proteins, such as the above-mentioned milk protein preparations, will be of particular importance.

11.2 Nutritional value of yeast biomass proteins

Protein is an essential and major building block of all live cells and molecules responsible for enzymatic metabolic processes. Yeast biomass can consist of high-protein concentration (30%–65%) in dry weight, depending on the yeast species and growth conditions, being great nonconventional sources of amino acids and other nutrients (Ganeva et al., 2020; Ma et al., 2023; Onofre et al., 2017). Yeast proteins are made of 18 amino acids, of which 40%–48% are nutritionally important exogenous amino acids. The most representative amino acids are glutamic acid, glutamine, alanine, isoleucine, leucine, valine, threonine, and lysine (two last limited in plant proteins), with lower content of sulfur amino acids—methionine and cysteine (Ganeva et al., 2020; Ma et al., 2023; Millward, 2012; Onofre et al., 2017). However, the Springer Proteissimo™ 101 yeast protein preparation available on the market (Biospringer—Lesaffre), containing at least 75% protein, is characterized by a complete, balanced composition of exogenous amino acids, containing significantly more phenylalanine and tyrosine, lysine, leucine, valine, isoleucine, threonine, histidine, and tryptophan with a comparable content of methionine and cytosine to the WHO/FAO/UNU 2007 protein standard with defined reference amino acid profile (Springer Proteissimo™ 102 - Biospringer Yeast protein, n.d.; Pacheco et al., 1997). Other examples of yeast protein preparations available on the market, are AngeoPro F80 (Angel Yeast, China) or ProteVin (Nextferm, Izrael). It was observed that various yeast protein preparations exhibited protein nutritive values in the range of 70% to 90% of casein (Pacheco et al., 1997).

Nutritional value of yeast biomass depends on metabolic phase, essential amino acid composition, and bioavailability (Pacheco et al., 1997). The amino acid composition of yeast protein can be modeled by the growth conditions of these microorganisms. Above facts makes yeast biomass a very valuable, high-quality, alternative source of protein, important in maintaining growth and supporting human and animal health. However, in order for these novel protein sources to be broadly used in future food products, their digestion and bioavailability must be assessed, as well as their safety for consumers must be clearly demonstrated. In particular, mycoproteins may become novel allergens triggering adverse reactions and, therefore, a comprehensive allergenicity risk assessment is needed but also toxicity assessment of emergent proteins is of utmost importance, especially when nonconventional species are proposed as a source (Scientific Foresight Unit (STOA) EPRS, 2024).

Yeasts are heterotrophic microorganisms. They use organic substances of animal and plant origin but also inorganic nitrogen compounds. Different relatively abundant wastes streams from agrifood industry can be used as and low-cost feedstock source of carbon and nitrogen for wide range of yeast species cultivation directed to protein synthesis, e.g., soy molasses, soybean wastewater, corn steep liquor, sugarcane molasses, potato starch wastewater, cheese whey, wheat and rice bran and husk, apple pomace, banana waste, pineapple waste, cauliflower waste, orange peel, wheat straw, sweet orange, paper mill waste, cassava waste, wood shavings, sawdust, corn cobs, sugar beet pulp, coconut waste, grape waste, mango waste, lignocellulosic biomass (cellulose and hemicellulose waste), and other. Some of the raw materials need to be hydrolyzed by physical, chemical, and enzymatic methods before being used as substrates for yeasts (Rasool et al., 2023). Their use does not compete with food or feed crop production but must comply with relevant regulatory standards, including those related to food and feed safety or environmental protection. There are different factors which influence on production efficiency of single cell proteins like substrate and media composition used for cultivation (C and N sources), producing microorganism, size and age of inoculum, as well as environmental factors like pH, temperature, and degree of culture aeration, time of cultivation as well as the method of conducting cultivation on a biofermentor scale (Lapeña et al., 2020). The biomass productivity has a crucial meaning in the economic aspect of yeast protein production. Hence, nonconventional yeast, which is often characterized by higher biomass productivity compared to conventional ones, can become a very efficient source of protein. Many yeasts can grow both aerobically and under oxygen limitation. In the presence of oxygen, they convert sugars to CO_2, energy, and biomass, while in anaerobic conditions, alcoholic fermentation occurs, leading to not efficient biomass growth. In Crabtree-positive yeast aerobic fermentation can occur at high sugar concentrations lowering biomass productivity. Therefore, in yeast biomass propagation, the supply of oxygen and balanced availability of carbon source is necessary for optimum biomass and protein production (Kayikci et al., 2015; Zitomer & Lowry, 1992). Different yeast species can use for growth ethanol or methanol. These forms of carbon can be obtained as a result of chemical catalysis of carbon dioxide, supported by electrical energy. This process is analogous to photosynthesis but the energy produced in the chloroplasts of plants is replaced by electrical energy, e.g., solar or wind, and the product of carbon dioxide transformation is not glucose, but alcohol. The resulting product can be used as a carbon source for yeast to synthesize food for humans, animal consumption, and aquaculture (Bell et al., 2022). Bell et al. (2022) indicated that the yeast *Saccharomyces cerevisiae* VITF1 pre-adapted to efficient growth on ethanol as a carbon source in a minimal medium, using classical selection techniques, produced high-quality, food-grade, single-cell protein. After 24 h of cultivation of discussed strain on the medium with ethanol as a sole carbon source and urea as nitrogen source, lacking vitamins and other organic micronutrients (e.g., inositol), the final cell density was approx. 104 g per L with the protein content of 32.5%, 38% of dietary fiber, 10% of lipids, and 12.7% of trehalose. The noted biomass yield was approx. 0.45 g yeast per g of ethanol and productivity approx.1.53 g yeast per L per h. It was calculated that 151 g of the protein per day would provide the recommended daily intakes (RDI) for all essential amino acids, while the whole biomass (108 kg) produced during 24 h hours cultivation at 3000 L fermentor scale would theoretically fulfill RDI for vitamins from B group of 240 adults. The electro–microbial process of carbon dioxide utilization could therefore generate of complex, organic food on demand with potentially high yields, enabling food production in climatically challenged environments, ex. deserts.

The spent biomass of *Saccharomyces* sp. yeast after alcoholic fermentation is also an very good source of protein, supplementing macro- and micronutrient and vitamin B-complex at the same time (Pacheco et al., 1997).

Yeast protein can effectively substitute conventional plant origin proteins in animal, including aquaculture, feeding (Lapeña et al., 2020). For example, after substitution of soybean meal or barley in dairy cow diets by *Cyberlindnera jadinii* yeast as a protein source no adverse effect on milk production and milk composition have been observed (Kidane et al., 2022). The use of yeast biomass as a full-value feed is important not only because of the possibility of limiting the feeding of plants to animals, and thus increasing the availability of agricultural areas for growing plants for human consumption, but it can also be a solution for areas with a challenging climate for agriculture, with low availability of typical grassland and limited potential to grow crops for animals. However, long-term effects of yeast supplementation, in combination with varying silage qualities, is important to adequately describe effects of the diet on milk production and milk composition according to quoted data.

11.3 Methods of obtaining yeast protein isolates

Yeast cells are covered by thick cell walls. Yeast cell wall polymers, such as mannoproteins and β(1,3/1,6)-glucan, are considered to be dietary fibers that are not subject to the action of animal digestive enzymes but they can be digested by the enzymes of beneficial microorganisms living in the digestive tract, acting prebiotically. In the process of producing yeast-derived protein preparations, the cell wall is disintegrated or enzymatically digested, while the cytoplasm components are releasing outside the cell (Ganeva et al., 2020). The necessity of disintegrating yeast cells in order to increase the bioavailability of intracellular protein for animals is a certain limitation, increasing the costs of the process. On the other hand, the conditions of yeast cultivation can influence the composition of cell wall polymers, increasing or decreasing the content of mannoproteins, β-glucans, and chitin and changing the chemical structure of these polymers, and thus probably their digestibility and therapeutic properties. Mentioned cell wall polymers act as prebiotics, antimicrobial substances and improve immune function and these fiber components may have important value as food ingredients (Bastos et al., 2015; Bastos et al., 2022; Bzducha-Wróbel et al., 2018; Bzducha-Wróbel et al., 2024). Therefore, appropriately cultivated yeast biomass can be both a source of protein and perform therapeutic functions in the prevention of animal diseases caused by pathogenic microorganisms. These are important properties in the context of the need to reduce the use of antibiotics in animal treatment, in line with the objectives of the From Farm to Fork strategy arising from the European Green Deal (Bzducha-Wróbel et al., 2024).

The factor which negatively influences on yeast protein nutritional value for human is the high content of nucleic acids, particularly RNA. The nucleic acid content of yeast cells can be between 6 to 15 grams per 100 grams of crude protein and their content must be reduced to approx. 2 g/ 100 g of dry biomass before a substantial amount of the protein is used for human nutrition. In humans, purines are metabolized to uric acid, in the absence of the uricase that converts uric acid to allantoin eliminated through urine. Overproduction of uric acid and its increase in serum is inversely associated with different disease states (Ganeva et al., 2020; Pacheco et al., 1997). However, nutritional studies have proved that dietary nucleotides have also multiple valuable physiological and biological functions, including stimulation of positive intestinal flora, maintaining normal growth and development in infants, modulating immune function, gastrointestinal growth and development, reducing the incidence of diarrheal episodes, and hepatic morphology and function (Ding et al., 2021).

To obtain yeast protein derivatives with lower nucleic acid content different methods can be employed. The intact yeast cell consist of soluble cytoplasmic proteins (approx. 40%), ribosomal proteins (approx. 40%) and proteins that build or are bounded to cell wall and/or cell membranes (approx. 20%). Yeast protein solubilization occurs in alkali condition at pH 12. The process of pH decreasing cause the precipitation of protein fractions at their isoelectric point (the range of pH 4.5–6) for most of the cytoplasmic proteins in yeast. During the process of yeast biomass disruption the cytoplasmic proteins become complexed with nucleic acids and mostly precipitate in the isoelectric point pH (4.2) as a nucleoprotein complex. Some 50%–60% of this nucleoprotein may contain 15–20% of nucleic acid. Different chemical solutions have been proposed in an attempt to separate protein from yeast biomass with lower nucleic acid contents, e.g. application of a chaotropic anion (e.g., NaClO4) at 0.5 M that was effective in reducing the RNA concentration in the yeast biomass from 7.5% to 2.2% in the isoelectric precipitate. By weakening the hydrogen-bonded structure of water chaotropic anions alter the interfacial properties in protein association and reduce the RNA content of the isolated proteins. The content of RNA may be also lowered by phosphorylation. Phosphorous oxychloride (POCl3) at pH 9.0 is proposed to be used for yeast nucleoprotein phosphorylation of epsilon-amino groups what result in 90% reduction in the RNA content of iso-electrically precipitated protein (Ma et al., 2023; Onofre et al., 2017; Pacheco et al., 1997). However it should be noted that chemical treatment, often applied in combination with heating, may cause denaturation and amino acid degradation and formation of toxic artefacts, e.g., lysinoalanine, which decreases the nutritive properties of the proteins. Utilization of exogenous nucleases for RNA degradation is limited because of the economic aspects, while autolysis promotes protein hydrolysis (Ganeva et al., 2020).

Yeast protein concentrates consist of approx. one and a half to two times more protein comparing to yeast biomass with lower nucleic acid content (Pacheco et al., 1997). Ganeva et al. (2020) noted that pulsed electric field (PEF) treatment, the nonthermal method for extraction, could be applied for efficient yeast protein releasing and production of protein concentrates of *Saccharomyces* origin, as well as non-Saccharomyces yeast, e.g., *Hansenula polymorpha*. Discussed process coupled with ultrafiltration (10 kDa cutoff) allowed to yeast protein obtain concentrates with lower nucleic acid content (protein/ nucleic acid ratio=100/4.5) in comparison with cell lysates obtained by mechanical disintegration. High hydrostatic pressure can also be used for yeast protein extraction (Ganeva et al., 2020). Summarizing, the main drawbacks related to yeast protein large-scale production are efficiency of protein recovery and high level of nucleic acid content.

11.4 Technological properties of yeast protein for food production

Yeast origin single cell proteins are mainly used as protein components in feeds despite their nutritional, functional, and sensory benefits for various food applications. Yeast proteins present interesting technological properties that can be used for innovative food product development. Nowadays, their application is gaining attraction in food industry. Physicochemical properties of yeast proteins, connected with rheological properties of protein solution, are important for food product formulation and are affected by different factors, like protein molecular weight, composition, protein size and shape, level of hydration, intermolecular interaction, and hydrophobicity. Yeast proteins show good water solubility and water-holding capacity approx. 2.5–6.0 mL of water/g of protein, depending on the source of yeast and growth conditions (structure and isoelectric point of proteins), extraction conditions of proteins and final purity of the extract (Ma et al., 2023).

Yeast protein concentrates may be used as flavoring additive, being responsible for creation of fresh, meaty, and cheesy taste as well as umami and kokumi tastes that allows partial replacements for salt in the product. Many properties of yeast proteins makes it possible to limit the addition of other ingredients to food, and thus achieving products with a white label (Ma et al., 2023). Yeast proteins can be used for formulation of meat alternatives, seasonings, sauces, soups, dips or spreads what provides ideas for innovative products for flexitarian consumers. As meat extenders yeast proteins may improve the products' texture (bond strength and hardness) and dimensional stability, reduce cooking losses and costs in food processing (improved tenderness, juiciness, and stability) Proteins derived from *Cyberlindnera jadinii* have very interesting functional attributes for technological application being highly scalable, efficient, and sustainable, with economical production (Sousa-Silva et al., 2021).

11.5 Conclusions

Growing global demand for protein will probably not be able to meet sustainably by increasing plant biomass synthesis and meat and dairy production. New solutions are needed and yeast SCP as well as yeast protein isolates produced by fermentation are promising direction. Yeast protein provides a relatively small proportion of current human nutrition, being mostly used for animal purposes. Considering growing global demand for protein it is likely to make yeast proteins increasingly important in human nutrition. They are very good source of nutritionally important amino acids and other nutrients. The valorization of side streams agrifood wastes by using yeast to cultivate biomass rich in protein is possible. However, the use of nonconventional carbon sources requires the development of single cell protein production processes using yeasts other than *Saccharomyces*, with lower nutritional requirements and the ability to use sugars other than glucose for effective growth. Therefore the evaluation of the safety aspects of using nonconventional yeast in the production of nutritional quality proteins is very important issue as well as the high-protein products obtained in this way.

References

Bastos, R., Coelho, E., & Coimbra, M. A. (2015). Modifications of Saccharomyces pastorianus cell wall polysaccharides with brewing process. *Carbohydrate Polymers, 124*, 322–330. https://doi.org/10.1016/j.carbpol.2015.02.031, http://www.elsevier.com/wps/find/journaldescription.cws_home/405871/description#description.

Barnett, J. A. (2004). A history of research on yeasts 8: Taxonomy. *Yeast (Chichester, England), 21*(14), 1141–1193. https://doi.org/10.1002/yea.1154.

Bastos, R., Oliveira, P. G., Gaspar, V. M., Mano, J. F., Coimbra, M. A., & Coelho, E. (2022). Brewer's yeast polysaccharides—a review of their exquisite structural features and biomedical applications. *Carbohydrate Polymers*.

Bell, P. J. L., Paras, F. E., Mandarakas, S., Arcenal, P., Robinson-Cast, S., Grobler, A. S., & Attfield, P. V. (2022). An electro–microbial process to uncouple food production from photosynthesis for application in space exploration. Life (Chicago, Ill.: 1978).

Brainy Insights (2022). Yeast Protein Market Size by Type (Conventional and Organic), Application (Food, Medicine, Feed and Others), Regions, Global Industry Analysis, Share, Growth, Trends, and Forecast 2022 to 2030, Report ID: TBI-13114. https://www.thebrainyinsights.com/report/yeast-protein-market-13114?srsltid=AfmBOorz8XFa9awoqwXr4kKYBVruqihKLhNto5XmhLuhFxYi5yJdUpd.

Bzducha-Wróbel, A., Błażejak, S., Kieliszek, M., Pobiega, K., Falana, K., & Janowicz, M. (2018). Modification of the cell wall structure of Saccharomyces cerevisiae strains during cultivation on waste potato juice water and glycerol towards biosynthesis of functional polysaccharides. *Journal of Biotechnology, 281*, 1–10. https://doi.org/10.1016/j.jbiotec.2018.06.305.

Bzducha-Wróbel, A., Farkaš, P., Bieliková, S., Čížová, A., & Sujkowska-Rybkowska, M. (2024). How do the carbon and nitrogen sources affect the synthesis of β-(1,3/1,6)-glucan, its structure and the susceptibility of Candida utilis yeast cells to immunolabelling with β-(1,3)-glucan monoclonal antibodies? *Microbial Cell Factories*.

Carbon Foot Print Cofalec. https://cofalec.com/en/carbon-footprint/.

Choi, K. R., Yu, H. E., & Lee, S. Y. (2022). Microbial food: Microorganisms repurposed for our food. *Microbial Biotechnology, 15*(1), 18–25. https://doi.org/10.1111/1751-7915.13911, http://onlinelibrary.wiley.com/journal/10.1111/(ISSN)1751-7915.

Comitini, F., Canonico, L., Agarbati, A., & Ciani, M. (2023). Probiotic function of non-saccharomyces yeasts: New insights in agri-food industry. *Microorganisms*.

Ding, T., Song, G., Liu, X., Xu, M., & Li, Y. (2021). Nucleotides as optimal candidates for essential nutrients in living organisms: A review. *Journal of Functional Foods, 82*, 104498. https://doi.org/10.1016/j.jff.2021.104498.

Epsztein, J. (2022). Environmental benefits of eating mycoprotein. *Nature, 605*, 34–35.

Fasolin, L. H., Pereira, R. N., Pinheiro, A. C., Martins, J. T., Andrade, C. C. P., Ramos, O. L., & Vicente, A. A. (2019). Emergent food proteins – Towards sustainability, health and innovation. *Food Research International, 125*, 108586. https://doi.org/10.1016/j.foodres.2019.108586.

Ganeva, V., Angelova, B., Galutzov, B., Goltsev, V., & Zhiponova, M. (2020). Extraction of proteins and other intracellular bioactive compounds from Baker's yeasts by pulsed electric field treatment. *Frontiers in Bioengineering and Biotechnology, 8*, 552335. https://doi.org/10.3389/fbioe.2020.552335.

Geijer, C., Ledesma-Amaro, R., & Tomas-Pejo, E. E. (2022). Unraveling the potential of non-conventional yeasts in biotechnology. *FEMS Yeast Research.*

Graham, A. E., & Ledesma-Amaro, R. (2023). The microbial food revolution. *Nature Communications, 14*(1). https://doi.org/10.1038/s41467-023-37891-1, https://www.nature.com/ncomms/.

Herman, L., Chemaly, M., Cocconcelli, P. S., Fernandez, P., Klein, G., Peixe, L., Prieto, M., Querol, A., Suarez, J. E., Sundh, I., Vlak, J., & Correia, S. (2019). The qualified presumption of safety assessment and its role in EFSA risk evaluations: 15 years past. *FEMS Microbiology Letters*(1).

Huang, M., Bao, J., Hallström, B. M., Petranovic, D., & Nielsen, J. (2017). Efficient protein production by yeast requires global tuning of metabolism. *Nature Communications, 8*(1). https://doi.org/10.1038/s41467-017-00999-2, http://www.nature.com/ncomms/index.html.

Hugot, C., Poirier, M., Spatz, M., Costa, G. D., Michaudel, C., Lapiere, A., Danne, C., Martin, V., Langella, P., Sokol, H., Michel, M. L., Boyaval, P., & Richard, M. L. (2023). Cyberlindnera jadinii and Kluyveromyces lactis, two fungi used in food processes, have potential probiotic effects on gut inflammation. *Msystems, 8*(6). https://doi.org/10.1128/msystems.00841-23, https://journals.asm.org/doi/10.1128/msystems.00841-23.

Kidane, A., Gregersen Vhile, S., Ferneborg, S., Skeie, S., Olsen, M. A., Torunn Mydland, L., Øverland, M., & PrestlØkken, E. (2022). , Cyberlindnera jadinii yeast as a protein source in early- to mid-lactation dairy cow diets: Effects on feed intake, ruminal fermentation, and milk production. *Journal of Dairy Science, 105*(3), 2343–2353. https://doi.org/10.3168/jds.2021-20139, http://www.elsevier.com/wps/find/journaldescription.cws_home/721317/description#description.

Kayikci, Ö., Nielsen, J., & Bolotin-Fukuhara, M. (2015). Glucose repression in Saccharomyces cerevisiae. *FEMS Yeast Research, 15*(6). https://doi.org/10.1093/femsyr/fov068 fov068.

Lapeña, D., Olsen, P. M., Arntzen, M., Kosa, G., Passoth, V., Eijsink, V. G. H., & Horn, S. J. (2020). Spruce sugars and poultry hydrolysate as growth medium in repeated fed-batch fermentation processes for production of yeast biomass. *Bioprocess and Biosystems Engineering, 43*(4), 723–736. https://doi.org/10.1007/s00449-019-02271-x, https://rd.springer.com/journal/449.

Lynch, B., Wang, T., Tafazoli, T. S., & Ryder, J. (2023). Safety evaluation of oubli fruit sweet protein (brazzein) derived from Komagataella phaffii, intended for use as a sweetener in food and beverages. *Toxicology Research and Application,* 1–21.

Ma, J., Sun, Y., Meng, D., Zhou, Z., Zhang, Y., & Yang, R. (2023). Yeast proteins: The novel and sustainable alternative protein in food applications. *Trends in Food Science and Technology, 135*, 190–201. https://doi.org/10.1016/j.tifs.2023.04.003, http://www.elsevier.com/wps/find/journaldescription.cws_home/601278/description#description.

Millward, D. J. (2012). Amino acid scoring patterns for protein quality assessment. *British Journal of Nutrition, 108*(2), S31. https://doi.org/10.1017/S0007114512002462.

Onofre, S. B., Bertoldo, I. C., Abatti, D., & Refosco, D. (2017). Chemical composition of the biomass of Saccharomyces cerevisiae - (Meyen ex E. C. Hansen, 1883) yeast obtained from the beer manufacturing process. *International Journal of Environment, Agriculture and Biotechnology, 2*(2), 558–562. https://doi.org/10.22161/ijeab/2.2.2.

Pacheco, M. T. B., Caballero-Córdoba, G. M., & Sgarbieri, V. C. (1997). Composition and nutritive value of yeast biomass and yeast protein concentrates. *Journal of Nutritional Science and Vitaminology, 43*(6), 601–612. https://doi.org/10.3177/jnsv.43.601, http://www.jstage.jst.go.jp/browse/.

Rasool, K., Hussain, S., Shahzad, A., Miran, W., Mahmoud, K. A., Ali, N., & Almomani, F. (2023). Comprehensive insights into sustainable conversion of agricultural and food waste into microbial protein for animal feed production. *Reviews in Environmental Science and Biotechnology, 22*(2), 527–562. https://doi.org/10.1007/s11157-023-09651-6, https://www.springer.com/journal/11157.

Scientific Foresight Unit (STOA) EPRS. (2024). European Parliamentary Research Service Study Panel for the Future of Science and Technology. Alternative protein sources for food and feed. https://doi.org/10.2861/999488.

Sousa-Silva, M., Vieira, D., Soares, P., Casal, M., & Soares-Silva, I. (2021). Expanding the knowledge on the skillful yeast Cyberlindnera jadinii. *Journal of Fungi, 7*(1), 36. https://doi.org/10.3390/jof7010036.

Springer Proteissimo™ 102 - Biospringer Yeast protein, (n.d.). https://biospringer.com/en/springer-proteissimo/.

Turck, D., Castenmiller, J., Henauw, Hirsch-Ernst, K. I., Kearney, J., Maciuk, A., Mangelsdorf, I., McArdle, H. J., Naska, A., Pelaez, C., Pentieva, K., Siani, A., Thies, Tsabouri, S., Vinceti, M., Cubadda, F., Engel, K. H., Frenzel, T., Heinonen, M., ... van, al) Loveren, H. (2019). Safety of an extension of use of Yarrowia lipolytica yeast biomass as a novel food pursuant to Regulation (EU) 2015/2283. *EFSA Journal, 17.*

Zitomer, R. S., & Lowry, C. V. (1992). Regulation of gene expression by oxygen in Saccharomyces cerevisiae. *Microbiological Reviews, 56*(1), 1–11. https://doi.org/10.1128/mmbr.56.1.1-11.1992.

Oleochemicals production from microalgae and yeast

12

Mihai Lucian Lomanar[1,2], *Rudolphus Antonius Timmers*[3], *Marcela Levío-Raimán*[1,2], *Cristina González-Fernández*[1,2], *Antonio David Moreno*[4], *and María José Negro*[4]

[1]Department of Chemical Engineering and Environmental Technology, School of Industrial Engineering, University of Valladolid, Valladolid, Spain, [2]Institute of Sustainable Processes, Valladolid, Spain, [3]Area of Circular Economy, CARTIF Technology Centre, Boecillo, Valladolid, Spain, [4]Advanced Biofuels and Bioproducts Unit, Department of Energy, Research Center for Energy, Environment and Technology (CIEMAT), Madrid, Spain

12.1 Introduction

12.1.1 Importance of the microbial oils in the industry

Fossil fuels—including coal, oil, and natural gas—have been the world's primary source of energy for over a century and nowadays they account for over 80% of the global energy sources (Neupane, 2023). The transport sector accounts for the most significant oil consumption in 2023 (about 60%). World oil demand is forecast from 2023 to 2030 to rise about 0.4%, supported by increased jet fuel and feedstock from petrochemical sector (IEA, 2024). In addition to the increasing demand, another concern about fossil fuels is the greenhouse gas emissions they have associated. Gases released (mostly carbon dioxide) to the environment when fossil fuels are burned are the major causes of deforestation, ozone layer depletion, and global warming. To diminish fossil fuel's carbon fingerprint and mitigate their effects related to climate change, new renewable energy sources are needed, and environmental policies must be changed (Neupane, 2023).

One promising alternative to fossil fuels is oleochemicals. Those molecules are composed of lipids present in animal or vegetable fats and exhibit a high energy. The energy produced when oleochemicals are burned is very similar—up to 91%—to fossil fuels (Gayathri et al., 2023). In the last decades, many companies have been developing products based on plant and animal fats and the production of oleochemical seeds has been constantly growing since then. However, lipid production yields from these resources are too low to meet global fuel needs. Alternatively, the oleaginous microorganisms should be an interesting source of oils because of their fast growth rates, high lipid content, ability to thrive in diverse environmental conditions, and lack of competition with the food sector (Szczepańska et al., 2022).

Microbial oils (MO), also called single-cell oils (SCOs), are lipids produced under some metabolic stress conditions by oleaginous microorganisms, including some species of bacteria, fungi, yeast, and microalgae. It should be highlighted that these

Eukaryotic Microorganisms as Sources of Bioproducts. DOI: https://doi.org/10.1016/B978-0-443-30188-9.00009-3

microorganisms are characterized by a lipid content higher than 20% of their dry weight. In comparison with plants or animals, the main advantage of MO is that microorganisms display a faster growth rate and that they can accumulate up to 60–70% of their dry cell weight (Tomás-Pejó et al., 2023). In addition, their flexible metabolism allows them to use a wide range of carbon sources to grow or to accumulate energy for stressful situations. It should be highlighted that these microorganisms cannot only grow in suspended cultivation but can be also cultured in solid-state fermentation mode. This is important as products attained in this kind of fermentation are normally produced at lower cost, mainly related to the lower energy and water consumption requirements.

Besides the technological aspect, industries have an increasing interest in the large-scale production of MO. Nowadays, the actual production cost of MO is higher than plant or animal oil because of the feedstock's cost and the need to maintain axenic conditions. On the other hand, the prices of vegetables and animal oils range between 0.30 $ to over 100 $/kg (Szczepańska et al., 2022). Therefore, the identification of microorganisms capable of producing high lipids content with structure and composition like that of high-value fats, and subsequent large-scale production can present an enormous market interest (Szczepańska et al., 2022).

12.1.2 Why to choose microalgae and yeast for microbial oil production?

Microalgae and yeast offer several benefits for microbial oil production compared with other microorganisms such as bacteria and fungi. Bacteria usually do not accumulate significant quantities of triacylglycerols (TAG) compared with microalgae and yeast (Dias et al., 2019). Also, fungi have lower lipid accumulation yields respect to yeast, and their FA profiles are different (Rajak et al., 2022). The following section describes the advantages of using oleaginous microalgae and yeast for oil production.

12.1.2.1 Oleaginous microalgae

Microalgae are also an interesting option for lipid production due to their higher lipid accumulation ability, faster growth rate, and higher photosynthetic efficiency in comparison to plants. Microalgae capture CO_2 and convert it into organic compounds via the Calvin cycle in photosynthesis process; this carbon fixed is subsequently utilized to generate acetyl-CoA, a key building block for FA synthesis (Sun et al., 2019). Besides, some nutrients limitation (nitrogen and phosphorus) and the supplementation of several cations stimulates the lipid accumulation in microalgae (Hess et al., 2018). Phosphorus limitation changes the profile of fatty acids (FAs) (Yun et al., 2019). On another hand, focusing on productivities some studies have reported microalgae lipid productivities around 136,900 L ha^{-1} $year^{-1}$, which are several times higher than those achieved by oil palm plantations (22,780 L ha^{-1} $year^{-1}$) (Tibocha-Bonilla et al., 2018). Another advantage of using microalgae as a source of oils is that they can be cultivated on nonarable land, and they do not require herbicides or pesticides like oilseed crops (Zhu et al., 2022). Given their photosynthetic ability, microalgae can use CO_2 from different industrial waste streams, rendering this biomass particularly beneficial as it also contributes for the elimination of greenhouse gas emissions (Hoyos et al., 2024). Among the oleaginous microalgae, the most common strains include *Chlorella* sp., *Neochloris oleoabundans*, *Cylindrotheca* sp., *Euglena gracilis*, *Spirulina maxima*, *Navicula* sp., *Scenedesmus* sp., and *Dunaliella* sp. Their lipid levels range between 20% and 50% of total cell dry weight, although there are some species like *Botryococcus braunii* whose lipid content can reach up to 80% (Papone et al., 2016). For all these reasons, microalgae are also a suitable option for oil production.

12.1.2.2 Oleaginous yeasts

Yeasts have evolved to colonize a huge diversity of environments and develop specific adaptations for surviving in each one of them. One of these adaptations is the ability to accumulate at least 20% intracellular oil of their dry weight, the so-called oleaginous microorganisms (Žganjar et al., 2024). In the presence of an excess of carbon and after exhaustion of an essential nutrient in the medium (usually nitrogen) (Tomás-Pejó et al., 2023), oleaginous yeasts can redirect the carbon flux from energy production to lipid synthesis. These unique microorganisms exhibit advantages for lipid production over other oleaginous microorganisms, such as filamentous fungi or microalgae, due to their unicellular form, short duplication times, capability to use a wide variety of raw materials to grow, and their easy cultivation in large fermenters (Patel et al., 2020).

So far, about 70 of the 1600 species of oleaginous yeasts have been identified and characterized, and the number is still growing each year. The most studied species belong to the division *Basidiomycota* (e.g., *Cutaneotrichosporon curvatum*, *Trichosporon oleaginosus*, *Rhodotorula toruloides*) and *Ascomycota*; subdivision *Saccharomycotina* (e.g., *Yarrowia lipolytica*, *Lipomyces starkeyi*). Some *Candida* species (*C. viswanathii*, *C. tropicalis*, *C. oleophila*) also exhibit oleaginous phenotypes. Depending on the species, the lipid content varies and can reach up to 70% of cell dry weight. Most of these species can use low-cost waste materials including sugar beet molasses, waste cooking oil, waste paper, and agroindustrial wastes, which makes them suitable platforms to use in waste management applications. In addition, most of these species are susceptible to improvement through genetic modification and are also suitable for large-scale fermentation to increase lipid production. The lipid content and FA composition differ from one yeast species to the other, but the main FA present in MO are palmitic (C16:0), palmitoleic (C16:1), stearic (C18:0), oleic (C18:1), and linoleic (C18:2) acids (Szczepańska et al., 2022).

12.1.3 Applications of microbial oils

MO can be used in different applications depending on the FA compositions. MO rich in TAG can be used to produce biofuels such as biodiesel. The unsaturated FA can be transesterified to produce fatty acid alkyl esters (FAAE), which have similar properties to petroleum-based fuels (de Oliveira et al., 2020). Given the current environmental situation and the nonstopping depletion of fossil fuels, biodiesel seems a suitable alternative due to its renewability, biodegradability, and nontoxicity (Bharathiraja et al., 2017). Other biofuels that have been gaining interest in recent years are sustainable aviation fuels (SAF). SAF can substitute traditional aviation kerosene without any modification to the current aircraft or fuel infrastructure, while also decreasing CO_2 emissions by 55%–92% (Song et al., 2023). Starting from lipids as raw materials for SAF production, the most common route is the Hydroprocessed Esters and Fatty Acids Synthetic Paraffinic Kerosene (HEFA-SPK). Within this route, triglycerides, the dominant component of lipids, are subjected to a catalytic hydroteatment and further to a deoxygenation process to obtain bioaviation fuels. For more information on this topic, a recent review has been released (Song et al., 2023).

Another promising application is to use microbial lipids as feedstock in chemical industries to produce different oleochemicals. Oleochemicals can be used as a substitute for petroleum-based products known as petrochemicals. Traditional oleochemicals are obtained from animal and vegetable fats, but MO can contribute to the high demand for sustainable feedstocks. The most relevant oleochemicals include FA, FAME, fatty alcohols, and fatty amines and can be used in a wide variety of industries such as soaps, surfactants, detergents, lubricants, polymers, cosmetic additives, and many more (Greses et al., 2024). In the pharmaceutical industry, MO can act as retardants for dental cements, and for the preparation of liniments, demulcents, emollients, or laxatives (Bharathiraja et al., 2017).

MO can also serve as stabilizing agents, thickening agents, water-binding agents, emulsifiers, and many more, which are used in food processing operations (Bharathiraja et al., 2017). Among FA, there is a special type very important for the maintenance of physiological functions in the human body, the so-called polyunsaturated fatty acids (PUFAs). This group includes several classes such as omega-3 (EPA and DHA) and omega-6. PUFAs of these classes play important roles as structural components of membrane phospholipids and as precursors of the eicosanoids, which are hormone-like substances involved in inflammatory reactions. Humans lack the ability to synthesize these PUFAs, and thereby, should be included in diets in an external input mode (Ochsenreither et al., 2016). The best source of PUFAs is fish oil. Yet, some limitations concerning such as their odor, low stability, and the presence of contaminants including heavy metals and other pollutants, limit their use as food additives (Bellou et al., 2016). Because of that, oleaginous microorganisms are regarded as an alternative source of PUFAs (Galán et al., 2019).

Other applications of MO in the industry are the formulation of aquafeeds for aquaculture industry. The most frequently used species in aquaculture are *Chlorella*, *Tetraselmis*, *Isochrysis*, *Pavlova*, *Phaeodactylum*, *Chaetoceros*, *Nannochloropsis*, *Skeletonema*, and *Thalassiosira* (Shah et al., 2018). Microalgae such as *Dunaliella salina*, *Haematococcus pluvialis*, and *Spirulina* are also used as a source of natural pigments for the culture of prawns, salmonid fish, and ornamental fish. In addition, the genera *Arthrospira* and *Chlorella* are used also to feed many types of animals: cats, dogs, aquarium fish, ornamental birds, horses, poultry, cows, and breeding bulls (Remize et al., 2021).

Recently, there has been growing interest in producing unusual FA with significant industrial value, especially essential in medicine, nutrition, and cosmetic industries. Some studies have been focused on biology techniques and gene editing technologies to investigate the ability of *Yarrowia lipolytica* to produce unusual fatty acids, including ricinoleic acid through metabolic engineering (Shah et al., 2022).

12.2 Biology of microalgae and yeasts

Understanding oleaginous microorganisms' metabolism is crucial to knowing the key reactions needed to produce lipids and the possibilities to improve the process of MO production, maximize yields and minimize costs in the industry.

12.2.1 Microalgae metabolism

Microalgae are a group of eukaryotic autotrophic microorganisms that can convert inorganic carbon (CO_2 from the atmosphere) into organic compounds using the energy from sunlight through photosynthesis (Fig. 12.1). Although there are some multicellular species, most of them are unicellular or form filaments. Microalgae have colonized a wide range of niches, not just aquatic but also terrestrial. According to scientific literature, there are around 50,000 microalgae species but only 30,000 have been studied and analyzed (Morales et al., 2021). Their ability to survive under harsh conditions is responsible for their actual diversity and ability to regulate lipid metabolism depending on the environmental conditions (Chen & Wang, 2021).

Microalgae can produce different types of metabolites both primary (lipids, carbohydrates, and proteins) and secondary (vitamins, pigments, carotenoids, waxes, ketones, or lectins). Proteins are synthetized mainly during the growth stage while

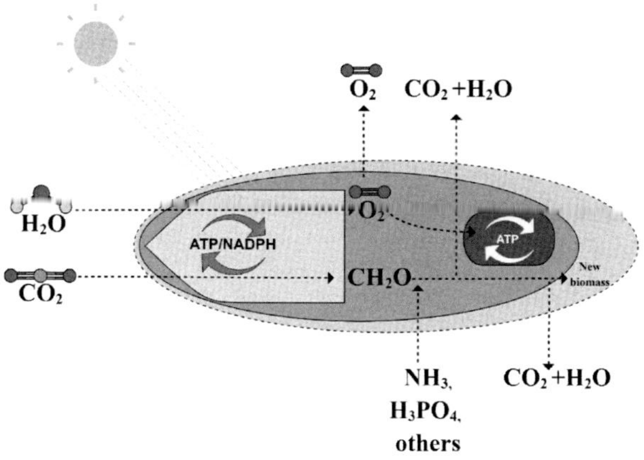

Figure 12.1 Microalgae photosynthesis process (Janssen, 2016).
Source: Adapted from Janssen, M. (2016). Microalgal photosynthesis and growth in mass culture. *Advances in Chemical Engineering, 48*, 185–256. https://doi.org/10.1016/BS.ACHE.2015.11.001.

lipids and carbohydrates are synthesized mainly as energy storage molecules during nutrient deficiency or light limitation (Vuppaladadiyam et al., 2018). Different lipid groups exist, such as glycerolipids, sterols, hydrocarbons, and waxes. Glycerolipids consist of glycerol combined with one, two, or three FA groups attached. Glycerolipids can also be divided based on specific functions: storage and structural lipids. Among structural lipids, membrane lipids usually contain two fatty acid groups and have a polar group bound to glycerol structure. According to their carbon number, FA can be classified into two main groups. FA, having 14–20 carbons, are mainly used to produce biofuels while PUFAs with more than 20 carbon atoms are mostly used as health food supplements (Sun et al., 2019). Indeed, some PUFAs present in microalgae lipids are rarely found in plant oils, such as C16:2 (hexadecadienoic acid), C16:3 (hexadecatrienoic acid), C16:4 (hexadecatetraenoic acid) and EPA (Morales et al., 2021).

Microalgae can synthesize lipids via two different pathways: fatty acid biosynthesis or TAG biosynthesis pathway. The fatty acid biosynthesis pathway, as depicted in Fig. 12.2, takes place primarily in the chloroplasts and produces saturated and monounsaturated fatty acids (MUFAs). This pathway begins with the conversion of acetyl-CoA to malonyl-CoA catalyzed by acetyl-CoA carboxylase (ACCase), which is the rate-limiting step in fatty acid synthesis. Malonyl-CoA is then transferred to an acyl carrier protein (ACP) catalyzed by malonyl-CoA transacylase (MAT). After that, malonyl-CoA is elongated in a reaction cycle catalyzed by the fatty acid synthetase (FAS) complex with repeated rounds of condensation by β-ketoacyl-ACP synthase (KAS), reduction by β-ketoacyl-ACP reductase (KAR), dehydration by β-hydroxyacyl-ACP dehydrase (HAD), and repeated reduction by enoyl-ACP reductase (EAR) to form SFA such as palmitic acid (C16:0) or stearic acid (C18:0). The last

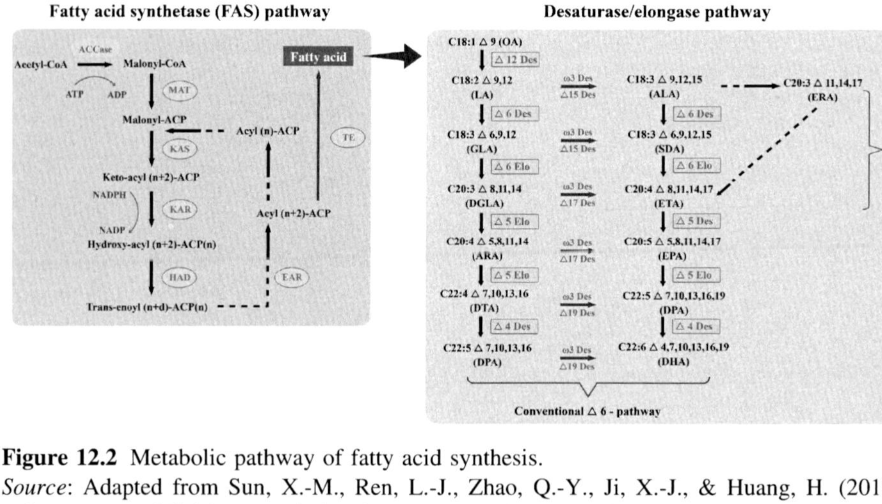

Figure 12.2 Metabolic pathway of fatty acid synthesis.
Source: Adapted from Sun, X.-M., Ren, L.-J., Zhao, Q.-Y., Ji, X.-J., & Huang, H. (2019). Enhancement of lipid accumulation in microalgae by metabolic engineering. *Biochimica et Biophysica Acta (BBA)—Molecular and Cell Biology of Lipids, 1864*(4), 552–566. https://doi.org/10.1016/j.bbalip.2018.10.004.

step of the fatty acid synthesis is the hydrolysis of acyl-ACP catalyzed by acyl-ACP thioesterases (TE) and the release of FA into the cytosol, which is ultimately incorporated into glycerolipids (Sun et al., 2019).

Fatty acid elongation begins with stearic acid in mitochondria in a reaction that involves several cycles of condensation, reduction, dehydration, and reduction and is catalyzed by three enzymes: enoyl-CoA hydratase (ECH), trans-2-enoyl CoA reductase (TER), and palmitoyl-protein thioesterase (PPT). The introduction of double bonds to produce unsaturated FA occurs via elongation and desaturation reactions catalyzed by fatty acid elongases (FAE) and fatty acid desaturases (FAD) (Su et al., 2021).

TAG synthesis on the other way takes place via the Kennedy pathway in endoplasmic reticulum and chloroplasts and involves three major steps. The first one is the acylation of glycerol-3-phosphate (G3P) by the acyl-CoA: glycerol-*sn*-3-phosphate acyl-transferase (GPAT) to produce lysophosphatidate (LPA). Then, LPA is condensed with another acyl-CoA, catalyzed by the lysophosphatidate acyl-transferase (LPAAT), to produce phosphatidate (PA), which is further dephosphorylated by the phosphatidic acid phosphatase (PAP) to generate diacylglycerol (DAG). Finally, DAG is used by diacylglycerol acyltransferase (DGAT) to synthesize TAG, which is deposited in the cytosol as lipid droplets (De Bhowmick et al., 2015; Sun et al., 2019).

12.2.2 Yeast metabolism

Among the yeast species, some of them have evolved to store high amounts of intracellular lipids under certain conditions, such as nitrogen or phosphorus limitations (Morales-Palomo et al., 2023; Tomás-Pejó et al., 2023). These are known as oleaginous yeasts, and they represent 11% approximately of all yeast species. This group has members in different clades, but the most important are the phylum Ascomycota with 26 genera, and the phylum Basidiomycota with 32 genera (Salvador-López et al., 2022). Despite the huge spectrum of species, the research and industrial applications are focused only on a few species, principally *Yarrowia lipolytica* or *Rhodotorula toruloides* because of their easy manipulation and growth (Salvador-López et al., 2022).

In oleaginous yeasts, the accumulated lipids are primarily TAGs and steryl esters (SEs). In most cases, the TAG content of the lipid in oleaginous yeasts is around 80% to 90% (w/w). Several factors such as yeast species, environmental conditions, or the election of carbon sources can influence lipid composition (Abeln & Chuck, 2021). In the case of a nonoleaginous yeast like *Saccharomyces cerevisiae*, lipids droplets contained 51.2% (w/w) TAGs and 44.4% (w/w) SEs (Leber et al., 1994). In comparison, in an oleaginous yeast like *Yarrowia lipolytica*, the TAG fraction represents 84.5% (w/w) while SEs represent 7.8% (w/w) (Leber et al., 1994). The main FA present in yeast lipids droplets are oleic (C18:1), palmitic (C16:0), linoleic (C18:2), and stearic (C18:0) acids (Abeln & Chuck, 2021).

Yeast can produce TAGs via two different pathways: the anabolic de novo or ex novo pathway. De novo synthesis occurs when there is a high C/N ratio and nitrogen deficiency in the growth medium. This pathway involves the formation of acetyl-CoA resulting from the inhibition of the Krebs cycle (Fabiszewska et al., 2019). Nitrogen depletion activates adenosine monophosphate deaminase (AMPD), which catalyzes the conversion of AMP to inosine monophosphate (IMP) and ammonia, so the cells can receive some nitrogen. The low levels of AMP have a negative impact on the Krebs cycle resulting in the inhibition of isocitrate dehydrogenase (IDH) and the accumulation of isocitrate, which is further balanced with citrate before being exported outside the mitochondria. Citrate is cleaved into acetyl-CoA and oxaloacetate (OOA) in a reaction catalyzed by the ATP-citrate lyase (ACL), which is a key enzyme for lipogenesis in oleaginous yeasts. After that, the acetyl-CoA is transformed into TAGs following the Kennedy pathway, which is stored as lipid droplets in the cytosol (Robles-Iglesias et al., 2023). Alternatively, the ex novo synthesis, on the other hand, implies the incorporation of FA and glycerol, which are intermediate products of fatty acid β-oxidation, into TAG molecules in media containing hydrophobic carbon sources (Ratledge, 2002).

12.3 Cultivation of microalgae and yeasts

12.3.1 *Microalgae cultivation*

Since microalgae are photosynthetic microorganisms, light plays an important role during lipid accumulation. In general, it has been seen that over-illumination favors MO production, but depending on the light intensity, the lipid profile is different. Under high light-intensity conditions, PUFAs might get damaged as a consequence of reactive oxygen species generation. The result is an increase in neutral lipid content and a decrease in polar lipid content (He et al., 2015). The optimum light intensity for most microalgae species is within the range of 200–400 μmol m^{-2} s^{-1} (Shekh et al., 2022). Nzayisenga et al. (2020) observed that *Scenedesmus obliquus* doubled FA content from 5.8% to 11.6% when using 300 μmol m^{-2} s^{-1}. Another study showed that the neutral lipid content of *Chlorella* sp. and *Monoraphidium* sp. increased when using 400 μmol m^{-2} s^{-1} (He et al., 2015). Light cycles have also an influence on lipid profile. Wahidin et al. (2013) tested different light/dark cycles for *Nannochloropsis* sp. at 100 μmol m^{-2} s^{-1} and observed that the highest lipid content (31.3% w/w) was obtained with 18 h light: 6 h dark cycle. Because microalgae chlorophyll absorbs red and blue light, providing light radiation at those wavelength regions could boost photosynthetic activity and lipid productivity. Blue light can enhance the synthesis of carbonic anhydrase and ribulose biphosphate carboxylase, enzymes involved in MO production (Ananthi et al., 2021).

The temperature has also a significant impact on the metabolism of microalgae, including growth and oil production. Although many species can accumulate lipids between 15 and 40°C, their optimal range is reported to be between 20 and 30°C. It has been seen that lower temperatures could boost the accumulation of unsaturated lipids, while higher temperatures could increase saturated lipids yields (Chaisutyakorn et al., 2018). In line with this, *Monoraphidium* sp. was able to grow well and produce lipids under 25–35°C, but when shifting the temperature to 40°C, the biomass decreased as well as lipid productivity (Wu et al., 2013). Operational temperature can also affect the lipid profile. Wei et al. (2015) observed that an increase in temperature from 20 to 30°C produced lower neutral lipids and PUFA and higher saturated FA and MUFAs in *Tetraselmis subcodiformis*.

Besides carbon, nitrogen, and phosphorus are the two other macronutrients needed for microalgal growth. It has been reported that nitrogen limitation can enhance cellular lipid content by redirecting carbon flux from protein to fatty acid synthesis (Hess et al., 2018). However, nitrogen deprivation also decreases growth rate and photosynthetic activity, and due to this tradeoff, lipid productivity is significantly lower (Griffiths et al., 2014). This strategy has been largely explored to obtain high lipids production yields. Latsos et al. (2020) observed an increase in lipid content in *Rhodomonas* sp. from 9 to 30% when cultured under nitrogen starvation conditions. Similar trend was observed for *Monoraphidium* sp. QLY-1 that increased lipid content from 22.5% to 44.8% under nitrogen limitation (Zhao et al., 2016). Likewise, phosphorus starvation has also been used to increase oil accumulation. Phosphorus inhibits ADP-glucose pyrophosphorylase, which is involved in starch synthesis, and consequently, metabolism flux is shifted towards lipid production (Shi et al., 2020). Yang et al. (2018) studied *Chlamydomonas reinhardtii* under nitrogen and phosphorus deprivation conditions and observed a 104.7% increase in FA compared with the control treatment. Yet, biomass growth reduction is commonly associated to an increase of lipid content. In the study of Almutairi (2020), reducing phosphorus concentration by 50%, increased the lipid content of *Dunaliella salina* from 5.9% to 16.4%. When the phosphorus was completely removed from the medium, the lipid content reached 24.9%. In both cases, the biomass was reduced by more than 30% compared with the control. Therefore, it is necessary to choose wisely nitrogen and phosphorus concentrations that can provide a good growth rate and lipid accumulation.

pH is another crucial parameter when culturing microalgae because variations in pH can affect biomass growth and lipid production. The optimum pH for most microalgal species lies between 7 and 9. Microalgae tend to increase medium pH due to their photosynthetic activity. pH values higher or lower than the optimum pH of the microalgae will cause a reduction in the photosynthesis rate, and thus, in the growth and lipid accumulation (Khoo et al., 2023; Sakarika & Kornaros, 2016). Species like *Chlorella* sp., *Nannochloropsis salina*, and *Pavlova lutheri* growing at pH of 8 were found to present a lipid content ranging 23–35% (Bartley et al., 2014; Shah et al., 2014). Another strategy to increase lipid content in microalgal cells is salinity stress. Under NaCl stress, some amino acids (proline, homoserine) and carbohydrates (sucrose, trehalose) increase, which reduces the pyruvate concentration, and thus TAG synthesis (Bromke et al., 2013). The membrane lipid reduces the TAG content to resist the osmotic pressure in the cells, and the profile of lipids shifts to polar lipids instead of TAG (Chen et al., 2008). Many microalgae have been studied under high salt concentrations. In the study of Yun et al. (2019), *Chlorella vulgaris YH703* was able to tolerate 500 mM NaCl and to increase the lipid content from 12.7% in the control to 24.5%.

Microalgae can be cultivated under four different types of cultivation conditions: photoautotrophic, heterotrophic, mixotrophic, and photoheterotrophic (Fig. 12.3). Photoautotrophic cultivation, the most frequent, takes place when microalgae use light and inorganic carbon to form chemical energy through photosynthesis. This type of cultivation can reduce CO_2 levels, but the microalgae cultivation systems must be closer to industries that supply high concentrations of CO_2. Some microalgae present a heterotrophic metabolism and they can use organic carbon sources to grow. Using this cultivation mode can increase biomass and productivity. Xu et al. (2006) obtained a 40% increase in lipid content using *C. photothecoides* after shifting the culture conditions from phototrophic to heterotrophic. There is a wide diversity of organic carbon sources that can be assimilated by the microalgae including glucose, fructose, lactose, glycerol, or acetate. For instance, *C. protothecoides* was able to produce 2 g $L^{-1}d^{-1}$ of biomass and 932 mg $L^{-1}d^{-1}$ of lipid using maize powder hydrolysate (Xu et al., 2006). The main drawback is that the use of organic pure compounds increases the total costs of cultivation due to their expensive price. Moreover, carbon sources based on sugars have a higher risk of contamination (Bora et al., 2024). For these reasons, many studies have tried to use low-cost substrates including industrial effluents as carbon sources. In mixotrophic cultivation, microalgae use simultaneously both inorganic (CO_2) and organic compounds to grow. This strategy allows the microalgae to live in phototrophic and heterotrophic conditions and it has the potential to increase growth rate, biomass, and lipid accumulation, even in the absence of light (Patel et al., 2021). Photoheterotrophic cultivation is very similar to mixotrophic, but the main difference is that the first one requires both light and organic substrates at the same time, while mixotrophic mode can still work with only organic materials. Despite the advantages of these two later strategies, they have been poorly used to produce microalgal oils (Chen et al., 2011).

Another important aspect that must be considered when working with microalgae is the type of production system used. The selection of the cultivation system will depend on the species, the availability of nutrients, and the purpose of the biomass produced (Mehariya et al., 2021). Choosing between one or other system will have a significant impact on the productivity of MO. There are different technologies to grow microalgae, but all of them are mainly classified into two groups: open systems and closed systems (Russell et al., 2022). Open systems are the most commonly used industrially due to their cheaper construction, maintenance, and operational costs.

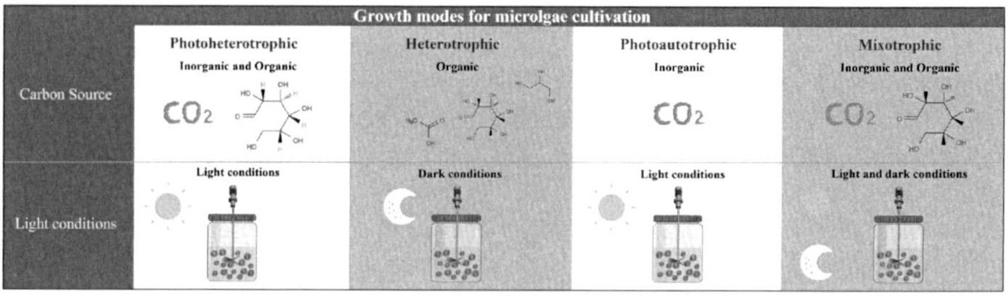

Figure 12.3 Different growth modes for microalgae cultivation.

They are also easier to scale up and have larger production capacities compared with closed systems. In these systems, microalgae grow using natural conditions, which allows them to have a high solar radiation availability. However, due to its open nature, it is difficult to control some environmental parameters such as temperature, light intensity, pH, and dissolved oxygen concentration, which might result in a weak microalgae growth. Moreover, these systems are more susceptible to seasonal changes and contamination from bacteria or other algae species. Another disadvantage is that due to the low concentration of CO_2 in the atmosphere (0.03–0.06%), there is a poor mass transfer, and the growth of microalgae can be slow (Mata et al., 2010). Therefore, only certain microalgae species can be successfully cultured using this system. Some examples include *Spirulina* and *Dunaliella*, which can withstand high alkalinity and high salinity, respectively (Jeevanandam et al., 2020).

Closed systems, also known as photobioreactors (PBRs), have gained attention in the last decades due to their potential to obtain higher biomass productivity, better control over culture conditions (pH, light, and temperature), and lower contamination risks compared to open systems. They also prevent evaporation and CO_2 losses (Paul et al., 2021). One of the biggest advantages of PBRs is that they allow higher volumetric productivities for the capture of light energy, and thus it might be expected that microalgae cultivated in PBRs can reach higher oil yield (Koru & Delice, 2014). PBRs can have different shapes, such as tubular, reactors, vertical columns, and flat plates. A vertical configuration allows better gas–liquid mass transfer and is less energy demanding, while the horizontal type is easier to scale up, but requires extensive use of land. Although having a high surface-to-volume ratio, the performance of flat plates PBRs showed lower yields of biomass per area compared to tubular PBRs due to shorter light penetration depths and a subsequential photo-inhibition of microalgal growth (Jeevanandam et al., 2020). Despite all these advantages, PBRs have some challenges that need to be addressed such as overheating, bio-fouling, oxygen accumulation, scale-up difficulties, and higher operational and building costs than open systems (Mata et al., 2010). Table 12.1 provides some examples of microalgae cultivated on different systems and the lipid content obtained in each case.

Hybrid systems have also been proposed to overcome problems encountered in open and closed-culture systems. This system combines an open raceway pond and a closed PBR and it is designed to reduce costs and increase biomass productivity. Additionally, this two-stage culture might also enhance lipid accumulation. The first phase takes place in a PBR, where the parameters directly related to microalgae growth are optimized to produce high-density biomass. In the second phase, microalgae collected from the PBR is used to seed the ponds. In this second stage, microalgae are cultured under certain environmental stresses, such as nitrogen or phosphorus depletion to favor lipid accumulation (Kadir et al., 2018). This system was used by Huntley & Redalje (2007) to produce astaxanthin and oil from *Haematococcus pluvialis* and they obtained an annual average of 10–24 ton ha^{-1} of lipids per year depending on the climatic conditions and 76 ton ha^{-1} when using *C. vulgaris*. Rodolfi et al. (2009) also used this system to grow *Nannochloropsis* sp. and their experiments showed that this microalga has the potential to produce 20 ton ha^{-1} of lipids in the Mediterranean climate and more than 30 ton ha^{-1} in sunny tropical areas.

Table 12.1 Examples of microalgae cultured on different bioreactors.

Strain	Bioreactor type	Carbon source	Bioreactor conditions(pH, temperature (T), light)	Lipid yield (% dry weight)	References
Chlorella vulgaris	Biofilm reactor	Wallnut shell	- pH 7.0–7.5 - T: 25 °C - Photoperiod 12:12 h light/dark cycle.	28.9	Zou et al., 2021
Scenedesmus quadricauda	Fluidized bed photobioreactor	Anaerobic digestate	- pH 7.0 - T: 25°C - Photoperiod 12:12 h light/dark cycle.	60.4	Xu et al. (2024)
Trebouxiophyceae	Continuous photobioreactor	Synthetic medium (Modified f/2 medium)	- pH 7.5 - T: 30°C - Continuous light	17–31	Policastro et al. (2024)
Chlorella sorokiniana	Tubular photobior-eactors	Synthetic medium (Modified Chu 13 medium)	- pH variable - T: 25–30°C - Continuous light	55.1	Montoya-Vallejo et al. (2023)

12.3.2 Yeast cultivation

In lipid production using oleaginous yeasts, cell cultivation has a central role in the overall process performance and cost-competitiveness. This is due to the influence of the cultivation strategy in the final conversion yields and the costs related to the carbon source and other required nutrients. Glucose has been the preferred carbon source, as it can be efficiently converted into lipids, achieving high cell mass concentrations and lipid yields. Gallego-García et al. (2022) used a pure glucose solution to increase the C/N ratio when cultivating the yeast *Cutaneotrichosporon curvatus* CL6032 (former *Cryptococcus curvatus*) in pepper-derived medium to trigger lipid accumulation. Using the resulting experimental data and the software Aspen Plus, those researchers estimated that the carbon source might account for up to 94% of the total input costs. The high costs associated to glucose represent an important limitation to scale up these bioprocesses.

Most oleaginous yeast strains isolated and characterized to date have shown the capacity to grow and produce lipids on a wide range of substrates. This carbon source portfolio includes different monomeric hexose (glucose, mannose) and pentose (xylose, arabinose) sugars, di- and polysaccharides (sucrose, cellobiose, and starch), certain carbohydrate-derived compounds (N-acetylglucosamine, galacturonate, and levoglucosan), and volatile fatty acids (acetic acid, propionic acid, and butyric acid) (Poontawee et al., 2023). Chmielarz et al. (2021) used crude glycerol (a by-product from the biodiesel industry) and a hemicellulosic hydrolysate (a by-product from the pulp and paper industry) to test the growth of 27 different oleaginous yeast strains. Only 11 strains showed the ability to grow on crude glycerol: one strain of the yeast *Lipomyces starkeyi* and 10 other strains identified as *Rhodotorula* sp. However, when mixing crude glycerol and the hemicellulosic hydrolysate as cultivation media, only the strains belonging to the genera *Rhodotorula* showed glycerol conversion. These authors also observed a stimulating effect on lipid accumulation when mixing these carbon sources to cultivate the yeasts *Rhodotorula toruloides* CBS 14 and *Rhodotorula glutinis* CBS 3044, increasing lipid production rates and yields and thus, shortening fermentation time. Alternatively, the use of lignocellulosic biomass for microbial fermentation has received increased attention in recent years, as it is an abundant and cheap feedstock that does not compete with food crops. Lignocellulosic biomass entails carbohydrate polymers, being glucose and xylose the main sugar monomers in lignocellulosic feedstocks. Although an elevated number of oleaginous yeast strains are capable of utilizing both glucose and xylose from lignocellulosic hydrolysates, glucose is usually the preferred carbon source and xylose is only consumed after glucose depletion. Intending to reach higher cell biomass concentration and increase lipid yield from the substrate, several studies investigated oleaginous yeast strains capable of utilizing glucose and xylose simultaneously. Tanimura et al. (2018) isolated different novel yeast strains from plant and soil samples from Iriomote Island in Japan, which was named *Cystobasidium iriomotense*. These isolated strains showed simultaneous glucose and xylose assimilation in a 1:1 glucose:xylose medium, showing even higher xylose consumption rates when compared to glucose (2.10 g L^{-1} d^{-1} and 1.21 g L^{-1} d^{-1}, respectively). The strain

C. iriomotense IPM46-17 showed the highest potential from all isolates with a total intracellular lipid content of 33% and a lipid conversion yield of 0.06 g g^{-1} of consumed sugars. Higher intracellular lipid content and conversion yields have been reported with other yeast strains capable of simultaneous glucose/xylose consumption. Wang et al. (2020) isolated different yeast strains from the soil of Nanyang in China with the capacity of assimilating glucose and xylose simultaneously in glucose:xylose ratios of 2:1, 1:1, and 1:2, selecting the strain *Cutaneotrichosporon dermatis* ZZ-46 as the most promising candidate. This strain showed the capacity to accumulate lipids intracellularly up to 50% of its cell dry weight with conversion yields of 0.142 g g^{-1} from a medium containing 47 and 23 g L^{-1} of glucose and xylose, respectively.

Another important constraint during the production of microbial lipids from lignocellulosic biomass is the presence of several inhibitory compounds (e.g., weak acids, furan derivatives, and phenolic compounds) that are generated during the biomass processing step. To increase the accessibility of hydrolytic enzymes to the carbohydrates contained in lignocellulose, a pretreatment process under harsh conditions is needed. The inhibitors released during this process stage directly affect yeast cells, highly influencing the overall process performance. In general, microorganisms are capable of transforming/tolerating, to a greater or lesser extent, these inhibitory compounds. Using metabolic engineering and/or evolutionary engineering techniques, new strains of oleaginous yeasts have been developed with higher tolerance to lignocellulosic inhibitors when compared to their respective native strains. Park et al. (2023) observed an improvement in cell growth rates in the presence of inhibitors (2.5 times higher than the parental strain) in *R. toruloides* AD16 strain obtained after evolutionary engineering. Antunes et al. (2024) recently reported that the improved phenotype of an evolved *R. toruloides* IST536 MM15 against lignocellulose-derived inhibitors can be attributed to changes in gene expression of genes associated with cell biogenesis, stress response, and cell membrane integrity and surface remodeling. In addition to evolutionary engineering, metabolic engineering is another approach to increase cell robustness towards the inhibitory compounds. Lyu et al. (2021) integrated the manganese peroxidase and versatile peroxidase enzymes into *R. toruloides* NP11 genome to cope with phenolic compounds. The modified strain showed higher growth rates and increased the final cell concentration and lipid content by 30% and 25%, respectively. Notwithstanding, it is important to highlight that the use of strains obtained by metabolically and evolutionary engineering methods usually results in extended process times and cell energy expenditure, which ultimately impacts the overall conversion yields.

Volatile fatty acids (VFAs) produced in anaerobic fermentation of organic waste have also received increased attention as alternative substrates for microbial lipid production. Bettencourt et al. (2020) reported about 45% (w/w) of intracellular lipid content and VFAs conversion yields of 0.25 g g^{-1} using the yeasts *Apiotrichum brassicae* and *Pichia kudriavzevii* cultivated in a dark fermentation effluent filtrate. VFAs are directly converted to acetyl-CoA in the cytoplasm, thus shortening the metabolic pathways and increasing the conversion yield. However, depending on the concentration and distribution of acids, VFAs-containing media may result in yeast growth inhibition. Hexanoic acid is the acid with the highest inhibitory potential and the most difficult to assimilate for yeasts (Llamas et al., 2020). Similar to increasing the tolerance to lignocellulose-derived inhibitors, the use of adaptive laboratory evolution (ALE) represents an attractive strategy to increase the tolerance of yeast strains to VFAs. Morales-Palomo et al. (2024) subjected the yeast *Yarrowia lipolytica* ACA DC 50109 to evolutionary engineering in the presence of increasing concentrations of caproic acid, resulting in an evolved strain with higher tolerance to this organic acid and with the capacity to accumulate a higher lipid content. The improved phenotype was attributed to the upregulation of the carbon catabolite repression and the YALI0 D17864g gene.

In addition to using different carbon sources, oleaginous yeasts can be cultivated under different processes and strategies to improve overall conversion yields and/or volumetric productivities (Poontawee et al., 2023). In this context, high cell densities are needed to meet these requirements and to facilitate downstream processing. As previously mentioned, nutrient-limiting conditions and high C/N ratios are required to trigger lipid production in yeast cells. Nitrogen, sulfur, and phosphorus limitations during yeast cultivation are the main factors used to stimulate lipogenesis. The formation of cell biomass is however hindered under nutrient-limiting conditions, which may also cause cell growth inhibition. As an alternative to the batch processes, fed-batch, continuous, and two-stage cultivation strategies have been investigated to improve substrate utilization (Fig. 12.4), favoring productivity values and/or reaching higher final lipids production yields (Rakicka et al., 2015).

In fed-batch cultivation mode, a fresh substrate is fed to the culture medium at different time points with no medium removal. Due to the dilution effect after substrate addition, it is essential to use concentrated substrate solutions to avoid reducing cell density and to keep nutrient concentrations at adequate levels (Rakicka et al., 2015). During fed-batch cultivations, the substrate can be added using a pulse-feeding or a continuous strategy. Independently of the feeding mode, fed-batch cultivation allows for reaching higher concentrations of cell biomass and lipids (Abeln & Chuck, 2019). For instance, Ayadi et al. (2016) increased the cell biomass concentration of *Candida viswanathii* from 17 g L^{-1} to 21 g L^{-1} dry weight by changing the cultivation mode from a batch to a fed-batch strategy, respectively. These authors also observed an increase in the intracellular lipid content from 33% to 50% and a 2-fold increase in the final lipid concentration (5.6 g L^{-1} *vs.* 10.5 g L^{-1}). Substrate addition during fed-batch cultivation processes can be designed based on nutrient concentrations (e.g., the carbon source, C/N ratio), the fermentation time, and the dissolved oxygen content. The concentration of the carbon source in the medium, usually considering a concentration lower than 10 g L^{-1}, is one of the main approaches that rules substrate feeding during MO production in fed-batch (Maina et al., 2017). As mentioned above, the increase of cell biomass concentration in fed-batch cultures is dependent on C/N ratio. (Capusoni et al., 2017) increased the cell biomass concentration of *Trichosporon oleaginosus* up to 132 g L^{-1} dry weight and 55% lipid (w/w) with a continuous feeding strategy to maintain the C/N ratio.

The feeding strategy can be also designed considering the relative dissolved oxygen levels. The dissolved oxygen content increases in response to carbon source

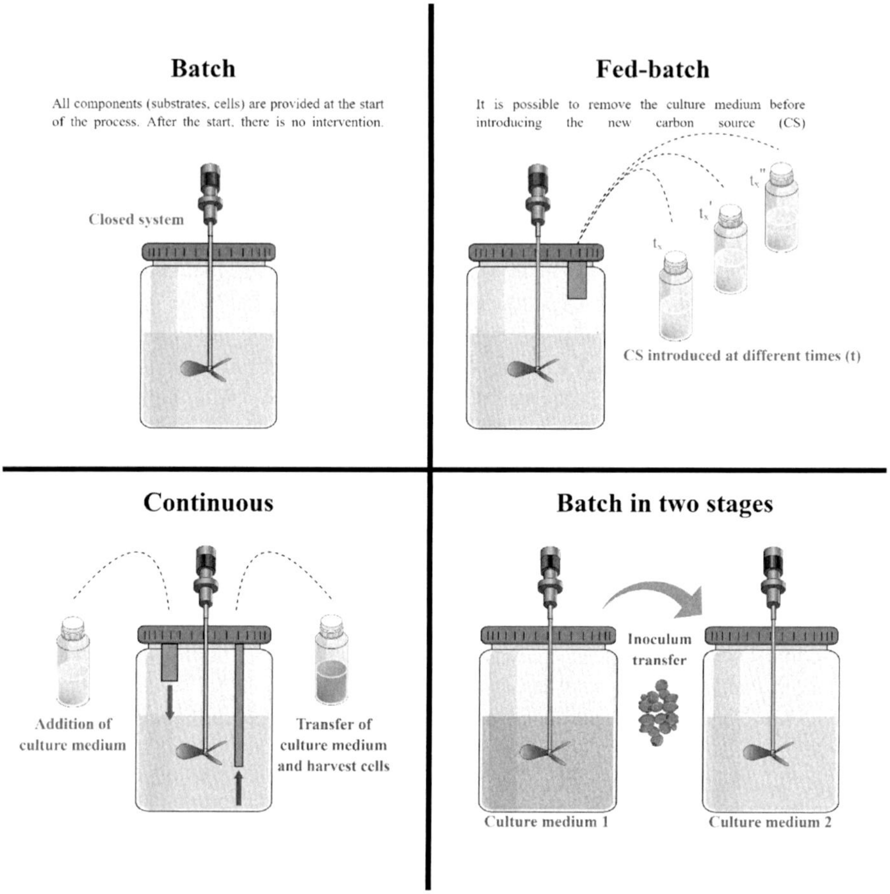

Figure 12.4 Different cultivation strategies for yeast.

depletion due to the lower metabolic activity of the fermenting microorganism (Lv et al., 2020). In this context, the dissolved oxygen levels can be maintained with continuous feeding after substrate depletion, balancing oxygen consumption and supply. Finally, the substrate can be also fed at different time points. Karamerou et al. (2016) designed a fed-batch process by adding fresh substrate every 24 h to culture *Rhodotorula glutinis* in a glycerol-based medium. This strategy allowed reaching almost a 2-fold and 1.5-fold higher concentration of cell biomass (9.4 g L^{-1} *vs*. 5.3 g L^{-1}) and lipid concentration (2.6 g L^{-1} *vs*. 1.7 g L^{-1}) when compared to the batch process, respectively.

The continuous operational mode has been also studied for MO production by yeast. This system provides fresh carbon and nitrogen sources at the same time as collecting products and cells. The main parameter influencing final cell biomass and lipid concentration is the dilution rate. Continuous cultures keep constant growth and substrate uptake rates, while differences in the nitrogen concentration modulate lipid production, allowing increasing productivity values (Gallego-García et al., 2023). In contrast, these cultivation systems may suffer from biofilm formation and media sedimentation in the long term.

Two-stage cultivation processes are also considered an effective strategy to boost lipid production (Qian et al., 2021). This process first targets growing yeast cells and producing cell biomass under nutrient-rich conditions. Cells are then subjected to nutrient-limiting conditions to trigger intracellular lipid accumulation. Combining different cultivation modes (fed-batch, continuous) usually benefits the two-stage approach, especially during the lipid accumulation phase. Fu et al. (2018) reported lipid yields of 0.32 $g\ g^{-1}$ and lipid content of 56.4% when working under a high-cell-density continuous cultivation system using a membrane cell recycling approach and a two-stage nutrient limitation strategy.

12.4 Microbial lipids production by coculture microalgae and oleaginous yeast

The coculture between microalgae and oleaginous yeast can be an interesting option for reaching higher biomass and lipid yields in comparison to individual pure cultures. Microalgae and yeast as coculture or mixed cultures can develop complementary metabolism, promoting the gas balance, metabolite exchanges, and adjusting the pH of the medium resulting in an enhanced lipid accumulation (Dias et al., 2023). Recent studies have demonstrated a mutualistic relation in microalgae–yeast cocultures due to the interchanging carbon and nitrogen where fermentation of sugar-based sources by yeast produces carbon dioxide (CO_2), which is taken up by microalgae to produce O_2 in photosynthesis (Das et al., 2022; Dias et al., 2019), as depicted in Fig. 12.5. This interaction reduces the CO_2 produced by yeast and remediates the loss of carbon generated in the process by yeast respiration. To stimulate cell growth and lipid accumulation in the co-culture microalgae-yeast, the main factors to control are the O_2/CO_2 equilibrium, pH balance, nutrient concentrations, and substrate exchange in the mixed culture system (Dias et al., 2022; Ray et al., 2022).

The exchange between O_2 and CO_2 generated by mixed microalgae–yeast culture may mitigate the stresses caused by CO_2 in the yeast and O_2 in the microalgae, resulting in a balance and improvement in respiration and photosynthesis processes (Ray et al., 2022). As a consequence, symbiotic microalgae–yeast cultures can remediate or solve the low growth rates displayed by pure cultures when cells synthesize lipids (Dias et al., 2019). The pH of the medium can be self-regulated due to the metabolite exchange, where the yeasts synthesize organic acids (reducing the pH) and the microalgae form OH^- due to the consumption the CO_2 (increasing the pH) (Rajak et al., 2022). In this sense, it is not necessary to control the pH using chemicals or equipment, and thereby, the cost involved in the production process can be decreased.

Table 12.2 summarizes the advantages and disadvantages of using coculture microalgae yeast for lipid production.

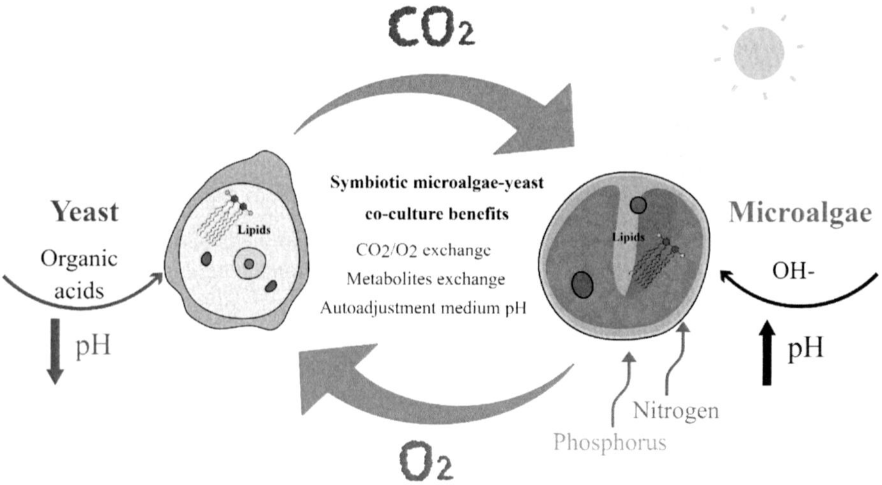

Figure 12.5 Synergistic interaction between microalga–yeast coculture for intracellular lipid accumulation.

Table 12.2 Advantages and disadvantages of microalgae–yeast coculture for lipid production.

Advantages	Disadvantages
Complementary metabolism	Inefficient light penetration under high-concentration yeast cells
Cell stress reduced	Preparation of independent inoculum
Auto-adjustment of the pH medium	Contamination risks
Balance and O_2/CO_2 exchange	Additional supplementation of carbon or nitrogen sources
Metabolites exchange (e.g., organics acids)	
Higher lipid productivity than pure cultures	
Higher cell densities	
Efficient biomass recovery	

The common carbon sources to cultivate microalgae and yeast strains are sugar-based substrates such as hexose and pentose, which stimulate lipids accumulation triggered by the depletion of essential nutrients in the culture medium (Rajpurohit & Eiteman, 2022). Nevertheless, the economic viability of lipids production depends on both cell performance and substrate cost. Therefore, different low-cost substrates have been used to grow symbiotic microalgae–yeast coculture using agro- and food industries effluents/wastes, reaching values of lipid content between 12 and 26.3% (w/w) (Dias et al., 2022; Liu et al., 2018; Zeng et al., 2018). Another important aspect is the biomass recovery from microalgae–yeast coculture. Biomass recovery can be enhanced given the flocculation properties of some yeast strains such as *Cryptococcus, Yarrowia, Trichosporonoides*, and *Lipomyces*, reducing the harvesting cost (de Andrade et al., 2022).

12.5 Microalgal and yeast cell disruption and lipid extraction

MO are produced intracellularly, making biomass harvesting and lipid extraction crucial steps in downstream processing. These technologies have been reviewed in detail by Nunes et al. (2024), highlighting major benefits and constraints for each technology.

Before lipid extraction, cells must be collected from culture media. This harvesting step may account for up to 30% of the total production costs (Barta et al., 2021). Harvesting alternatives include mechanical (filtration, centrifugation, sedimentation), chemical (coagulation, flocculation), biological (bio-flocculation), and electrical-based (electro-flocculation) methods or combinations between them. Filtration and centrifugation techniques have been commonly used for collecting cell biomass for almost all microbial species, allowing high recovery yields. In contrast, these processes show high capital and operational costs due to high energy demand and the need to face clogging and fouling issues.

After cell harvesting, cells must be disrupted to release the intracellular content and subsequently extracted for lipid recovery. For optimal lipid extraction, several procedures have been developed for cell conditioning before the lysis processes (Khot et al., 2020). Knowing the composition and chemical properties of the cell wall is therefore of utmost importance to develop appropriate cell extraction and lysis protocols. Among the methods developed for this step, cell disruption by cold or heat, chemical agents (detergents, acids, alkali), the use of hydrolytic enzymes, and other physical/mechanical processes including bead grinding, pressure extrusion, and sonication are promising methods. The advantages and disadvantages of cell disruption methods can be found in Table 12.3. After cell disruption, lipids are recovered by a liquid–liquid extraction step with solvents (e.g., methanol, chloroform, or petroleum ether). In addition, high-speed homogenization, pulsed electric field, and microwave have been also applied to promote cell lysis in microalgal biomass (Liu et al., 2016). Mechanical cell disruption using abrasion, high pressures, and/or ultrasound are commonly favored methods due to their scaling-up potential and the benefit of avoiding the need to add chemicals during the process, which may end up in the final product. One of the most common methods used for mechanical disruption is bead milling. This method, however, is highly energy-intensive and may damage the final product. Moreover, the method is strain-specific as different cell walls are targeted (Vasconcelos et al., 2018).

Table 12.3 Advantages and disadvantages of cell disruption methods.

Method	Aspect	Advantages	Disadvantages
Mechanical	Bead Milling	– High disruption efficiency and product recovery – Effective against rigid cell walls – Fast operation time – Low operational cost compared to chemical or enzymatic methods – Possibility of continuous mode operation – Easily scale-up	– Possible product degradation – High energy demand – Non-selective release of intracellular products
	High-pressure homogenization	– Effective against rigid cell walls – Transfer of intracellular products – Low solvent requirements – Low-temperature extraction – Suitable for medium to large-scale use	– Possible product degradation – Formation cell debris may cause agglomeration – High energy demand – High capital and maintenance cost – Non-selective release of intracellular products
	Ultrasonication	– High reproducibility – Easily optimized process – Rapid process – Low maintenance cost of equipment	– Poor selectivity (negatively affects product recovery) – Lipid degradation – Not very efficient against rigid cell walls – Requires a very long time – High energy consumption
Nonmechanical	Electrical treatment	– It can be combined with other treatment methods – Easily scale-up – High energetic efficiency	– The solution must be free of ions – Treatment requires pre-washing and deionization to increase the electrical resistance of the surrounding media

(Continued)

Table 12.3 (Continued)

Method	Aspect	Advantages	Disadvantages
		– Fast operation time	– Varying energy consumption – Varying disruption efficiency – High initial capital investment-temperature extraction
	Enzymatic treatment	– High selectivity – Mild operating conditions – Low energy requirements – Low capital investment – Non-aggressive physical conditions – Use of water as a green solvent	– High cost of enzymes – Lower extraction efficiency compared to solvent extraction – The addition of certain enzymes may complicate downstream processing – Possible loss of valuable by-product by conversion – Higher time requires Enzyme cocktail formulation – Difficult to scale up
	Chemical treatment	– Mild operating conditions – Cost-effective – Can be applied to wet biomass – Readily available chemicals – Easily scale-up	– Product quality affected by oxidation or disruption agent contamination – Varying efficiency and selectivity – High cost of some chemicals

Source: Based on Lopes da Silva et al. (2023), Nunes et al. (2024), and Saini et al. (2021).

Nonmechanical cell disruption methods include electric disruption, enzymatic treatment, and chemical lysis. Compared to mechanical cell disruption methods that seek to completely break down cells, these processes promote gentle cell lysis by perforating the cell wall instead. Although enzyme costs represent a great bottleneck for scaling up, enzymatic treatment is considered an attractive cell disruption method due to its selectivity and the possibility of combining it with other techniques. Protease, lipase, glucanase, glucosidase, mannanase, chitinase, and cellulases are major enzymes activities required for cell lysis (Selvakumar & Sivashanmugam, 2019). However, further optimization of the enzyme cocktail and activities is needed, since lipid recovery yields are lower than those obtained with other conventional methods. Kruger et al. (2018) reported a 70% lipid recovery yield from the yeast *Lipomyces starkeyi* by using two enzyme preparations containing glucanase, mannanase, chitinase, and protease. Higher recovery yields of up to 90% were, however, obtained with an acid lysis protocol. Finally, chemical cell lysis involves the use of acids and alkali solutions, solvents (acetone, ethanol, methanol, dimethyl sulfoxide, and ethyl acetate), surfactants, antibiotics, ionic liquids, and deep eutectic solvents. The chemical treatment process usually yields high lipid recovery yield (>80%), requires low energy inputs, and it is easy to scale up. In contrast, these treatments can affect the product quality, resulting in more downstream steps (Gautério et al., 2023).

Different cell extraction methods have been also developed for lipid recovery. These methods mainly include solvent-based and mechanical-based processes. The most common lipid extraction processes are the Folch method and the Bligh and Dyer method (Nunes et al., 2024). Although less sensitive when compared to other methods, the Folch method is a fast and easy procedure that uses a 2:1 chloroform-methanol mixture, resulting in two phases, with lipids being accumulated in the organic phase. A similar approach is followed by the Bligh and Dyer lipid extraction method but using a 1:2 chloroform-methanol mixture and different solvent:cell ratios. In addition to these two processes, lipid extraction with n-hexane is also a common procedure, typically applied for oil recovery from plant seeds. As an alternative to these methods, extraction protocols with supercritical CO_2, bio-based solvents (terpenes, ethyl lactate, and methyl soyate), ionic liquids, and mechanical extraction methods (expeller press) have been also investigated (Barta et al., 2021). In addition, the Soxhlet method using methanol/hexane as solvent is a widely used method for lipid recovery in the laboratory (Khot et al., 2020; Vasconcelos et al., 2019). After lipid recovery, the presence of non-lipid contaminants can be removed by including a washing step of the solvent extract with an aqueous saline solution. Finally, the organic solvent is removed to yield the purified lipid solution.

12.6 Conclusions

Oil-based chemistry arises as a promising alternative to petroleum for the production of fuels and chemicals MO production using microalgae and yeasts holds tremendous potential for sustainable biofuel production and beyond. However, it still faces challenges related to cost, scalability, and process optimization. As a promising alternative, the cocultivation of microalgae and yeast is presented. Their complementarities in nutrient requirements, gas exchanges, and physical behaviors might be key to advancement in MO. Advances in biotechnological tools, waste valorization, and integrated biorefinery approaches can also significantly improve the prospects for large-scale adoption.

Acknowledgments

This work has been supported by the project OILI (CNS2022–135848) and LILIS (EUR2022–134030) funded by MICINN/AEI_NEXT GENERATION-UE and grant PID2020–119403RB-C22 funded by y MICIU/AEI/10.13039/501100011033.

References

Abeln, F., & Chuck, C. J. (2019). Achieving a high-density oleaginous yeast culture: Comparison of four processing strategies using *Metschnikowia pulcherrima*. *Biotechnology and Bioengineering, 116*(12), 3200–3214. https://doi.org/10.1002/bit.27141.

Abeln, F., & Chuck, C. J. (2021). The history, state of the art and future prospects for oleaginous yeast research. *Microbial Cell Factories, 20*(1), 221. https://doi.org/10.1186/s12934-021-01712-1.

Almutairi, A. W. (2020). Effects of nitrogen and phosphorus limitations on fatty acid methyl esters and fuel properties of *Dunaliella salina*. *Environmental Science and Pollution Research, 27*, 32296–32303. https://doi.org/10.1007/s11356-020-08531-8/Published Renewable Energy and Water Sustainability.

Ananthi, V., Brindhadevi, K., Pugazhendhi, A., & Arun, A. (2021). Impact of abiotic factors on biodiesel production by microalgae. *Fuel, 284*.

Antunes, M., Mota, M. N., & Sá-Correia, I. (2024). Cell envelope and stress-responsive pathways underlie an evolved oleaginous *Rhodotorula toruloides* strain multi-stress tolerance. *Biotechnology for Biofuels and Bioproducts, 17*(1), 71. https://doi.org/10.1186/s13068-024-02518-0.

Ayadi, I., Kamoun, O., Trigui-Lahiani, H., Hdiji, A., Gargouri, A., Belghith, H., & Guerfali, M. (2016). Single cell oil production from a newly isolated *Candida viswanathii* Y-E4 and agro-industrial by-products valorization. *Journal of Industrial Microbiology and Biotechnology, 43*(7), 901–914. https://doi.org/10.1007/s10295-016-1772-4.

Barta, D. G., Coman, V., & Vodnar, D. C. (2021). Microalgae as sources of omega-3 polyunsaturated fatty acids: Biotechnological aspects. *Algal Research, 58*, 102410. https://doi.org/10.1016/j.algal.2021.102410.

Bartley, M. L., Boeing, W. J., Dungan, B. N., Holguin, F. O., & Schaub, T. (2014). pH effects on growth and lipid accumulation of the biofuel microalgae *Nannochloropsis salina* and invading organisms. *Journal of Applied Phycology, 26*(3), 1431–1437. https://doi.org/10.1007/s10811-013-0177-2.

Bellou, S., Triantaphyllidou, I.-E., Aggeli, D., Elazzazy, A. M., Baeshen, M. N., & Aggelis, G. (2016). Microbial oils as food additives: recent approaches for improving microbial oil production and its polyunsaturated fatty acid content. *Current Opinion in Biotechnology, 37*, 24–35. https://doi.org/10.1016/j.copbio.2015.09.005.

Bettencourt, S., Miranda, C., Pozdniakova, T. A., Sampaio, P., Franco-Duarte, R., & Pais, C. (2020). Single cell oil production by oleaginous yeasts grown in synthetic and waste-derived volatile fatty acids. *Microorganisms, 8*(11), 1809.

Bharathiraja, B., Sridharan, S., Sowmya, V., Yuvaraj, D., & Praveenkumar, R. (2017). Microbial oil – A plausible alternate resource for food and fuel application. *Bioresource Technology, 233*, 423–432. https://doi.org/10.1016/j.biortech.2017.03.006.

Bora, A., Thondi Rajan, A. S., Ponnuchamy, K., Muthusamy, G., & Alagarsamy, A. (2024). Microalgae to bioenergy production: Recent advances, influencing parameters, utilization of wastewater – A critical review. *Science of the Total Environment, 946*.

Bromke, M. A., Giavalisco, P., Willmitzer, L., & Hesse, H. (2013). Metabolic Analysis of Adaptation to Short-Term Changes in Culture Conditions of the Marine Diatom *Thalassiosira pseudonana*. *PLoS One, 8*(6). https://doi.org/10.1371/journal.pone.0067340.

Capusoni, C., Rodighiero, V., Cucchetti, D., Galafassi, S., Bianchi, D., Franzosi, G., & Compagno, C. (2017). Characterization of lipid accumulation and lipidome analysis in the oleaginous yeasts *Rhodosporidium azoricum* and *Trichosporon oleaginosus*. *Bioresource Technology, 238*, 281–289. https://doi.org/10.1016/j.biortech.2017.03.188.

Chaisutyakorn, P., Praiboon, J., & Kaewsuralikhit, C. (2018). The effect of temperature on growth and lipid and fatty acid composition on marine microalgae used for biodiesel production. *Journal of Applied Phycology, 30*, 37–45. https://doi.org/10.1007/s10811-017-1186-3/Published 9th Asia Pacific Conference on Algal Biotechnology.

Chen, C. Y., Yeh, K. L., Aisyah, R., Lee, D. J., & Chang, J. S. (2011). Cultivation, photobioreactor design and harvesting of microalgae for biodiesel production: A critical review. *Bioresource Technology, 102*(1), 71–81. https://doi.org/10.1016/j.biortech.2010.06.159.

Chen, G. Q., Jiang, Y., & Chen, F. (2008). Salt-induced alterations in lipid composition of diatom *Nitzschia laevis* (Bacillariophyceae) under heterotrophic culture condition. *Journal of Phycology, 44*(5), 1309–1314. https://doi.org/10.1111/j.1529-8817.2008.00565.x.

Chen, H., & Wang, Q. (2021). Regulatory mechanisms of lipid biosynthesis in microalgae. *Biological Reviews, 96*(5), 2373–2391. https://doi.org/10.1111/brv.12759.

Chmielarz, M., Blomqvist, J., Sampels, S., Sandgren, M., & Passoth, V. (2021). Microbial lipid production from crude glycerol and hemicellulosic hydrolysate with oleaginous yeasts. *Biotechnology for Biofuels, 14*(1), 65. https://doi.org/10.1186/s13068-021-01916-y.

Das, P. K., Rani, J., Rawat, S., & Kumar, S. (2022). Microalgal Co-cultivation for Biofuel Production and Bioremediation: Current Status and Benefits. *BioEnergy. Research; a Journal of Science and its Applications, 15*, 1–26. https://doi.org/10.1007/s12155-021-10254-8/Published.

de Andrade, F. P., de Farias Silva, C. E., Medeiros, J. A., Vieira, R. C., de Sá Filho, M. L. F., & Santos, G. K. S. (2022). Consortium between microalgae and other microbiological groups: A promising approach to emphasise the sustainability of open cultivation systems for wastewater treatment. *Journal of Water Process Engineering, 50*. Elsevier Ltd.

De Bhowmick, G., Koduru, L., & Sen, R. (2015). Metabolic pathway engineering towards enhancing microalgal lipid biosynthesis for biofuel application—A review. *Renewable and Sustainable Energy Reviews, 50*, 1239–1253. https://doi.org/10.1016/j.rser.2015.04.131.

de Oliveira, V. F., Parente, E. J. S., Manrique-Rueda, E. D., Cavalcante, C. L., & Luna, F. M. T. (2020). Fatty acid alkyl esters obtained from babassu oil using C1–C8 alcohols and process integration into a typical biodiesel plant. *Chemical Engineering Research and Design, 160*, 224–232. https://doi.org/10.1016/j.cherd.2020.05.028.

Dias, C., Reis, A., Santos, J. A. L., Gouveia, L., & Lopes da Silva, T. (2022). Primary brewery wastewater as feedstock for the yeast *Rhodosporidium toruloides* and the microalga *Tetradesmus obliquus* mixed cultures with lipid production. *Process Biochemistry, 113*, 71–86. https://doi.org/10.1016/j.procbio.2021.12.019.

Dias, C., Santos, J. A. L., Reis, A., & Lopes da Silva, T. (2023). The Use of Oleaginous Yeasts and Microalgae Grown in Brewery Wastewater for Lipid Production and Nutrient Removal: A Review. *Waste and Biomass Valorization, 14*, 1799–1822. Springer Science and Business Media B.V.

Dias, C., Santos, J., Reis, A., & Lopes da Silva, T. (2019). Yeast and microalgal symbiotic cultures using low-cost substrates for lipid production. *Bioresource Technology Reports, 7*. Elsevier Ltd.

Fabiszewska, A., Misiukiewicz-Stępień, P., Paplińska-Goryca, M., Zieniuk, B., & Białecka-Florjańczyk, E. (2019). An insight into storage lipid synthesis by *Yarrowia lipolytica* yeast relating to lipid and sugar substrates metabolism. *Biomolecules, 9*(11). https://doi.org/10.3390/biom9110685.

Fu, R., Fei, Q., Shang, L., Brigham, C. J., & Chang, H. N. (2018). Enhanced microbial lipid production by *Cryptococcus albidus* in the high-cell-density continuous cultivation with membrane cell recycling and two-stage nutrient limitation. *Journal of Industrial Microbiology and Biotechnology, 45*(12), 1045–1051. https://doi.org/10.1007/s10295-018-2081-x.

Galán, B., Santos-Merino, M., Nogales, J., de la Cruz, F., & García, J. L. (2019). Microbial Oils as Nutraceuticals and Animal Feeds. In H. Goldfine (Ed.). *Health Consequences of Microbial Interactions with Hydrocarbons, Oils, and Lipids* (pp. 1–45). Springer International Publishing. https://doi.org/10.1007/978-3-319-72473-7_34-1.

Gallego-García, M., Susmozas, A., Moreno, A. D., & Negro, M. J. (2022). Evaluation and identification of key economic bottlenecks for cost-effective microbial oil production from fruit and vegetable residues. *Fermentation, 8*(7).

Gallego-García, M., Susmozas, A., Negro, M. J., & Moreno, A. D. (2023). Challenges and prospects of yeast-based microbial oil production within a biorefinery concept. *Microbial Cell Factories, 22*(1), 246. https://doi.org/10.1186/s12934-023-02254-4.

Gautério, G. V., da Silva, R. M., Karraz, F. C., Coelho, M. A. Z., Ribeiro, B. D., & Lemes, A. C. (2023). Cell disruption and permeabilization methods for obtaining yeast bioproducts. *Cleaner Chemical Engineering, 6*, 100112. https://doi.org/10.1016/j.clce.2023.100112.

Gayathri, R., Ranjitha, J., & Shankar, V. (2023). Oleochemical Resources for Biodiesel Production. In *Biodiesel Technology and Applications*. Wiley, 311–340. https://doi.org/10.1002/9781119724957.ch11.

Greses, S., Llamas, M., Morales-Palomo, S., González-Fernández, C., & Tomás-Pejó, E. (2024). Microbial Production of Oleochemicals. In V. Bisaria (Ed.). *Handbook of Biorefinery Research and Technology: Production of Biofuels and Biochemicals* (pp. 427–448). Springer Nature Singapore. https://doi.org/10.1007/978-981-97-7586-6_17.

Griffiths, M. J., Van Hille, R. P., & Harrison, S. T. L. (2014). The effect of nitrogen limitation on lipid productivity and cell composition in *Chlorella vulgaris*. *Applied Microbiology and Biotechnology, 98*(5), 2345–2356. https://doi.org/10.1007/s00253-013-5442-4.

He, Q., Yang, H., Wu, L., & Hu, C. (2015). Effect of light intensity on physiological changes, carbon allocation and neutral lipid accumulation in oleaginous microalgae. *Bioresource Technology, 191*, 219–228. https://doi.org/10.1016/j.biortech.2015.05.021.

Hess, S. K., Lepetit, B., Kroth, P. G., & Mecking, S. (2018). Production of chemicals from microalgae lipids – status and perspectives. *European Journal of Lipid Science and Technology, 120*.

Hoyos, E. G., Kuri, R., Toda, T., & Muñoz, R. (2024). Innovative design and operational strategies to improve CO_2 mass transfer during photosynthetic biogas upgrading [Article]. *Bioresource Technology, 391*, 129955. Article https://doi.org/10.1016/j.biortech.2023.129955.

Huntley, M. E., & Redalje, D. G. (2007). CO2 mitigation and renewable oil from photosynthetic microbes: A new appraisal. *Mitigation and Adaptation Strategies for Global Change, 12*(4), 573–608. https://doi.org/10.1007/s11027-006-7304-1.

IEA. (2024). Oil 2024. International Energy Agency (IEA). Paris. https://www.iea.org/reports/oil-2024

Janssen, M. (2016). Microalgal Photosynthesis and Growth in Mass Culture. *Advances in Chemical Engineering, 48*, 185–256. https://doi.org/10.1016/BS.ACHE.2015.11.001.

Jeevanandam, J., Harun, M. R., Lau, S. Y., Sewu, D. D., & Danquah, M. K. (2020). Microalgal biomass generation via electroflotation: A cost-effective dewatering technology. In Applied Sciences (Switzerland), 10, 1–24.

Kadir, W.N.A., Lam, M.K., Uemura, Y., Lim, J.W., & Lee, K.T. (2018). Harvesting and pre-treatment of microalgae cultivated in wastewater for biodiesel production: A review. In Energy Conversion and Management, 171, 1416–1429.

Karamerou, E. E., Theodoropoulos, C., & Webb, C. (2016). A biorefinery approach to microbial oil production from glycerol by *Rhodotorula glutinis*. *Biomass and Bioenergy, 89*, 113–122. https://doi.org/10.1016/j.biombioe.2016.01.007.

Khoo, K.S., Ahmad, I., Chew, K.W., Iwamoto, K., Bhatnagar, A., & Show, P.L. (2023). Enhanced microalgal lipid production for biofuel using different strategies including genetic modification of microalgae: A review. In Progress in Energy and Combustion Science, 96.

Khot, M., Raut, G., Ghosh, D., Alarcón-Vivero, M., Contreras, D., & Ravikumar, A. (2020). Lipid recovery from oleaginous yeasts: Perspectives and challenges for industrial applications. *Fuel, 259*, 116292. https://doi.org/10.1016/j.fuel.2019.116292.

Koru, E., & Delice, G. (2014). Microalgae and their Cultivation System to Produce Biodiesel. International Burch University. BU Journal of. Science and Technology, 2, 73-84 .

Kruger, J. S., Cleveland, N. S., Yeap, R. Y., Dong, T., Ramirez, K. J., Nagle, N. J., ... Biddy, M. J. (2018). Recovery of fuel-precursor lipids from oleaginous yeast. *ACS Sustainable Chemistry & Engineering, 6*(3), 2921–2931. https://doi.org/10.1021/acssuschemeng.7b01874.

Latsos, C., van Houcke, J., & Timmermans, K. R. (2020). The Effect of Nitrogen Starvation on Biomass Yield and Biochemical Constituents of Rhodomonas sp. *Frontiers in Marine Science, 7*. https://doi.org/10.3389/fmars.2020.563333.

Leber, R., Zinser, E., Zellnig, G., Paltauf, F., & Daum, G. (1994). Characterization of lipid particles of the yeast, *Saccharomyces cerevisiae. Yeast (Chichester, England), 10*(11), 1421–1428. https://doi.org/10.1002/yea.320101105.

Liu, D., Ding, L., Sun, J., Boussetta, N., & Vorobiev, E. (2016). Yeast cell disruption strategies for recovery of intracellular bio-active compounds—A review. *Innovative Food Science & Emerging Technologies, 36*, 181–192. https://doi.org/10.1016/j.ifset.2016.06.017.

Liu, L., Chen, J., Lim, P. E., & Wei, D. (2018). Enhanced single cell oil production by mixed culture of *Chlorella pyrenoidosa* and *Rhodotorula glutinis* using cassava bagasse hydrolysate as carbon source. *Bioresource Technology, 255*, 140–148. https://doi.org/10.1016/j.biortech.2018.01.114.

Llamas, M., Magdalena, J. A., González-Fernández, C., & Tomás-Pejó, E. (2020). Volatile fatty acids as novel building blocks for oil-based chemistry via oleaginous yeast fermentation. *Biotechnology and Bioengineering, 117*(1), 238–250. https://doi.org/10.1002/bit.27180.

Lopes da Silva, T., Fontes, A., Reis, A., Siva, C., & Gírio, F. (2023). Oleaginous Yeast Biorefinery: Feedstocks, Processes. *Techniques, Bioproducts. Fermentation, 9*(12).

Lv, P. J., Qiang, S., Liu, L., Hu, C. Y., & Meng, Y. H. (2020). Dissolved-oxygen feedback control fermentation for enhancing β-carotene in engineered *Yarrowia lipolytica. Scientific Reports, 10*(1), 17114. https://doi.org/10.1038/s41598-020-74074-0.

Lyu, L., Chu, Y., Zhang, S., Zhang, Y., Huang, Q., Wang, S., & Zhao, Z. K. (2021). Engineering the oleaginous yeast *Rhodosporidium toruloides* for improved resistance against inhibitors in biomass hydrolysates [Brief Research Report]. *Frontiers in Bioengineering and Biotechnology, 9*. https://doi.org/10.3389/fbioe.2021.768934.

Maina, S., Pateraki, C., Kopsahelis, N., Paramithiotis, S., Drosinos, E. H., Papanikolaou, S., & Koutinas, A. (2017). Microbial oil production from various carbon sources by newly isolated oleaginous yeasts. *Eng Life Sci, 17*(3), 333–344. https://doi.org/10.1002/elsc.201500153.

Mata, T. M., Martins, A. A., & Caetano, N. S. (2010). Microalgae for biodiesel production and other applications: A review. *Renewable and Sustainable Energy Reviews, 14*(1), 217–232. https://doi.org/10.1016/j.rser.2009.07.020.

Mehariya, S., Goswami, R. K., Karthikeysan, O. P., & Verma, P. (2021). Microalgae for high-value products: A way towards green nutraceutical and pharmaceutical compounds. *Chemosphere, 280*. https://doi.org/10.1016/j.chemosphere.2021.130553.

Montoya-Vallejo, C., Guzmán Duque, F. L., & Quintero Díaz, J. C. (2023). Biomass and lipid production by the native green microalgae *Chlorella sorokiniana* in response to nutrients, light intensity, and carbon dioxide: experimental and modeling approach. *Frontiers in Bioengineering and Biotechnology, 11*. https://doi.org/10.3389/fbioe.2023.1149762.

Morales, M., Aflalo, C., & Bernard, O. (2021). Microalgal lipids: A review of lipids potential and quantification for 95 phytoplankton species. *Biomass and Bioenergy, 150*, 106108. https://doi.org/10.1016/j.biombioe.2021.106108.

Morales-Palomo, S., Navarrete, C., Martínez, J. L., González-Fernández, C., & Tomás-Pejó, E. (2024). Transcriptomic profiling of an evolved *Yarrowia lipolytica* strain: tackling hexanoic acid fermentation to increase lipid production from short-chain fatty acids. *Microbial Cell Factories, 23*(1), 101. https://doi.org/10.1186/s12934-024-02367-4.

Morales-Palomo, S., Tomás-Pejó, E., & González-Fernández, C. (2023). Phosphate limitation as crucial factor to enhance yeast lipid production from short-chain fatty acids. *Microbial Biotechnology, 16*(2), 372–380. https://doi.org/10.1111/1751-7915.14197.

Neupane, D. (2023). Biofuels from Renewable Sources, a Potential Option for Biodiesel Production. *Bioengineering, 10*(1).

Nunes, D. D., Pillay, V. L., Van Rensburg, E., & Pott, R. W. M. (2024). Oleaginous microorganisms as a sustainable oil source with a focus on downstream processing and cost-lowering production strategies: A review. *Bioresource Technology Reports, 26*, 101871. https://doi.org/10.1016/j.biteb.2024.101871.

Nzayisenga, J. C., Farge, X., Groll, S. L., & Sellstedt, A. (2020). Effects of light intensity on growth and lipid production in microalgae grown in wastewater. *Biotechnology for Biofuels, 13*(1). https://doi.org/10.1186/s13068-019-1646-x.

Ochsenreither, K., Glück, C., Stressler, T., Fischer, L., & Syldatk, C. (2016). Production strategies and applications of microbial single cell oils [Review]. *Frontiers in Microbiology, 7*. https://doi.org/10.3389/fmicb.2016.01539.

Papone, T., Kookkhunthod, S., Paungbut, M., & Leesing, R. (2016). Producing of microbial oil by mixed culture of microalgae and oleaginous yeast using sugarcane molasses as carbon substrate. *Journal of Clean Energy Technologies, 4*(4), 4.

Park, G. W., Shin, S., Kim, S. J., Lee, J.-S., Moon, M., & Min, K. (2023). Rice straw-derived lipid production by HMF/furfural-tolerant oleaginous yeast generated by adaptive laboratory evolution. *Bioresource Technology, 367*, 128220. https://doi.org/10.1016/j.biortech.2022.128220.

Patel, A., Karageorgou, D., Rova, E., Katapodis, P., Rova, U., Christakopoulos, P., & Matsakas, L. (2020). An overview of potential oleaginous microorganisms and their role in biodiesel and omega-3 fatty acid-based industries. *Microorganisms, 8*(3), 434.

Patel, A., Singhania, R., Sim, S., & Dong, C. (2021). Recent advancements in mixotrophic bioprocessing for production of high value microalgal products. *Bioresource Technology, 320*. https://doi.org/10.1016/j.biortech.2020.124421.

Paul, S., Bera, S., Dasgupta, R., Mondal, S., & Roy, S. (2021). Review on the recent structural advances in open and closed systems for carbon capture through algae. *Energy Nexus, 4*. https://doi.org/10.1016/j.nexus.2021.100032.

Policastro, G., Ebrahimi, S., Weissbrodt, D. G., Fabbricino, M., & van Loosdrecht, M. C. M. (2024). Selecting for a high lipid accumulating microalgae culture by dual growth limitation in a continuous bioreactor. *Science of the Total Environment, 912*. https://doi.org/10.1016/j.scitotenv.2023.169213.

Poontawee, R., Lorliam, W., Polburee, P., & Limtong, S. (2023). Oleaginous yeasts: Biodiversity and cultivation. *Fungal Biology Reviews, 44*, 100295. https://doi.org/10.1016/j.fbr.2022.11.003.

Qian, X., Zhou, X., Chen, L., Zhang, X., Xin, F., Dong, W., ... Jiang, M. (2021). Bioconversion of volatile fatty acids into lipids by the oleaginous yeast *Apiotrichum porosum* DSM27194. *Fuel, 290*, 119811. https://doi.org/10.1016/j.fuel.2020.119811.

Rajak, R. C., Rajlakshmi, Saravanabhupathy, S., & Banerjee, R. (2022). Microbial lipids production by oleaginous yeasts. In *Biomass, Biofuels, Biochemicals: Microbial Lipids - Processes, Products, and Innovations*. Elsevier, 161–189. https://doi.org/10.1016/B978-0-323-90631-9.00001-6.

Rajpurohit, H., & Eiteman, M. A. (2022). Nutrient-Limited Operational Strategies for the Microbial Production of Biochemicals. *Microorganisms, 10*.

Rakicka, M., Lazar, Z., Dulermo, T., Fickers, P., & Nicaud, J. M. (2015). Lipid production by the oleaginous yeast *Yarrowia lipolytica* using industrial by-products under different culture conditions. *Biotechnology for Biofuels, 8*(1), 104. https://doi.org/10.1186/s13068-015-0286-z.

Ratledge, C. (2002). Regulation of lipid accumulation in oleaginous micro-organisms. *Biochemical Society Transactions, 30*(Pt 6), 1047–1050. https://doi.org/10.1042/bst0301047.

Ray, A., Nayak, M., & Ghosh, A. (2022). A review on co-culturing of microalgae: A greener strategy towards sustainable biofuels production. *Science of the Total Environment, 802*.

Remize, M., Brunel, Y., Silva, J. L., Berthon, J. Y., & Filaire, E. (2021). Microalgae n-3 PUFAs Production and Use in Food and Feed Industries. *Marine Drugs, 19*(2). https://doi.org/10.3390/md19020113.

Robles-Iglesias, R., Naveira-Pazos, C., Fernández-Blanco, C., Veiga, M. C., & Kennes, C. (2023). Factors affecting the optimisation and scale-up of lipid accumulation in oleaginous yeasts for sustainable biofuels production. *Renewable and Sustainable Energy Reviews, 171*. Elsevier Ltd.

Rodolfi, L., Zittelli, G. C., Bassi, N., Padovani, G., Biondi, N., Bonini, G., & Tredici, M. R. (2009). Microalgae for oil: Strain selection, induction of lipid synthesis and outdoor mass cultivation in a low-cost photobioreactor. *Biotechnology and Bioengineering, 102*(1), 100–112. https://doi.org/10.1002/bit.22033.

Russell, C., Rodriguez, C., & Yaseen, M. (2022). High-value biochemical products & applications of freshwater eukaryotic microalgae. *Science of the Total Environment, 809*. Elsevier B.V.

Saini, R. K., Prasad, P., Shang, X., & Keum, Y.-S. (2021). Advances in Lipid Extraction Methods—A Review. *International Journal of Molecular Sciences, 22*(24).

Sakarika, M., & Kornaros, M. (2016). Effect of pH on growth and lipid accumulation kinetics of the microalga *Chlorella vulgaris* grown heterotrophically under sulfur limitation. *Bioresource Technology, 219*, 694–701. https://doi.org/10.1016/j.biortech.2016.08.033.

Salvador-López, J. M., Vandeputte, M., & Van Bogaert, I. N. A. (2022). Oleaginous yeasts: Time to rethink the definition? *Yeast (Chichester, England), 39*(11-12), 553–606. https://doi.org/10.1002/yea.3827.

Selvakumar, P., & Sivashanmugam, P. (2019). Ultrasound assisted oleaginous yeast lipid extraction and garbage lipase catalyzed transesterification for enhanced biodiesel production. *Energy Conversion and Management, 179*, 141–151. https://doi.org/10.1016/j.enconman.2018.10.051.

Shah, A. M., Yang, W., Mohamed, H., Zhang, Y., & Song, Y. (2022). Microbes: A Hidden Treasure of Polyunsaturated Fatty Acids. *Front Nutr, 9*, 827837. https://doi.org/10.3389/fnut.2022.827837.

Shah, M. R., Lutzu, G. A., Alam, A., Sarker, P., Kabir Chowdhury, M. A., Parsaeimehr, A., ... Daroch, M. (2018). Microalgae in aquafeeds for a sustainable aquaculture industry. *Journal of Applied Phycology, 30*(1), 197–213. https://doi.org/10.1007/s10811-017-1234-z.

Shah, S. M. U., Che Radziah, C., Ibrahim, S., Latiff, F., Othman, M. F., & Abdullah, M. A. (2014). Effects of photoperiod, salinity and pH on cell growth and lipid content of Pavlova lutheri. *Annals of Microbiology, 64*(1), 157–164. https://doi.org/10.1007/s13213-013-0645-6.

Shekh, A., Sharma, A., Schenk, P. M., Kumar, G., & Mudliar, S. (2022). Microalgae cultivation: photobioreactors, CO_2 utilization, and value-added products of industrial importance. *Journal of Chemical Technology and Biotechnology, 97*, 1064–1085. John Wiley and Sons Ltd.

Shi, T. Q., Wang, L. R., Zhang, Z. X., Sun, X. M., & Huang, H. (2020). Stresses as First-Line Tools for Enhancing Lipid and Carotenoid Production in Microalgae. *In Frontiers in Bioengineering and Biotechnology, 8*. Frontiers Media S.A.

Song, M., Zhang, X., Chen, Y., Zhang, Q., Chen, L., Liu, J., & Ma, L. (2023). Hydroprocessing of lipids: An effective production process for sustainable aviation fuel. *Energy, 283*, 129107. https://doi.org/10.1016/j.energy.2023.129107.

Su, H., Feng, J., Lv, J., Liu, Q., Nan, F., Liu, X., & Xie, S. (2021). Molecular mechanism of lipid accumulation and metabolism of oleaginous *Chlorococcum sphacosum* GD from soil under salt stress. *Int J Mol Sci, 22*(3). https://doi.org/10.3390/ijms22031304.

Sun, X.-M., Ren, L.-J., Zhao, Q.-Y., Ji, X.-J., & Huang, H. (2019). Enhancement of lipid accumulation in microalgae by metabolic engineering. *Biochimica et Biophysica Acta (BBA) - Molecular and Cell Biology of Lipids, 1864*(4), 552–566. https://doi.org/10.1016/j.bbalip.2018.10.004.

Szczepańska, P., Hapeta, P., & Lazar, Z. (2022). Advances in production of high-value lipids by oleaginous yeasts. *Critical Reviews in Biotechnology, 42*(1), 1–22. https://doi.org/10.1080/07388551.2021.1922353.

Tanimura, A., Sugita, T., Endoh, R., Ohkuma, M., Kishino, S., Ogawa, J., ... Takashima, M. (2018). Lipid production via simultaneous conversion of glucose and xylose by a novel yeast, *Cystobasidium iriomotense*. *PLoS One, 13*(9), e0202164. https://doi.org/10.1371/journal.pone.0202164.

Tibocha-Bonilla, J. D., Zuñiga, C., Godoy-Silva, R. D., & Zengler, K. (2018). Advances in metabolic modeling of oleaginous microalgae. *Biotechnology for Biofuels, 11*(1), 241. https://doi.org/10.1186/s13068-018-1244-3.

Tomás-Pejó, E., Morales-Palomo, S., & González-Fernández, C. (2023). *Cutaneotrichosporon curvatum* and *Yarrowia lipolytica* as key players for green chemistry: efficient oil producers from food waste via the carboxylate platform [Article]. *Bioengineered, 14*(1), Article 2286723. https://doi.org/10.1080/21655979.2023.2286723.

Vasconcelos, B., Teixeira, J. C., Dragone, G., & Teixeira, J. A. (2018). Optimization of lipid extraction from the oleaginous yeasts *Rhodotorula glutinis* and *Lipomyces kononenkoae*. *AMB Express, 8*(1), 126. https://doi.org/10.1186/s13568-018-0658-4.

Vasconcelos, B., Teixeira, J. C., Dragone, G., & Teixeira, J. A. (2019). Oleaginous yeasts for sustainable lipid production—from biodiesel to surf boards, a wide range of "green" applications. *Applied Microbiology and Biotechnology, 103*(9), 3651–3667. https://doi.org/10.1007/s00253-019-09742-x.

Vuppaladadiyam, A. K., Prinsen, P., Raheem, A., Luque, R., & Zhao, M. (2018). Microalgae cultivation and metabolites production: a comprehensive review. *Biofuels, Bioproducts and Biorefining, 12*(2), 304–324. https://doi.org/10.1002/bbb.1864.

Wahidin, S., Idris, A., & Shaleh, S. R. M. (2013). The influence of light intensity and photoperiod on the growth and lipid content of microalgae *Nannochloropsis* sp. *Bioresource Technology, 129*, 7–11. https://doi.org/10.1016/j.biortech.2012.11.032.

Wang, L., Wang, D., Zhang, Z., Cheng, S., Liu, B., Wang, C., ... Guo, S. (2020). Comparative glucose and xylose coutilization efficiencies of soil-isolated yeast strains identify *Cutaneotrichosporon dermatis* as a potential producer of lipid. *ACS Omega, 5*(37), 23596–23603. https://doi.org/10.1021/acsomega.0c02089.

Wei, L., Huang, X., & Huang, Z. (2015). Temperature effects on lipid properties of microalgae *Tetraselmis subcordiformis* and *Nannochloropsis oculata* as biofuel resources. *Chinese Journal of Oceanology and Limnology, 33*(1), 99–106. https://doi.org/10.1007/s00343-015-3346-0.

Wu, L. F., Chen, P. C., & Lee, C. M. (2013). The effects of nitrogen sources and temperature on cell growth and lipid accumulation of microalgae. *International Biodeterioration and Biodegradation, 85*, 506–510. https://doi.org/10.1016/j.ibiod.2013.05.016.

Xu, H., Miao, X., & Wu, Q. (2006). High quality biodiesel production from a microalga *Chlorella protothecoides* by heterotrophic growth in fermenters. *Journal of Biotechnology, 126*(4), 499–507. https://doi.org/10.1016/j.jbiotec.2006.05.002.

Yang, L., Chen, J., Qin, S., Zeng, M., Jiang, Y., Hu, L., ... Wang, J. (2018). Growth and lipid accumulation by different nutrients in the microalga *Chlamydomonas reinhardtii*. *Biotechnology for Biofuels, 11*(1). https://doi.org/10.1186/s13068-018-1041-z.

Xu, H., Wang, H., Wang, X., Tang, Z., Chen, X., Chen, Y., ... Wang, H. (2024). Fluidized bed photobioreactor based on diatomite powder and high light intensity improved microalgae harvesting, nutrient removal and lipid accumulation: Performance and microscopic mechanism. *Water Research, 264*. https://doi.org/10.1016/j.watres.2024.122172.

Yun, C. J., Hwang, K. O., Han, S. S., & Ri, H. G. (2019). The effect of salinity stress on the biofuel production potential of freshwater microalgae *Chlorella vulgaris* YH703. *Biomass and Bioenergy, 127*. https://doi.org/10.1016/j.biombioe.2019.105277.

Zeng, Y., Xie, T., Li, P., Jian, B., Li, X., Xie, Y., & Zhang, Y. (2018). Enhanced lipid production and nutrient utilization of food waste hydrolysate by mixed culture of oleaginous yeast *Rhodosporidium toruloides* and oleaginous microalgae *Chlorella vulgaris*. *Renewable Energy, 126*, 915–923. https://doi.org/10.1016/j.renene.2018.04.020.

Žganjar, M., Ogrizović, M., Matul, M., Čadež, N., Gunde-Cimerman, N., González-Fernández, C., ... Petrovič, U. (2024). High-throughput screening of non-conventional yeasts for conversion of organic waste to microbial oils via carboxylate platform. *Scientific Reports, 14*(1). https://doi.org/10.1038/s41598-024-65150-w.

Zhao, Y., Li, D., Ding, K., Che, R., Xu, J. W., Zhao, P., ... Yu, X. (2016). Production of biomass and lipids by the oleaginous microalgae *Monoraphidium* sp. QLY-1 through heterotrophic cultivation and photo-chemical modulator induction. *Bioresource Technology, 211*, 669–676. https://doi.org/10.1016/j.biortech.2016.03.160.

Zhu, J., Guo, B., Qie, F., Li, X., Zhao, X., Rong, J., & Zong, B. (2022). A sustainable integration of removing CO_2/NO_x and producing biomass with high content of lipid/protein by microalgae. *Journal of Energy Chemistry, 73*, 13–25. https://doi.org/10.1016/j.jechem.2022.04.008.

Zou, X., Xu, K., Chang, W., Qu, Y., & Li, Y. (2021). A novel microalgal biofilm reactor using walnut shell as substratum for microalgae biofilm cultivation and lipid accumulation. *Renewable Energy, 175*, 676–685. https://doi.org/10.1016/j.renene.2021.04.122.

Section 2

Some key microalgae and yeast species for bioproduction

Debaryomyces hansenii and its role on mineral nutrition of dicot plants 13

Carlos Lucena[1], and José Ramos[2]
[1]Department of Agronomy, ceiA3 Campus of International Excellence in Agri-Food of Rabanales, María de Maeztu Unit of Excellence 2021-2024, University of Córdoba, Córdoba, Spain, [2]Department of Agriculture Chemistry, Edaphology and Microbiology, ceiA3 Campus of International Excellence in Agri-Food of Rabanales, University of Córdoba, Córdoba, Spain

Every day, humanity needs to produce more food to satisfy the increasing population, but this has to be done in a sustainable way. In relation to the plant mineral nutrition, two aspects that can contribute to this goal are the development of varieties more efficient in nutrient acquisition and a better management of the rhizosphere, the soil volume influenced by the root (Pii et al., 2015; Romera et al., 2019). Among the essential mineral nutrients required by plants, iron (Fe) represents one of the major constraints for crop productivity worldwide. Fe is a very abundant element in most soils but its acquisition for plants is problematic because their low solubility, mainly in calcareous soils, abundant in the World (Marschner, 2012; Vélez-Bermúdez & Schmidt, 2023). To facilitate their acquisition, plants develop several responses in their roots, all of them are switched off once Fe have been acquired in enough quantity. Their regulation is not totally understood but in dicot plants it is known that some substances act as activators of the responses, like ethylene and nitric oxide (Romera et al., 2019). In the last years, it has been found that some microorganisms, some of them with the capacity to elicit induced systemic resistance (ISR), can facilitate the acquisition of Fe. In this case, ISR-eliciting microorganisms can trigger certain Fe deficiency responses, likely due to the involvement of similar signaling pathways in both processes (ISR and Fe deficiency responses) (Pieterse et al., 2014; Verbon et al., 2017).

The demand for food is increasing due to gradual growth of the world's population which is expected to reach 9,000 million inhabitants by 2050 (FAO, 2015). The pressure that modern agriculture puts on crops over the last decades has led to an abuse of agrochemicals, mainly Fe chelates, as Fe deficiency is a major agronomic problem causing a significant decrease in crop yield and quality (Romera et al., 2019). Moreover, their use has become a serious environmental problem and has raised the cost of production owing to the high price of these products. One of the most sustainable alternatives to the use of chemical fertilizers is the employment of soil microorganisms as biofertilizers. Some results obtained with the yeast specie *D. hansenii* on Fe nutrition in dicot plants, have beenreally awesome. These results have been recently published (Lucena et al., 2021), showing for the first time the possibility of using yeast species as possible Fe biofertilizers. Authors have been

Eukaryotic Microorganisms as Sources of Bioproducts. DOI: https://doi.org/10.1016/B978-0-443-30188-9.00022-6

pioneer in the study of yeasts as inducers of Fe deficiency responses in plants (Lucena et al., 2021; Sevillano-Caño, García, et al., 2024). To date, there are no other studies on the roles played by yeasts on the Fe nutrition of plants. This opens an innovative, sustainable and environmentally friendly way to solve Fe deficiency associated problems by using yeasts as Fe biofertilizers.

13.1 Potential of *D. hansenii* as Fe biofertilizer and biopesticide in dicot plants

Yeasts, like other beneficial microorganisms that interact with plants, either through the roots or leaves, can positively influence plant growth, facilitate nutrient acquisition, and strengthen their defenses against pathogens. Despite this, they have been much less studied in these aspects compared to rhizobacteria, mycorrhizal fungi, and other rhizospheric fungi (Fernández-San Millán et al., 2020; Raklami et al., 2024). The reasons for their positive effects on plants are also not well understood, but they are often related to their ability to produce certain substances, such as hormones, chelating agents, and antimicrobial compounds. Hormones, such as auxins, gibberellins, or ethylene, can alter root morphology and physiology; chelating agents, such as organic acids and siderophores, contribute to the solubilization of phosphate, iron, zinc, and other nutrients; and antimicrobial substances restrict the growth of pathogens (Fernández-San Millán et al., 2020; Jamal et al., 2021; Liao et al., 2024; Ling et al., 2020; Lonhienne et al., 2014; Prabina et al., 2019; Raklami et al., 2024; Sobrino-Plata et al., 2022). In certain cases, yeasts can induce systemic defense responses in the plants they colonize, likely through ISR (Lee et al., 2017), and also responses to deficiencies in some nutrients, such as Fe (Lucena et al., 2021). Lucena et al. (2021), in a preliminary study on the influence of yeasts on Fe nutrition in dicotyledonous plants, investigated the effect of three yeasts [*Saccharomyces cerevisiae*, *Hansenula polymorpha, and D. hansenii*] on the induction of responses to Fe deficiency. The results showed that two of them, *H. polymorpha* and *D. hansenii*, significantly induced physiological and morphological responses to Fe deficiency (Lucena et al., 2021). The two yeast species selected for the authors belong to the family Saccharomycetaceae. Both yeasts divide by budding, and neither forms "mycelium" nor behaves as a pathogen.

H. polymorpha (now called *Ogataea angusta*) is a ubiquitous microorganism that can be isolated from many different habitats, such as fruit juices or the insect gut. However, this yeast is particularly abundant in soils. It is a methylotrophic organism, capable of growing on methanol and metabolizing a wide range of carbohydrates as carbon sources. It is also noteworthy that *H. polymorpha* can utilize nitrate and nitrite as its sole nitrogen sources (Siverio, 2002), making it an excellent model organism for studying nitrate acquisition processes. Another key feature, especially important for this proposal, is its thermotolerant nature. Its optimal growth temperature is around 37°C, and it tolerates temperatures close to 50°C (Cabeç-Silva & Madeira-Lopes, 2007). Thus, considering its adaptability to high temperatures, *H. polymorpha* could play a crucial role in helping higher plants adapt to future environmental conditions associated with climate change.

D. hansenii is an organism that belongs to the so-called nonconventional group of yeasts. It can be found in very diverse habitats such as sea water, soil, or salty food. It is a nonpathogenic yest and from a taxonomical point of view it is an ascomycetous yeast in the Order *Saccharomycetales* and in the Family *Saccharomycetaceae* (Kreger-van Rij, 1984). Phylogenetically is more related to some *Candida* species than to the model yeast *Saccharomyces cerevisiae*.

D. hansenii has some peculiarities that it is important to keep in mind when working with this organism: *D. hansenii* has the highest coding capacity among yeasts, amounting to 79.2% of the genome with a putative number of 6906 detected CDS. This yeast belongs to the ambiguous CUG decoding group of yeasts, in which CUG codon can be ambiguously translated, mainly as serine but also as leucine, although in a minor percentage (Prista et al., 2016; Santos et al., 2011).

Moreover, strains belonging to that specie are very heterogeneous both genetically and biochemically and, consequently, they can show wide phenotypic differences. Research groups have mainly focussed on the study of its salt-loving character and its very wide biotechnological potential.

The salt tolerant behavior of *D. hansenii* has been studied during decades (Norkrans & Kylin, 1969) and although multiple salt tolerance determinants previously identified in *S. cerevisiae* have been found in *D. hansenii*, their real contribution to the whole process remains obscure. Several differential determinants between the two yeast species have been reported, such as the existence of ion transporters absent in *Saccharomyces* (Hak1) (Martínez et al., 2011; Prista et al., 2007), the high salt tolerance of some *D. hansenii* enzymes (Hal2 or Ppz1) (Aggarwal & Mondal, 2006), specific changes in *D. hansenii* plasma membrane composition in response to salt (Turk et al., 2007) or the ability to transport and sequester sodium into the vacuole (Herrera-Cruz & Simmen, 2017; Montiel & Ramos, 2007). In addition, cellular adaptations of the metabolism to the presence of salt may be involved in the halotolerant/halophilic behavior of *D. hansenii* (Ruiz-Pérez et al., 2023). Most probably there is not a single reason to explain why *D. hansenii* happily grows in high salt and the final reason is a combination of the different existing determinants that fit in a magnificent puzzle.

Recently, Nimsi et al. (2024) have provided the first insights into the multifarious plant growth-promoting traits of manglicolous yeasts and their role in the promotion of the growth of *Rhizophora mucronata*. The mangrove yeasts displayed varied plant growth-promoting traits, including phosphate solubilization, IAA, siderophore, ammonia, catalase, and hydrolytic enzyme production. Growth in seedlings of *R. mucronata* was stimulated by both phyllospheric and rhizospheric yeast species. Among the 215 screened isolates *D. hansenii* PV 23, *Candida tropicalis*, P 9, and *Aureobasidium melanogenum* KV 35 were found to promote the growth of *Rhizophora* seedlings in a pot experiments. Based on the results obtained from potted plants with P 9, PV 23, and KV 35 both individually and as consortia promoted the growth of mangrove seedlings of *R. mucronata*, the authors proposed to test the effect of the yeast in the field as well (Nimsi et al., 2024).

In addition to its beneficial role in plant growth and nutrition, *D. hansenii* serves as a biocontrol agent against various bacterial and fungal diseases (Liao et al., 2024; Medina-Córdova et al., 2016). Moreover, it is an endophytic yeast capable of colonizing leaves, roots, and other plant organs (Kaur et al., 2020; Ling et al., 2020). Endophytic microorganisms offer several advantages, such as better protection against biotic and abiotic stresses and better adaptation to the plant they colonize (Ling et al., 2020; Mercado-Blanco & JJ Lugtenberg, 2014). *D. hansenii* is adapted to adverse conditions such as salinity and high pH (Prista et al., 2005), like those found in calcareous soils. These characteristics contribute to their efficacy and competitiveness under natural conditions, where their favorable effects on plants in calcareous soil are observed. These findings were confirmed in a subsequent study with *D. hansenii*, which also demonstrated its ability to improve the growth of cucumber plants, their flowering, and their nutrition of Fe and other nutrients when grown in calcareous soil (Sevillano-Caño, García, et al., 2024) (Fig. 13.1).

In summary, the ability of the yeast *D. hansenii* to induce responses to Fe deficiency; their growth-promoting effect on dicotyledonous plants; the possibility of being used as biocontrol agents and their easy handling as a lyophilized product make them ideal candidates for use as biofertilizer and biopesticide in sustainable agriculture, less dependent on synthetic fertilizers and pesticides (Kaur et al., 2020; Lucena et al., 2021, Sevillano-Caño, García, et al., 2024).

Figure 13.1 Effect of the yeast *D. hansenii* (Dh) on the growth at 20 days after inoculation (left) and flowering at 28 days after inoculation (right) of cucumber plants grown in nonsterile calcareous soil. After germination in the dark within papers moistened with water during 4 days, seedlings were transferred to a plastic conteiners with nutrient solution for 10 days. Finally, plants were individually transplanted into pots with 500 g of nonsterile calcareous soil as shown in Fig. 13.1.
Source: From Sevillano-Caño, J., García, M. J., Córdoba-Galván, C., Luque-Cruz, C., Agustí-Brisach, C., Lucena, C., & Romera, F. J. (2024). Exploring the role of debaryomyces hansenii as Biofertilizer in iron-deficient environments to enhance plant nutrition and crop production sustainability. *International Journal of Molecular Sciences*, *25*(11), 5729.

13.2 Fe nutrition of plants

Fe is very abundant in most soils, mainly as Fe^{3+}, although its availability to plants is low, especially in calcareous soils (Briat, Dubos, et al., 2015), very abundant around the World. Based on the mechanisms used by plant roots to facilitate mobilization and uptake of Fe, plants are classified into Strategy I species (dicots and nongrass monocots) and Strategy II species (grasses) (Kobayashi & Nishizawa, 2012; Lucena et al., 2015).

It is important to note that this book chapter focuses on highlighting the role that *D. hansenii* plays in Fe nutrition in dicot plants.

Dicots (Strategy I), such as Arabidopsis and tomato, need to reduce Fe^{3+} to Fe^{2+}, by means of a ferric reductase (encoded by *FRO2* in Arabidopsis) at the root surface, prior to its subsequent uptake through a Fe^{2+} transporter (encoded by *IRT1* in Arabidopsis; Kobayashi & Nishizawa, 2012; Lucena et al., 2015). Grasses (Strategy II), like barley, release PS (PhytoSiderophores) from their roots, which form stable Fe^{3+}-chelates. These Fe^{3+}-chelates (Fe^{3+}-PS) are then taken up by specific epidermal root cell plasma membrane transporters (Kobayashi & Nishizawa, 2012).

When grown under Fe deficiency, Strategy I species develop several physiological and morphological responses, mainly in roots, known as Fe deficiency responses and aimed to facilitating Fe mobilization and uptake (Kobayashi & Nishizawa, 2012; Brumbarova et al., 2015; Lucena et al., 2015). Among the physiological responses are: an enhanced ferric reductase activity (due to higher expression of the *FRO* gene; Brumbarova et al., 2015; Lucena et al., 2015); an enhanced Fe^{2+} uptake capacity (due to higher expression of the *IRT1* gene; Brumbarova et al., 2015; Lucena et al., 2015); the acidification of the rhizosphere [due to higher expression of *HA* (H^+-ATPase) genes; (Waters et al., 2007)]; and an increase of the synthesis and release of organic acids (i.e., citrate, malate), phenols, and flavins to the medium (Kabir et al., 2012; Tsai & Schmidt, 2017a; Tsai et al., 2018; Zamioudis et al., 2014). The acidification facilitates the solubilization of Fe and the functioning of the ferric reductase which has an optimum pH around 5.0 (Waters et al., 2007). Organic acids, phenols, and flavins can act as chelating and/or reducing agents for Fe in the soil or inside the plant (Durrett et al., 2007; Tsai & Schmidt, 2017b). Among the morphological responses are: development of subapical root hairs, of cluster roots and of transfer cells, all of them aimed to increase the surface of contact with the soil (Lucena et al., 2015; Romera et al., 2017). Both physiological and morphological responses are located in the subapical regions of the roots (Lucena et al., 2015; Molnár et al., 2023). The activation of the Fe deficiency responses is not fully understood but in the last years several transcription factors (TFs) that participate in the up-regulation of most of their associated genes have been found (Brumbarova et al., 2015; Gao et al., 2019; Kobayashi & Nishizawa, 2012; Schwarz & Bauer, 2020). In Arabidopsis, the master regulator of most of these genes is FIT (bHLH29), homolog of the tomato FER gene (Lucena et al., 2015). The FIT regulatory network comprises other bHLH TFs of the Ib subgroup, such as bHLH38, bHLH39, bHLH100, and bHLH101. All of them have redundant functions and can interact with FIT to form heterodimers that activate the expression of the Fe acquisition genes *FRO2* and *IRT1* (Brumbarova et al., 2015; Gao et al., 2019; Schwarz & Bauer, 2020). *FIT/FER* is induced in roots in response to Fe deficiency while the other Ib bHLH genes cited above are induced in both roots and leaves in response to Fe deficiency (Brumbarova et al., 2015 and references therein). FIT also controls MYB72, a TF essential for growth of plants on low Fe conditions and for ISR development (Palmer et al., 2013; Zamioudis et al., 2014, 2015). Some of the above mentioned

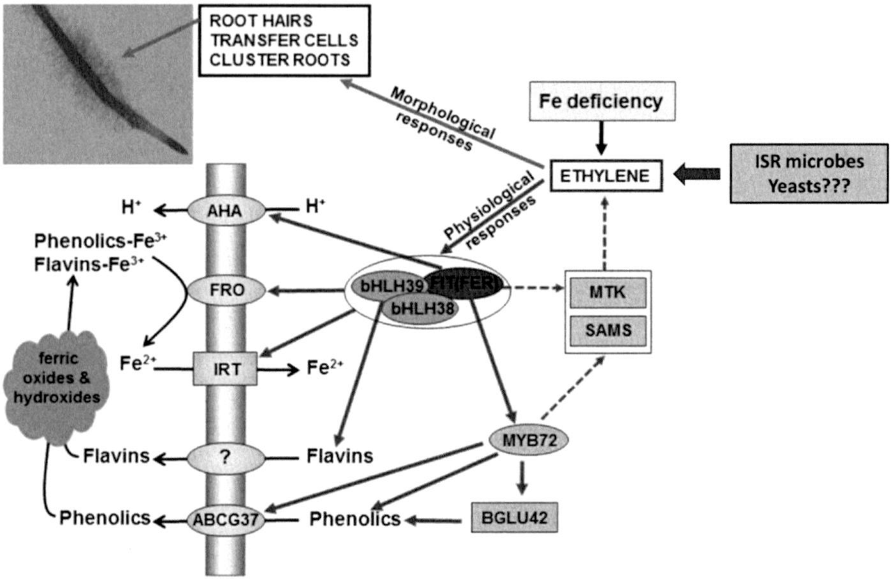

Figure 13.2 Possible pathways involved in the activation of Fe responses under Fe deficiency and upon colonization of roots by ISR microbes and yeasts. Overview of the role of ethylene on the regulation of morphological and physiological responses to Fe deficiency in Strategy I species. Ethylene, through the activation of the transcription factors FIT (FER), bHLH38, and bHLH39, can up-regulate the expression of *FRO* (ferric reductase), *IRT* (iron transporter), and flavin synthesis genes, thus increasing ferric reductase activity, Fe^{2+} uptake and flavin synthesis. Similarly, ethylene, through FIT (FER), can up-regulate *AHA* (H^{+}-ATPase) genes, thus causing acidification, and activate the MYB72 transcription factor, which in turn up-regulates genes related to phenolics synthesis. Moreover, MYB72 activates the b-glucosidase BGLU42 and the phenolic efflux transporter ABCG37, both being implicated in the secretion of phenolic compounds. Ethylene has also been implicated in the development of different morphological responses, such as subapical root hairs, root epidermal transfer cells, and cluster roots. For the development of these morphological responses, FIT(FER) could indirectly act by affecting ethylene synthesis, through the up regulation of *MTK* and *SAMS*. Based on Lucena et al. (2015); Zamioudis et al. (2014, 2015); Liu, Karemera, et al. (2017); Liu, Li, et al., (2017).
Source: From Lucena, C., Romera, F. J., García, M. J., Alcántara, E., & Pérez-Vicente, R. (2015). Ethylene participates in the regulation of Fe deficiency responses in strategy I plants and in rice. *Frontiers in Plant Science*, *6*(November). https://doi.org/10.3389/fpls.2015.01056.

genes, like *FIT*, are subjected to post-transcriptional regulation, related to the internal Fe concentration, in addition to their transcriptional regulation. This implies that sometimes their mRNAs are up-regulated but their related proteins are not (Connolly et al., 2003; Lingam et al., 2011; Meiser et al., 2011) (Fig. 13.2).

Under Fe-deficient conditions, Strategy II species greatly increase the production and release of PS, the number of Fe^{3+}-PS transporters and develop other physiological and regulatory responses (Kobayashi & Nishizawa, 2012). Rice, traditionally considered a Strategy II species (Kobayashi & Nishizawa, 2012), presents some characteristics of Strategy I species, such as enhanced Fe^{2+} uptake through an IRT1-Fe^{2+} transporter. For this reason, some authors consider it as a plant species that uses a combined strategy (Núñez-Cano et al., 2023; Ricachenevsky & Sperotto, 2014).

13.3 Regulation of Fe deficiency responses in dicot plants

Once Fe has been acquired in enough quantity by the plant, its signaling response is switched off. Since Fe is very abundant in soils but with low availability, the responses of dicot plants to favor its acquisition are subject to a complex regulatory machinery. Although not totally known, in dicot plants, it has been found that some hormones and signaling substances act as activators of most of the physiological and morphological responses to Fe deficiency, like ET, auxin, and NO, and some of them as repressors, like Fe itself (Buet et al., 2019; Galatro et al., 2020; García et al., 2010, 2011, 2015, 2021; Lucena et al., 2015, 2018; Meng et al., 2012; Wang, Li, et al., 2010; Wang, Tang, et al., 2010; Romera et al., 2017, 2021). The simultaneous implication of ET, auxin, and NO is not unexpected, since all of them are closely interrelated (Romera et al., 2011, 2017). All the above-mentioned substances increase their synthesis and/or sensitivity in Fe-deficient roots although their precise roles and interactions are not totally known (Buet et al., 2019; Galatro et al., 2020; García et al., 2010, 2011; Graziano & Lamattina, 2007; Huang et al., 2018; Lucena et al., 2015, 2018, 2021; Romera et al., 2011, 2017, 2021). By contrast to these activating signals, other ones have been implicated in the suppression of the responses, like cytokinins, some phloem Fe-related signals or miRNAs (García et al., 2013, 2018, 2021; Lucena et al., 2018; Mendoza-Cózatl et al., 2014; Zhai et al., 2014). Since ET and NO play a key role in the regulation of Fe deficiency responses, and they probably act downstream of most other signaling substances (Romera et al., 2011, 2017), both of them will be described in more detail.

ET is synthetized from the methionine amino acid, through a pathway including SAMS (SAM Synthetase), ACS (ACC Synthase), and ACO (ACC Oxidase; Sauter et al., 2013):

***SAMS ACS ACO* Methionine → SAM → ACC → Ethylene**

Although ethylene's mode of action is not fully understood, a linear canonical signaling pathway has been proposed in Arabidopsis (Binder, 2020; Dubois et al., 2018; Shakeel et al., 2013):

ET—‖ ET receptors → CTR1—‖ EIN2 → EIN3/EILs → ERFs → ET responses

In this pathway, CTR1 is a kinase, EIN2 is a transmembrane protein located in the endoplasmic reticulum membrane, and EIN3, EILs, and ERFs are TFs (Binder, 2020; Dubois et al., 2018; Shakeel et al., 2013). Mutants of CTR1 present constitutive activation of ET signaling, while mutants of EIN2 and EIN3 display reduced sensitivity to ET (Shakeel et al., 2013). The ERF ("ET response factor") TFs act downstream of EIN3 to activate or repress ET-responsive genes (Binder, 2020; Dubois et al., 2018). The real pathway is probably more complex than the linear pathway depicted above (Binder, 2020; Shakeel et al., 2013). Further studies are still necessary.

Fe deficiency influence both ET synthesis and signaling (García et al., 2010, 2015; Lucena et al., 2015; Nagarajan & Smith, 2012; Song & Liu, 2015; Romera et al., 2017). The implication of ET in the activation of physiological and morphological responses to Fe deficiency in dicot (Strategy I) plants, and in the up-regulation of their associated genes, like *FIT*, *FRO2*, *IRT1*, and many others, is supported by many experimental results (reviewed in Lucena et al., 2015; Romera et al., 2017) (Fig. 13.2). As more recent evidence, it has been shown that EIN3 and EIL1 interact with MED16 (Mediator) to form a complex involved in the transcription of *FIT*, the key TF of Fe acquisition genes (Yang et al., 2014). Moreover, Lingam et al. (2011) found that EIN3 and EIL1 also participate in the posttranscriptional regulation of FIT. In Strategy II plants, the role of ET has been less studied but some experimental results also support a key role for ET in the regulation of several Fe-related genes in rice (Wu et al., 2011), which uses a combined Strategy I-Strategy II (Ricachenevsky & Sperotto, 2014).

Besides, there are also many experimental results showing that ET plays an important role in the activation of physiological and morphological responses to phosphorus (P) deficiency, and in the up-regulation of their associated genes, like *PHR1*, *PHT1s*, *PAPs*, and others (Lei et al., 2011; Nagarajan & Smith, 2012; Neumann, 2016; Song & Liu, 2015). Moreover, EIN3 has also been implicated in the expression of *PHR1*, the key TF of P acquisition genes (Liu, Liu, et al., 2017).

In relation to NO, it should be noted that both ET and NO are closely interrelated (Romera et al., 2011, 2017). In fact, ET can enhance NO synthesis in the subapical regions of roots, while NO can activate both ET synthesis and signaling (García et al., 2011, 2021). Both ET and NO greatly activate the expression of Fe acquisition genes and also the development of several morphological responses, like root hairs and cluster roots, in dicot plants (García et al., 2010, 2011; Graziano & Lamattina, 2007; Lucena et al., 2015; Meng et al., 2012; Neumann, 2016; Wang, Li, et al., 2010; Wang, Tang, et al., 2010).

NO can interact with glutathione (GSH) to generate S-nitrosoglutathione (GSNO), a nitrosothiol compound which is the main NO reservoir in plants (Buet et al., 2019). GSNO concentration is regulated by the GSNO reductase, whose expression (the *GSNOR1* gene) increases in both Fe-deficient and P-deficient Arabidopsis roots (García et al., 2018). The roles of NO and GSNO would not be exactly the same since NO concentration increases in Fe-deficient roots while GSNO concentration decreases (García et al., 2011, 2018; Kailasam et al., 2018; Shanmugam et al., 2015). Although NO has been involved in the regulation of morphological responses to P deficiency (Meng et al., 2012; Wang, Li, et al., 2010; Wang, Tang, et al., 2010), its role, and the one of GSNO/GSNOR, have not been studied yet in relation to the physiological responses to P deficiency (García et al., 2021).

Both ET and NO greatly activate the expression of Fe acquisition genes in dicot plants grown with low levels of Fe (or without Fe) but have much less effect in plants grown with high levels of Fe (García et al., 2011; Graziano & Lamattina, 2007). All these results suggest that the up-regulation of Fe acquisition genes does not solely depend on hormones and signaling molecules (such as ET or NO), that would act as activators, but also on the internal Fe content of plants, that would act as a repressor (García et al., 2011, 2013, 2018; Romera et al., 2011). However, different results suggest that total Fe in roots is not the repressor of Fe acquisition genes but a Fe-related signal moving from shoots to roots through the phloem, probably a peptide (García et al., 2013, 2018; Mendoza-Cózatl et al., 2014; Zhai et al., 2014). In this sense, in the last years, it has been found that some peptides, called "Iron Man," are involved in the regulation of Fe deficiency responses (Gautam et al., 2021; Grillet et al., 2018). As with Fe acquisition genes, ET alone is not enough to induce the expression of P acquisition genes (Lei et al., 2011; Nagarajan & Smith, 2012). This suggests that, besides ET, other specific signals would participate in the regulation of the responses to this deficiency (Lei et al., 2011). Again, a parallelism is found in the regulation of the responses to both deficiencies (Fe and P), suggesting the existence of specific repressor signals that would interact with ET/NO. In the P case, a signal that could interact with ET/NO in the regulation of P deficiency responses could be the miRNA399 (Buhtz et al., 2010), which interacts with the PHO2 conjugase to inhibit it (Bari et al., 2006). This is supported by some results showing that the *pho2* mutant presents higher ACC (ethylene precursor) content in roots than the wild-type when grown under P sufficient conditions (Lucena et al., 2019). Another question is whether ET regulates P and Fe deficiency responses through the same transduction pathway or not. The results obtained with different ET signaling mutants suggest that it does not. For example, the upregulation of the *PHT1;4* (*PT2*) gene, induced under P deficiency, is drastically impaired in the Arabidopsis ET insensitive *ein2* mutant (Lei et al., 2011) while the one of the *IRT1* gene, induced under Fe deficiency, is not (Angulo et al., 2021; García et al., 2010).

13.4 Induced systemic resistance (ISR)

Similarly to the responses to abiotic stresses, such as those induced under Fe deficiency, plants respond to biotic stresses provoked by biological agents, like pathogens or insects (Martínez-Medina et al., 2017; Pieterse et al., 2014; Verbon et al., 2017). Some of these responses are located but others are systemic, spreading far from the attacked organ and inducing defensive responses in the entire plant (Pieterse et al., 2014; Verbon et al., 2017). Among the systemic responses are the systemic acquired resistance (SAR) and the induced systemic resistance (ISR) (Pieterse et al., 2014; Verbon et al., 2017).

SAR is originated when pathogens infect a part of the plant while ISR is strikingly induced by beneficial rhizobacteria, like *Pseudomonas simiae* (syn. *P. fluorescens*), *Paenibacillus polimyxa,* and *Azospirillum brasilense*; or rhizofungi, like *Trichoderma* spp., *Glomus* spp. (syn. *Rhizophagus* spp.), or *Fusarium oxysporum* (nonpathogenic races) (Alabouvette & Olivain, 2002; Alabouvette et al., 2009; Alizadeh et al., 2013; Martínez-Medina et al., 2017; Patil et al., 2011; Pieterse et al., 2012, 2014; Pii et al., 2016; Verbon et al., 2017; Zamioudis et al., 2014, 2015; Zhang et al., 2009; Zhao et al., 2014). SAR is characterized by increased levels of SA, which activates the expression of a large set of Pathogenesis-Related (PR) genes, involved in defense responses (Pieterse et al., 2012, 2014).

By contrast to SAR, ISR is generally mediated by an SA-independent pathway where JA and ET (ET) are the central players, and typically functions without PR gene activation (Pieterse et al., 2012, 2014). The ways beneficial rhizospheric microbes elicit ISR is not totally understood but several substances produced by these microbes, and that interact with the plants, have been proposed as elicitors. Among them, there are volatile organic compounds (VOCs; Martínez-Medina et al., 2017; Orozco-Mosqueda et al., 2013; Sharifi & Ryu, 2018; Zamioudis et al., 2015; Zhang et al., 2007, 2009), siderophores, which are Fe chelating agents released by the bacteria to further acquire Fe from the medium (Aznar et al., 2015; Lemanceau et al., 2009), and hormones either produced by the microbes or generated upon interaction of microbe elicitors with the plants. Among the hormones implicated in the ISR, JA, ET, auxin, and NO play a key role (Acharya et al., 2011; Knoester et al., 1999; Martínez-Medina et al., 2017; Nascimento et al., 2018; Nie et al., 2017; Pieterse et al., 2014; Shoresh et al., 2005; Stringlis et al., 2018; Van der Ent, 2008; Zhang et al., 2007). Some years ago, it was found that the *MYB72* gene, encoding a TF, was greatly induced in Arabidopsis roots upon treatment with *Pseudomonas simiae* (Van der Ent, 2008). Arabidopsis *myb72* mutant plants cannot develop ISR. This suggests that this TF plays a key role in the transduction pathway leading to (Segarra et al., 2009; Van der Ent, 2008; Zamioudis et al., 2014, 2015).

13.5 Possible relationship of ISR, the yeast specie *D. hansenii* and Fe deficiency responses in dicot plants

In the previous works, it has been demonstrated that ISR eliciting microorganisms can induce Fe deficiency responses in dicot plants, such as enhanced ferric reductase activity, acidification of the rhizosphere, release of phenols and development of root hairs; and the expression of genes associated with these responses, like *FIT*, *MYB72*, *FRO2*, *IRT1*, *HA*, and others (Martínez-Medina et al., 2017; Pii et al., 2016; Verbon et al., 2019; Zamioudis et al., 2014, 2015; Zhang et al., 2009; Zhao et al., 2014; Zhou, Ma, et al., 2016). As a consequence, these microorganisms can improve the Fe nutrition of plants (Zhou, Guo, et al., 2016). It is remarkable that the root-specific MYB72 TF that plays a key role in the onset of ISR (Van der Ent, 2008; Verbon et al., 2017; Zamioudis et al., 2014), also plays a key role in the adaptive responses of plants to Fe deficiency (García et al., 2011; Lucena et al., 2015; Palmer et al., 2013;

Verbon et al., 2017; Zamioudis et al., 2014, 2015). This indicates that MYB72 is a node of convergence between ISR and Fe deficiency responses (Pieterse et al., 2014; Romera et al., 2019; Verbon et al., 2017; Zamioudis et al., 2014). Moreover, both ISR and Fe responses are regulated by similar hormones and signaling substances, like ET, auxin, and NO (Acharya et al., 2011; Elías et al., 2018; García et al., 2010; Lucena et al., 2015; Romera et al., 2019; Stringlis et al., 2018), indicating that their regulation pathways overlap. In relation to *MYB72*, there is evidence that both ET and NO activate its expression in the same way that the one of *FIT*, since *MYB72* is a FIT-dependent gene (García et al., 2010, 2011; Lucena et al., 2015).

In relation to FO12 (a nonpathogenic race of *Fusarium oxysporum*), Romera´s research group has found that it can induce several Fe deficiency responses, and regulate the genes associated, in dicot plants like cucumber or Arabidopsis (Núñez-Cano et al., 2023; Sevillano-Caño, Núñez-Cano, et al., 2024). FO12 also improves the Fe nutrition of several dicot plants when applied to calcareous soils where the plants are grown. That fungal race has also the ability to protect olive trees against a serious pathogen like Verticillium (Varo et al., 2016), which supports that FO12 could act through the activation of ISR (Pieterse et al., 2014; Romera et al., 2019).

The effects of FO12 have also been studied in rice plants and preliminary results have shown that it can induce the release of phytosiderophores (PS), and some associated genes, a typical response of Strategy II plants to Fe deficiency (Kobayashi & Nishizawa, 2012; Núñez-Cano et al., 2023).

Similarly to the results obtained with FO12, Lucena et al. (2021) have found in preliminary experiments that the application of the yeast specie *D. hansenii* to cucumber plants grown in nutrient solution also induce Fe deficiency responses, such as enhanced ferric reductase activity, acidification, development of root hairs and the expression of Fe deficiency responses genes (Fig. 13.3). To the best of our knowledge, this is the first time that yeasts are related to the induction of Fe deficiency responses, and this phenomenon opens the way to the use of this yeast as biofertilizers, even in saline soils because *D. hansenii* tolerates high levels of salinity (Prista et al., 2005).

ISR eliciting microorganisms have been more extensively studied in their relation to Fe deficiency responses and Fe nutrition of dicot (Strategy I) plants but hardly in grasses (Strategy II; Balmer et al., 2013). Similarly, there are some studies relating ISR eliciting microorganisms to P solubilization (Grady et al., 2016; Li et al., 2015) but rarely with P deficiency responses. However, several responses to P deficiency are similar to the ones induced under Fe deficiency: acidification, release of organic acids and phenols, development of root hairs, and ET and NO are common signals in all these processes (P deficiency responses, Fe deficiency responses, and ISR). Moreover, PHR1, a key TF in the regulation of P deficiency responses, is a direct regulator of a functionally relevant set of plant immune genes (Castrillo et al., 2017). Consequently, it is possible that ISR eliciting microorganisms could also induce some P deficiency responses (Fig. 13.4).

The form in which plants acquire P from the soil is as inorganic phosphate (Pi). The Pi concentration in soils is frequently below the one necessary for optimal maintenance of crops, due to several Pi properties, such as its low mobility and

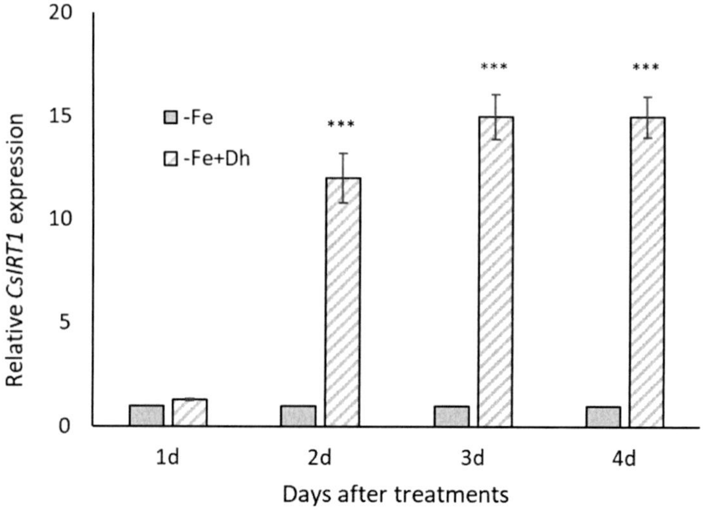

Figure 13.3 Time course of *CsIRT1* expression in cucumber plants along the four determination days of the experiment. The inoculation with *D. hansenii* (Dh) was carried out the same day Fe deficiency treatment was applied. After germination in the dark within papers moistened with water during 4 days, cucumber seedlings were inserted in plastic lids and held in the holes of a thin polyurethane raft floating on aerated nutrient solution containing 10 μM Fe-EDDHA during 10 days. After that, seedlings were inoculated by root immersion in an adjusted cell suspension to reach a final concentration of 10^7 cells ml^{-1} for 30 minutes. After inoculation, plants were transferred to the following treatments: Fe 40 μM or -Fe. Each treatment had its corresponding control. Data of *CsIRT1* expression represent the mean ± SE of three independent biological replicates and two technical replicates. Within each day, bars with *** indicate significant differences ($p < 0.001$) in relation to the –Fe treatment according to the Dunnett's test.
Source: From Lucena, C., Alcalá-Jiménez, M. T., Romera, F. J., & Ramos, J. (2021). Several yeast species induce iron deficiency responses in cucumber plants (Cucumis sativus L.). *Microorganisms*, 9, 2603. https://doi.org/10.3390/microorganisms9122603.

solubility, and its adsorption to Fe oxides (García et al., 2021). Pi is absorbed from the soil trough phosphate transporters, encoded by PHT1 like genes, which are induced by P deficiency and have a great potential to improve P acquisition (García et al., 2021; Karthikeyan et al., 2002; Lucena et al., 2018; Nagarajan & Smith, 2012). In Arabidopsis, the main phosphate transporters implicated in its acquisition from the medium are PHT1;1 (also named PT1) and PHT1;4 (also named PT2; Crombez et al., 2019; Lucena et al., 2018, 2019; Nagarajan & Smith, 2012). Both transporters are located in subapical root hairs (Karthikeyan et al., 2002).

When grown under P deficiency, plants also develop several physiological and morphological responses, mainly in roots, known as P deficiency responses and aimed to facilitating P mobilization and uptake. Most of these responses are similar to the ones developed under Fe deficiency in dicot plants (García et al., 2021; Lucena et al., 2018). Among the physiological responses are: an enhanced expresion of Pi

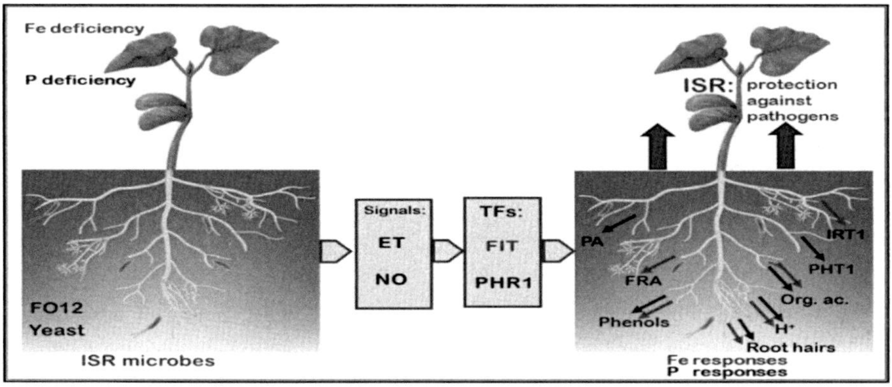

Figure 13.4 Working model to explain the relationship between ISR and Fe and P deficiency responses in dicot plants. Upon Fe or P deficiency, plants induce several responses in their roots, like enhanced expression of Fe (IRT1) or P (PHT1) transporters, the acidification of the rhizosphere, enhanced ferric reductase activity (FRA), enhanced phosphatase activity (PA), release of phenols and organic acids, and development of root hairs. Some of these responses are also induced upon colonization of roots by ISR eliciting microorganisms (Yeasts?). In Fe deficiency, P deficiency and ISR have been implicated common signals, like ethylene (ET) and nitric oxide (NO). In blue, deficiency P responses; in red, Fe deficiency responses.
Source: Modified from Pieterse, C. M., Zamioudis, C., Berendsen, R. L., Weller, D. M., Van Wees, S. C., & Bakker, P. A. (2014). Induced systemic resistance by beneficial microbes. *Annual Review of Phytopathology*, *52*(1), 347–375.

transporters (due to higher expression of the *PHT1* genes); and enhanced phosphatase activity (due to higher expression of *PAP* genes, like *PAP7* or *PAP17* also named *ACP5*; Lei et al., 2011; Lucena et al., 2018, 2019; Venuti et al., 2019); the acidification of the rhizosphere [due to higher expression of *HA* (H^+-ATPase) genes]; and an increase of the synthesis and release of organic acids (i.e., citrate, malate), and phenols to the medium (García et al., 2021; Karthikeyan et al., 2002; Lucena et al., 2019; Nagarajan & Smith, 2012; Venuti et al., 2019). The acidification facilitates the solubilization of Pi, organic acids release Pi from Pi-metal compounds through the chelation of the metal, and phosphatases release Pi from organic compounds containing it (García et al., 2021). Among the morphological responses are: alteration of the root system architecture, increase of the root/shoot ratio, development of subapical root hairs, of cluster roots and of transfer cells, all of them aimed to increase the surface of contact with the soil (García et al., 2021; Lucena et al., 2018; Neumann, 2016). The activation of the P deficiency responses is not fully understood but several transcription factors (TFs) that participate in the up-regulation of most of their associated genes are already known. In Arabidopsis, the master TF is *PHR1*, a MYB TF which interacts with other related TFs, like *PHL1*, to activate the expression of P starvation genes, like *PHT1;1*, *PHT1;4,* and *PAP17* (Briat, Rouached, et al., 2015; Bustos et al., 2010; Nagarajan & Smith, 2012; Rubio, 2001; Sun et al., 2016). Results related to the transcriptional regulation of *PHR1* present some controversy

(Huang et al., 2018; Rubio, 2001; Sega & Pacak, 2019) since hile some authors have found *PHR1* upregulation under P deficiency (Huang et al., 2018) but some others did not (Rubio, 2001; Sega & Pacak, 2019, and references therein).

13.6 The latest achievements

Previous studies highlighted the ability of the yeast *D. hansenii* to induce Fe deficiency responses (Lucena et al., 2021). However, its capacity to compete with other microorganisms in the rhizosphere and to promote plant growth under such conditions remained unexplored until the study by Sevillano-Caño, García, et al. (2024). For their research, the authors used cucumber seedlings grown for 14 days in a hydroponic system with continuously aerated nutrient solution. The seedlings were then transplanted into pots containing calcareous soil, characterized by its low Fe and P availability. Experiments were conducted using either sterile soil (autoclaved twice) or non-sterile soil. Plants were inoculated once or twice with CBS767 by irrigating with 100 mL of adjusted cell suspension to reach a final concentration of 10^8 cells g^{-1} of soil: (1) at transplant and (2) 20 d after transplant.

The validation of this yeast's competitiveness in the rhizosphere marked a pivotal milestone towards its potential application as an Fe biofertilizer, serving as the primary focus of that research.

The findings from Sevillano-Caño, García, et al. (2024) clearly demonstrate that *D. hansenii* promotes plant growth by enhancing nutrient absorption in calcareous soils under the experimental conditions in which the experiments were conducted by the authors. In all cases, the nutrients measured in inoculated plants showed significant increases compared to uninoculated controls. In addition to promoting growth, *D. hansenii* also stimulated flower development, which could translate into higher productivity in cucumber plants. This positions *D. hansenii* as a promising candidate for boosting agricultural productivity in alignment with the Green Deal's sustainability goals.

Additionally, research by Sánchez-Rodríguez et al. (2013, 2014) on calcareous soils with plant species sensitive to Fe chlorosis revealed that P fertilization can alter Fe availability, exacerbating Fe chlorosis in sensitive plants. P often forms complexes with soil components, notably Fe oxides, which act as surfaces for P adsorption (Ziegler, 2016). Interestingly, coumarins, released in response to Fe deficiency, help release P from these Fe oxides, improving P availability for plants (García et al., 2021). The regulation of BGLU42 and PDR9, key enzymes in coumarin biosynthesis, is controlled by FIT through MYB72 a critical point where Fe deficiency responses and ISR pathways converge. This approach leverages the potential of beneficial microorganisms to enhance the availability of both Fe and P, without negatively impacting either nutrient. Based on the aforementioned, it could be stated that *D. hansenii* might play a key role in improving Fe and P nutrition in close connection with ISR. However, this assertion still needs to be confirmed through field experiments, as it is based on results obtained under controlled growth chamber conditions.

Despite the inverse correlation between P, Fe, and manganese (Mn), inoculation with *D. hansenii* significantly improved the uptake of all these elements in soil. This improvement highlights the direct influence of *D. hansenii* on nutrient absorption, emphasizing its role in enhancing the acquisition of essential minerals (Sevillano-Caño et al., 2024).

In addition to promoting Fe uptake (Lucena et al., 2021), *D. hansenii* also triggered the expression of several genes related to ethylene (ET) biosynthesis (*ACO1*), ET signaling (*EIN2* and *EIN3*), and salicylic acid (SA) biosynthesis (*PAL*) in cucumber plants grown in calcareous soils inoculated with *D. hansenii* CBS767. These findings suggest that *D. hansenii*, alongside other microorganisms, could induce both induced systemic resistance (ISR) and systemic acquired resistance (SAR) (Sevillano-Caño, García, et al., 2024). This is consistent with results published by García-Espinoza, Quesada-Moraga, et al. (2023), who demonstrated that inoculating cucumber and melon plants with the entomopathogenic fungus *Metarhizium brunneum* Petch strain EAMa 01/58-Su led to the upregulation of genes involved in ET synthesis and signaling, as well as SA and jasmonic acid (JA) biosynthesis genes (*PAL* and *LOX*). This gene expression enhancement was associated with lethal and sublethal effects on *Spodoptera littoralis*. Importantly, their study showed that ISR-SAR induction was not necessarily tied to endophytic colonization, although it had significant effects on insect pest fitness (García-Espinoza, García, et al., 2023).

A recent study by Lu & Liang (2023) revealed that Fe deficiency-induced ET synthesis confers resistance to *Botrytis cinerea* in *Arabidopsis thaliana*. This suggests that *D. hansenii* could "prime" plants, putting them in a state of readiness that enhances their defense against pathogens and insect pests. This priming effect has been observed in response to rhizosphere microbes, entomopathogenic fungi (EPF), and pathogens (Pieterse et al., 2014). However, Sevillano-Caño, García, et al. (2024) were the first ones describing this phenomenon in response to yeast.

The increased expression of ET-related genes may partially explain the higher number of flowers observed in plants inoculated with *D. hansenii*, as ET is known to regulate flowering (Maric, 2023; Wang, Li, et al., 2010; Wang, Tang, et al., 2010). While the exact mechanism by which ET influences flower development is still not fully understood, its involvement in this process is well-accepted. However, reports on ethylene's role in flowering are somewhat contradictory (Maric, 2023). Some studies suggest that ET delays flowering in species such as *Arabidopsis thaliana* and morning glory (*Pharbitis nil*), while others show that exogenous ET promotes flowering in pineapple (*Ananas comosus*) and urn plants (*Aechmea fasciata*) (Iqbal et al., 2017). In cucumbers, ET has been shown to promote female flower development (Wang, Li, et al., 2010; Wang, Tang, et al., 2010), which aligns with our findings, though Sevillano-Caño, García, et al. (2024) assessed the total number of flowers without differentiating their sex.

In conjunction, the results shown by Sevillano-Caño, García, et al. (2024) evidence the great ability of *D. hansenii* to improve nutrient acquisition and plant growth in Fe-deficient environments. Besides this, *D. hansenii* has also demonstrated its capability to induce systemic responses despite not being an endophytic microorganism. Although further studies are still necessary, that results would indicate that this yeast could induce ISR and SAR. However, as this aspect is explored further and elucidated, it is crucial to consider that the inclusion of *D. hansenii* as a viable candidate for use as a biofertilizer hinges predominantly on its demonstrated ability to compete effectively in the rhizosphere, although all the progress presented so far is still preliminary and must be validated in the field.

13.7 Future perspectives

Fe deficiency is one of the more important plant nutrition disorders worldwide. It takes place mainly in calcareous soils which represents 30% of the arable soils. Fe deficiency leads to reduced crop yield and significant economic loss. In EEUU, the yield reduction in soybean grown in calcareous soils is estimated at $120 million per year (Lauter et al., 2020). In Spain, for example, due to the high incidence of calcareous soils, the Fe deficiency problem is very extended and affect several important crops, like peach, citrus and olive trees. About 70% of the olive trees in Spain are grown in calcareous soils. In general, growers invest large amounts of money trying to solve the Fe deficiency problem (Martín-Sanz et al., 2018). Besides the high economic losses, plant Fe deficiency entails an important human health hazard in many underdeveloped countries, whose populations acquire Fe mainly from plants. As we have previously mentioned, World population is estimated to increase up to 9.000 million of persons by 2050 (FAO, 2015). This will require a substantial increase in crop production to satisfy the food demand (Cieschi et al., 2019). Any crop yield increase will require the finding of solutions to the problem of Fe deficiency, and this should be done in a sustainable way with the environment.

Nowadays, the most frequent ways to solve nutrient deficiencies and to fight against pests are the abusive use of fertilizers and pesticides. For Fe deficiency, the most common fertilizers are Fe chelates, generally very expensive. One alternative is the use of more efficient varieties and rootstocks (Romera et al., 2019). However, this is not always enough, and other additional and complementary solutions are required. In this sense, the use of rhizospheric microbes to optimize crop nutrition and protection is a promising strategy. Within this context, the scientific community focuses its efforts on develop a new and environment friendly Fe biofertilizer to solve Fe deficiency in calcareous soils (Aparicio et al., 2023).

A really good option could be to optimize *D. hansenii* use as a Fe biofertilizer. To get a Fe fertilizer based on a yeast would be a great achievement that would allow us to minimize the cost associated to Fe chelates use and it would mean a great advance to a more sustainable agriculture. Under the results shown, a yeast specie *D. hansenii* is considered as a possible Fe biofertilizer for the first time all over the world (Lucena et al., 2021; Sevillano-Caño et al., 2024).

All the results described by Lucena et al. (2021) and Sevillano-Caño et al. (2024) suggest that *D. hansenii* could have a big potential as a Fe biofertilizer. As it mentioned previously, right now, the most frequent way to control Fe deficiency, very abundant in calcareous soils, is the application of Fe chelates, which are generally very expensive and, in addition, produce a detrimental effect for the environment (Briat, Dubos, et al., 2015). One alternative is the use of Fe-efficient plant genotypes, but this is not always enough in certain soils (Jin et al., 2014; Pii et al., 2015). Consequently, the application of *D. hansenii* could be a complementary practice to the use of Fe-efficient genotypes, avoiding the abusive use of Fe-chelates and probably of other fertilizers. It should be noted that some of the Fe deficiency responses, like the acidification of the rhizosphere or the development of subapical root hairs, are also induced under other nutrient deficiencies, like P, Zn or Mn deficiency, and can favor the acquisition of these other nutrients besides Fe (Duffner et al., 2012; García et al., 2015; Liu, Liu, et al., 2017). In fact, results obtained by Sevillano-Caño, García, et al. (2024) with cucumber plants inoculated or not with *D. hansenii*, and grown in a calcareous soil, showed higher Fe, P, and Mn concentrations in leaves of the inoculated plants.

The exact mechanism by which *D. hansenii* induces Fe deficiency responses remains unclear, though preliminary results suggest it shares traits with ISR-eliciting microorganisms (Sevillano-Caño, García, et al., 2024). Nevertheless, its strong capacity to trigger these responses, ease of management, and ability to thrive in saline and calcareous soils make it an ideal candidate for use as an Fe biofertilizer. One of the key challenges for the scientific community is to optimize the field application of *D. hansenii* to maximize crop yields under adverse conditions, such as high-salinity soils. Future research should focus on studying and comparing different inoculum concentrations in various soil types to identify the optimal dosage for each plant species and soil condition. Besides, more targeted future studies are necessary, especially on *D. hansenii*´s interaction with changing environment conditions with the aim of confirming viability and effectiveness as Fe biofertilizer of this yeast.

Based on the progress made so far regarding the role of *Debaryomyces hansenii* in Fe nutrition and plant growth, the next step could be to achieve its industrial use, developing a new Fe biofertilizer based in the D. *hansenii* yeast to solve Fe deficiency in calcareous soils and to improve plant nutrition in general evaluating the viability of several *D. hansenii* formulates. For example, after the lyophilization process or preparing its a carrier-based formulate that helps it remain viable in the soil or on the leaf after application. Another step could be to optimize the use of *D. hansenii* as a Fe biofertilizer in horticultural species and in woody species in two different kind of soils, calcareous, and saline soils. The first one could be selected since Fe deficiency is a very frequent problem in calcareous soils, and *D. hansenii* can favor Fe nutrition. Saline soil could be chosen because *D. hansenii* is a tolerant yeast to high salt concentrations. For these reasons, the next step must be to study and compare different inoculum concentrations in both kind of soils to find the optimum dose to each plant species and kind of soil.

References

Acharya, K., Chandra, S., Chakraborty, N., & Acharya, R. (2011). Nitric oxide functions as a signal in induced systemic resistance. *Archives of Phytopathology and Plant Protection, 44*, 1335–1342. https://doi.org/10.1080/03235408.2010.496552.

Aggarwal, M., & Mondal, A. K. (2006). Role of N-terminal hydrophobic region in modulating the subcellular localization and enzyme activity of the bisphosphate nucleotidase from Debaryomyces hansenii. *Eukaryotic Cell, 5*(2), 262–271.

Alabouvette, C., & Olivain, C. (2002). Modes of action of non-pathogenic strains of Fusarium oxysporum in controlling Fusarium wilts. *Plant Protection Science, 38*, 195–199.

Alabouvette, C., Olivain, C., Migheli, Q., & Steinberg, C. (2009). Microbiological control of soil-borne phytopathogenic fungi with special emphasis on wilt-inducing Fusarium oxysporum. *The New Phytologist, 184*(3), 529–544. https://doi.org/10.1111/j.1469-8137.2009.03014.x.

Alizadeh, H., Behboudi, K., Ahmadzadeh, M., Javan-Nikkhah, M., Zamioudis, C., Pieterse, C. M., & Bakker, P. A. (2013). Induced systemic resistance in cucumber and Arabidopsis thaliana by the combination of Trichoderma harzianum Tr6 and Pseudomonas sp. Ps14. *Biological Control, 65*(1), 14–23.

Angulo, M., García, M. J., Alcántara, E., Pérez-Vicente, R., & Romera, F. J. (2021). Comparative study of several Fe deficiency responses in the Arabidopsis thaliana ethylene insensitive mutants ein2-1 and ein2-5. *Plants, 10*(2), 262.

Aparicio, M. A., Lucena, C., García, M. J., Ruiz-Castilla, F. J., Jiménez-Adrián, P., López-Berges, M. S., Prieto, P., Alcántara, E., Pérez-Vicente, R., Ramos, J., & Romera, F. J. (2023). The nonpathogenic strain of fusarium oxysporum FO12 induces Fe deficiency responses in cucumber (Cucumis sativus L.) plants. *Planta, 257*, 50.

Aznar, A., Chen, N. W. G., Thomine, S., & Dellagi, A. (2015). Immunity to plant pathogens and iron homeostasis. *Plant Science (Shannon, Ireland), 240*, 90–97. https://doi.org/10.1016/j.plantsci.2015.08.022.

Balmer, D., Planchamp, C., & Mauch-Mani, B. (2013). On the move: Induced resistance in monocots. *Journal of Experimental Botany, 64*, 1249–1261. https://doi.org/10.1093/jxb/ers248.

Bari, R., Datt Pant, B., Stitt, M., & Scheible, W. R. (2006). PHO2, microRNA399, and PHR1 define a phosphate-signaling pathway in plants. *Plant Physiology, 141*(3), 988–999.

Binder, B. M. (2020). Ethylene signaling in plants. *Journal of Biological Chemistry, 295*(22), 7710–7725.

Briat, J. F., Rouached, H., Tissot, N., Gaymard, F., & Dubos, C. (2015). Integration of P, S, Fe, and Zn nutrition signals in Arabidopsis thaliana: Potential involvement of Phosphate Starvation Response 1 (PHR1). *Frontiers in Plant Science, 6*, 290.

Briat, J. F., Dubos, C., & Gaymard, F. (2015). Iron nutrition, biomass production, and plant product quality. *Trends in Plant Science, 20*, 33–40. https://doi.org/10.1016/j.tplants.2014.07.005.

Brumbarova, T., Bauer, P., & Ivanov, R. (2015). Molecular mechanisms governing arabidopsis iron uptake. *Trends in Plant Science, 20*, 124–133.

Buet, A., Galatro, A., Ramos-Artuso, F., & Simontacchi, M. (2019). Nitric oxide and plant mineral nutrition: Current knowledge. *Journal of Experimental Botany, 70*(17), 4461–4476.

Buhtz, A., Pieritz, J., Springer, F., & Kehr, J. (2010). Phloem small RNAs, nutrient stress responses, and systemic mobility. *BMC Plant Biology, 10*, 1–13.

Bustos, R., Castrillo, G., Linhares, F., Puga, M. I., Rubio, V., Pérez-Pérez, J., & Paz-Ares, J. (2010). A central regulatory system largely controls transcriptional activation and repression responses to phosphate starvation in Arabidopsis. *PLoS Genetics, 6*(9), e1001102.

Cabeç-Silva, & Madeira-Lopes (2007). Occurrence of two maximum temperatures for growth in yeasts. *Zeitschrift fur Allgemeine Mikrobiologie, 11*, 59–61.

Castrillo, G., Teixeira, P. J. P. L., Paredes, S. H., Law, T. F., De Lorenzo, L., Feltcher, M. E., & Dangl, J. L. (2017). Root microbiota drive direct integration of phosphate stress and immunity. *Nature, 543*(7646), 513–518.

Cieschi, M. T., Polyakov, A. Y., Lebedev, V. A., Volkov, D. S., Pankratov, D. A., Veligzhanin, A. A., & Lucena, J. J. (2019). Eco-friendly iron-humic nanofertilizers synthesis for the prevention of iron chlorosis in soybean (Glycine max) grown in calcareous soil. *Frontiers in Plant Science, 10*, 413.

Connolly, E. L., Campbell, N. H., Grotz, N., Prichard, C. L., & Guerinot, M. L. (2003). Overexpression of the FRO2 ferric chelate reductase confers tolerance to growth on low iron and uncovers posttranscriptional control. *Plant Physiology, 133*(3), 1102–1110.

Crombez, H., Motte, H., & Beeckman, T. (2019). Tackling plant phosphate starvation by the roots. *Developmental Cell, 48*(5), 599–615.

Dubois, M., Van den Broeck, L., & Inzé, D. (2018). The pivotal role of ethylene in plant growth. *Trends in Plant Science*. https://doi.org/10.1016/j.tplants.2018.01.003.

Duffner, A., Hoffland, E., & Temminghoff, E. J. (2012). Bioavailability of zinc and phosphorus in calcareous soils as affected by citrate exudation. *Plant and Soil, 361*(1), 165–175. https://doi.org/10.1007/s11104-012-1273-9.

Durrett, T. P., Gassmann, W., & Rogers, E. E. (2007). The FRD3-mediated efflux of citrate into the root vasculature is necessary for efficient iron translocation. *Plant Physiology, 144*, 197–205. https://doi.org/10.1104/pp.107.097162.

Elías, J. M., Guerrero-Molina, M. F., Martínez-Zamora, M. G., Díaz-Ricci, J. C., & Pedraza, R. O. (2018). Role of ethylene and related gene expression in the interaction between strawberry plants and the plant growth-promoting bacterium Azospirillum brasilense. *Plant Biology, 20*(3), 490–496.

FAO. (2015). Sustainable development goals helpdesk. http://www.fao.org/sustainable-development-goals/overview/fao-and-post2015/sustainable-agriculture/es/.

Fernández-San Millán, A., Farran, I., Larraya, L., Ancin, M., Arregui, L. M., & Veramendi, J. (2020). Plant growth-promoting traits of yeasts isolated from Spanish vineyards: Benefits for seedling development. *Microbiological Research, 237*, 126480.

Galatro, A., Ramos-Artuso, F., Luquet, M., Buet, A., & Simontacchi, M. (2020). An update on nitric oxide production and role under phosphorus scarcity in plants. *Frontiers in Plant Science, 11*, 413.

Gao, F., Robe, K., Gaymard, F., Izquierdo, E., & Dubos, C. (2019). The transcriptional control of iron homeostasis in plants: A tale of bHLH transcription factors? *Frontiers in Plant Science, 10*, 6.

García, M. J., Corpas, F. J., Lucena, C., Alcántara, E., Pérez-Vicente, R., Zamarreño, Á. M., & Romera, F. J. (2018). A shoot Fe signaling pathway requiring the OPT3 transporter controls GSNO reductase and ethylene in Arabidopsis thaliana roots. *Frontiers in Plant Science, 9*, 1325.

García, M. J., Lucena, C., & Romera, F. J. (2021). Ethylene and nitric oxide involvement in the regulation of Fe and P deficiency responses in dicotyledonous plants. *International Journal of Molecular Sciences, 22*(9), 4904.

García, M. J., Lucena, C., Romera, F. J., Alcantara, E., & Pérez-Vicente, R. (2010). Ethylene and nitric oxide involvement in the up-regulation of key genes related to iron acquisition and homeostasis in Arabidopsis. *Journal of Experimental Botany, 61*(14), 3885–3899.

García, M. J., Romera, F. J., Lucena, C., Alcántara, E., & Pérez-Vicente, R. (2015). Ethylene and the regulation of physiological and morphological responses to nutrient deficiencies. *Plant Physiology, 169*(1), 51–60.

García, M. J., Romera, F. J., Stacey, M. G., Stacey, G., Villar, E., Alcántara, E., & Pérez-Vicente, R. (2013). Shoot to root communication is necessary to control the expression of iron-acquisition genes in Strategy I plants. *Planta, 237*, 65–75.

García, M. J., Suárez, V., Romera, F. J., Alcántara, E., & Pérez-Vicente, R. (2011). A new model involving ethylene, nitric oxide and Fe to explain the regulation of Fe-acquisition genes in Strategy I plants. *Plant Physiology and Biochemistry, 49*(5), 537–544.

García-Espinoza, F., García, M. J., Quesada-Moraga, E., & Yousef-Yousef, M. (2023). Entomopathogenic fungus-related priming defense mechanisms in cucurbits impact Spodoptera littoralis (Boisduval) fitness. *Applied and Environmental Microbiology, 89*(8), e00940 23.

García-Espinoza, F., Quesada-Moraga, E., García del Rosal, M. J., & Yousef-Yousef, M. (2023). Entomopathogenic fungi-mediated solubilization and induction of Fe related genes in melon and cucumber plants. *Journal of Fungi, 9*(2), 258.

Gautam, C. K., Tsai, H. H., & Schmidt, W. (2021). IRONMAN tunes responses to iron deficiency in concert with environmental pH. *Plant Physiology, 187*(3), 1728–1745.

Grady, E. N., MacDonald, J., Liu, L., Richman, A., & Yuan, Z. C. (2016). Current knowledge and perspectives of Paenibacillus: A review. *Microbial Cell Factories, 15*, 203. https://doi.org/10.1186/s12934-016-0603-7.

Graziano, M., & Lamattina, L. (2007). Nitric oxide accumulation is required for molecular and physiological responses to iron deficiency in tomato roots. *The Plant Journal: For Cell and Molecular Biology, 52*, 949–960.

Grillet, L., Lan, P., Li, W., Mokkapati, G., & Schmidt, W. (2018). IRON MAN is a ubiquitous family of peptides that control iron transport in plants. *Nature Plants, 4*(11), 953–963.

Herrera-Cruz, M. S., & Simmen, T. (2017). Of yeast, mice and men: MAMs come in two flavors. *Biology Direct, 12*, 1–21.

Huang, K. L., Ma, G. J., Zhang, M. L., Xiong, H., Wu, H., Zhao, C. Z., & Ren, F. (2018). The ARF7 and ARF19 transcription factors positively regulate PHOSPHATE STARVATION RESPONSE1 in Arabidopsis roots. *Plant Physiology, 178*(1), 413–427.

Iqbal, N., Khan, N. A., Ferrante, A., Trivellini, A., Francini, A., & Khan, M. I. R. (2017). Ethylene role in plant growth, development and senescence: Interaction with other phytohormones. *Frontiers in Plant Science, 8*, 475.

Jamal, A., Farhat, H., Urooj, F., Rahman, A., Irfan, M., & Ehteshamul-Haque, S. (2021). Characterization of endophytic yeast and its suppressive effect on root rotting fungi of tomato under neem cake soil amendment. *Egyptian Journal of Biological Pest Control, 31*, 1–12.

Jin, C. W., Ye, Y. Q., & Zheng, S. J. (2014). An underground tale: Contribution of microbial activity to plant iron acquisition via ecological processes. *Annals of Botany, 113*, 7–18.

Kabir, A. H., Paltridge, N. G., Able, A. J., Paull, J. G., & Stangoulis, J. C. R. (2012). Natural variation for Fe-efficiency is associated with up-regulation of Strategy I mechanisms and enhanced citrate and ethylene synthesis in *Pisum sativum* L. *Planta, 235*, 1409–1419. https://doi.org/10.1007/s00425-011-1583-9.

Kailasam, S., Wang, Y., Lo, J. C., Chang, H. F., & Yeh, K. C. (2018). S-nitrosoglutathione works downstream of nitric oxide to mediate iron deficiency signaling in Arabidopsis. *The Plant Journal: For Cell and Molecular Biology, 94*, 157–168. https://doi.org/10.1111/tpj.13850.

Karthikeyan, A. S., Varadarajan, D. K., Mukatira, U. T., D'Urzo, M. P., Damsz, B., & Raghothama, K. G. (2002). Regulated expression of Arabidopsis phosphate transporters. *Plant Physiology, 130*(1), 221–233.

Kaur, J., Anand, V., Srivastava, S., Bist, V., Tripathi, P., Naseem, M., & Srivastava, S. (2020). Yeast strain Debaryomyces hansenii for amelioration of arsenic stress in rice. *Ecotoxicology and Environmental Safety, 195*, 110480.

Knoester, M., Pieterse, C. M. J., Bol, J. F., & Van Loon, L. C. (1999). Systemic resistance in *Arabidopsis* induced by rhizobacteria requires ethylene-dependent signaling at the site of application. *Molecular Plant-Microbe Interactions: MPMI, 12*, 720–727. https://doi.org/10.1094/MPMI.1999.12.8.720.

Kobayashi, T., & Nishizawa, N. K. (2012). Iron uptake, translocation, and regulation in higher plants. *Annual Review of Plant Biology, 63*(1), 131–152.

Kreger-van Rij, N. J. W. (Ed.). (1984). *The yeasts - a taxonomic study* (3rd ed.). Amsterdam: Elsevier Sci. Publ.

Lauter, A. N. M., Rutter, L., Cook, D., O'Rourke, J. A., & Graham, M. A. (2020). Examining short-term responses to a long-term problem: RNA-seq analyses of iron deficiency chlorosis tolerant soybean. *International Journal of Molecular Sciences, 21*(10), 3591.

Lee, G., Lee, S. H., Kim, K., & Ryu, C. (2017). Foliar application of the leaf-colonizing yeast Pseudozyma churashimaensis elicits systemic defense of pepper against bacterial and viral pathogens. *Scientific Reports, 7*(1), 39432.

Lei, M., Zhu, C., Liu, Y., Karthikeyan, A. S., Bressan, R. A., Raghothama, K. G., & Liu, D. (2011). Ethylene signalling is involved in regulation of phosphate starvation-induced gene expression and production of acid phosphatases and anthocyanin in Arabidopsis. *New Phytologist, 189*(4), 1084–1095.

Lemanceau, P., Expert, D., Gaymard, F., Bakker, P. A. H. M., & Briat, J. F. (2009). Role of iron in plant-microbe interactions. *Advances in Botanical Research, 51*, 491–549. https://doi.org/10.1016/S0065-2296(09)51012-9.

Li, R.-X., Cai, F., Pang, G., Shen, Q.-R., Li, R., & Chen, W. (2015). Solubilisation of phosphate and micronutrients by *Trichoderma harzianum* and its relationship with the promotion of tomato plant growth. *PLoS One, 10*, e0130081. https://doi.org/10.1371/journal.pone.0130081.

Liao, Q., Zhao, Y., Wang, Z., Yu, L., Su, Q., Li, J., & Liu, J. (2024). Kiwifruit resistance to gray mold is enhanced by yeast-induced modulation of the endophytic microbiome. *Science of The Total Environment, 932*, 173109.

Ling, L., Tu, Y., Ma, W., Feng, S., Yang, C., Zhao, Y., & Zhang, J. (2020). A potentially important resource: Endophytic yeasts. *World Journal of Microbiology and Biotechnology, 36*, 1–7.

Lingam, S., Mohrbacher, J., Brumbarova, T., Potuschak, T., Fink-Straube, C., Blondet, E., & Bauer, P. (2011). Interaction between the bHLH transcription factor FIT and ETHYLENE INSENSITIVE3/ETHYLENE INSENSITIVE3-LIKE1 reveals molecular linkage between the regulation of iron acquisition and ethylene signaling in Arabidopsis. *The Plant Cell, 23*(5), 1815–1829.

Liu, M., Liu, X. X., He, X. L., Liu, L. J., Wu, H., Tang, C. X., & Jin, C. W. (2017). Ethylene and nitric oxide interact to regulate the magnesium deficiency-induced root hair development in Arabidopsis. *The New Phytologist, 213*, 1242–1256. https://doi.org/10.1111/nph.14259.

Liu, W., Karemera, N. J. U., Wu, T., Yang, Y., Zhang, X., Xu, X., et al. (2017). The ethylene response factor AtERF4 negatively regulates the iron deficiency response in *Arabidopsis thaliana*. *PLoS One, 12*, e0186580. https://doi.org/10.1371/journal.pone.0186580.

Liu, W., Li, Q., Wang, Y., Wu, T., Yang, Y., Zhang, X., et al. (2017). Ethylene response factor AtERF72 negatively regulates *Arabidopsis thaliana* response to iron deficiency.

Biochemical and Biophysical Research Communications, 491, 862–868. https://doi.org/10.1016/j.bbrc.2017.04.014.

Lonhienne, T., Mason, M. G., Ragan, M. A., Hugenholtz, P., Schmidt, S., & Paungfoo-Lonhienne, C. (2014). Yeast as a biofertilizer alters plant growth and morphology. *Crop Science, 54*(2), 785–790.

Lu, C. K., & Liang, G. (2023). Fe deficiency-induced ethylene synthesis confers resistance to Botrytis cinerea. *New Phytologist, 237*(5), 1843–1855.

Lucena, C., Alcalá-Jiménez, M. T., Romera, F. J., & Ramos, J. (2021). Several Yeast Species Induce Iron Deficiency Responses in Cucumber Plants (Cucumis sativus L.). *Microorganisms, 9*, 2603. https://doi.org/10.3390/microorganisms9122603.

Lucena, C., Porras, R., García, M. J., Alcántara, E., Pérez-Vicente, R., Zamarreño, Á. M., & Romera, F. J. (2019). Ethylene and phloem signals are involved in the regulation of responses to Fe and P deficiencies in roots of strategy I plants. *Frontiers in Plant Science, 10*, 1237.

Lucena, C., Porras, R., Romera, F. J., Alcántara, E., García, M. J., & Pérez-Vicente, R. (2018). Similarities and differences in the acquisition of Fe and P by dicot plants. *Agronomy, 8*(8), 148.

Lucena, C., Romera, F. J., García, M. J., Alcántara, E., & Pérez-Vicente, R. (2015). Ethylene participates in the regulation of Fe deficiency responses in Strategy I plants and in rice. *Frontiers in Plant Science, 6*, 1056. https://doi.org/10.3389/fpls.2015.01056.

Maric, A. (2023). Beyond the genetics of flowering: Integration of ethylene signaling and histone methylation controls flowering time. *Plant Physiology, 192*, 2224–2226.

Marschner, P. (2012). *Marschner's Mineral Nutrition of Higher Plants.* (3rd edition). London: Academic Press.

Martín-Sanz, J. P., Valverde-Asenjo, I., de Santiago-Martín, A., Quintana-Nieto, J. R., González-Huecas, C., López-Lafuente, A. L., & Diéguez-Antón, A. (2018). Enzyme activity indicates soil functionality affectation with low levels of trace elements. *Environmental Pollution, 243*, 1861–1866.

Martínez, T., Bellincontro, A., Peinado, R. A., Mauricio, J. C., Mencarelli, F., & Moreno, J. J. (2011). Discrimination of sweet wines partially fermented by two osmo-ethanol-tolerant yeasts by gas chromatographic analysis and electronic nose. *Food Chemistry, 127*(3), 1391–1396.

Martínez-Medina, A., Van Wees, S. C. M., & Pieterse, C. M. J. (2017). Airborne signals from *Trichoderma* fungi stimulate iron uptake responses in roots resulting in priming of jasmonic acid dependent defences in shoots of *Arabidopsis thaliana* and *Solanum lycopersicum. Plant, Cell & Environment, 40*, 2691–2705. https://doi.org/10.1111/pce.13016.

Medina-Córdova, N., López-Aguilar, R., Ascencio, F., Castellanos, T., Campa-Córdova, A. I., & Angulo, C. (2016). Biocontrol activity of the marine yeast Debaryomyces hansenii against phytopathogenic fungi and its ability to inhibit mycotoxins production in maize grain (Zea mays L.). *Biological Control, 97*, 70–79.

Meiser, J., Lingam, S., & Bauer, P. (2011). Posttranslational regulation of the iron deficiency basic helix-loop-helix transcription factor FIT is affected by iron and nitric oxide. *Plant Physiology, 157*(4), 2154–2166.

Mendoza-Cózatl, D. G., Xie, Q., Akmakjian, G. Z., Jobe, T. O., Patel, A., Stacey, M. G., & Schroeder, J. I. (2014). OPT3 is a component of the iron-signaling network between leaves and roots and misregulation of OPT3 leads to an over-accumulation of cadmium in seeds. *Molecular Plant, 7*(9), 1455–1469.

Meng, Z. B., Chen, L. Q., Suo, D., Li, G. X., Tang, C. X., & Zheng, S. J. (2012). Nitric oxide is the shared signalling molecule in phosphorus-and iron-deficiency-induced formation of cluster roots in white lupin (Lupinus albus). *Annals of Botany, 109*(6), 1055–1064.

Mercado-Blanco, J., & JJ Lugtenberg, B. (2014). Biotechnological applications of bacterial endophytes. *Current Biotechnology, 3*(1), 60–75.

Molnár, Z., Solomon, W., Mutum, L., & Janda, T. (2023). Understanding the mechanisms of Fe deficiency in the rhizosphere to promote plant resilience. *Plants, 12*(10), 1945.

Montiel, V., & Ramos, J. (2007). Intracellular Na+ and K+ distribution in Debaryomyces hansenii. Cloning and expression in Saccharomyces cerevisiae of DhNHX1. *FEMS Yeast Research, 7*(1), 102–109.

Nagarajan, V. K., & Smith, A. P. (2012). Ethylene´s role in phosphate starvation signaling more than just a root growth regulator. *Plant & Cell Physiology, 53*, 277–286.

Nascimento, F. X., Rossi, M. J., & Glick, B. R. (2018). Ethylene and 1-aminocyclopropane-1-carboxylate (ACC) in plant–bacterial interactions. *Frontiers in Plant Science, 9*, 114. https://doi.org/10.3389/fpls.2018.00114.

Neumann, G. (2016). The role of ethylene in plant adaptations for phosphate acquisition in soils – A review. *Frontiers in Plant Science, 6*, 1224.

Nie, P., Li, X., Wang, S., Guo, J., Zhao, H., & Niu, D. (2017). Induced Systemic Resistance against *Botrytis cinerea* by *Bacillus cereus* AR156 through a JA/ET- and NPR1-dependent signaling pathway and activates PAMP-triggered immunity in *Arabidopsis. Frontiers in Plant Science, 8*, 238. https://doi.org/10.3389/fpls.2017.00238.

Nimsi, K. A., Arya, H., Manjusha, K., & Kathiresan, K. (2024). Multifarious plant growth-promoting traits of mangrove yeasts: growth enhancement in mangrove seedlings (Rhizophora mucronata) for conservation. *Archives of Microbiology, 206*(4), 192.

Norkrans, B., & Kylin, A. (1969). Regulation of the potassium to sodium ratio and of the osmotic potential in relation to salt tolerance in yeasts. *Journal of Bacteriology, 100*(2), 836–845.

Núñez-Cano, J., Romera, F. J., Prieto, P., García, M. J., Sevillano-Caño, J., Agustí-Brisach, C., & Lucena, C. (2023). Effect of the Nonpathogenic Strain Fusarium oxysporum FO12 on Fe Acquisition in Rice (Oryza sativa L.). *Plants. Plants, 12*(17), 3145.

Orozco-Mosqueda, M. C., Velázquez-Becerra, C., Macías-Rodríguez, L. I., Santoyo, G., Flores-Cortez, I., Alfaro-Cuevas, R., et al. (2013). *Arthrobacter agilis* UMCV2 induces iron acquisition in *Medicago truncatula* (Strategy I plant) in vitro via dimethylhexadecylamine emission. *Plant and Soil, 362*, 51–66. https://doi.org/10.1007/s11104-012-1263-y.

Palmer, C. M., Hind, M. N., Schmidt, H., Clemens, S., & Guerinot, M. L. (2013). MYB10 and MYB72 are required for growth under iron-limiting conditions. *PLoS Genetics, 9*, e1003953. https://doi.org/10.1371/journal.pgen.1003953.

Patil, S., Sriram, S., Savitha, M. J., & Arulmani, N. (2011). Induced systemic resistance in tomato by non-pathogenic Fusarium species for the management of Fusarium wilt. *Archives of Phytopathology and Plant Protection, 44*, 1621–1634. https://doi.org/10.1080/03235408.2010.526774.

Pieterse, C. M., Zamioudis, C., Berendsen, R. L., Weller, D. M., Van Wees, S. C., & Bakker, P. A. (2014). Induced systemic resistance by beneficial microbes. *Annual Review of Phytopathology, 52*(1), 347–375.

Pieterse, C. M. J., Van der Does, D., Zamioudis, C., Leon-Reyes, A., & Van Wees, S. C. M. (2012). Hormonal modulation of plant immunity. *Annual Review of Cell and Developmental Biology, 28*, 489–521. https://doi.org/10.1146/annurev-cellbio-092910-154055.

Pii, Y., Marastoni, L., Springeth, C., Fontanella, M. C., Beone, G. M., Cesco, S., et al. (2016). Modulation of Fe acquisition process by *Azospirillum brasilense* in cucumber plants. *Environmental and Experimental Botany, 130*, 216–225. https://doi.org/10.1016/j.envexpbot.2016.06.011.

Pii, Y., Mimmo, T., Tomasi, N., Terzano, R., Cesco, S., & Crecchio, C. (2015). Microbial interactions in the rhizosphere: Beneficial influences of plant growth-promoting rhizobacteria on nutrient acquisition process. A review. *Biology and Fertility of Soils, 51*, 403–415.

Prabina, B. J., Kumutha, K., Anandham, R., & Durga, P. (2019). Isolation and characterization of multifunctional yeast as plant probiotics for better crop nutrition in pulses. *International Journal of Current Microbiology and Applied Sciences, 8*(01), 2711–2718.

Prista, C., Gonzalez-Hernandez, J. C., Ramos, J., & Loureiro-Dias, M. C. (2007). Cloning and characterization of two K+ transporters of Debaryomyces hansenii. *Microbiology (Reading, England), 153*(9), 3034–3043.

Prista, C., Loureiro-Dias, M. C., Montiel, V., García, R., & Ramos, J. (2005). Mechanisms underlying the halotolerant way of Debaryomyces hansenii. *FEMS Yeast Research, 5*(8), 693–701.

Prista, C., Michán, C., Miranda, I., & Ramos, J. (2016). The halotolerant *Debaryomyces hansenii*, the Cinderella of non-conventional yeasts. *Yeast (Chichester, England), 33*(10), 523–533. https://doi.org/10.1002/yea.3177.

Raklami, A., Babalola, O. O., Jemo, M., & Nafis, A. (2024). Unlocking the plant growth-promoting potential of yeast spp.: Exploring species from the Moroccan extremophilic environment for enhanced plant growth and sustainable farming. *FEMS Microbiology Letters, 371*, fnae015.

Ricachenevsky, F. K., & Sperotto, R. A. (2014). There and back again, or always there? The evolution of rice combined strategy for Fe uptake. *Frontiers in Plant Science, 5*, 189.

Romera, F. J., García, M. J., Alcántara, E., & Pérez-Vicente, R. (2011). Latest findings about the interplay of auxin, ethylene and nitric oxide in the regulation of Fe deficiency responses by Strategy I plants. *Plant Signaling & Behavior, 6*(1), 167–170.

Romera, F. J., García, M. J., Lucena, C., Martínez-Medina, A., Aparicio, M. A., Ramos, J., & Pérez-Vicente, R. (2019). Induced systemic resistance (ISR) and Fe deficiency responses in dicot plants. *Frontiers in Plant Science, 10*, 287.

Romera, F. J., Lucena, C., García, M. J., Alcántara, E., & Pérez-Vicente, R. (2017). The role of ethylene and other signals in the regulation of Fe deficiency responses by dicot plants. *Stress Signaling in Plants: Genomics and Proteomics Perspective, 2*, 277–300.

Romera, F. J., Lucena, C., García, M. J., Alcántara, E., Angulo, M., Aparicio, M.Á., & Pérez-Vicente, R. (2021). Plant hormones and nutrient deficiency responses. *Hormones and Plant Response*, 29–65.

Rubio, V., et al. (2001). A conserved MYB transcription factor involved in phosphate starvation signaling both in vascular plants and in unicellular algae. *Genes & Development, 15*, 2122–2133.

Ruiz-Pérez, F. S., Ruiz-Castilla, F. J., Leal, C., Martínez, J. L., & Ramos, J. (2023). Sodium and lithium exert differential effects on the central carbon metabolism of Debaryomyces hansenii through the glyoxylate shunt regulation. *Yeast (Chichester, England), 40*(7), 265–275.

Sánchez-Rodríguez, A. R., Canasveras, J. C., del Campillo, M. C., Barrón, V., & Torrent, J. (2013). Iron chlorosis in field grown olive as affected by phosphorus fertilization. *European Journal of Agronomy, 51*, 101–107.

Sánchez-Rodríguez, A. R., Del Campillo, M. C., Torrent, J., & Jones, D. L. (2014). Organic acids alleviate iron chlorosis in chickpea grown on two P-fertilized soils. *Journal of Soil Science and Plant Nutrition, 14*(2), 292–303.

Santos, A., Navascués, E., Bravo, E., & Marquina, D. (2011). Ustilago maydis killer toxin as a new tool for the biocontrol of the wine spoilage yeast Brettanomyces bruxellensis. *International Journal of Food Microbiology, 145*(1), 147–154.

Sauter, M., Moffatt, B., Saechao, M. C., Hell, R., & Wirtz, M. (2013). Methionine salvage and S-adenosylmethionine: Essential links between sulfur, ethylene and polyamine biosynthesis. *The Biochemical Journal, 451*, 145–154. https://doi.org/10.1042/BJ20121744.

Schwarz, B., & Bauer, P. (2020). FIT, a regulatory hub for iron deficiency and stress signaling in roots, and FIT-dependent and -independent gene signatures. *Journal of Experimental Botany, 71*, 1694–1705.

Sega, P., & Pacak, A. (2019). Plant PHR transcription factors: Put on a map. *Genes, 10*, 1018.

Segarra, G., Van der Ent, S., Trillas, I., & Pieterse, C. M. J. (2009). MYB72, a node of convergence in induced systemic resistance triggered by a fungal and a bacterial beneficial microbe. *Plant Biology, 11*, 90–96. https://doi.org/10.1111/j.1438-8677.2008.00162.x.

Sevillano-Caño, J., García, M. J., Córdoba-Galván, C., Luque-Cruz, C., Agustí-Brisach, C., Lucena, C., & Romera, F. J. (2024). Exploring the role of debaryomyces hansenii as biofertilizer in iron-deficient environments to enhance plant nutrition and crop production sustainability. *International Journal of Molecular Sciences, 25*(11), 5729.

Sevillano-Caño, J., Núñez-Cano, J., Prieto, P., Trapero, A., Sánchez-Rodríguez, A. R., & Agustí-Brisach, C. (2024). The endophytic role and growth-promoting ability of the nonpathogenic strain Fusarium oxysporum FO12 in herbaceous hosts. *Scientia Horticulturae, 332*, 113220.

Shakeel, S. N., Wang, X., Binder, B. M., & Schaller, G. E. (2013). Mechanisms of signal transduction by ethylene: Overlapping and non-overlapping signalling roles in a receptor family. *AoB Plants, 5*, plt010.

Shanmugam, V., Wang, Y. W., Tsednee, M., Karunakaran, K., & Yeh, K. C. (2015). Glutathione plays an essential role in nitric oxide-mediated irondeficiency signaling and iron-deficiency tolerance in Arabidopsis. *The Plant Journal: For Cell and Molecular Biology, 84*, 464–477. https://doi.org/10.1111/tpj.13011.

Sharifi, R., & Ryu, C. M. (2018). Sniffing bacterial volatile compounds for healthier plants. *Current Opinion in Plant Biology, 44*, 88–97. https://doi.org/10.1016/j.pbi.2018.03.004.

Shoresh, M., Yedidia, I., & Chet, I. (2005). Involvement of jasmonic acid/ethylene signaling pathway in the systemic resistance induced in cucumber by *Trichoderma asperellum* T203. *Phytopathol, 95*, 76–84. https://doi.org/10.1094/PHYTO-95-0076.

Siverio, J. M. (2002). Assimilation of nitrate by yeasts. *FEMS Microbiology Reviews, 26*, 277–284.

Sobrino-Plata, J., Martínez-Arias, C., Ormeño-Moncalvillo, S., Fernández, I., Collada, C., Gil, L., & Martín, J. A. (2022). No priming, just fighting—endophytic yeast attenuates the defense response and the stress induced by Dutch elm disease in Ulmus minor Mill. *Tree Physiology, 42*(10), 2086–2099.

Song, G., & Liu, S. (2015). Phosphorus speciation and distribution in surface sediments of the Yellow Sea and East China Sea and potential impacts on ecosystem. *Acta Oceanologica Sinica, 34*, 84–91.

Stringlis, I. A., Proietti, S., Hickman, R., Van Verk, M. C., Zamioudis, C., & Pieterse, C. M. J. (2018). Root transcriptional dynamics induced by beneficial rhizobacteria and microbial immune elicitors reveal signatures of adaptation to mutualists. *The Plant Journal: For Cell and Molecular Biology, 93*, 166–180. https://doi.org/10.1111/tpj.13741.

Sun, L., Song, L., Zhang, Y., Zheng, Z., & Liu, D. (2016). Arabidopsis PHL2 and PHR1 act redundantly as the key components of the central regulatory system controlling transcriptional responses to phosphate starvation. *Plant Physiology, 170*(1), 499–514.

Tsai, H. H., & Schmidt, W. (2017a). Mobilization of iron by plant-borne coumarins. *Trends in Plant Science, 22*, 538–548. https://doi.org/10.1016/j.tplants.2017.03.008.

Tsai, H. H., & Schmidt, W. (2017b). One way. Or another? Iron uptake in plants. *The New Phytologist, 214*, 500–505. https://doi.org/10.1111/nph.14477.

Tsai, H. H., Rodríguez-Celma, J., Lan, P., Wu, Y. C., Vélez-Bermúdez, I. C., & Schmidt, W. (2018). Scopoletin 8-Hydroxylase-mediated fraxetin production is crucial for iron mobilization. *Plant Physiology, 177*, 194–207. https://doi.org/10.1104/pp.18.00178.

Turk, M., Abramović, Z., Plemenitaš, A., & Gunde-Cimerman, N. (2007). Salt stress and plasma-membrane fluidity in selected extremophilic yeasts and yeast-like fungi. *FEMS Yeast Research, 7*(4), 550–557.

Van der Ent, S. (2008). Transcriptional regulators of rhizobacteria induced systemic resistance. PhD Thesis. Utrecht, The Netherlands.

Varo, A., Moral, J., Lozano-Tóvar, M. D., & Trapero, A. (2016). Development and validation of an inoculation method to assess the efficacy of biological treatments against Verticillium wilt in olive trees. *Biocontrol, 61*, 283–292. https://doi.org/10.1007/s10526-015-9710-3.

Vélez-Bermúdez, I. C., & Schmidt, W. (2023). Plant strategies to mine iron from alkaline substrates. *Plant and Soil, 483*, 1–25.

Venuti, S., Zanin, L., Marroni, F., Franco, A., Morgante, M., Pinton, R., & Tomasi, N. (2019). Physiological and transcriptomic data highlight common features between iron and phosphorus acquisition mechanisms in white lupin roots. *Plant Science, 285*, 110–121.

Verbon, E. H., Trapet, P. L., Kruijs, S., Temple-Boyer-Dury, C., Rouwenhorst, T. G., & Pieterse, C. M. (2019). Rhizobacteria-mediated activation of the Fe deficiency response in Arabidopsis roots: Impact on Fe status and signaling. *Frontiers in Plant Science, 10*, 909.

Verbon, E. H., Trapet, P. L., Stringlis, I. A., Kruijs, S., Bakker, P. A. H. M., & Pieterse, C. M. J. (2017). Iron and Immunity. *Annual Review of Phytopathology, 55*, 355–375.

Wang, B. L., Tang, X. Y., Cheng, L. Y., Zhang, A. Z., Zhang, W. H., Zhang, F. S., & Shen, J. B. (2010). Nitric oxide is involved in phosphorus deficiency-induced cluster-root development and citrate exudation in white lupin. *New Phytologist, 187*(4), 1112–1123.

Wang, D. H., Li, F., Duan, Q. H., Han, T., Xu, Z. H., & Bai, S. N. (2010). Ethylene perception is involved in female cucumber flower development. *The Plant Journal, 61*(5), 862–872.

Waters, B. M., Lucena, C., Romera, F. J., Jester, G. G., Wynn, A. N., Rojas, C. L., et al. (2007). Ethylene involvement in the regulation of the H^+-ATPase *CsHA1* gene and of the new isolated ferric reductase *CsFRO1* and iron transporter *CsIRT1* genes in cucumber plants. *Plant Physiology and Biochemistry: PPB / Societe Francaise de Physiologie Vegetale, 45*, 293–301. https://doi.org/10.1016/j.plaphy.2007.03.011.

Wu, Z., Ren, H., McGrath, S. P., Wu, P., & Zhao, F. J. (2011). Investigating the contribution of the phosphate transport pathway to arsenic accumulation in rice. *Plant Physiology, 157*(1), 498–508.

Yang, Y., Ou, B., Zhang, J., Si, W., Gu, H., Qin, G., et al. (2014). The Arabidopsis Mediator subunit MED16 regulates iron homeostasis by associating with EIN3/EIL1 through subunit MED25. *The Plant Journal: For Cell and Molecular Biology, 77*, 838–851. https://doi.org/10.1111/tpj.12440.

Zamioudis, C., Hanson, J., & Pieterse, C. M. (2014). β-Glucosidase BGLU 42 is a MYB 72-dependent key regulator of rhizobacteria-induced systemic resistance and modulates iron deficiency responses in A rabidopsis roots. *New Phytologist, 204*(2), 368–379.

Zamioudis, C., Korteland, J., Van Pelt, J. A., van Hamersveld, M., Dombrowski, N., Bai, Y., et al. (2015). Rhizobacterial volatiles and photosynthesis-related signals coordinate MYB72 expression in *Arabidopsis* roots during onset of induced systemic resistance and iron-deficiency responses. *The Plant Journal: For Cell and Molecular Biology, 84*, 309–322. https://doi.org/10.1111/tpj.12995.

Zhai, Z., Gayomba, S. R., Jung, H. I., Vimalakumari, N. K., Piñeros, M., Craft, E., & Vatamaniuk, O. K. (2014). OPT3 is a phloem-specific iron transporter that is essential for systemic iron signaling and redistribution of iron and cadmium in Arabidopsis. *The Plant Cell, 26*(5), 2249–2264.

Zhang, H., Kim, M. S., Krishnamachari, V., Payton, P., Sun, Y., Grimson, M., et al. (2007). Rhizobacterial volatile emissions regulate auxin homeostasis and cell expansion in *Arabidopsis*. *Planta, 226*, 839–851. https://doi.org/10.1007/s00425-007-0530-2.

Zhang, H., Sun, Y., Xie, X., Kim, M. S., Dowd, S. E., & Paré, P. W. (2009). A soil bacterium regulates plant acquisition of iron via deficiency inducible mechanisms. *The Plant Journal: for Cell and Molecular Biology, 58*, 568–577. https://doi.org/10.1111/j.1365-313X.2009.03803.x.

Zhao, L., Wang, F., Zhang, Y., & Zhang, J. (2014). Involvement of *Trichoderma asperellum* strain T6 in regulating iron acquisition in plants. *Journal of Basic Microbiology, 54*, S115–S124. https://doi.org/10.1002/jobm.201400148.

Zhou, C., Guo, J., Zhu, L., Xiao, X., Xie, Y., Zhu, J., et al. (2016). *Paenibacillus polymyxa* BFKC01 enhances plant iron absorption via improved root systems and activated iron acquisition mechanisms. *Plant Physiology and Biochemistry: PPB / Societe Francaise de Physiologie Vegetale, 105*, 162–173. https://doi.org/10.1016/j.plaphy.2016.04.025.

Zhou, C., Ma, Z., Xiao, X., Xie, Y., Zhu, J., & Wang, J. (2016). Potential enhancement of plant iron assimilation by microbial-induced root exudation of phenolic compounds. *Research & Reviews: Journal of Botanical Sciences, 5*, 34–37.

Ziegler, V. L. (2016). Exploration of the use of treatment wetlands as a nutrient management strategy in Wisconsin. The Nature Conservancy. Natural Infrastructure Fellow.

Scheffersomyces stipitis: development of cell factories beyond ethanol production

14

Valeria Mapelli, Vittorio Senatore, Immacolata Serra, and Paola Branduardi
Department of Biotechnology and Biosciences, University of Milano-Bicocca, Milan, Italy

14.1 Introduction

Species now assigned to *Scheffersomyces* were originally placed in the genus *Pichia* because of phenotypic similarities, such as the formation of hat-shaped ascospores and inability to assimilate nitrate as a sole source of nitrogen. A phylogenetic analysis of *Pichia* species based on analysis of D1/D2 domain of large subunit (LSU) rRNA gene sequences showed that the *Scheffersomyces* species were distantly related to *Pichia membranifaciens* that was the type species of the genus *Pichia* (Kurtzman & Robnett, 1998). As a result from (Kurtzman et al., 2011) analysis, the ascosporic species of the *Pichia stipitis* clade (including *P. stipitis*, *P. segobiensis*, and *P. spartinae*) were transferred to the genus *Scheffersomyces*, with *S. stipitis* being the type species of the genus.

S. stipitis was originally isolated from insects' larvae (Nardi et al., 2006), in particular from the hindgut of larvae and adults of the passalid beetle (*Odontotaenius disjunctus)* that usually live inside logs or stumps and is known to degrade white-rotted hardwood (Suh et al., 2003). The importance of the symbiosis between this yeast and wood degrading insects becomes extremely clear when considering the enzymatic activities that *S. stipitis* is endowed with. In fact, sequencing of the genome (Jeffries et al., 2007) has revealed that *S. stipitis* produces a number of cellulases, mannanases, xylanase, and other hydrolytic enzymes that liberate wooden sugars and hence are necessary for the survival of wood degrading insects (Nardi et al., 2006).

The capacity of *S. stipitis* to consume and ferment wooden sugars has long arisen the interest for this yeast in biotechnological applications aiming at using the sugar streams from lignocellulose materials as source of sugars for fermentation-based production processes. While, as all other yeasts, *S. stipitis* can consume glucose, this yeast has been noticed for its natural and great capacity to consume and efficiently metabolize sugars deriving from the hemicellulose fraction of lignocellulose, in particular, xylose, galactose, mannose, and also oligosaccharides, such as cellobiose, mannan, and xylan oligomers (Jeffries & Van Vleet, 2009). The capacity to metabolize sugar oligomers represents a great advantage for fermentation-based processes using pretreated lignocellulosic biomass, as mild acidic pre-treatments can be used preventing the formation of inhibitory compounds and leaving solubilized hemicellulose under the form of bioavailable sugar oligomers.

Eukaryotic Microorganisms as Sources of Bioproducts. DOI: https://doi.org/10.1016/B978-0-443-30188-9.00021-4

The natural capacity of *S. stipitis* to metabolize and ferment wood-derived sugars has been the main driver, within a biobased economy frame, to develop strains and processes aimed at biotechnological productions from renewable lignocellulosic sources. *S. stipitis* has been used directly in production processes or its genes have been introduced in other microorganisms to plug existing metabolic pathways to xylose (or other hemicellulose sugars) consumption. Production of second-generation ethanol has long been the main focus of scientific research and applications. However, in this chapter, we aim at emphasizing the characterization and use of *S. stipitis* in fermentation processes intended to target products different from ethanol, on which extensive and exhaustive literature reviews are already available. At the same time, we try to give a complete picture by highlighting the latest advances in metabolism characterization and development of molecular tools that ease the historically challenging genetic and metabolic engineering process in *S. stipitis*.

14.2 *Scheffersomyces stipitis* strains

The most studied and characterized *S. stipitis* strain is CBS 6054 isolated from passalid beetle hindgut; this is the strain on which most of scientific research work has been performed. Experimental work on this strain revealed the peculiar respiratory capacity thanks to the presence of an alternative respiration system donating electrons directly to O_2 from ubiquinone (Jeppsson et al., 1995), and to the presence of NADH dehydrogenase Complex I within the respiratory electron transport chain (Shi et al., 2002). The genome sequence of CBS 6054 was published in 2007 (Jeffries et al., 2007) revealing the presence of eight chromosomes and the proximity of *S. stipitis* to *Debaryomyces hansenii*, sharing with it 151 gene families that were not found in the other compared yeast genomes. Genome sequencing was of course of great importance to describe and identify the genes responsible for several metabolic traits that were observed and characterized before the availability of the genome sequence. In particular, the coding sequence of numerous genes for endoglucanases and β-glucosidases, along with xylanase, mannanase and chitinase activities was revealed. Based on the CBS 6054 genome sequence, a genome-scale metabolic model has been reconstructed (Balagurunathan et al., 2012) and allowed to identify in silico the bottlenecks in *S. stipitis* metabolic network for xylose uptake and nucleotide cofactor recycling (Shi et al., 2002).

Other *S. stipitis* strains often reported in scientific research studies are the CBS 5773 (Grootjen et al., 1990) and CBS 5776 (Tran & Chambers, 1986). A systematic and detailed comparison between CBS 5773 and CBS 6054, quantitatively describing the process for the conversion of xylose to ethanol was performed (Krahulec et al., 2012) showing that previously suggested differences between the two strains were mostly related to differences in fermentation conditions. In fact, both strains could convert xylose to ethanol with a yield of 0.41 g/g. However, no comparative genetic information between these strains is available to exhaustively define the differences among the different strains.

An important genetic feature of *S. stipitis* is its belonging to the so-called CUG(Ser1) clade (Krassowski et al., 2018), that is that clade of yeasts using an alternative codon system substituting serine for leucine at CUG codons. Genome sequencing of *S. stipitis* revealed that CUG usage is particularly frequent in genes located at chromosome termini (Jeffries et al., 2007). In addition, a study comparing 27 different *S. stipitis* isolates demonstrated that *S. stipitis* has an intrinsically plastic genome with different isolates having distinct chromosome organization (Vega-Estévez et al., 2021), that is possibly associated with fitness benefits coming from adaptation to hostile environments.

14.3 Development of synthetic biology and metabolic engineering tools for *Scheffersomyces stipitis*

Challenges in genetic engineering of the so-called non-conventional yeasts have been exhaustively reviewed by Löbs et al. (2017) and are also applicable to *S. stipitis*. One of the challenges in the transformation and selection of *S. stipitis* is the natural resistance of this yeast to the antibiotic geneticin (up to 2 mg/mL), that is typically used for yeast transformant selection (Ho et al., 1991). Therefore, research studies focused on the search of additional selectable markers conferring resistance to antibiotics. A ShBle selectable marker conferring resistance to phleomycin D1 (also known as zeocin) has been used upon substitution of the CUG codons with other codons specifying leucine. In fact, in times when gene synthesis and codon optimization were not as developed as today, the CUG(Ser1) genetic feature of *S. stipitis* represented a big challenge for genetic and metabolic engineering of this yeast (Laplaza et al., 2006). Other genes conferring resistance to antibiotics can be effectively used in *S. stipitis* upon codon optimization adapting the genetic sequence to the CUG(Ser1) codon usage (Matanović et al., 2022). The most common antibiotic-resistance markers are reported in Table 14.1.

An additional obstacle to easily introduce or delete specific genetic functions is the scarce (Lu et al., 1998) or null homologous recombination (HR) process as the main DNA repair system. Contrary to the well-known yeast *Saccharomyces cerevisiae*, which shows high HR rates, the so-called nonconventional yeasts preferentially use nonhomologous end joining (NHEJ) for the repair of double strand DNA breaks. Due to the scarce capacity of HR, a limited number of auxotrophic *S. stipitis* strains have been generated by the deletion of specific genes. Auxotrophic strains were obtained for uracil (*ura3*) (Laplaza et al., 2006; Lu et al., 1998), leucine (*leu2*) (Lu et al., 1998), histidine (*his3–1*), and tryptophan (*trp5–10*) biosynthesis (Piontek et al., 1998) (Table 14.1), allowing for the selection of transformed cells with vectors harboring the respective complementation markers (i.e., prototrophic genes).

When a gene of interest must be integrated into the genome, a random integration occurs, hence limiting the control of the integration event. This aspect is typically addressed by deleting the *KU70* or *KU80* gene. Ku is a dimeric protein complex composed of Ku70 and Ku80 subunits that bind to double-strand break (DSB) ends to facilitate NHEJ-mediated genome repair (Di Primio et al., 2005). Deletion of *KU*

Table 14.1 Antibiotic resistance markers and auxotrophic strains for engineering of *Scheffersomyces stipitis*.

Antibiotic-resistance marker genes	Antibiotic	Concentration of antibiotic in YPD medium	Reference	Prototrophic markers	Auxotrophic strain (mutation)	References
ble^{R}	phleomycin D1[a]	100 µg/mL	Laplaza et al. (2006), and Matanović et al. (2022)	*URA3*	*S. stipitis* UC7 (*ura3-3* or *ura3-1*)[b]	Laplaza et al. (2006), Lu et al. (1998), and Yang et al. (1994)
nat^{R}	nourseothricin	50 µg/mL	Matanović et al. (2022)	*LEU2*	*S. stipitis* LU75 (*leu2Δ-1*)[b]	Lu et al. (1998)
hyg^{R}	hygromycin	600 µg/mL	Matanović et al. (2022)	*HIS3, TRP5*	*S. stipitis* PJH53 (*trp5–10, his3-1*)[c]	Piontek et al. (1998)

[a] Zeocin is the trade name.
[b] Wild type strain, CBS 6054.
[c] Mutant obtained via chemical mutagenesis from wild type strain, CBS5773.

genes impairs NHEJ and consequently reinforces HR. In *S. stipitis,* deletion of *KU80* gene considerably increased the frequency of homologous integration; however, it resulted in a remarkable decrease of the transformation efficiency (Maassen et al., 2008), very likely because most transformants remain abortive due to an inefficient HR pathway. An efficient transformation protocol has been later developed for *ku70Δ* and *ku80Δ* single mutants (Di Primio et al., 2005) and the *ku70Δku80Δ* double mutant strain, that was obtained via CRISPR mediated genome editing technology (Cao et al., 2018). This *ku70Δku80Δ* strain was used as a platform to assess the enhancement of the HR-mediated genome editing and the generation of a *trp1* or *ade2* auxotrophic strains by providing specific guide RNA and a double strand DNA fragment featured by different base pairs (bp) length of the homology arms (HA). Overall, it was shown that very similar HR efficiencies (73%–83%) could be obtained among the different HA lengths, and HA of 500 bp and 1 kb exhibited higher transformation efficiencies compared to the 50 bp HA. In addition, despite a much lower transformation efficiency of the *ku70Δku80Δ*, compared to its parental UC7 (*ura3-3*) strain, HR was the dominant double strand repair system in *ku70Δku80Δ* after Cas9 targeting of the selected loci (Cao et al., 2018). Transformation and HR efficiency of the *kuΔ* mutants generated with CRISPR-Cas9 are reported in Table 14.2.

Table 14.2 CRISPR-Cas9-based tools for generation of engineered *Scheffersomyces stipitis* strains.

CRISPR-Cas9 tool for gene deletion			
Host strain	**Target gene**	**KO and HR efficiency**	**Transformation efficiency**
S. stipitis UC7 *ku70Δ*	*KU80*	78%	n.a.
S. stipitis UC7 *ku80Δ*	*KU70*	100%	n.a.
S. stipitis UC7 *ku70Δ, ku80Δ*	*TRP1*	71%[a]	5%
S. stipitis UC7 *ku70Δ, ku80Δ*	*ADE2*	64%[b]	1%
CRISPR-Cas9 tool for gene repression			
Host strain	**Integrated cassette**	**Cas9 variants**	**Gene repression**
S. stipitis UC7 *ku70Δ, ku80Δ, trp1Δ*	p*TEF1-eGFP*-t*PGK1*[c]	dCas9 (D10A, H840A)	17%
S. stipitis UC7 *ku70Δ, ku80Δ, trp1Δ*		dCas9(D10A, H840A)-Mxi1[d]	32%

[a] HR efficiency was calculated using 500 bp homology arms.
[b] HR efficiency was calculated using 50 bp homology arms.
[c] The expression cassette was integrated in the genome (*trp1Δ::eGFP*).
[d] Cas9 fused to transcription repressor, Mxi1 (Gilbert et al., 2013).
Source: Adapted from Cao, M., Gao, M., Ploessl, D., Song, C., & Shao, Z. (2018). CRISPR–mediated genome editing and gene repression in *Scheffersomyces stipitis*. *Biotechnology Journal*, *13*(9). https://doi.org/10.1002/biot.201700598.

The CRISPR-based gene knockout system used for the generation of the mutant strains described above, has been developed specifically for *S. stipitis* (Cao, Gao, et al., 2017). As other yeasts species, *S. stipitis* does not have the naturally born 2μ high-copy number plasmid, typically present in *S. cerevisiae*; therefore, an exogenous plasmid can be recognized and maintained by *S. stipitis* only if harboring an autonomous replicating sequence (ARS) that has been first identified and isolated by Yang et al. (1994). In addition, a centromere (CEN) has been shown crucial to guarantee direct and stable segregation of an episomal plasmid. The success of the CRISPR system in *S. stipitis* relied on the identification and use of the CEN5 centromeric DNA as the partitioning element that was integrated in the plasmid carrying the Cas9 gene (Cao, Gao, et al., 2017). Expressing the CRISPR components on the newly developed and stable plasmid led to an improvement of the gene knockout efficiency from $< 1\%$ to more than 80% based on indel mutations. Based on this approach, a library-based strategy that integrates in silico chromosome scanning and high-throughput functional screening enabled the isolation of all eight *S. stipitis* centromeres (Cao, Seetharam, et al., 2017); thus, promoting the generation of a panel of vectors differing for the specific CEN sequence. For gene repression or gene knockdown, a CRISPR interference (CRISPRi) tool specific for *S. stipitis* has also been developed using the mutant dCas9 (Dominguez et al., 2016), and tested on the expression levels of the reporter gene coding for GFP, integrated in the genome (Cao, Seetharam, et al., 2017) of the *trp1Δ* strain (Cao et al., 2018), resulting in gene repression rate of 17%–32% (Table 14.2).

Advances in the use of CRISPR-Cas9 system resulted also in the development of a reusable CRISPR-Cas9 based bioluminescent detectable and selectable reporter system, namely, BLINCAR (Bio Luminescent Indicator Nullified by Cas9 Actuated Recombination). Such a system was established for introducing and/or removing genetic elements to/from wild type *S. stipitis* (Reichard et al., 2023). BLINCAR is a two-component system based on the use of click beetle green luciferase CBG99 gene as bioluminescence reporter gene. The first component delivers an antibiotic selectable bioluminescent element that can be used to introduce an exogenous gene or can be used to delete native elements. The second component is an episomal plasmid carrying Cas9 gene and sgDNAs targeting the selectable bioluminescent element for its removal. The loss of bioluminescence serves as a quick and reliable screening method to identify the desired transformants.

14.4 An overview on carbon and nitrogen metabolism of *Scheffersomyces stipitis*: lessons from ethanol production showed the way to medium and process development for aerobic processes

Sugar metabolism in *S. stipitis* has long been focused on its natural ability to consume and metabolize the pentose sugar xylose (du Preez & Prior, 1985; Skoog & Hahn-Hägerdal, 1989; Toivola et al., 1984). Reduction of xylose to xylulose happens via the activity of xylose reductase (XR) producing xylitol, and xylitol dehydrogenase (XDH), ultimately producing xylulose (Barnett, 1976). Xylulose enters the pentose phosphate pathway after being phosphorylated by a xylulose kinase (XK) and then the C3 and C6 compounds enter glycolysis. As for long time the main interest has been on the conversion of xylose into ethanol, several studies were conducted to clarify how the oxygenation levels in *S. stipitis* fermentation processes affected ethanol formation from xylose. In fact, *S. stipitis* is a Crabtree negative yeast (Gancedo & Serrano, 1989) that does not produce ethanol if it grows under fully aerobic conditions, regardless of the level and flux of sugar(s) supplied. On the other hand, oxygen plays an important role for the consumption of xylose, as under anaerobic conditions xylose is not consumed by yeasts (Ligthelm et al., 1988; Skoog & Hahn-Hägerdal, 1989) and xylose uptake rate was shown to increase with increasing oxygenation of the cultures (Skoog & Hahn-Hägerdal, 1990). Notably, oxygen partial pressure plays a crucial role also in the uptake of glucose in *S. stipitis*, with anaerobic conditions causing drastic decrease of glucose uptake rate and ethanol formation (Skoog et al., 1992); while increasing oxygen partial pressure leads to the onset of respiro-fermentative metabolism on glucose (Klinner et al., 2005). Therefore, when the target compound is ethanol, oxygen availability is essential, but the oxygenation levels and the oxygen transfer rate must be carefully tuned, as an excess of available oxygen favors cell growth and is detrimental to ethanol formation (Silva et al., 2012).

The metabolic shift occurring in *S. stipitis* when coping with reduced oxygen uptake is linked to a shift in redox balance. A study taking advantage of a modified and improved genome-scale metabolic model demonstrated that the shift of reducing power from NADPH to NADH was crucial in directing the overall shift in the metabolic state (Hilliard et al., 2018). In *S. stipitis*, NADPH is the favored cofactor for XR; while XDH depends strictly on NAD^+, hence causing redox imbalance within the xylose catabolic pathway and limiting growth performance and ethanol production on xylose. This model clarified that when the oxygen uptake rate decreases, less ATP is produced via the electron transport chain and therefore, less NADH is oxidized via this pathway. To compensate for this condition, XR shifts its preference from NADPH to NADH, restoring the cofactor balance of the XR and XDH reactions, hence reducing the need to shuttle NADH to the mitochondria. As consequence of the XR cofactor shift, less NADPH is necessary to metabolize xylose, leading to a reduced flux through the oxidative route of the pentose phosphates pathway and to the channeling of carbon to ethanol (Hilliard et al., 2018).

In view of the use of hemicellulose rich biomass for the development of biorefineries based on *S. stipitis* metabolic activities, the metabolism of hemicellulose sugars other than xylose has also been investigated. In particular, the pentose sugar arabinose cannot be consumed by *S. stipitis* under anaerobic conditions, while arabinose uptake occurs under microaerobic and aerobic conditions. However, when aiming at ethanol production, the idea of exploiting also arabinose is not a real advantage for the process, as in a multi-sugar medium arabinose uptake occurs after consumption of glucose, xylose and part of the produced ethanol (Delgenes et al., 1988).

Based on this evidence, the full exploitation of hemicellulose sugars in *S. stipitis* fermentations can occur only in aerated processes. A complete and quantitative analysis on how the sugars deriving from lignocellulose biomass has been conducted (du Preez et al., 1986), confirming that the presence of glucose represses the consumption of other sugars and showing that *S. stipitis* can metabolize most of the sugars present in hemicellulose and that arabinose and rhamnose could be used for biomass formation but were not fermented into ethanol. Ethanol yield on cellobiose is comparable to the ethanol yield achieved on galactose; however, the specific productivity on cellobiose was found to be very low ($q = 0.05$ g/g·h) compared to the one on galactose ($q = 0.18$ g/g·h) (du Preez et al., 1986). The quantitative details on the different sugars are summarized in Table 14.3.

Overall, the studies aiming at exploiting the natural metabolic capacities of *S. stipitis* for the bio-based production of ethanol were of crucial importance also for elucidating the basis of the carbon metabolism of this yeast. However, most probably due to the complex process conditions necessary to achieve competitive ethanol yield and productivity, *S. stipitis* never succeeded as primary ethanol producer and the transfer of valuable metabolic activities native of *S. stipitis* to *S. cerevisiae* was the preferred route, as reviewed by (Ruchala et al., 2020). Nonetheless, the potential of *S. stipitis* for other applications is still being explored and one example is reported in Rodriguez-Ocasio et al. (2022), where the ability of *S. stipitis* and other nonconventional yeasts to grow on several lipids and hydrocarbon compounds has been tested for possible environment remediation applications. This comparative study used several model compounds, including alcohols, acids, diacids, alkanes, alkenes, and esters. *S. stipitis* was shown to be able to grow on several of the tested compounds, interestingly showing that it was not negatively affected by fatty alcohols, that instead had an inhibitory effect for other yeasts. The use of N-acetyl-glucosamine (NAG) as carbon source was characterized in *S. stipitis*, NAG being the monomeric constituent

Table 14.3 Quantitative parameters of *S. stipitis* consumption and fermentation of single hexose and pentose sugars.

Sugar	μ_{max} (h^{-1})[a]	$Y_{EtOH/S}$ (g/g)[b]	$Y_{X/S}$ (g/g)[c]
Glucose	0.23	0.40	0.18
Mannose	0.18	0.36	0.13
Galactose	0.20	0.32	0.19
Xylose	0.22	0.39	0.16
Arabinose	0.14	0	0.32
Rhamnose	0.16	0	0.40
Cellobiose	0.21	0.32	0.34

[a] maximum specific growth rate.
[b] Ethanol yield coefficient, g ethanol·(g substrate utilized)$^{-1}$.
[c] Cell yield coefficient, g biomass·(g substrate utilized)$^{-1}$.
Source: Adapted from du Preez, J. C., Bosch, M., & Prior, B. A. (1986). The fermentation of hexose and pentose sugars by *Candida shehatae* and *Pichia stipitis*. *Applied Microbiology and Biotechnology*, *23*(3–4), 228–233. https://doi.org/10.1007/BF00261920.

of chitin, which is the most abundant nitrogen-containing organic compound in nature (Inokuma et al., 2018). The metabolomic study revealed that in the presence of NAG as sole carbon source and under microaerophilic conditions, ammonia was generated upon deamination of NAG and that the levels of aromatic, branched chain, and sulfur-containing amino acids and adenine, guanine, and cytosine nucleotides were the highest in NAG-grown cells. Based on this preliminary study, specific processes could be designed to establish production of specific target products starting from chitin-derived NAG (Inokuma et al., 2018).

In a similar way as NAG, it has been shown that amino acids, such as glutamate, aspartate, and proline can work as both C- and N-source in *S. stipitis* (Freese et al., 2011). Glutamate dehydrogenase 2 (Gdh2) enzymatic activity has been shown to be responsible for the catabolism of glutamate as carbon source; Gdh2 catalyses the deamination of glutamate to α-ketoglutarate, that is channeled to the TCA cycle. This reaction also leads to the release of NH_4^+ and the generation of one NADH equivalent per mole of glutamate. Using glutamate as nitrogen source instead of NH_4^+ in a glucose containing medium did not affect the specific growth rate and resulted in a higher biomass yield on glucose, indicating that glutamate was not used only as N-source but contributed also as carbon source. This effect was also reflected on the higher ethanol accumulation, possibly due to an increased NADH production linked to the activity of Gdh2, and to a greater availability of glucose for cytosolic metabolic reactions, as the TCA cycle is fueled by α-ketoglutarate deriving from glutamate (Freese et al., 2011). This impact of amino acid supplementation specifically characterized for glutamate could be linked to the positive effect that is reported in connection with the supplementation of yeast extract (Silva et al., 2012), that positively affects ethanol accumulation and yield upon xylose consumption.

As for the study of central carbon metabolism and oxygen uptake, also nitrogen metabolism in *S. stipitis* has been mostly studied in relation to ethanol production. However, not many studies have focused on medium optimization and nitrogen metabolism for fully aerobic processes, wherein the formation of biomass or growth-related products is the target. Medium optimization for cell biomass and folate production in the presence of both glucose and xylose showed that optimizing the amount of salts and trace element was key to improve biomass formation. In addition, the sole substitution of NH_4^+ with urea allowed the complete consumption of both sugars (Mastella et al., 2022). The positive effect of urea was shown also in Silva et al. (2012) and although the metabolic details still need to be explained, a possible model has been proposed (Mastella et al., 2022), as represented in Fig. 14.1.

Assuming that nitrogen uptake in *S. stipitis* works as in *S. cerevisiae*, when ammonium sulfate is used as nitrogen source, NH_4^+ uptake occurs via facilitated diffusion through an ammonium permease, coded by *MEP1* and *MEP2* genes that are also present in *S. stipitis* (Jeffries et al., 2007). In order to maintain homeostasis and regulate cytosolic pH, the H^+-ATP-dependent pump Pma1 (Cueto-Rojas et al., 2017), also present in *S. stipitis*, must be active; thus, resulting in the consumption of 1 ATP every NH_4^+ ion entering the cell. When urea is the nitrogen source, the uptake in *S. stipitis* occurs via Dur3 and Dur4 permeases via symport mechanism with H^+. After the cellular uptake, urea amidolysis happens via the action of Dur1 and Dur2

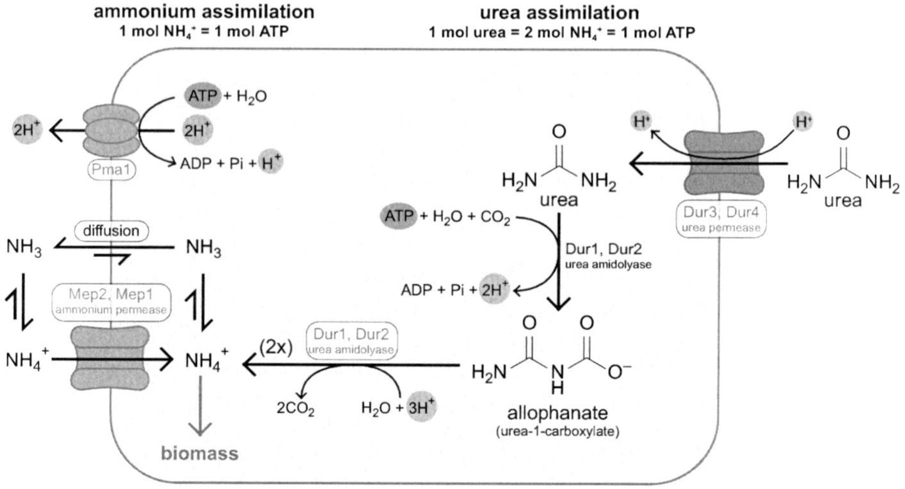

Figure 14.1 Putative pathways for ammonium and urea uptake in *S. stipitis*. Ammonium (left) is internalized in the cell by facilitated diffusion via ammonium permease; to maintain the potential difference across the plasma membrane the H^+-ATPase pump Pma1 pumps a proton out of the cell, contributing to acidification of the growth medium. Urea (right) is internalized through a symport system with an H+ ion (Dur3,4), and it is converted into ammonium and CO_2 via urea amidolyase (Dur1,2).
Source: Modified from Mastella, L., Senatore, V. G., Guzzetti, L., Coppolino, M., Campone, L., Labra, M., Beltrani, T., & Branduardi, P. (2022). First report on Vitamin B_9 production including quantitative analysis of its vitamers in the yeast *Scheffersomyces stipitis*. *Biotechnology for Biofuels and Bioproducts*, *15*(1). https://doi.org/10.1186/s13068-022-02194-y.

(Roon & Levenberg, 1972), which catalyze the overall conversion of urea into 2 CO_2 and 2 NH_4^+ with the consumption of 1 ATP and 1 CO_2. Interestingly, the H+ imported by the permeases and the 2 H^+ generated by the carboxylation reaction are all used in the decarboxylation step, resulting in a neutral H^+ balance and therefore eliminating the need of 1 ATP to maintain intracellular pH and homeostasis (Fig. 14.1). Overall, the use of urea as nitrogen source is energetically favorable, as it generates two moles of NH_4^+ with the consumption of only 1 ATP, that is 50% of the energy required for ammonium uptake.

14.5 Industrial scalability challenges for the development of *S. stipitis* cell factories

As we have described above, despite *S. stipitis* natural ability to metabolize and ferment xylose and other sugars deriving from lignocellulosic biomasses, this yeast has never become the workhorse for second generation bioethanol production. The biggest limitations that prevented the development of efficient and cost-effective processes are linked to the low tolerance to ethanol, as compared to *S. cerevisiae* (du Preez et al., 1989; Meyrial et al., 1995; Shin et al., 2019) and to inhibitors deriving from lignocellulose biomass. Evolutionary engineering, random mutagenesis, and genome shuffling have been the most common approaches to promote adaptation to inhibitors or ethanol. Repetitive cultures and xylose fed continuous cultures were used to challenge *S. stipitis* toward increasing concentrations of inhibitors or ethanol (Slininger et al., 2015). The improved features of the evolved strain growing in enzyme hydrolysates of pretreated switchgrass included reduced lag phase, increased fermentation rates (up to 40 g/L ethanol in about 167 hours) along with higher ethanol tolerance. *S. stipitis* strains with improved tolerance to inhibitors in pretreated hardwood were also selected following UV mutagenesis (Bajwa et al., 2009); the selected strains could ferment most of the sugars present in hardwood spent sulfite liquor, showing a better xylose fermentation ability and higher tolerance to the inhibitors compared to the parental strain.

In addition to the naturally lower fermentation capacity of sugars compared to *S. cerevisiae* (Shin et al., 2019), the need of aerobic conditions to promote the consumptions of pentose sugars has greatly hampered the development and scale up of *S. stipitis* cell factories for bioethanol production. However, the deeper understanding of *S. stipitis* physiology, metabolism and nutritional needs have led to research studies aiming at developing fermentation processes that could take advantage of the metabolic traits that specifically distinguish *S. stipitis* from other yeasts, projecting metabolic engineering and process development beyond ethanol.

14.6 The potential of *Scheffersomyces stipitis* beyond ethanol production

Thanks to the native ability of *S. stipitis* to metabolize xylose (and other lignocellulose derived sugars) and to produce ethanol with high yields, most of the scientific work has focused for long time primarily on the optimization of second-generation bioethanol production with *S. stipitis*. However, regardless of the microorganism used, second-generation bioethanol production has very high intrinsic costs linked to biomass pretreatment and the saccharification steps. In addition, *S. stipitis* has low tolerance for ethanol, lower ethanol productivity compared to *S. cerevisiae*, and aerobic conditions are required for growth. Considering all these aspects, bioethanol production processes using *S. stipitis* (or other yeasts) were never brought to industrial production scale (Kwak et al., 2019; Zha et al., 2021). Nonetheless, given the interest in establishing biobased production processes from hemicellulose deriving sugars, the potential of *S. stipitis* has been—and is currently being—explored for production of other organic compounds via biobased and fermentation-based processes. In the following sections we make an overview of *S. stipitis* engineered strains and *S. stipitis*-based processes that have been studied for production of compounds different from ethanol, exploiting the metabolic capacity of *S. stipitis* under fully aerobic conditions (Fig. 14.2).

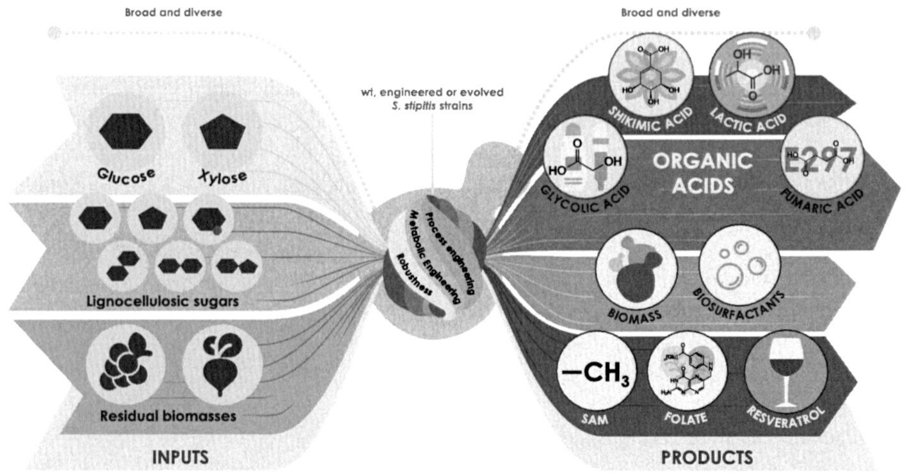

Figure 14.2 Overview of *S. stipitis* applications for fermentation-derived products other than ethanol. Graphical representations of products that have been shown in scientific research studies as products derived from *S. stipitis* fermentation processes under aerobic conditions and from pure sugars or sugar streams from renewable biomass. Organic acids are mostly derived from engineered *S. stipitis* strains, except for glycolic acid that has been produced via a bioconversion process from ethylene glycol and a wild type strain. Added value products, such as folate and resveratrol have also been shown as products from *S. stipitis*; wild type *S. stipitis* was also shown to accumulate folate under controlled fermentation conditions using glucose-xylose mix, while resveratrol was produced via an engineered strain, exploiting the high flux through shikimate. Minor products that are not further discussed in the text are S-adenosyl-methionine (SAM), reported as accumulated metabolite in a *S. stipitis* strain selected upon random mutagenesis, and a biosurfactant molecule with larvicide activity produced by a wild type *S. stipitis* strain (NRRL Y-7124) from sugarcane bagasse hemicellulosic hydrolysate.

14.7 Productive pathways from erythrose-4-P and shikimate metabolic nodes

As previously reported, *S. stipitis* metabolism is featured by a high flux toward erythrose-4-P (E4P) (Jeffries et al., 2007) that is the precursor for the natural biosynthesis of aromatic compounds including aromatic amino acids that are naturally synthesized from the intermediate shikimate and, later in the pathway, chorismate (Fig. 14.3A). Shikimate is a very interesting molecule as it is the precursor of muconic acid, that can be used for the synthesis of nylon and of other molecules that are otherwise derived from fossil sources, such as vanillin (Brochado et al., 2010). Nonetheless, shikimate is also the precursor in the biosynthetic pathway of complex aromatic molecules, such as flavonoids or alkaloids (Galanie et al., 2015; Koopman et al., 2012). Thanks to the interest in shikimate production and to the availability of molecular tools, *S. cerevisiae* has been considered as preferential host for shikimate production. However, E4P flux in *S. cerevisiae* is much lower (Feng & Zhao, 2013; Fiaux et al., 2003) compared to the flux

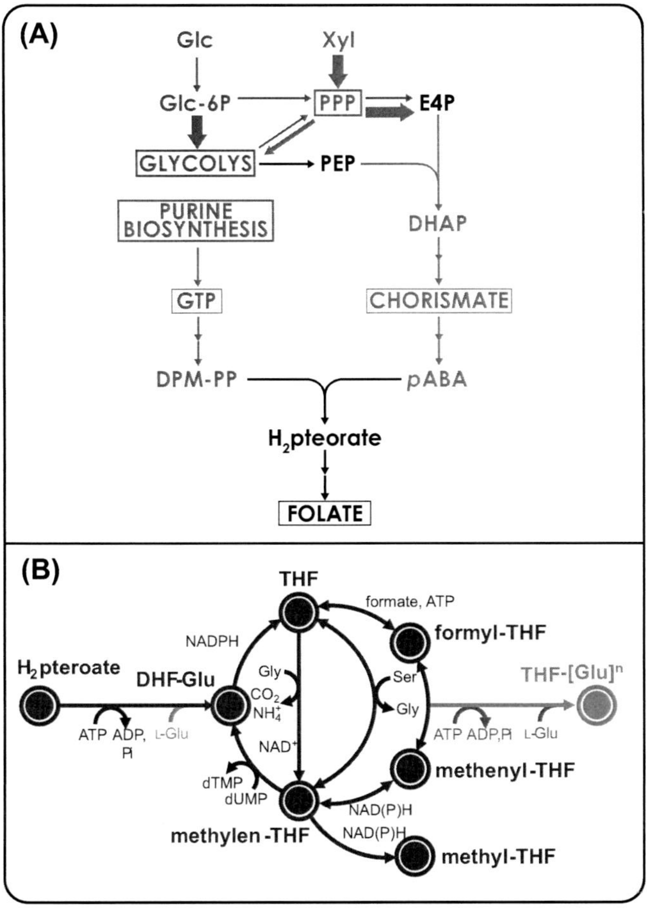

Figure 14.3 Biosynthesis of folate from sugars and the shikimate metabolic pathway. (A) schematic representation of the glucose and xylose metabolic pathways, highlighting the carbon flux toward shikimate and chorismate. (B) Representation of the pathway in which the different vitamers of folate are produced. *Glc*, glucose; *Xyl*, xylose; *Glc-6P*, glucose 6-phosphate; *PPP*, pentose phosphate pathway; *E4P*, erythrose 4-phospate; *PEP*, phosphoenolpyruvate; *DHAP*, 3-deoxy-d-arabin o-heptulosonate-7-phosphate; *DPM-PP*, (7,8-dihydropterin-6-yl) methyl diphosphate; *pABA*, *para*-aminobenzoic acid; *H_2-pteroate*, dihydropteroate; *THF*, tetrahydrofolate; *DHF*, dihydrofolate; *Gly*, glycine; *Ser*, serine; *Glu*, glutamate.

Source: Adapted from Mastella, L., Senatore, V. G., Guzzetti, L., Coppolino, M., Campone, L., Labra, M., Beltrani, T., & Branduardi, P. (2022). First report on Vitamin B_9 production including quantitative analysis of its vitamers in the yeast *Scheffersomyces stipitis*. *Biotechnology for Biofuels and Bioproducts*, *15*(1). https://doi.org/10.1186/s13068-022-02194-y.

observed in *S. stipitis*, thus representing a difficult bottleneck to overcome. Therefore, a shikimate producing strain was developed by overexpressing the minimal set of genes of the shikimate pathway, that is, *TKT1* (transketolase in the pentose phosphate pathway (PPP)), *ARO4* (3-deoxy-D-arabino-heptulosonate-7-phosphate (DHAP) synthase), and

ARO1 (pentafunctional enzyme converting DAHP to 5-enolpyruvyl-3-shikimate phosphate (EPSP)). Mutant variants of Aro4 ($Aro4_{K220L}$) and Aro1 ($Aro1_{D900A}$) were exploited to increase the accumulation of shikimate. The variant $Aro4_{K220L}$ is insensitive to feedback inhibition from phenylalanine and tyrosine, thus allowing an increased production of DHAP. The $Aro1_{D900A}$ variant allows the conversion of DHAP to shikimate but prevents the further conversion to shikimate-3P. The overexpression of the genes coding for Tkt1, $Aro4_{K220L}$, and $Aro1_{D900A}$ was carefully designed by a thorough analysis and selection of the best promoters and terminators under different cultivation conditions (Gao et al., 2017). The set of promoters resulting in the best shikimate titers was composed by *pADH1*, *pPIR1*, and *pUAG* controlling the expression of *TKT1*, $ARO4_{K220L}$, and $ARO1_{D900A}$, respectively. The maximum titers (2.83–2.93 g/L of shikimate) were achieved under oxygen rich conditions in the presence of only glucose or the mix of glucose and xylose.

Further improvement of the shikimate production in *S. stipitis* should involve tuning of the promoter strength and combination, and increasing the precursor availability, phosphoenol pyruvate (PEP) being one of the most important targets. Two main strategies are envisioned (Gao et al., 2017) to increase PEP availability: (1) replacing the promoter of pyruvate kinase (PYK) with an inducible one, (2) overexpressing PEP carboxylase (*PPCK*) and pyruvate carboxylase (PC) to recirculate pyruvate to PEP. Both strategies can now rely on the developed CRISPR-Cas9 system that facilitate the otherwise challenging path towards *S. stipitis* genetic engineering (Cao et al., 2018; Cao, Seetharam, et al., 2017).

A strain with increased flux to shikimate synthesis would be desirable not only for shikimate production but also to be used as chassis strain for the biosynthesis of more complex molecules deriving from the conversion of shikimate. One example is the synthesis of folate (Fig. 14.3B), that represent the active form of folic acid (vitamin B_9) upon the action of dihydrofolate reductase (Ohrvik & Witthoft, 2011). The wild type *S. stipitis* CBS 6054 was tested for its ability to produce folate under different growth conditions. In the process of medium and growth condition optimization, it became clear that a fully aerobic condition was necessary, with oxygen limitation shifting the metabolism toward ethanol production and reduced cell biomass (Mastella et al., 2022). Surprisingly, in this study, the complete consumption of the provided sugars (20 g/L glucose or 20 g/L xylose) could be achieved only if urea was used as nitrogen source and the highest amount of folates was obtained when glucose and xylose were provided as a mix. In addition, it was notable that the highest folate production occurred during the xylose consumption phase; thus, proving that a high flux through E4P is favorable for folate biosynthesis. The maximum folate production was obtained in fully aerobic controlled bioreactors, reaching 192.7 ± 46.9 μg folate/g CDW. Speciation analysis of the total folate revealed that tetrahydrofolate (THF) was the main folate form present (76%), followed by 5-methyl-THF and 5-formyl-THF. With a titer of folates equal to 3.36 mg/L of cultivation, this study reports the highest titer obtained by fermentation of nonengineered microorganism cultivated in minimal medium with xylose as one of supplied sugars. Ideally, a *S. stipitis* engineered strain for the increase of folates precursors, that is, shikimate and chorismate, could lead to a further improved folate production.

The importance of xylose as carbon source when the aim is to maximize the folate production is highlighted by comparing the above-mentioned study with a study conducted by the same research group, measuring folate in *S. stipitis* grown in media based on residual biomass of different origin (Mastella et al., 2023). Upon characterization of the soluble sugars in the pretreated biomass, no xylose was detected and the maximum titer of folate reported was only 188.2 ± 24.86 μg/L in e medium containing sugar beet molasses.

Another high-added value product synthesized in *S. stipitis* exploiting the pathway downstream of shikimate is resveratrol that has tyrosine as precursor (Kobayashi et al., 2021). Also, in this study, the feedback insensitive $Aro4_{K220L}$ variant of Aro4 was used, along with a mutant chorismate mutase from *S. stipitis* $Aro7_{G139S}$ insensitive to chorismate feedback inhibition. The exogenous genes for resveratrol biosynthesis derived from the plants *Herpetosiphon aurantiacus*, *Arabidopsis thaliana*, and *Vitis vinifera*. *S. stipitis* cultivations were conducted only in shake flasks and resveratrol production was measured in the presence of different sugars. The highest resveratrol titers were achieved when the supplied sugars were cellobiose or sucrose with the final titers of resveratrol equal to 529.8 and 668.6 mg/L, respectively. It is also important to notice that ethanol was not detected only in the cultivations supplemented with cellobiose or sucrose. While we cannot evaluate the oxygenation level of the media, the rate of sugar consumption was much lower for cellobiose and sucrose, compared to the other sugars, possibly limiting the overflow towards the fermentation pathways. Intracellular metabolite profile revealed an altered profile of the metabolites involved in the glycolysis and shikimate pathways, depending on the sugar present in the medium. During disaccharides fermentations, the accumulation of PEP was higher than those observed in maltose and glucose fermentations, possibly suggesting that conversion of PEP to pyruvic acid by pyruvate kinase (Pyk) was strongly inhibited under cellobiose and sucrose fermentations. In addition, the detailed metabolite profile analysis showed that tyrosine levels were higher in cells grown on cellobiose or sucrose, compared to cells grown on glucose or maltose.

Overall, these studies indicate the need to better clarify the interdependence between sugars provided and their internal flux redistribution in redirecting C flux towards final products, to maximize the potential of this yeast as cell factory. An implementation of the metabolic model currently available could be of support for predicting further engineering and optimization (Fig. 14.3).

14.8 Organic acids from glycolysis and TCA cycle: lactic acid and fumaric acid

Among organic acid of industrial interest, lactic acid has long been receiving attention in the field of industrial biotechnology, as it can be used in manufacture of diverse products, including the production of the biodegradable polymer polylactic acid (PLA) and textile fibers. One of the advantages to produce lactic acid in yeasts is linked to the tolerance of most yeast species to acidic pH.

The interest to produce bio-based lactic acid using the sugar stream from lignocellulosic biomass led to the development of engineered *S. stipitis* strains that would allow production of lactic acid from xylose. A *S. stipitis* strain able to produce lactate was generated by integrating the L-lactate dehydrogenasae (*ldhL*) gene from *Lactobacillus helveticus* (Ilmén et al., 2007), under the control of the endogenous *ADH1* promoter. Cultivations of the engineered strain on a single sugar, that is, xylose or glucose, or a mix of the two sugars showed that xylose was consumed at a lower rate than glucose and that most of the xylose was converted to lactate, with a yield of about 0.60 g lactate/g xylose, versus a yield of 0.44 g lactate/g of glucose. The reported study (Ilmén et al., 2007) was far from achieving an optimized *S. stipitis* strain or lactic acid production, as yields could be improved by deleting the pathway to ethanol production. In addition, the titers and yields reported were obtained only in noncontrolled fermentation conditions, where dissolved oxygen was not tuned and the need to buffer the cultivation medium and neutralize lactic acid with calcium carbonate was necessary to obtain the reported yields. Although further steps could have been taken to generate an improved *S. stipitis* strain and a better characterized process, this study remains the only one aiming at L-lactic acid production in *S. stipitis*.

Genetic and metabolic engineering of *S. stipitis* were applied for production of fumaric acid from xylose in an anaerobic process (Wei et al., 2015). Fumaric acid is naturally occurring in living organisms as part of the TCA cycle and the industrial demand of such compound is high, as it is used for the synthesis of diverse polymers and as precursor of succinate or maleic acid (Roa Engel et al., 2008). Metabolic engineering of *S. stipitis* for fumaric acid accumulation consisted of introducing and overexpressing the heterologous reductive pathway of the fungus *Rhizopus oryzae*, that is a typical industrial producer of fumaric acid via fermentation (Roa Engel et al., 2008; Wang et al., 2015). While the titers reached with a mutant *R. oryzae* were up to 56 g/L, transferring the fumaric acid biosynthetic route to *S. stipitis* would have allowed bypassing the process challenges linked to the complex morphology of the *Rhizopus* spp. and a xylose-based production having a maximum theoretical yield of 1.67 mol fumaric acid/mol xylose (Wei et al., 2015). Three genes from the best performing *R. oryzae* strain were introduced in *S. stipitis*, namely *PYC1*, *MDH*, and *FUM1*. In spite of an increase of fumaric acid production (1.86 g/L) compared to the wild type strain, a further increase of fumaric acid could be obtained by deletion of the native fumarase genes and by the overexpression of the fumaric acid transporter from *Schyzosaccharomyces pombe* (*MAE1* gene). In a batch fermentation process under anaerobic conditions and with xylose as the only carbon source, the maximum titer reached was 4.67 g/L, with ethanol as the main by-product and a yield of fumaric acid on xylose equal to 0.12 (mol/mol), much lower than the maximum theoretical yield.

14.9 Processes for *S. stipitis* biomass production and bioconversion reactions

A few research studies have focused on whole cell conversion of substrates with *S. stipitis*. In particular *S. stipitis* has been studied for whole-cell catalyst, exploiting the native xylose reductase (XR) enzymatic activity for the reduction of *o*-chloroacetophenone to (*S*)-1-(2-chlorophenyl)ethanol, that is used in the preparation of chemotherapeutics (Gruber et al., 2013). The NAD(P)H-dependent reduction was characterized using the purified XR, the crude extract and the whole-cell conversion of the substrate. The catalytic efficiency (k_{cat}/k_M) of the purified enzyme was calculated as 228 M^{-1} s^{-1} and whole-cell conversion using xylose-grown cells showed the ability to convert 100 mM of substrate to 15 mM of product. While the enzymatic activity was shown to be quite stable, the limited conversion of the substrate was possibly related to sensitivity of XR towards the formed product. This hypothesis was confirmed by the fact that product removal allowed further conversion. However, the latter was tested only by exploiting the XR activity of *Candida tenuis*, that in the same study showed higher potential for this kind of application.

Another bioconversion that has been shown possible using *S. stipitis* is the conversion of ethylene glycol (EG) to glycolic acid (GA), a small two-carbon α-hydroxy acid with a widespread application in various industries (Salusjärvi et al., 2019). EG has a growing market value, with the personal care segment holding the largest revenue share. EG is one of the products resulting from polyethylene terephthalate (PET) enzymatic depolymerization, which is an emerging strategy to recycle PET (Weiland et al., 2024) by breaking it down and then valorize the resulting products.

Conversion of EG to GA was evaluated in a screening study including eight different yeast genera, wherein *S. stipitis* CBS 6054 stood out with a production of 112 mM GA from 150 mM EG supplied as carbon source in complex medium (Senatore et al., 2024). *S. stipitis* was then used for whole-cell bioconversion process in controlled bioreactors; cell biomass accumulation was performed in batch mode with glucose as carbon source and upon complete glucose consumption EG was added and completely depleted from the cultivation broth after 120 hours. Conversion of EG to GA started immediately after the addition of EG to the culture and after 120 hours, the conversion yield was calculated as 0.77 mol GA/mol EG, featured by a volumetric productivity of 190.41 mg GA L^{-1} h^{-1}. No by-products were observed during the bioconversion; however, given the overall production of CO_2 (80 mmol), it was possible to hypothesize that the fraction of EG not converted to GA could have been shunted to glyoxylate and then channeled to TCA, with subsequent CO_2 production. This study is certainly seminal for the use of *S. stipitis* as whole-cell catalyst for the valorization of EG via GA production and further process improvement could be made to optimize the process, both by engineering *S. stipitis* for a more effective conversion and by using sugar stream from waste biomass for the initial step of cell biomass accumulation.

When whole-cell conversion is a solution for the synthesis of a specific product, increasing the productivity and the yield of biomass is certainly an aspect to consider. On this aspect, a mutant strain of *S. stipitis* with high biomass yield was designed applying system metabolic engineering guided by flux balance analysis and then characterized. Flux balance analysis predicted four target genes, whose overexpression would have led to maximizing biomass yield (Unrean et al., 2016). In particular, the candidate genes were *GPD1* (Glycerol-3-phosphate dehydrogenase), *ZWF1* (Glucose 6-phosphate dehydrogenase), *TKL1* (Transketolase) and *CIT1* (Citrate synthase). The biomass yield predicted following the overexpression of the four target genes was 0.49 g/g glucose, with biomass production efficiency per oxygen uptake equal to 0.08 g CDW/$mmol_{O2}$. All the single mutants expressed the target gene under the control of the constitutive promoter *pTEF1*. All mutants showed a decreased ethanol yield and increased biomass yield, compared to the control strain, with the strain overexpressing *ZWF1* reaching a biomass yield of 0.49 g/g glucose, that is a 44% yield increase compared to the wild type strain. Regardless of the specific gene overexpressed, all the mutants showed higher ATP and NADPH availability. The *ZWF1* mutant had the highest NADPH availability, suggesting a redirection of fluxes towards the pentose phosphate pathway. In addition, a comparative flux distribution showed that the fluxes through the TCA cycle reactions were increased in all the mutants, thus leading to the biosynthesis of several precursors of lipid and biomass synthesis. Also fluxes towards oxygen uptake and oxidative phosphorylation were higher in the mutants, hence enhancing oxygen utilization and ATP generation. Further improvement of the mutants was obtained via directed evolution, using subsequent batch cultivations. All the evolved mutants reached a biomass yield of 0.49 g/g and showed a 10%–22% higher specific growth rate compared to the parental strains and a faster glucose consumption rate. The strains developed in this study could possibly be used as platform for high-cell-density fermentations as well as for metabolic engineering of pathways that require precursors and cofactors available at higher levels in these mutants.

14.10 Conclusions

Most of the scientific work on *S. stipitis* focused for long time exclusively on the optimization of the process conditions and parameters for the production of ethanol from sugars deriving form hemicellulose, that is, xylose, arabinose, cellobiose, and other oligosaccharides. This work was seminal and essential to elucidate the metabolic properties of *S. stipitis*. Although bioethanol production with *S. stipitis* was never realized due to several drawbacks linked to the conditions required by this yeast, the fruits of the scientific investigations could be applied to other biotechnological processes. On one hand, the enzymatic activities (Kwak et al., 2019; Zha et al., 2021) native of *S. stipitis* enabling the efficient consumption of xylose were transferred to other microorganisms to achieve xylose-based processes. On the other hand, the Crabtree negative metabolism and the natural capacity to catabolize diverse sugars are properties that make *S. stipitis* interesting for the development of second generation processes exploiting the sugar streams from lignocellulose. Although *S. stipitis* has not been widely considered yet as one of the main potential industrial biotechnology microbial producers, the definition of genome scale metabolic models, along with the development of genetic engineering tools make this yeast a valuable host for industrial biotechnology processes and biomanufacturing, as we have tried to outline in our review work.

Acknowledgments and funding

This work was supported by the National Center 5 "National Biodiversity Future Center" Award Number: Project code CN_00000033, Concession Decree No. 1034 of 17 June 2022 adopted by the Italian Ministry of University and Research, CUP H43C22000530001, Project title "National Biodiversity Future Center – NBFC. Funders: Project funded under the National Recovery and Resilience Plan (NRRP), Mission 4 Component 2 Investment 1.4—Call for tender No. 3138 of 16 December 2021, rectified by Decree n.3175 of 18 December 2021 of Italian Ministry of University and Research funded by the European Union – NextGenerationEU, and under the European COST programme, Action No CA18229 Yeast4Bio (Non-Conventional Yeasts for the Production of Bioproducts).

References

Bajwa, P. K., Shireen, T., D'Aoust, F., Pinel, D., Martin, V. J. J., Trevors, J. T., & Lee, H. (2009). Mutants of the pentose-fermenting yeast Pichia stipitis with improved tolerance to inhibitors in hardwood spent sulfite liquor. *Biotechnology and Bioengineering, 104*(5), 892–900. https://doi.org/10.1002/bit.22449.

Balagurunathan, B., Jonnalagadda, S., Tan, L., & Srinivasan, R. (2012). Reconstruction and analysis of a genome-scale metabolic model for Scheffersomyces stipitis. *Microbial Cell Factories, 11*(1). https://doi.org/10.1186/1475-2859-11-27.

Barnett, J. A. (1976). The utilization of sugars by yeasts. *Advances in Carbohydrate Chemistry and Biochemistry, 32*(C), 125–234. https://doi.org/10.1016/S0065-2318(08)60337-6.

Brochado, A. R., Matos, C., Møller, B. L., Hansen, J., Mortensen, U. H., & Patil, K. R. (2010). Improved vanillin production in baker's yeast through in silico design. *Microbial Cell Factories, 9*. https://doi.org/10.1186/1475-2859-9-84, http://www.microbialcellfactories.com/content/9/1/84.

Cao, M., Gao, M., Lopez-Garcia, C. L., Wu, Y., Seetharam, A. S., Severin, A. J., & Shao, Z. (2017). Centromeric DNA facilitates nonconventional yeast genetic engineering. *ACS Synthetic Biology, 6*(8), 1545–1553. https://doi.org/10.1021/acssynbio.7b00046, http://pubs.acs.org/journal/asbcd6.

Cao, M., Gao, M., Ploessl, D., Song, C., & Shao, Z. (2018). CRISPR–mediated genome editing and gene repression in Scheffersomyces stipitis. *Biotechnology Journal, 13*(9). https://doi.org/10.1002/biot.201700598.

Cao, M., Seetharam, A. S., Severin, A. J., & Shao, Z. (2017). Rapid isolation of centromeres from Scheffersomyces stipitis. *ACS Synthetic Biology, 6*(11), 2028–2034. https://doi.org/10.1021/acssynbio.7b00166.

Cueto-Rojas, H. F., Milne, N., Helmond, W., Pieterse, M. M., Maris, A. J. A., Daran, J. M., & Wahl, S. A. (2017). Membrane potential independent transport of NH3 in the absence of ammonium permeases in Saccharomyces cerevisiae. *BMC Systems Biology, 11*(1). https://doi.org/10.1186/s12918-016-0381-1, http://www.biomedcentral.com/bmcsystbiol/.

Delgenes, Moletta, & Navarro (1988). Fermentation of d-xylose, d-glucose and l-arabinose mixture by Pichia stipitis Y 7124: Sugar tolerance. *Applied Microbiology and Biotechnology, 29*(2-3), 155–161. https://doi.org/10.1007/BF01982895.

Di Primio, C., Galli, A., Cervelli, T., Zoppè, M., & Rainaldi, G. (2005). Potentiation of gene targeting in human cells by expression of Saccharomyces cerevisiae Rad52. *Nucleic Acids Research, 33*(14), 4639–4648. https://doi.org/10.1093/nar/gki778.

Dominguez, A. A., Lim, W. A., & Qi, L. S. (2016). Beyond editing: Repurposing CRISPR-Cas9 for precision genome regulation and interrogation. *Nature Reviews. Molecular Cell Biology, 17*(1), 5–15. https://doi.org/10.1038/nrm.2015.2.

du Preez, J. C., Bosch, M., & Prior, B. A. (1986). The fermentation of hexose and pentose sugars by Candida shehatae and Pichia stipitis. *Applied Microbiology and Biotechnology, 23*(3-4), 228–233. https://doi.org/10.1007/BF00261920.

du Preez, J. C., & Prior, B. A. (1985). A quantitative screening of some xylose-fermenting yeast isolates. *Biotechnology Letters, 7*(4), 241–246. https://doi.org/10.1007/bf01042370.

du Preez, J. C., van Driessel, B., & Prior, B. A. (1989). Ethanol tolerance of Pichia stipitis and Candida shehatae strains in fed-batch cultures at controlled low dissolved oxygen levels. *Applied Microbiology and Biotechnology, 30*(1), 53–58. https://doi.org/10.1007/BF00255996.

Feng, X., & Zhao, H. (2013). Investigating glucose and xylose metabolism in Saccharomyces cerevisiae and Scheffersomyces stipitis via 13 C metabolic flux analysis. *AIChE Journal, 59*(9), 3195–3202. https://doi.org/10.1002/aic.14182.

Fiaux, J., Çakar, Z. P., Sonderegger, M., Wüthrich, K., Szyperski, T., & Sauer, U. (2003). Metabolic-flux profiling of the yeasts Saccharomyces cerevisiae and Pichia stipitis. *Eukaryotic Cell, 2*(1), 170–180. https://doi.org/10.1128/EC.2.1.170-180.2003.

Freese, S., Vogts, T., Speer, F., Schäfer, B., Passoth, V., & Klinner, U. (2011). C- and N-catabolic utilization of tricarboxylic acid cycle-related amino acids by *Scheffersomyces stipitis* and other yeasts. *Yeast, 28*(5), 375–390. https://doi.org/10.1002/yea.1845.

Galanie, S., Thodey, K., Trenchard, I. J., Interrante, M. F., & Smolke, C. D. (2015). Complete biosynthesis of opioids in yeast. *Science (New York, N.Y.), 349*(6252), 1095–1100. https://doi.org/10.1126/science.aac9373, http://www.sciencemag.org/content/349/6252/1095.full.pdf.

Gancedo, C., & Serrano, R. (1989). Energy-yielding metabolism. *The Yeast, 3*, 205–259.

Gao, M., Cao, M., Suástegui, M., Walker, J., Quiroz, N. R., Wu, Y., Tribby, D., Okerlund, A., Stanley, L., Shanks, J. V., & Shao, Z. (2017). Innovating a nonconventional yeast platform for producing shikimate as the building block of high-value aromatics. *ACS Synthetic Biology, 6*(1), 29–38. https://doi.org/10.1021/acssynbio.6b00132, http://pubs.acs.org/journal/asbcd6.

Gilbert, L. A., Larson, M. H., Morsut, L., Liu, Z., Brar, G. A., Torres, S. E., Stern-Ginossar, N., Brandman, O., Whitehead, E. H., Doudna, J. A., Lim, W. A., Weissman, J. S., & Qi, L. S. (2013). XCRISPR-mediated modular RNA-guided regulation of transcription in eukaryotes. *Cell, 154*(2). https://doi.org/10.1016/j.cell.2013.06.044.

Grootjen, D. R. J., van der Lans, R. G. J. M., & Luyben, K. C. A. M. (1990). Effects of the aeration rate on the fermentation of glucose and xylose by Pichia stipitis CBS 5773. *Enzyme and Microbial Technology, 12*(1), 20–23. https://doi.org/10.1016/0141-0229(90)90174-O.

Gruber, C., Krahulec, S., Nidetzky, B., & Kratzer, R. (2013). Harnessing Candida tenuis and Pichia stipitis in whole-cell bioreductions of o -chloroacetophenone: Stereoselectivity, cell activity, in situ substrate supply and product removal. *Biotechnology Journal, 8*(6), 699–708. https://doi.org/10.1002/biot.201200322.

Hilliard, M., Damiani, A., He, Q. P., Jeffries, T., & Wang, J. (2018). Elucidating redox balance shift in Scheffersomyces stipitis' fermentative metabolism using a modified genome-scale metabolic model. *Microbial Cell Factories, 17*(1). http://www.microbialcellfactories.com/home/, https://doi.org/10.1186/s12934-018-0983-y, http://www.microbialcellfactories.com/home/.

Ho, N. W. Y., Petros, D., & Deng (1991). Genetic transformation of xylose-fermenting yeast Pichia stipitis - Scientific note. *Applied Biochemistry and Biotechnology, 28-29*(1), 369–375. https://doi.org/10.1007/BF02922616.

Ilmén, M., Koivuranta, K., Ruohonen, L., Suominen, P., & Penttilä, M. (2007). Efficient production of l -lactic acid from xylose by pichia stipitis. *Applied and Environmental Microbiology, 73*(1), 117–123. https://doi.org/10.1128/aem.01311-06.

Inokuma, K., Matsuda, M., Sasaki, D., Hasunuma, T., & Kondo, A. (2018). Widespread effect of N-acetyl-d-glucosamine assimilation on the metabolisms of amino acids, purines, and pyrimidines in Scheffersomyces stipitis. *Microbial Cell Factories, 17*(1). https://doi.org/10.1186/s12934-018-0998-4.

Jeffries, T. W., Grigoriev, I. V., Grimwood, J., Laplaza, J. M., Aerts, A., Salamov, A., Schmutz, J., Lindquist, E., Dehal, P., Shapiro, H., Jin, Y. S., Passoth, V., & Richardson, P. M. (2007). Genome sequence of the lignocellulose-bioconverting and xylose-fermenting yeast Pichia stipitis. *Nature Biotechnology, 25*(3), 319–326. https://doi.org/10.1038/nbt1290.

Jeffries, T. W., & Van Vleet, J. R. H. (2009). Pichia stipitis genomics, transcriptomics, and gene clusters. *FEMS Yeast Research, 9*(6), 793–807. https://doi.org/10.1111/j.1567-1364.2009.00525.x.

Jeppsson, H., Alexander, N. J., & Hahn-Hagerdal, B. (1995). Existence of cyanide-insensitive respiration in the yeast Pichia stipitis and Its possible influence on product formation during xylose utilization. *Applied and Environmental Microbiology, 61*(7), 2596–2600. https://doi.org/10.1128/aem.61.7.2596-2600.1995.

Klinner, U., Fluthgraf, S., Freese, S., & Passoth, V. (2005). Aerobic induction of respiro-fermentative growth by decreasing oxygen tensions in the respiratory yeast Pichia stipitis. *Applied Microbiology and Biotechnology, 67*(2), 247–253. https://doi.org/10.1007/s00253-004-1746-8.

Kobayashi, Y., Inokuma, K., Matsuda, M., Kondo, A., & Hasunuma, T. (2021). Resveratrol production from several types of saccharide sources by a recombinant Scheffersomyces stipitis strain. *Metabolic Engineering Communications, 13*, e00188. https://doi.org/10.1016/j.mec.2021.e00188.

Koopman, F., Beekwilder, J., Crimi, B., van Houwelingen, A., Hall, R. D., Bosch, D., van Maris, A. J. A., Pronk, J. T., & Daran, J. M. (2012). De novo production of the flavonoid naringenin in engineered Saccharomyces cerevisiae. *Microbial Cell Factories, 11*. https://doi.org/10.1186/1475-2859-11-155, http://www.microbialcellfactories.com/content/11/1/155. Netherlands.

Krahulec, S., Kratzer, R., Longus, K., & Nidetzky, B. (2012). Comparison of Scheffersomyces stipitis strains CBS 5773 and CBS 6054 with regard to their xylose metabolism: Implications for xylose fermentation. *MicrobiologyOpen, 1*(1), 64–70. https://doi.org/10.1002/mbo3.5.

Krassowski, T., Coughlan, A. Y., Shen, X. X., Zhou, X., Kominek, J., Opulente, D. A., Riley, R., Grigoriev, I. V., Maheshwari, N., Shields, D. C., Kurtzman, C. P., Hittinger, C. T., Rokas, A., & Wolfe, K. H. (2018). Evolutionary instability of CUG-Leu in the genetic code of budding yeasts. *Nature Communications, 9*(1). http://www.nature.com/ncomms/index.html, https://doi.org/10.1038/s41467-018-04374-7, http://www.nature.com/ncomms/index.html.

Kurtzman, C. P., Kurtzman, S., & Suzuki, M. (2011). *The Yeasts*. United States: Elsevier, 773–777. http://doi.org/10.1016/B978-0-444-52149-1.00065-3, http://www.sciencedirect.com/science/book/9780444521491 2.

Kurtzman, C. P., & Robnett, C. J. (1998). Identification and phylogeny of ascomycetous yeasts from analysis of nuclear large subunit (26S) ribosomal DNA partial sequences. *Antonie Van Leeuwenhoek, 73*(4), 331–371. https://doi.org/10.1023/A:1001761008817. International Journal of General and Molecular Microbiology.

Kwak, S., Jo, J. H., Yun, E. J., Jin, Y. S., & Seo, J. H. (2019). Production of biofuels and chemicals from xylose using native and engineered yeast strains. *Biotechnology Advances, 37*(2), 271–283. Available from https://doi.org/10.1016/j.biotechadv.2018.12.003, www.elsevier.com/inca/publications/store/5/2/5/4/5/5/index.htt.

Laplaza, J. M., Torres, B. R., Jin, Y. S., & Jeffries, T. W. (2006). Sh ble and Cre adapted for functional genomics and metabolic engineering of Pichia stipitis. *Enzyme and Microbial Technology, 38*(6), 741–747. https://doi.org/10.1016/j.enzmictec.2005.07.024.

Ligthelm, M. E., Prior, B. A., & du Preez, J. C. (1988). The oxygen requirements of yeasts for the fermentation of d-xylose and d-glucose to ethanol. *Applied Microbiology and Biotechnology, 28*(1), 63–68. https://doi.org/10.1007/BF00250500.

Löbs, A.-K., Schwartz, C., & Wheeldon, I. (2017). Genome and metabolic engineering in non-conventional yeasts: Current advances and applications. *Synthetic and Systems Biotechnology, 2*(3), 198–207. https://doi.org/10.1016/j.synbio.2017.08.002.

Lu, P., Davis, B. P., Hendrick, J., & Jeffries, T. W. (1998). Cloning and disruption of the β-isopropylmalate dehydrogenase gene (LEU2) of Pichia stipitis with URA3 and recovery of the double auxotroph. *Applied Microbiology and Biotechnology, 49*(2), 141–146. https://doi.org/10.1007/s002530051150.

Maassen, N., Freese, S., Schruff, B., Passoth, V., & Klinner, U. (2008). Nonhomologous end joining and homologous recombination DNA repair pathways in integration mutagenesis in the xylose-fermenting yeast Pichia stipitis. *FEMS Yeast Research, 8*(5), 735–743. https://doi.org/10.1111/j.1567-1364.2008.00383.x.

Mastella, L., Senatore, V., Beltrani, T., & Branduardi, P. (2023). Scheffersomyces stipitis ability to valorize different residual biomasses for vitamin B 9 production. *Microbial Biotechnology, 16*(2), 392–403. https://doi.org/10.1111/1751-7915.14177.

Mastella, L., Senatore, V. G., Guzzetti, L., Coppolino, M., Campone, L., Labra, M., Beltrani, T., & Branduardi, P. (2022). First report on Vitamin B9 production including quantitative analysis of its vitamers in the yeast Scheffersomyces stipitis. *Biotechnology for Biofuels and Bioproducts, 15*(1). https://biotechnologyforbiofuels.biomedcentral.com/, https://doi.org/10.1186/s13068-022-02194-y, https://biotechnologyforbiofuels.biomedcentral.com/.

Matanović, A., Arambašić, K., Žunar, B., Štafa, A., Miklenić, M. S., Šantek, B., & Svetec, I.-K. (2022). Toolbox for genetic transformation of non-conventional saccharomycotina yeasts: High efficiency transformation of yeasts belonging to the Schwanniomyces genus. *Journal of Fungi, 8*(5). https://doi.org/10.3390/jof8050531.

Meyrial, Delgenes, Romieu, Moletta, & Gounot (1995). Ethanol tolerance and activity of plasma membrane ATPase in Pichia stipitis grown on d-xylose or on d-glucose. *Enzyme and Microbial Technology, 17*(6), 535–540. https://doi.org/10.1016/0141-0229(94)00065-Y.

Nardi, J. B., Bee, C. M., Miller, L. A., Nguyen, N. H., Suh, S. O., & Blackwell, M. (2006). Communities of microbes that inhabit the changing hindgut landscape of a subsocial beetle. *Arthropod Structure and Development, 35*(1), 57–68. https://doi.org/10.1016/j.asd.2005.06.003.

Ohrvik, V. E., & Witthoft, C. M. (2011). Human folate bioavailability. *Nutrients, 3*(4), 475–490. https://doi.org/10.3390/nu3040475.

Piontek, Hagedorn, Hollenberg, Gellissen, & Strasser (1998). Two novel gene expression systems based on the yeasts Schwanniomyces occidentalis and Pichia stipitis. *Applied Microbiology and Biotechnology, 50*(3), 331–338. https://doi.org/10.1007/s002530051300.

Reichard, W. D., Smith, S. E., & Robertson, J. B. (2023). BLINCAR: A reusable bioluminescent and Cas9-based genetic toolset for repeatedly modifying wild-type Scheffersomyces stipitis. *mSphere, 8*(4). https://doi.org/10.1128/msphere.00224-23.

Roa Engel, C. A., Straathof, A. J. J., Zijlmans, T. W., Van Gulik, W. M., & Van Der Wielen, L. A. M. (2008). Fumaric acid production by fermentation. *Applied Microbiology and Biotechnology, 78*(3), 379–389. https://doi.org/10.1007/s00253-007-1341-x.

Rodriguez-Ocasio, E., Khalid, A., Truka, C. J., Blenner, M. A., & Jarboe, L. R. (2022). Survey of nonconventional yeasts for lipid and hydrocarbon biotechnology. *Journal of Industrial Microbiology and Biotechnology, 49*(4). https://doi.org/10.1093/jimb/kuac010.

Roon, R. J., & Levenberg, B. (1972). Urea amidolyase. I. Properties of the enzyme from Candida utilis. *Journal of Biological Chemistry, 247*(13), 4107–4113.

Ruchala, J., Kurylenko, O. O., Dmytruk, K. V., & Sibirny, A. A. (2020). Construction of advanced producers of first- and second-generation ethanol in Saccharomyces cerevisiae and selected species of non-conventional yeasts (Scheffersomyces stipitis, Ogataea polymorpha). *Journal of Industrial Microbiology and Biotechnology, 47*(1), 109–132. Available from https://doi.org/10.1007/s10295-019-02242-x, www.springerlink.com/app/home/journal.asp.

Salusjärvi, L., Havukainen, S., Koivistoinen, O., & Toivari, M. (2019). Biotechnological production of glycolic acid and ethylene glycol: Current state and perspectives. *Applied Microbiology and Biotechnology, 103*(6), 2525–2535. https://doi.org/10.1007/s00253-019-09640-2.

Senatore, V. G., Milanesi, R., Masotti, F., Maestroni, L., Pagliari, S., Cannavacciuolo, C., Campone, L., Serra, I., & Branduardi, P. (2024). Exploring yeast biodiversity and process conditions for optimizing ethylene glycol conversion into glycolic acid. *FEMS Yeast Research, 24*. https://doi.org/10.1093/femsyr/foae024, http://femsyr.oxfordjournals.org/.

Shi, N. Q., Cruz, J., Sherman, F., & Jeffries, T. W. (2002). SHAM-sensitive alternative respiration in the xylose-metabolizing yeast Pichia stipitis. *Yeast (Chichester, England), 19*(14), 1203–1220. https://doi.org/10.1002/yea.915.

Shin, M., Kim, J.-w, Ye, S., Kim, S., Jeong, D., Lee, D. Y., Kim, J. N., Jin, Y.-S., Kim, K. H., & Kim, S. R. (2019). Comparative global metabolite profiling of xylose-fermenting Saccharomyces cerevisiae SR8 and Scheffersomyces stipitis. *Applied Microbiology and Biotechnology, 103*(13), 5435–5446. https://doi.org/10.1007/s00253-019-09829-5.

Silva, J. P. A., Mussatto, S. I., Roberto, I. C., & Teixeira, J. A. (2012). Fermentation medium and oxygen transfer conditions that maximize the xylose conversion to ethanol by Pichia stipitis. *Renewable Energy, 37*(1), 259–265. https://doi.org/10.1016/j.renene.2011.06.032.

Skoog, K., & Hahn-Hägerdal, B. (1989). Intermediary metabolite concentrations in xylose fermentingCandida tropicalis at varying oxygen limitations. *Biotechnology Techniques, 3*(1), 1–6. https://doi.org/10.1007/bf01876212.

Skoog, K., & Hahn-Hägerdal, B. (1990). Effect of oxygenation on xylose fermentation by *Pichia stipitis. Applied and Environmental Microbiology, 56*(11), 3389–3394. https://doi.org/10.1128/aem.56.11.3389-3394.1990.

Skoog, K., Jeppsson, H., & Hahn-Hägerdal, B. (1992). The effect of oxygenation on glucose fermentation with pichia stipitis - Scientific note. *Applied Biochemistry and Biotechnology, 34-35*(1), 369–375. https://doi.org/10.1007/BF02920560.

Slininger, P. J., Shea-Andersh, M. A., Thompson, S. R., Dien, B. S., Kurtzman, C. P., Balan, V., Da Costa Sousa, L., Uppugundla, N., Dale, B. E., & Cotta, M. A. (2015). Evolved strains of Scheffersomyces stipitis achieving high ethanol productivity on acid- and base-pretreated biomass hydrolyzate at high solids loading. *Biotechnology for Biofuels, 8*(1). https://doi.org/10.1186/s13068-015-0239-6.

Suh, S. O., Marshall, C. J., McHugh, J. V., & Blackwell, M. (2003). Wood ingestion by passalid beetles in the presence of xylose-fermenting gut yeasts. *Molecular Ecology, 12*(11), 3137–3145. https://doi.org/10.1046/j.1365-294X.2003.01973.x.

Toivola, A., Yarrow, D., van den Bosch, E., van Dijken, J. P., & Scheffers, W. A. (1984). Alcoholic fermentation of d -Xylose by yeasts. *Applied and Environmental Microbiology, 47*(6), 1221–1223. https://doi.org/10.1128/aem.47.6.1221-1223.1984.

Tran, A. V., & Chambers, R. P. (1986). Ethanol fermentation of red oak acid prehydrolysate by the yeast Pichia stipitis CBS 5776. *Enzyme and Microbial Technology, 8*(7), 439–444. https://doi.org/10.1016/0141-0229(86)90154-7.

Unrean, P., Jeennor, S., & Laoteng, K. (2016). Systematic development of biomass overproducing Scheffersomyces stipitis for high-cell-density fermentations. *Synthetic and Systems Biotechnology, 1*(1), 47–55. https://doi.org/10.1016/j.synbio.2016.01.006.

Vega-Estévez, S., Armitage, A., Bates, H. J., Harrison, R. J., & Buscaino, A. (2021). The genome of the CTG(Ser1) yeast Scheffersomyces stipitis is plastic. *mBio, 12*(5). https://doi.org/10.1128/mBio.01871-21, https://journals.asm.org/doi/10.1128/mBio.01871-21.

Wang, G., Huang, D., Li, Y., Wen, J., & Jia, X. (2015). A metabolic-based approach to improve xylose utilization for fumaric acid production from acid pretreated wheat bran by Rhizopus oryzae. *Bioresource Technology, 180*, 119–127. https://doi.org/10.1016/j.biortech.2014.12.091.

Wei, L., Liu, J., Qi, H., & Wen, J. (2015). Engineering Scheffersomyces stipitis for fumaric acid production from xylose. *Bioresource Technology, 187*, 246–254. https://doi.org/10.1016/j.biortech.2015.03.122.

Weiland, F., Kohlstedt, M., & Wittmann, C. (2024). Biobased de novo synthesis, upcycling, and recycling — the heartbeat toward a green and sustainable polyethylene terephthalate industry. *Current Opinion in Biotechnology, 86*, 103079. https://doi.org/10.1016/j.copbio.2024.103079.

Yang, V. W., Marks, J. A., Davis, B. P., & Jeffries, T. W. (1994). High-efficiency transformation of Pichia stipitis based on its URA3 gene and a homologous autonomous replication sequence, ARS2. *Applied and Environmental Microbiology, 60*(12), 4245–4254. https://doi.org/10.1128/aem.60.12.4245-4254.1994.

Zha, J., Yuwen, M., Qian, W., & Wu, X. (2021). Yeast-based biosynthesis of natural products from xylose. *Frontiers in Bioengineering and Biotechnology, 9*. https://doi.org/10.3389/fbioe.2021.634919.

Yarrowia lipolytica as cell factory for bioproduction

15

Aleksandra Maria Mirończuk[1], *Dorota Alicja Rzechonek*[2], *and Mateusz Szczepańczyk*[1]
[1]Laboratory for Biosustainability, Institute of Environmental Biology, Wrocław University of Environmental and Life Sciences, Wroclaw, Poland, [2]Department of Life Sciences (LIFE), University of Technology, Göteborg, Sweden

15.1 Introduction

Yarrowia lipolytica is an aerobic yeast best known as a model organism for metabolism of lipids. The oleaginous characteristics are reflected in the species name, as the yeast was often isolated from environments rich in fats and proteins. The first record dates back to 1928, but the species was described then as *Mycotorula lipolytica*. The genus affiliation was changed several times. For a long time, the teleomorph (sexual state) could not be determined, thus species was classified in *Candida* genus, together with other anamorph yeast. After identification of the teleomorph form, it was reclassified as *Endomycopsis* and then *Saccharomycopsis* (Barth & Gaillardin, 1996). A new genus *Yarrowia* was created in 1980 (Al-Feel et al., 1992; van der Walt and von Arx, 1980). For many years, *Y. lipolytica* was its only representative, but since 2013, the situation has been rapidly changing. Due to popularization of sequencing in taxonomy, it is easier to classify new isolates; moreover, several other *Candidas* were reclassified to *Yarrowia* genus (Michely et al., 2013), which currently include 14 species (NCBI Taxonomy Browser, accessed in September 2024). The first isolations of *Y. lipolytica* were associated with food products, but over time, the species began to be found in a wide variety of environments such as soil (strain ATCC 48436), oil-contaminated soil (A-101 and WSH-Z06), marine water (NCIM 3589), sewage (W29), corn-processing plant (ATCC 18942 and YB-419), and intestine or feces of vertebrates (SWJ-1b) or insects (Madzak, 2021). *Y. lipolytica* is an obligatory aerobe but can survive in the digestive system for some time (Madzak et al., 2023). This diversity of habitats is attributed to yeast's ability to utilize unusual carbon sources and its resistance to adverse conditions and pollution.

A characteristic feature of *Y. lipolytica* is dimorphism. The oval yeast form is the most common; however, there is a possibility of switching to more complex structures such as elongated cells, pseudohyphae or even hyphae. This diversity of forms translates into a varied appearance of colonies on agar media, which range from smooth and shiny to matte or rippled with an irregular shape (see Fig. 15.1) (Barth & Gaillardin, 1996). In most of the typical liquid media, there is a mixture of the prevalent yeast form and occasional micelles (Fig. 15.2). Change of proportion

Eukaryotic Microorganisms as Sources of Bioproducts. DOI: https://doi.org/10.1016/B978-0-443-30188-9.00016-0

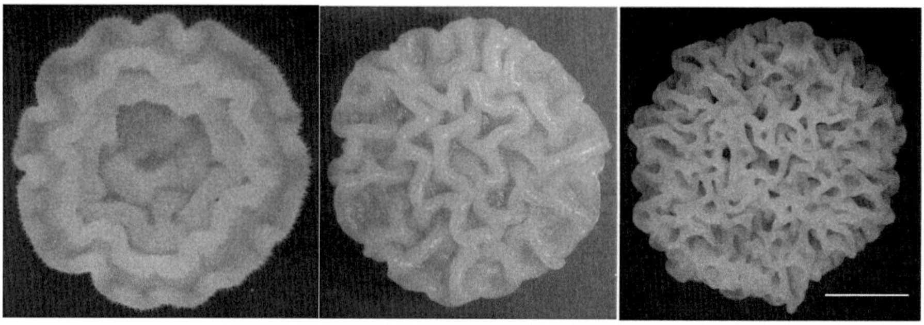

Figure 15.1 Colony morphology of different *Y. lipolytica* strains; left panel: *Y. lipolytic*a A 101; center: *Y. lipolytica* K1; right panel: *Y. lipolytica* MK1. Stereomicroscope pictures taken 4 days after inoculation. The bar indicates 1 cm.

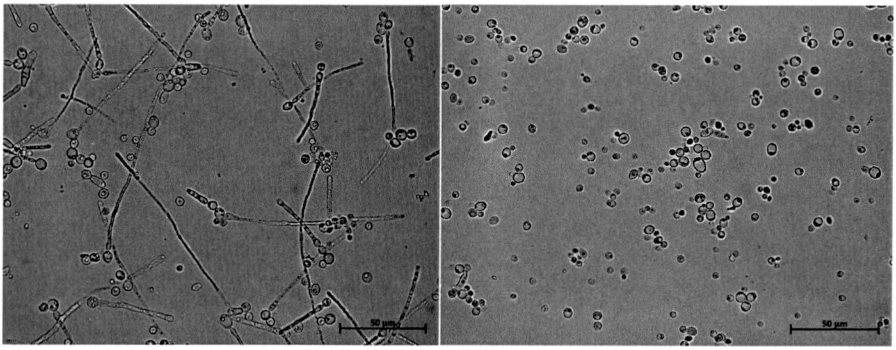

Figure 15.2 Dimorphic nature of *Y. lipolytica*
Source: Adopted from Mirończuk et al. (2019). *Microb Cell Fact*, *18*(1), 176.

toward micelles might be associated with more challenging environmental condition, certain carbon sources, presence of heavy metals or insufficient oxygenation (Celinska, 2022; Ruiz-Herrera & Sentandreu, 2002); however, factors triggering filamentation can vary between different. The change in growth form can also be a result of spontaneous mutations or a side effect of intentional genetic modifications, e.g., disrupting signaling pathways in the cell (Rzechonek et al., 2023). Increased filamentation is a potential problem—elongated cells may be more vulnerable to damage in the reactor. The possibility of a change in growth form could also complicate measurements based on turbidimetric (OD_{600}). For this reason, modifications that prevent filamentous growth, such as deletion of MHY1 gene (Konzock & Norbeck, 2020) are introduced in some research.

Y. lipolytica is a Crabtree-negative yeast (Barth & Gaillardin, 1996), which means that it carries out no alcoholic fermentation during the oxidation of carbon sources. It does not produce ethanol. Instead, the flow of carbon can be directed toward (1) biomass through accumulation of lipids or protein or (2) production of polyols and intermediates of citric acid cycle. This offers very different possibilities from the processes performed by Crabtree-positive yeasts such as *S. cerevisiae*. The potential for alternative applications was quickly recognized. However, the use of the microorganism in the food industry requires assurance that it does not pose a risk to human health.

Many processes carried out by *Y. lipolytica* have GRAS status (Generally Recognized As Safe) (Groenewald et al., 2014). The status is granted to products added to food by the United States Food and Drug Administration. The first GRAS involving *Y. lipolytica* was given in 1977 for the process of citric acid production (21 CFR 173.165). The European Food Safety Authority also define microorganisms with Qualified Presumption of Safety (QPS). Notably, *Y. lipolytica* was granted QPS status for the consumption of whole yeast biomass (Efsa Panel on Nutrition, et al., 2020), and not just purified compounds, produced by the yeast (Koutsoumanis et al., 2022).

The high potential for industrial applications was a good impetus for the development of tools for genetic modification of *Y. lipolytica*. The first transformation system was developed as early as 1985, and the first sequenced genome was available in 2004 (the E150 strain also referred to as CLIB122) (Dujon et al., 2004). Most strains are haploid (Madzak, 2021), which facilitates modifications. When starting to work on metabolic engineering of *Y. lipolytica*, its characteristic properties must be considered. A major problem is the dominance of DNA repair by nonheterologous recombination (NHEJ). This complicates the introduction of mutations, especially if genome editing tool is based solely on homologous recombination. For deletion or overexpression cassettes to be incorporated into the desired sequence, it is necessary to flank them with homologous fragments of at least 500 bp and preferably 1000 bp in length (Ji et al., 2020). Improving recombination efficiency while shortening homologous fragments is possible by deleting ku70/ku80 (Kretzschmar et al., 2013; Verbeke et al., 2013) or introducing elements of the homologous recombination mechanism from *S. cerevisiae* such as *Sc*Rad52 (Ji et al., 2020). Alternatives to the use of homologous recombination are genome editing methods adapted for *Y. lipolitca* based on CRISPR-Cas9 or TALENs nucleases (Shi et al., 2018). Large chromosome rearrangements also exist between *Y. lipolytica* strains (Casaregola et al., 2000). This should be considered especially when preparing cassettes for strains other than the available reference genomes.

Currently, there is a wide range of molecular biology tools adapted for *Y. lipolytica*. Available DNA assembly systems are Gateway, BioBricks, Gibson, Golden Gate, and OE-PCR. For gene overexpression, both native and hybrid promoters can be used. They can be constitutive or induced by environmental factors such as the presence of alkanes and fatty acids (FAs) or erythritol (Larroude et al., 2018).

Biomass characteristics of *Y. lipolytica* is depended on the strain and the cultivation conditions. It contains high content of lipid (up to 40%), proteins (up to 38%), macroelements, such as calcium, copper, phosphorus, potassium, and zinc and microelements such as chromium, iodine, and selenium (Jach & Malm, 2022; Juszczyk et al., 2013; Yang et al., 2022).

15.2 Substrates for Yarrowia lipolytica

15.2.1 Hydrophobic carbon sources

15.2.1.1 Hydrophobic carbon sources

The ability to utilize a variety of hydrophobic carbon sources is the most characteristic feature of *Y. lipolytica*. Access to poorly water-soluble compounds is enabled by the secretion of extracellular surfactants, emulsifiers or lipases, and presence of transport proteins in the cell membrane (Fukuda, 2023). *Y. lipolytica* are also known for lipid accumulation, which is particularly pronounced during growth on hydrophobic carbon sources.

Metabolism of n-alkanes is possible through hydroxylation to fatty alcohols via cytochromes P450ALK, followed by conversion to FAs (Michely et al., 2013). There are 12 *ALK* genes encoding cytochromes P450ALK in the genome of *Y. lipolytica* genome, and two of them, *AKL1* and *ALK2*, are crucial for the assimilation of alkanes (Fukuda, 2013). The ability to use alkanes as a carbon source very early on attracted the attention of researchers. In the 1960s, British Petroleum was developing technologies to use *Y. lipolytica* to produce single-cell proteins (SCPs) from crude oil and ammonia as a nitrogen source. The product was commercialized as Toprina G and was sold as livestock feed, but this production lost profitability because of the oil crisis in 1970s (Madzak, 2021). Currently, the ability to utilize n-alkanes is being explored more in terms of bioremediation of soil and wastewaters (Zinjarde et al., 2014).

The breakdown of triglycerides (TGA) is possible thanks to lipases, that hydrolyze the ester bond between FAs and glycerol. There are 16 genes encoding lipases in the *Y. lipolytica* genome, although not all are expressed. The first papers dedicated to the characterization of these enzymes date back to 1948, and since then they have been the subject of hundreds of publications. Some lipases are secreted (LIP2, LIP7, and LIP8), so that the breakdown of triglycerides occurs outside the cell, and free FAs can be transported across the cell membrane (Fickers et al., 2011). Importantly, although in nature lipases catalyze the breakdown of triglycerides, under the right conditions they can catalyze the degradation of other substrates as well, examples of which can be found in the later paragraphs.

15.2.2 Hydrophilic carbon sources

Most strains of *Y. lipolytica* can naturally utilize relatively few hydrophilic carbon sources compared to other yeast species. The most preferred substrate is glycerol, which is utilized before sugars or other polyols (Lubuta et al., 2019) and inhibit transcription of enzymes involved in metabolism of n-alkanes (Fukuda, 2013). It is in contrast with most yeasts that primarily utilize glucose. Glycerol assimilation can be accelerated by overexpression of *GUT1* encoding glycerol kinase (Mirończuk et al., 2016). Another polyol that can be utilized is erythrol. The use of erythrol as a carbon source is not applicable to industrial processes, but research on this process has led to the development of synthetic promoters induced by the presence of erythrol (Vidal et al., 2023). A polyol commonly found in nature, and therefore a potentially good carbon source for industrial applications, is mannitol. *Y. lipolytica* can utilize it in the presence of glucose (Dobrowolski et al., 2022); however, genetic modifications are required to make the process efficient. The ability of *Y. lipolytica* to grow on sugar carbon sources is limited. Hexoses like glucose, fructose, and mannose can be efficiently utilized. Glucose is preferred over other sugars and polyols, except for glycerol. Its presence in the environment does not inhibit the transcription of genes involved in the utilization of n-alkanes (Fukuda, 2013). The usage of galactose (hexose) and xylose (pentose) is more challenging. In both cases, genes encoding enzymes necessary for utilization have been found in the genome of *Y. lipolytica* (Lazar et al., 2015; Spagnuolo et al., 2018), but even so, these sugars cannot be used as the sole carbon source. In the case of galactose, utilization is possible in the presence of glucose. Growth on galactose as the only carbon source was achieved by overexpression of native *ylGAL1* encoding galactokinase (Lazar et al., 2015).

Much effort has been devoted to improving the utilization of xylose, given its huge prevalence in nature—as a component of hemicelluloses. Individual strains of *Y. lipolityca* have been reported as capable of utilizing xylose as the sole source of carbon (Madzak, 2021), but the results of various studies are inconsistent (Spagnuolo et al., 2018). Efficient utilization has been achieved through genetic modifications combined with laboratory adaptations. The problem has been approached in several ways, involving overexpression of heterologous or native genes (*ylXDH*, *ylXKS,* and *ylXYR2*) (Spagnuolo et al., 2018).

Metabolic engineering was also necessary to utilize the disaccharide sucrose. *Y. lipolytica* lacks the enzyme invertase, which allows sucrose to be split into glucose and fructose. This problem was solved by introducing the heterologous gene *scSUC2*. This was one of the earliest modifications *Y. lipolytica* was subjected to, and sc*SUC2* was used as a selection marker for genetic modifications (Nicaud et al., 1989). Today, this modification is present in many laboratory strains (Madzak, 2021). Another promising substrate is acetate. Most *Y. lipolytica* strains can use it as a carbon source, but the limiting factor is its toxicity to the cells. Toxicity tolerance ranges from 4 to 10 g/L and is dependent on strain and the pH of the media (Robles-Iglesias et al., 2023).

15.2.3 Resilience to challenging conditions enables the usage of crude and waste carbon sources

Y. lipolytica can withstand many adverse environmental conditions, which is highly important for industrial applications. First, it can grow in a wide pH range, from 2.5 to even 9 for some strains (Celinska, 2022; Egermeier et al., 2017; Sekova et al., 2021). Performing cultures at a low pH makes it easier to avoid bacterial contamination. However, a change in pH can result in the production of different compounds. When growing on glycerol, pH below 4.5 leads to a predominant formation of polyols, while pH range of 4.5–7.5 promotes the production of citric and isocitric acid (Egermeier et al., 2017). In nitrogen-limited conditions, acetic pH might also promote lipid synthesis from glucose and glycerol (Dobrowolski et al., 2020;

Zhang et al., 2019). Another factor often posing a problem for yeast is osmotic stress generated by the presence of salts or high concentration of substrates. In the case of *Y. lipolytica*, cultures can be conducted even with 200 g/L glycerol. Salinity also does not significantly impair growth; the most halotolerant strains can grow even in the presence of 90 g/L NaCl (Andreishcheva et al., 1999). Moreover, osmotic pressure is necessary for performing some processes, such as erythrol production. Another type of hazard is the presence of heavy metals such as chromium, mercury, cadmium, nickel, zinc, lead, or uranium. *Y. lipolytica* can grow in an environment contaminated by heavy metals and even absorb them. However, the tolerable concentrations vary between strains (Bankar et al., 2018; Kolhe et al., 2022). Finally, there are also hazardous organic compounds. *Y. lipolytica* was reported to degrade phenol (J et al., 2024), nitro compound 2,4,6-trinitrotoluene (TNT), and halogenated compounds: 1-chlorohexadecane, 1-chlorooctadecane, 2-bromopropane, 1-bromobutane, 1,5-dibromopentane, and 1- bromodecane (Bankar et al., 2009).

Resistance to a large spectrum of impurities is important for industrial applications. It offers the possibility of utilizing substrates that are less purified or even waste from other processes. Cheaper substrates increase the profitability of production, but this is not the only benefit. In some cases, the purification of wastewater is the main purpose of applying *Y. lipolytica*. The ability to remediate contaminated water and soil is important for reducing pressure on the environment and making industrial processes more sustainable.

A very promising substrate is crude glycerol, which is a by-product of processes such as the production of soap, stearin and, above all, biodiesel. The biodiesel market is constantly growing, which increases the availability of crude glycerol. Depending on the technologies used, the resulting glycerol can contain varying amounts of impurities and salts (18). Nevertheless, it can still be used by *Y. lipolytica* to produce additional bio-based chemicals, such as erythritol, lipids, organic acids, phospholipids, isopropanol (Fickers et al., 2011; Fukuda 2013, 2023; Zinjarde et al., 2014). A very big group of waste are used cooking oils and animal fats. It is estimated that 15%–20% of produced cooking oils become a waste during its manufacture of food processing (Thushari & Babel, 2022). They consist mostly of triglycerides, what makes it a good substrate for *Y. lipolytica*. Research into the use of these waste oils focuses on the production of value-added organic compounds like terpenes, terpenoids, or wax ester which often involves additional metabolic engineering (Soong et al., 2023).

There is also research on the use of hydrolysates of plant biomass, especially lignocellulose. Hydrolysis with enzymes or chemical pretreatment (acids, alkaline, or organic solvents) allows the breakdown of polysaccharides into components assimilable by *Y. lipolytica*. Chemical pretreatment of lignocellulose is cheaper and more efficient butresults in the formation of inhibitory compounds such as five-hydroxymethylfurfural (HMF), furfural, formic acid, or levulinic acid (Konzock et al., 2021). Several strategies are currently being considered to address this problem, such as detoxification of the substrate (Zainuddin et al., 2022), metabolic engineering, and adaptive laboratory evolution (Drzymala-Kapinos et al., 2022; Zhou et al., 2021). Some recently isolated strains are naturally more resistant to the inhibitors present in lignocellulose hydrolysates (Walker et al., 2018). Algae, in contrast to vascular plants, do not contain lignocellulose, so they can be a less challenging material. Hydrolysates of brown algae have been successfully tested as a substrate for lipid production (Dobrowolski et al., 2022).

Y. lipolytica can also grow in wastewater from food industries such as fish processing or edible oil milling. In this case, reducing the environmental burden of wastewater is just as important as producing value-added compounds like microbial proteins and lipids. The growth of yeast biomass reduces chemical oxygen demand (COD) by recovering nutrients, absorbing nitrates, phosphates, sulfites, and removing some of the phenolic compounds (Hamimed et al., 2021). Food industry wastewater might be also highly saline, and *Y. lipolytica* treatment enables desalination by adsorption of crystals on the surface of the cells (Hamimed et al., 2022). It was also shown that *Y. lipolytica* can grow on seawater-based medium with crude and pure glycerol and produces lipids (Dobrowolski et al., 2019). Some wastewaters, such as from olive oil mill, may not contain enough carbon sources, which in some studies has been supplemented by the addition of glucose or glycerol (Sarris et al., 2017, 2019).

15.3 *Yarrowa lipolytica* as a microbial cell factory

15.3.1 Enhancing natural production capacities

Y. lipolytica is an obligatory aerobe, thus much of the carbon flow is directed to the tricarboxylic acid cycle (also named TCA or Krebs cycle). This provides a good basis to produce its intermediates. The TCA cycle consist of eight reactions resulting in oxidation of acetyl-CoA molecule that enter the cycle (Fig. 15.3). This is coupled with generation of a reducing power in the form of NADH or $FADH_2$, later used to produce ATP. Intermediates of the TCA can be also used as precursors for production of some amino acids. When essential nutrients such as nitrogen are limited in the environment, the cells are unable to further divide, and energy requirements decrease. However, if there is still access to the abundant carbon source, it is assimilated and converted to acetyl-CoA. When there is too much acetyl-CoA to be processed by TCA cycle, the intermediates of TCA accumulate and might be exported from the cell. Other possibilities for utilizing the excess carbon source include the production of lipids or polyols, but the final direction of carbon flow is affected by many factors, such as pH, salinity, type of carbon source or redox balance.

15.3.1.1 Citric acid

Citric acid is an important food additive due to its acidity regulation and antioxidant properties. The possibility of its production using yeast quickly attracted the attention of researchers—it was the first process performed by *Y. lipolityca* for which GRAS status was granted. Citric acid is formed in the first reaction of the TCA cycle as a result of the condensation of acetyl-CoA and oxaloacetate and citrate synthase (CIT1). There are several factors that contribute to accumulation of citrate and its

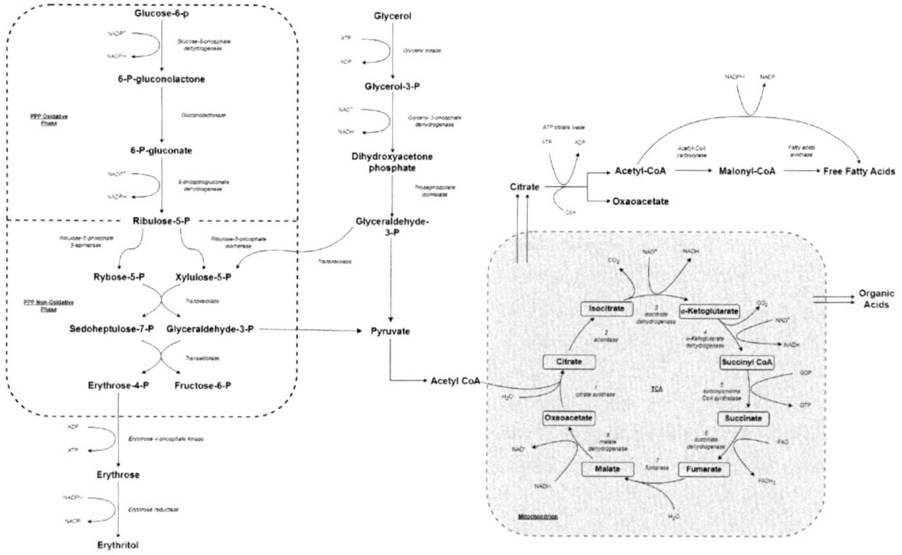

Figure 15.3 Schematic metabolic pathways in *Y. lipolytica*. Production of erythritol, fatty acids and organic acids from glycerol.

subsequent transport out of the mitochondria and export from the cell. The main cause is a low ratio of nitrogen or other minerals necessary for growth to abundant carbon sources. Factors that further promote redirection of carbon flow to citrate production are: pH in the range of 4.5–6 (Egermeier et al., 2017; Zhang et al., 2019) and availability of thiamine, that should be supplemented to the media (100–200 μg/L) (Kamzolova, 2023).

Citric acid can be produced from almost all carbon sources assimilated by *Y. lipolytica*, but the highest production yield is achieved for hydrophobic compounds such as oils (rapeseed, sunflower, or corn kernel) and alkanes, for which it can exceed the value of 1 g/g of substrate (Kamzolova, 2023). Exceeding the theoretical maximum yield is because hydrophobic substrates are rich in highly reduced carbon (high carbon-to-oxygen ratio), while a significant part of the mass of the obtained product is due to the presence of oxygen atoms in the molecule of citric acid. For optimal production, the C/N ratio should be selected for each substrate type.

Despite the large natural potential, there is still room for improving production through substrate optimization and metabolic engineering. The main problem of citric acid production by *Y. lipolytica* is formation of isocitric acid as a by-product. Conversion of citric acid to isocitric acid is the second reaction of the TCA cycle, carried out by the enzyme aconitase. Citrate and isocitrate are difficult to separate. The ratio of their production depends on the used substrates: citric acid is a main product obtained from glycerol and glucose, while usage of hydrophobic substrates leads to formation of more equal amounts of both products (Kamzolova, 2023; Papanikolaou et al., 2008). A solution to this problem may be the introduction of additional copies of the enzyme isocitrate lyase (ICL1), which breaks down isocitrate into glyoxylate and succinate (Forster et al., 2007). Another solution is the deletion of the *PHD1* gene encoding 2-methyl-citrate dehydratase, which results in the accumulation of 2-methyl-citrate that inhibit aconitase activity (Papanikolaou et al., 2013).

Other modifications that improves citric acid production include overexpression of YlYHM2, which encodes the major mitochondrial citrate transporter (Yuzbasheva et al., 2019). Some modifications not directly related to TCA cycle can unexpectedly increase citric acid production, for example, overexpression of Yl*AMPD*, encoding adenosine monophosphate deaminase (Yuzbasheva et al., 2019). Overexpression of glycerol kinase encoded by *GUT1*, and glycerol-3P-dehydrogenase encoded by *GUT2* not only increases the synthesis of citric acid from glycerol but also enables to the carry out the process at pH 3, much lower than the typical optimum for citric acid production (Rzechonek, 2019).

15.3.1.2 α-Ketoglutarate acid and pyruvate

Another reaction of TCA is the decarboxylation of isocitrate to α-ketoglutarate by α-ketoglutarate dehydrogenase. Interest in the production of this compound stems from its possible applications in cosmetics, dietetics, and as a building block to produce more complex organic compounds. So far, no biological method of producing α-ketoglutarate has been commercialized (Guo et al., 2016). There are several strains of *Y. lipolytica* producing significant amounts of α-ketoglutaric acid. As in the case of citric acid production, the necessary conditions are an excess of carbon sources upon nitrogen limitation. The highest yield productivity is also achieved for hydrophobic carbon sources, for example, 0.95 g/g from rapeseed oil (Kamzolova & Morgunov, 2013). The important difference, however, is that to redirect metabolism toward α-ketoglutaric acid production, thiamine limitation is necessary. Thiamine is a cofactor of ketoglutarate dehydrogenase, catalyzing a subsequent reaction of the TCA cycle which creates succinyl-CoA from α-ketoglutarate. Lack of thiamine results in the accumulation of α-ketoglutarate (Kamzolova, 2023). However, the initiation of production by thiamine limitation involves the formation of another product, pyruvic acid. Pyruvate is a product of glycolysis that is converted to acetyl-CoA by pyruvate dehydrogenase. Pyruvate dehydrogenase, like ketoglutarate dehydrogenase, needs thiamine as a cofactor. Therefore, thiamine limitation causes accumulation of both ketoglutarate and pyruvate. Pyruvate is also a valuable compound, and there are papers devoted to optimizing its production by *Y. lipolytica* (Cybulski et al., 2019; Kamzolova & Morgunov, 2018). The problem, however, is that pyruvate and ketoglutarate are produced simultaneously, and their separation can be costly. So, the main effort in optimizing the production of ketoglutarate and pyruvate is to switch the flow of carbon to only one of these compounds. The ratio of each product may depend on the substrate—the use of hydrophobic carbon sources, allows to omit the pyruvate formation step, thus promoting the synthesis of α-ketoglutarate. However, the production of α-ketoglutarate is not limited only to hydrophobic substrates, as glycerol and crude glycerol were also tested (Morgunov et al., 2013). Other aspects of optimizing production relate to the concentration of thiamine, and the pH of the media that for α-ketoglutarate is 3.5–4 (Cybulski et al., 2019; Kamzolova & Morgunov, 2018; Morgunov et al., 2013). To further reduce the by-product formation, selected strains were also subjected to metabolic engineering (Guo et al., 2016).

15.3.1.3 Lipid biosynthesis

Due to many unique natural features mentioned above, yeast *Y. lipolytica* become an attractive candidate for industrial production. The genome sequence has been published in 2004 (Dujon et al., 2004), and since this time many works have been done to improve its capabilities for enhanced production. By modification of the fermentation conditions such as pH or carbon source, it was possible to stimulate a production of the desired product such as citric acid or erythritol (polyol, a natural sweetener). But first, *Y. lipolytica* became a model organism for studying lipogenesis in eukaryotic cells. It was observed that under conditions of nitrogen limitation with high access of carbon, this yeast naturally accumulates a huge quantities of lipids in lipids bodies (Papanikolaou & Aggelis, 2002). *Y. lipolytica* accumulates mostly triacylglycerols (TAG) (45%–55% w/w) (Papanikolaou et al., 2001) and it contains saturated (SFA), monounsaturated (MUFA), and polyunsaturated (PUFA) FAs (Papanikolaou et al., 2002). The FA profile might vary and it depend on the applied substrate, but the most abundant FA are oleic (C18:1) and linoleic acids (C18:2) (Dobrowolski et al., 2016; Papanikolaou & Aggelis, 2002). In *Y. lipolytica*, the biosynthesis of TAG during the de novo synthesis follows the Kennedy pathway (Kennedy, 1962). During this process, yeast uses hydrophilic substrate such as glucose, organic acids, or polyols to produce the major precursors acetylo-CoA and glycerol-3-phosphate (G3P). As its name suggests, *Y. lipolytica* uses lipids as a carbon source and has developed several strategies to achieve this goal. It is well known for its extracellular lipolytic and proteolytic activities (Barth & Gaillardin, 1996), it produces biosurfactants that reduce the hydrophilicity of the substrate (T et al., 2016). As mentioned above, to utilize triglycerides as a carbon source, its secretes lipases that hydrolyze TGA into glycerol and free FAs. In its genome, 16 lipase-encoding genes have been identified (Fickers et al., 2005; Meunchan et al., 2015), including extracellular lipases Lip2, Lip7, and Lip8, with different substrate specifications. Lipase Lip2 hydrolyzes a long-chain FAs esters (Fickers et al., 2011), whereas Lip7 and Lip8 medium-long (Fickers et al., 2005). All this makes *Y. lipolytica* capable to synthetize own lipids during ex novo synthesis. During ex novo synthesis, a hydrophobic substrates are used such as FAs, alkanes, oils or TAG to synthetize FAs and glycerol, to reassemble them inside the cells into TAGs (Papanikolaou & Aggelis, 2003). Nitrogen limitation in the environment causes activation of adenosine monophosphate (AMP) deaminase (AMPD), which converts AMP into inosine monophosphate (IMP) and ammonia. Next, the citric acid is accumulated in the mitochondria and consequently is transported into the cytosol by malate/citrate transferase. In the cytosol, citrate is cleaved by ATP-citrate lyase (ACL1 and ACL2) into cytosolic acetyl-CoA and oxaloacetate. In the next step, acetylo-CoA is catalyzed by acetyl-CoA carboxylase (ACC) into malonyl-CoA (Al-Feel et al., 1992). Elongation of FAs requires NADPH as a reducing cofactor and the oxidative pentose phosphate pathway is a primary source of this cofactor (Wasylenko et al., 2015). The FA chain length is usually 16 or 18 carbon atoms they are directed into endoplasmic reticulum (ER) for further elongation and desaturation (Jenni et al., 2007). To form TGAs, three FAs link with one glycerol 3-phosphate (G3P), which is formed mainly by the reduction of dihydroxyacetone phosphate (DHAP). G3P is acylated by glycerol phosphate acyltransferase (SCT1) to lysophosphatidic acid (LPA) (Beopoulos et al., 2008), which is then further acylated by LPA acyltransferase (SLC1) to phosphatidic acid (PA). PA is then dephosphorylated by PA phosphohydrolase (PAP) to release diacylglycerol (DAG). In the final step, is acylated either by DAG acyltransferase (Athenstaedt, 2011) (DGA1 with acyl-CoA as an acyl donor) or by phospholipid DAG acyltransferase (LRO1 with glycerophospholipids as an acyl donor) to produce TAG (Dobrowolski & Mirończuk, 2020). These processes occur in the ER or on the lipid body (LB) surface, where the relevant enzymes have been located (Athenstaedt et al., 2006).

Under starvation conditions *Y. lipolytica* utilizes storage material, consequently the degradation of FA occurs in peroxisomes following the β-oxidation pathway. In this process, the FA backbone is shortened, resulting in the release of one molecule of acetyl-CoA. In the first step, six POX genes encoding acetyl-CoA are involved. Each of enzymes shows different chain-length preferences, deletion of six of them results in lack of capability for lipids biodegradation (Beopoulos et al., 2009). Two next steps are carried out by one gene MFE, therefore mutation in it, results in increased lipid accumulation (Blazeck et al., 2014; Dulermo & Nicaud, 2011). The last step is catalyzed by gene POT1 encoding thiolase.

Since the metabolisms of lipids had been well studied, the different strategies have been developed to increase lipid biosynthesis by *Y. lipolytica*. One of them is overexpressing genes involved in lipogenesis, such as DGA1 and DGA2 (Beopoulos et al., 2012; Tai & Stephanopoulos, 2013). Co-expression of DGA1 and ACC resulted in increased lipid content in dry biomass to 67% within 120 hours of cultivation (Tai & Stephanopoulos, 2013). In addition, the strategy "pull and push" was also used with overexpression of SCD (delta-stearoyl-Co-adesaturase) what allowed for rapid cell growth and high carbon o lipid conversion (more than 80% of theoretical maximal yield) (Qiao et al., 2015). Deletion of the genes responsible for lipid degradation, also improved their biosynthesis, deletion of β oxidation pathway, PEX1 6, Pex10, and MFE1 to prevent peroxisome biogenesis and β-oxidation (Blazeck et al., 2014; Dulermo & Nicaud, 2011). Interestingly, it has been found that the restoration of a complete leucine biosynthesis pathway increases lipid accumulation (Blazeck et al., 2014), what was later confirmed by other study (Kerkhoven et al., 2017). Also nonmetabolic engineering techniques have been employed to increase lipid production by the yeast. By establishing a high C/N ratio and using a suitable substrate, it is possible to modulate not only the lipid content but also the FA profile. It has been shown by using a mixture of glucose and saturated free FAs. Cultures performed on mixtures of stearin

(ranging 7.0–10.5 g/l) and glucose (ranging from 10.0 to 31.5 g/l). During the growth on glucose, mainly C18:1 was produced 53.1% (wt/wt in total lipid pool) whereas production on glucose(20.5 g/l)/sterine (10.5 g/l) resulted in 6.9% of C18:1 (Papanikolaou et al., 2006). Another techniques improving FAs synthesis are laboratory evolution strategies. These process allowed increasing after 77 generations accumulation up to 44% w/w of lipid, which was 30% higher than that of the starting strain (Daskalaki et al., 2019). What is important to notice, a FA profile might be modified by the applied substrate (Cordova et al., 2022; Dobrowolski et al., 2016; Papanikolaou & Aggelis, 2010) or metabolic engineering (Blazeck et al., 2014; Dobrowolski et al., 2020; Konzock et al., 2022; Lin et al., 2024; Tang et al., 2024; Xue et al., 2013). Overexpression of SCT1 gene resulted in tenfold improvement (>20%) in linoleic acid synthesis (C18:2) in a shake-flask experiment during growth on crude glycerol (Dobrowolski et al., 2020). Heterologous overexpression of gene clusters (BGCs) encoding polyketide synthase-like PUFA synthases from myxobacteria, allowed for increased biosynthesis of omega-3-LC-PUFAs eicosapentaenoic acid (EPA), docosapentaenoic acid (DPA), and docosahexaenoic acid (DHA) by the engineered *Y. lipolytica* starins (Gemperlein et al., 2019). Also bioproduction of storage lipids in *Y. lipolytica* mimicking cocoa butter have been established by optimization of fermentation production (Papanikolaou et al., 2003) or by metabolic engineering (Konzock et al., 2022). In the first study, application glycerol and stearin resulted in higher lipid content that usage glucose and stearin. The pool of the synthetized lipids contained significant amounts of stearic acid (50%–70%) and lower ones of palmitic (15%–20%), oleic (7%–20%), and linoleic (2%–7%) acids (Papanikolaou et al., 2003). It was shown that *Y. lipolytica* metabolisms might be directed by using mixtures of low-cost saturated fats, glycerol, and glucose as cosubstrates, in order to accumulate lipids with predetermined composition, e.g., cocoa-butter equivalents. Employing different substrates such as crude glycerol, a side-waste product from the various industries, also results in profiling a FA composition. Using waste glycerol from the biodiesel industry (87% of glycerol) resulted in significant high production of PUFA (more than 20% in FA pool) (Dobrowolski et al., 2016). In another study, the profile of FA have been changed by replacement of the native Δ9 FA desaturase (Ole1p) with homologs from *Arxula adeninivorans* (AaOLE1), *Gloeophyllum trabeum* (GtOLE1), or *Rhodotorula toruloides* (RtOLE1) and changing the expression of both Ole1p and the Δ12 FA desaturase (Fad2p). Also cultivation on low concentration of the volatile FAs (2.5 g/L) derived from food wastes, increases the tolerance in the high temperature and acidic environment of *Y. lipolytica* strain CICC 31596. In cultures at higher temperatures (38°C) or without pH adjustment, the content of unsaturated FAs in lipids decreased and that of saturated FAs increased (Gao et al., 2017).

In summary, the combination of the metabolic engineering of *Y. lipolytica* and process optimization may contribute to the establishment of microbial cell factories for the sustainable production of food oils and biodiesel.

15.3.1.4 Polyols production

Polyols, the so-called sweet alcohols, are the group of organic compound containing multiple hydroxyl groups, are used by microorganisms as a osmoprotectants or as a storage material. *Y. lipolytica* produces polyols such as mannitol, arabitol but mainly erythritol (Rzechonek et al., 2018). This phenomenon occurs under high osmotic pressure, as was proved by increased concentration of glycerol correlated with increasing osmotic pressure, and production of erythritol (Yang et al., 2014). Unlike glycerol produced by *Saccharomyces cerevisiae* is not stored in the cell but is secreted into the environment. Erythritol is a four-carbon polyols due to its sweet taste is used as a natural sweetener. This compound has many advantages, it does not affect the insulin level in blood and the highest no-effect dose for causing diarrhea among the polyols (Mirończuk et al., 2015). It was shown that in yeast *Y. lipolytica*, the production of erythritol occurs via the pentose phosphate pathway (Fig. 15.3) from glycerol or glucose (Mirończuk et al., 2017) and the final step, reduction of erythrose into erythritol, is catalyzed by erythritol reductase (Janek et al., 2017; Szczepańczyk et al., 2022; Szczepańczyk et al., 2023). To increase the production level of erythritol, genes potentially involved in its synthesis were overexpressed, genes encoding: transketolase (*TKL1*), transaldolase (*TAL1*), glucose-6-phosphate dehydrogenase (*ZWF1*), and 6-phosphogluconate dehydrogenase (*GND1*) (Mirończuk et al., 2017). The transketolase (*TKL1*) have been identified as a key gene, and its overexpression results in significant increasing of erythritol production by the yeast (Mirończuk et al., 2017; Carly et al., 2017). Moreover, the overexpression of TKL1 allows for efficient growth under dissolved oxygen (DO) limitation in the medium. In addition, to increase erythritol biosynthesis from glycerol, two genes involved in its metabolism were overexpressed, *GUT1* encoding glycerol kinase and *GUT2* encoding glycerol dehydrogenase (Mirończuk et al., 2016). In this process, the carbon flux were pushed toward enhanced glycerol metabolism and by modifying the fermentation conditions, the biosynthesis of the desired product was consequently enhanced. Strains with overexpression of *GUT1/GUT2* produced high quantities of erythritol or citric acid, depending on the media used (Mirończuk et al., 2017). Moreover, this strain was also capable to produce citric acid at low pH 3.0, which had not previously been observed in the wild types (Wojtatowicz & Rymowicz, 1991). Production on industrial scale causes some bottlenecks such as requirement for DO in medium or foam forming, what might cause a technical problems. In order to solve the problem, overexpression of all identified before genes have been done in the yeast using adopted CRISPR-Cas9-based system (Yang et al., 2024). Authors were able to produce 256 g/L erythritol from glucose in a 5 L bioreactor and a productivity of 1.78 g/L/h. Production of erythritol was also established at high temperature, by overexpressing RSP5 gene from *Saccharomyces cerevisia*e encoding ubiquitin ligase. The engineered strain HCY118 retained an efficient erythritol production capacity at 33°C (Wang et al., 2020).

Together with the high production of erythritol, *Y. lipolytica* can use this polyol as a carbon source, which is undesirable in industrial production. Some strains lacks this feature, such as *Y. lipolytica Wratislavia* K1 (Mirończuk et al., 2014; Rymowicz et al., 2009) but most of them can utilize it. Identification of the erythritol utilization factor (EUF1) (Rzechonek et al., 2017), a transcriptional factor responded for activation of the genes localized on chromosome F on the so-called the erythritol utilization cluster (Mirończuk et al., 2018), allowed for solving the issue of the erythritol utilization. Deletion of each the genes, localized in the cluster, results in impaired or lack of erythritol utilization (Carly et al., 2017, 2018; Mirończuk et al., 2018; Rzechonek et al., 2017). Deletion of *EYD1* (Carly et al., 2018) or *EYK1* (Carly et al., 2017) genes makes *Y. lipolytica* unable to utilize erythritol. Deletion of *EYI1*, which encodes an isomerase, results in impaired erythritol utilization (Mirończuk et al., 2018). Moreover, an overexpression of each gene involved in erythritol utilization cluster, results in enhanced erythritol utilization.

15.3.1.5 Heterologous protein production

The employing of *Y. lipolytica* as a host for heterologous expression was established in the 1980s thanks to the effective method of its transformation (Madzak, 2021). Since this time, more than 150 protein have been proved to be successfully produced by this yeast (Madzak, 2015). Since *Y. lipolytica* lacks a 1,3-mannosyltransferase it has a similar glycolization profile of the proteins to the humans cells, thus it might be used for humanized protein biosynthesis (Kim et al., 2015). As mentioned above, a wide range of the molecular tools allowed but also unique natural capabilities make this microorganism a convenient host. Many studies have been done to improve heterologous gene expression, such as creating hybrid and inducible promoters (Blazeck et al., 2011, 2013; Vidal et al., 2023), searching for novel secretory tags (Celinska et al., 2018), secretory helpers (Korpys-Wozniak et al., 2021) and recombinant protein expression (Yu et al., 2023).

Y. lipolytica was as a host for overexpression of cutinases from *Trichoderma reesei, Fusarium solani pisi,* and PETase from *Ideonella sakaiensis* (Kosiorowska et al., 2022, 2022) for successful PET biodegradation directly in the culture media. In *Y. lipolytica* the capsid protein of red-spotted grouper nervous necrosis virus (RGNNV-CP) was overexpressed demonstrating high potential for the *Y. lipolytica* expression system as a platform for recombinant vaccine production (Luu et al., 2017).

Other examples of employing this yeast as a host are described in the following section.

15.3.2 Synthetic biology and new metabolic pathways

Y. lipolytica emerged as a chassis for biosynthesis of many industrially important compounds. To date a large amount of compounds was successfully produced in this organism via genetic engineering of the new metabolic pathways, usually coupled with the improvement of the already existing pathways (Zhang et al., 2020, 2022). That includes lipid synthesis pathway, mevalonate pathway and Krebs cycle. This strategy increases the amount of precursor compounds required for the synthesis of novel, desired chemicals. The advantages of *Y. lipolytica* that made it a viable alternative of environmentally friendly cell factory can also be disadvantageous in some cases. As the yeast possess the capability to utilize a wide array of compounds, the intermediates of the planned pathways can be incorporated in the metabolic cycle. However, the ever developing genetic engineering tools allow for the more precise and complicated genetic manipulations while implementing the new metabolic pathways for sustainable production of value-added products (Zhang et al., 2020).

A group of compounds which can be produced by *Y. lipolytica* in large quantities, allowing for potential industrial application, are terpenoids. Those compounds have a wide variety of applications, from food industry to medicine, therefore the need to obtain an alternative source to produce terpenoids in a more sustainable way is a motivating factor to modify microorganisms to increase their production (Yang et al., 2016). It is especially important as terpenoids are currently derived from plants, which makes the process not viable in a long term. *Y. lipolytica* constitutes a good chassis for industrial production of terpenoids, as this yeast species produce an abundance of the precursor molecules, such as acetyl-CoA, malonyl-CoA, as well as cofactors in a form of NADPH, along other, already mentioned advantages of this yeast specie, like relatively cheap and easy cultivation and access to many genetic engineering tools optimized for this organism (Cao et al., 2016).

Synthesis of terpenoids is strongly regulated by the presence of their precursor, acetyl-CoA. However, this compound is also essential for the synthesis of FAs, which *Y. lipolytica* produces in high quantities. Therefore there is a need to modify the pathways of lipid synthesis, allowing for the shift of the carbon flux towards terpenoids instead of lipids. One of the strategies of improving terpenoid synthesis is the increase of β-oxidation, which is responsible for breaking down FAs, resulting in the increase of the amount of acetyl-CoA in the cells. The impact of β-oxidation on terpenoid synthesis was analyzed by Hanko et al. (2018) and Liu et al. (2020) and is one of the most common strategies implemented in the terpenoid production design in *Y. lipolytica* (Hanko et al., 2018; Liu et al., 2020).

Terpenoid synthesis in *Y. lipolytica* occurs in the cytosol in mevalonate (MVA) pathway in which the first step is a condensation of two acetyl-CoA molecules to produce acetoacetyl-CoA. Next steps are responsible for further modifications leading to other precursor molecules, such as geranyl diphosphate (GPP) or farnesyl diphosphate (FPP). The pathway can therefore be modified depending on the desirable final product of the synthesis. Terpenoids are then stored in the lipid compartment within the cell or transported to the extracellular matrix. It is necessary to take this into consideration, while designing the pathway for the product, since the transport of terpenoids is not yet fully understood and could result in a lower than desirable titer of the compound of interest.

The metabolic engineering of the MVP pathway to increase the production of terpenoids often involves upregulation of the 3-Hydroxy-3-methylglutaryl-CoA reductase (HMG), responsible for reduction of HMG-CoA to mevalonate. This strategy is a popular approach, with many reported improvements in the overall terpenoid production yields. The upregulation of HMG is often coupled with other modifications of the indigenous genes of MVP pathway of *Y. lipolytica*, especially two other genes, which were described as limiting in terms of the rate and regulation throughout the pathway: isopentenyl-diphosphate δ-isomerase (IDI) and hydroxy-methylglutaryl-CoA synthase (ERG). ERG is an important enzyme when it comes to the synthesis of triterpenoids, while IDI is responsible for maintaining the balance between isopentenyl pyrophosphate (IPP) and dimethylallyl pyrophosphate (DMAPP) (Yang et al., 2016).

The strategy of overexpression of HMG and ERG was used for example in creation of limonene synthesis pathway. Limonene is a monoterpene, which is used in food industry thanks to its flavoring properties and smell properties—limonene is responsible for the citrus smell in the fruits. It can be extracted from the orange peels (Park et al., 2024). As *Y. lipolytica* produces one of the precursors of limonene, it was a good host to carry out the research. The precoursors, isopentyl diphosphate (IPP) and its isomer dimethylallyl diphosphate (DMAPP) are derived from the MVA pathway. Both of those compounds can be converted to neryl diphosphate (NPP) by neryl diphosphate synthase 1 (NDS1). This gene was introduced *to the genome of Y. lipolytica* from *Solanum lycopersicum after the codon-optimization. The next step requires the presence of d-limonene synthase, which can be heterologously expressed from Agastache rugosa. Additional modifications in the pathways can further increase the yield of limonene synthesis, achieving 23.56 mg/L of limonene in the rich medium* (Cao et al., 2016). *Y. lipolytica can also act as a host for synthesis of more complex terpenoids, such as colorants:* β-carotene and lycopene.

High demand of β-carotene was the driving force behind creation of recombinant production in several microorganisms. Despite the lack of ability to produce β-carotene, *Y. lipolytica* remains one of the most promising hosts. In the base of this phenomenon lays the fact that both β-carotene and lipids, produced in large amounts in *Y. lipolytica*, share the same precursor-acetyl-CoA (Larroude et. al, 2018). Genetic engineering towards increased lipid production leads to the metabolic flux towards mevalonate pathway, which can later be altered to produce β-carotene. The strain introduced in 2017 by Larrude et al. *Y. lipolytica* W29 (ATCC20460) was used for further modifications, including addition of geranylgeranyl pyrophosphate synthase (GGS1) gene cassette, phytoene synthase/lycopene cyclase (CarRP) and phytoene dehydrogenase (CarB), each gene under its own promoter. The last modification resulted in addition of hydroxymethylglutaryl-CoA reductase (HMG1) gene cassette. The construction was carried out by Golden Gate Assembly (Larroude et al., 2018), obtaining 1.5 g/L of the product in the shake flask study on minimal medium containing 6% of glucose. β-carotene can also be produced from lycopene. To achieve that, a lycopene cyclase is necessary, as it does not naturally occur in *Y. lipolytica*. Genetic engineering, resulting in expression of CarRA, encoding phytoene cyclase/lycopene cyclase from *B. trispora* allowed for conversion of lycopene to β-carotene (Liu et al., 2021), with the total production reaching 1.7 g/L in minimal medium containing 20 g/L of glucose, as well as the synthesis of lycopene precoursor, phytoene. Further research by other groups focused on optimization and increase of β-carotene production as well as additional modifications to achieve production of other compounds (Jacobsen et al., 2020; Lu et al., 2020), such as astaxanthin.

On a large-scale astaxanthin is obtained by chemical synthesis; however, the impurity and presence of other chiral forms of the compound makes it useless for human consumption. An alternative for chemical synthesis is production by natural microalgae *Haematococcus pluvialis* (Oslan, 2021) but this process results in limited amounts of astaxanthin over a long period of time and large open areas are needed. To increase the production, noncarotenogenic organisms were selected as potential hosts of production. *Y. lipolytica* became an attractive target for modification due to its high natural production of carotenoids precursors. Early research by Kildegaard et al. (2017) focused on screening 14 different strains of *Y. lipolytica* to determine the best host for further modifications (Kildegaard et al., 2017). The criteria was genetic amenability as well as higher β-carotene titers. To achieve β-carotene production two genes from red yeast *Xanthophyllomyces dendrorhous* were expressed in *Y. lipolytica*: bi-functional phytoene synthase/lycopene cyclase (crtYB) and phytoene desaturase (crtI). To induce the production of astaxanthin in yeast, the isoprenoid pathway was altered. The main steps that increased the flux of the pathway was overexpression of 3-hydroxy-3-methyl-glutarylecoenzyme A reductase (HMG1) and heterologous expression of geranylgeranyl diphosphate synthase (crtE). crtE gene expressed in *Y. lipolytica*, similarly to crtI and crtYB, was obtained from *X. dendrorhous*. This modification increased the overall production of β-carotene and allowed for the next modifications in order to direct the products into astaxanthin production, which requires two oxidation reactions. Due to a lack of astaxanthin biosynthetic genes in *Y. lipolytica*, heterological expression of bacterial genes was required. crtW (β-carotene ketolase from *Paracoccus sp.* N81106) and crtZ (β-carotene hydroxylase from *P. ananatis*) genes were used to create the astaxanthin biosynthesis pathway. HPLC analysis allowed detection of astaxanthin in the engineered strains, however, many intermediates were observed as well. Further modifications in the number of copies of astaxanthin biosynthetic genes allowed to push the flux towards astaxanthin, significantly reducing the amount of intermediates. Multiple integrations of both genes allowed it to produce 54.6 mg/L (3.5 mg/g DCW) in 96 deep-well cultivation experiments (Kildegaard et al., 2017).

Alongside β-carotene and lycopene, other colorants were successfully produced by modified *Y. lipolytica* strains. A significant progress was made in synthesis of betalains, mainly betanin and isobetanin, which naturally occur in beetroot. The low content of those compounds in the biomass prompted the research to implement the microbial cell factories, as opposed to production in plants, to produce them more sustainably. As *Y. lipolytica* naturally does not possess the capabilities to produce betalains, the metabolic engineering was implemented. The strategy followed three essential steps: improving the precursor (L- thyrosine) availability, shifting the flux towards 4,5-seco-DOPA and reduction of betanin degradation by deletions of beta-glucosidases genes. Tyrosine hydroxylases and DOPA-4,5-extradiol dioxygenase from different organisms were screened for the most optimal combination (Thomsen et al., 2023).

Overexpression of tyrosine hydroxylase, which leads to higher production of 3,4-dihydroxy-L-phenylalanine (L-DOPA) was the first step towards production of betalains in *Y. lipolytica*. Its further oxidation results in 4,5-seco-DOPA synthesis, which in turn undergoes cyclization and condensation with betalamic acid to produce betanidin. Introduced betanidin glycosyltransferase (GT) allowed for attachment of a glycosyl group to betanidin, yielding betanin (betanidin-5-*O*-β-glucoside), a much more stable compound (Thomsen et al., 2023). As a result, of beta-glucosidases knockouts, the colorant is not degraded in the cells. The final yield of betanin reached 1,271 ± 141 mg/l in a fed-batch cultivation.

Colorants are certainly not the only compounds produced in *Y. lipolytica*. The increase in demand of pheromones and hormones, following the research about their novel applications, lead to the development of strategies to produce them more sustainably in yeast. The metabolic engineering allowed for synthesis of phytophormones, such as abscisic acid and giberelins.

The production of abscisic acid (ABA) is possible in a phatogenic fungus *Botrytis cinerea*, however its cultivation is difficult and the genetic engineering tools are not well established (Wang et al., 2024). *Y. lipolytica* is a perfect host for ABA synthesis, because as mentioned above, the production of Acetyl-CoA is high, and the strains with modified mevalonate pathway were already created and tested. Since production of ABA requires farnesyl diphosphate, produced in MVP, the pathway for ABA synthesis needs to be established from this point forward. The genes for abscisic acid synthesis were introduced from *Botrytis cinerea*. First α-ionylideneethane synthases and terpene synthase, which allows for synthesis of α-ionylideneethane. It is further modified to α-ionylideneacetic acid by cytochrome P450 (BcABA1p) and then oxygenated by a different cytochrome P450 (BcABA2p) to 1,4-*trans*-dihydroxy-α-ionylideneacetic acid. The dehydrogenase (BcABA4p) then allows for biosynthesis of abscisic acid. After further modification and optimization, which included overexpression of plant ABA transporter, the created strain was able to produce 263.5 mg/L of ABA (Arnesen et al., 2022), which is significant, yet still lower than the production by *B. cinerea*, which was reported at 2 g/L. The easier cultivation of *Y. lipolytica* and the potential for further improvement in the titer of ABA can be an important factor in possible change of the host organism for synthesis of this phytohormone.

Similarly, the group of plant hormones gibberelins was successfully produced in *Y. lipolytica*. The synthesis of those compounds is dependent on the presence of the precursor geranylgeranyl diphosphate, which is also produced in the MVP pathway. Additional genes, responsible for further conversion leading to final production of gibberelins, necessary for the creation of the pathway were expressed from *Fusarium fujikuroi* or *Arabidopsis thaliana*. And while the final titers of the gibberelins were not as high as for the industrially utilized fungus *Fusarium fujikuroi* (3.9 g/L G_3) as compared to 17.29 mg/L (G_4) for *Y. lipolytica*, the sheer capability to produce the compounds in an easier way is a promising for the further implementation of the strains on an industrial scale (Kildegaard et al., 2021).

Plant hormones are not the only industrially important active compounds derived from different organisms that are produced in *Y. lipolytica*. A strategy for pest control by mating disruption by sex pheromones arose as a viable option for crops protection. High cost of those compounds lead to the creation of *Y. lipolytica* strains able to produce sex pheromones of few months that are destructive to crop plants. The precursors of Z7–12:OH, Z9–12:OH, and Z7–14:OH, which after further modifications constitute a large percentage of the sex pheromones of the soybean looper (*Chrysodeixis includens*), the olive moth (*Prays olea*) and the grape berry moth *Paralobesia viteana* were successsfully produced in genetically engineered *Y.*

lipolytica strains. The process of pheromones production in *Y. lipolytica* is possible thanks to its ability to produce large quantities of lipids, which in turn can be further processed into FAs alcohols, via FAs desaturation, reduction, chain-shortening, alcohol acetylation, or oxidation (Petkevicius et al., 2022).

The strategy was implemented to produce (Z)-11-tetradecenol (Z11–14:OH), part of the complex group of pheromones of the European borer *Ostrinia nubilalis* (Petkevicius et al., 2021). The production of the precoursor for further synthesis of other compounds (C14; myristic acid) was improved in the parent strain by introduction of the mutation in the α-subunit of the FA synthase. Additional genes, which are required for the production of specific fatty alcohols, were expressed from different species of *Lepidoptera*. Those included fatty acyl-CoA desaturases (FAD) and fatty acyl-CoA reductases (FAR). Additional copies of the genes from *Lobesia botrana* and *Helicoverpa armigera,* respectively, lead to further improvement in the titer of Z11–14:OH (Petkevicius et al., 2021).

Excessive FA synthesis and known pathways in *Y. lipolytica* allows also for possible modifications of the FA synthesis profiles, depending on the demand. *Y. lipolytica* was used as a chassis for the synthesis of human milk fat substitute, which requires specific FA production and their composition resemblance to human milk fat. After modifications resulting in lipid higher production of triacylglycerols in the cells, including overexpressing acyl-CoA diacylglycerol acyltransferase (DGA2) and glycerol-3-phosphate dehydrogenase (GPD1), as well as disruption of β-oxidation, the strains were further modified with a gene from *Chlamydomonas reinhardtii.* The gene encoding lysophosphatidic acid acyltransferase allows for the modifications resulting in esterification of the palmitic acid, which does not naturally undergo this reaction in *Y. lipolytica* to form esterified to the middle position (sn-2) (Bhutada et al., 2022).

15.4 Commercialization of the results

Y. lipolytica is used as a chassis for synthesis of many compounds, from naturally produced lipids, polyols, and organic acids to dyes, bioactive compounds, and pheromones. Some of the research remains in the initial laboratory phase, where the production titers are not high enough to facilitate the industrial scale production, while other require optimization and more testing to replace the currently used method of production. However, there are already companies that offer products derived from *Y. lipolytica* that are on the markets worldwide. That was possible due to optimizations of the processes and approval by the federal organizations in different countries. One of the most important regulations regarding the usage of *Y. lipolytica* biomass for human consumption was announced in 2023, when the European Food Safety Authority (EFSA) concluded, that the yeast biomass is "safe under the proposed conditions of use" and is not nutritionally disadvantageous. That opened the door for the possible supplementation of food products with the *Y. lipolytica* biomass. The yeast biomass was, however, accepted for animal consumption as an addition to fodder, as the research demonstrated *Y. lipolytica* biomass possesses nutritional and probiotic properties in mammals, fish, birds, mollusks, and crustaceans (Guardiola et al., 2021).

The story of *Y. lipolytica* biomass as a source of SCPs from petroleum products dates back to the 1970s. British Petroleum (BP) created an industrial size plant that had an annual SCP production capabilities estimated to be up to 100,000 tons. *Y. lipolytica* biomass produced on the waste materials was aimed to be used as an addition to the animal feed. This plant, however, among others, was abandoned due to environmental concerns and high-input costs (Groenewald et al., 2014). This setback did not stop the entrepreneurs to take advantage of the capabilities of *Y. lipolytica*. Currently, the yeast biomass is produced in a Polish company Equinox, part of the Onesano S.A., however, the feedstock for the production changed and it is no longer petroleum products, as it was in the plant of BP. The portfolio of this group offers a wide range of specialized supplements for horses with the addition of biomass of *Y. lipolytica* strain A-101 (Wojtatowicz & Rymowicz, 1991) The biomass used in those supplements is obtained from a variety of different cultivation parameters, based on the desired profile of production, resulting in a different concentrations of microelements, amino acids, and lipids. The high content of B group vitamins is also an advantage. The company established the production line in 2011 and is capable to produce up to 600 tons of the yeast biomass supplements each year from glycerol. The brand expanded and now offers also supplements for other animals, including dogs and cats, advertising the positive impact of their product on digestion, immune system, and regeneration after the exertion of the animals.

Y. lipolytica is a natural producer of high quantities of FAs, therefore a lot of research was conducted to both boost the production and change the profile of produced FAs to obtain the desired product. DuPont de Nemorus Inc. (USA) focused on the production of eicosapentaenoic acid (EPA), which is an omega-3 FA. This ω-3 FA is naturally found in fish and marine plankton, but it is expected that the supply will not meet the demand, therefore genetic engineering tools were used to modify *Y. lipolytica* strains for production of this compound. The obtained strains, with the production of EPA exceeding 25% of the overall production profile were patented in the United States (patent number: US7932077B2). The modification of the strains included different desaturases, elongases and acyltransferases and the biomass rich in omega-3 FAs was also used as a fodder for sustainable production of Verlasso™ salmon. Despite the initial hype, as this product was not only more environmentally friendly than the extraction from fish oils but also could be used in the vegan and vegetarian diets, the final commercial product by DuPont, New Harvest™ EPA oil, appears to no longer be on the market. The data about the sales and production scale is also not available.

The capabilities of *Y. lipolytica* strains to produce various sweeteners was also used to obtain a commercially available product for food and beverage producers. Avansya, a joined venture of Evolva and Cargill, created a strain of *Y. lipolytica* producing large quantities of steviol glycosides, which could replace a current method of obtaining this sweetener from stevia leafs, making the process more sustainable. Eversweet™ is a product currently available in the United States (commercialized in 2019), Mexico, Canada, and other major markets, and soon can be present in the European Union, as the EFSA panel on Food Additive and Flavourings gave a positive opinion on the product in 2024, citing the lack of cells and DNA in the sweetener (Cargill, 2024).

Engineered strains of *Y. lipolytica* are also used to produce large quantities of insect pheromones. The start-up BioPhero (currently acquired by the FMC) established by Irina Borodina specializes in the production of those compounds, in hopes to use them to crop protection by disruption of insects mating. Due to the limited capabilities of production and the high cost, the pheromones are currently used on high-value plants, such as grapes, but the estimates suggest that the pheromones produced by modified *Y. lipolytica* could be used on up to 100 million hectares of crops, making it an important strategy for large-scale plants protection.

Commercialization of the results is a long and expensive process. Multiple regulations, especially involving the usage of genetically modified microorganisms, limits the abilities to push the products on the market. Positive decisions on the usage of *Y. lipolytica* biomass in human nutrition and the products of *Y. lipolytica* metabolism as food additives paves the way for more products for more commercially available products. That is important mostly due to the more frequent usage of this yeast as a chassis for the synthesis of different compounds, which have an industrial significance.

15.5 Conclusions

Sustainability requires finding new solutions to existing, well-established industrial processes. The production of biodiesel, the biodegradation of plastics, and the synthesis of new pharmaceuticals and chemicals could be implemented by microbial cell factories instead of traditional industry. The bottleneck in these processes is not only finding a suitable host that will allow for efficient, ecological but also economical production of the desired product. *Y. llipolytica* with all unique features is a promising, still underestimated microorganism, that might meet the expectations. By employing *Y. lipolytica* for the production of FAs, which after the esterification process can be used as a substrate for biodiesel production, we are protecting the natural environment and finding a solution to the depletion of fossil fuels. By using waste, renewable feedstocks, to synthetize a value-added bioproducts, we contribute to economical production and sustainable development. For this reason, *Y. lipolytica* should be investigated further, as expanding our knowledge will provide insights into further possibilities.

References

Al-Feel, W., Chirala, S. S., & Wakil, S. J. (1992). Cloning of the yeast FAS3 gene and primary structure of yeast acetyl-CoA carboxylase. *Proceedings of the National Academy of Sciences of the United States of America, 89*(10), 4534–4538.

Andreishcheva, E. N., et al. (1999). Adaptation to salt stress in a salt-tolerant strain of the yeast Yarrowia lipolytica. *Biochemistry (Mosc), 64*(9), 1061–1067.

Arnesen, J. A., et al. (2022). Production of abscisic acid in the oleaginous yeast Yarrowia lipolytica. *FEMS Yeast Research, 22*(1).

Athenstaedt, K. (2011). YALI0E32769g (DGA1) and YALI0E16797g (LRO1) encode major triacylglycerol synthases of the oleaginous yeast Yarrowia lipolytica. *Biochimica et Biophysica Acta, 1811*(10), 587–596.

Athenstaedt, K., et al. (2006). Lipid particle composition of the yeast Yarrowia lipolytica depends on the carbon source. *Proteomics, 6*(5), 1450–1459.

Bankar, A., et al. (2018). Morphological response of Yarrowia lipolytica under stress of heavy metals. *Canadian Journal of Microbiology, 64*(8), 559–566.

Bankar, A. V., Kumar, A. R., & Zinjarde, S. S. (2009). Environmental and industrial applications of Yarrowia lipolytica. *Applied Microbiology and Biotechnology, 84*(5), 847–865.

Barth, G., & Gaillardin, C. (1996). Yarrowia lipolytica. In K. Wolf (Ed.). *Nonconventional yeasts in biotechnology: A handbook* (pp. 313–388). Springer.

Beopoulos, A., et al. (2008). Control of lipid accumulation in the yeast Yarrowia lipolytica. *Applied and Environmental Microbiology, 74*(24), 7779–7789.

Beopoulos, A., Chardot, T., & Nicaud, J. M. (2009). Yarrowia lipolytica: A model and a tool to understand the mechanisms implicated in lipid accumulation. *Biochimie, 91*(6), 692–696.

Beopoulos, A., et al. (2012). Identification and characterization of DGA2, an acyltransferase of the DGAT1 acyl-CoA:diacylglycerol acyltransferase family in the oleaginous yeast Yarrowia lipolytica. New insights into the storage lipid metabolism of oleaginous yeasts. *Applied Microbiology and Biotechnology, 93*(4), 1523–1537.

Bhutada, G., et al. (2022). Production of human milk fat substitute by engineered strains of Yarrowia lipolytica. *Metabolic Engineering Communications, 14*, e00192.

Blazeck, J., et al. (2011). Tuning gene expression in Yarrowia lipolytica by a hybrid promoter approach. *Applied and Environmental Microbiology, 77*(22), 7905–7914.

Blazeck, J., et al. (2013). Generalizing a hybrid synthetic promoter approach in Yarrowia lipolytica. *Applied Microbiology and Biotechnology, 97*(7), 3037–3052.

Blazeck, J., et al. (2014). Harnessing Yarrowia lipolytica lipogenesis to create a platform for lipid and biofuel production. *Nature Communications, 5*, 3131.

Cao, X., et al. (2016). Metabolic engineering of oleaginous yeast Yarrowia lipolytica for limonene overproduction. *Biotechnology for Biofuels, 9*, 214.

Cargill. (2024). How is EverSweet® stevia sweetener made.

Carly, F., et al. (2017). Enhancing erythritol productivity in Yarrowia lipolytica using metabolic engineering. *Metabolic Engineering, 42*, 19–24.

Carly, F., et al. (2017). Identification and characterization of EYK1, a key gene for erythritol catabolism in Yarrowia lipolytica. *Applied Microbiology and Biotechnology, 101*(17), 6587–6596.

Carly, F., et al. (2018). Identification and characterization of EYD1, encoding an erythritol dehydrogenase in Yarrowia lipolytica and its application to bioconvert erythritol into erythrulose. *Bioresource Technology, 247*(Supplement C), 963–969.

Casaregola, S., et al. (2000). Genomic exploration of the hemiascomycetous yeasts: 17. Yarrowia lipolytica. *FEBS Letters, 487*(1), 95–100.

Celinska, E. (2022). "Fight-flight-or-freeze" - how Yarrowia lipolytica responds to stress at molecular level? *Applied Microbiology and Biotechnology, 106*(9-10), 3369–3395.

Celinska, E., et al. (2018). Robust signal peptides for protein secretion in Yarrowia lipolytica: identification and characterization of novel secretory tags. *Applied Microbiology and Biotechnology, 102*(12), 5221–5233.

Cordova, L. T., Palmer, C. M., & Alper, H. S. (2022). Shifting the distribution: modulation of the lipid profile in Yarrowia lipolytica via iron content. *Applied Microbiology and Biotechnology, 106*(4), 1571–1581.

Cybulski, K., et al. (2019). Production of pyruvic acid from glycerol by Yarrowia lipolytica. *Folia Microbiol (Praha), 64*(6), 809–820.

Daskalaki, A., et al. (2019). Laboratory evolution strategies for improving lipid accumulation in Yarrowia lipolytica. *Applied Microbiology and Biotechnology, 103*(20), 8585–8596.

Dobrowolski, A., & Mirończuk, A. M. (2020). The influence of transketolase on lipid biosynthesis in the yeast Yarrowia lipolytica. *Microbial Cell Factories, 19*(1), 138.

Dobrowolski, A., et al. (2016). Efficient conversion of crude glycerol from various industrial wastes into single cell oil by yeast Yarrowia lipolytica. *Bioresource Technology, 207*, 237–243.

Dobrowolski, A., et al. (2019). Lipid Production From Waste Materials in Seawater-Based Medium by the Yeast Yarrowia lipolytica. *Frontiers in Microbiology, 10*, 547.

Dobrowolski, A., et al. (2020). Production of tailor-made fatty acids from crude glycerol at low pH by Yarrowia lipolytica. *Bioresource Technology, 314*, 123746.

Dobrowolski, A., Nawijn, W., & Mirończuk, A. M. (2022). Brown seaweed hydrolysate as a promising growth substrate for biomass and lipid synthesis of the yeast yarrowia lipolytica. *Frontiers in Bioengineering and Biotechnology, 10*, 944228.

Drzymala-Kapinos, K., Mirończuk, A. M., & Dobrowolski, A. (2022). Lipid production from lignocellulosic biomass using an engineered Yarrowia lipolytica strain. *Microbial Cell Factories, 21*(1), 226.

Dujon, B., et al. (2004). Genome evolution in yeasts. *Nature, 430*(6995), 35–44.

Dulermo, T., & Nicaud, J. M. (2011). Involvement of the G3P shuttle and beta-oxidation pathway in the control of TAG synthesis and lipid accumulation in Yarrowia lipolytica. *Metabolic Engineering, 13*(5), 482–491.

Efsa Panel on Nutrition, N. F., et al. (2020). Safety of chromium-enriched biomass of Yarrowia lipolytica as a novel food pursuant to Regulation (EU) 2015/2283. *EFSA J, 18*(3), e06005.

Egermeier, M., et al. (2017). Metabolic Flexibility of Yarrowia lipolytica Growing on Glycerol. *Frontiers in Microbiology, 8*, 49.

T, F., et al. (2016). *Uptake and Assimilation of Hydrophobic Substrates by the Oleaginous Yeast Yarrowia lipolytica.*

Fickers, P.,, et al. (2005). Identification and characterisation of LIP7 and LIP8 genes encoding two extracellular triacylglycerol lipases in the yeast Yarrowia lipolytica. *Fungal Genetics and Biology, 42*(3), 264–274.

Fickers, P., Marty, A., & Nicaud, J. M. (2011). The lipases from Yarrowia lipolytica: genetics, production, regulation, biochemical characterization and biotechnological applications. *Biotechnology Advances, 29*(6), 632–644.

Forster, A., et al. (2007). Overexpression of the ICL1 gene changes the product ratio of citric acid production by Yarrowia lipolytica. *Applied Microbiology and Biotechnology, 77*(4), 861–869.

Fukuda, R. (2013). Metabolism of hydrophobic carbon sources and regulation of it in n-alkane-assimilating yeast Yarrowia lipolytica. *Bioscience, Biotechnology, and Biochemistry, 77*(6), 1149–1154.

Fukuda, R. (2023). Utilization of n-alkane and roles of lipid transfer proteins in Yarrowia lipolytica. *World Journal of Microbiology and Biotechnology, 39*(4), 97.

Gao, R., et al. (2017). Oleaginous yeast Yarrowia lipolytica culture with synthetic and food waste-derived volatile fatty acids for lipid production. *Biotechnology for Biofuels, 10*, 247.

Gemperlein, K., et al. (2019). Polyunsaturated fatty acid production by Yarrowia lipolytica employing designed myxobacterial PUFA synthases. *Nature Communications, 10*(1), 4055.

Groenewald, M., et al. (2014). Yarrowia lipolytica: safety assessment of an oleaginous yeast with a great industrial potential. *Critical Reviews in Microbiology, 40*(3), 187–206.

Guardiola, F. A., Esteban, M. A., & Angulo, C. (2021). Yarrowia lipolytica, health benefits for animals. *Applied Microbiology and Biotechnology, 105*(20), 7577–7592.

Guo, H., et al. (2016). Applying pathway engineering to enhance production of alpha-ketoglutarate in Yarrowia lipolytica. *Applied Microbiology and Biotechnology, 100*(23), 9875–9884.

Hamimed, S., et al. (2021). High-performance biological treatment of tuna wash processing wastewater using Yarrowia lipolytica. *Environmental Science and Pollution Research International, 28*(2), 1545–1554.

Hamimed, S., et al. (2022). Bio-nanocrystallization of NaCl using saline wastewaters through biological treatment by Yarrowia lipolytica. *Environmental Technology & Innovation, 26*, 102338.

Hanko, E. K. R., et al. (2018). Engineering beta-oxidation in Yarrowia lipolytica for methyl ketone production. *Metabolic Engineering, 48*, 52–62.

J, B.-S., et al. (2024). Isolation and Characterization of Yarrowia lipolytica YQ22 from Diesel Samples for Phenol Biodegradation. *Environmental Research, Engineering and Management, 80*(1), 32–38.

Jach, M. E., & Malm, A. (2022). Yarrowia lipolytica as an Alternative and Valuable Source of Nutritional and Bioactive Compounds for Humans. *Molecules, 27*(7).

Jacobsen, I. H., Ledesma-Amaro, R., & Martinez, J. L. (2020). Recombinant beta-Carotene Production by Yarrowia lipolytica - Assessing the Potential of Micro-Scale Fermentation Analysis in Cell Factory Design and Bioreaction Optimization. *Frontiers in Bioengineering and Biotechnology, 8*, 29.

Janek, T., et al. (2017). Characterization of erythrose reductase from Yarrowia lipolytica and its influence on erythritol synthesis. *Microbial Cell Factories, 16*(1), 118.

Jenni, S., et al. (2007). Structure of fungal fatty acid synthase and implications for iterative substrate shuttling. *Science, 316*(5822), 254–261.

Ji, Q., et al. (2020). Improving the homologous recombination efficiency of Yarrowia lipolytica by grafting heterologous component from Saccharomyces cerevisiae. *Metabolic Engineering Communications, 11*, e00152.

Juszczyk, P., et al. (2013). Biomass production by novel strains of Yarrowia lipolytica using raw glycerol, derived from biodiesel production. *Bioresource Technology, 137*, 124–131.

Kamzolova, S. V. (2023). A Review on Citric Acid Production by Yarrowia lipolytica Yeast: Past and Present Challenges and Developments. *Processes, 11*(12), 3435.

Kamzolova, S. V., & Morgunov, I. G. (2013). alpha-Ketoglutaric acid production from rapeseed oil by Yarrowia lipolytica yeast. *Applied Microbiology and Biotechnology, 97*(12), 5517–5525.

Kamzolova, S. V., & Morgunov, I. G. (2018). Biosynthesis of pyruvic acid from glycerol-containing substrates and its regulation in the yeast Yarrowia lipolytica. *Bioresource Technology, 266*, 125–133.

Kennedy, E. P. (1962). The biosynthesis of phosphatides and triglycerides. *Deutsche Medizinische Wochenschrift, 87*, 99–102.

Kerkhoven, E. J., et al. (2017). Leucine Biosynthesis Is Involved in Regulating High Lipid Accumulation in Yarrowia lipolytica. *mBio, 8*(3).

Kildegaard, K. R., et al. (2017). Engineering of Yarrowia lipolytica for production of astaxanthin. *Synthetic and Systems Biotechnology, 2*(4), 287–294.

Kildegaard, K. R., et al. (2021). Tailored biosynthesis of gibberellin plant hormones in yeast. *Metabolic Engineering, 66*, 1–11.

Kim, H., Yoo, S. J., & Kang, H. A. (2015). Yeast synthetic biology for the production of recombinant therapeutic proteins. *FEMS Yeast Research, 15*(1), 1–16.

Kolhe, N., et al. (2022). A comprehensive assessment of Yarrowia lipolytica and its interactions with metals: Current updates and future prospective. *Biotechnology Advances, 59*, 107967.

Konzock, O., & Norbeck, J. (2020). Deletion of MHY1 abolishes hyphae formation in Yarrowia lipolytica without negative effects on stress tolerance. *PLoS ONE, 15*(4), e0231161.

Konzock, O., Zaghen, S., & Norbeck, J. (2021). Tolerance of Yarrowia lipolytica to inhibitors commonly found in lignocellulosic hydrolysates. *BMC Microbiology, 21*(1), 77.

Konzock, O., et al. (2022). Altering the fatty acid profile of Yarrowia lipolytica to mimic cocoa butter by genetic engineering of desaturases. *Microbial Cell Factories, 21*(1), 25.

Korpys-Wozniak, P., Kubiak, P., & Celinska, E. (2021). Secretory helpers for enhanced production of heterologous proteins in Yarrowia lipolytica. *Biotechnology Reports (Amst), 32*, e00669.

Kosiorowska, K. E., et al. (2022). Metabolic engineering of Yarrowia lipolytica for poly (ethylene terephthalate) degradation. *The Science of the Total Environment, 831*, 154841.

Kosiorowska, K. E., et al. (2022). Production of PETase by engineered Yarrowia lipolytica for efficient poly(ethylene terephthalate) biodegradation. *The Science of the Total Environment, 846*, 157358.

Koutsoumanis, K., et al. (2022). Update of the list of QPS-recommended biological agents intentionally added to food or feed as notified to EFSA 15: suitability of taxonomic units notified to EFSA until September 2021. *EFSA J, 20*(1), e07045.

Kretzschmar, A., et al. (2013). Increased homologous integration frequency in Yarrowia lipolytica strains defective in non-homologous end-joining. *Current Genetics, 59*(1-2), 63–72.

Larroude, M., et al. (2018). Synthetic biology tools for engineering Yarrowia lipolytica. *Biotechnology Advances, 36*(8), 2150–2164.

Larroude, M., et al. (2018). A synthetic biology approach to transform Yarrowia lipolytica into a competitive biotechnological producer of beta-carotene. *Biotechnology and Bioengineering, 115*(2), 464–472.

Lazar, Z., et al. (2015). Awakening the endogenous Leloir pathway for efficient galactose utilization by Yarrowia lipolytica. *Biotechnology for Biofuels, 8*, 185.

Lin, L., et al. (2024). Harnessing oleaginous yeast to produce omega fatty acids. *Trends in Biotechnology, 42*(11), 1335–1338.

Liu, G. S., et al. (2020). The yeast peroxisome: A dynamic storage depot and subcellular factory for squalene overproduction. *Metabolic Engineering, 57*, 151–161.

Liu, L., et al. (2021). Elevated beta-Carotene Production Using Codon-Adapted CarRA&B and Metabolic Balance in Engineered Yarrowia lipolytica. *Frontiers in Microbiology, 12*, 627150.

Lu, Y., et al. (2020). A modular pathway engineering strategy for the high-level production of beta-ionone in Yarrowia lipolytica. *Microbial Cell Factories, 19*(1), 49.

Lubuta, P., et al. (2019). Investigating the Influence of Glycerol on the Utilization of Glucose in Yarrowia lipolytica Using RNA-Seq-Based Transcriptomics. *G3 (Bethesda), 9*(12), 4059–4071.

Luu, V. T., et al. (2017). Development of recombinant Yarrowia lipolytica producing virus-like particles of a fish nervous necrosis virus. *Journal of Microbiology, 55*(8), 655–664.

Madzak, C. (2015). Yarrowia lipolytica: recent achievements in heterologous protein expression and pathway engineering. *Applied Microbiology and Biotechnology, 99*(11), 4559–4577.

Madzak, C. (2021). Yarrowia lipolytica Strains and Their Biotechnological Applications: How Natural Biodiversity and Metabolic Engineering Could Contribute to Cell Factories Improvement. *J Fungi (Basel), 7*(7).

Madzak, C., et al. (2023). Study of the persistence and dynamics of recombinant mCherry-producing Yarrowia lipolytica strains in the mouse intestine using fluorescence imaging. *Microbial Biotechnology, 16*(3), 618–631.

Meunchan, M., et al. (2015). Comprehensive Analysis of a Yeast Lipase Family in the Yarrowia Clade. *PLoS ONE, 10*(11), e0143096.

Michely, S., et al. (2013). Comparative physiology of oleaginous species from the Yarrowia clade. *PLoS ONE, 8*(5), e63356.

Mirończuk, A. M., et al. (2014). Enhanced production of erythritol by Yarrowia lipolytica on glycerol in repeated batch cultures. *Journal of Industrial Microbiology and Biotechnology, 41*(1), 57–64.

Mirończuk, A. M., et al. (2018). A Role of a Newly Identified Isomerase From Yarrowia lipolytica in Erythritol Catabolism. *Frontiers in Microbiology, 9*, 1122.

Mirończuk, A. M., et al. (2015). Newly isolated mutant of Yarrowia lipolytica MK1 as a proper host for efficient erythritol biosynthesis from glycerol. *Process Biochemistry, 50*(1), 61–68.

Mirończuk, A. M., et al. (2016). A novel strain of Yarrowia lipolytica as a platform for value-added product synthesis from glycerol. *Biotechnology for Biofuels, 9*(1), 1–12.

Mirończuk, A. M., Biegalska, A., & Dobrowolski, A. (2017). Functional overexpression of genes involved in erythritol synthesis in the yeast Yarrowia lipolytica. *Biotechnol Biofuels, 10*, 77.

Morgunov, I. G., Kamzolova, S. V., & Samoilenko, V. A. (2013). Enhanced alpha-ketoglutaric acid production and recovery in Yarrowia lipolytica yeast by effective pH controlling. *Applied Microbiology and Biotechnology, 97*(19), 8711–8718.

Nicaud, J. M., Fabre, E., & Gaillardin, C. (1989). Expression of invertase activity in Yarrowia lipolytica and its use as a selective marker. *Current Genetics, 16*(4), 253–260.

Oslan, S. N. H., et al. (2021). Haematococcus pluvialis as a Potential Source of Astaxanthin with Diverse Applications in Industrial Sectors: Current Research and Future Directions. *Molecules, 26*(21).

Papanikolaou, S., & Aggelis, G. (2002). Lipid production by Yarrowia lipolytica growing on industrial glycerol in a single-stage continuous culture. *Bioresource Technology, 82*(1), 43–49.

Papanikolaou, S., & Aggelis, G. (2003). Modeling lipid accumulation and degradation in Yarrowia lipolytica cultivated on industrial fats. *Current Microbiology, 46*(6), 398–402.

Papanikolaou, S., & Aggelis, G. (2010). Yarrowia lipolytica: A model microorganism used for the production of tailor-made lipids. *European Journal of Lipid Science and Technology, 112*(6), 639–654.

Papanikolaou, S., et al. (2001). Kinetic profile of the cellular lipid composition in an oleaginous Yarrowia lipolytica capable of producing a cocoa-butter substitute from industrial fats. *Antonie van Leeuwenhoek, 80*(3-4), 215–224.

Papanikolaou, S., et al. (2002). Single cell oil production by Yarrowia lipolytica growing on an industrial derivative of animal fat in batch cultures. *Applied Microbiology and Biotechnology, 58*(3), 308–312.

Papanikolaou, S., et al. (2003). Accumulation of a cocoa-butter-like lipid by Yarrowia lipolytica cultivated on agro-industrial residues. *Current Microbiology, 46*(2), 124–130.

Papanikolaou, S., et al. (2006). Influence of glucose and saturated free-fatty acid mixtures on citric acid and lipid production by Yarrowia lipolytica. *Current Microbiology, 52*(2), 134–142.

Papanikolaou, S., et al. (2008). Citric acid production by Yarrowia lipolytica cultivated on olive-mill wastewater-based media. *Bioresource Technology, 99*(7), 2419–2428.

Papanikolaou, S., et al. (2013). Importance of the methyl-citrate cycle on glycerol metabolism in the yeast Yarrowia lipolytica. *Journal of Biotechnology, 168*(4), 303–314.

Park, Y. K., et al. (2024). Efficient synthesis of limonene production in Yarrowia lipolytica by combinatorial engineering strategies. *Biotechnol Biofuels Bioprod, 17*(1), 94.

Petkevicius, K., et al. (2021). Biotechnological production of the European corn borer sex pheromone in the yeast Yarrowia lipolytica. *Biotechnology Journal, 16*(6), e2100004.

Petkevicius, K., et al. (2022). Biosynthesis of insect sex pheromone precursors via engineered beta-oxidation in yeast. *FEMS Yeast Research, 22*(1).

Qiao, K., et al. (2015). Engineering lipid overproduction in the oleaginous yeast Yarrowia lipolytica. *Metabolic Engineering, 29*, 56–65.

Robles-Iglesias, R., et al. (2023). Integrated fermentative process for lipid and beta-carotene production from acetogenic syngas fermentation using an engineered oleaginous Yarrowia lipolytica yeast. *Bioresource Technology, 389*, 29815.

Ruiz-Herrera, J., & Sentandreu, R. (2002). Different effectors of dimorphism in Yarrowia lipolytica. *Archives of Microbiology, 178*(6), 477–483.

Rymowicz, W., Rywinska, A., & Marcinkiewicz, M. (2009). High-yield production of erythritol from raw glycerol in fed-batch cultures of Yarrowia lipolytica. *Biotechnology Letters, 31*(3), 377–380.

Rzechonek, D. A., et al. (2017). EUF1 - a newly identified gene involved in erythritol utilization in Yarrowia lipolytica. *Scientific Reports, 7*(1), 12507.

Rzechonek, D. A., et al. (2018). Recent advances in biological production of erythritol. *Critical Reviews in Biotechnology, 38*(4), 620–633.

Rzechonek, D. A., et al. (2019). Aseptic production of citric and isocitric acid from crude glycerol by genetically modified Yarrowia lipolytica. *Bioresource Technology, 271*, 340–344.

Rzechonek, D. A., Szczepanczyk, M., & Mironczuk, A. M. (2023). Mutation in yl-HOG1 represses the filament-to-yeast transition in the dimorphic yeast Yarrowia lipolytica. *Microbial Cell Factories, 22*(1), 155.

Sarris, D., et al. (2017). Production of added-value metabolites by Yarrowia lipolytica growing in olive mill wastewater-based media under aseptic and non-aseptic conditions. *Engineering in Life Sciences, 17*(6), 695–709.

Sarris, D., et al. (2019). Production of Added-Value Chemical Compounds through Bioconversions of Olive-Mill Wastewaters Blended with Crude Glycerol by a Yarrowia lipolytica Strain. *Molecules, 24*(2).

Sekova, V. Y., et al. (2021). Proteomics Readjustment of the Yarrowia lipolytica Yeast in Response to Increased Temperature and Alkaline Stress. *Microorganisms, 9*(12).

Shi, T. Q., et al. (2018). Advancing metabolic engineering of Yarrowia lipolytica using the CRISPR/Cas system. *Applied Microbiology and Biotechnology, 102*(22), 9541–9548.

Soong, Y. V., et al. (2023). Using oils and fats to replace sugars as feedstocks for biomanufacturing: Challenges and opportunities for the yeast Yarrowia lipolytica. *Biotechnology Advances, 65*, 108128.

Spagnuolo, M., et al. (2018). Alternative Substrate Metabolism in Yarrowia lipolytica. *Frontiers in Microbiology, 9*, 1077.

Szczepanczyk, M., et al. (2022). The role of YALI0B07117g and other erythrose reductse homologs in yeast Yarrowia lipolytica. In *FEBS OPEN BIO*. 111 RIVER ST, HOBOKEN 07030-5774, NJ USA: WILEY.

Szczepańczyk, M., et al. (2023). In-depth analysis of erythrose reductase homologs in Yarrowia lipolytica. *Scientific Reports, 13*(1), 9129.

Tai, M., & Stephanopoulos, G. (2013). Engineering the push and pull of lipid biosynthesis in oleaginous yeast Yarrowia lipolytica for biofuel production. *Metabolic Engineering, 15*, 1–9.

Tang, Y. X., et al. (2024). Engineering Yarrowia lipolytica for sustainable Cis-13, 16-docosadienoic acid production. *Bioresource Technology, 406*, 130978.

Thomsen, P. T., et al. (2023). Beet red food colourant can be produced more sustainably with engineered Yarrowia lipolytica. *Nature Microbiology, 8*(12), 2290–2303.

Thushari, I., & Babel, S. (2022). Comparative study of the environmental impacts of used cooking oil valorization options in Thailand. *Journal of Environmental Management, 310*, 114810.

Verbeke, J., Beopoulos, A., & Nicaud, J. M. (2013). Efficient homologous recombination with short length flanking fragments in Ku70 deficient Yarrowia lipolytica strains. *Biotechnology Letters, 35*(4), 571–576.

Vidal, L., et al. (2023). Bidirectional hybrid erythritol-inducible promoter for synthetic biology in Yarrowia lipolytica. *Microbial Cell Factories, 22*(1), 7.

Walker, C., et al. (2018). Draft Genome Assemblies of Five Robust Yarrowia lipolytica Strains Exhibiting High Lipid Production, Pentose Sugar Utilization, and Sugar Alcohol Secretion from Undetoxified Lignocellulosic Biomass Hydrolysates. *Microbiology Resource Announcements, 7*(12).

van der Walt, J. P., & von Arx, J. A. (1980). The yeast genus Yarrowia gen. nov. *Antonie van Leeuwenhoek, 46*(6), 517–521.

Wang, N., et al. (2020). Metabolic engineering of Yarrowia lipolytica for thermoresistance and enhanced erythritol productivity. *Biotechnology for Biofuels, 13*, 176.

Wang, Y., et al. (2024). Engineering strategies for enhanced 1', 4'-trans-ABA diol production by Botrytis cinerea. *Microbial Cell Factories, 23*(1), 185.

Wasylenko, T. M., Ahn, W. S., & Stephanopoulos, G. (2015). The oxidative pentose phosphate pathway is the primary source of NADPH for lipid overproduction from glucose in Yarrowia lipolytica. *Metabolic Engineering, 30*, 27–39.

Wojtatowicz, M., & Rymowicz, W. (1991). *Effect of inoculum on kinetics and yield of citric acids production on glucose by Yarrowia lipolytica A-101, 41*(2), 137–143.

Xue, Z., et al. (2013). Production of omega-3 eicosapentaenoic acid by metabolic engineering of Yarrowia lipolytica. *Nature Biotechnology, 31*(8), 734–740.

Yang, L. B., et al. (2014). A novel osmotic pressure control fed-batch fermentation strategy for improvement of erythritol production by Yarrowia lipolytica from glycerol. *Bioresource Technology, 151*, 120–127.

Yang, R., et al. (2022). Two-stage fermentation enhanced single-cell protein production by Yarrowia lipolytica from food waste. *Bioresource Technology, 361*, 127677.

Yang, S., et al. (2024). Systematic metabolic engineering of Yarrowia lipolytica for the enhanced production of erythritol. *Bioresource Technology, 391*(Pt A), 129918.

Yang, X., et al. (2016). Heterologous production of alpha-farnesene in metabolically engineered strains of Yarrowia lipolytica. *Bioresource Technology, 216*, 1040–1048.

Yu, S., et al. (2023). Construction and testing of Yarrowia lipolytica recombinant protein expression chassis cells based on the high-throughput screening and secretome. *Microbial Cell Factories, 22*(1), 185.

Yuzbasheva, E. Y., et al. (2019). The mitochondrial citrate carrier in Yarrowia lipolytica: Its identification, characterization and functional significance for the production of citric acid. *Metabolic Engineering, 54*, 264–274.

Zainuddin, M. F., et al. (2022). Production of single cell oil by Yarrowia lipolytica JCM 2320 using detoxified desiccated coconut residue hydrolysate. *PeerJ, 10*, e12833.

Zhang, S., et al. (2019). pH selectively regulates citric acid and lipid production in Yarrowia lipolytica W29 during nitrogen-limited growth on glucose. *Journal of Biotechnology, 290*, 10–15.

Zhang, X. K., et al. (2020). Metabolic engineering of beta-carotene biosynthesis in Yarrowia lipolytica. *Biotechnology Letters, 42*(6), 945–956.

Zhang, Y., et al. (2022). Engineering thermotolerant Yarrowia lipolytica for sustainable biosynthesis of mannitol and fructooligosaccharides. *Biochemical Engineering Journal, 187*, 108604.

Zhou, L., et al. (2021). Combined adaptive evolution and transcriptomic profiles reveal aromatic aldehydes tolerance mechanisms in Yarrowia lipolytica. *Bioresource Technology, 329*, 124910.

Zinjarde, S., et al. (2014). Yarrowia lipolytica and pollutants: Interactions and applications. *Biotechnology Advances, 32*(5), 920–933.

Section 3

Up-scaling, industrial application and process assessments

Commercial applications for yeast-based technologies

16

Mustafa Türker, Ümit Cem Derman, and Filiz Alemdar
Pak Gıda Üretim ve Pazarlama A.Ş., Kocaeli, Turkey

16.1 Introduction

Fermentation is a natural process that has been used for thousands of years by humans to produce a variety of foods and beverages. Egyptian and Mesopotamian civilizations leavened dough and fermented beer and wine before spreading to China, Europe, and India. The earliest known records of yeast risen bread come from Ancient Egypt in 1300–1500 BC and China in 500–300 BC (Lahue et al., 2020). Thus, yeasts, particularly those of the genus *Saccharomyces*, have been exploited by humans for millennia. Until the 19th century, yeasts were used without knowledge of their biological nature. The nature of fermentation, the degradation of sugar into alcohol, and the role of oxygen during fermentation were only understood in the nineteenth century. Since then, there has been an exponential increase in our understanding of yeasts and their biology, leading to our current understanding of nutrition, metabolism, molecular genetics, and cell biology (Barnett & Barnett, 2011; Money, 2018).

Twentieth-century developments led to a modern yeast manufacturing industry (Frey, 1930). The uses of yeasts have been expanded to the production of biofuels, food and feed supplements, probiotics, nutrients for fermentations, vitamins, enzymes, biochemical precursors, and applications as cell factories for manufacturing chemicals, pharmaceuticals, and vaccines (Türker, 2014). *Saccharomyces cerevisiae* and certain other yeast species are arguably the most important group of biotechnological organisms. Yeasts are the major producer of biotechnology products worldwide, exceeding the production, capacity, and economic revenues, of any other group of industrial microorganisms (Attfield, 2023).

Yeasts are a heterogeneous group of single-celled fungi recognized for their crucial contributions to various industries, particularly in food and beverage manufacturing. Currently, there are approximately 2000 recognized yeast species listed on theyeasts.org and the estimated total number is around 200,000 (Boekhout et al., 2022) and more than 30 yeast species have reached the industrial scale. These organisms play a crucial role in fermentation, by transforming sugars into alcohol and carbon dioxide. The investigation and application of different yeast species are continuously creating new opportunities in sustainable production methods, making significant contributions to the progress of food, pharmaceutical, and environmental applications (Boekhout et al., 2022; Tullio, 2022). Yeasts are an important source of enzymes for traditional markets, including food, feed, and agriculture, as well as specific biocatalysts, which are increasingly used in pharmaceutical and fine chemical

Eukaryotic Microorganisms as Sources of Bioproducts. DOI: https://doi.org/10.1016/B978-0-443-30188-9.00011-1

synthesis. Enhanced utilization of yeasts as sources of native and heterologous enzymes, proteins, and lipids in the pharmaceutical, chemical, and food industries are leading to new products such as chiral intermediates for the production of pharmaceuticals, fine chemicals, alternative food proteins, and microbial oils (Attfield, 2023; De Brabander et al., 2023; Knychala et al., 2024).

Among other yeast species, industrial-scale production of *S. cerevisiae* biomass exceeds that of any other propagated microorganism by over 100-fold, and it is currently produced over 1.5 million tons per year and is expected to reach over 2 million tonnes (dry basis) by 2026. The economic value of yeasts is predicted to reach USD 9 billion by 2026 and 13.3 Billion by 2031 (Yeast Market Outlook 2031, 2020). However, yeasts underpin far greater economic value based on their applications. Baked goods are an essential part of the human diet and baker's yeast *Saccharomyces cerevisiae* is a very important microorganism in the bakery industry. The market's global value for bakery products reached USD 498 billion in 2022. The global alcoholic beverages markets (Beer, Wine, Vodka, Gin, Tequila, Rum, Whiskey, and Brandy) reached USD 1,108 billion in 2020, among which beer and wine are the most popular. The global beer market was valued at USD 605 billion in 2020 and is projected to reach USD 685 billion, and the global wine market was USD 340 billion in 2020 and is estimated to reach USD 420 billion, by 2025. The global biofuels market was valued at nearly USD 110 billion in 2021 and is projected to increase by USD 201.2 billion until 2030 (Attfield, 2023; Niego et al., 2023).

16.2 Raw materials

Industrial production of yeast biomass requires low-cost substrates to be readily available in high volumes. Substrates must provide high concentrations of carbon compounds that can be thoroughly converted to biomass. Residual unmetabolized carbon and nitrogen compounds are undesirable because they contribute to biological and chemical oxygen demands of effluent from yeast manufacturing plants. Sucrose-rich sugarcane and beet molasses, and high glucose content starch-based syrups fit these requirements and have long been the favored substrates across the yeast manufacturing industry (Chen & Chiger, 1979; Trivedi et al., 1986). However, yeast producers have to compete with other high-volume users of these substrates, including the growing bioethanol industry. Alternative substrates are continuously sought as full or partial replacements of molasses and syrups. Two attractive options are stillage derived from ethanol distillations, and glycerol obtained from transesterification of triglycerides in biodiesel production. Stillages contain glycerol, acetate, and lactate, which are aerobic substrates for *Saccharomyces* and some other yeast species, but they also contain residual solids that need to be removed (as is the case with molasses, which has to be clarified before use). The potential use of stillage and glycerol as substrates for propagating *Saccharomyces* has been demonstrated with engineered and adapted strains (Mattenberger et al., 2017). Other yeast species in Table 16.1 can efficiently use glycerol to make valuable products such as enzymes, recombinant proteins, lipids, and biochemicals. The dairy industry generates

Table 16.1 Commercially important GRAS[a] yeast species and their physiological properties (Dalbanjan et al., 2024; De Brabander et al., 2023; Johnson, 2013; Karim et al., 2024; Kosaka et al., 2022; Lehnen et al., 2019; Löbs et al., 2017; Park & Ledesma-Amaro, 2023; Patra et al., 2021; Pobiega et al., 2024; Sousa-Silva et al., 2021).

Yeasts	Crabtree effect	Oxygen metabolism	Optimal growth temperature	Native carbohydrate substrates
Saccharomyces cerevisiae	Positive	Facultative anaerobic	30°C–32°C	Glucose, fructose, maltose, sucrose, glycerol, ethanol, VFAs
Pichia pastoris	Negative	Facultative anaerobic	28°C–30°C	Glucose, glycerol, ethanol, methanol
Yarrowia lipolytica	Negative	Obligate aerobic	30°C–34°C	Alkanes, alkenes, fatty acids, triacylglycerides, glucose, fructose, and glycerol; and organic acids, including acetate, lactate, citrate, and succinate,
Kluyveromyces lactis	Negative	Obligate aerobic	25°C–28°C	Glucose, fructose, xylose, glycerol, ethanol, lactose
Kluyveromyces marxianus	Negative	Obligate aerobic	40°C (even up to 47°C)	Glucose, fructose, xylose, sucrose, lactose, cellobiose, glycerol, ethanol, lactose
Hansenula polymorpha	Negative	Facultative methylo-trophic	20°C–45°C (even higher than 50°C)	Methanol, glucose, maltose, sucrose, glycerol, xylitol, xylose
Candida utilis	Negative	Crabtree negative	19°C–40°C	Glucose, arabinose, sucrose, raffinose, D-xylose, alcohols, acetaldehyde, lactate, succinate, citrate, D-gluconate, and various nitrogen sources nitrate, ammonium, amino acids, VFAs

[a] GRAS: Generally recognized as safe.

significant quantities of high-strength lactose and protein-containing waste stream called cheese whey (CW), and 9 L of CW is released for every 1 kg of cheese produced. (Türker et al., 2022). Whey is a good substrate to grow *Kluyveromyces* and has also been demonstrated as a possible partial replacement of molasses in the cultivation of baker's yeast after hydrolyzing lactose into glucose and galactose. Heavy steep water (HSW) from corn processing contains amino acids, glucose, and organic acids and can be used by *Saccharomyces* and other yeast species as carbon and nitrogen sources.

Nonfood lignocellulosic biomass is considered a potential source of new substrates. Lignocellulose consists of cellulose, hemicellulose, and lignin, which have to be hydrolyzed to release free sugars. However, considerable economic and technical challenges must be overcome for them to be routinely available at an industrial scale for the propagation of yeasts. Released sugars are mixtures of hexoses (mostly glucose) and pentoses (xylose and arabinose), but inhibitory compounds, such as furan aldehydes, weak acids, and phenolics, are also generated during hydrolysis. *Saccharomyces cerevisiae* can naturally utilize the hexose sugars but does not readily assimilate pentose sugars. Therefore, substantial effort has gone into engineering strains of the yeast, both to enable the metabolism of xylose and arabinose, and to withstand the toxicity of hydrolyzates. However, other biotechnologically important yeast species given in Table 16.1 can use pentose sugars and glycerol to make a range of compounds (Ceaser et al., 2024; Sharma et al., 2020).

16.3 Yeast biomass for food, beverages, and biofuels

16.3.1 Baker's yeast

The production and processing of yeast *Saccharomyces cerevisiae* at an industrial scale involves well-established fermentation techniques using bubble column fermenters. Potential improvements in cell yields and product performance continue to be made through modified process design and control systems. Details of manufacturing procedures vary according to the design of fermentors and other equipment as well as the type of yeast product being produced. All fermentations are aerobic to maximize cell yield on substrate. Ten tonnes of sugar supports the production of close to five tonnes of yeast biomass. (Chen & Chiger, 1979). Early-stage propagations up to 15 m^3 are batch cultures. Subsequent fermentations to 150–200 m^3 are fed-batch, with the largest and final fermentation being called the "commercial" fermentation. The fed-batch technique was developed a century ago to overcome ethanol formation for Crabtree-positive yeasts, such as *Saccharomyces cerevisiae*, to redirect the carbon source to biomass formation rather than to ethanol. Crabtree-negative yeasts can be cultivated in batch mode, but fed-batch mode can still be used to achieve high-cell densities >60 g/L. The biomass is then separated from the spent culture medium by centrifugation and counter-current washing to produce "cream yeast" (cell density 180–200 g dry cell weight/L), which is stored refrigerated in stirred tanks. Cream yeast may be formulated with polyols and gums, to both preserve fermentative activity and maintain cells in suspension without continuous stirring. Filter presses or rotary vacuum filters are used to further remove extracellular water and produce biomass with 28%–30% yeast solids. This biomass is formulated as crumble, or compressed yeasts, which are stored and transported under refrigeration. Cream, crumble, compressed and liquid yeasts are termed "fresh" or "wet." Alternatively, pressed or filtered yeast is fed through a screw conveyor and sprayed with an emulsifying mixture.

The drying, with varying times between 30 and 200 minutes, reduces the moisture content from 65%–70% to 4%–6% to improve its shelf life. Emulsified yeast is then extruded to form noodles that are commonly dried in batch or continuous fluidized bed, tunnel, and rotolouver dryers. Of these drying processes, fluidized-bed drying has become the most popular because it causes less damage to yeast cells, and thus maintains their viability. Depending on the drying rate, granules are either in the oval-shaped granular form, called quick-rise yeast or produced in small spherical or ball forms (Beker & Rapoport, 2005; Chen & Chiger, 1979; Frey, 1930; Mattenberger et al., 2017; Türker et al., 2006).

The storage method is very important for maintaining the stability and fermentative activity of active dry yeast. When dry yeast products are stored in the presence of oxygen or air, their colony-forming unit (CFU) is reduced. Therefore, the use of vacuum-packaging, or storing in packaging material with inert gas, will help ensure optimum stability (Yüzgec et al., 2008). Heated air is blown through the yeast particles to remove water, producing biomass with about 5% moisture content. Dried yeast is vacuum-packed in heat-sealed, nitrogen or carbon dioxide flushed foil packs, forming hard air-tight bricks.

Among the most critical phenotypes for improvement are *S. cerevisiae* strains that tolerate high levels of sucrose, endure freeze-thaw stress, rapidly utilize maltose, and produce high levels of carbon dioxide. Baking procedures subject yeasts to several stresses. Doughs can contain up to 30% sucrose per weight of flour, which exerts severe osmotic stress on *S. cerevisiae* and other component organisms. Baker's yeast strains must efficiently utilize maltose, which is derived from flour in low-sugar doughs. *S. cerevisiae* strains that have high sucrose tolerance, and also rapidly use maltose, have been genetically developed and are used commercially. One of the most desired properties of baker's yeast strains is a high fermentation rate. Free sugars in dough comprise mainly sucrose, glucose, fructose, and maltose, which are sequentially consumed due to transport and catabolite repression regulatory mechanisms. Amylases in the dough release maltose from starch, but many strains of baker's yeast poorly utilize maltose, primarily due to the repression of maltose utilization by other sugars through catabolite repression. Overcoming this bottleneck has long been a desired trait to enhance baking productivity (Struyf et al., 2017; Trivedi et al., 1986).

The popularity of frozen doughs, particularly sweet frozen doughs, has increased markedly in recent years. Enhancing the freeze tolerance, and freeze-thawing survival of yeasts, is an attribute that would benefit the production and quality of these bakery products, and cryo-resistant baker's yeast strains have been developed. *Torulaspora delbrueckii* is highly tolerant to freezing and freeze-thawing, and this yeast has been considered a primary yeast or adjunct in frozen-dough products. Propionate resistance is also an important property of bakery strains during the production of packaged baked goods (Attfield, 1997; Attfield, 2023; Deparis et al., 2017; Randez-Gil et al., 1999; Zhou, Semumu, & Gamero, 2021).

The "conventional" baker's yeast, *Saccharomyces cerevisiae*, is the most used industrial workhorse of all time. Other yeast species principally contribute to the leavening of flour dough and the aroma of bread. Around 30 different species have been isolated. The genera *Saccharomyces* (*S. cerevisiae*), *Kazachstania* (*K. humilis*, *K. exigua*), *Wickerhamomyces* (*W. anomalus*), *Pichia* (*P. kudriavzevii*), and *Torulaspora* (*T. delbrueckii*) are the most isolated and geographically widespread in Asia, Europe, as well as in Africa and Australia in certain cases (Carbonetto et al., 2018). The increasing number of studies highlight the potential of nonconventional yeasts to develop new bakery products, with novel aroma profiles. Some of these yeasts present several attributes interesting for the bakery industry, such as osmotolerance, halotolerance, or thermotolerance, among others, in addition to the ability to synthesize a huge range of aroma compounds that can result in higher aroma complexity (Aslankoohi et al., 2016). These species exhibit great potential and an untapped source for product diversification and innovation. However, further studies are needed to adjust the different manufacturing processes as well as to demonstrate the qualified presumption of safety (QPS) or GRAS (Generally recognized as safe) status of most of them, such that they can be freely used in the food industry (Zhou, Semumu, & Gamero, 2021).

The global yeast market size is estimated to reach about USD 6 billion by 2025, USD 9 billion by 2026, and 13.3 Billion by 2031 (Yeast Market Outlook 2031, 2020). Major global players in baker's yeast manufacturing include Lesaffre (France), AB Mauri (UK); Lallemand (Canada), Angel Yeast (China), and Pakmaya (Türkiye). It is estimated that together with around 20 local manufacturers, there are approximately 150 baker's yeast manufacturing plants in the world. In addition to baker's yeast, they also mass cultivate other live yeast biomass used for yeast extract and other yeast derivatives, wine and beer yeast, bioethanol yeast, mineral enriched yeast, and yeast species used in animal feed and agricultural applications. World yeast production is estimated at around 1.5 to 2 million tons of dry yeast, 80%–90% of which is probably baker's yeast.

16.3.2 Brewery yeasts

Generally, *Saccharomyces cerevisiae* strains dominate the industrial production of fermented beverages, such as beer, wine, and distilled spirits. In brewing, yeasts are conventionally classified into two groups: (a) ale, that is top fermenting (*Saccharomyces cerevisiae*), and (b) lager, that is bottom fermenting (*Saccharomyces pastorianus* var. *carlsbergensis*) yeasts. There are many separate strains of brewing yeast, and it is well known that the type of yeast strain used in beer production significantly affects fermentation time and the chemical composition of the produced beer (Bamforth, 2017). At the end of fermentation, yeast is separated from the rest of the contents by flocculation. A small amount of yeast from the previous fermentation is used to start the next fermentation, in a process known as refermentation. Although the recycling of brewer's yeast in the new fermentation cycle is a common practice, the number of reuses is limited to maintain the quality of the beer (Zeko-Pivač et al., 2023). The application of non-*Saccharomyces* yeasts in beer, as a natural tool for innovation, has attracted great interest from both researchers and commercial brewers. *Brettanomyces* species are the most relevant nonconventional yeasts involved in the production of sour beers, mainly in Lambic beers. For instance, Belgian Sour Beer uses the yeast *Brettanomyces bruxellensis on* a commercial scale (Capece et al., 2018). Among customer demand for more diversity in the beer market, requests for low-alcohol (0.5–1.2% *v/v*) and nonalcoholic beers (<0.5% *v/v*) are constantly increasing. The use of yeasts with limited abilities to utilize wort sugars, that is, maltose-negative yeasts, but producing typical concentrations of aroma compounds, is a useful alternative to produce low-alcohol beers that still retain some of the aromatic complexity of standard beers. A strain of *Saccharomycodes ludwigii* has been commercially used for this purpose for many years; it is a good example of the novel use of non-*Saccharomyces* yeasts in the brewing industry. As a result, a higher diversity of non-*Saccharomyces* yeasts for beer production is expected to appear on the market in the coming years. According to available public data, 1.88 billion hectoliters of beer were produced worldwide in 2023 (Brauwelt, 2024), from which we can estimate the exogenous yeast consumption of approximately 30,000 to 60,000 tons per year (Capece et al., 2018; Miguel et al., 2022; Simões et al., 2023; Zeko-Pivač et al., 2023).

16.3.3 Wine yeasts

Similar to brewing, the yeasts carry out the alcoholic fermentation (AF), that is, conversion of sugars to ethanol. However, complex ranges of primary and secondary yeast metabolites are responsible for deeply influencing wine quality. Following AF, a secondary bioconversion of enological interest can take place in wine. It is called malolactic fermentation (MLF), the decarboxylation of malic acid to lactic acid by lactic acid bacteria (LAB). *Saccharomyces cerevisiae*, and more rarely *Saccharomyces bayanus*, can complete enological conditions, meaning to consume all sugar initially present in grape must. However, the MLF and AF are driven simultaneously. *Saccharomyces cerevisiae* and *Oenococcus oeni* are generally recognized as the main microorganisms responsible for the AF and the MLF, respectively. However, the spontaneous fermentations in the traditional winemaking process do not rely only on these two species and high concentrations of enological yeasts that do not belong to the *Saccharomyces* genera are present. For this reason, this heterogeneous group of microorganisms is generally mentioned as non-*Saccharomyces*. Non-*Saccharomyces* species drive the first steps of the AF, followed by the dominance of *S. cerevisiae*, which is responsible for the core and final phases of the AF. Traditionally, non-*Saccharomyces* yeasts have been considered as contaminants due to the production of undesirable metabolites by many of the species currently known. This aspect has been changing in recent years due to the growing interest in certain strains that contribute to metabolites that positively impact wine (Roudil et al., 2019; Vejarano & Gil-Calderón, 2021). In the last 20 years, this scenario has resulted in the search for and selection of new strains by the scientific community, as evidenced in the high number of publications related to non-*Saccharomyces* yeasts of enological interest as well as in the development and market launch of commercial products based on these selected non-*Saccharomyces* strains (Comitini et al., 2021). Based on the market research, it is estimated that about 42 commercial products based on 12 non-*Saccharomyces* yeasts, and in combination with *Saccharomyces cerevisiae* strains, are available for winemaking (Roudil et al., 2019; Vejarano & Gil-Calderón, 2021). The dominant species are *Torulaspora delbrueckii*, *Lachancea thermotolerans*, and *Metschnikowia pulcherrima*. A recent OIV Report gives an estimate of world wine production of approximately 244 million hectoliters for 2023 (World Wine Production Outlook OIV First Estimates, 2023). Based on world wine production, it is estimated that the exogenous yeast consumption in the wine industry is approximately 5000 to 10,000 tons per year.

16.3.4 Biofuels

Bioethanol has been attracting increasing attention as a sustainable alternative to fossil fuels, for many years. Depending on the feedstock, it is classified into four generations. Presently, the predominant form of bioethanol is categorized as the first generation, produced from sugar-based materials, like sugarcane or starch-rich biomass such as corn, potatoes, cereals, and grains. The used yeast ferments sugars into ethanol, as in beer and wine fermentations. Starch-based bioethanol production consists of three significant steps: (1) liquefaction of starch, using alpha-amylase at a temperature about 30°C to smaller oligo-saccharides or polysaccharides, followed by (2) saccharification, to depolymerize polysaccharides to simpler fermentable sugars, like glucose, using glucoamylase and, finally, (3) fermentation of hydrolysate (refers to any product of the hydrolysis), where yeast is involved in the breakdown of sugars to bioethanol (Da Silva et al., 2022; Jain & Kumar, 2024).

Second-generation bioethanol uses low-cost lignocellulosic feedstocks and can be widely adopted. However, despite the technological development in second-generation bioethanol, the cost of processing is still high, making second-generation bioethanol uncompetitive with gasoline. The method of producing bioethanol from the second generation feedstock is quite complex compared to the first generation, comprising several stages such as physical-chemical pretreatment to breakdown structure, hydrolysis of pretreated biomass rich in cellulosic content (yielding simple sugars), hydrolysate fermentation (utilizing suitable bioethanol-producing strains that convert both pentose and hexose sugars) (Ceaser et al., 2024; Da Silva et al., 2022; Sharma et al., 2020; Walker & Walker, 2018).

Current bioethanol production mainly comes from the first-generation route like corn, sugarcane, and wheat, and global production has increased by 16.5 folds since 2008 and has reached 110 million m^3 (29.5 billion gallons) in 2023 from 6.7 million m^3 in 2008 (Akita & Matsushika, 2024; RFA Ethanol Industry Outlook, 2024). The United States has led the world in ethanol output for close to two decades. The production of 60 million m^3 (15.6 billion gallons) in 2023, nearly twice that of Brazil and the United States together with Brazil provides approximately 80% of world bioethanol production, followed by European Union (4.9%), India (4.8%), China (3.2%), and the rest of the world (6.5%) (RFA Ethanol Industry Outlook, 2024). The second-generation bioethanol is slowly taking off and their contribution to overall bioethanol production is still limited and close to 1% (Ceaser et al., 2024; RFA Ethanol Industry Outlook, 2024). There are currently more than 30 commercial full-scale or demonstration scale units (some of which may not be operational) producing bioethanol from the second-generation feedstocks (Ewing et al., 2022; Jain & Kumar, 2024; Jansen et al., 2017; Padella et al., 2019).

Industry practice for average exogenous yeast consumption for bioethanol fermentations is approximately 0.10 kg dry yeast/100 L ethanol. As a result, it can be estimated that the maximum size of the bioethanol yeast market is expected to be around 110,000 tons of dry yeast per year for 2023. The actual figure would be much less than this figure, owing to much less exogenous yeast used in molasses-based bioethanol. *K. marxianus* is also used for bioethanol production from lactose-containing whey streams and was first introduced on a commercial scale by the Carbery Group in Ireland in 1978, and subsequently adopted by dairies in the United States and New Zealand (Löbs et al., 2016, 2017; Varela et al., 2017).

Isobutanol and *n*-butanol are the most studied biofuel targets among advanced alcohols and esters due to their high energy contents and properties that are more similar to those of petrol than those of ethanol. Since butanol is less hygroscopic and has a longer chain length than ethanol, it can be blended up to 85% with gasoline. The genetically modified (GM) strains of *S. cerevisiae* are now being commercially exploited for the production of these alcohols from first generation feedstock using a yeast-based process similar to bioethanol (Buijs et al., 2013; Gevo Isobutanol, 2024). At dry-mill corn ethanol plants, the corn mash stream used for ethanol fermentation can also be used for isobutanol fermentation. Ethanol and isobutanol must then go through additional steps at a biorefinery, where they can be refined into sustainable aviation fuel using three steps including alcohol dehydration, oligomerization, and hydrogenation. These steps are already in use on a commercial scale (Gevo WP Aviation Fuel, 2022). Isobutanol production is being commercialized by three companies: Butalco, Butamax, and Gevo (Buijs et al., 2013). Together with the Indian bioenergy company Praj Industries, Gevo is also developing and commercializing the production of isobutanol from sugarcane and sugarbeet-derived feedstocks (Ewing et al., 2022).

Native strains of *S. cerevisiae* are widely used as industrial microorganism in the bioethanol industry; however, new GM yeast strains have been introduced with new properties for the first and second generation feedstocks. For the first-generation feedstocks, glucoamylase genes have been expressed in the yeast to reduce enzyme costs during the saccharification process. In addition to secreting glucoamylase, some commercial strains possess alternative pathways to reduce glycerol formation, with higher ethanol yield. The GM strains for lignocellulosic feedstocks can ferment C6 and C5 compounds (i.e., glucose and xylose) in lignocellulosic feedstocks (Da Silva et al., 2022; Walker & Walker, 2018). In the corn-based process, the yeast is used once and recovered with dried distiller's grains solubles (DDGSs), which is eventually used as animal feed, whereas in the molasses-based process, the yeast is recovered at the end of the process and recycled back into the process reducing exogenous yeast supply compared to the corn-based process (Shurson, 2017).

16.4 Yeasts for human and animal nutrition

Single-cell proteins (SCPs) are derived from microorganisms such as fungi and bacteria. These organisms can convert a wide range of substrates, including agricultural byproducts, industrial waste and carbon dioxide, into biomass enriched in proteins. The microbial species must be GRAS to be used in food systems, and most of the currently discussed microbial species hold GRAS certification (Dalbanjan et al., 2024; Koukoumaki et al., 2024). Today, several SCP processes using microbial species are approved for the production of SCP, and their distribution is shown in Fig. 16.1. Among them are *Saccharomyces cerevisiae* (known as Baker's or Brewer's yeast), *Candida utilis* or *Cyberlindnera jadinii* (known as Torula yeast), *Yarrowia lipolytica, Kluyveromyces marxianus, Kluyveromyces lactis, Schwanniomyces occidentalis, Torulaspora delbrueckii, Zygosaccharomyces rouxii, Metschnikowia pulcherrima,* and *Starmerella bombicola* (Dalbanjan et al., 2024; Koukoumaki et al., 2024). These yeast species have GRAS status and are mentioned in the QPS list notified to European Food Safety Authority (EFSA). They have been used in a wide range of fermentations, such as in bread, bakery products, and alcoholic beverages, as nutritional yeast (wine and beer), dairy product fermentations, bioethanol production, soy sauce and miso production, flavoring agents in various foods, and biocontrol agent in agriculture (Dalbanjan et al., 2024; Koutsoumanis et al., 2022; Sobhi et al., 2023). Commercially viable SCP products must meet a set of quality criteria, comprising nutritional value (protein content, bioavailability), safety (toxin-free, low

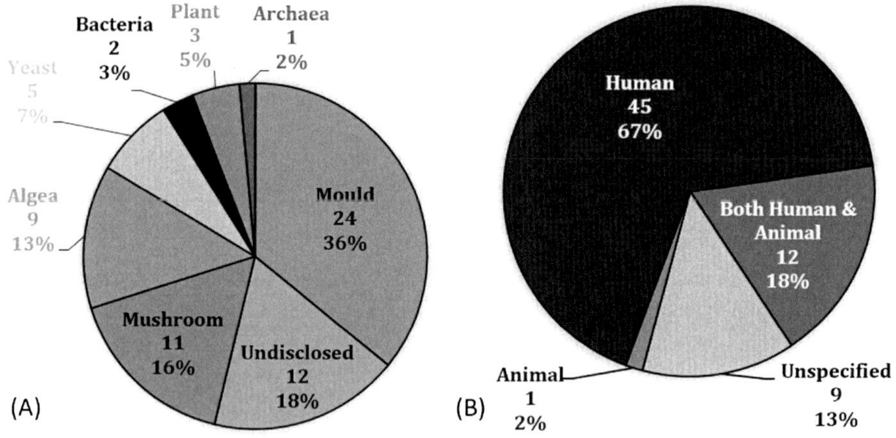

Figure 16.1 Commercial scale SCP; (A) distribution of microorganisms, (B) application (Good Food Institute, 2024; Graham & Ledesma-Amaro, 2023; Linder, 2023). (Note: Proteins from precision fermentations are not included in these figures).

nucleic acid content), and consumer acceptance (taste, texture). On the other hand, smart process engineering should enable technically feasible large-scale solutions with low production costs (Ye et al., 2024).

SCP had been the first high point during World War I in Germany, when baker's yeast *S. cerevisiae* was grown on molasses to replace up to 60% of imported protein and fossil fuels. Similarly, mass production of *Candida utilis* from paper manufacturing waste in World War II efficiently helped to supply starving populations with protein (Ye et al., 2024). The British Petroleum (BP) Company developed, in the 1950s, the alkane-degrading *Y. lipolytica*, for the production of SCP from crude oil for livestock feeding (Toprina G). Over the coming years, BP developed two microbial systems, *Candida tropicalis* and *Yarrowia lipolytica*, using gas oil and n-alkanes as substrates, respectively. For both organisms, pilot and demonstration plants were built, for *C. tropicalis* in Lavera, France (Toprina L) and *Y. lipolytica* (Toprina G) in the United Kingdom. These processes were later discontinued due to the sharp increase in raw material prices with the 1973 oil crisis (Groenewald et al., 2014; Tullio, 2022).

US-based company Arbiom constructed a commercial scale SCP plant based on *Candida utilis* (Torula yeast) using lignocellulosic residues. The plant, located in France, will deliver 10,000 tons of high-quality and sustainable protein ingredients for SylPro for animal feed (pet-food, piglets, and feed for salmonoids and other aqua species) and Yusto for human nutrition to replace animal and plant-based proteins (Arbiom, 2024; He, Tang, et al., 2024).

Currently, DuPont uses *Y. lipolytica* to produce the nutritional supplement Newharvest™ EPA oil, and its by-product biomass is used as salmon feed, Verlasso®. Polish company Onesano (previously Skotan SA) started the production of SCP (edible protein for humans and animals) from waste glycerol, and registered the feed product in the EU (Groenewald et al., 2014; Madzak, 2021; Onesano, 2024). The product portfolio contains ultrapure, highly concentrated omega-3, 6, and 9 acids, in the form of ethyl esters, which ensures high bioavailability and bioretention in the human body and products with a unique combination of ingredients, combining the postbiotic *Yarrowia lipolytica* enriched food with minerals (Onesano, 2024). In addition to oils and carotenoids, which are the most common *Yarrowia* products for human use, the American-based Nucelis also offers a protein-rich Yarrowia Flour (Groenewald et al., 2014; Nucelis, 2024; Nyyssölä et al., 2022; Tullio, 2022).

Probiotic yeast *Saccharomyces boulardii*, genetically close to *Saccharomyces cerevisiae*, and originally isolated from fruit in Indochinahas been studied more extensively than other probiotic yeast strains and it has been used for the treatment of intestinal diseases in children and adults since 1950 (McFarland, 2010). The optimal growth temperature of *S. boulardii* parallels the human host temperature (37°C) and this yeast is also resistant to very high temperatures (keeping 65% viability after 1 hour at 52°C). This yeast generally appears to be safe for most healthy individuals, and it may be effective in addressing issues like antibiotic-associated diarrhea, traveler's diarrhea, acute infectious diarrhea, and even inflammatory bowel disease. Recent advancements in genetic engineering techniques are opening up possibilities for developing next-generation probiotic yeasts, with enhanced probiotic properties or targeted functionalities (Fernández-Pacheco et al., 2021; Shruthi et al., 2022; Tullio, 2024). The studies with non-*Saccharomyces* yeasts have shown numerous beneficial effects, such as antimicrobial activity, improvement of gastrointestinal functions and antioxidant activity, among others. The potential candidates from various yeast genera like *Debaryomyces*, *Kluyveromyces*, *Yarrowia*, *Pichia*, and *Torulaspora* have been isolated from fermented foods, traditional beverages, human microbiota, and natural sources and have been under intensive investigation (Fernández-Pacheco et al., 2021; Shruthi et al., 2022; Tullio, 2024). *Saccharomyces boulardii* is widely available in the market, supplied by various companies. Polish company Onesano (previously Skotan SA) is developing *Y. lipolytica* for prebiotic and probiotic use in feed and food (Groenewald et al., 2014; Madzak, 2021; Onesano, 2024).

Yeast and yeast derivatives have been used for animal feed, in various forms, for more than a century. However, due to government restrictions and the elimination of the use of growth promoting antibiotics in animal feed, there is a growing interest in using alternative products (including yeast products) to promote animal health and growth performance benefits (Shurson, 2018). Furthermore, feed ingredients produced from yeast-based (or other) fermentation processes (e.g., distillers dried grains with solubles, nominate DDGS, brewery, and citric acid) in animal feeds have dramatically increased in recent years (Deuchande et al., 2024; Shurson, 2017; Thayer et al., 2023; Zeko-Pivač et al., 2023). With the global transition from a fossil economy to a bioeconomy, increases in the diversity and volume of chemicals and products derived from cell metabolism will take place. As a consequence, the generation of industrial spent biomass will also increase (Stikane et al., 2022). Therefore, many types of feed additives and feed ingredients contain, yeast in various forms, but animal nutritionists are often uncertain about the unique differences among these products and their potential roles in animal nutrition and health.

As a feed additive for livestock and poultry breeding, active dry yeast should exhibit higher microbial activity, enabling it to play a better role in regulating intestinal flora, inhibiting the propagation of harmful bacteria, improving the activity of intracellular enzymes and promoting the digestion and absorption of nutrients by livestock and poultry. Dry yeast is comprised of $>10^{10}$ live yeast cells, as colony forming units (CFUs), per gram for animal feed (Broadway et al., 2015; Fathima et al., 2023; Pang et al., 2003; Schlabitz et al., 2022). Many animal feeds are manufactured and fed in pellet forms. The pelleting process uses steam (moist heat) to condition the feed prior to forcing through a pellet die. Moist heat is considered to be one of the most effective methods for killing microorganisms because it denatures enzyme systems and their metabolic activity within the organism, which inhibits their survival. Therefore, when adding active dry yeast products to animal feeds, there is concern that their viability will be reduced, or potentially eliminated, rendering them ineffective as probiotics. The resistance of yeast cells to heat treatment is much lower as compared to bacterial cells. Dry, active *Saccharomyces cerevisiae* (5% moisture) survival was unaffected when heated at 85°C for 8 minutes, but when the moisture was increased to 19%, only 0.1% survived. Similar results were also reported when evaluating pelleting conditions on *Saccharomyces cerevisiae*, *Kluyveromyces marxianus,* and *Candida utilis* survival (Shurson, 2018). Studies have determined viable yeast cell counts under different moisture and temperature conditions, and moist heat and dehydration were shown to be the primary factors causing yeast cell death. These results suggest that the survival of unprotected yeast cells would be expected to be low after the typical pelleting process. Therefore, the measurement of both cell viability and cell vitality must be considered when evaluating the functionality of yeast cells in pelleted feeds, and the advantages and disadvantages of different methods of evaluating viability and vitality have been compared (Kwolek-Mirek & Zadrag-Tecza, 2014; Shurson, 2018).

16.5 Yeasts for plant nutrition and protection

Biostimulants are a general category of natural substances or microorganisms that are applied to plants to improve their nutritional efficiency, tolerance to abiotic stress, and/or quality traits of crops, regardless of their nutrient content. Biofertilizers, a subcategory of biostimulants, can be defined as microbial inoculants, in which active or inactive formulations of beneficial microorganisms can improve the nutritional efficiency of plants. Biofertilizers enhance soil productivity by fixing down atmospheric nitrogen and solubilizing different nutrients, thereby exerting direct or indirect beneficial effects on crop growth and yield through different mechanisms. (Allouzi et al., 2022). The use of biofertilizers raised crop yields by around 10%–40%, which increased the content of proteins, essential amino acids, vitamins, and nitrogen fixation (Bhardwaj et al., 2014; Naik et al., 2019; Nunes et al., 2024).

There are several works focused on bacteria and fungi as biofertilizers (Nunes et al., 2024) but the literature on yeasts and their potential ability to safely promote plant growth has been gaining particular attention in recent years. The yeasts promote plant growth and increase crop yields by providing soluble nutrients to plants, producing organic acids and phytohormones. In addition, the yeasts act as biocontrol agents to protect the plants through their high antifungal activity and lower insecticidal and herbicidal activity. They also act as protective agents against extreme environmental factors by activating defense mechanisms, as shown in Fig. 16.2 (Hernández-Fernández et al., 2021; Lonhienne et al., 2014; Nimsi et al., 2023).

Considerable progress has been made in research showing that plant growth can be enhanced by yeasts. Some of their positive effects could be due to the provision of soluble nutrients and the production of different phytohormones and enzymes that directly contribute to plant growth, acting as biostimulants (Hernández-Fernández et al., 2021).

Plant growth-promoting microorganisms (PGPMs) have several beneficial effects on plants due to both direct and indirect mechanisms. The largest yeast populations are generally found in the rhizosphere rather than in the soil. In fact, a growing number of studies indicate that plant root growth may be favored by yeasts in the rhizosphere, although a wide diversity of soil yeasts has also been investigated for their potential as biofertilizers (Hernández-Fernández et al., 2021; Naik et al., 2019; Nimsi et al., 2023). The yeast *C. tropicalis* was able to solubilize $Ca_3(PO_4)_2$, in the

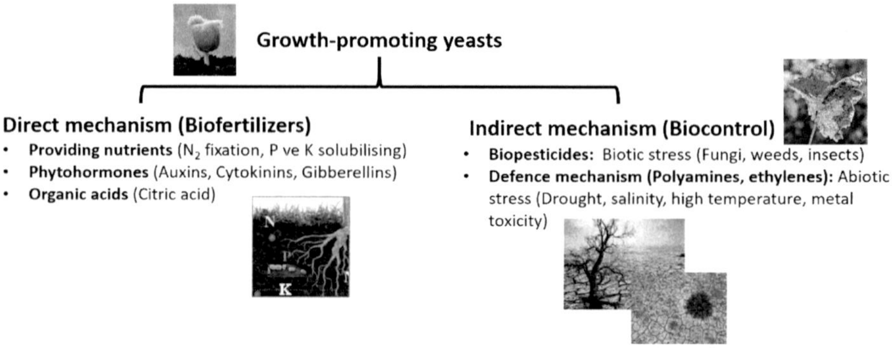

Figure 16.2 Direct and indirect mechanisms of growth-promoting yeasts. Yeast provides essential nutrients, phytohormones, and organic acids as biofertilizer, and produce defensive chemicals to protect plants from harmful organisms as biocontrol agent.

same way as *Rhodotorula* sp. and *Lachancea thermotolerans can* provide soluble inorganic P with the consequent pH reduction. Moreover, it has also been demonstrated the ability of some soil strains belonging to *Yarrowia lipolytica* and *Saccharomyces cerevisiae* to solubilize inorganic phosphate compounds by producing citric acid (Hernández-Fernández et al., 2021).

Phytohormones are organic compounds that are responsible for plant growth. Yeasts can produce some plant growth-promoting phytohormones. Phytohormones, such as abscisic acid (ABA), auxins, cytokinins, and gibberellins (GA), control the growth and development of root, shoot and fruit in a plant, and mitigate the stress on the plant. The yeast cells, when used in combination with the phytohormones, promote better plant growth than when used alone (Nimsi et al., 2023).

The regulation and control of the growth of undesired microorganisms could be exploited through different mechanisms. The modes of action usually involved include (1) competition for nutrients and space; (2) production of toxins; (3) secreted enzymes; (4) production of volatile organic compounds (VOCs); (5) parasitism; (6) induced systemic resistance. In this way, yeasts act as biocontrol agents through an antagonistic action against pathogenic microorganisms that indirectly affect plant growth, acting mainly as a biofungicide (Agirman et al., 2023; He, Tang, et al., 2024; Hernández-Fernández et al., 2021). Many studies have shown that some yeasts could represent an alternative against fungal infections. It is well known that fungi are more proficient in colonization and competition for space and nutrients than other groups of microbes that exist on the surface of fruits and vegetables. Furthermore, yeasts do not usually produce toxic secondary metabolites such as antibiotics, allergenic spores, or mycotoxins in their inhibitory activity, so they can be used as an effective biocontrol agent (Agirman et al., 2023; Freimoser et al., 2019; He, Degraeve, et al., 2024; Hernández-Fernández et al., 2021).

Among the commercialized products, it should be highlighted that the commercial biofertilizer BioGro 2 consists of a strain cocktail of bacteria and the yeast *C. tropicalis* (Nguyen et al., 2017). This yeast strain was initially selected for its

Table 16.2 Some commercially available bioprotective products developed to control postharvest diseases in fruits and vegetables (Agirman et al., 2023; Ali et al., 2024; He, Degraeve, et al., 2024; Zhang et al., 2020).

Product	Yeast	Target pathogens	Application field	Manufacturer
Aspire®	*Candida oleophila*	*Botrytis, Penicillium, Rhizopus, Monilinia*	Citrus fruits, pome fruits, apple, peach	Ecogen, USA
Candifruit®	*Candida sake*	*Penicillium, Botrytis, Rhizopus, Aspergillus*	Pome, grapevine	IRTA/Sipcam-Inagra, Spain
Nexy®	*Candida oleophila*	*Penicillium, Botrytis,*	Pome, citrus, banana	Belgium
Blossom protect®	*Aureobasidium pullulans*	*Botrytis, Penicillium, Monilinia, Erwinia amylovora*	Pome fruits (apples, pears, quince)	Bioferm, Austria
Boni Protect®	*Aureobasidium pullulans*	*Botrytis, Penicillium, Monilinia*	Pome fruits	IOBC-WPRS, Germany
Botector®	*Aureobasidium pullulans*	*Botrytis*	Grape, strawberry, tomato, apples, pears, other fruits	Nufarm, Australia
YieldPlus®	*Cryptococcus albidus*	*Botrytis, Penicillium, Mucor*	Pome fruits, citrus, peal, apple	Canada
Shemer®	*Metschnikowian fructocola*	*Penicillium, Botrytis, Rhizopus, Aspergillus*	Pome, strawberry, grape, stone fruit, citrus, sweet potato, peach, pepper	The Netherlands
Noli®	*Metschnikowian fructocola*	*Botrytis, Monilinia, Penicillium, Rhizopus, Aspergillus*	Strawberry, blueberry, grape, stone fruit, soft fruits	The Netherlands
Pantovital®	*Pantoea agglomerans*	*Botrytis, Monilinia, Penicillium, Rhizopus*	Pome, citrus fruits	Spain
Romeo®	*Saccharomyces cerevisiae* (cell wall)	*Botrytis, Erysiphales*	Lettuce, tomato, strawberry, cucumber, grapevines	France

ability to solubilize insoluble PO_4 in agar medium, but it has also been shown to have a set of properties that promote rice plants' growth. The application of this biofertilizer is effective in reducing the use of conventional N fertilizers by up to 52% on farms, without decreasing rice yields (Rose et al., 2014). On the other hand, there is NutriSmart®, a biofertilizer composed of a group of *S. cerevisiae* yeast strains that fix atmospheric nitrogen, break down phosphate rock into soluble phosphate and enhance the amount of exchangeable potassium in the soil. Due to the unique characteristics of yeasts to decompose complex organic matter, various wastes, including animal manure, compost, and sludge, are often used as substrates. Once applied to the soil, yeasts decompose the residues and release essential elements, such as N, P, and K. Thus, crop residues can be used to produce these biofertilizers by entering a sustainable recycling loop. Moreover, it has been demonstrated that it increases soil microbial activity while allowing to reduce the use of synthetic fertilizers by 20%–30% (Pang et al., 2003).

The use of yeast as a biopesticide in postharvest disease management holds great potential due to its effectiveness, safety, and sustainability. However, despite the extensive research conducted on yeasts as biopesticides, the commercialization of an antagonistic yeast is a long and costly process requiring extensive testing of toxicology and biocontrol efficacy under commercial conditions (Agirman et al., 2023; Zhang et al., 2020). Several yeast-based commercial biopesticides, including registered and nonregistered products, are worldwide applied against the post-harvested fungal pathogen of fruits, against most important fungal pathogens such as *Penicillium digitatum, P. expansum, P. italicum, Rhizopus stolonifer, Botrytis cinerea*, and *Fusarium* species, and there are around 11 products that have achieved commercial success in the market (Table 16.2). Various factors, such as a deficiency in mechanistic understanding, challenges and expenses associated with product registration, a scarcity of partnerships or consortia with the necessary expertize, and perceived limited commercial potential, likely contribute to the apparent challenges in the development of yeast-based plant protection products. Nonetheless, ongoing research efforts in this area aim to develop consistent and reliable yeast-based biocontrol products, having the potential to significantly contribute to the reduction of postharvest food losses and enhance food safety (Agirman et al., 2023; He, Degraeve, et al., 2024; Hernández-Fernández et al., 2021; Zhang et al., 2020).

16.6 Conclusion

The industrial production of yeast biomass started in the last century and is currently a technologically optimized process generating a product that offers a large number of applications. This review attempts to provide an overview of the many current and future applications of yeasts, from the production of bakery, biofuel, alcoholic beverages and nutrition, to biocontrol to combat phytopathogens, which already reached commercial scale. Commercial scale production of *S. cerevisiae* biomass exceeds that of any other propagated microorganism by several hundred-fold, approximately 80%–90% of which is estimated to be used in the bakery industry and the remaining is distributed other industries. However, traditionally used in food fermentations, the impact of yeasts, including non-*Saccharomyces* yeasts is far-reaching and penetrates all aspects of our life in size and multitude. Yeast-based bioeconomy underpins far greater monetary value based on their applications. More than 30 new yeast species have entered into commercial production and their impacts are expanding, ranging from traditional and large-volume productions to high-value-added products. In addition, these yeast species, built using synthetic biology tools, have been used to produce biopharmaceutical proteins, enzymes, and alternative proteins, such as insulin to the development of safe vaccines, and also for the production of food additives to the recent cow-free milk.

References

Agirman, B., Carsanba, E., Settanni, L., & Erten, H. (2023). Exploring yeast-based microbial interactions: The next frontier in postharvest biocontrol. *Yeast, 40*(10), 457–475. https://doi.org/10.1002/yea.3895.

Akita, H., & Matsushika, A. (2024). A short review of second-generation isobutanol production by SHF and SSF. *Applied Biosciences, 3*(3), 296–309. https://www.mdpi.com/2813-0464/3/3/20.

Ali, A., Ölmez, F., Zeshan, M. A., Mubeen, M., Iftikhar, Y., Sajid, A., Abid, M., Kumar, A., Kumar Divvela, P., & Kumar Solanki, M. (2024). Yeast-based solutions in controlling plant pathogens. *Biocatalysis and Agricultural Biotechnology, 58*. https://doi.org/10.1016/j.bcab.2024.103199.

Allouzi, M. M. A., Allouzi, S. M. A., Keng, Z. X., Supramaniam, C. V., Singh, A., & Chong, S. (2022). Liquid biofertilizers as a sustainable solution for agriculture. *Heliyon, 8*(12). https://doi.org/10.1016/j.heliyon.2022.e12609.

Arbiom. (2024). https://arbiom.com/.

Aslankoohi, E., Herrera-Malaver, B., Rezaei, M. N., Steensels, J., Courtin, C. M., & Verstrepen, K. J. (2016). Non-Conventional Yeast Strains Increase the Aroma Complexity of Bread. *PLOS ONE, 11*(10), e0165126. https://doi.org/10.1371/journal.pone.0165126.

Attfield, P. V. (1997). Stress tolerance: The key to effective strains of industrial baker's yeast. *Nature Biotechnology, 15*(13), 1351. https://doi.org/10.1038/nbt1297-1351 11357.

Attfield, P. V. (2023). Crucial aspects of metabolism and cell biology relating to industrial production and processing of Saccharomyces biomass. *Critical Reviews in Biotechnology, 43*(6), 920–937. https://doi.org/10.1080/07388551.2022.2072268.

Bamforth (2017). Progress in brewing science and beer production. *Annual Review of Chemical and Biomolecular Engineering. 8*, 161–1176. https://doi.org/10.1146/annurev-chembioeng-060816-101450.

Barnett, J. A., & Barnett, L. (2011). In J. A. Barnett, & L. Barnett (Eds.). *Yeast research a historical review*. Washington, DC: ASM Press.

Beker, M. J., & Rapoport, A. I. (2005). Biotechnology methods conservation of yeasts by dehydration. Springer, Berlin, Heidelberg. Biotechnology Methods Advances in Biochemical Engineering/Biotechnology, 35.

Bhardwaj, D., Ansari, M. W., Sahoo, R. K., & Tuteja, N. (2014). Biofertilizers function as key player in sustainable agriculture by improving soil fertility. *Microbial Cell Factories, 13*, 1–10. https://doi.org/10.1186/1475-2859-13-66.

Boekhout, T., Amend, A. S., El Baidouri, F. G., Toni, G., József, M., Moritz R., V., Tan, C. S., Turchetti, B., Vu, D., Wang, Q. M., & Yurkov, A. (2022). Trends in yeast diversity discovery. *Fungal Diversity, 114*(1), 491–537. https://doi.org/10.1007/s13225-021-00494-6.

De Brabander, P., Uitterhaegen, E., Delmulle, T., De Winter, K., & Soetaert, W. (2023). Challenges and progress towards industrial recombinant protein production in yeasts: A review. *Biotechnology Advances, 64*. https://doi.org/10.1016/j.biotechadv.2023.108121.

Brauwelt. (2024). https://brauwelt.com/.

Broadway, P. R., Carroll, J. A., & Sanchez, N. C. B. (2015). Live yeast and yeast cell wall supplements enhance immune function and performance in food-producing livestock: A review. *Microorganisms, 3*(3), 417–427. https://doi.org/10.3390/microorganisms3030417 Immunity Livestock Metabolism Yeast.

Buijs, N. A., Siewers, V., & Nielsen, J. (2013). Advanced biofuel production by the yeast saccharomyces cerevisiae. *Current Opinion in Chemical Biology, 17*(3), 480–488. https://doi.org/10.1016/j.cbpa.2013.03.036.

Capece, A., Romaniello, R., Siesto, G., & Romano, P. (2018). Conventional and non-conventional yeasts in beer production. *Fermentation, 4*.

Carbonetto, B., Ramsayer, J., Nidelet, T., Legrand, J., & Sicard, D. (2018). Bakery yeasts, a new model for studies in ecology and evolution. *Yeast, 35*(11), 591–603. https://doi.org/10.1002/yea.3350.

Ceaser, R., Montané, D., Constantí, M., & Medina, F. (2024). Current progress on lignocellulosic bioethanol including a technological and economical perspective. In *Environment, development and sustainability*. Springer Science and Business Media B.V. 10.1007/s10668-024-04792-2.

Chen, S. L., & Chiger, M. (1979). Production of baker's yeast. In *Comprehensive biotechnology*. Oxford, U.K.: Pergamon Press 3.

Comitini, F., Agarbati, A., Canonico, L., & Ciani, M. (2021). Yeast interactions MDPI Yeast interactions and molecular mechanisms in wine fermentation: A comprehensive review. *International Journal of Molecular Sciences, 22*(14). https://doi.org/10.3390/ijms22147754.

Dalbanjan, N. P., Eelager, M. P., & Narasagoudr, S. S. (2024). Microbial protein sources: A comprehensive review on the potential usage of fungi and cyanobacteria in sustainable food systems. *Food and Humanity, 3*, 100366. https://doi.org/10.1016/j.foohum.2024.100366, https://linkinghub.elsevier.com/retrieve/pii/S2949824424001411.

Deparis, Q., Claes, A., Foulquié-Moreno, M. R., & Thevelein, J. M. (2017). Engineering tolerance to industrially relevant stress factors in yeast cell factories. *FEMS Yeast Research, 17*(4). https://doi.org/10.1093/femsyr/fox036.

Deuchande, T., Fundo, J. F., Rodrigues, D., Pintado, M. E., & Amaro, A. L. (2024). Characterization of spent yeast streams from precision fermentations driven by genetically engineered yeasts: Valorization potential for a circular bioeconomy. *Waste and Biomass Valorization, 15*(6), 3553–3567. https://doi.org/10.1007/s12649-023-02394-z.

Ewing, T. A., Nouse, N., van Lint, M., van Haveren, J., Hugenholtz, J., & van Es, D. S. (2022). Fermentation for the production of biobased chemicals in a circular economy: A perspective for the period 2022–2050. *Green Chemistry, 24*(17), 6373–6405. https://doi.org/10.1039/D1GC04758B.

Fathima, S., Shanmugasundaram, R., Sifri, M., & Selvaraj, R. (2023). Yeasts and yeast-based products in poultry nutrition. *Journal of Applied Poultry Research, 32*(2), 100345. https://doi.org/10.1016/j.japr.2023.100345, https://www.sciencedirect.com/science/article/pii/S105661712300017X.

Fernández-Pacheco, P., Pintado, C., Pérez, A., & Briones Arévalo-Villena, M. (2021). Potential probiotic strains of saccharomyces and non-saccharomyces: Functional and biotechnological characteristics. *Journal of Fungi, 7*(3), 1–18. https://doi.org/10.3390/jof7030177.

Freimoser, F. M., Rueda-Mejia, M. P., Tilocca, B., & Migheli, Q. (2019). Biocontrol yeasts: Mechanisms and applications. *World Journal of Microbiology and Biotechnology, 35*(10), 154. https://doi.org/10.1007/s11274-019-2728-4.

Frey, C. N. (1930). History and development of the mordern yeast industry. *Industrial and Engineering Chemistry, 22*(11), 1154–1161.

Gevo Isobutanol. (2024). https://gevo.com/. 2024.

Gevo WP Aviation Fuel. (2022). https://gevo.com/product/sustainable-aviation-fuel/.

Good Food Institute. (2024). https://gfi.org/.

Graham, A. E., & Ledesma-Amaro, R. (2023). The microbial food revolution. *Nature Communications, 14*(1). https://doi.org/10.1038/s41467-023-37891-1.

Groenewald, M., Boekhout, T., Neuvéglise, C., Gaillardin, C., Van Dijck, P. W. M., & Wyss, M. (2014). Yarrowia lipolytica: Safety assessment of an oleaginous yeast with a great industrial potential. *Critical Reviews in Microbiology, 40*(3), 187–206. https://doi.org/10.3109/1040841X.2013.770386.

He, J. T., Tang, M., Zhong, F. F., Deng, J. L., Wen, Zhang, L., Lin, Q. L., Xia, X., Li, J., & Guo, T. (2024). Current trends and possibilities of typical microbial protein production approaches: A review. *Critical Reviews in Biotechnology*. https://doi.org/10.1080/07388551.2024.2332927.

He, Y., Degraeve, P., & Oulahal, N. (2024). Bioprotective yeasts: Potential to limit postharvest spoilage and to extend shelf life or improve microbial safety of processed foods. *Heliyon, 10*(3). https://doi.org/10.1016/j.heliyon.2024.e24929.

Hernández-Fernández, M., Cordero-Bueso, G., Ruiz-Muñoz, M., & Cantoral, J. M. (2021). Culturable yeasts as biofertilizers and biopesticides for a sustainable agriculture: A comprehensive review. *Plants, 10*(5). https://doi.org/10.3390/plants10050822.

Jain, S., & Kumar, S. (2024). A comprehensive review of bioethanol production from diverse feedstocks: Current advancements and economic perspectives. *Energy, 296*. https://doi.org/10.1016/j.energy.2024.131130.

Jansen, M. L. A., Bracher, J. M., Papapetridis, I., Verhoeven, M. D., de Bruijn, H., de Waal, P. P., van Maris, A., J.A. Klaassen, P., & Pronk, J. T. (2017). Saccharomyces cerevisiae strains for second-generation ethanol production: From academic exploration to industrial implementation. *FEMS Yeast Research, 17*(5). https://doi.org/10.1093/femsyr/fox044.

Johnson, E. A. (2013). Biotechnology of non-Saccharomyces yeasts - The ascomycetes. *Applied Microbiology and Biotechnology, 97*(2), 503–517. https://doi.org/10.1007/s00253 012 4197 y.

Karim, A. A., Martínez-Cartas, M. L., & Cuevas-Aranda, M. (2024). Fermentation of sugar by thermotolerant hansenula polymorpha yeast for ethanol production. *Fermentation, 10*(5). https://doi.org/10.3390/fermentation10050260.

Knychala, M. M., Boing, L. A., Ienczak, J. L., Trichez, D., & Stambuk, B. U. (2024). Precision fermentation as an alternative to animal protein, a review. *Fermentation, 10*(6). https://doi.org/10.3390/fermentation10060315.

Kosaka, T., Tsuzuno, T., Nishida, S., Pattanakittivorakul, S., Murata, M., Miyakawa, I., Lertwattanasakul, N., Limtong, S., & Yamada, M. (2022). Distinct metabolic flow in response to temperature in thermotolerant Kluyveromyces marxianus. *Applied and Environmental Microbiology, 88*(6). https://doi.org/10.1128/AEM.02006-21.

Koukoumaki, D. I., Tsouko, E., Papanikolaou, S., Ioannou, Z., Diamantopoulou, P., & Sarris, D. (2024). Recent advances in the production of single cell protein from renewable resources and applications. *Carbon Resources Conversion, 7*(2). https://doi.org/10.1016/j.crcon.2023.07.004.

Koutsoumanis, K., Allende, A., Alvarez-Ordóñez, A., Bolton, D., Bover-Cid, S., Chemaly, M., Davies, R., De Cesare, A., Hilbert, F., Lindqvist, R., Nauta, M., Peixe, L., Ru, G., Simmons, M., Skandamis, P., Suffredini, E., Cocconcelli, P. S., Fernández Escámez, P. S., Prieto-Maradona, M., ... Herman, L. (2022). Update of the list of QPS-recommended biological agents intentionally added to food or feed as notified to EFSA 15: Suitability of taxonomic units notified to EFSA until September 2021. *EFSA Journal, 20*(1). https://doi.org/10.2903/j.efsa.2022.7045.

Kwolek-Mirek, M., & Zadrag-Tecza, R. (2014). Comparison of methods used for assessing the viability and vitality of yeast cells. *FEMS Yeast Research, 14*(7), 1068–1079. https://doi.org/10.1111/1567-1364.12202.

Lahue, C., Madden, A. A., Dunn, R. R., & Smukowski Heil, C. (2020). History and domestication of saccharomyces cerevisiae in bread baking. *Frontiers in Genetics, 11*. https://doi.org/10.3389/fgene.2020.584718.

Lehnen, M., Ebert, B. E., & Blank, L. M. (2019). Elevated temperatures do not trigger a conserved metabolic network response among thermotolerant yeasts. *BMC Microbiology, 19*(1). https://doi.org/10.1186/s12866-019-1453-3.

Linder, T. (2023). Beyond agriculture—how microorganisms can revolutionize global food production. *ACS Food Science and Technology, 3*(7), 1144–1152. https://doi.org/10.1021/acsfoodscitech.3c00099.

Lonhienne, T., Mason, M. G., Ragan, M. A., Hugenholtzc, P., Schmidt, S., & Paungfoo-Lonhienne, C. (2014). Yeast as a biofertilizer alters plant growth and morphology. *Crop Science, 54*(2), 785–790. https://doi.org/10.2135/cropsci2013.07.0488.

Löbs, A. K., Schwartz, C., & Wheeldon, I. (2017). Genome and metabolic engineering in non-conventional yeasts: Current advances and applications. *Synthetic and Systems Biotechnology, 2*(3). https://doi.org/10.1016/j.synbio.2017.08.002.

Löbs, A.-K., Lin, J.-L., Cook, M., & Wheeldon, I. (2016). High throughput, colorimetric screening of microbial ester biosynthesis reveals high ethyl acetate production from Kluyveromyces marxianus on C5, C6, and C12 carbon sources. *Biotechnology Journal, 11*(10), 1274–1281. https://doi.org/10.1002/biot.201600060.

Madzak, C. (2021). Yarrowia lipolytica engineering as a source of microbial cell factories. In *Microbial cell factories engineering for production of biomolecules*. Elsevier, 345–380. 10.1016/B978-0-12-821477-0.00009-X.

Mattenberger, F., Sabater-Muñoz, B., Hallsworth, J. E., & Fares, M. A. (2017). Glycerol stress in Saccharomyces cerevisiae: Cellular responses and evolved adaptations. *Environmental Microbiology, 19*(3), 990–1007. https://doi.org/10.1111/1462-2920.13603.

McFarland, L. V. (2010). Systematic review and meta-analysis of Saccharomyces boulardii in adult patients. *World Journal of Gastroenterology, 16*(18), 2202. https://doi.org/10.3748/wjg.v16.i18.2202.

Miguel, G. A., Carlsen, S., Arneborg, N., Saerens, S. M. G., Laulund, S., & Knudsen, G. M. (2022). Non-Saccharomyces yeasts for beer production: Insights into safety aspects and considerations. *International Journal of Food Microbiology, 383*. https://doi.org/10.1016/j.ijfoodmicro.2022.109951.

Money, N. P. (2018). *The rise of yeast how the sugar fungus shaped civilization*. Great Clarendon Street, Oxford, ox2 6dp, United Kingdom: Oxford University Press.

Naik, K., Mishra, S., Srichandan, H., Singh, P. K., & Sarangi, P. K. (2019). Plant growth promoting microbes: Potential link to sustainable agriculture and environment. *Biocatalysis and Agricultural Biotechnology, 21*. https://doi.org/10.1016/j.bcab.2019.101326.

Nguyen, T. H., Phan, T. C., Choudhury, A. T. M. A., Rose, M. T., Deaker, R. J., & Kennedy, I. R. (2017). *BioGro: A plant growth-promoting biofertilizer validated by 15 Years' research from laboratory selection to rice farmer's fields of the Mekong delta agro-environmental sustainability*. Cham: Springer International Publishing, 237–254. 10.1007/978-3-319-49724-2_11.

Niego, A. G. T., Lambert, C., Mortimer, P., Thongklang, N., Rapior, S., Grosse, M., Schrey, H., Charria-Girón, E., Walker, A., Hyde, K. D., & Stadler, M. (2023). The contribution of fungi to the global economy. *Fungal Diversity, 121*(1), 95–137. https://doi.org/10.1007/s13225-023-00520-9.

Nimsi, K. A., Manjusha, K., Kathiresan, K., & Arya, H. (2023). Plant growth-promoting yeasts (PGPY), the latest entrant for use in sustainable agriculture: A review. *Journal of Applied Microbiology, 134*(2). https://doi.org/10.1093/jambio/lxac088.

Nucelis. (2024). https://www.nucelis.com/products.php?%20product=oils#circles.

Nunes, P. S. O., Lacerda-Junior, G. V., Mascarin, G. M., Guimarães, R. A., Medeiros, F. H. V., Arthurs, S., & Bettiol, W. (2024). Microbial consortia of biological products: Do they have a future? *Biological Control, 188*. https://doi.org/10.1016/j.biocontrol.2024.105439.

Nyyssölä, A., Suhonen, A., Ritala, A., & Oksman-Caldentey, K. M. (2022). The role of single cell protein in cellular agriculture. *Current Opinion in Biotechnology, 75*. https://doi.org/10.1016/j.copbio.2022.102686.

Onesano. (2024). https://onesano.pl/.

Padella, M., O'Connell, A., & Prussi, M. (2019). What is still limiting the deployment of cellulosic ethanol? Analysis of the current status of the sector. *Applied Sciences (Switzerland), 9*(21). https://doi.org/10.3390/app9214523.

Pang, S. F., Lui, A., Goold, G., Wong, W. L. S.C. E., Kwok, I., & Cheung, L. (2003). NutriSmart: A fertiliser capable of re-establishing the sustainibility of ecosystems and enhancing the productivitiy of farmland. In *Proceeding of the 11th Australian Agronomy Conference VIC*, Australia.

Park, Y.-K., & Ledesma-Amaro, R. (2023). What makes Yarrowia lipolytica well suited for industry? *Trends in Biotechnology, 41*(2), 242–254. https://doi.org/10.1016/j.tibtech.2022.07.006.

Patra, P., Das, M., Kundu, P., & Ghosh, A. (2021). Recent advances in systems and synthetic biology approaches for developing novel cell-factories in non-conventional yeasts. *Biotechnology Advances, 47*. https://doi.org/10.1016/j.biotechadv.2021.107695.

Pobiega, K., Sękul, J., Pakulska, A., Latoszewska, M., Michońska, A., Korzeniowska, Z., Macherzyńska, Z., Pląder, M., Duda, W., Szafraniuk, J., Kufel, A., Dominiak, Ł., Lis, Z., Kłusek, E., Kozicka, E., Wierzbicka, A., Trusińska, M., Rybak, K., Kot, A. M., & Nowacka, M. (2024). Fungal proteins: Sources, production and purification methods, industrial applications, and future perspectives. *Applied Sciences, 14*(14), 6259. https://doi.org/10.3390/app14146259.

Randez-Gil, F., Sanz, P., & Prieto, J. A. (1999). Engineering baker's yeast: Room for improvement. *Trends in Biotechnology, 17*(6), 237–244. https://doi.org/10.1016/s0167-7799(99)01318-9.

RFA Ethanol Industry Outlook. (2024). https://d35t1syewk4d42.cloudfront.net/file/2661/2024%20Ethanol%20Industry%20Outlook.pdf.

Rose, M. T., Phuong, T. L., Nhan, D. K., Cong, P. T., Hien, N. T., & Kennedy, I. R. (2014). Up to 52% N fertilizer replaced by biofertilizer in lowland rice via farmer participatory research. *Agronomy for Sustainable Development, 34*(4), 857–868. https://doi.org/10.1007/s13593-014-0210-0.

Roudil, L., Russo, P., Berbegal, C., Albertin, W., Spano, G., & Capozzi, V. (2019). Non-Saccharomyces commercial starter cultures: Scientific trends, recent patents and innovation in the wine sector. *Recent Patents on Food, Nutrition & Agriculture, 11*(1), 27–39. https://doi.org/10.2174/2212798410666190131103713.

Schlabitz, C., Neutzling Lehn, D., & Volken de Souza, C. F. (2022). A review of Saccharomyces cerevisiae and the applications of its byproducts in dairy cattle feed: Trends in the use of residual brewer's yeast. *Journal of Cleaner Production, 332*. https://doi.org/10.1016/j.jclepro.2021.130059.

Sharma, B., Larroche, C., & Dussap, C. G. (2020). Comprehensive assessment of 2G bioethanol production. *Bioresource Technology, 313*. https://doi.org/10.1016/j.biortech.2020.123630.

Shruthi, B., Deepa, N., Somashekaraiah, R., Adithi, G., Divyashree, S., & Sreenivasa, M. Y. (2022). Exploring biotechnological and functional characteristics of probiotic yeasts: A review. *Biotechnology Reports, 34*. https://doi.org/10.1016/j.btre.2022.e00716.

Shurson, G. C. (2017). The role of biofuels coproducts in feeding the world sustainably. *Annual Review of Animal Biosciences, 5*, 229–254. https://doi.org/10.1146/annurev-animal-022516-022907.

Shurson, G. C. (2018). Yeast and yeast derivatives in feed additives and ingredients: Sources, characteristics, animal responses, and quantification methods. *Animal Feed Science and Technology, 235*. https://doi.org/10.1016/j.anifeedsci.2017.11.010.

Da Silva, F. F., De Souza, É. S., Carneiro, L. M., Alves Silva, J. P., De Souza, J. V., & Braga Da Batista, S. J. (2022). Current ethanol production requirements for the yeast Saccharomyces cerevisiae. *International Journal of Microbiology, 2022*. https://doi.org/10.1155/2022/7878830 Hindawi Limited.

Simões, J., Coelho, E., Magalhães, P., Brandão, T., Rodrigues, P., Teixeira, J. A., & Domingues, L. (2023). Exploiting non-conventional yeasts for low-alcohol beer production. *Microorganisms, 11*(2). https://doi.org/10.3390/microorganisms11020316.

Sobhi, M., Zakaria, E., Zhu, F., Liu, W., Aboagye, D., Hu, X., Cui, Y., & Huo, S. (2023). Advanced microbial protein technologies are promising for supporting global food-feed supply chains with positive environmental impacts. *Science of the Total Environment, 894*. https://doi.org/10.1016/j.scitotenv.2023.165044.

Sousa-Silva, M., Vieira, D., Soares, P., Casal, M., & Soares-Silva, I. (2021). Expanding the knowledge on the skillful yeast cyberlindnera jadinii. *Journal of Fungi, 7*(1), 1–19. https://doi.org/10.3390/jof7010036.

Stikane, A., Dace, E., & Stalidzans, E. (2022). Closing the loop in bioproduction: Spent microbial biomass as a resource within circular bioeconomy. *New Biotechnology, 70*, 109–115. https://doi.org/10.1016/j.nbt.2022.06.001.

Struyf, N., Van der Maelen, E., Hemdane, S., Verspreet, J., Verstrepen, K. J., & Courtin, C. M. (2017). Bread dough and baker's yeast: An uplifting synergy. *Comprehensive Reviews in Food Science and Food Safety, 16*(5), 850–867. https://doi.org/10.1111/1541-4337.12282.

Thayer, M. T., Garcia, R. M., Duttlinger, A. W., Mahoney, J. A., Schinckel, A. P., Asmus, M. D., Jones, D. B., Dunn, J. L., & Richert, B. T. (2023). Feeding a whole-cell inactivated Pichia guilliermondii yeast to gestating and lactating sows in a commercial production system. *Translational Animal Science, 7*(1). https://doi.org/10.1093/tas/txac160.

Trivedi, N. B., Jacobson, G. K., Tesch, W., & Friend, J. P. (1986). Bakers' yeast. *Critical Reviews in Biotechnology, 4*(1), 75–109. https://doi.org/10.3109/07388558609150791.

Tullio, V. (2022). Yeast genomics and its applications in biotechnological processes: What is our present and near future? *Journal of Fungi, 8*(7). https://doi.org/10.3390/jof8070752.

Tullio, V. (2024). Probiotic yeasts: A developing reality. *Journal of Fungi*. https://doi.org/10.20944/preprints202405.1641.v1.

Türker, M. (2014). Yeast biotechnology: Diversity and applications. In *Proceedings of 27th VH Yeast Conference* (pp. 4–15). Advances in Science and Industrial Productions of Baker's Yeast, İstanbul, Türkiye İstanbul.

Türker, M., Kanarya, A., Yüzgeç, U.ğur, Kapucu, H., & Şenalp, Z. (2006). Drying of baker's yeast in batch fluidized bed. *Chemical Engineering and Processing: Process Intensification, 45*(12), 1019–1028. https://doi.org/10.1016/j.cep.2006.01.016.

Türker, M., Selimoğlu, S. M., & Taşpınar-Demir, H. (2022). Waste(water) to feed protein: Effluent characteristics, protein recovery, and single cell protein production from food industry waste streams.

Varela, J. A., Gethins, L., Stanton, C., Ross, P., & Morrissey, J. P. (2017). Applications of Kluyveromyces marxianus in biotechnology. In *Yeast diversity in human welfare*. Singapore: Springer Singapore, 439–453. 10.1007/978-981-10-2621-8_17.

Vejarano, R., & Gil-Calderón, A. (2021). Commercially available non-saccharomyces yeasts for winemaking: Current market, advantages over saccharomyces, biocompatibility, and safety. *Fermentation, 7*(3). https://doi.org/10.3390/fermentation7030171.

Walker, G. M., & Walker, R. S. K. (2018). Enhancing yeast alcoholic fermentations. *Advances in Applied Microbiology, 105*, 87–129. https://doi.org/10.1016/bs.aambs.2018.05.003 9780128151815.

World Wine Production Outlook OIV First Estimates. (2023). https://www.oiv.int/sites/default/files/documents/OIV_World_Wine_Production_Outlook_2023.pdf.

Ye, L., Bogicevic, B., Bolten, C. J., & Wittmann, C. (2024). Single-cell protein: overcoming technological and biological challenges towards improved industrialization. *Current Opinion in Biotechnology, 88*. https://doi.org/10.1016/j.copbio.2024.103171.

Yeast Market Outlook, 2031. (2020). https://www.transparencymarketresearch.com/yeast-market.html.

Yüzgec, U.ğur, Türker, M., & Becerikli, Y. (2008). Modelling spatial distributions of moisture and quality during drying of granular baker's yeast. *Canadian Journal of Chemical Engineering, 86*(4), 725–738. https://doi.org/10.1002/cjce.20089.

Zeko-Pivač, A., Habschied, K., Kulisic, B., Barkow, I., & Tišma, M. (2023). Valorization of spent Brewer's yeast for the production of high-value products, materials, and biofuels and environmental application. *Fermentation, 9*(3). https://doi.org/10.3390/fermentation9030208.

Zhang, X., Li, B., Zhang, Z., Chen, Y., & Tian, S. (2020). Antagonistic yeasts: A promising alternative to chemical fungicides for controlling postharvest decay of fruit. *Journal of Fungi, 6*(3). https://doi.org/10.3390/jof6030158.

Zhou, N., Semumu, T., & Gamero, A. (2021). Non-conventional yeasts as alternatives in modern baking for improved performance and aroma enhancement. *Fermentation, 7*(3). https://doi.org/10.3390/fermentation7030102.

Yeast-derived products and their commercial applications

17

Mustafa Türker, Ümit Cem Derman, and Filiz Alemdar
Pak Gıda Üretim ve Pazarlama A.Ş., Kocaeli, Turkey

17.1 Introduction

Microbial biomass is an integral part of any microbial bioproduction process and biocatalyst where live microorganisms are used to transform organic substrates into useful products in a process called "fermentation." The biomass itself can be the product including baker's yeast, probiotic yeast, dried yeast, and single-cell protein. However, most industrial bioproducts are compounds secreted outside cells including alcohols, organic acids, enzymes, proteins, and biochemicals, and then purified from cultivation media. Thus, product recovery often starts with cell separation from the liquid broth, and the remaining cells called spent microbial biomass (SMB) are a common bulk fermentation residue. Traditionally, most research attention has been directed toward valorization of spent brewer's yeast to produce nutritional yeast and yeast components for the food and feed industry (Jaeger et al., 2020; Oliveira, Ferreira, et al., 2022; Oliveira, Odila Pereira, et al., 2022). With the global transition to bioeconomy, significant market growth is expected for both bulk and specialty bioproducts. The global bio-based chemicals market is expected to increase and as a result, the generation of industrial SMB will also increase (Deuchande et al., 2024; Stikane et al., 2022).

Microbial biomass consists of insoluble cell walls containing a high proportion (15%–30% of the dry cell mass) of polysaccharides, mainly β-glucan, mannan oligosaccharides, and chitin, and remaining soluble fractions mainly containing proteins, nucleotides, several bioactive compounds, and minerals. All these components of yeast cells can be extracted and separated into specialty yeast products using cell disruption and extraction techniques. These specialty products have several applications in food, healthcare, cosmetic, biotechnology and feed industries, replacing animal origin compounds with much lower footprints (Dong et al., 2024, Liu et al., 2016; Tao et al., 2023).

The production of most natural products from plant extraction, biochemicals, and proteins also faces challenges in environments, due to worsening climatic conditions. Therefore, there is a growing interest in the green production of target products with renewable raw materials. Commercial applications of enzymes are now growing in diverse sectors like food, agro, biofuel, pharmaceutical, textile, leather, cosmetics, waste management, and so on. Most of the enzymes are produced at an industrial scale from fungi and bacteria (Fasim et al., 2021), but enzymes from yeasts are steadily growing as a consequence of the robustness of yeasts in large-scale fermentations. Yeast cell factories, together with new synthetic biology tools, provide

Eukaryotic Microorganisms as Sources of Bioproducts. DOI: https://doi.org/10.1016/B978-0-443-30188-9.00007-X

a promising alternative due to fast growth, clean process, and no competition in land and food sector. In recent decades, yeast cell factories have been applied in large-scale production of numerous products and there are many commercial examples (De Brabander et al., 2023; Ewing et al., 2022; Matassa et al., 2016).

17.2 Yeast-derived products

Yeast proteins are traditionally used in the form of yeast extract, which is safe and nutritious, and is now considered a natural, high-quality product capable of meeting diverse food flavor requirements and supplying essential dietary nutrients. Yeast extract is usually defined as the water-soluble extract produced from yeast waste streams, including baker's yeast, brewer's yeast, *Candida utilis*, *Candida tropicalis,* and *Kluyveromyces marxianus*, following disruption of the cell membrane (Lukondeh et al., 2003; Tao et al., 2023; Timira et al., 2024; Tullio, 2022). Several cell disruption techniques are used to disintegrate yeast cells; autolysis, plasmolysis, enzymatic lysis, and physical methods. After the yeast cells are disrupted, cell wall debris can be removed by centrifugation, and further processing such as treatments with acids, bases, salts, or heat can achieve the filtration and purification of yeast proteins shown in Fig. 17.1 (Felix Jacob et al., 2019). Typically, the released yeast proteins can be extracted by alkaline solubilization and isoelectric point precipitation methods for further use (Felix Jacob et al., 2019; Timira et al., 2024; Va Gautério et al., 2023).

Yeast cell wall polysaccharides mainly consist of *β*-glucans and mannans as well as small amounts of chitins and glycogens. The acid-base method was most commonly used in the preparation of β-glucan and mannan. Subsequent extractions are then carried out by hot alkali treatment using 3% NaOH or 1 M NaOH, which is the condition often preferred to cold alkali to solubilize β-glucans. Glucan fractions

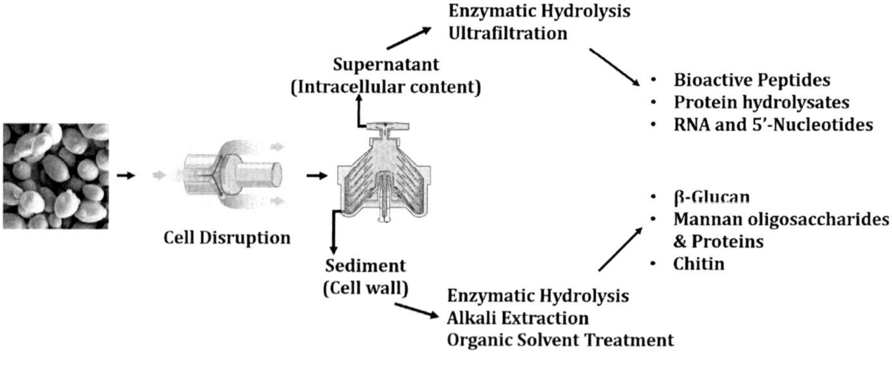

Figure 17.1 Steps for the extraction of protein hydrolysates, bioactive peptides, *β*-Glucan, nucleotides, and MOS.
Cell disruption techniques are used to disintegrate yeast cells, further processing such as treatments with acids, bases, salts, or heat can achieve the filtration and purification of yeast derivatives.

are therefore defined according to their solubility under alkaline conditions after extraction. Alkali-insoluble glucans have higher molecular weight (MW) than alkali-soluble glucans (Liu et al., 2021; Va Gautério et al., 2023). To obtain a mannoproteins-rich fraction (i.e., mannan-protein complexes), the alkali-soluble fraction is further treated with cold organic solvents such as ethanol or methanol to precipitate mannoproteins (Schiavone et al., 2023).

The global specialty yeast market size and share revenue were valued at approximately 2.9 billion USD in 2020. Furthermore, it is projected to reach 4.0–4.3 billion USD by 2026, with a 7.9% compound annual growth rate (CAGR) from 2021 to 2026 (Baek et al., 2024; Patterson et al., 2023). The increased demand for natural food ingredients is expected to amplify the specialty yeast industry's market growth. The yeast market can be further divided by type: yeast extract, yeast β-glucan, yeast autolysates, and hydrolyzed yeast; species: *Saccharomyces cerevisiae*, *Candida utilis,Pichia pastoris,* and *Kluyveromyces*; and application segment: feed, food, savory snacks, cheese, bakery, confectionary, ready-to-eat, functional foods, biofuels, bioethanol, and beverages (Baek et al., 2024; Specialty Yeast Market, 2024).

17.2.1 Yeast protein hydrolysates

Yeast protein-rich extracts have also been used as human dietary supplements, flavor enhancers, or additives in fermentation processes and animal feed. Some of the protein-rich products currently on the market are derived from yeast [*S. cerevisiae* and *Candida utilis* (Torula)] obtained from different extraction processes (hydrolysis and autolysis) and/or spray-dried. Beyond the high-protein content, these extracts are also rich in amino acids and bioactive peptides, among others, which are responsible for some of the bioactive properties observed. In fact, except for being used as protein sources, several of the remaining applications of these extracts (e.g., flavor enhancement) are dependent on the smaller molecules present (amino acids), and the choice of commercializing the whole protein extract is mainly economic (Oliveira et al., 2022; Pereira et al., 2021).

The amino acids content in protein biomass among yeast species is comparable with a high content of lysine, the most limiting amino acid in wheat and other cereals given in Table 17.1. The essential amino acids in yeast proteins contain more than 40% of the total amino acids, and the ratio of essential to nonessential amino acids ranges from 0.86 to 0.91, which is in accordance with and exceeds the Food & Agriculture Organization (FAO)/World Health Organization (WHO) standards of 40% and 0.6 respectively (Jaeger et al., 2020; Ma, Xia, et al., 2023; Pobiega et al., 2024; Tomé & Extracts, 2021). One hundred grams of yeast biomass contains a mean of 45–55 g of protein which corresponds exactly to almost 100% of recommended daily intake for adults (50 g). Yeast protein also contains all essential amino acids in the required quantity according to FAO recommendations, making yeast SCP a complete protein. Therefore, yeast biomass provides a good quality protein with a healthy balance of amino acids including complete exogenous ones, being adequate for human and animal consumption (Bratosin et al., 2021; Jach et al., 2022; Prakash et al., 2024; Tao et al., 2023).

Table 17.1 Average amino acid content of proteins of some yeast species in comparison to plant and animal proteins (g amino acids per 100 g of protein) (Food & Agriculture Organization of the United Nations World Health Organization United Nations University, 2007; Jach et al., 2022; Karimi et al., 2019; Landi et al., 2021; Molfetta et al., 2022; Pobiega et al., 2024; Razzaq et al., 2020).

Amino acids	***S. cerevisiae***	***K. marxianus***	***C. utilis***	***Y. lipolytica***	**Soybean**	**Wheat**	**Cow milk**	**Egg**	**FAO recommendations for essential amino acids (g per 100 g protein for adults)**
Isoleucine	4.4	6.2	4.8	4.4	5.3	3.3	4.0	6.8	3.0
Leucine	9.1	9.6	7.1	6.8	7.7	6.7	8.8	9.0	5.9
Valine	8.0	5.3	5.5	5.3	5.3	4.4	4.8	7.4	3.9
Lysine	6.6	8.2	5.1	7.0	6.4	2.8	7.8	6.3	4.5
Phenylalanine	6.7	5.6	4.1	4.0	5.0	4.5	4.7	6.3	3.8
Tyrosine			2.0		3.7	3.6	5.0	2.0	
Methionine	2.7	2.5	2.4	1.2	1.3	2.5	2.9	3.2	1.6
Cysteine		1.1	1.6	1.1	1.9	1.5	0.9	0.9	0.6
Tryptophan			3.9		1.4	1.1		2.9	0.6
Threonine	5.2	5.1	4.1	4.8	4.0	2.9	4.9	5.0	2.3
Alanine	3.9	3.6		8.0	5.0		3.3		
Arginine	6.7	6.0	3.2	4.8	7.4	4.8	3.3	1.2	
Aspartic acid	8.2				1.3		7.5		
Asparagine	2.9								
Glutamic acid	5.3				19.0		21.7		
Glutamine	4.3								
Glycine	3.2	2.7		4.6	4.5		1.8		
Histidine	1.7	2.7	1.6	2.6	2.6	1.6	3.7	0.4	1.5
Proline	0.8				5.3		6.6		
Serine	3.0				5.8		6.0		

Yeast extract-based seasonings were investigated as salt replacements, showing their potential to replace salt in foods while maintaining the original taste and nutritional value. Yeast extract can also be useful in yogurt fermentation, which in combination with added yeast extract containing low-molecular-weight peptides promotes the growth of beneficial bacteria and shortens the fermentation time by 21%. The addition of β-glucan from yeast extract to yogurt increased its thickness and improved its sensory evaluation (Tao et al., 2023; Tomé & Extracts, 2021). These foods are manufactured in many countries under different names but with quite similar characteristics: Vegemite and Promite (Australia), Marmite (United Kingdom and New Zealand), Vitam-R (Germany), and Cenovis (France and Switzerland) (Tullio, 2022). In recent years, a new food additive has been developed, that is, yeast extract rich in polyphosphate, which contains a high-energy phosphoric anhydride bond and is an effective meat preservative, which can keep fresh meat stable, soft, and juicy. In particular, meat that has become tough can be restored to its original tender state by treatment with the high-phosphate extract (Demling et al., 2024).

Baker's/brewery yeast is known to be a good source of nucleotides due to its high content of RNA, which makes up 8%–12% of dry matter. Moreover, certain nucleotides, especially 5'-monophosphates, are recognized as flavor enhancers. During autolysis, endogenous enzymes break down nucleic acids into nucleotides, producing 5'-nucleotides (71%–88% RNA). They can be partially hydrolyzed by the addition of enzymes, usually with 5'-phosphodiesterase (Sombutyanuchit et al., 2001; Zhao et al., 2005). Different processing methods can result in a wide range of flavor compounds in yeast as a result of interactions between amino acids, nucleotides, carbohydrates, and peptides. Yeast extracts rich in 5'-nucleotides are widely used in the food industry, mimicking meat or "umami" flavors. These flavors are created by the synergistic action of 5'-nucleotides, glutamic acid, and cysteine. The capacity of glutamate to improve taste increases by 10–15 times if 5'-nucleotides are added (Belem & Lee, 1998; Sombutyanuchit et al., 2001; Zhao et al., 2005). Nucleotides have a positive effect on the digestion and absorption of nutrients in the gut microbiota. Therefore, the supplementation of nucleotides in animal feed has positively affected intestinal morphology and function, immune response, the composition of intestinal microbiota, liver function, and morphology as well as growth performance in animals (Fathima et al., 2023; Shurson, 2018). High-purity RNA has several applications in the fermentation, pharmaceutical, and cosmetics industries. Japanese company Nippon Paper Industries Co. Ltd. produces high-purity RNA from *Candida utilis* (Torula yeast) cultured on wood sugar using unique extraction and isolation techniques (Nippon Paper Industries, 2024). Yeast extract contains abundant amino acids, vitamins, nucleosides, polypeptides, and minerals and is, therefore, an ideal nutrient growth medium for both laboratory and industrial microbial fermentation, especially for auxotrophic strains. The peptides and amino acids in yeast extract are also the main factors in the growth of lactic acid bacteria and can promote up to 60% of biomass. Overall, yeast extract is inexpensive and often has an ideal nutrient composition for use by the fermentation industry (e.g., growth of lactic acid bacteria); however, careful selection of a suitable extract is required for each microorganism to optimize the fermentation (Jacob et al., 2019; Jaeger et al., 2020; Tao et al., 2023).

Yeast extract has beneficial biological effects on the skin, such as moisturizing the skin, promoting cell renewal, slowing skin aging, and speeding up wound healing. It is a well-established cosmetic ingredient to achieve the desired cosmetic functions when combined with other substances, such as vitamins, moisturizers, and antioxidants. Yeast extract degrades and removes melanin from human skin and can be used in skin-lightening formulations (Tao et al., 2023).

Yeast extracts have found applications in the medical and healthcare fields, because of their biological activities, high nutritional content, activities in managing and preventing human diseases, and improving dysfunction of the intestinal microbial balance. Commonly used antiinflammatory and antibacterial treatments usually contain yeast extract. Yeast extract can be used to treat skin diseases, and their efficacy of yeast extracts was mainly attributed to the flavonoids, dextran, amino acids, and vitamins present in yeast extract (Tadioto et al., 2023; Tao et al., 2023). Bioactive peptides become popular over the years as they have demonstrated important biological benefits in the digestive, immune, cardiovascular, and nervous human systems. Although many commercial peptides are chemically synthesized, they can also be obtained from yeast. The recovery of this fermentation by-product for the production of functional ingredients is an important step in the increasing demand to implement and promote a circular economy-based industry (Oliveira et al., 2022).

17.2.2 Bioactive peptides

Bioactive peptides can be found in protein-rich extracts and several studies reported their antihypertensive, antioxidant, antimicrobial, and antidiabetic effects (Oliveira et al., 2022). Yeast-originated peptides can be categorized into three different groups, including (1) bioactive peptides produced through autolysis or hydrolysis of yeast cells which are responsible for functional properties of yeast extract, (2) bioactive peptides produced by yeast cells during the fermentation process and are responsible for the functional properties of yeast fermented products, and (3) antimicrobial peptides released by yeast cells during the growth phase (Doolam et al., 2024; Mirzaei et al., 2021). For example, after the hydrolysis by pepsin, trypsin, alcalase, flavorase, and chymotrypsin, yeasts can produce yeast polypeptides with angiotensin-converting enzyme inhibitor (ACE-I) activity and antioxidant peptides (Ma, Sun, et al., 2023).

Yeast bioactive peptides can be obtained by physical, chemical, or enzymatic extraction processes in which proteins are hydrolyzed and released from cell membranes, followed by a purification process. Membrane filtration technology has been widely used for the separation and concentration of baker's or brewer's yeast proteins and bioactive peptides, since it can be operated under gentle conditions, allowing cell debris removal with low energy consumption and without compromising molecular structures, and consequently, their biological activities. Depending on protein and peptides MW and the intended application, different membrane filtration cut-offs and materials can be used in order to obtain the desired product (Marson et al., 2022; Oliveira et al., 2024; Oliveira et al., 2022). Frequently, this technique is applied after yeast protein extraction or prior to other protein purification methods, such as chromatography (Liu et al., 2016; Oliveira et al., 2022; Wei et al., 2024).

The glutathione (GSH) tripeptide, formed by a thiol of cysteine, glycine, and glutamine, is a major example of these products and protective functions in eukaryotes and in most prokaryotes. The primary function of GSH is to maintain the redox potential of the cell, which is directly connected to GSH concentration, and to prevent cellular damage caused by reactive oxygen species or toxic heavy metals. Therefore, GSH has a wide range of applications in food, beverages, and healthcare as active pharmaceutical ingredients. The fed-batch procedure with yeasts of the genera *Saccharomyces* and *Candida* is still a common method for industrial production. Japanese company (Kohjin Life Sciences, 2024) produces GSH from Torula yeast (*Candida utilis*) (Kohjin Life Sciences, 2024). The GSH and GSH-enriched yeast cells are also used to improve the technical properties of food or beverage products, e.g., as aroma stabilizers, flavor enhancers, or baking agents. In the latter case, GSH-enriched yeasts are used to alter disulfide bonds in the protein network of wheat dough with the effect of dough weakening and modified baking properties (Raza et al., 2020). GSH is available in the market in the form of GSH-enriched yeast cells or its derivatives and crystalline form. Annually, more than 200 tons of pure crystalline GSH are produced worldwide. Due to its antioxidant character, it is widely used in the pharmaceutical, cosmetic, beverage, food, and feed industries with a market value of around 317 M$ in 2020 (Schmacht et al., 2017). Riboflavin is also an important bioactive compound produced by yeasts. Capable of preventing deficiency symptoms such as dermatitis, this vitamin also plays a key role in the food industry. Currently, the yeast *Meyerozyma guilliermondii* stands out in the large-scale production of riboflavin, with a market value of around 400 M$ in 2023 (Tadioto et al., 2023; Timira et al., 2024). The high incidence and prevalence of metabolic diseases have valorized the peptide therapeutics market, valued at $25 billion in 2018, and which is expected to increase at a CAGR of 7.9% from 2019 to 2027 (Oliveira, et al., 2022; Wei et al., 2024).

17.2.3 Yeast polysaccharides

17.2.3.1 β-glucans

Yeast cell wall glucan and mannan have been widely studied and gained commercial interest for their extensive industrial applications. Although food and feed industries initially use yeast polysaccharides as dietary fibers, emulsifying agents, or fat replacers, there is an increasing interest in potential pharmaceutical and biomedical applications, with an emerging focus on immunology, tissue engineering, vaccines, and drug delivery (Freitas et al., 2015). β-glucan exhibits diverse health functions, including immunomodulatory, cholesterol-lowering, anti-inflammatory, antidiabetic, anticancer, and prebiotic effects. These attributes have sparked intense interest in exploring β-glucan's potential in various clinical investigations, shedding light on its efficacy in managing conditions such as diabetes, cardiovascular diseases, and gastrointestinal disorders (Bastos et al., 2022; Schiavone et al., 2023). The immune activity of β-glucans can be increased by simple chemical (phosphorylation or carboxylation) and physical (purification and drying) modifications. The combination of carboxylation and air drying is known to result in high-quality extraction and increased immune activity of β-glucans from yeast. In animal species, the consumption of β-glucan plays an important role in the adsorption of mycotoxins that accumulate in animal feed. Yeast β-glucans have helped to enhance the immune defenses of fish (Machuca et al., 2022). In addition, depending on the β-glucan nature, β-glucan intake decreases plasma cholesterol and immune system stimulation. Many mycotoxins cause varying degrees of toxicity to humans and livestock. The structures of β-glucan and mycotoxin complement each other and form a chemical complex with the hydroxyl group of the mycotoxin, thereby reducing toxicity (Baek et al., 2024; Fathima et al., 2023; Pang et al., 2022; Singla et al., 2024; Zeko-Pivač et al., 2023).

In the cosmetic industry, a β-D-glucan polysaccharide was applied in the formulations of several products, such as body lotions, sunscreen creams, facial cleansers, toners, eye creams, shampoo, body wash, and hand-care creams. This bioproduct is mainly added for its properties as a moisturizing agent and for conferring oxidation and radiation resistance. β-glucan contains beneficial cosmeceutical properties including antioxidants, moisture retention and skin regeneration, that can be applied in skincare especially for sensitive skin (Natakankitkul et al., 2015; Sousa-Silva et al., 2021).

β-glucan is not widely used in the food industry, which is due to the insufficient awareness of manufacturers regarding its functional and technological properties (Baek et al., 2024; Mykhalevych et al., 2022). Generally, β-glucan is associated with a biologically active additive that is used for therapeutic purposes or as part of various diets for people with diabetes, obesity, and cardiovascular diseases. This makes sense because there have been a lot more clinical studies on the chemical makeup of β-glucan and how it affects the way the human body works than on how it is used in food technology. β-glucans are used in the food industry as thickeners, water-retention or oil-binding agents, and emulsion stabilization agents. Yeast β-glucans can be used to partially replace fat in mayonnaise, resulting in increased moisture content with a significant reduction in caloric value. However, they can cause negative sensory properties in mayonnaise. Enrichment of yogurt with β-glucan can greatly affect the gel formation, which significantly reduces the fermentation time and allows for the potential use of β-glucan as a thickener (Mykhalevych et al., 2022; Rose, 2023). Techniques such as homogenization and spray drying after autolysis and extraction of spent brewers yeast (SBY) increase the β-glucan yield. It is possible to achieve higher viscosity, water retention capacity, and emulsion stabilization ability compared to commercial products. The biological properties of β-glucan depend on the source (fungi, bacteria, algae, or cereals), structure, degree of solubility, formation of conformation in solution, MW and polymer charge (Baek et al., 2024; Mykhalevych et al., 2022). The global β-glucan market was assessed to be worth 340 million USD in 2018 and is predicted to generate a revenue of approximately 657 million USD by 2027, which includes the dairy industry as one of the branches that need β-glucan as a technological ingredient (Mykhalevych et al., 2022). However, the yeast β-glucan market was assessed to be worth 174.2 million USD in 2021 (Niego et al., 2023).

17.2.3.2 Mannan-oligosaccharides

Mannan is a polysaccharide, or mannose polymer ($C_6H_{12}O_6$), the generic name for polysaccharide moiety of glycoprotein and their structure can vary significantly in terms of MW, degree of branching, and the presence of side chains. More highly branched mannans might exhibit enhanced interactions with specific receptors on immune cells, potentially leading to stronger or more targeted immune responses (Baek et al., 2024). Mannan is a bioactive polysaccharide prominent in various sectors due to its exploitable biodegradable properties. Furthermore, its amphiphilic nature serves as a bioemulsifier and is also used as a bioactive material in health-related applications. For example, mannans conjugated to vaccine preparations are already in clinical practice. Mannan extracted from yeast has received increasing attention because the yeast extract sector produces extensive cell wall by-products, establishing a cheap source for obtaining mannan. In addition, long mannan chains are hydrolyzed into short chains known as mannan-oligosaccharide (MOS) (Schiavone et al., 2023; Timira et al., 2024).

Mannan or OS is regularly used in the feed industry. For instance, livestock that consume MOS from an early stage exhibit remarkably improved immune systems. In addition, mannan extracts have also been studied for their beneficial effects on humans (Baek et al., 2024; Schiavone et al., 2023; Timira et al., 2024). They are recognized as valuable biologically active raw materials for the pharmaceutical and cosmetology industries. Nontraditional immunostimulants demonstrating antimicrobial, antiviral, and antitumor activities. Their antimicrobial, antiviral, antitumor, antimutagenic, and antioxidant properties have been confirmed and their effectiveness in wound healing has been proven. For example, mannan aids probiotic bacteria survival in yogurt, while a clinical study reported yeast mannan supplements improved gut microbiota and skin dryness in healthy subjects (Snyman et al., 2021).

Yeast mannoproteins stabilize the yeast cell wall, promote yeast cell development and survival, and affect wine and beer flavor, aroma, and mouthfeel. Mannoprotein is commonly utilized in the food industry as a biological emulsifier that can replace synthetic agents and stabilizers. Besides their functional qualities, yeast mannoproteins may improve gut health and immunological function. Yeast mannoproteins, isolated from yeast cells by acid hydrolysis or enzymatic digestion, are utilized as additives in the food and beverage industries to improve product quality (Baek et al., 2024; Snyman et al., 2021; Timira et al., 2024).

Mannan oligosaccharides significantly improve the production performance of broiler chickens, characterized by improved body weight gain and feed conversion ratio. In breeders, MOS supplementation improves sperm density, hatchability, and maternal antibody titers in progeny. Mannoproteins can also adsorb mycotoxins by as much as 21%. Hence, mannoproteins can be used as a feed additive to decrease the mycotoxin levels in poultry (Fathima et al., 2023; Pang et al., 2022). Spring et al. (2015) reviewed results from 733 published trials that evaluated the effects of feeding MOS to animals and various aquaculture species and showed generally positive improvements in body weight gain and feed conversion while reducing mortality. These beneficial responses are a result of MOS binding and limiting the colonization of pathogens in the gastrointestinal tract (Shurson, 2018; Spring et al., 2015).

17.2.3.3 Chitin

Chitin is a polymer of glucosamine found in the cell walls of fungi and yeast, but its concentration is much less than glucans and mannans. This biopolymer has versatile applications in the biomedical, food industry, and agriculture due to its biocompatibility, biodegradability, strong antibacterial effect, and nontoxicity (Abo Elsoud & El Kady, 2019; Freitas et al., 2015). However, its effects on animal nutrition and health are not well studied (Shurson, 2018).

17.3 Spent microbial biomass

SMB is the second major by-product derived from fermentation processes, representing about 15%–20% of the total residues generated. The quantity of this residual biomass is increasing as the biobased economy is expanding. Dried yeast, brewer's dried yeast, brewer's liquid yeast, Torula dried yeast, *Yarrowia lipolytica,* and whey yeast are considered to be nutritional yeast products because they consist of yeast biomass of dead yeast cells and are used for their nutritional value and source of biochemicals. Brewers spent yeast (BSY) is one of the major brewing by-products, with 2–4 kg of spent yeast (SY) waste produced per 100 L of beer, generally recovered for reuse in various applications as the platforms of synthetic biotechnology evolve, the production of a myriad of molecules through fermentation increases, and consequently, the side streams generated in these processes are also foreseen to increase in the coming years. The amount of SMB formed depends on the bioproduct and the fermentation process. (Oliveira et al., 2022). Brewer's yeast is commonly used in the animal feed industry as a specialty amino acid, vitamin, and mineral supplement. According to the available public data, 1.86 billion hectoliters of beer were produced worldwide in 2021 and 1.88 billion hectoliters in 2023 (Brauwelt, 2024). Therefore, BSY production in 2023 can be estimated between 376,000 and 752,000 tons. The Global SBY market was estimated at USD 1290 million in 2021, with the average CAGR of 6.12% from 2021 to 2027 reaching USD 1842 million. Considering this, BSY holds a significant economic potential. The trend of increasing valorization of BSY is confirmed by the growing number of reviews on this topic (Bastos et al., 2022; Doolam et al., 2024; Łukaszewicz et al., 2024; Oliveira et al., 2022; Zeko-Pivač et al., 2023). Torula yeast (*Candida utilis*) also contains high concentrations of protein, minerals, and vitamins, and it is used as an additive to a wide variety of processed foods. Historically, Torula yeast was also used in animal feeds, but today, most of it is used by food manufacturers. Although whey yeast was commonly available and used in the feed industry for many years, very little is produced and available today.

Similar to brewing, wine and bioethanol industries also generate spent microbial yeast biomass as a by-product. A recent OIV report gives an estimate of world wine production of approximately 244 million hectoliters for 2023 (World Wine Production Outlook, 2023). Assuming similar biomass yields, the amount of wine SY can be estimated in the range of 49,000–98,000 tons per annum which is significantly lesser than BSY. Similarly, spent yeast biomass generated in the bioethanol industry is recovered within dried distiller grains solubles (DDGS) in corn-based processes and used in animal feed (Shurson, 2017), while the yeast biomass is recycled back to the process after acid treatment to control bacterial contamination in molasses-based alcohol production (Stikane et al., 2022).

Currently, valuable biomolecules are sustainably produced through fermentation using genetically modified microorganisms. However, these industries also generate spent yeast as a by-product, and their potential need to be evaluated. It can be observed that the production of specialty bioproducts, such as vitamins, protein, and antibiotics, generates higher amounts of SMB per unit weight of product than bulk chemicals such as alcohols, organic acids, and amino acids. Thus, the highest SMB production amounts were estimated to be 110 kg per kg of vitamins, while the lowest was 0.1 kg per kg of organic acids. To close the sustainability loop in bioproduction, it is essential to characterize the composition and define valorization strategies for such waste streams (Deuchande et al., 2024; Stikane et al., 2022). The gross composition of six different SMBs derived from genetically engineered *Saccharomyces cerevisiae* strains has been analyzed and it was observed that slight differences in the fermentation conditions led to significant differences in yeast composition by the end of the fermentation process which may lead to different valorization strategies in integrated biorefineries (Deuchande et al., 2024). Spent yeast *Pichia guilliermondii* generated from citric acid production has been commercially used for animal feed by US-based company Archer Daniels and Midland (ADM; Thayer et al., 2023). Astaxanthin is a red–orange carotenoid belonging to the xanthophyll family. Although the natural sources of astaxanthin include bacteria, fungi, algae, crustaceans, and certain fishes, important fungal astaxanthin producers are the yeasts *Phaffia rhodozyma* and *Xanthophyllomyces dendrorhous*. The market price of astaxanthin ranges from USD 2500 to USD 7000/kg with a global market value of USD 1.94 billion in 2022 (Niego et al., 2023). Although the natural sources of astaxanthin such as yeasts are more expensive than the synthesized carotenoids, their bioavailability may be higher because it is related to the organic matrix (Shurson, 2018).

17.4 Yeast cultures or fermentates

Yeast cultures, also called fermentates, are unique among yeast products because they contain a combination of yeast biomass and fermentation metabolites produced during specific fermentation processes, including (1) bioactive peptides produced through autolysis or hydrolysis of yeast cells which are responsible for functional properties of yeast extract, (2) bioactive peptides produced by yeast cells during the fermentation process and are responsible for the functional properties of yeast fermented products, and (3) antimicrobial peptides released by yeast cells during the growth phase. While they contain some residual viable yeast cells, they are not a substantial source of yeast biomass. To produce yeast cultures, a specific culture media is inoculated with live yeast cells and allowed to ferment under specific conditions, upon which the entire fermented media is subsequently dried. As yeast cells ferment the sugars present in the culture media, they produce a wide variety of metabolic products including peptides, alcohols, esters, and organic acids. However, the composition of the various metabolites produced depends on the composition of the fermentation media and conditions (Doolam et al., 2024; Mirzaei et al., 2021). Therefore, yeast cultures or yeast fermentates are unique food and feed additives that contain several undefined metabolites that may have beneficial nutritional and health effects for animals as well as for humans but are not considered to be a nutrient source in animal feeds. The Diamond V (owned by Cargill) commercialized a brand of unique yeast (*Saccharomyces cerevisiae*) fermentation product EpiCor® for improved human and animal health worldwide to support their immune and gut health (Evans et al., 2012; Jensen et al., 2015; Schauss et al., 2012).

Yeast culture supplementation improves weight gain, feed conversion ratio, serum lysozyme concentration, and antibody titer in broiler chickens. Supplementation of the soluble fraction of yeast culture has antiinflammatory effects in vitro. The fermentation times and growth media can alter the beneficial effects of yeast cultures (Fathima et al., 2023; Schlabitz et al., 2022). Adding appropriate yeast culture to the diet significantly improved the digestibility of calcium and phosphorus for broilers. More importantly, yeast culture increased the humoral and cellular immunity of broilers and alleviated the inhibitory effect of coccidiosis on the performance of broilers. The stress ability of weaning piglets was enhanced after yeast culture supplementation, while the number of panting or dying piglets was decreased. The activities of xylanase, endoglucanase, fibrinoglycogenase, and cellulase in the rumen of beef cattle were significantly increased by adding yeast culture to the feeding diet, which indicated that yeast culture improved the digestibility of ruminants by increasing the enzyme activity in the rumen (Mirzaei et al., 2021; Pang et al., 2022). There are a lot of dairy products available in our markets, such as cheese, yogurt, kefir, koumiss, and fermented milks in which yeast–bacteria interactions create unique and different flavors. There are many yeast species actively involved in association with bacterial strains in the development of desirable and sensory properties in these products such as *K. marxianus*/*Candida kefyr*, *K. lactis*, *S. cerevisiae*, *G. candidum*, and *D. hansenii*. They have shown some therapeutic properties as a natural probiotic beverage (Nandy & Srivastava, 2018).

17.5 Specialty yeast products

Some specific yeast products are also being used in a variety of food and feed industries. Several specialty yeast products are used in various amounts and for unique nutritional applications in the feed industry including irradiated yeast, selenium yeast, chromium yeast, and Phaffia yeast. The yeast biomass contains trace minerals, such as calcium, copper, iron, phosphorus, potassium, magnesium, manganese, sodium, selenium, chromium, and zinc. Industrial trace-element-enriched yeast is mainly produced by adding mineral salts of trace elements to the fermentation medium to complete the enrichment of trace elements in the metabolic process of yeast (Mapelli et al., 2011). Yeast contains ergosterol which can be converted to vitamin D2 (ergocalciferol) when irradiated with ultraviolet light. At one time, irradiated yeast was an important source of vitamin D activity, but less expensive synthetic vitamin D3 (cholecalciferol) has replaced it in the market (Hohman et al., 2011; Shurson, 2018).

Selenium deficiency causes many diseases for both humans and animals and is a global issue. At present, there are many reports on selenium-rich yeast and it has a unique role in regulating animal metabolism, improving animal health, and increasing serine content in meat and eggs compared with inorganic feeding. Selenium yeast is commercially produced and marketed as a highly bioavailable form of Se (selenomethionine) and has unique effects on metabolism, improves the health status of animals, and increases Se content in meat and eggs compared to feeding inorganic forms (Ferrari et al., 2023; Jach et al., 2022). Although brewer's yeast contains 1–2 mg Se/kg, some commercial high selenium yeasts may contain up to 2000 mg Se/kg. Selenium yeast is a successful case of *S. cerevisiae*-enriching trace elements. Selenium-enriched yeast is widely used not only as a selenium nutritional supplement but also in the food field to produce selenium-enriched milk, selenium-enriched chicken, selenium-enriched eggs, and so on. Chromium yeast contains trivalent chromium in combination with biologically active peptides, amino acids, and niacin, which seems to function as a "glucose tolerance factor" acting as a physiological potentiator of insulin action to improve carbohydrate metabolism (Sun et al., 2022).

17.6 Enzymes from yeasts

The last few decades have seen a drastic surge in commercial applications of enzymes in diverse sectors like food, agro, biofuel, pharmaceutical, textile, leather, cosmetics, waste management, and so on. Nearly, 75% of all the enzymes produced at an industrial scale are hydrolases. The global demand and trade for industrial enzymes is growing steadily. Approximately, 60% of industrial enzymes are produced from fungi, 24% from bacteria, 4% from yeast, and the rest 10% from plant and animal sources. Nearly, 90% of commercial enzymes are made by recombinant microorganisms (Binati et al., 2021; Fasim et al., 2021). Only a limited number of yeast enzymes are currently used as bulk enzymes in commodity processes (Johnson, 2013). Komagataella phaffii is used for most of the yeast-based enzyme productions; Roche (Mannheim, Germany) produces recombinant trypsin for the digestion of proteins. Phytex, LLC/United Animal Health (Sheridan, USA) produces recombinant phytase to be used as an animal feed additive. The Nitrate Elimination Co. (Lake Linden, USA) produces Superior Stock recombinant nitrate reductase to treat and test water for enzyme-based products. Verenium/DSM (Heerlen, Netherlands) produces purifine (a recombinant phospholipase C) to degum high phosphorus oils (Barone et al., 2023). *Candida antarctica* is used for its lipase enzyme in the polymer industry for ring-opening polymerization of lactones, carbonates, and polycondensation, pulp and paper industry for pitch control, and textile industry for removing the size lubricants and denim finishing (Singh et al., 2016).

Saccharomyces species are used for producing enzymes such as invertase from *S. cerevisiae* by Biocon Ltd. (Bangalore, India) and lactase from S. fragilis (Fasim et al., 2021; Nandy & Srivastava, 2018). *K. lactis* is widely used for the industrial production of β-galactosidase and recombinant chymosin. The β-galactosidase is used to produce lactose-free dairy products and prebiotic galacto-oligosaccharides. Several commercial β-galactosidase are available from *K. lactis*, such as Maxilact® from DSM (the Netherlands), Lactozym® from Novozymes (Denmark), and HA-Lactase from Chr. Hansen (Denmark) (Zolnere & Ciprovica, 2017). Recombinant bovine chymosin from *K. lactis* is an important protein for cheese production and shows a higher specific activity than traditional rennet. Other commercially relevant proteins produced using *K. lactis* include inulinase, phospholipase B, chitinase, xylanase, and the sweet-tasting protein brazzein. Kluyveromyces lactis is also used for the manufacture of infant nutrition products and single-cell proteins (Belem & Lee, 1998; Binati et al., 2021).

In addition, several yeast enzymes or whole-cell yeasts have found application in the production of high-value specialized fine chemicals and for the biotransformation of pharmaceutical intermediates (Pscheidt & Glieder, 2008; Qiu et al., 2023). Regarding enzymes for industrial applications, USD 6.3 billion and an annual growth rate (CAGR) of 4.7% were predicted in 2021; their global market is expected to grow from USD 6.3–6.4 billion in 2021 to USD 8.7 billion within 2026, with CAGR of 6.3% for the years between 2021 and 2026 (Barone et al., 2023; Yang et al., 2024).

17.7 Food ingredients and biochemicals from yeasts

Currently, the production of most natural products either from fossil-based resources or from plant extraction causes big challenges due to environmental conditions. The green production of target products with renewable raw materials has opened up a new way for sustainable development. In this regard, yeast cell factories provide a promising alternative due to fast growth, simple raw materials, clean processes, and much less land, water, and CO2 footprint. In the recent decades, microbial cell factories have been applied in large-scale production of numerous yeast-based products that are listed in Table 17.2.

US-based company Amyris reported the development of yeast strains, for β-farnesene production. Farnesene is used as a flavoring agent and found to be an important component of the plant's essential oil with healthcare and cosmetic applications (Rautela et al., 2024). Amyris produced β-farnesene commercially at a facility in Brotas, Brazil, using sugarcane-derived sugar as a feedstock. Amyris also focused on and commercialized the production of smaller volumes of specialty products for the health, personal care, and flavor and fragrance markets; rebaudioside M (Reb M, terpenoid-based steviol glycoside), a sweetener destined for the food industry; and fragrances used in cosmetics, including: manool (MN, labdane diterpenoid), patchouli-type oil (PAT, sesquiterpene alcohol), sclareol (SCL, diterpene alcohol) and sandalwood-type oil (SDW, sesquiterpene alcohol), Sclereol, hemisqualane, bisabolol, cannabigerol (Amyris, Biotechnology Company, 2024). A major product candidate from the pharma sector is artemisinin, an antimalaria drug that is traditionally derived from plants. Sanofi/Amyris launched large-scale industrial artemisinin production and entered the market based on the original development of the cell factory by Amyris (Deuchande et al., 2024; Jullesson et al., 2015; Rautela et al., 2024).

Table 17.2 Commercial use of yeast cell factories for the production of some biochemicals (Becker et al., 2015; Chai et al., 2022; Cost Efficiencies In Lactic Acid Production, 2024; Crater et al., 2016; Deuchande et al., 2024; Ewing et al., 2022; Groenewald et al., 2014; Jullesson et al., 2015; Kumar et al., 2024; Madzak, 2021; Morrissey et al., 2017; Skoog et al., 2018; Li & Mupondwa, 2021; Madzak, 2021; Nandy & Srivastava, 2018; Rautela et al., 2024; Thayer et al., 2023; Van Dien, 2013).

Microorganisms (Yeast)	Substrate	Product	Company (Manufacturer)
Saccharomyces cerevisiae	Glucose, glycerol	β-farnesene	Amyris
Saccharomyces cerevisiae	Glucose, glycerol	Isoprene	Amyris
Saccharomyces cerevisiae	Glucose	Rebaudioside M, manool, patchouli-type oil sclareol and sandalwood-type oil, Sclereol, hemisqualane, bisabolol, cannabigerol	Amyris
Saccharomyces cerevisiae	Glucose	Artemisinin	Sanofi, Amyris
Yeast		Nootkatone, valencene	Isobionics
Saccharomyces cerevisiae (GM)		Vanillin, nootkatone, valencene, L-arabinosa, resveratrol	Evolva
Kluyveromyces marxianus		2-phenylethanol (2-PE)	Puris, Lasaffre
Yeast	Glucose	Lactic acid	Cargill
Saccharomyces cerevisiae or Pichia kudriavzevii	Glucose	Malonic acid (propanedioic acid)	Lygos
Saccharomyces cerevisiae (acid-tolerant recombinant)	Corn-starch derived sugars	Succinic acid (butanedioic acid)	Reverdia
Candida krusei (acid-tolerant yeast strain recombinant)	Corn syrup	Succinic acid (butanedioic acid)	BioAmber, Cargill, LCY Biosciences
Yeast	Fatty acids	Adipic acid (hexanedioic acid)	Verdezyne
Candida tropicalis	Fatty acids	Dodecanedioic acid	Verdezyne
Yarrowia lipolytica	Alkanes	Citric acid	Archer Daniels Midland Company
Yarrowia lipolytica		Erythritol	Baolingbao Biology Co., Ltd

In the food additives, UK-based company Isobionics produced the citrus flavors nalencene and nootkatone based on engineered yeast cells (Jullesson et al., 2015). Swiss company Evolva successfully commercialized vanillin (flavoring), nootkatone and valencene (aroma compounds), L-arabinose (sugar substitute), and resveratrol (functional ingredient) produced using different recombinant *S. cerevisiae* strains as the microbial chassis (Chai et al., 2022; Nandy & Srivastava, 2018). The flavor molecule 2-phenylethanol (2-PE) has a rose-like smell and has a range of applications in the cosmetics and perfume industry to the food and beverage industry. The two companies, Puris (South Africa) and Lasaffre (France) commercially produce 2-PE from the yeast Kluyveromyces marxianus by fermentation technology (Morrissey et al., 2017). Other applications in the domain of food industry include erythritol production. Yarrowia lipolytica has been chosen as the production host for erythritol. Since 2003, Baolingbao Biology Co., Ltd., from Shandong, China, produces erythritol with *Y. lipolytica*, for addition to foods as a nutritive sweetener, flavor enhancer, formulation aid, humectant, stabilizer and thickener, sequestrant, and texturizer (Groenewald et al., 2014; Madzak, 2021).

Probably the best-known organic acid that is dommercially produced by fermentation is lactic acid using bacterial and yeast strains. Lactic acid currently finds application as a food additive and cosmetics ingredient. Lactic acid is also of significant interest as the precursor to polylactic acid (PLA), a biobased and biodegradable polymer. US-based company Cargill (formerly NatureWorks) developed a yeast strain that required significant metabolic engineering to produce lactic acid for their PLA polymer (Cost Efficiencies In Lactic Acid Production, 2024; Van Dien, 2013). Succinic acid (SA) is in the list of top platform chemicals that are obtainable from biomass by the US Department of Energy and has several applications in food and beverages, polymers, paints, and pharmaceutical industries. The commercial production of bio-based SA has already been proven by several companies, using recombinant bacterial (*E. coli* and *B. succiniciproducens*) and acid-tolerant yeast strains. The first commercial plant for bio-based SA was opened in 2012 for the production of succinic acid by acid-tolerant recombinant *Saccharomyces cerevisia* from corn-starch derived sugars in Cassano Spinola, Italy, by Reverdia. BioAmber opened the first biobased succinic acid demonstration plant in Pomacle, France, in 2010, with a capacity of 3 kt per annum. BioAmber later applied an engineered, acid-tolerant yeast strain recombinant *Candida krusei* in a biobased succinic acid plant with a capacity of 30 kt per annum in Sarnia, Canada (Becker et al., 2015; Ewing et al., 2022; Kumar et al., 2024; Li & Mupondwa, 2021). The dicarboxylic acid dodecanedioic acid is applied in the production of coatings, adhesives, and engineering plastics such as polyamide nylon and is also a potential monomer for the production of aliphatic polyesters. The US-based company Verdezyne also invested in the development of biobased dodecanedioic acid using engineered strains of the yeast *Candida tropicalis* and constructed the first commercial plant in Malaysia. However, the operation of the plant was discontinued in 2018 due to economic reasons (Ewing et al., 2022). The industrial use of wild-type (or traditionally improved) *Y. lipolytica* strains remains relevant nowadays, notably for citric acid production by Archer Daniels Midland Company (ADM, United States) (Crater et al., 2016; Madzak, 2021; Madzak, 2021). Most probably, ADM replaced the *Y. lipolytica* strain with yeast Pichia guilliermondii for citric acid production (Thayer et al., 2023).

17.8 Conclusion

Microbial biomass is a product from which several components are derived including yeast proteins, nucleotides, biopeptides, and other functional compounds or it can be used as baker's yeast, probiotic yeast, dried yeast, and single-cell protein. The development and utilization of yeast proteins made from yeast biomass is not new and large-scale production is carried out around the world. Although the development of new, high value applications for yeast proteins is advancing, most of the production is still used in relatively low-value applications, such as animal feed and microbial culture. The application of yeast proteins in nutritional supplements, medicine, and cosmetics is still limited, and considerable further development is needed to maximize the high-value application potential of yeast proteins. Yeast polysaccharides have shown extensive bioactivity and have potential applications in many fields, which have attracted wide attention. However, they have not been fully exploited at a commercial scale yet. Yeast polysaccharides have antioxidant properties, stimulate immunity, promote intestinal health, adsorb mycotoxins, promote wound healing, and reduce blood sugar and lipids. In addition, yeast polysaccharides have interesting functional properties. Most industrial bioproducts are compounds secreted outside cells including alcohols, organic acids, enzymes, proteins, and biochemicals and then purified from cultivation media. Commercial scale production of yeast biomass and derived products are expanding and are penetrating in all aspects of our daily life.

References

Abo Elsoud, M. M., & El Kady, E. M. (2019). Current trends in fungal biosynthesis of chitin and chitosan. *Bulletin of the National Research Centre, 43*(1), 59. https://doi.org/10.1186/s42269-019-0105-y, https://doi.org/10.1186/s42269-019-0105-y.

Amyris, Biotechnology Company. (2024). https://amyris.com/.

Baek, K. R., Rani, R. S., Kim, S. J., & Oh, S. S. (2024). Yeast cell wall mannan structural features, biological activities, and production strategies. *Heliyon, 10*(6). https://doi.org/10.1016/j.heliyon.2024.e27896.

Barone, G. D., Emmerstorfer-Augustin, A., Antonino, B., Pisano, I., Coccetti, P., Mapelli, V., & Camattari, A. (2023). Industrial production of proteins with *Pichia pastoris—Komagataella phaffii*. *Biomolecules, 13*(3), 441. https://doi.org/10.3390/biom13030441.

Bastos, R., Oliveira, P. G., Gaspar, V. M., Mano, J. F., Coimbra, M. A., & Coelho, E. (2022). Brewer's yeast polysaccharides — A review of their exquisite structural features and biomedical applications. *Carbohydrate Polymers, 277*, 118826. https://doi.org/10.1016/j.carbpol.2021.118826.

Becker, J., Anna, L., Jonathan, F., & Christoph, W. (2015). Top value platform chemicals: Bio-based production of organic acids. *Current Opinion in Biotechnology, 36*, 168–175. https://doi.org/10.1016/j.copbio.2015.08.022.

Belem, & Lee (1998). Production of bioingredients from Kluyveromyces marxianus grown on whey: An alternative. *Critical Reviews in Food Science and Nutrition, 38*(7), 565–598. https://doi.org/10.1080/10408699891274318.

Binati, R. L., Salvetti, E., Bzducha-Wróbel, A., Bašinskienė, L., Cizeikienė, D., Bolzonella, D., & Felis, G. E. (2021). Non-conventional yeasts for food and additives production in a circular economy perspective. *FEMS Yeast Research, 21*(7), foab052. https://doi.org/10.1093/femsyr/foab052.

Bratosin, B., Constantin, D., Sorina, V., & Dan, C. (2021). Single cell protein: A potential substitute in human and animal nutrition. *Sustainability (Switzerland), 13*(16), 9284. https://doi.org/10.3390/su13169284.

Brauwelt. (2024). 2024 9 17. https://brauwelt.com/.

Chai, K. F., Ng, K. R., Samarasiri, M., & Chen, W. N. (2022). Precision fermentation to advance fungal food fermentations. *Current Opinion in Food Science, 47*, 100881. https://doi.org/10.1016/j.cofs.2022.100881.

Cost Efficiencies In Lactic Acid Production. (2024). 2024 9 18 https://www.cargill.com/about/research/fermentation-case-study.

Crater, J., Galleher, C., & Lievense, J. (2016). Consultancy on large-scale submerged aerobic cultivation process design – Final Technical report: February 1, 2016. June 30, 2016 http://www.nrel.gov/publications.

De Brabander, P., Uitterhaegen, E., Delmulle, T., De, W., & Karel, S. W. (2023). Challenges and progress towards industrial recombinant protein production in yeasts: A review. *Biotechnology Advances, 64*, 108121. https://doi.org/10.1016/j.biotechadv.2023.108121.

Demling, P., Baier, M., Deitert, A., Fees, J., & Blank, L. M. (2024). Biotechnological polyphosphate as an opportunity to contribute to the circularization of the phosphate economy. *Current Opinion in Biotechnology, 87*, 103107. https://doi.org/10.1016/j.copbio.2024.103107.

Deuchande, T., Fundo, J. F., Rodrigues, D., Pintado, M. E., & Amaro, A. L. (2024). Characterization of spent yeast streams from precision fermentations driven by genetically engineered yeasts: Valorization potential for a circular bioeconomy. *Waste and Biomass Valorization, 15*(6), 3553–3567. https://doi.org/10.1007/s12649-023-02394-z.

Dong, W., Li, Y., Xue, S., Wen, F., Meng, D., Zhang, Y., & Yang, R. (2024). Yeast polysaccharides: The environmentally friendly polysaccharides with broad application potentials. *Comprehensive Reviews in Food Science and Food Safety, 23*(5), e70003. https://doi.org/10.1111/1541–4337.70003.

Doolam, B., Mishra, B., Surabhi, D., Mandal, S. K., Sada, S., Reddy, N. R., Panda, J., Rustagi, S., Mishra, A. K., & Mohanta, Y. K. (2024). A systematic review of potential bioactive compounds from *Saccharomyces cerevisiae*: Exploring their applications in health promotion and food development. *Environment, Development and Sustainability, 27*, 2945–2982. https://doi.org/10.1007/s10668-024-04969-9.

Evans, M., Reeves, S., & Robinson, L. E. (2012). A dried yeast fermentate prevents and reduces inflammation in two separate experimental immune models. *Evidence-based Complementary and Alternative Medicine, 2012*, 973041. https://doi.org/10.1155/2012/973041.

Ewing, T. A., Nouse, N., Van Lint, M., Van Haveren, J., Hugenholtz, J., & Van Es, D. S. (2022). Fermentation for the production of biobased chemicals in a circular economy: A perspective for the period. *Green Chemistry, 24*(17), 6373–6405. https://doi.org/10.1039/d1gc04758b.

Fasim, A., More, V. S., & More, S. S. (2021). Large-scale production of enzymes for biotechnology uses. *Current Opinion in Biotechnology, 69*, 68–76. https://doi.org/10.1016/j.copbio.2020.12.002.

Fathima, S., Shanmugasundaram, R., Sifri, M., & Selvaraj, R. (2023). Yeasts and yeast-based products in poultry nutrition. *Journal of Applied Poultry Research, 32*(2), 100345. https://doi.org/10.1016/j.japr.2023.100345.

Felix Jacob, F., Hutzler, M., & Methner, F.-J. (2019). Comparison of various industrially applicable disruption methods to produce yeast extract using spent yeast from top-fermenting beer production: Influence on amino acid and protein content. *European Food Research and Technology, 245*(1), 95–109. https://doi.org/10.1007/s00217-018-3143-z.

Ferrari, L., Cattaneo, D. M. I. R., Abbate, R., Manoni, M., Ottoboni, M., Luciano, A., Von Holst, C., & Pinotti, L. (2023). Advances in selenium supplementation: From selenium-enriched yeast to potential selenium-enriched insects, and selenium nanoparticles. *Animal Nutrition, 14*, 193–203. https://doi.org/10.1016/j.aninu.2023.05.002.

Food & and Agriculture Organization of the United Nations World Health Organization United Nations University (2007). *Protein and amino acid requirements in human nutrition report of a joint FAO/WHO/UNU expert consultation.* World Health Organization Geneva. https://iris.who.int/handle/10665/43411.

Freitas, F., Roca, C., & Reis, M. A. M. (2015). Fungi as sources of polysaccharides for pharmaceutical and biomedical applications. In *Handbook of polymers for pharmaceutical technologies (vol. 3).* Wiley, 61–103. https://doi.org/10.1002/9781119041450.ch3.

Groenewald, M., Boekhout, T., Neuvéglise, C., Gaillardin, C., Van Dijck, P. W. M., & Wyss, M. (2014). Yarrowia lipolytica: Safety assessment of an oleaginous yeast with a great industrial potential. *Critical Reviews in Microbiology, 40*(3), 187–206. https://doi.org/10.3109/1040841X.2013.770386.

Hohman, E. E., Martin, B. R., Lachcik, P. J., Gordon, D. T., Fleet, J. C., & Weaver, C. M. (2011). Bioavailability and efficacy of vitamin D2 from UV-irradiated yeast in growing, vitamin D-deficient rats. *Journal of Agricultural and Food Chemistry, 59*(6), 2341–2346. https://doi.org/10.1021/jf104679c.

Jach, M. E., Serefko, A., Ziaja, M., & Kieliszek, M. (2022). Yeast protein as an easily accessible food source. *Metabolites, 12*(1), 63. https://doi.org/10.3390/metabo12010063.

Jacob, F. F., Hutzler, M., Michel, M., & Methner, F. J. (2019). Boosting yeast propagation via yeast extract supplementation. *BrewingScience, 72*(9/10), 157–167. https://doi.org/10.23763/BrSc19–22jacob.

Jaeger, A., Arendt, E. K., Zannini, E., & Sahin, A. W. (2020). Brewer's Spent Yeast (BSY), an underutilized brewing by-product. *Fermentation, 6*(4), 123. https://doi.org/10.3390/fermentation6040123.

Jensen, G. S., Carter, S. G., Reeves, S. G., Robinson, L. E., & Benson, K. F. (2015). Anti-inflammatory properties of a dried fermentate in vitro and in vivo. *Journal of Medicinal Food, 18*(3), 378–384. https://doi.org/10.1089/jmf.2013.0158.

Johnson, E. A. (2013). Biotechnology of non-Saccharomyces yeasts - The ascomycetes. *Applied Microbiology and Biotechnology, 97*(2), 503–517. https://doi.org/10.1007/s00253-012-4497-y.

Jullesson, D., David, F., Pfleger, B., & Nielsen, J. (2015). Impact of synthetic biology and metabolic engineering on industrial production of fine chemicals. *Biotechnology Advances, 33*(7), 1395–1402. https://doi.org/10.1016/j.biotechadv.2015.02.011.

Karimi, S., Soofiani, N. M., Lundh, T., Mahboubi, A., Kiessling, A., & Taherzadeh, M. J. (2019). Evaluation of filamentous fungal biomass cultivated on vinasse as an alternative nutrient source of fish feed: Protein, lipid, and mineral composition. *Fermentation, 5*(4), 99. https://doi.org/10.3390/fermentation5040099.

Kohjin Life Sciences. (2024). https://www.kohjinls.com/.

Kumar, V., Kumar, P., Maity, S. K., Agrawal, D., Narisetty, V., Jacob, S., Kumar, G., Bhatia, S. K., Kumar, D., & Vivekanand, V. (2024). Recent advances in bio-based production of top platform chemical, succinic acid: An alternative to conventional chemistry. *Biotechnology for Biofuels and Bioproducts, 17*(1), 72. https://doi.org/10.1186/s13068-024-02508-2.

Landi, N., Ragucci, S., & Di Maro, A. (2021). Amino acid composition of milk from cow, sheep and goat raised in ailano and valle agricola, two localities of 'alto casertano' (Campania region). *Foods, 10*(10), 2431. https://doi.org/10.3390/foods10102431.

Li, X., & Mupondwa, E. (2021). Empirical analysis of large-scale bio-succinic acid commercialization from a technoeconomic and innovation value chain perspective: BioAmber biorefinery case study in Canada. *Renewable and Sustainable Energy Reviews, 137*, 110587. https://doi.org/10.1016/j.rser.2020.110587.

Liu, D., Ding, L., Sun, J., Boussetta, N., & Vorobiev, E. (2016). Yeast cell disruption strategies for recovery of intracellular bio-active compounds — A review. *Innovative Food Science and Emerging Technologies, 36*, 181–192. https://doi.org/10.1016/j.ifset.2016.06.017.

Liu, Y., Wu, Q., Wu, X., Algharib, S. A., Gong, F., Hu, J., Luo, W., Zhou, M., Pan, Y., Yan, Y., Yuan, P., & Yulian, W. (2021). Structure, preparation, modification, and bioactivities of β-glucan and mannan from yeast cell wall: A review. *International Journal of Biological Macromolecules, 173*, 445–456. https://doi.org/10.1016/j.ijbiomac.2021.01.125.

Łukaszewicz, M., Leszczyński, P., Jabłoński, S. J., & Kawa-Rygielska, J. (2024). Potential Applications of Yeast Biomass Derived from Small-Scale Breweries. *Applied Sciences (Switzerland), 14*(6). https://doi.org/10.3390/app14062529.

Lukondeh, T., Ashbolt, N. J., & Rogers, P. L. (2003). Evaluation of Kluyveromyces marxianus as a source of yeast autolysates. *Journal of Industrial Microbiology and Biotechnology, 30*(1), 52–56. https://doi.org/10.1007/s10295-002-0008-y.

Ma, C., Xia, S., Song, J., Hou, Y., Hao, T., Shen, S., Li, K., Xue, C., & Jiang, X. (2023). Yeast protein as a novel dietary protein source: Comparison with four common plant proteins in physicochemical properties. *Current Research in Food Science, 7*, 100555. https://doi.org/10.1016/j.crfs.2023.100555.

Ma, J., Sun, Y., Meng, D., Zhou, Z., Zhang, Y., & Yang, R. (2023). Yeast proteins: The novel and sustainable alternative protein in food applications. *Trends in Food Science and Technology, 135*, 190–201. https://doi.org/10.1016/j.tifs.2023.04.003.

Machuca, C., Méndez-Martínez, Y., Reyes-Becerril, M., & Angulo, C. (2022). Yeast β-glucans as fish immunomodulators: A review. *Animals, 12*(16), 2154. https://doi.org/10.3390/ani12162154.

Madzak, C. (2021). Yarrowia lipolytica strains and their biotechnological applications: How natural biodiversity and metabolic engineering could contribute to cell factories improvement. *Journal of Fungi, 7*(7), 548. https://doi.org/10.3390/jof7070548.

Madzak, C. (2021). *Yarrowia lipolytica* engineering as a source of microbial cell factories. In V. Singh (Ed.). *Microbial cell factories engineering for production of biomolecules* (pp. 345–380). Elsevier. https://doi.org/10.1016/B978-0-12-821477-0.00009-X.

Mapelli, V., Hillestrøm, P. R., Kápolna, E., Larsen, E. H., & Olsson, L. (2011). Metabolic and bioprocess engineering for production of selenized yeast with increased content of seleno-methylselenocysteine. *Metabolic Engineering, 13*(3), 282–293. https://doi.org/10.1016/j.ymben.2011.03.001.

Marson, G. V., Lacour, S., Dupas Hubinger, M., & Belleville, M. P. (2022). Serial fractionation of spent brewer's yeast protein hydrolysate by ultrafiltration: A peptide-rich product with low RNA content. *Journal of Food Engineering, 312*, 110737. https://doi.org/10.1016/j.jfoodeng.2021.110737.

Matassa, S., Boon, N., Pikaar, I., & Verstraete, W. (2016). Microbial protein: Future sustainable food supply route with low environmental footprint. *Microbial Biotechnology, 9*(5), 568–575. https://doi.org/10.1111/1751-7915.12369.

Mirzaei, M., Shavandi, A., Mirdamadi, S., Soleymanzadeh, N., Motahari, P., Mirdamadi, N., Moser, M., Subra, G., Alimoradi, H., & Goriely, S. (2021). Bioactive peptides from yeast:

A comparative review on production methods, bioactivity, structure-function relationship, and stability. *Trends in Food Science and Technology, 118*, 297–315. https://doi.org/10.1016/j.tifs.2021.10.008.

Molfetta, M., Morais, E. G., Barreira, L., Luigi Bruno, G., Porcelli, F., Dugat-Bony, E., Bonnarme, P., & Minervini, F. (2022). Protein sources alternative to meat: State of the art and involvement of fermentation. *Foods, 11*(14), 2065. https://doi.org/10.3390/foods11142065.

Morrissey, J. P., Varela, J. A., Gethins, L., Stanton, C., & Ross, P. (2017). Applications of *Kluyveromyces marxianus* in biotechnology. In T. Satyanarayana, & G. Kunze (Eds.). *Yeast diversity in human welfare* (pp. 439–453). Springer. https://doi.org/10.1007/978-981-10-2621-8_17.

Mykhalevych, A., Polishchuk, G., Nassar, K., Osmak, T., & Buniowska-Olejnik, M. (2022). β-Glucan as a techno-functional ingredient in dairy and milk-based products—A review. *Molecules (Basel, Switzerland), 27*(19), 6313. https://doi.org/10.3390/molecules27196313.

Nandy, S. K., & Srivastava, R. K. (2018). A review on sustainable yeast biotechnological processes and applications. *Microbiological Research, 207*, 83–90. https://doi.org/10.1016/j.micres.2017.11.013.

Natakankitkul, S., Homdok, P., Wandee, P., Krisdaphong, T., & Toida, T. (2015). Development of skincare cosmetic from yeast beta-glucans. *Thai Journal of Pharmaceutical Sciences, 40*, 9–12. https://www.researchgate.net/publication/305432797.

Niego, A. G. T., Lambert, C., Mortimer, P., Thongklang, N., Rapior, S., Grosse, M., Schrey, H., Charria-Girón, E., Walker, A., Hyde, K. D., & Stadler, M. (2023). The contribution of fungi to the global economy. *Fungal Diversity, 121*(1), 95–137. https://doi.org/10.1007/s13225-023-00520-9.

Nippon Paper Industries. (2024). http://www.npchem.co.jp/.

Oliveira, A. S., Ferreira, C., Pereira, J. O., Ptado, M. E., & Carvalho, A. P. (2022). Spent brewer's yeast (*Saccharomyces cerevisiae*) as a potential source of bioactive peptides: An overview. *International Journal of Biological Macromolecules, 208*, 1116–1126. https://doi.org/10.1016/j.ijbiomac.2022.03.094.

Oliveira, A. S., Odila Pereira, J., Ferreira, C., Faustino, M., Durão, J., Pintado, M. E., & Carvalho, A. P. (2022). Peptide-rich extracts from spent yeast waste streams as a source of bioactive compounds for the nutraceutical market. *Innovative Food Science and Emerging Technologies, 81*, 103148. https://doi.org/10.1016/j.ifset.2022.103148.

Oliveira, A. S., Odila Pereira, J., Ferreira, C., Faustino, M., Durão, J., Margarida Pereira, A., Maria Oliveira, C., Pintado, M. E., & Carvalho, A. P. (2022). Spent yeast valorization for food applications: Effect of different extraction methodologies. *Foods, 11*(24), 4002. https://doi.org/10.3390/foods11244002.

Oliveira, A. S., Margarida Pereira, A., Ferreira, C. M. H., Odila Pereira, J., Amorim, M., Faustino, M., Durão, J., Pintado, M. E., & Carvalho, A. P. (2024). Purification of bioactive peptides from spent yeast autolysates. *Food and Bioproducts Processing, 143*, 45–53. https://doi.org/10.1016/j.fbp.2023.10.010.

Pang, Y., Zhang, H., Wen, H., Wan, H., Wu, H., Chen, Y., Li, S., Zhang, Sun, L., Xiaojie, L., & Bichen, L. X. (2022). Yeast probiotic and yeast products in enhancing livestock feeds utilization and performance: An overview. *Journal of Fungi, 8*(11), 1191. https://doi.org/10.3390/jof8111191.

Patterson, R., Rogiewicz, A., Kiarie, E. G., & Slominski, B. A. (2023). Yeast derivatives as a source of bioactive components in animal nutrition: A brief review. *Frontiers in Veterinary Science, 9*, 1067383. https://doi.org/10.3389/fvets.2022.1067383.

Pereira, P. R., Freitas, C. S., & Paschoalin, V. M. F. (2021). Saccharomyces cerevisiae biomass as a source of next-generation food preservatives: Evaluating potential proteins as a source of antimicrobial peptides. *Comprehensive Reviews in Food Science and Food Safety, 20*(5), 4450–4479. https://doi.org/10.1111/1541-4337.12798.

Pobiega, K., Sękul, J., Pakulska, A., Latoszewska, M., Michońska, A., Korzeniowska, Z., Macherzyńska, Z., Pląder, M., Duda, W., Szafraniuk, J., Kufel, A., Dominiak, Ł., Lis, Z., Kłusek, E., Kozicka, E., Wierzbicka, A., Trusińska, M., Rybak, K., Kot, A. M., ... Proteins, F. (2024). Sources, production and purification methods, industrial applications, and future perspectives. *Applied Sciences, 14*(14), 6259. https://doi.org/10.3390/app14146259.

Prakash Dalbanjan, N., Eelager, M. P., & Narasagoudr, S. S. (2024). Microbial protein sources: A comprehensive review on the potential usage of fungi and cyanobacteria in sustainable food systems. *Food and Humanity, 3*, 100366. https://doi.org/10.1016/j.foohum.2024.100366.

Pscheidt, B., & Glieder, A. (2008). Yeast cell factories for finechemical and API production. *Microbial Cell Factories, 7*, 25. https://doi.org/10.1186/1475-2859-7-25.

Qiu, Y., Lei, P., Wang, R., Sun, L., Luo, Z., Li, S., & Xu, H. (2023). Kluyveromyces as promising yeast cell factories for industrial bioproduction: From bio-functional design to applications. *Biotechnology Advances, 64*, 108125. https://doi.org/10.1016/j.biotechadv.2023.

Rautela, A., Chatterjee, R., Yadav, I., & Kumar, S. (2024). A comprehensive review on engineered microbial production of farnesene for versatile applications. *Journal of Environmental Chemical Engineering, 12*(2), 112398. https://doi.org/10.1016/j.jece.2024.112398.

Raza, A., Song, H., Raza, J., Li, P., Li, K., & Yao, J. (2020). Formation of beef-like odorants from glutathione-enriched yeast extract: Via Maillard reaction. *Food and Function, 11*(10), 8583–8601. https://doi.org/10.1039/d0fo01946a.

Razzaq, Z. U., Iqbal Khan, M. K., Maan, A. A., & Rahman, S. U. (2020). Characterization of single cell protein from Saccharomyces cerevisiae for nutritional, functional and antioxidant properties. *Journal of Food Measurement and Characterization, 14*(5), 2520–2528. https://doi.org/10.1007/s11694-020-00498-x.

Rose, H. (2023). Beta-glucans: Structure and utilization in fermented milk: A mini review. *International Journal of Food and Fermentation Technology, 13*(1). https://doi.org/10.30954/2277-9396.01.2023.6.

Schauss, A. G., Glavits, Endres, J., Jensen, G. S., & Clewell, A. (2012). Safety evaluation of a proprietary food-grade, dried fermentate preparation of saccharomyces cerevisiae. *International Journal of Toxicology, 31*(1), 34–45. https://doi.org/10.1177/1091581811425195.

Schiavone, M. F., Jean, M., Zerbib, D., & Capp, J. P. (2023). Emerging relevance of cell wall components from non-conventional yeasts as functional ingredients for the food and feed industry. *Current Research in Food Science, 7*, 100603. https://doi.org/10.1016/j.crfs.2023.100603.

Schlabitz, C., Neutzling, L., Daniel, V., de, S., & Claucia, F. (2022). A review of Saccharomyces cerevisiae and the applications of its byproducts in dairy cattle feed: Trends in the use of residual brewer's yeast. *Journal of Cleaner Production, 332*, 130059. https://doi.org/10.1016/j.jclepro.2021.130059.

Schmacht, M., Lorenz, E., & Senz, M. (2017). Microbial production of glutathione. *World Journal of Microbiology and Biotechnology, 33*(6), 106. https://doi.org/10.1007/s11274-017-2277-7.

Shurson, G. C. (2017). The role of biofuels coproducts in feeding the world sustainably. *Annual Review of Animal Biosciences, 5*, 229–254. https://doi.org/10.1146/annurev-animal-022516-022907.

Shurson, G. C. (2018). Yeast and yeast derivatives in feed additives and ingredients: Sources, characteristics, animal responses, and quantification methods. *Animal Feed Science and Technology, 235*, 76. https://doi.org/10.1016/j.anifeedsci.2017.11.01060.

Singh, R., Kumar, M., Mittal, A., & Mehta, P. K. (2016). Microbial enzymes: Industrial progress in 21st century. *3 Biotech, 6*(2), 174. https://doi.org/10.1007/s13205-016-0485-8.

Singla, A., Gupta, O. P., Sagwal, V., Kumar, A., Patwa, N., Mohan, N., Ankush, Kumar, D., Vir, O., Singh, J., Kumar, L., Lal, C., & Singh, G. (2024). Beta-glucan as a soluble dietary fiber source: Origins, biosynthesis, extraction, purification, structural characteristics, bioavailability, biofunctional attributes, industrial utilization, and global trade. *Nutrients, 16*(6), 900. https://doi.org/10.3390/nu16060900.

Skoog, E., Ho Shin, J., Saez-Jimenez, V., Mapelli, V., & Olsson, L. (2018). Biobased adipic acid – The challenge of developing the production host. *Biotechnology Advances, 36*(8), 2248–2263. https://doi.org/10.1016/j.biotechadv.2018.10.012.

Snyman, C., Nguela, J. M., Sieczkowski, N., Marangon, M., & Divol, B. (2021). Optimised extraction and preliminary characterisation of mannoproteins from non-saccharomyces wine yeasts. *Foods, 10*(5), 924. https://doi.org/10.3390/foods10050924.

Sombutyanuchit, Suphantharika, & Verduyn (2001). Preparation of 5′-GMP-rich yeast extracts from spent brewer's yeast. *World Journal of Microbiology and Biotechnology, 17*(2), 163–168. https://doi.org/10.1023/A:1016686504154.

Sousa-Silva, M., Vieira, D., Soares, P., Casal, M., & Soares-Silva, I. (2021). Expanding the knowledge on the skillful yeast *Cyberlindnera jadinii*. *Journal of Fungi, 7*(1), 1–19. https://doi.org/10.3390/jof7010036.

Specialty Yeast Market. (2024). https://www.marketsandmarkets.com/PressReleases/specialty-yeast.asp.

Spring, P., Wenk, C., Connolly, A., & Kiers, A. (2015). A review of 733 published trials on Bio-Mos®, a mannan oligosaccharide, and Actigen®, a second generation mannose rich fraction, on farm and companion animals. *Journal of Applied Animal Nutrition, 3*, e8. https://doi.org/10.1017/jan.2015.6.

Stikane, A., Dace, E., & Stalidzans, E. (2022). Closing the loop in bioproduction: Spent microbial biomass as a resource within circular bioeconomy. *New Biotechnology, 70*, 109–115. https://doi.org/10.1016/j.nbt.2022.06.001.

Sun, J., Xu, S., Du, Y., Yu, K., Jiang, Y., Weng, H., & Yuan, W. (2022). Accumulation and enrichment of trace elements by yeast cells and their applications: A critical review. *Microorganisms, 10*(9), 1746. https://doi.org/10.3390/microorganisms10091746.

Tadioto, V., Giehl, A., Cadamuro, R. D., Guterres, I. Z., dos Santos, A. A., Bressan, S. K., Werlang, L., Stambuk, B. U., Fongaro, G., Silva, I. T., & Alves, S. L. (2023). Bioactive compounds from and against yeasts in the one health context: A comprehensive review. *Fermentation, 9*(4), 363. https://doi.org/10.3390/fermentation9040363.

Tao, Z., Yuan, H., Liu, M., Liu, Q., Zhang, S., Liu, H., Jiang, Y., Huang, D., & Wang, T. (2023). Yeast extract: Characteristics, production, applications and future perspectives. *Journal of Microbiology and Biotechnology, 33*(1), 151–166. https://doi.org/10.4014/jmb.2207.07057.

Thayer, M. T., Garcia, R. M., Duttlinger, A. W., Mahoney, J. A., Schinckel, A. P., Asmus, M. D., Jones, D. B., Dunn, J. L., & Richert, B. T. (2023). Feeding a whole-cell inactivated *Pichia guilliermondii* yeast to gestating and lactating sows in a commercial production system. *Translational Animal Science, 7*(1), txac160. https://doi.org/10.1093/tas/txac160.

Timira, V., Chen, X., Zhou, P., Wu, J., & Wang, T. (2024). Potential use of yeast protein in terms of biorefinery, functionality, and sustainability in food industry. *Comprehensive Reviews in Food Science and Food Safety, 23*(3), e13326. https://doi.org/10.1111/1541-4337.13326.

Tomé, D., & Extracts, Y. (2021). Nutritional and Flavoring Food Ingredients. *ACS Food Science and Technology, 1*(4), 487–494. https://doi.org/10.1021/acsfoodscitech.0c00131.

Tullio, V. (2022). Yeast genomics and its applications in biotechnological processes: What is our present and near future? *Journal of Fungi, 8*(7), 752. https://doi.org/10.3390/jof8070752.

Va Gautério, G., da Silva, R. M., Karraz, F. C., Zarur Coelho, M. A., Ribeiro, B. D., & Lemes, A. C. (2023). Cell disruption and permeabilization methods for obtaining yeast bioproducts. *Cleaner Chemical Engineering, 6*, 100112. https://doi.org/10.1016/j.clce.2023.100112 100112.

Van Dien, S. (2013). From the first drop to the first truckload: Commercialization of microbial processes for renewable chemicals. *Current Opinion in Biotechnology, 24*(6), 1061–1068. https://doi.org/10.1016/j.copbio.2013.03.002.

Wei, G., Shang, W., Xie, Z., Zhang, M., Dan, M., Zhao, G., & Wang, D. (2024). Unlocking high-value components from Brewer's spent yeast for innovative food applications. *Food Bioscience, 59*, 104047. https://doi.org/10.1016/j.fbio.2024.104047.

World Wine Production Outlook. (2023). https://www.oiv.int/sites/default/files/documents/OIV_World_Wine_Production_Outlook_2023.pdf.

Yang, S., Song, L., Wang, J., Zhao, J., Tang, H., & Bao, X. (2024). Engineering Saccharomyces cerevisiae for efficient production of recombinant proteins. *Engineering Microbiology, 4*(1), 100122. https://doi.org/10.1016/j.engmic.2023.100122.

Zeko-Pivač, A., Habschied, K., Kulisic, B., Barkow, I., & Tišma, M. (2023). Valorization of spent Brewer's yeast for the production of high-value products, materials, and biofuels and environmental application. *Fermentation, 9*(3), 208. https://doi.org/10.3390/fermentation9030208.

Zhao, J., Graham, F., Zhao, H., & Fleet G H, J. (2005). Degradation of RNA during the autolysis of Saccharomyces cerevisiae produces predominantly ribonucleotides. *Journal of Industrial Microbiology & Biotechnology, 32*, 415–438.

Zolnere, K., & Ciprovica, I. (2017). The comparison of commercially available ß-Galactosidases for dairy industry: Review. *Research for Rural Development, 1*, 215–222. https://doi.org/10.22616/RRD.23.2017.032.

18 Upscale and economics aspects in the use of microalgae and macroalgae for biofuels production

Jorge Aburto
Bioenergy consultant, Mexico City, Mexico

18.1 Introduction

The steadily increase of human activities since the industrial revolution in the middle XIX centaury till today have caused the incremental use of fossil sources for energy, food, materials, chemical generation, among other needs of contemporary societies. Hence, the atmospheric concentration of CO_2 has raised from 280 to 420 ppm in such time (National Oceanic and Atmospheric Administration NOAA, 2024), which is the most likely reason of the rise of average earth's surface temperature and climate change (Intergovernmental Panel on Climate Change IPCC, 2024). The transport, energy, industrial, animal breeding, and agriculture sectors contribute at different extents to such CO_2 emissions. The international community reached a consensus to assume the reduction of greenhouse gases (GHG) at the 21th Conference of the Parties (COP21) in Paris (Consulted on internet in November 2024, 2015), to disrupt economic growth from gas emission generation, mainly CO_2 y CH_4, through the Intended National Determined Contributions (INDC).

In addition, the incremental activity of industrial agriculture has significantly increased the use of agrochemicals, mainly fertilizers, which some quantities are released to the aquifers, lakes, rivers, and finally the oceans. This nutrient enrichment of waters causes its eutrophication, i.e., it favors the growth of eukaryotic organisms like microalgae and seaweed (macroalgae) bloom and a low oxygen content which may reduce fish and other species habitats. Eutrophication causes a disruption on microalgae ecology that reduces some microalgae species and increases the density of toxic microalgae species (Ha et al., 2015). In Mexico, there is a huge strain diversity along the country such as *Spirulina* sp., *Anabaena* sp., *Oscillatoria* sp., *Desmodesmus* sp., *Scenedesmus* sp., *Chlorella* sp., among others, which can be used to treat residual wastewaters or bioproducts. By instance, Spirulina was collected in Texcoco Lake in Central Mexico by Mexica people, and it is still cultivated for nutritional supplement in present Mexico (Sosa-Hernández et al., 2019).

On the other side, macroalgae bloom in the Atlantic Ocean (Sargassum sea) is responsible of massive yearly arrivals to the Caribbean, Central America countries, and Mexico coasts with serious environmental, health, economic, and touristic impacts. The most common species are *Sargassum fluitants* and *Sargassum natans* (hereafter sargassum) that belong to the family *Sargassaceae*, order *Fucales*, class *Phaeophyceae* of *Eucarya* dominion (López-Contreras et al., 2022). Even if millions

Eukaryotic Microorganisms as Sources of Bioproducts. DOI: https://doi.org/10.1016/B978-0-443-30188-9.00023-8

of tons of Sargassum arrive to the mentioned coasts, Mexico informed just the collection of 1233 ton in 2019 (Anuario estadístico de acuacultura y pesca Consulted in internet in November, 2019). In the case of Sargassum, the issue is located mainly in Quintana Roo State coasts, Southern Mexico. Several initiatives have offshore and onshore collected the sargassum which is further transported to landfills or transformed to biogas, bricks or alginate production.

Therefore, contemporary societies are facing the increase of greenhouse emissions in the atmosphere but also the pollution of soil and water as seen before. A holistic approach to such problem is using alternative fuels with lower carbon fingerprint derived from microalgae and macroalgae (Sargassum), which may also help to decrease its environmental impacts. Indeed, they may be both transformed into fuels like ethanol, biodiesel and renewable diesel, biogas, sustainable aviation fuel (SAF), biocompounds, among others (Anto et al., 2020; Bhushan et al., 2020).

Since some Mexican agencies like Aeropuertos y Servicios Auxiliares (ASA) and Agencia Federal de Aviación Civil (AFAC) altogether with the Mexican Aviation Airline Chamber (CANAERO) are committed to the mitigation of such CO_2 emissions, the microalgae and sargassum valorization into fuels may be an attractive approach. Therefore, the research and technology developments of microalgae and sargassum-based biofuels emerged as attractive solutions to the need of reduction of CO_2 emissions of aviation industry but also to create value from microalgae farms and to minimize the impact of sargassum arrivals and related environmental, health, economic, social, and touristic issues.

Hence, such challenge was approached using the TEA for the up-scaling of microalgae and sargassum conversion to biofuels, i.e., biochar, renewable diesel, and SAF. The scientific and technological performance achieved by this approach helps to guide further research and technology development, but also, to get an understanding of how the latter affects the economic and financial dimensions to move forward into the market. Finally, the TEA of up-scaling of microalgae and sargassum based biofuels is discussed.

18.2 Criteria for techno-economic analysis

The TEA is an important decision tool that drives the conception, development, scaling-up, and commercialization of technologies in terms of three main dimensions: technology, economy, and finances (Fig. 18.1). At earlier to medium fundamental and applied research (1<TRL<5), it gives insights to address scientific, technical, and initial financial questions. Further technology development (5<TRL<9) requires information that guides decisions for near-to-market technologies (U.S. Energy Information Administration (USEIA), 2015; Kargbo et al., 2021; NASA, 2012; National Renewable Energy Laboratory (NREL), 2011). Hence, the basics of TEA are presented and applied to the fuel production from Chlorella powder and Sargassum pellets.

The TEA may be easily explained in terms of such three dimensions: (1) the scientific knowledge applied to solve questions as well as technical, social, environmental, and market issues. Here, a specific technology is assessed in terms of conversion and yield

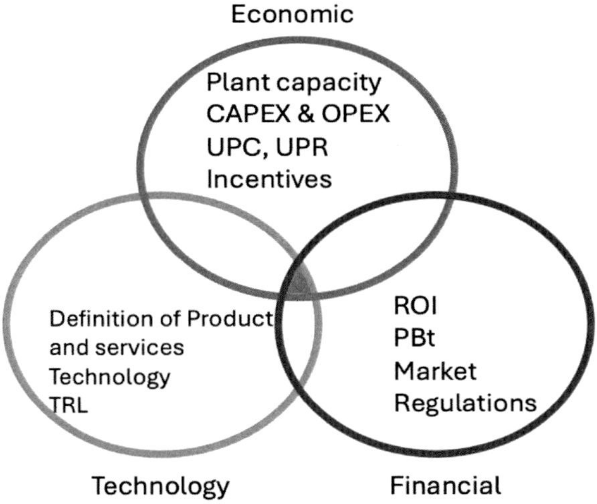

Figure 18.1 Dimensions of technology, economy, and finances to be considered in techno-economic analysis. The circles represent every dimension: economic, technology, and financial. No Permission Required.

using kinetic and/or stoichiometric equilibria and modeled using a process simulator. Then, the (2) economic (CAPEX, OPEX, plant capacity, UPC, UPR, incentives) and (3) financial (ROI, PBt, market, regulations) dimensions are evaluated. Moreover, the Chorella powder production as well as Sargassum pellet supply chain must also be considered (Lo et al., 2021; Mungodla et al., 2019) but such subject is out of the scope of present work.

Hence, the TEA allows to define the conceptual engineering considering inside or outside battery limits (ISBL, OSBL) including the scheme of main unit operations, input and output mass flows, chemical reactions, the engineering involving equipment, material properties, and operation conditions in every unit operation and energy requirements such as steam and electricity. Today, there are process simulators that allow such design like Superpro Designer® from Intelligen Inc., Batch Plus® from Aspen Technology, Inc., (Petrides & Siletti, 2004; Petrides et al., 1989; Shanklin et al., 2001); the BEFS Cogeneration tool (Food and Agriculture Organization FAO, 2024); IMP Bio2Energy® (Bco Cluster, 2021); and IMPES® from the Mexican Petroleum Institute, among others. In many cases, the simulators lack on information concerning physicochemical properties of biomolecules, kinetics, thermodynamics, and stoichiometry of chemical and thermochemical reactions as well as engineering data of bioprocesses, which uncertainty must be approach through experimental data validation.

Moreover, the economic and financial analysis must consider the price of all raw materials, operational, and capital costs and to determine the UPC and UPR values as well as the capacity plant. Sensitivity analysis of financial parameters is a main activity where minimum acceptable plant capacity and return of inversion and PBt are calculated.

18.3 Valorization of microalgae (*Chlorella* sp.) and macroalgae (*Sargassum* sp.) into biofuels

The use of native and unmodified microorganisms as phototropic microalgae to produce biofuels corresponds to the known third biofuel generation. These microorganisms produce biomass and oils through heterotrophic fermentation or autotrophic CO_2 assimilation, respectively. Such biomass and/or oils can be further converted by conventional or advanced processing into biofuels. The microalgae are cultivated in open ponds or closed bioreactors to produced biomass or extracts of different components as oils and pigments (Fig. 18.2). Moreover, microalgae are harvested and processed into microalgal powder for food, feed, nutraceulticals, or pharmaceuticals products (Qin et al., 2023). *Chlorella* sp. is a very versatile strain that can be cultivated both in open and closed systems and has raised attention because a high oil titer as 50% w/w may be attained under stressing nutrient conditions (Guccione et al., 2014; Thirugnanasambantham et al., 2020).

Besides, the valorization of *Sargassum* sp. arrivals to the coasts must include an integral pathway considering upstream activities as (1) the collecting of sargassum, (2) its stabilization and sizing and downstream processing, as (3) its further processing to valuable services and/or products (Fig. 18.3). Upstream activities will not be considered further on this work, but includes:

1. Collecting of Sargassum:
   a. Using specialized boats to collect Sargassum offshore and,
   b. Using manual and/or machinery to collect it at the beach.

Figure 18.2 Up- and downstream schematic processing of microalgae. Scheme process of microalgae.
No Permission Required.

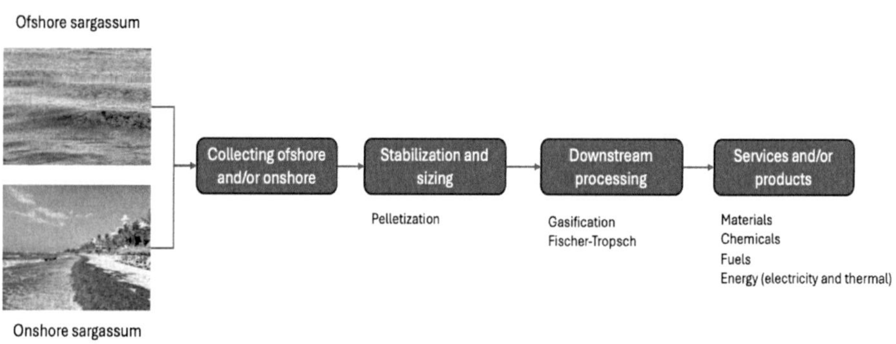

Figure 18.3 Up- and down-stream processing of *Sargassum* sp. Scheme process of *Sargassum* sp.
No Permission Required.

2. Stabilization and sizing:
   a. Stabilization through washing, drying, torrefaction, among others,
   b. Sizing using cutting, grinding, pelletization, or other sizing technology.

Downstream processing of microalgae as *Chlorella* sp. and *Sargassum* sp. in terms of chemicals, fuels, and energy would be very versatile as many robust and mature technologies are available such as deconstruction, anaerobic digestion, solid and liquid fermentation, hydrothermal liquefaction, combustion, pyrolysis, and gasification. Any of these technologies can produce either materials as biochar, energy services as heat and electricity, chemicals as polysaccharides, saccharides, fine chemicals, and methane or fuels as methane, renewable diesel, and SAF. An environmental solid and socially responsible technological approach must consider local and regional needs to supply the more adequate proposal solution. We have identified a need and opportunity for sustainable biofuels in the international airports of Cancún and Tulum, located in Quintana Roo State in Mexico, to comply to IATA and CORSIA statements concerning the off trade of CO_2 emissions of international flights. Therefore, we propose here the valorization of both *Chlorella* sp. and *Sargassum* sp. to produce SAF and renewable diesel as discussed below.

18.4 Techno-economic analysis of the use of microalgae (*Chlorella* sp.) and macroalgae (*Sargassum* sp.) for biofuels

Hence, Chlorella powder and Sargassum pellets can be valorized through its gasification to biochar and biosyngas and further upgrading of the latter to methane, renewable diesel, and SAF (Fig. 18.4). The schematic process includes the (1) gasification, (2) integrated gas-to-liquids reaction (Fischer-Tropsch) and hydrotreatment, (3) off gas separation and drying, and (4) the separation of products by distillation: renewable diesel and SAF.

The process simulation of gasification of Chlorella powder and Sargassum pellets starts by defining its ultimate and proximate analysis as well as a common and its simulated composition of obtained syngas through gasification (Table 18.1).

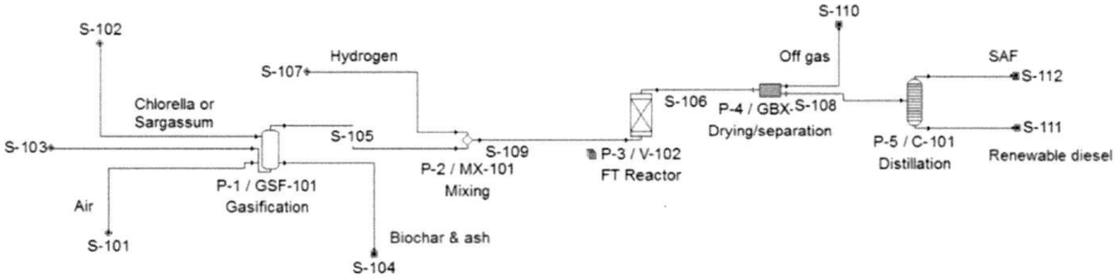

Figure 18.4 Biofuels production process from Chlorella powder and Sargassum pellets. Scheme process.
No Permission Required.

Table 18.1 Ultimate and proximate analysis of chlorella powder and sargassum pellets and its syngas common and simulated composition obtained by gasification.

Ultimate analysis (%)	**C**	**H**	**O**	**N**	**S**		**References**
Chlorella powder	52.35	7.79	27.55	6.74	0.57		Cordoba-Perez and de Lasa (2021)
Sargassum pellets	29.36	4.67	29.77	3.13	1.29		Velázquez-Hernández & Aguillón-Martínez (2023)
Proximate analysis (%)	**Moisture**	**Fixed carbon**	**Volatile Matter**	**Ash**	**HHV (MJ/kg)**	**LHV (MJ/kg)**	
Chlorella powder	10	72	13	5.0	21.18	19.21	Almeida et al., (2017); Chen & Torii (2015)
Sargassum pellets	6.62	13.58	52.33	31.78	12.68	9.29	Velázquez-Hernández & Aguillón-Martínez (2023)
Syngas experimental composition (%)	**H_2**	**CO**	**CO_2**	**CH_4**			
Chlorella powder	18	15	13	4			Raheem et al. (2017)
Sargassum pellets	6.9	17.2	11.8	1.4			Velázquez-Hernández & Aguillón-Martínez (2023)
Syngas simulated composition (%)	**H_2**	**CO**	**CO_2**	**CH_4**			
Chlorella powder	12.65	12.08	12.39	18			
Sargassum pellets	12.56	11.98	12.29	3.04			

Table 18.2 Main simulation parameters used for the gasification of Chlorella powder and Sargassum pellets in Superpro Designer®.

Parameter	Chlorella powder	Sargassum pellets
Equilibrium ratio	0.36%	0.72%
Carbon conversion	90%	90%
Gasification temperature	512°C	501°C
Operation pressure	1 bar	1 bar
Heat losses	5%	5%

No Permission Required.

The gasification process was simulated employing only air as gasifying agent and the parameters were set to get a common syngas composition (Table 18.1). The parameters set were a 5% heat loss and a conversion of 90% (Table 18.2) which allowed a syngas composition closed to data reported in literature.

The process simulator SuperPro Designer® uses an equilibrium reaction model that considers the following chemical reactions:

Gasification:

$$C + H_2O \rightarrow CO + H_2$$

Boudouard:

$$C + CO_2 \rightarrow 2CO$$

Methanation:

$$C + H_2 \rightarrow CH_4$$

$$CH_xO_yN_wS_z + lH_2O + m(O_2 + N_2) \rightarrow n_{CO}CO + n_{H_2}H_2 + n_{CH_4}CH_4 + n_{CO_s}CO_2 + n_{H_2O}H_2O + n_CC + zH_2S + \left(\frac{w}{2} + rm\right)N_2$$

Here, x, y, w, and z are atoms of H, O, N, and S per atoms of C, respectively. The coefficients l and m are the initial number of moles of H_2O and O_2 per atom of C, $_r$ is the N_2/O_2 ratio of air, n_{CO}, n_{CO_2}, n_{H_2}, n_{CH_4}, and n_{H_2O} are the moles of products of CO, CO_2, H_2, n_{CH_4}, and n_{H_2O} per atom of C, and n_C is the moles of unreacted C per atom of C. The composition of the produced syngas from the simulation was validated with the literature (Table 18.1; Velázquez-Hernández & Aguillón-Martínez, 2023).

Now, the produced syngas is transformed to liquids as linear n-paraffins which are the mainly components of renewable diesel; through the Fischer-Tropsch and isomerization reactions (Hu et al., 2012). Here, we have only considered the synthesis of hexadecane to simplify the respective simulation in Superpro designer®.

Fischer-Tropsch:

$$16CO + 33H_2 \rightarrow C_{16}H_{34} + 16H_2$$
$$16CO_2 + 49H_2 \rightarrow C_{16}H_{34} + 32H_2$$

Isomerization reaction:

$$C_{16}H_{34} \rightarrow iso - C_{16}H_{34}$$

The technological assessment of renewable diesel and SAF production from chlorella powder and sargassum pellets requires to define the dimensions in Figs. 18.2 and 18.3, which are case specific and presented in Table 18.2. The inside battery limit (ISBL) process scheme and simulation, stoichiometric reactions and products yield need to be defined. Then, all streams as feedstocks, products, and energy requirements must be defined to calculate the mass and energy balances and establish the plant capacity and yields of products and coproducts. Most advanced biofuel technologies as Hydrotreated esters and fatty acids (HEFA) and Alcohol-to-jet (ATJ; TRL=9) are based on the use of plant oils, greases, cooking waste oil, and CO_2 rich streams or ethanol, respectively. The gasification and Fisher-Tropsch process (GFT) are very well-known technologies (Hu et al., 2012) but the use of Chlorella powder and Sargassum pellets has still to be further studied. Moreover, the costs associated with the production, processing, and transport (Lo et al., 2021) of such feedstocks to the conversion plant and fuels to supplying site must be considered. HEFA and ATJ technologies are already commercial when compared to the GFT production from Chlorella powder and Sargassum pellets (Table 18.3), but their feedstock availability is a serious issue when we considered the fuels supply to the aviation industry but more importantly, the solution approach to the Sargassum arrivals problem to the Quintana Roo State coasts. Plant-based oils, animal greases, cooking waste oils, CO_2 streams, and ethanol are currently the more practicable source for fuel production, but attention is needed with respect to the food-water-energy nexus, the land use change (LUC), and use of agrochemicals that impact severely the environmental and may affect fuel production sustainability (Amezcua-Allieri et al., 2019; Martinez-Hernandez, Magdaleno Molina, et al., 2019; Martinez-Hernandez, Ramírez-Verduzco, et al., 2019). Chlorella powder is produced in open or closed reactor with high biomass or oil titer with a low land use but harvesting and concentration is still a issue to be resolved. Besides, the valorization of sargassum to produce fuels through GFT process presents some issues related to high ash content due to marine salt content, collection, intermediate TRL, and scaling-up which may increase associated costs (Table 18.3).

Concerning to the economic issues, Chlorella powder (CPP) and Sargassum pellet prices (SPP) impact both unit production cost (UPC) as well as return of inversion (ROI, %: Fig. 18.5). The UPC value considers both the production costs for renewable diesel as well as for SAF.

The CPP range was set from 50 to 250 USD/ton considering an average price of 152 USD/ton of Chlorella powder to be competitive with crude oil (Sosa-Hernández et al., 2019), and it does not consider cultivation, harvesting, and drying. The UPC varies from 1.8 to 2.5 USD/kg with a fixed unit production revenue (UPR) of 2.55 USD/kg and with a decrease in ROI from 30% to 10% (Fig. 18.5A). In the case of Sargassum pellet costs, the variation of SPP in a range from 25 to 150 USD/ton augments the UPC value from ca. 1.92 to 2.75 USD/kg of product and diminishes the ROI value from 21% to 3% (Fig. 18.5B). A financial attractive ROI of 10% will need

Table 18.3 Technological characteristics of GFT, HEFA, and ATJ for fuels.

Technology	Feedstock	Products	Risks	Benefits
Gasification and Fischer-Tropsch (GFT)	• Chlorella powder • Sargassum pellets	• Biochar • Methane • Renewable diesel • SAF	• Low to high ash content • Offshore and onshore collection • Intermediate TRL • Scaling-up	• Solution to economic and environmental issues • No LUC issues • Local employment • Supply of biofuels for aviation industry
HEFA	• Soybean • Sunflower • Corn • Oil palm • Coconut • Cooking waste oil	• Renewable propane • Renewable diesel • SAF	• Feedstock availability • Food-water-energy nexus • Land use change (LUC) • Agrochemicals	• Production of biofuels, bio-based chemicals, and electric and steam power • TRL 9: Commercial
ATJ	CO_2 streams	• Renewable diesel • SAF	• Availability of CO_2 streams • Long distances from CO_2 generation to fuel production and supply	• Production of low carbon fuels and chemicals • TRL 9: Commercial

No Permission Required.

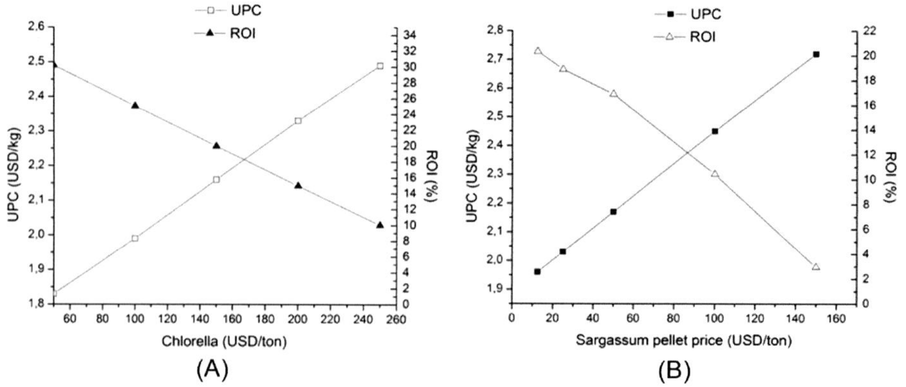

Figure 18.5 Variation of UPC and ROI with Chlorella powder (A) and Sargassum pellet price (B) with respect to the processing capacity. Variation of UPC and ROI with processing capacity. No Permission Required.

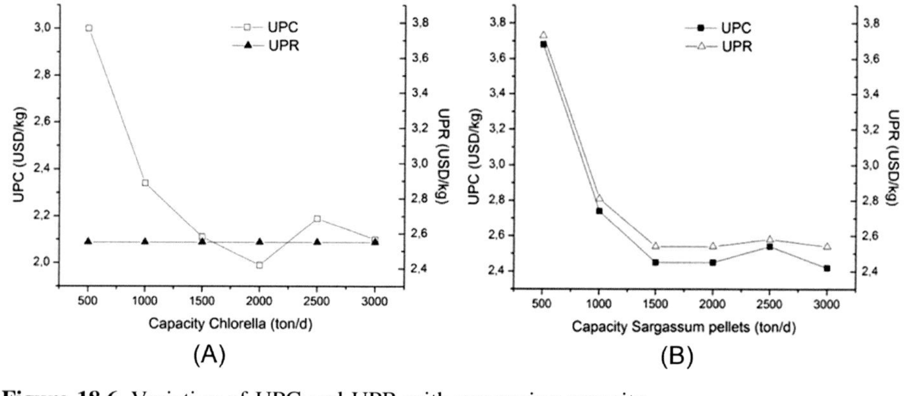

Figure 18.6 Variation of UPC and UPR with processing capacity. No Permission Required.

a SPP of at least 80 USD/ton and an UPC value of 2.3 USD/kg. The SPP considers all activities from offshore and/or onshore collection as well as the transport to the washing, grinding, and sizing facility. The feasibility of such SPP value is not consider in this work, but it serves as a comparison value to future entrepreneurs. By instance, the SAF's UPC for the HEFA and the ATJ technologies were recently reported: 1.09 and 5.06 USD/kg (1.36 and 6.33 USD/L; Aburto et al., 2025), respectively. We observed that the Chlorella powder and Sargassum pellets might be competitive with both HEFA and ATJ technologies, even if they have already a TRL value of 9, i.e., they are already on commercial operation.

In addition, the price of jet fuel from fossil oil in international markets is about 0.70 USD/kg (Consulted online in November, 2024), i.e., three times less than the

UPC obtained in this work. With respect to the processing capacity for renewable diesel and SAF and a CPP or SPP of 100 USD/ton, the UPC and UPR values diminish from 500 to 1500 ton/d and becomes asymptotic at capacities larger than 1500 ton/d (Fig. 18.6). The UPR values diminish from about 3.7 or 3.0 to 2.4 or 2.5 USD/kg for Chlorella powder or Sargassum pellets, respectively. It is important to notice that the UPR value is composed of to product prices: SAF and renewable diesel. In this study, we fixed them at 1.1 and 1.5 USD/kg, respectively. Hence, the market price of SAF obtained from Chlorella powder or Sargassum pellets is slightly higher that the fossil jet fuel.

Concerning CAPEX and OPEX, they both increase steadily from 500 to 2000 ton/d capacity of Chorella powder or Sargassum pellets (Fig. 18.7). Beyond 2000 ton/d processing capacity, the need for additional FT reactors increases severely both expenses but UPC and UPR remained constant as seen before. This means that CAPEX and OPEX may still be optimized to each processing capacity of Chlorella powder or Sargassum pellets, but this was not considered on this work.

Furthermore, the production of biochar, renewable diesel, and SAF increases steadily with processing capacity of chlorella powder or sargassum pellets (Fig. 18.8). In the case of the former, renewable diesel, SAF and biochar production may reach 5112 and 6377 B/d as well as 5889 kg/h, respectively. Now, the use of sargassum pellets allows a high production of 4182, 5217 B/d, and 4487 kg/h. The conversion of Chlorella powder results in a higher production of liquids and solid biofuels when compared with Sargassum pellets and present better financial values as seen above. Now, biochar production may help to further decarbonize the aviation sector using as a fertilizer carrier or another industrial sectors if used to produce electric or steam power.

The sargassum arrivals to the Quintana Roo State in Southern Mexico affects severely the tourism, economy, and environment as present earlier. The State has two main international airports located at two of main touristic cities: Cancún and Tulum (open in 2024), which received flights from around the world but intensively from

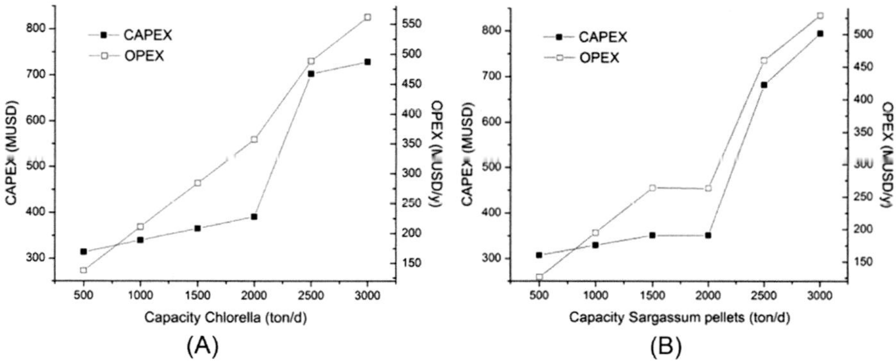

Figure 18.7 Sensibility of CAPEX and OPEX values with respect to processing capacity process of Chlorella powder (A) and Sargassum pellets (B). Sensibility of CAPEX and OPEX values with processing capacity.
No Permission Required.

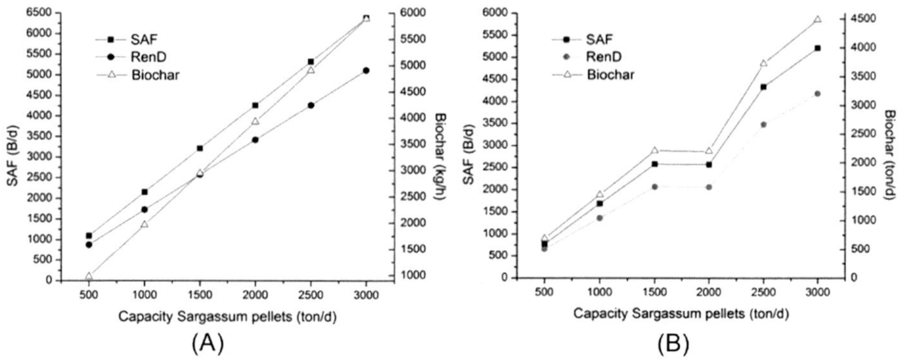

Figure 18.8 Sensibility of SAF, renewable diesel, and biochar production with respect to processing capacity process of Chlorella powder (A) and Sargassum pellets (B). Sensibility of biofuel production with processing capacity.
No Permission Required.

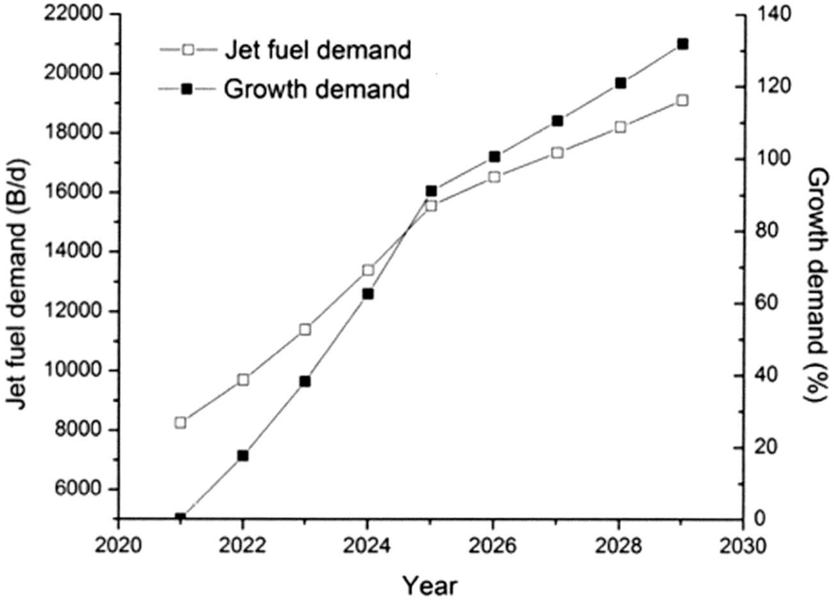

Figure 18.9 Trends in jet fuel demand and growth demand in Cancun airport from 2021 to 2029 (Aeropuertos, Servicios Auxiliares, 2024).
No Permission Required.

USA and Europe, and the operating international airlines are committed to the use of a certain blend of SAF with fossil jet fuel to mitigate their CO_2 emissions. Cancun airport reported a daily jet fuel demand of 11,393 B/d in 2023 and a growth of 62.5%

with respect to 2021, indicating the recovery from the pandemia economy shutdown and an intensive jet fuel demand growth is observed between 2021 and 2029 (Aeropuertos, Servicios Auxiliares, 2024; Fig. 18.9). In our study, Chlorella powder or Sargassum pellets may provide about 1093 to 6377 or 750 to 5000 B/d of SAF within a processing capacity of 500–3000 ton/d, respectively; which is very attractive to decarbonize the aviation international flights in coming years.

18.5 Conclusions

Bioproducts from Eucaryotic organisms as microalgae or macroalgae, and in particular *Chlorella* sp. or *Sargassum* sp., may contribute to the mitigation of CO_2 emissions from the aviation sector, diminish beach arrivals and related tourism and economic issues, create social and economic value from such feedstocks and contribute to the mitigation of environmental impacts.

Eukaryotic organisms as *Chlorella* sp. and *Sargassum* sp. may be transformed into valuable bioproducts as biofuels like renewable diesel, SAF, and biochar through gasification, Fischer-Tropsch, and isomerization. The TEA of the up-scaling to produce such fuels showed that the UPC varies from 1.8 to 2.5 or 1.92 to 2.75 USD/kg for Chlorella powder or Sargassum pellets, respectively. The UPR reached a value around 2.5 USD/kg but with a SAF price established at 1.1 USD/kg, slightly higher than the 0,7 USD/kg of the fossil jet fuel. SAF production from chlorella powder and sargassum pellets might be competitive with commercial technologies as HEFA and ATJ, that use plant oils or ethanol as feedstocks.

Concerning SAF production, we observed that neither feedstock can account for the jet fuel demand (ca. 1440 B/d in 2024) by its own at a capacity of 500 ton/d, but the parallel production of SAF from Chlorella powder and Sargassum pellets may afford a combined production of 1900 B/d in 2030 using a capacity plant of 1000 ton/d for each feedstock.

The technology of gasification and Fischer-Tropsch requires still further study and development to reach a TRL value of 9, but this study showed that both Eukaryote may have the potential to supply the SAF demand in 2030 to the Cancún and Tulum airports in Quintana Roo State, Mexico.

References

Aburto, J., Martínez-Hernández, E., & Castillo-Landero, A. (2025). Is Sustainable Aviation Fuel Production Through Hydroprocessing of Esters and Fatty Acids (HEF and Alcohol-to-Jet (ATJ) Technologies Feasible in Mexico? *Sustainability, 17*(4), 1584. https://doi.org/10.3390/su17041584.

Aeropuertos, Servicios Auxiliares, Consulted on internet on. (2024). ASA.

Almeida, H. N., Calixto, G. Q., Chagas, B. M. E., Melo, D. M. A., Resende, F. M., Melo, M. A. F., & Braga, R. M. (2017). Characterization and pyrolysis of Chlorella vulgaris and Arthrospira platensis: potential of bio-oil and chemical production by Py-GC/MS analysis. *Environmental Science and Pollution Research, 24*(16), 14142–14150. https://doi.org/10.1007/s11356-017-9009-2.

Amezcua-Allieri, M. A., Martínez-Hernández, E., Anaya-Reza, O., Magdaleno-Molina, M., Melgarejo-Flores, L. A., Palmerín-Ruiz, M. E., Eguía-Lis, J. A. Z., Rosas-Molina, A., Enríquez-Poy, M., & Aburto, J. (2019). Techno-economic analysis and life cycle assessment for energy generation from sugarcane bagasse: Case study for a sugar mill in Mexico. *Food and Bioproducts Processing, 118*, 281–292. https://doi.org/10.1016/j.fbp.2019.09.014.

Anto, S., Mukherjee, S. S., Muthappa, R., Mathimani, T., Deviram, G., Kumar, S. S., Verma, T. N., & Pugazhendhi, A. (2020). Algae as green energy reserve: Technological outlook on biofuel production. *Chemosphere, 242*. https://doi.org/10.1016/j.chemosphere.2019.125079.

Anuario estadístico de acuacultura y pesca Consulted in internet in November. (2019). Comisión Nacional de Acuacultura y Pesca (CONAPESCA), Comisión Nacional de Acuacultura y Pesca (CONAPESCA). 2019.

Bcs Cluster. (2021).

Bhushan, S., Kalrac, A., Simsek, H., Kumard, G., & Prajapatia, S. K. (2020). Current trends and prospects in microalgae-based bioenergy production. *J. Environ.Chem.Engin. 8*(5), 104025.

Chen, A.-P., & Torii, S. (2015). Characterized the Microalgae (Chlorella and Spirulina) and Macro Algae by Using TGA and Bomb Calorific Meter for the Biomass Energy Application. *Journal of Mechanics Engineering and Automation, 5*(12). https://doi.org/10.17265/2159-5275/2015.12.007.

Consulted on internet in November 2024. (2015). United Nations. Paris Agreement, United Nations. Paris Agreement. (2015).

Consulted online in November. (2024). International Air Transport Association (IATA), International Air Transport Association (IATA). (2024).

Cordoba-Perez, M., & de Lasa, H. (2021). CO2-Derived Carbon Capture Using Microalgae and Sodium Bicarbonate in a PhotoBioCREC Unit: Kinetic Modeling. *Processes, 9*(8), 1296. https://doi.org/10.3390/pr9081296.

Food and Agriculture Organization (FAO). (2024). BEFS Cogeneration tool.

Guccione, A., Biondi, N., Sampietro, G., Rodolfi, L., Bassi, N., & Tredici, M.R. (2014). Chlorella for protein and biofuels: from strain selection to outdoor cultivation in a Green Wall Panel photobioreactor. *Biotechnology for Biofuels*, *7*, 84. http://www.biotechnologyforbiofuels.com/content/7/1/84.

Ha, N. T. H., Dao, P. T. B., Loan, D. T., Thanh, N. N., Tuan, N. D., Mai, P. T., & Khoa, T. D. (2015). Microalgae as indicators of eutrophication in Hoan Kiem lake, Hanoi, Vietnam. *Tạp chí Công nghệ Sinh học, 13*(2), 355–365.

Hu, J., Yu, F., & Lu, Y. (2012). Application of fischer-tropsch synthesis in biomass to liquid conversion. *Catalysts, 2*(2), 303–326. https://doi.org/10.3390/catal2020303.

Intergovernmental Panel on Climate Change (IPCC). 2021. (2024). Climate change 2021: Summary for all.

Kargbo, H., Harris, J. S., & Phan, A. N. (2021). Drop-in" fuel production from biomass: Critical review on techno-economic feasibility and sustainability. *Renewable and Sustainable Energy Reviews, 135*, 110168.

Lo, S. L. Y., How, B. S., Leong, W. D., Teng, S. Y., Rhamdhani, M. A., & Sunarso, J. (2021). Techno-economic analysis for biomass supply chain: A state-of-the-art review. *Renewable and Sustainable Energy Reviews, 135*. https://doi.org/10.1016/j.rser.2020.110164.

López-Contreras, A. M., Núñez Valenzuela, P., Celis García, B., Driegen, J., Huerta Lwanga, E., Domin, P., Polett Gurrola, M., Rosas-Luis, R., Verde Gómez, Y., & de Vrije, T. (2022). Sargassum in Mexico: From environmental problem to valuable resource. WUR report 2319. https://doi.org/10.18174/574423.

Martinez-Hernandez, E., Magdaleno Molina, M., Melgarejo Flores, L. A., Palmerín Ruiz, M. E., Zermeño EguiaLis, J. A., Rosas Molina, A., Aburto, J., & Amezcua-Allieri, M. A. (2019). Energy-water nexus strategies for the energetic valorization of orange peels based on techno-economic and environmental impact assessment. *Food and Bioproducts Processing, 117*, 380–387. https://doi.org/10.1016/j.fbp.2019.08.002.

Martinez-Hernandez, E., Ramírez-Verduzco, L. F., Amezcua-Allieri, M. A., & Aburto, J. (2019). Process simulation and techno-economic analysis of bio-jet fuel and green diesel production—Minimum selling prices. *Chemical Engineering Research and Design, 146*, 60–70. https://doi.org/10.1016/j.cherd.2019.03.042.

Mungodla, S. G., Linganiso, L. Z., Mlambo, S., & Motaung, T. (2019). Economic and technical feasibility studies: technologies for second generation biofuels. *Journal of Engineering, Design and Technology, 17*(4), 670–704. https://doi.org/10.1108/JEDT-07-2018-0111.

NASA. (2012). Technology Readiness Level. Consulted on June 2021: https://www.nasa.gov/directorates/heo/scan/engineering/technology/technology_readiness_level.

National Oceanic and Atmospheric Administration (NOAA). (2024).

National Renewable Energy Laboratory (NREL). (2011). Process Design and Economics for Biochemical Conversion of Lignocellulosic Biomass to Ethanol Dilute-Acid Pretreatment and Enzymatic Hydrolysis of Corn Stover. (2011). Task No. BB07.2410.

Petrides, D., Cooney, C. L., Evans, L. B., Field, R. P., & Snoswell, M. (1989). Bioprocess simulation: An integrated approach to process development. *Computers & Chemical Engineering, 13*(4-5), 553–561. https://doi.org/10.1016/0098-1354(89)85038-0.

Petrides, D. P., Siletti, C. A. (2004). The role of process simulation and scheduling tools in the development and manufacturing of biopharmaceuticals. Proceedings - Winter Simulation Conference 2046–2051. United States Volume 2.

Qin, S., Wang, K., Gao, F., Ge, B., Cui, H., & Li, W. (2023). Biotechnologies for bulk production of microalgal biomass: from mass cultivation to dried biomass acquisition. *Biotechnology for Biofuels and Bioproducts, 16*(1). https://doi.org/10.1186/s13068-023-02382-4.

Raheem, A., Dupont, V., Channa, A. Q., Zhao, X., Vuppaladadiyam, A. K., Taufiq Yap, Y. H., Zhao, M., & Harun, R. (2017). Parametric Characterization of Air Gasification of Chlorella vulgaris Biomass. *Energy Fuels, 31*, 2959–2969. https://doi.org/10.1021/acs.energyfuels.6b03468.

Shanklin, T., Roper, K., Yegneswaran, P. K., & Marten, M. R. (2001). Selection of bioprocess simulation software for industrial applications. *Biotechnology and Bioengineering, 72*(4), 483–489. https://doi.org/10.1002/1097-0290(20010220)72:4<483::AID-BIT1010>3.0.CO;2-3.

Sosa-Hernández, J. E., Romero-Castillo, K. D., Parra-Arroyo, L., Aguilar-Aguila-Isaías, M. A., García-Reyes, I. E., Ahmed, I., Parra-Saldivar, R., Bilal, M., & Iqbal, H. M. N. (2019). Mexican microalgae biodiversity and state of the art extraction strategies to meet sustainable circular economy challenges: High-value compounds and their applied perspectives. *Marine Drugs, 17*(3). https://doi.org/10.3390/md17030174.

Thirugnanasambantham, R., Elango, T., & Elangovan, K. (2020). Chlorella vulgaris sp. microalgae as a feedstock for biofuel. *Materials Today: Proceedings, 33*, 3182–3185. https://doi.org/10.1016/j.matpr.2020.04.106.

U.S. Energy Information Administration (USEIA). (2015). Technical Economic Analysis Guide - DRAFT. (2015). U.S. Department of Energy. Consulted online in June 2021: https://www.eia.gov/outlooks/documentation/workshops/pdf/tea_guide_071015_draft.pdf.

Velázquez-Hernández, A., & Aguillón-Martínez, J. E. (2023). Aspen plus simulation of sargassum for quality synthesis gas. *Heliyon, 9*(7), e17731. https://doi.org/10.1016/j.heliyon.2023.e17731.

Index

Note: Page numbers followed by "*f*" and "*t*" refer to figures and tables, respectively.

H

Q

R